APPLIED REGRESSION ANALYSIS, LINEAR MODELS, AND RELATED METHODS

APPLIED REGRESSION ANALYSIS, LINEAR MODELS, AND RELATED METHODS

John Fox

For information address:

SAGE Publications, Inc.
2455 Teller Road
Thousand Oaks, California 91320
E-mail: order@sagepub.com

SAGE Publications Ltd.
6 Bonhill Street
London EC2A 4PU
United Kingdom

SAGE Publications India Pvt. Ltd.
M-32 Market
Greater Kailash I
New Delhi 110 048 India

Printed in the United States of America

Library of Congress Cataloging-in-Publication Data

Fox, John, 1947–
Applied regression analysis, linear models, and related models / author, John Fox.
p. cm.
Includes bibliographical references and index.
ISBN 0-8039-4540-X (alk. paper)
1. Regression analysis. 2. Linear models (Statistics) 3. Social sciences—Statistical methods. I. Title.
HA31.3.FR69 1997
300′.01′519536—dc21 96-38084

10 9 8 7

Acquiring Editor: C. Deborah Laughton
Editorial Assistant: Eileen Carr
Production Editor: Astrid Virding
Production Assistant: Denise Santoyo
Book Designer: Ravi Balasuriya
Typesetter: Technical Typesetting, Inc.
Cover Designer: Lesa Valdez
Print Buyer: Anna Chin

For Bonnie and Jesse

Contents

Preface

Linear models, their variants, and extensions are among the most useful and widely used statistical tools for social research. This book aims to provide an accessible, in-depth, modern treatment of regression analysis, linear models, and closely related methods.

The book should be of interest to students and researchers in the social sciences. Although the specific choice of methods and examples reflects this readership, I expect that the book will prove useful in other disciplines that employ linear models for data analysis, and in courses on applied regression and linear models where the subject matter of applications is not of special concern.

This book is a revision of my 1984 text *Linear Statistical Models and Related Methods*. In revising the book, I have freely incorporated material from my 1991 monograph on *Regression Diagnostics*.

The new title of the text reflects a change in organization and emphasis: I have thoroughly reworked the book, removing some topics and adding a variety of new material. Even more fundamentally, however, the book has been extensively rewritten. It is a new and different book.

I have endeavored, in particular, to make the text as accessible as possible. With the exception of three chapters, several sections, and a few shorter passages, the prerequisite for reading the book is a course in basic applied statistics that covers the elements of statistical data analysis and inference.

Many topics (e.g., logistic regression in Chapter 15) are introduced with an example that motivates the statistics or (as in the case of bootstrapping, in Chapter 16) by appealing to familiar material. The treatment of regression analysis starts (in Chapter 2) with an elementary discussion of nonparametric regression, developing the notion of regression as a conditional average—in the absence of restrictive assumptions about the nature of the relationship between the dependent and independent variables. This approach begins closer to the data than the traditional starting point of linear least-squares regression, and should make readers skeptical about glib assumptions of normality, constant variance, and so on.

More difficult chapters and sections are marked with asterisks. These parts of the text can be omitted without loss of continuity, but they provide greater understanding and depth, along with coverage of some topics that depend on more extensive mathematical or statistical background. I do not, however, wish to exaggerate the background that is required for this "more difficult" material: All that is necessary is some exposure to matrices, elementary linear algebra, and elementary differential calculus. Appendices to the text provide additional background material.

All chapters end with summaries, and most include recommendations for additional reading. Summary points are also presented in boxes interspersed with the text.

Synopsis

Part I

The first part of the book consists of preliminary material:

Chapter 1 discusses the role of statistical data analysis in social science, expressing the point of view that statistical models are essentially descriptive, not direct (if abstract) representations of social processes. This perspective provides the foundation for the data-analytic focus of the text.

Chapter 2 introduces the notion of regression analysis as tracing the conditional distribution of a "dependent" variable as a function of one or several "independent" variables. This idea is initially explored "nonparametrically," in the absence of a restrictive statistical model for the data.

Chapter 3 describes a variety of graphical tools for examining data. These methods are useful both as a preliminary to statistical modeling and to assist in the diagnostic checking of a model that has been fit to data.

Chapter 4 discusses variable transformation as a solution to several sorts of problems commonly encountered in data analysis, including skewness, nonlinearity, and nonconstant spread.

Part II

The second part, on linear models fit by the method of least squares, comprises the heart of the book:

Chapter 5 discusses linear least-squares regression. Linear regression is the prototypical linear model, and its direct extension is the subject of Chapters 7 through 10.

Chapter 6, on statistical inference in regression, develops tools for testing hypotheses and constructing confidence intervals that apply generally to linear models. This chapter also introduces the basic methodological distinction between empirical and structural relationships—a distinction central to understanding causal inference in nonexperimental research.

Chapter 7 shows how "dummy variables" can be employed to extend the regression model to qualitative independent variables. Interactions among independent variables are introduced in this context.

Chapter 8, on analysis-of-variance models, deals with linear models in which all of the independent variables are qualitative.

*Chapter 9** develops the statistical theory of linear models, providing the foundation for much of the material in Chapters 5 through 8 along with some additional results.

*Chapter 10** applies vector geometry to linear models, allowing us literally to visualize the structure and properties of these models. Many topics are revisited from the geometric perspective, and central concepts—such as "degrees of freedom"—are given a natural and compelling interpretation.

Part III

The third part of the book describes "diagnostic" methods for discovering whether a linear model fit to data adequately represents the data. Methods are also presented for correcting problems that are revealed:

Chapter 11 deals with the detection of unusual and influential data in linear models.

Chapter 12 describes methods for diagnosing a variety of problems, including non-normally distributed errors, nonconstant error variance, and nonlinearity. Some more advanced material in this chapter discusses how the method of maximum likelihood can be employed for selecting transformations.

Chapter 13 takes up the problem of collinearity—the difficulties for estimation that ensue when the independent variables are highly correlated.

Part IV

The fourth part of the book discusses important extensions of linear least squares. In selecting topics, I was guided by the proximity of the methods to the general linear model and by the promise that these methods hold for data analysis in the social sciences. The methods described in this part of the text are (with the exception of logistic regression in Chapter 15) given introductory—rather than extensive—treatments. My aim in introducing these relatively advanced topics is (i) to provide enough information so that readers can begin to use these methods in their research; and (ii) to provide sufficient background to support further work in these areas should readers choose to pursue them:

*Chapter 14** describes several important direct extensions of linear least squares: Time series regression (and generalized least squares), where the observations are ordered in time, and the errors in the linear model are, consequently, not assumed to be independent; nonlinear models fit by least squares; robust estimation of linear models (i.e., using methods of estimation more resistant than least squares to unusual data); and nonparametric regression, in which the functional form of the relationship between the dependent and independent variables is not specified in advance.

Chapter 15 takes up the centrally important topic of linear-like models for qualitative and ordinal dependent variables, most notably, logit models (logistic regression). The chapter concludes with a brief introduction to generalized linear models, a grand synthesis encompassing linear least-squares regression, logistic regression, and a variety of related statistical models.

Chapter 16 discusses two broadly applicable techniques for assessing sampling variation: the "bootstrap" and cross-validation. The bootstrap is a computationally intensive simulation method for constructing confidence intervals and hypothesis tests. The bootstrap does not make strong distributional assumptions about the data, and can be made to reflect the manner in which the data were collected (e.g., in complex survey-sampling designs). Cross-validation is a simple method for drawing honest statistical inferences when—as is commonly the case—the data are employed both to select a statistical model and to estimate its parameters.

Appendices

Several appendices provide background, principally—but not exclusively—for the starred portions of the text:

Appendix A describes the notational conventions employed in the text.

*Appendix B** shows how vector geometry can be used to visualize key concepts of linear algebra. The material in this appendix is required for Chapter 10 and presupposes a basic knowledge of matrices and linear algebra.

*Appendix C** assumes an acquaintance with elementary differential calculus of one independent variable and shows how, employing matrices, differential calculus can be extended to several independent variables. This material is required for some starred portions of the text, for example, the derivation of least-squares and maximum-likelihood estimators.

Appendix D provides an introduction to the elements of probability theory and to basic concepts of statistical estimation and inference. A few, more demanding, parts of the appendix are starred. The background developed in this appendix is required for some of the material on statistical inference in the text.

Computing

Nearly all of the examples in this text employ real data from the social sciences, many of them previously analyzed and published. The exercises that involve data analysis also almost all use real data drawn from various areas of application. Most of the datasets are relatively small. I encourage readers to analyze their own data as well.

The datasets can be downloaded free of charge via the World Wide Web; point your web browser at `http://www.sagepub.com/authors_home_page.htm`. If you do not have access to the Internet, then you can write to me at the Department of Sociology, McMaster University, Hamilton, Ontario, Canada L8S 4L8, for information about obtaining the datasets on disk. Each dataset is associated with two files: The file with extension `.cbk` (e.g., `duncan.cbk`) contains a "codebook" describing the data; the file with extension `.dat` (e.g., `duncan.dat`) contains the data themselves. Smaller datasets are also presented in tabular form in the text.

I occasionally comment in passing on computational matters, but the book generally ignores the finer points of statistical computing in favor of methods that are computationally simple. I feel that this approach facilitates learning. Once basic techniques are absorbed, an experienced data analyst has recourse to carefully designed programs for statistical computations.

I think that it is a mistake to tie a general discussion of linear and related statistical models too closely to particular software. Although the marvelous proliferation of statistical software has routinized the computations for most of the methods described in this book, the workings of standard computer programs are not sufficiently accessible to promote learning. I, consequently, find it desirable, where time permits, to teach the use of a statistical computing environment as part of a course on applied regression and linear models.

For nearly 20 years, I used the interactive programming language APL in this role. More recently, I use Lisp-Stat. I particularly recommend the R-code program, written in Lisp-Stat, which accompanies Cook and Weisberg's (1994) fine book on regression graphics. Other programmable computing environments that are used for statistical data analysis include S, Gauss, Stata, Mathematica, and the interactive matrix language (IML) in SAS. Descriptions of these environments appear in a book edited by Stine and Fox (1996).

To Readers, Students, and Instructors

I have used the material in this book for two types of courses (along with a variety of short courses and lectures):

- I cover the unstarred sections of Chapters 1–8, 11–13, 15, and 16 in a one-semester (13-week) course for social-science graduate students (at McMaster University in Hamilton, Ontario) who have had (at least) a one-semester introduction to statistics at the level of Moore (1995). The outline of this course is as follows:

Week	*Topic*	*Readings*
1	Introduction to the course and to MS/Windows	Chap. 1
2	Introduction to regression, Lisp-Stat, and R-code	Chap. 2, App. A, D
3	Examining and transforming data	Chap. 3, 4
4	Linear least-squares regression	Chap. 5
5	Statistical inference for regression	Chap. 6
6	Dummy-variable regression	Chap. 7
7	Analysis of variance	Chap. 8
8	Diagnostics I: Unusual and influential data	Chap. 11
9	Diagnostics II: Nonlinearity and other ills	Chap. 12
10	Diagnostics III: Collinearity and variable selection	Chap. 13
11	Logit and probit models I	Chap. 15
12	Logit and probit models II	Chap. 15 (cont.)
13	Assessing sampling variation and review	Chap. 16

  The readings from this text are supplemented with parts of Cook and Weisberg's (1994) book on regression graphics, a paper by Tierney (1995) on Lisp-Stat, and several handouts on computing. Students complete required weekly homework assignments, which are mostly focused on data analysis. Homework is collected and corrected, but not graded. I distribute answers when the homework is collected, and take it up in class after it is corrected and returned. There are midterm and final take-home exams, also focused on data analysis.

- I used the material in Chapters 1–13 and 15, along with the appendices and basic introductions to matrices, linear algebra, and calculus, for a two-semester course for social-science graduate students (at York University in Toronto) with similar statistical preparation. For this second, more intensive, course, background topics (such as linear algebra) were introduced as required, and constituted about one-fifth of the course. The organization of the course was similar to the first one.

Both courses include some treatment of statistical computing, with more information on programming in the second course. For students with the requisite mathematical and statistical background, it should be possible to cover the whole text in a reasonably paced two-semester course.

In learning statistics, it is important for the reader to participate actively, both by working though the arguments presented in the book, and—even more importantly—by applying methods to data. Statistical data analysis is a craft and, like any craft, it requires practice. Reworking examples is a good place to start, and I have presented illustrations in such a manner as to make reanalysis and further analysis possible. Where possible, I have relegated formal "proofs" and derivations to exercises, which, nevertheless, typically provide some guidance to the reader. I believe that this type of material is best learned constructively.

As well, including too much algebraic detail in the body of the text invites readers to lose the statistical forest for the mathematical trees. You can decide for yourself (or your students) whether or not to work the theoretical exercises. It is my experience that some people feel that the process of working through derivations cements their understanding of the statistical material, while others find this activity tedious and pointless. Some of the theoretical exercises, marked with asterisks, are comparatively difficult. (Difficulty is assessed relative to the material in the text, so the threshold is higher in starred sections and chapters.)

In preparing the data-analytic exercises, I have tried to find datasets of some intrinsic interest that embody a variety of characteristics. You can safely assume, for example, that datasets for exercises in Chapter 11 include unusual data. In many instances, I try to supply some direction in the data-analytic exercises, but—like all real data analysis—these exercises are fundamentally open ended. It is, therefore, important for instructors to set aside time to discuss data-analytic exercises in class, both before and after students tackle them. Although students often miss important features of the data in their initial analyses, this experience—properly approached and integrated—is an unavoidable part of learning the craft of data analysis.

A few exercises, marked with pound signs (#) are meant for "hand" computation. Hand computation (i.e., with a calculator) is tedious, and is practical only for unrealistically small problems, but it sometimes serves to make statistical procedures more concrete. Similarly, despite the emphasis in the text on analyzing real data, a small number of exercises generate simulated data to clarify certain properties of statistical methods.

I struggled with the placement of cross references to exercises and to other parts of the text, trying brackets [too distracting!], marginal boxes (too imprecise), and finally settling on traditional footnotes.[1] I suggest that you ignore both the cross references and the other footnotes[2] on the first reading of the text.

---

[1] Footnotes are a bit awkward, but you do not have to read them.

[2] Footnotes other than cross references generally develop small points and elaborations.

Finally, a word about style: I try to use the first person singular—"I"—when I express opinions. "We" is reserved for you—the reader—and me.

Acknowledgments

Many individuals have helped me in the preparation of this book.

I am grateful to Georges Monette of York University, to Bob Stine of the University of Pennsylvania, and to two anonymous reviewers, for insightful comments and suggestions.

Mike Friendly of York University provided detailed comments, corrections, and suggestions on almost all of the text.

A number of friends and colleagues donated their data for illustrations and exercises—implicitly subjecting their research to scrutiny and criticism.

Several individuals contributed to this book indirectly by helpful comments on its predecessor (Fox, 1984), both before and after publication: Ken Bollen, Gene Denzel, Shirley Dowdy, Paul Herzberg, and Doug Rivers.

Edward Ng capably assisted in the preparation of some of the figures that appear in the book.

C. Deborah Laughton, my editor at Sage Publications, has been patient and supportive throughout the several years that I have worked on this project.

I am also in debt to the students at York University and McMaster University, and to participants at the Inter-University Consortium for Political and Social Research Summer Program, all of whom were exposed to various versions and portions of this text and who have improved it through their criticism, suggestions, and—occasionally—informative incomprehension.

Finally, I am grateful to York University for providing me with a sabbatical-leave research grant during the 1994–95 academic year, when much of the text was drafted.

If, after all of this help, deficiencies remain, then I alone am at fault.

John Fox
Toronto, Canada

PART I

Preliminaries

1

Statistics and Social Science

The social world is exquisitely complex and rich. From the improbable moment of birth, each of our lives is governed by chance and contingency. The statistical models typically used to analyze social data—and, in particular, the models considered in this book—are, by contrast, ludicrously simple. How can simple statistical models help us to understand a complex social reality?

This is a book on data analysis and statistics, not on the philosophy of the social sciences. I shall, therefore, address this question, and related issues, very briefly here. Nevertheless, I feel that it is useful to begin with a consideration of the role of data analysis in the larger process of social research. You need not agree with the point of view that I express in this chapter in order to make productive use of the statistical tools presented in the remainder of the book; but the emphasis and specific choice of methods in this text partly reflect the ideas in this chapter. You may wish to reread this material after you study the methods described in the sequel.

1.1 Statistical Models and Social Reality

As I said, social reality is complex: Consider how my income is "determined." I am a relatively well paid professor at a Canadian university, teaching sociology and statistics. That the billiard ball of my life would fall into this particular pocket was, however, hardly predictable 30 years ago, when I was attending a science high school in New York City. My subsequent decision to study sociology at New York's City College (after several other majors), my interest in statistics (the consequence of a course taken without careful consideration in my senior year), my decision to attend graduate school in sociology at the University of Michigan (one of several more or less equally attractive possibilities), and the opportunity and desire to move to Canada (the vote to hire me at the University of Alberta was, I later learned, very close) are all events that could easily have occurred differently.

I do not mean to imply that personal histories are completely capricious, unaffected by social structures of race, ethnicity, class, gender, and so on, just that they are not in detail determined by these structures. That social structures—and other sorts of systematic factors—condition, limit, and encourage specific events is clear from each of the illustrations in the previous paragraph, and in fact makes sense of the argument for the statistical analysis of social data presented below. To take a particularly gross example: The public high school that I attended admitted its students by competitive examination, but no young women could apply.

Each of these precarious occurrences clearly affected my income, as have other events—some significant, some small—too numerous and too tedious to mention, even if I were aware of them all. If, for some perverse reason, you were truly interested in my income (and, perhaps, in other matters more private), you could study my biography, and through that study arrive at a detailed (if inevitably incomplete) understanding. It is clearly impossible, however, to pursue this strategy for many individuals or, more to the point, for individuals in general.

Nor is an understanding of income in general inconsequential, because income inequality is an important feature of our society. If such an understanding hinges on a literal description of the process by which each of us receives an income, then the enterprise is clearly hopeless. We might, alternatively, try to capture significant features of the process in general without attempting to predict the outcome for specific individuals. One could draw formal analogies (largely unproductively, I expect) to chaotic physical processes, such as the determination of weather and earthquakes.

Concrete mathematical theories purporting to describe social processes sometimes appear in the social sciences (e.g., in economics and in some areas of psychology), but they are relatively rare.[1] If a theory, like Newton's laws of motion, is mathematically concrete then, to be sure, there are difficulties in applying and testing it; but, with some ingenuity, experiments and observations can be devised to estimate the free parameters of the theory (a gravitational constant, for example), and to assess the fit of the theory to the resulting data.

In the social sciences, verbal theories abound. These social theories tend to be vague, elliptical, and highly qualified. Often, they are, at least partially, a codification of "common sense." I believe that vague social theories are potentially useful abstractions for understanding an intrinsically complex social reality, but how can such theories be linked empirically to that reality?

A vague social theory may lead us to expect, for example, that racial prejudice is the partial consequence of an "authoritarian personality," which, in turn, is a product of rigid child rearing. Each of these terms requires elaboration and

---

[1] The methods for fitting nonlinear models described in Section 14.2 are sometimes appropriate to the rare theories in social science that are mathematically concrete.

procedures of assessment or measurement. Other social theories may lead us to expect that higher levels of education should be associated with higher levels of income, perhaps because the value of labor power is enhanced by training, or, alternatively, because occupations requiring higher levels of education are of greater functional importance. In either case, we need to consider how to assess income and education, how to examine their relationship, and what other factors need to be included.[2]

Statistical models of the type considered in this book are grossly simplified descriptions of complex social reality. Imagine that we have data from a social survey of a large sample of employed individuals. Imagine further, anticipating the statistical methods described in subsequent chapters, that we regress these individuals' income on a variety of putatively relevant characteristics, such as their level of education, gender, race, region of residence, and so on. We recognize that a model of this sort will fail to account perfectly for individuals' incomes, so our model includes a "residual," meant to capture the component of income unaccounted for by the systematic part of the model, which incorporates the "effects" on income of education, gender, and so forth.

The residuals for our model are likely very large. Even if the residuals were small, however, we should still need to consider the relationships among our social "theory," the statistical model that we have fit to the data, and the social "reality" that we seek to understand. Social reality, along with our methods of observation, produce the data, our theory aims to explain the data, and the model to describe them.

I believe that a statistical model cannot, and is not literally meant to, capture the social process by which incomes are "determined." As I argued above, individuals receive their incomes as a result of their almost unimaginably complex personal histories. No regression model, not even one including a residual, can reproduce this process: It is not as if my income is partly determined by my education, gender, race, and so on, and partly by the detailed trajectory of my life. It is, therefore, not sensible, at the level of real social processes, to relegate chance and contingency to a random term that is simply added to the systematic part of a statistical model. The unfortunate tendency to reify statistical models—to forget that they are descriptive summaries, not literal accounts of social processes—can only serve to discredit quantitative data analysis in the social sciences.

Nevertheless, and despite the rich chaos of individuals' lives, social theories imply a structure to income inequality. Statistical models are capable of capturing and describing that structure, or at least significant aspects of it. Moreover, social research is often motivated by questions rather than by hypotheses: Has income inequality between men and women changed recently? Is there a relationship between public concern over crime and the level of crime? Data analysis can help to answer these questions, which frequently are of practical—as well as theoretical—concern. Finally, if we proceed carefully, data analysis can assist us in the discovery of social facts that initially escape our hypotheses and questions.

It is, in my view, a paradox that the statistical models that are at the heart of most modern quantitative social science are at once taken too seriously and not seriously enough by many practitioners of social science. On the one

[2] See Section 1.2.

hand, social scientists write about simple statistical models as if they were direct representations of the social processes that they purport to describe. On the other hand, there is frequently a failure to attend to the descriptive accuracy of these models.

As a shorthand, reference to the "effect" of education on income is innocuous. That the shorthand often comes to dominate the interpretation of statistical models is reflected, for example, in much of the social science literature that employs structural-equation models (sometimes called "causal models"). There is, I believe, a valid sense in which income is "affected" by education, because the complex real process by which individuals" incomes are determined is partly conditioned by their levels of education, but—as I have argued above—one should not mistake the model for the process.[3]

Although statistical models are very simple in comparison with social reality, they typically incorporate strong claims about the descriptive pattern of data. These claims rarely reflect the substantive social theories, hypotheses, or questions that motivate the use of the statistical models, and they are very often wrong. For example, it is common in social research to assume a priori, and without reflection, that the relationship between two variables, such as income and education, is linear. Now, we may well have good reason to believe that income tends to be higher at higher levels of education, but there is no reason to suppose that this relationship is linear. Our practice of data analysis should reflect our ignorance as well as our knowledge.

A statistical model is of no practical use if it is an inaccurate description of social reality, and we shall, therefore, pay close attention to the descriptive accuracy of statistical models. Unhappily, the converse is not true, for a statistical model may be descriptively accurate but of little practical use; it may even be descriptively accurate but substantively misleading. We shall explore these issues briefly in the next two sections, which tie the interpretation of statistical models to the manner in which data are collected.

> With few exceptions, statistical data analysis describes the outcomes of real social processes and not the processes themselves. It is therefore important to attend to the descriptive accuracy of statistical models, but to refrain from reifying them.

abstract vs. concrete

[3] There is the danger here of simply substituting one term ("conditioned by") for another ("affected by"), but the point is deeper than that: Education affects income because the choices and constraints that partly structure individuals' lives change systematically with their level of education. Many highly paid occupations in our society are closed to individuals who lack a university education, for example. To recognize this fact, and to examine its descriptive reflection in a statistical summary, is different from claiming that a university education literally adds an increment to individuals' incomes.

1.2 Observation and Experiment

It is common for (careful) introductory accounts of statistical methods (e.g., Freedman et al., 1991; Moore and McCabe, 1993) to distinguish strongly between observational and experimental data. According to the standard distinction, causal inferences are justified (or, at least, more certain) in experiments, where the independent variables (i.e., the possible "causes") are under the direct control of the researcher; causal inferences are particularly compelling in a randomized experiment, in which the values of independent variables are assigned by some chance mechanism to experimental units. In nonexperimental research, in contrast, the values of the independent variables are observed—not assigned—by the researcher, along with the value of the dependent variable (the "effect"), and causal inferences are not justified (or, at least, are less certain). I believe that this account is essentially correct, but requires qualification and elaboration.

To fix ideas, let us consider the data summarized in Table 1.1, drawn from a paper by Greene and Shaffer (1992) on Canada's refugee determination process. This table shows the outcome of 608 cases, filed in 1990, in which refugee claimants who were turned down by the Immigration and Refugee Board asked the Federal Court of Appeal for leave to appeal the board's determination. In each case, the decision to grant or deny leave to appeal was made by a single judge. It is clear from the table that the 12 judges who heard these cases differed widely in the percentages of cases that they granted leave to appeal. Employing a standard significance test for a contingency table (a chi-square test of independence), Greene and Shaffer calculated that a relationship as strong as the one in the table will occur by chance alone about two times in 100,000. These data became the basis for a court case contesting the fairness of the Canadian refugee determination process.

If the 608 cases had been assigned at random to the judges, then the data would constitute a natural experiment, and we could unambiguously conclude that the large differences among the judges reflect differences in their propensities to grant leave to appeal.[4] The cases were, however, assigned to the judges not randomly but on a rotating basis, with a single judge hearing all of the cases that arrived at the court in a particular week. In defending the current refugee determination process, expert witnesses for the Crown argued that the observed differences among the judges might therefore be due to factors that systematically differentiated the cases that different judges happened to hear.

It is possible, in practice, to "control" statistically for such extraneous "confounding" factors as may explicitly be identified,[5] but it is not, in principle, possible to control for all relevant factors, because we can never be certain that all relevant factors have been identified. Nevertheless, I would argue, the data in Table 1.1 establish a prima facie case for systematic differences in the judges' propensities. Careful researchers control statistically for potentially rel-

---

[4] Even so, this inference is not reasonably construed as a representation of the cognitive process by which judges arrive at their determinations. Following the argument in the previous section, it is unlikely that we could ever trace out that process in detail; it is quite possible, for example, that a specific judge would make different decisions faced with the same case on different occasions.

[5] See the further discussion of the refugee data in Exercise 15.11 and in Section 16.2.

TABLE 1.1 Percentages of Refugee Claimants in 1990 Who Were Granted or Denied Leave to Appeal a Negative Decision of the Canadian Immigration and Refugee Board, Classified by the Judge Who Heard the Case.

Judge	*Leave Granted?*		*Total*	*Number of Cases*
	Yes	*No*		
Pratte	9	91	100	57
Linden	9	91	100	32
Stone	12	88	100	43
Iacobucci	12	88	100	33
Décary	20	80	100	80
Hugessen	26	74	100	65
Urie	29	71	100	21
MacGuigan	30	70	100	90
Heald	30	70	100	46
Mahoney	34	66	100	44
Marceau	36	64	100	50
Desjardins	49	51	100	47
All judges	25	75	100	608

Source: Adapted from Greene and Shaffer (1992, Table 1) *International Journal of Refugee Law*, by permission of Oxford University Press.

evant factors that they can identify; cogent critics demonstrate that an omitted confounding factor accounts for the observed association between judges and decisions, or at least argue persuasively that a specific factor may be responsible for this association—they do not simply maintain the abstract possibility that such a factor may exist.

What makes an omitted factor "relevant" in this context?[6]

1. The omitted factor must influence the dependent variable. For example, if the gender of the refugee applicant has no impact on the judges' decisions, then it is irrelevant to control statistically for gender.
2. The omitted factor must be related as well to the independent variable that is the focus of the research. Even if the judges' decisions are influenced by the gender of the applicants, the relationship between outcome and judge will be unchanged by controlling for gender (e.g., by looking separately at male and female applicants) unless the gender of the applicants is also related to judges—that is, unless the different judges heard cases with substantially different proportions of male and female applicants.

The strength of randomized experimentation derives from the second point: If cases were randomly assigned to judges, then there would be no systematic tendency for them to hear cases with differing proportions of men and women—or, for that matter, with systematic differences of any kind.

[6] These points are developed more formally in Sections 6.3 and 9.6.

It is, however, misleading to conclude that causal inferences are completely unambiguous in experimental research, even within the bounds of statistical uncertainty (expressed, for example, in the *p*-value of a statistical test). Although we can unambiguously ascribe an observed difference to an experimental manipulation, we cannot unambiguously identify that manipulation with the independent variable that is the focus of our research.

In a randomized drug study, for example, in which patients are prescribed a new drug or an inactive placebo, we may establish with virtual certainty that there was greater average improvement among those receiving the drug, but we cannot be sure that this difference is due (or solely due) to the active ingredient in the drug. Perhaps the experimenters inadvertently conveyed their enthusiasm for the drug to the patients who received it, influencing the patients' responses; or perhaps the bitter taste of the drug subconsciously convinced these patients of its potency.

Experimenters try to rule out alternative interpretations of this kind by following careful experimental practices, such as "double-blind" delivery of treatments (neither the subject nor the experimenter knows whether the subject is administered the drug or the placebo), and by holding constant potentially influential factors deemed to be extraneous to the research (the taste, color, shape, etc., of the drug and placebo are carefully matched). One can never be certain, however, that all relevant factors are held constant in this manner. Although the degree of certainty achieved is typically much greater in a randomized experiment than in an observational study, the distinction is less clear-cut than it at first appears.

> Causal inferences are most certain—if not completely definitive—in randomized experiments, but observational data can also be reasonably marshaled as evidence of causation. Good experimental practice seeks to avoid confounding experimentally manipulated independent variables with other factors that can influence the dependent variable. Sound analysis of observational data seeks to control statistically for potentially confounding factors.

In subsequent chapters, we shall have occasion to examine observational data on the prestige, educational levels, and income levels of occupations. It will materialize that occupations with higher levels of education tend to have higher prestige, and that occupations with higher levels of income also tend to have higher prestige. The income and educational levels of occupations are themselves positively related. As a consequence, when education is controlled statistically, the relationship between prestige and income grows smaller; likewise, when income is controlled, the relationship between prestige and education grows smaller. In neither case, however, does the relationship disappear.

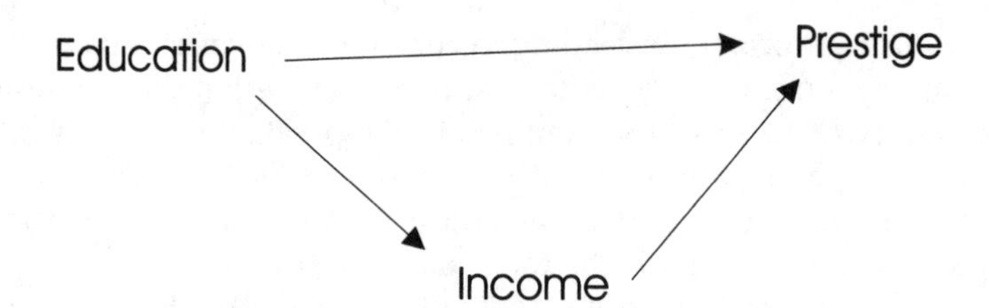

Figure 1.1. Simple "causal model" relating education, income, and prestige of occupations. Education is a common prior cause of both income and prestige; income intervenes causally between education and prestige.

How are we to understand the pattern of statistical associations among the three variables? It is helpful in this context to entertain an informal "causal model" for the data, as in Figure 1.1. That is, the educational level of occupations influences (potentially) both their income level and their prestige, while income potentially influences prestige. The association between prestige and income is "spurious" (i.e., not causal) to the degree that it is a consequence of the mutual dependence of these two variables on education; the reduction in this association when education is controlled represents the removal of the spurious component. In contrast, the causal relationship between education and prestige is partly mediated by the "intervening variable" income; the reduction in this association when income is controlled represents the articulation of an "indirect" effect of education on prestige (i.e., through income).

In the former case, we partly *explain away* the association between income and prestige: Part of the relationship is "really" due to education. In the latter case, we partly *explain* the association between education and prestige: Part of the relationship is mediated by income.

> In analyzing observational data, it is important to distinguish between a factor that is a common prior cause of an independent and dependent variable and a factor that intervenes causally between the two.

Causal interpretation of observational data is always risky, particularly—as here—when the data are cross-sectional (i.e., collected at one point in time) rather than longitudinal (where the data are collected over time). Nevertheless, it is usually impossible, impractical, or immoral to collect experimental data in the social sciences,[7] and longitudinal data are often hard to come by. Moreover, the essential difficulty of causal interpretation in nonexperimental investigations—

[7] Experiments with human beings also frequently distort the processes that they purport to study: Although it might well be possible, for example, to recruit judges to an experimental study of judicial decision-making, the artificiality of the situation could easily affect their simulated decisions. Even if the study entailed real judicial judgments, the mere act of observation might influence the judges' decisions.

due to potentially confounding factors that are left uncontrolled—applies to longitudinal as well as to cross-sectional observational data.

The notion of "cause" and its relationship to statistical data are notoriously difficult ideas. A relatively strict view requires an experimentally manipulable independent variable, at least one that is manipulable in principle. This is a particularly sticky point because, in social science, many independent variables are intrinsically not subject to direct manipulation, even in principle. Thus, for example, according to the strict view, gender cannot be considered a cause of income, even if it can be shown (perhaps after controlling for other determinants of income) that men and women systematically differ in their incomes, because an individual's gender cannot be changed.[8]

I believe that treating nonmanipulable independent variables, such as gender, as potential causes is, at the very least, a useful shorthand. Men earn higher incomes than women be*cause* women are (by one account) concentrated into lower-paying jobs, work fewer hours, are directly discriminated against, and so on (see, e.g., Ornstein, 1983). Explanations of this sort are perfectly reasonable, and are subject to statistical examination; the sense of "cause" here may be weaker than the narrow one, but it is nevertheless useful.

> It is overly restrictive to limit the notion of statistical causation to independent variables that are manipulated experimentally, to independent variables that are manipulable in principle, or to data that are collected over time.

1.3 Populations and Samples

Statistical inference is typically introduced in the context of random sampling from an identifiable population. There are good reasons for stressing this interpretation of inference—not the least of which are its relative concreteness and clarity—but the application of statistical inference is, at least arguably, much broader, and it is certainly broader in practice.

Take, for example, a prototypical experiment, in which subjects are assigned values of the independent variables at random: Inferences may properly be made to the hypothetical population of random rearrangements of the subjects, even when these subjects are not sampled from some larger population. If, for example, we find a highly "statistically significant" difference between two experimental groups of subjects in a randomized experiment, then we can

[8] This statement is, of course, arguable: There are historically many instances in which individuals have changed their gender, for example by disguise, not to mention surgery. Despite some fuzziness, however, I believe that the essential point—that some independent variables are not (normally) subject to manipulation—is valid.

be sure, with practical certainty, that the difference was due to the experimental manipulation. The rub here is that our interest almost surely extends beyond this specific group of subjects to some larger—often ill-defined—population.

Even when subjects in an experimental or observational investigation are literally sampled at random from a real population, we usually wish to generalize beyond that population. There are exceptions—election polling comes immediately to mind—but our interest is seldom confined to the population that is directly sampled. This point is perhaps clearest when no sampling is involved—that is, when we have data on every individual in a real population.

Suppose, for example, that we examine data on population density and crime rates for all large U. S. cities and find only a weak association between the two variables. Suppose further that a standard test of statistical significance indicates that this association is so weak that it easily could have been the product of "chance."[9] Is there any sense in which this information is interpretable? After all, we have before us data on the entire population of large U.S. cities at a particular historical juncture.

Because our interest inheres not directly—at least not exclusively—in these specific cities, but in the complex social processes by which density and crime are determined, we can reasonably imagine a different outcome. Were we to replay history conceptually, we would not observe precisely the same crime rates and population density statistics, dependent as these are on a myriad of contingent and chancy events; indeed, if the ambit of our conceptual replay of history is sufficiently broad, the identities of the cities themselves might change. (Imagine, for example, that Henry Hudson had not survived his trip or, if he survived it, that the capitol of the United States had remained in New York. Less momentously, imagine that Fred Smith had not gotten drunk and killed a friend in a brawl, reducing the number of homicides in New York by one.) It is, in this context, reasonable to draw statistical inferences to the process that produced the currently existing population. Similar considerations arise in the analysis of historical statistics, for example of time series data.[10]

Much interesting data in the social sciences—and elsewhere—are collected haphazardly. The data constitute neither a sample drawn at random from a larger population, nor a coherently defined population. Experimental randomization provides a basis for making statistical inferences to the population of rearrangements of a haphazardly selected group of subjects, but that is in itself cold comfort. For example, an educational experiment is conducted with students recruited from a school that is conveniently available. We are interested in drawing conclusions about the efficacy of teaching methods for students in general, however, not just for the students who participated in the study.

Haphazard data are also employed in many observational studies—for example, volunteers are recruited from among university students to study the association between eating disorders and overexercise. Once more, our interest transcends this specific group of volunteers.

To rule out haphazardly collected data would be a terrible waste; it is, instead, prudent to be careful and critical in the interpretation of the data. We should try, for example, to satisfy ourselves that our haphazard group does not

[9] Cf. the critical discussion of crime and population density in Freedman (1975).

[10] See Section 14.1 for a discussion of regression analysis with time series data.

differ in presumably important ways from the larger population of interest, or to control for factors thought to be relevant to the phenomena under study.

Statistical inference can speak to the internal stability of patterns in haphazardly collected data and—most clearly in experimental data—to causation. Generalization from haphazardly collected data to a broader population, however, is inherently a matter of judgment.

> Randomization and good sampling design are desirable in social research, but they are not prerequisites for drawing statistical inferences. Even when randomization or random sampling is employed, we typically want to generalize beyond the strict bounds of statistical inference.

1.4 Summary

- With few exceptions, statistical data analysis describes the outcomes of real social processes and not the processes themselves. It is therefore important to attend to the descriptive accuracy of statistical models, but to refrain from reifying them.
- Causal inferences are most certain—if not completely definitive—in randomized experiments, but observational data can also be reasonably marshaled as evidence of causation. Good experimental practice seeks to avoid confounding experimentally manipulated independent variables with other factors that can influence the dependent variable. Sound analysis of observational data seeks to control statistically for potentially confounding factors.
- In analyzing observational data, it is important to distinguish between a factor that is a common prior cause of an independent and dependent variable and a factor that intervenes causally between the two.
- It is overly restrictive to limit the notion of statistical causation to independent variables that are manipulated experimentally, to independent variables that are manipulable in principle, or to data that are collected over time.
- Randomization and good sampling design are desirable in social research, but they are not prerequisites for drawing statistical inferences. Even when randomization or random sampling is employed, we typically want to generalize beyond the strict bounds of statistical inference.

1.5 Recommended Reading

- Chance and contingency are recurrent themes in Stephen Gould's fine essays on natural history; see, in particular, Gould (1989). Gould's work has strongly influenced the presentation in Section 1.1.

- The legitimacy of causal inferences in nonexperimental research is a hotly debated topic. Perhaps the most vocal current critic of the use of observational data is David Freedman. See, for example, Freedman's (1987) critique of structural-equation modeling in the social sciences and the commentary that follows it.
- The place of sampling and randomization in statistical investigations has also been widely discussed and debated in the literature on research design. The classic presentation of the issues in Campbell and Stanley (1963) is still worth reading, as is Kish (1987). In statistics, these themes are reflected in the distinction between model-based and design-based inference (see, e.g., Koch and Gillings, 1983) and in the notion of superpopulation inference (see, e.g., Thompson, 1988).
- Achen (1982) argues eloquently for the descriptive interpretation of statistical models, illustrating his argument with effective examples.

2

What Is Regression Analysis?

Figure 2.1 is a *scatterplot* showing the relationship between income (in thousands of dollars) and formal education (in years) for the population of employed Canadians. Unlike most of the examples in this book, the "data" in Figure 2.1 are fictitious: There are, of course, millions of employed Canadians, but only 2600 points in the plot. The line in the plot shows the mean value of income for each level of education.

Regression analysis,[1] broadly construed, traces the distribution of a dependent variable (denoted by Y)—or some characteristic of this distribution (such as its mean)—as a function of one or more independent variables ($X_1, \ldots, X_k$):

$$p(y|x_1, \ldots, x_k) = f(x_1, \ldots, x_k) \qquad [2.1]$$

Here, $p(y|x_1, \ldots, x_k)$ represents the probability (or, for continuous Y, the probability density) of observing the specific value y of the dependent variable, *conditional* upon a set of specific values $(x_1, \ldots, x_k)$ of the independent variables; $p(Y|x_1, \ldots, x_k)$ is the *probability distribution of* Y for these specific values of the X's. In Figure 2.1, for example, where there is only one X (education), $p(Y|x)$ represents the population distribution of incomes for all individuals who share the specific value x of education (e.g., 12 years).[2]

[1] See Exercise 5.2 for an explanation of the archaic term "regression."

[2] If the concept of (or notation for) a conditional distribution is unfamiliar, you should consult Appendix D, Section D.1 on elementary probability theory. Please keep in mind that a variety of background information is located in the appendices.

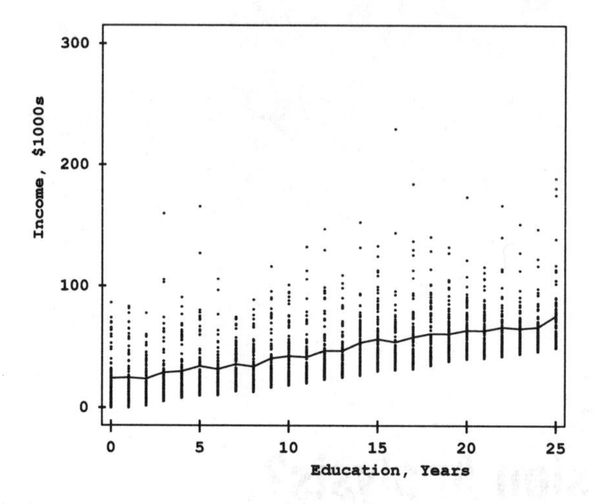

Figure 2.1. A scatterplot showing the relationship between income (in thousands of dollars) and education (in years) for the population of employed Canadians. The line connects the mean incomes at the different levels of education. The "data" are hypothetical.

The relationship of Y to the X's is of particular interest when we entertain the possibility that the X's affect Y, or—more weakly—when we wish to use the X's to predict the value of Y. Primarily for convenience of exposition, I shall use the term "regression analysis" to refer to those cases in which both Y and the X's are quantitative (as opposed to qualitative) variables.[3] This chapter introduces basic concepts of regression analysis in a very general setting, and explores some simple methods of regression analysis that make only very weak assumptions about the structure of the data.

> Regression analysis examines the relationship between a quantitative dependent variable Y and one or more quantitative independent variables, $X_1, \ldots, X_k$. Regression analysis traces the conditional distribution of Y—or some aspect of this distribution, such as its mean—as a function of the X's.

2.1 Preliminaries

Figure 2.2 illustrates the regression of a continuous Y on a single, discrete, X, which takes on the values $1, 2, 3, \ldots$. For concreteness, imagine (as in Figure 2.1) that Y is dollars of annual income, that X is years of formal education,

---

[3] Later in the book, we shall have occasion to consider statistical models in which the independent variables (Chapters 7 and 8) and the dependent variable (Chapter 15) are qualitative/categorical.

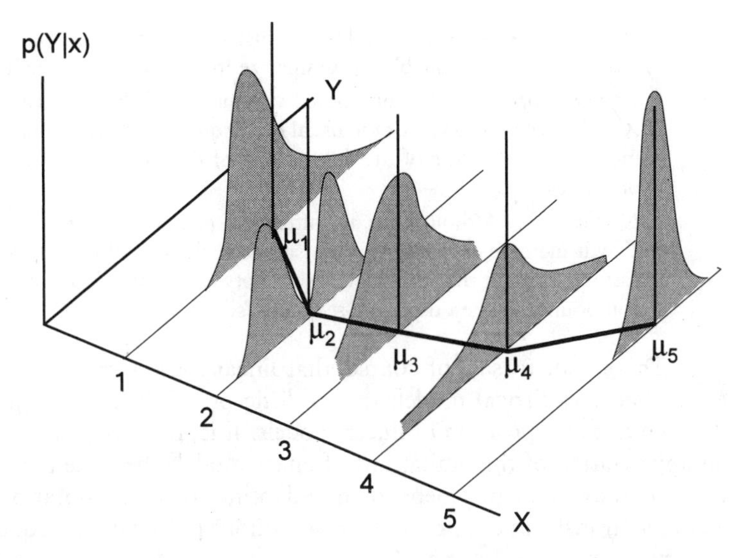

Figure 2.2. Population regression of Y on X. The conditional distribution of Y, $p(Y|x)$, is shown for each of a few values of X. The distribution of Y at $X = 1$ is positively skewed; at $X = 2$ it is bimodal; at $X = 3$ it is heavy tailed; the distribution at $x = 4$ has much greater spread than at $x = 5$. Notice that the conditional means of Y given X—$\mu_1, \ldots, \mu_5$—are not a linear function of X.

and that the figure shows the conditional distribution $p(Y|x)$ of income for some of the values of education.

Most discussions of regression analysis begin by assuming that the conditional distribution of the dependent variable, $p(Y|x_1, \ldots, x_k)$, is a normal distribution; that the variance of Y conditional on the X's is everywhere the same regardless of the specific values of $x_1, \ldots, x_k$; and that the expected value (the mean) of Y is a linear function of the X's:

$$\mu \equiv E(Y|x_1, \ldots, x_k) = \alpha + \beta_1 x_1 + \cdots + \beta_k x_k \qquad [2.2]$$

As we shall see,[4] these assumptions, along with independent random sampling, lead to linear least-squares estimation of model [2.2]. In this chapter, in contrast, we shall pursue the notion of regression with as few preconceived assumptions as possible.

Figure 2.2 illustrates why we should not be too hasty to make the assumptions of normality, equal variance, and linearity:

- *Skewness*. If the conditional distribution of Y is skewed, then the mean will not be a good summary of its center. This is the case as well in Figure 2.1, where the conditional distributions of income given education are all positively skewed.
- *Heavy tails*. If the conditional distribution of Y is substantially nonnormal—for example, heavy-tailed—then the sample mean will not be an efficient estimator of the center of the Y-distribution even when this distribution is symmetric.

[4] Chapter 6.

4 regression assumptions

- *Multiple modes.* If the conditional distribution of Y is multimodal, then it is intrinsically unreasonable to summarize its center with a single number.
- *Unequal spread.* If the conditional variance of Y changes with the values of the X's, then the efficiency of the usual least-squares estimates may be compromised; moreover, the nature of the dependence of the variance on the X's may itself be of interest.
- *Nonlinearity.* Although we are often in a position to suppose that the values of Y will increase or decrease with some X, there is almost never good reason to assume a priori that the relationship between Y and X is linear; this problem is compounded when there are several X's.[5]

This is not to say, of course, that linear regression analysis or, more generally, linear statistical models, are of little practical use. Much of this book is devoted to the exposition of linear models. It is, however, prudent to begin with an appreciation of the limitations of linear models, because their effective use in data analysis frequently depends on adapting to these limitations: We may, for example, transform data to make assumptions of normality, equal variance, and linearity more nearly correct.[6]

There are two additional advantages to approaching regression analysis from a general perspective: First, an appreciation for the practical difficulties of fitting the very general model [2.1] to data motivates the specification of more restrictive models, such as the usual linear regression model with normal errors. Second, modern methods of *nonparametric regression*, while not quite as general as model [2.1], are emerging as practical alternatives to the more traditional linear models.

The balance of the present chapter is devoted to an initial foray into the territory of nonparametric regression. I shall begin by taking a direct, or "naive" approach to the problem, and then shall improve this approach by local averaging and weighting. In the process, we shall encounter for the first time a number of recurring themes in this book, including the direct examination of data using graphical displays, smoothing to clarify patterns in data, and the detection and treatment of unusual data.[7]

2.2 Naive Nonparametric Regression

Imagine once more that we are interested in the relationship between income and education. Now, however, we do not have data for the whole population, but we have a very large sample—say, of one million employed Canadians. We could easily display the conditional distribution of income for each of the values of education $(0, 1, 2, \ldots, 25)$ that occur in our data, because (I assume) each value of education occurs many times.

Although income is (for practical purposes) a continuous variable, the large quantity of data makes it practical to display its conditional distribution using a histogram with narrow bars (each, say, \$1000 wide).[8] If, as is often the case,

[5] In Figure 2.1, in contrast, the conditional means for income are not a perfect linear function of education, but the relationship between income and education is close enough to linear to make a straight line a reasonable summary of the population.

[6] See Chapter 4.

[7] More sophisticated methods for nonparametric regression are discussed in Section 14.4.

[8] We shall explore other approaches to displaying distributions in Section 3.1.

our interest centers (to make a bad pun) on the average or typical value of income conditional on education, we could—in light of the large size of our dataset—estimate these conditional averages very accurately. The distribution of income given education is likely positively skewed, so it would be better to use conditional medians than conditional means as typical values; nevertheless, we shall, for simplicity, focus initially on the conditional means, $\overline{Y}|x$.[9]

Imagine now that X, along with Y, is a continuous variable. For example, X is the reported weight in kilograms for each of a sample of individuals, and Y is their measured weight, again in kilograms. We want to use reported weight to predict actual (i.e., measured) weight, and so we are interested in the mean value of Y as a function of X in the population of individuals from among whom the sample was randomly drawn:[10]

$$\mu = E(Y|x) = f(x) \qquad [2.3]$$

Even if the sample is large, replicated values of X will be rare because X is continuous.[11] In the absence of replicated X's, we cannot directly examine the conditional distribution of Y given X, and we cannot directly calculate conditional means. If we indeed have a large sample of individuals at our disposal, however, then we can dissect the range of X into many narrow intervals of reported weight, each interval containing many observations; within each such interval we can display the conditional distribution of measured weight and estimate the conditional mean of Y with great precision.

In very large samples, and when the independent variables are discrete, it is possible to estimate a regression by directly examining the conditional distribution of Y given the X's. When the independent variables are continuous, we can proceed similarly by dissecting the X's into a large number of narrow intervals.

If, as is more typical, we have a relatively small sample at our disposal, we have to make do with fewer intervals, each containing relatively few observations. This situation is illustrated in Figure 2.3, using data on reported and mea-

---

[9] We imagine that Figure 2.1 shows the *population* conditional means, $\mu|x$, the values that we now want to estimate from our sample.

[10] This is a peculiarly interesting problem in several respects: First, although it is more reasonable to suppose that actual weight "affects" the report than vice versa, our desire to use the report to predict actual weight (presumably because it is easier to elicit a verbal report than actually to weigh people) motivates treating measured weight as the dependent variable. Second, this is one of those comparatively rare instances in which a linear-regression equation is a natural specification, since if people are unbiased reporters of their weight, we should have $\mu = x$ (i.e., expected reported weight equal to actual weight). Finally, if people are accurate, as well as unbiased reporters of their weight, then the variance of Y given x should be very small.

[11] No numerical data are literally continuous, of course, since data are always recorded to some finite number of digits. This is why tied values are possible. The philosophical issues surrounding continuity are subtle but essentially irrelevant to us: For practical purposes, a variable is continuous when it takes on many different values.

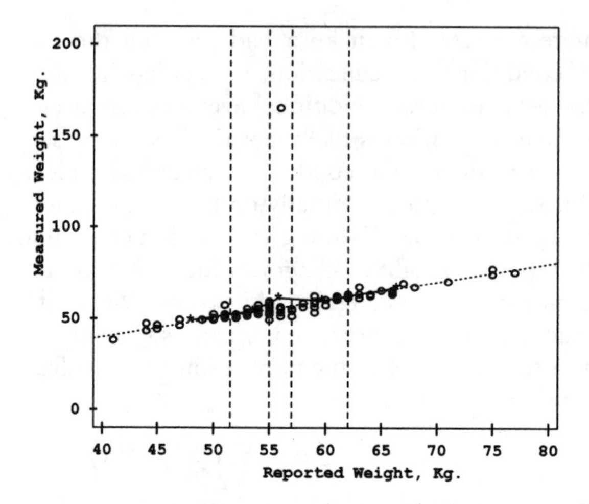

Figure 2.3. Naive nonparametric regression of measured weight on reported weight, each in kilograms. The range of reported weight has been dissected into five intervals (separated by broken lines), each containing about 20 observations. The solid line connects the averages of measured weight and reported weight in the five intervals, shown as asterisks. The dotted line around which the points cluster is $Y = X$.

sured weight for each of 101 Canadian women engaged in regular exercise.[12] A partially contrasting example, using the prestige and income levels of 102 Canadian occupations in 1971, is shown in Figure 2.4.[13]

The X-axes in Figures 2.3 and 2.4 are carved into five intervals (sometimes called "slices" or "strips"), each interval containing approximately 20 observations (the first and last intervals contain the extra observations). The "nonparametric regression line" displayed on each plot is calculated by connecting the points defined by the conditional dependent-variable means $\overline{Y}$ and the independent-variable means $\overline{X}$ in the five intervals.

Recalling our purpose, which is to estimate model [2.3], there are two sources of error in this simple procedure of dissecting and averaging:

- *Sampling error (variance).* The conditional sample means $\overline{Y}$ will, of course, change if we select a new sample (even if we could retain the same selection of x's). Sampling error is minimized by using a small number of relatively wide intervals, each with a substantial number of observations.
- *Bias.* Let x_i denote the center of the ith interval (here, $i = 1, \ldots, 5$), and suppose that the X-values are evenly spread in the interval. If the population regression

[12] These data were generously made available to me by Caroline Davis of York University, who used them as part of a larger study; see Davis (1990). The error in the data, described below, was located by Professor Davis. The 101 women were volunteers for the study, not a true sample from a larger population.

[13] The Canadian prestige data are described in Fox and Suschnigg (1989). Although there are many more occupations in the Canadian Census, these 102 do not constitute a random sample from the larger population of occupations. Treating the 102 occupations as a sample implicitly rests on the claim that they are "typical" of the population, at least with respect to the relationship between prestige and income—a problematic, if arguable, claim.

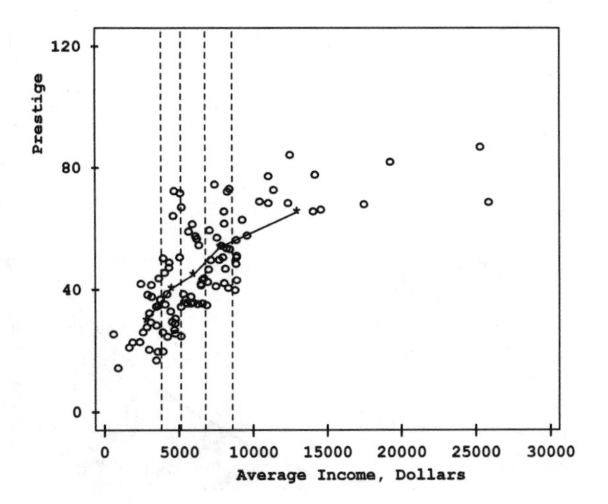

Figure 2.4. Naive nonparametric regression of occupational prestige on average income, for 102 Canadian occupations in 1971. The range of income has been dissected into five intervals, each containing about 20 observations. The line connects the average prestige and income scores in the five intervals, shown as asterisks.

curve $f(x)$ is nonlinear within the interval, then the average population value of Y in the interval is usually different from the value of the regression curve at the center of the interval, $f(x_i)$. This situation is illustrated in Figure 2.5. Bias is therefore minimized by making the class intervals as numerous and as narrow as possible.

As is typically the case in statistical estimation, reducing bias and reducing sampling variance work at cross purposes. Only if we select a very large sample can we have our cake and eat it, too—by constructing a very large number of narrow intervals, each with a large number of observations. This situation was, of course, our starting point.

The nonparametric regression lines in Figures 2.3 and 2.4 are also very crude: Although reported weights vary from about 40 to about 80 kg, we have evaluated the regression at only five points in this substantial range; likewise, income values for the 102 occupations vary from about $600 to about $26,000. Nevertheless, it is clear from Figure 2.3 that, except for one very discrepant data point,[14] the data are very close to the line $Y = X$; and it is clear from Figure 2.4 that while prestige appears to increase with income, the increase is nonlinear, with prestige values leveling off at relatively high income.

The opportunity that a very large sample presents to reduce both bias and variance suggests that naive nonparametric regression is, under very broad conditions, a consistent estimator of the population regression curve.[15] For, as the

[14] It seems difficult to comprehend how a 166-kg woman could have reported a weight of only 56 kg, but the solution to this riddle is simple: The woman's weight in kilograms and height in centimeters were accidentially switched when the data were entered into the computer.

[15] For example, we need to assume that the regression curve $\mu = f(X)$ is reasonably smooth and that the distribution of Y given x has finite variance (see Manski, 1991, for some details). We also should remember that the reasonableness of focusing on the mean μ depends on the symmetry—and unimodality—of the conditional distribution of Y.

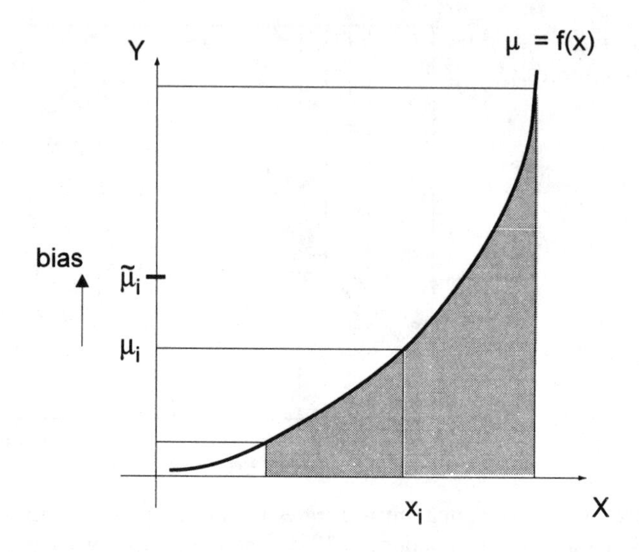

Figure 2.5. When the regression of Y on X is nonlinear in an interval centered at x_i, the average value of Y in the interval ($\tilde{\mu}_i$) can be a biased estimate of the regression curve at the center of the interval [$\mu_i = f(x_i)$].

sample size gets larger (i.e., as $n \to \infty$), we can ensure that the intervals grow successively narrower, yet each contains more data (e.g., by employing $\sqrt{n}$ intervals, each containing $\sqrt{n}$ observations). In the limit—never, of course, attained—we have an infinite number of intervals, each of zero width, and each containing an infinite number of observations. In this utopia, the naive nonparametric regression and the population regression curve coincide.

It may appear as if naive nonparametric regression—that is, dissecting and averaging—is a practical procedure in large datasets or when independent variables are discrete. Although this conclusion is essentially correct, it is instructive—and sobering—to consider what happens when there is more than one independent variable.

Suppose, for example, that we have three discrete independent variables, each with 10 values. There are, then, $10^3 = 1000$ combinations of values of the three variables, and within each such combination there is a conditional distribution of Y [i.e., $p(Y|x_1, x_2, x_3)$]. Even if the X's are uniformly and independently distributed—implying equal expected numbers of observations for each of the 1000 combinations—we should require a very large sample indeed to calculate the conditional means of Y with sufficient precision. The situation is even worse when the X's are continuous, because dissecting the range of each X into as few as 10 class intervals might introduce substantial bias into the estimation.

The problem of dividing the data into too many parts grows exponentially more serious as the number of X's increases. Statisticians therefore often refer to the intrinsic sparseness of multivariate data as the "curse of dimensionality." Moreover, the imaginary calculation on which the consistency of naive nonparametric regression is based—in which the number of independent variables remains the same as the sample size grows—is itself unrealistic, because we are apt in large samples to entertain more complex statistical models than in small samples.[16]

EXERCISES[17]

2.1 *Figure 2.4 illustrates how, when the relationship between Y and X is nonlinear in an interval, the average value of Y in the interval can be a biased estimate of $E(Y|x)$ at the center of the interval. Imagine that X-values are evenly distributed in an interval centered at x_i, and let $\mu_i \equiv E(Y|x_i)$.

(a) If the relationship between Y and X is *linear* in the interval, is the average value of Y a biased or an unbiased estimator of μ_i?

(b) Are there any circumstances under which the average Y in the interval is an unbiased estimator of μ_i if the relationship between Y and X is *nonlinear* in the interval?

2.2 #The data in Table 2.1 and `sahlins.dat` were compiled by Sahlins (1972) from information presented in Scudder's (1962) report on the Gwenba valley of Central Africa. The data describe agricultural production in Mazulu village. The independent variable (Consumers/Gardener) is the ratio of consumers to productive individuals in each of 20 households, making suitable adjustments for the consumption requirements of different household members. The dependent variable (Acres/Gardener) is a measure of domestic-labor intensity, based on the amount of land cultivated by each household. Think of Consumers/Gardener as representing the relative consumption needs of the household, and Acres/Gardener as representing how hard each productive individual in the household works. Sahlins was interested in production, consumption, and redistribution of the social product in "primitive" communities.

(a) Draw a scatterplot of Acres/Gardener (Y) versus Consumers/Gardener (X). What relationship, if any, do you discern in this plot—does the relationship appear to be linear or nonlinear; is it strong or weak? Is there anything else noteworthy about the data—for example, do any households appear to be unusual?

(b) Notice that the households are ordered by the values of Consumers/Gardener (X). Divide the 20 households into three groups, placing the first seven households in the first group, the next six in the second group, and the last seven in the third group. Calculate the mean Y and mean X in each of the three groups. Transfer these means to the scatterplot and connect them with a simple nonparametric regression line. Does the regression line help you to interpret the relationship between the variables?

---

[16] I am indebted to Robert Stine, of the University of Pennsylvania, for this insight.

[17] Recall that exercises intended for "hand" computation are marked with a pound sign (#). Relatively difficult exercises are marked with an asterisk (*).

TABLE 2.1 Data on Agricultural Production in Mazulu Village.

Household	*Consumers/ Gardener* X_i	*Acres/ Gardener* Y_i
1	1.00	1.71
2	1.08	1.52
3	1.15	1.29
4	1.15	3.09
5	1.20	2.21
6	1.30	2.26
7	1.37	2.40
8	1.37	2.10
9	1.43	1.96
10	1.46	2.09
11	1.52	2.02
12	1.57	1.31
13	1.65	2.17
14	1.65	2.28
15	1.65	2.41
16	1.66	2.23
17	1.87	3.04
18	2.03	2.06
19	2.05	2.73
20	2.30	2.36

Source of Data: Sahlins (1972, Table 3.1).

(c) Two of the households, one in the first group, one in the second (which ones?), stand out from the others. Recalculate the means in groups 1 and 2 omitting the outlying observations. How, if at all, do the new means differ from the original values?

2.3 #The data in Table 2.2 and `robey.dat` are drawn from a report by Robey et al. (1992) on the fertility decline in developing countries. The data show the total fertility rate (Y) and the percentage of married women aged 15 to 44 who use contraceptives (X) in each of 50 developing countries (along with the region of the world in which each country is located). The total fertility rate is the expected number of births to a woman who survives through her child-bearing years under current age-specific fertility rates. Repeat Exercise 2.2 for this dataset, using three groups of 17, 16, and 17 observations.

2.3 Local Averaging

Let us return to Figure 2.4, showing the naive nonparametric regression of occupational prestige on income. One problem with this procedure is that we have estimated the regression at only five points, a consequence of our desire to have relatively stable conditional averages, each based on a sufficient number of observations (here, 20). However, there is no intrinsic reason why we should restrict ourselves to partitioning the data by X-values into nonoverlapping strips.

TABLE 2.2 Fertility (Total Fertility Rate, TFR) and Contraception (Percentage of Women of Reproductive Age Who Practice Contraception) for 50 Developing Nations Around 1990

Nation	*TFR*	*Contraception*	*Nation*	*TFR*	*Contraception*
Africa			Sri Lanka	2.7	62
Botswana	4.8	35	Thailand	2.3	68
Burundi	6.5	9	Vietnam	3.9	53
Cameroon	5.9	16	*Latin America and Caribbean*		
Ghana	6.1	13	Belize	4.5	47
Kenya	6.5	27	Bolivia	4.9	32
Liberia	6.4	6	Brazil	3.6	66
Mali	6.8	5	Colombia	2.8	66
Mauritius	2.2	75	Costa Rica	3.6	70
Niger	7.3	4	Dominican Republic	3.3	56
Nigeria	5.7	6	Ecuador	3.8	53
Senegal	6.4	12	El Salvador	4.6	47
Sudan	4.8	9	Guatemala	5.6	23
Swaziland	5.0	21	Haiti	6.0	10
Tanzania	6.1	10	Jamaica	2.9	55
Togo	6.1	12	Mexico	4.0	55
Uganda	7.2	5	Panama	4.0	58
Zambia	6.3	15	Paraguay	4.6	48
Zimbabwe	5.3	45	Peru	3.5	59
Asia and Pacific			Trinidad-Tobago	3.1	54
Bangladesh	5.5	40	*Near East and North Africa*		
China	2.5	72	Egypt	4.6	40
India	4.3	45	Jordan	5.5	35
Indonesia	3.0	50	Morocco	4.0	42
Republic of Korea	1.7	77	Tunisia	4.3	51
Pakistan	5.2	12	Turkey	3.4	60
Philippines	4.3	34	Yemen	7.0	7

Source of Data: Robey et al. (1992).

We can allow X to vary continuously across the range of observed values, calculating the average value of Y within a moving *window* of fixed width centered at the current value x. Alternatively, we can employ a window of varying width, constructed to accommodate a fixed number of data values (say, m) that are the nearest X-neighbors to the x-value at the center of the window. As a practical matter, of course, we cannot perform these calculations at the uncountably infinite number of points produced by allowing X to vary continuously, but, using a computer, we can quickly calculate averages at a large number of x's. One attractive procedure, if the sample size n is not very large, is to evaluate the local average of Y in a neighborhood around each of the X-values observed in the data: $x_1, x_2, \ldots, x_n$.

> In smaller samples, local averages of Y can be calculated in a neighborhood surrounding each x-value.

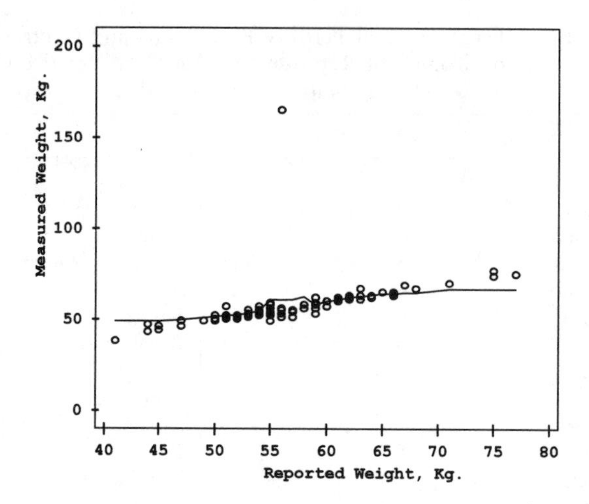

Figure 2.6. Nonparametric regression by local averaging for Davis's data on reported and measured weight. The data are shown by hollow dots, the local averages at each x-value by the line traced through the plot. Each local average is based on 20 of the 101 observations. Note the impact of the outlying observation on the averages that include it, and the flattening of the regression at the lowest and highest reported weights.

Figure 2.6 illustrates the process of local averaging for the reported-weight data, employing $m = 20$ observations for each local average. Similarly, Figure 2.7 shows the result of applying local averaging to the occupational prestige data, again using $m = 20$. Three defects of the procedure are apparent from these examples:

1. The first few local averages are identical to one another, as are the last few, flattening the regression line at extreme X-values.[18]
2. The line connecting the averages tends to be rough, because the average "jumps" up or down as observations enter and exit the moving window. (This roughness is more apparent in Figure 2.7, where the relationship between the two variables is weaker, than in Figure 2.6.)
3. Unusual data values, called *outliers*, unduly influence the average when they fall in the window (as in Figure 2.6). In regression analysis, an outlier is a value of Y that is substantially different from other dependent-variable values associated with similar X's.

A fully adequate solution to the first problem will have to await the more sophisticated tools for nonparametric regression discussed in Chapter 14. The second and third problems, however, can be addressed by weighting.

[18] Imagine, for example, that the x-values are evenly spaced and that m is 19. Let $x_{(1)}, x_{(2)}, \ldots, x_{(n)}$ represent these x-values ordered from smallest to largest. Then, the first 19 observations—the Y-values associated with $x_{(1)}, x_{(2)}, \ldots, x_{(19)}$ —would be used for the first 10 local averages, making these averages identical to one another. One solution is to employ "symmetric" neighborhoods around each $x_{(i)}$, with the same number of observations below and above the focal $x_{(i)}$, but this procedure implies using smaller and smaller neighborhoods as we approach the extreme values $x_{(1)}$ and $x_{(n)}$. For each extreme, for example, the symmetric neighborhood only includes the value itself.

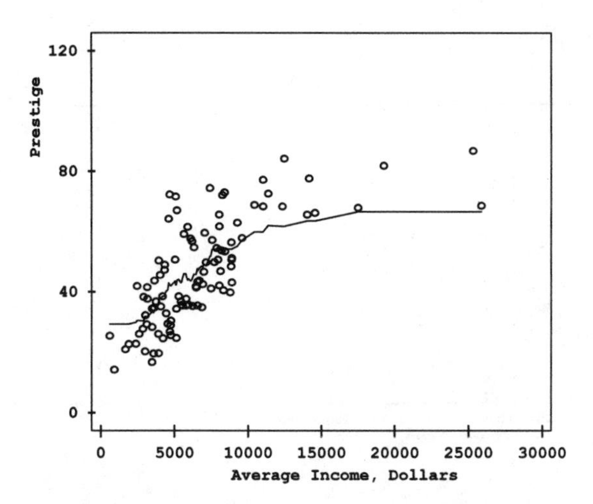

Figure 2.7. Nonparametric regression of occupational prestige on income, using local averages. Each average includes 20 of the 102 observations; the averages are connected by a line.

2.3.1 Weighted Local Averages*

Consider the m observations contained in the window centered at x_i. Let us denote the X-values for these observations $x_{i1}, x_{i2}, \ldots, x_{im}$; these are, recall, the m nearest neighbors of X_i. The corresponding Y-values are $Y_{i1}, Y_{i2}, \ldots, Y_{im}$; these are *not*, in general, the Y-values that are closest to Y_i—just the Y's associated with the X's that are the nearest neighbors of x_i.

We need a function that accords greatest weight to observations within the window that are closest to the central x_i than to those with X's that are farther away. Although there are many reasonable choices, the *tricube weight function* works well:

$$w_{ij} \equiv w_T(z_{ij}) = \begin{cases} (1 - |z_{ij}|^3)^3 & \text{for } |z_{ij}| < 1 \\ 0 & \text{for } |z_{ij}| \geq 1 \end{cases}$$

where, here,

$$z_{ij} = \frac{x_{ij} - x_i}{h_i}$$

and

$$h_i = \max |x_{ij} - x_i|$$

Thus, as shown in Figure 2.8 for the Canadian occupational prestige data, the weights are at their maximum value of 1 at the center of the window and fall gradually to 0 at its edges; h_i is the half-width of the window centered on x_i. The weighted-average Y-value at x_i—called the *fitted value* of Y—is then

$$\hat{Y}_i = \frac{\sum_{j=1}^{m} w_{ij} Y_{ij}}{\sum_{j=1}^{m} w_{ij}}$$

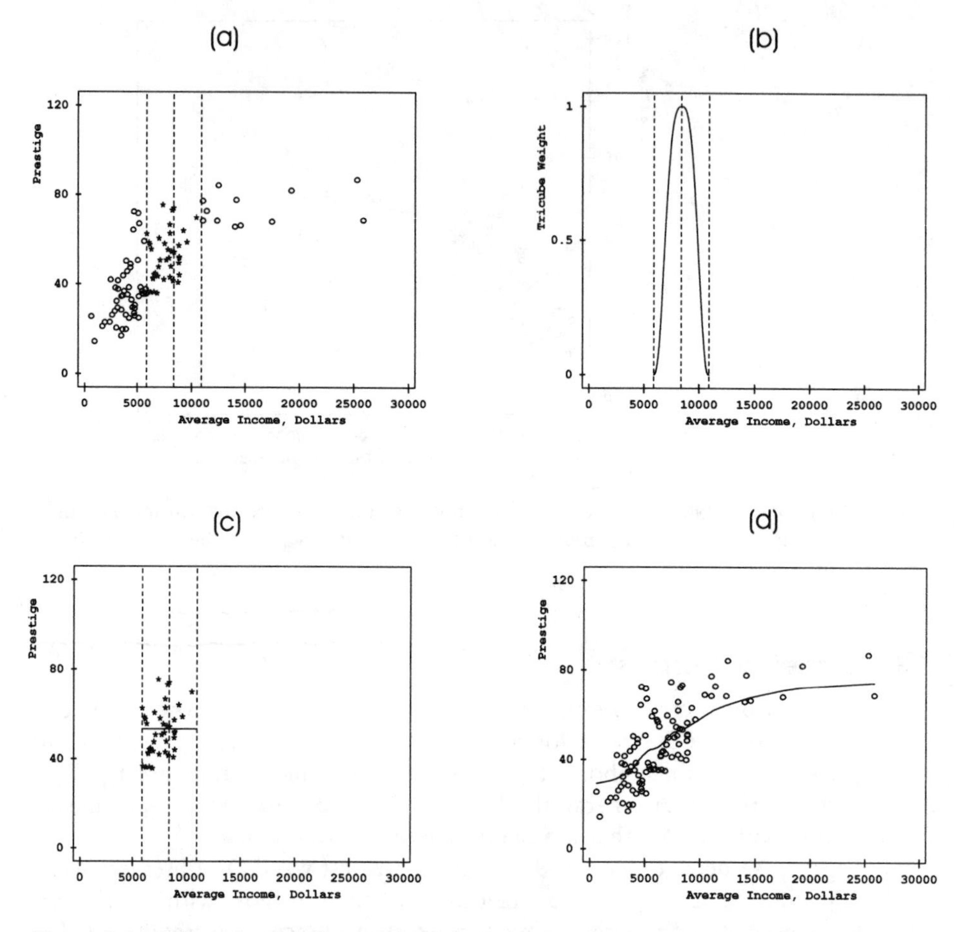

Figure 2.8. Calculating locally weighted averages for the data on the prestige and income of Canadian occupations. Panels (*a*), (*b*), and (*c*) illustrate the calculation of the locally weighted average at $x_{(80)} = 8403$: In panel (*a*), a window is defined enclosing the 40 nearest neighbors of $x_{(80)}$; panel (*b*) shows the tricube weight function centered at $x_{(80)}$; panel (*c*) shows the locally weighted average Y-value calculated at $x_{(80)}$. Panel (*d*) displays all 102 locally weighted averages connected by a line.

Source: Adapted with permission from Chambers et al. (1983, Figure 4.14).

The result of applying this procedure of locally weighted averaging to Davis's data is shown in Figure 2.9.[19]

Although locally weighted averaging produces a smoother regression, it is vulnerable to outliers, as is apparent in Figure 2.9. We can, however, discount outliers by extending the notion of weighting. Consider the following procedure:

1. Define the *residual* for each observation as the difference between the observed dependent-variable value and the fitted value:

$$E_i = Y_i - \hat{Y}_i$$

[19] Locally weighted averaging is also called *kernel smoothing*, and the local weight function is called the *kernel*. In Section 3.1.2, we shall use kernel estimation to smooth the histogram of a variable.

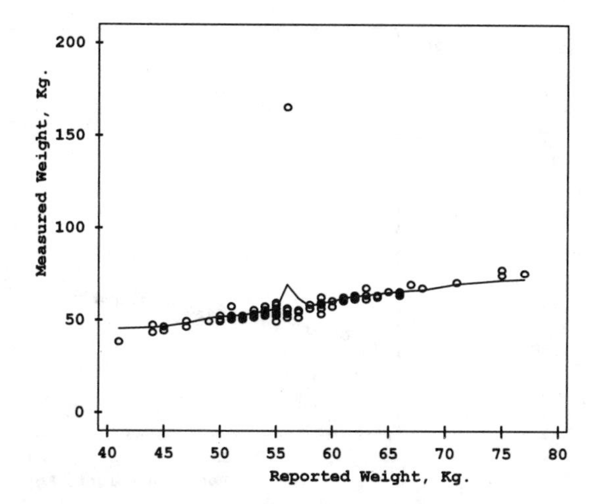

Figure 2.9. Locally weighted averages for the regression of measured weight on reported weight. Each average is based on 20 observations.

The division of data into fitted and residual components (here, $Y = \hat{Y} + E$) is a common feature of statistical modeling.

2. Calculate a measure of the *scale* of the residuals. One such measure is the standard deviation, but it is prudent to select a measure of scale that is less affected by outliers—for example, the median absolute deviation from the median residual:

$$\text{MAD} = \text{median}|E_i - \tilde{E}|$$

where

$$\tilde{E} = \text{median}(E_i)$$

3. Using an appropriate weight function, calculate weights that reflect the degree of outlyingness of the observations—that is, the relative sizes of their residuals. A good choice is the *bisquare* weight function:

$$v_i \equiv w_B(z_i) = \begin{cases} (1 - z_i^2)^2 & \text{for } |z_i| < 1 \\ 0 & \text{for } |z_i| \geq 1 \end{cases}$$

where, here,

$$z_i = \frac{E_i}{t \times \text{MAD}}$$

Thus, observations with residuals of 0 are accorded the maximum weight of 1, while those with absolute values exceeding t MADs are accorded zero weight. The multiplier t is called a "tuning constant"; $t = 6$ often works well.[20] Using these weights will make the regression *resistant* to outliers.

---

[20] The cutoff of six MADs is determined by appealing roughly to the setting of the normal distribution, where MAD/0.6745 estimates the population standard deviation σ. Six MADs correspond, therefore, to $6 \times 0.6745 \simeq 4$ standard deviations. The probability of observing a normally distributed random variable more than 4 standard deviations from the mean is very small, about 0.0001. It is common in reasoning about "robust" statistics to use the normal distribution as a point of reference. For a more detailed discussion of robust regression, see Section 14.3.

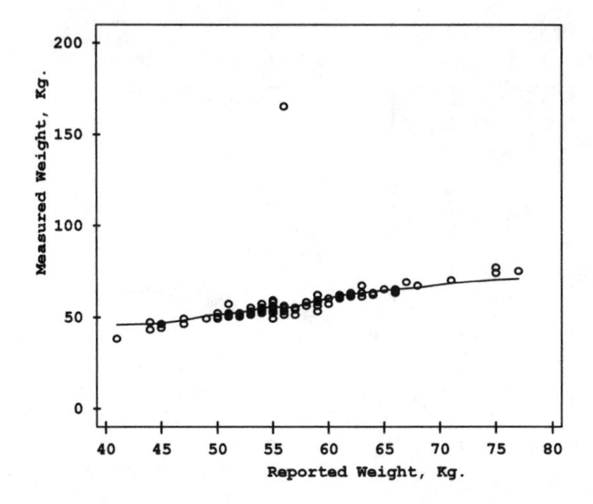

Figure 2.10. Robust regression for Davis's weight data by locally weighted averaging, downweighting outliers. Here, $m = 20$ observations are included in each local average, and the biweight tuning constant is $t = 10$. There is still some flattening of the regression curve at the lowest and highest X's.

4. Recalculate the local averages employing the resistance weights v_j (along with the local weight factors w_{ij}):

$$\hat{Y}'_i = \frac{\sum_{j=1}^{m} v_j w_{ij} Y_{ij}}{\sum_{j=1}^{m} v_j w_{ij}}$$

5. Steps 1–4 can be *iterated* (i.e., repeated) until fitted values (and hence the weights v_i) stabilize. Usually one or two iterations will suffice. The procedure is illustrated for Davis's weight data in Figure 2.10, using $t = 10$.

Local averages can be smoothed and made resistant to outliers by appropriate weighting.

EXERCISES

2.4 Returning to Sahlins's data on domestic-labor intensity in Mazulu village, described in Exercise 2.2, use a computer program to fit a nonparametric regression line to Sahlins's data using local averages and locally weighted averages. Then, fit a regression that downweights outlying observations. How, if at all, do the three regressions differ from one another? How would you summarize the relationship between Acres/Gardener and Consumers/Gardener?

TABLE 2.3 Moral Integration (Int.), Ethnic Heterogeneity (Het.), and Geographic Mobility (Mob.) for 43 U.S. Cities Around 1950

City	*Int.*	*Het.*	*Mob.*	*City*	*Int.*	*Het.*	*Mob.*
East				Columbus	8.0	27.4	25.0
Rochester	19.0	20.6	15.0	*West*			
Syracuse	17.0	15.6	20.2	Denver	13.9	13.0	34.5
Worcester	16.4	22.1	13.6	San Diego	12.5	15.9	49.8
Erie	16.2	14.0	14.8	Tacoma	10.9	17.8	31.2
Bridgeport	15.3	27.9	17.5	Spokane	9.6	12.3	38.9
Buffalo	15.2	22.3	14.7	Seattle	9.0	23.9	34.2
Reading	14.2	10.6	19.4	Portland OR	7.2	16.4	35.8
Trenton	13.0	32.5	15.8	*South*			
Baltimore	12.0	45.8	12.1	Richmond	10.4	65.3	24.9
Midwest				Houston	10.2	49.0	36.1
Milwaukee	15.8	17.4	17.6	Fort Worth	10.2	30.5	36.8
Dayton	14.3	23.7	23.8	Oklahoma City	9.7	20.7	47.2
Des Moines	14.1	12.7	31.9	Chattanooga	9.3	57.7	27.2
Cleveland	14.0	39.7	18.6	Nashville	8.6	57.4	25.4
Peoria	13.8	10.7	35.1	Birmingham	8.2	83.1	25.9
Wichita	13.6	11.9	42.7	Dallas	8.0	36.8	37.8
Grand Rapids	12.8	15.7	24.2	Louisville	7.7	31.5	19.4
Toledo	12.7	19.2	21.6	Jacksonville	6.0	73.7	27.7
South Bend	11.8	17.9	27.4	Memphis	5.4	84.5	26.7
Akron	11.3	20.4	22.1	Tulsa	5.3	23.8	44.9
Detroit	11.1	38.3	19.5	Miami	5.1	50.2	41.8
Flint	9.8	19.3	32.2	Atlanta	4.2	70.6	32.6
Indianapolis	8.8	29.2	23.1				

Source of Data: Angell (1951).

2.5 Repeat Exercise 2.4 for Robey et al.'s data on fertility and contraception, described in Exercise 2.3.

2.6 The data on 43 U.S. cities in Table 2.3 and `angell.dat` were employed by Angell (1951) in a multiple-regression analysis. Quite apart from their substantive content, these data are of interest because they represent an early use of multiple regression in sociology. Angell's index of "moral integration" (i.e., social integration) combines information on the incidence of crime with data on social welfare expenditures. Cities that have low crime rates and high levels of welfare effort are assigned high moral-integration scores. The ethnic-heterogeneity index is calculated from the proportions of nonwhites and foreign-born whites residing in the cities, and the mobility index is constructed from the proportions of residents moving into and out of the cities. Angell argues that the moral integration of cities should decrease with ethnic heterogeneity and geographic mobility. Employing the methods of this chapter, explore the relationship between moral integration and each of ethnic heterogeneity and geographic mobility. In each case, try to determine whether the regression is linear or nonlinear, and whether there are unusual data points.

2.4 Summary

- Regression analysis examines the relationship between a quantitative dependent variable Y and one or more quantitative independent variables, $X_1, \ldots, X_k$. Regression analysis traces the conditional distribution of Y—or some aspect of this distribution, such as its mean—as a function of the X's.
- In very large samples, and when the independent variables are discrete, it is possible to estimate a regression by directly examining the conditional distribution of Y given the X's. When the independent variables are continuous, we can proceed similarly by dissecting the X's into a large number of narrow intervals.
- In smaller samples, local averages of Y can be calculated in a neighborhood surrounding each x-value.
- Local averages can be smoothed and made resistant to outliers by appropriate weighting.

3

Examining Data

This chapter, on graphical methods for examining data, and the next, on transformations, represent a digression from the principal focus of the book. Nevertheless, the material here is important to us for two essential reasons: First, careful data analysis should begin with inspection of the data. You will find in this chapter simple methods for graphing univariate, bivariate, and multivariate data. Second, the techniques for examining and transforming data that are discussed in Chapters 3 and 4 will find direct application to the analysis of data using linear models.[1] Feel free, of course, to pass lightly over topics that are familiar.

To motivate the material in the chapter, and to demonstrate its relevance to the study of linear models, consider the four scatterplots shown in Figure 3.1. The data for these plots, given in Table 3.1, were cleverly contrived by Anscombe (1973) to illustrate the central role of graphical methods in data analysis: Anticipating the material in Chapters 5 and 6, the least-squares regression line and all other common regression "outputs"—such as the correlation coefficient, standard deviation of the residuals, and standard errors of the regression coefficients—are identical in the four datasets.

It is clear, however, that each graph tells a different story about the data. Of course, the data are simply made up, so we have to allow our imagination

[1] See, for example, the graphical regression "diagnostics" in Chapters 11 and 12.

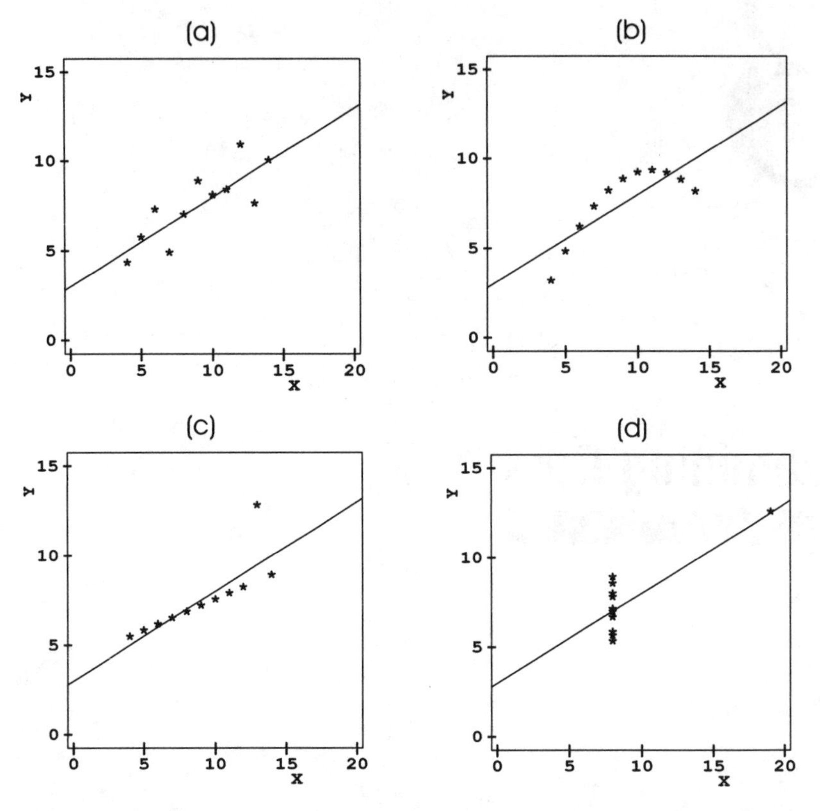

Figure 3.1. Four datasets, due to Anscombe (1973), with identical least-squares regressions. In (*a*), the linear regression is an accurate summary; in (*b*), the linear regression distorts the curvilinear relationship between Y and X; in (*c*), the linear regression is drawn toward an outlier; in (*d*), the linear regression "chases" the influential observation at the right.

Source: Reprinted with permission from *The American Statistician*.

some latitude:

- In Figure 3.1(*a*), the linear regression line is a reasonable descriptive summary of the tendency of Y to increase with X.
- In Figure 3.1(b), the linear regression fails to capture the clearly curvilinear relationship between the two variables; we would do much better to fit a quadratic function here,[2] that is, $Y = a + bX + cX^2$.
- In Figure 3.1(*c*), there is a perfect linear relationship between Y and X for all but one outlying data point. The least-squares line is pulled strongly toward the outlier, distorting the relationship between the two variables for the rest of the data. Perhaps the outlier represents an error in data entry or an observation that differs in some fundamental respect from the others. When we encounter an outlier in real data, we should look for an explanation.[3]

---

[2] Quadratic and other polynomial regression models are discussed in Section 14.2.1.
[3] Outlier detection in linear models is taken up in Section 11.3.

TABLE 3.1 Four Contrived Regression Datasets From Anscombe (1973)

$X_{a,b,c}$	Y_a	Y_b	Y_c	X_d	Y_d
10	8.04	9.14	7.46	8	6.58
8	6.95	8.14	6.77	8	5.76
13	7.58	8.74	12.74	8	7.71
9	8.81	8.77	7.11	8	8.84
11	8.33	9.26	7.81	8	8.47
14	9.96	8.10	8.84	8	7.04
6	7.24	6.13	6.08	8	5.25
4	4.26	3.10	5.39	19	12.50
12	10.84	9.13	8.15	8	5.56
7	4.82	7.26	6.42	8	7.91
5	5.68	4.74	5.73	8	6.89

- Finally, in Figure 3.1(d), the values of X are invariant (all are equal to 8), with the exception of one point (which has an X-value of 19); the least-squares line would be undefined but for this point—the line necessarily goes through the mean of the 10 Y's that share the value $X = 8$ and through the point for which $X = 19$. Furthermore, if this point were moved, then the regression line would chase it. We are usually uncomfortable having the result of a data analysis depend so centrally on a single influential observation.[4]

The essential point to be derived from Anscombe's "quartet" (so dubbed by Tufte, 1983) is that it is frequently helpful to examine data graphically. Important characteristics of data are often disguised by numerical summaries and—worse—the summaries can be fundamentally misleading. Moreover, directly examining the numerical data is often uninformative: Only in the fourth dataset is the problem immediately apparent upon inspection of the numbers.

> Statistical graphs are central to effective data analysis, both in the early stages of an investigation and in statistical modeling.

3.1 Univariate Displays

3.1.1 Histograms

Recall the Canadian data on occupational prestige, introduced in Chapter 2. In Figure 2.3, for example, we used a bivariate scatterplot to examine the relationship between rated prestige and average income for 102 occupations, but we did not stop to consider the univariate distributions of prestige and income.

Figure 3.2 shows two histograms for the distribution of the 102 income scores. I assume that the histogram is a familiar graphical display, so I shall offer

[4] Influential data are discussed in Chapter 11.

only the briefest of descriptions: We dissect the range of income into equal-width intervals (called "bins"); we count the number of observations falling into each interval; and we display these frequency counts in a bar graph.

Both histograms in Figure 3.2 use bins of width $1000; they differ in that the bins in Figure 3.2(*a*) start at 0 (i.e., 0–1000, 1000–2000, etc.), while those in Figure 3.2(*b*) start at 500 (i.e., 500–1500, 1500–2500, etc.). The two histograms for the income distribution are more similar than different—both, for example, show that the distribution of incomes is positively skewed—but they do differ in at least one respect: The distribution in Figure 3.2(*a*) appears more clearly bimodal than that in Figure 3.2(*b*), where we could easily imagine that the second mode is an artifact of the display rather than an intrinsic feature of the data.

Figure 3.3 shows an alternative form of histogram, called a stem-and-leaf display. The stem-and-leaf plot, introduced by John Tukey (1972, 1977), ingeniously employs the numerical data to form the bars of the histogram. You may also be familiar with the stem-and-leaf display. Here is a relatively compressed definition:

- Each data value is broken between two adjacent digits into a "stem" and a "leaf": In Figure 3.3, the break takes place between the 1000's and the 100's digits. For example, accountants earned on average $9271 (recall that the income data are from 1971!), which translates into the stem 9 and leaf 2:

$$9|271 \rightarrow 9|2$$

  Notice that 9271 is truncated to 9200 rather than rounded to 9300; this practice, suggested by Tukey, makes it easier to locate the original values in a list or table of the data.
- Stems (here, 0, 1, . . . , 14) are constructed to cover the data, implicitly defining a system of bins each of width $1000. Each leaf is placed to the right of its stem, and the leaves on each stem are then sorted into ascending order. We can produce a finer system of bins by dividing each stem into two parts (taking, respectively, leaves 0–4 and 5–9), or five parts (0–1, 2–3, 4–5, 6–7, 8–9); for the income data, two-part stems would correspond to bins of width $500 and five-part stems to bins of width $200. We could employ still finer bins by dividing stems from leaves between the 100's and 10's digits, but that would produce a display with more bins than observations. Similarly, a coarser division between the 10,000's and 1000's digits would yield only two stems—0 and 1.
- Unusually large values—"outliers"—are collected on a special "high" stem and displayed individually. Here, there are five occupations with unusually large incomes. Were there occupations with unusually small incomes, then these would be collected and displayed individually on a "low" stem.[5]
- The column of depths counts in toward the median from both ends of the distribution. The median is the observation at depth $(n + 1)/2$. If, as here, the number of observations is even, then the depth of the median has a fractional part $[(102 + 1)/2 = 51.5]$ and, by convention, we average the two middle observations (i.e., those at depth 51) to find the median. In Figure 3.3, there are two observations at stem 0, four at and below stem 1, and so on; there are seven observations (including the outliers) at and above stem 14, still seven at and

[5] The rule for identifying outliers is explained in Section 3.1.4 on boxplots.

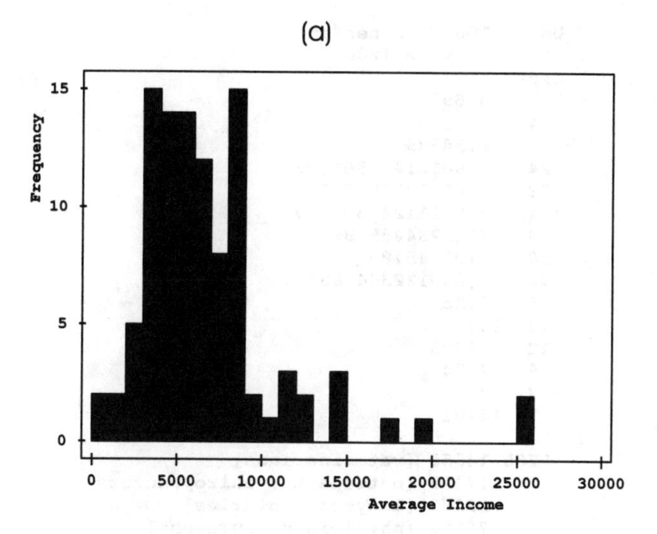

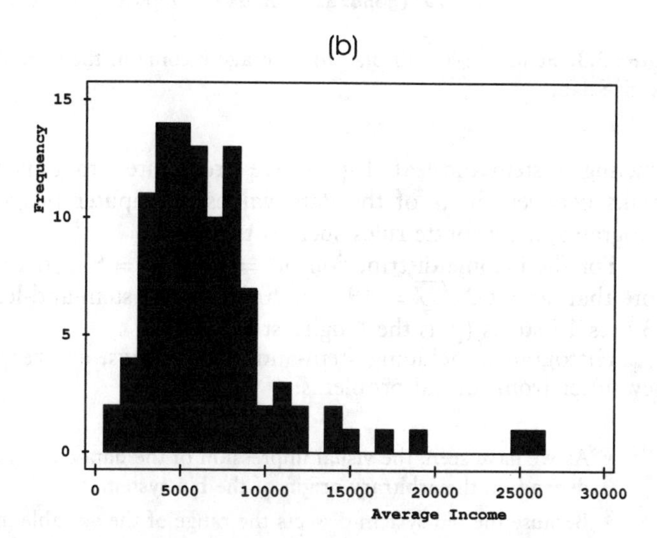

Figure 3.2. Distribution of average income for 102 occupations in the Canadian occupational prestige data. The histograms both use bins of width 1000; histogram (*a*) employs bins that start at 0, while (*b*) employs bins that start at 500.

above stem 13 (which contains no leaves), and so forth. The count at the stem containing the median is shown in parentheses—here, 14 at stem 5. Note that $38 + 14 + 50 = 102$.

In constructing histograms (including stem-and-leaf displays), we want enough bins to preserve some detail, but not so many that the display is too rough and dominated by sampling variation. Let n^* represent the number of nonoutlying observations. Then, for $n^* \leq 100$, using no more than about $2\sqrt{n^*}$ bins usually works well; for $n^* > 100$, we use no more than about $10 \times \log_{10} n^*$ bins. Of course, in constructing a histogram, we also want bins that start and end at "nice" numbers (e.g., 1000–2000 rather than 958.43–2104.57); in con-

```
Unit: 100     Lines/stem: 1
      1|2 <--> 1200
depth
  2     0|69
  4     1|68
  9     2|34589
 24     3|001114445667999
 38     4|00123345666777
(14)    5|00111245567899
 50     6|112344556899
 38     7|01445789
 30     8|000122344788888
 15     9|25
 13    10|4
 12    11|003
  9    12|34
  7    13|
  7    14|01

 HIGH:14558 [veterinarians]
      17498 [osteopaths, chiropractors]
      19263 [lawyers, notaries]
      25308 [physicians, surgeons]
      25879 [general managers, other senior officials]
```

Figure 3.3. Stem-and-leaf display for average income in the Canadian occupational prestige data.

structing a stem-and-leaf display, we are limited to bins that correspond to breaks between digits of the data values. Computer programs that construct histograms, incorporate rules such as these.

For the income distribution, $n^* = 102 - 5 = 97$, so we should aim for no more than about $2\sqrt{97} = 19.7 \simeq 20$ bins. The stem-and-leaf display in Figure 3.3 uses 15 stems (plus the "high" stem).

Histograms, including stem-and-leaf displays, are very useful graphs, but they suffer from several problems:

- As we have seen, the visual impression of the data conveyed by a histogram can depend on the arbitrary origin of the bin system.
- Because the bin system dissects the range of the variable into class intervals, the histogram is discontinuous (i.e., rough) even if, as in the case of income, the variable is continuous.
- The form of the histogram depends on the arbitrary width of the bins.
- Moreover, if we use bins that are narrow enough to capture detail where data are plentiful—usually near the center of the distribution—then they may be too narrow to avoid "noise" where data are sparse—usually in the tails of the distribution.

3.1.2 Density Estimation*

Nonparametric density estimation addresses the deficiencies of traditional histograms by averaging and smoothing, much in the manner of the nonparametric regression estimators introduced in the previous chapter. As the term implies, "density estimation" can be construed formally as an attempt to estimate the probability density function of a variable based on a sample, but it can also be thought of informally as a descriptive technique for smoothing histograms.

In fact, the histogram—suitably rescaled—is a particularly simple density estimator.[6] Imagine that the origin of the bin system is at x_0, and that each of the m bins has width $2h$; the end points of the bins are then at x_0, $x_0 + 2h$, $x_0 + 4h, \ldots, x_0 + 2mh$. An observation X_i falls in the jth bin if (by convention)

$$x_0 + 2(j-1)h \leq X_i < x_0 + 2jh$$

The histogram estimator of the density at any x-value located in the jth bin is based on the number of observations that fall in that bin:

$$\hat{p}(x) = \frac{\overset{n}{\underset{i=1}{\#}} [x_0 + 2(j-1)h \leq X_i < x_0 + 2jh]}{2nh}$$

where # is the counting operator.

We can dispense with the arbitrary origin x_0 of the bin system by counting locally within a continuously moving window of half-width h centered at x:

$$\hat{p}(x) = \frac{\overset{n}{\underset{i=1}{\#}} (x - h \leq X_i < x + h)}{2nh}$$

In practice, of course, we would use a computer program to evaluate $\hat{p}(x)$ at a large number of x-values covering the range of X. This "naive density estimator" (so named by Silverman, 1986) is equivalent to locally weighted averaging, using a rectangular weight function:

$$\hat{p}(x) = \frac{1}{nh} \sum_{i=1}^{n} W\left(\frac{x - X_i}{h}\right) \qquad [3.1]$$

where

$$W(z) = \begin{cases} \frac{1}{2} & \text{for } |z| < 1 \\ 0 & \text{otherwise} \end{cases}$$

a formulation that will be useful below when we consider alternative weight functions to smooth the density. The naive estimator is like a histogram that uses bins of width $2h$ but has no fixed origin.

An illustration, using average income from the Canadian occupational prestige data, appears in Figure 3.4, and reveals the principal problem with the naive estimator: Because the estimated density jumps up and down as observations enter and leave the window, the naive density estimator is intrinsically rough.

---

[6] Rescaling is required because a density function encloses a total area of 1. Histograms are typically scaled so that the height of each bar represents frequency, and thus the heights of the bars sum to the sample size n. If each bar spans a bin of width $2h$ (anticipating the notation below), then the total area enclosed by the bars is $n \times 2h$. Dividing the height of each bar by $2nh$ therefore produces the requisite rescaling.

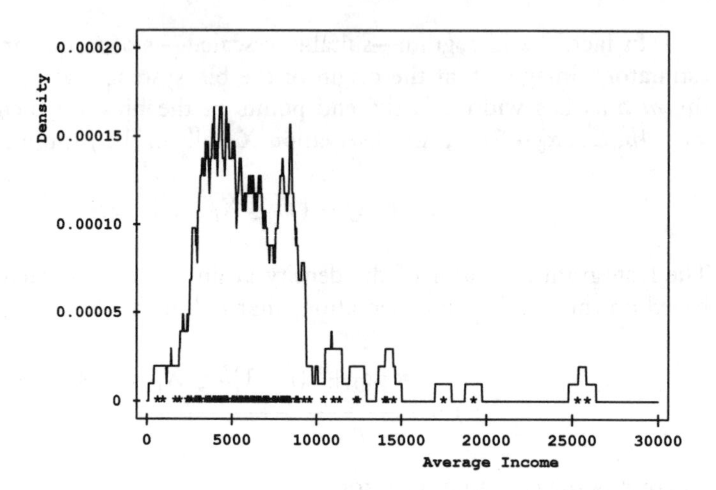

Figure 3.4. Naive density estimator for average income in the Canadian occupational prestige data, using a window half-width of $h = 500$. Note the roughness of the estimator. A "one-dimensional scatterplot" of the data values appears at the bottom of the graph.

The rectangular weight function $W(z)$ in Equation 3.1 is defined to enclose an area of $2 \times \frac{1}{2} = 1$, producing a density estimate that (as required) also encloses an area of 1. Any function that has this property—probability density functions are obvious choices—may be used as a weight function (called a "kernel"). Choosing a kernel that is smooth, symmetric, and unimodal smooths out the rough edges of the naive density estimator. This is the essential insight of kernel density estimation.

The general kernel density estimator is, then, given by

$$\hat{p}(x) = \frac{1}{nh} \sum_{i=1}^{n} K\left(\frac{x - X_i}{h}\right)$$

There are many reasonable choices of the kernel function $K(z)$, including the familiar standard normal density function, $\phi(z)$, which is what we shall use here. While the naive density estimator in effect sums suitably scaled rectangles centered at the observations, the more general kernel estimator sums smooth lumps. An example is shown in Figure 3.5, in which the kernel density estimator is given by the solid line.

Selecting the window width for the kernel estimator is primarily a matter of trial and error—we want a value small enough to reveal detail but large enough to suppress random noise. We can, however, look to statistical theory for rough guidance. If the underlying density that we are trying to estimate is normal with standard deviation σ, then (for the normal kernel) estimation is most efficient with the window half-width

$$h = 0.9\sigma n^{-1/5} \qquad [3.2]$$

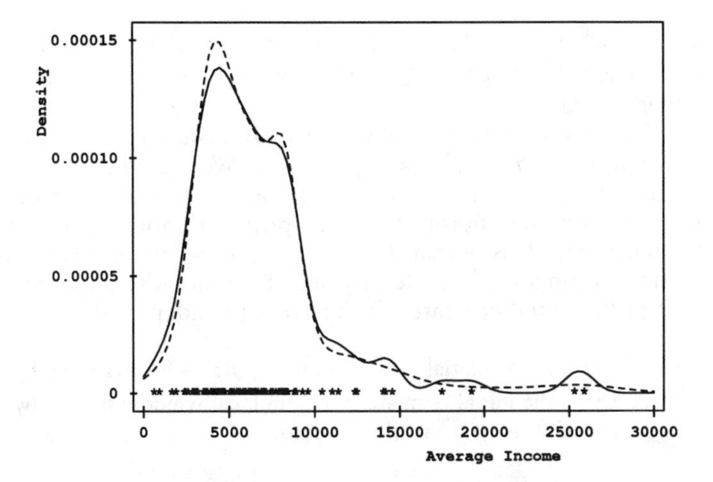

Figure 3.5. Kernel (solid line) and adaptive-kernel (broken line) density estimators for average income in the Canadian occupational prestige data, using a normal kernel and a window half-width of $h = 800$. Note the "images" of the normal kernel (i.e., the bumps) near the right of the display where data are sparse.

As is intuitively reasonable, the optimal window grows gradually narrower as the sample size is increased, permitting finer detail in large samples than in small ones.[7]

Although we might, by reflex, be tempted to replace the unknown σ in Equation 3.2 with the sample standard deviation S, it is prudent to be more cautious, for if the underlying density is sufficiently nonnormal, then the sample standard deviation may be seriously inflated. A common compromise is to use an "adaptive" estimator of spread:

$$A = \min\left(S, \frac{\text{hinge spread}}{1.349}\right)$$

The hinge spread is also called the interquartile range.[8] The factor 1.349 is the interquartile range of the standard normal distribution, making hinge spread/1.349 a robust estimator of σ in the normal setting.

One further caveat: If the underlying density is substantially nonnormal—in particular, if it is skewed or multimodal—then basing h on the adaptive estimator A generally produces a window that is too wide. A good procedure, then, is to start with

$$h = 0.9An^{-1/5}$$

and to adjust this value downwards until the resulting density plot becomes too rough. This is the procedure that was used to find the window width in Fig-

[7] Of course, if we really know that the density is normal, then it is even more efficient to estimate it parametrically by substituting the sample mean $\overline{X}$ and standard deviation S for μ and σ in the formula for the normal density, $p(x) = (2\pi\sigma^2)^{-1/2} \exp[-(x-\mu)^2/2\sigma^2]$.

[8] The hinges are simply defined sample quartiles: See the discussion of boxplots in Section 3.1.4 for definitions of the hinges and the hinge spread.

ure 3.5, where $S = 4246$ and hinge spread/1.349 = (8206 − 4075)/1.349 = 3062. Here, the "optimal" window width, $h = 0.9 \times 3062 \times 102^{-1/5} = 1093$ proved too large.

The kernel density estimator usually does a pretty good job, but the window half-width h remains a compromise: We would prefer a narrower window where data are plentiful (to preserve detail) and a wider one where data are sparse (to suppress noise). Because "plentiful" and "sparse" refer implicitly to the underlying density that we are trying to estimate, it is natural to begin with an initial estimate of the density, and to adjust the window half-width on the basis of the initial estimate. The result is the adaptive-kernel estimator:

1. Calculate an initial density estimate, $\tilde{p}(x)$—for example, by the kernel method.
2. Using the initial estimate, compute local window factors by evaluating the estimated density at the observations:

$$f_i = \left[\frac{\tilde{p}(X_i)}{\tilde{p}}\right]^{-1/2}$$

In this formula, $\tilde{p}$ is the geometric mean of the initial density estimates at the observations—that is,

$$\tilde{p} = \left[\prod_{i=1}^{n} \tilde{p}(X_i)\right]^{1/n}$$

(where the operator $\prod$ indicates continued multiplication). As a consequence of this definition, the f_i's have a product of 1, and hence a geometric mean of 1, ensuring that the area under the density estimate remains equal to 1.

3. Calculate the adaptive density estimator using the local window factors to adjust the width of the kernels centered at the observations:

$$\hat{p}(x) = \frac{1}{nh}\sum_{i=1}^{n}\frac{1}{f_i}K\left(\frac{x - X_i}{f_i h}\right)$$

Applying the adaptive kernel estimator to the income distribution produces the broken line in Figure 3.5. Notice that the two modes of the distribution are more clearly resolved and that the right tail of the distribution (where the data are sparse) is no longer "lumpy."

3.1.3 Quantile Comparison Plots

Quantile comparison plots are useful for comparing an empirical sample distribution with a theoretical distribution, such as the normal distribution. A strength of the display is that it does not require the use of arbitrary bins or windows: The continuity of continuous data is preserved.

Let $P(x)$ represent the theoretical cumulative distribution function (CDF) with which we want to compare the data; that is, $P(x) = \Pr(X \leq x)$. A simple (but, alas, flawed) procedure is to calculate the empirical cumulative distribution function (ECDF) for the observed data, which is simply the proportion of data below each value of x, as x moves continuously from left to right:

$$\hat{P}(x) = \frac{\overset{n}{\underset{i=1}{\#}}(X_i \leq x)}{n}$$

As illustrated in Figure 3.6, however, the ECDF is a "stair-step" function (each

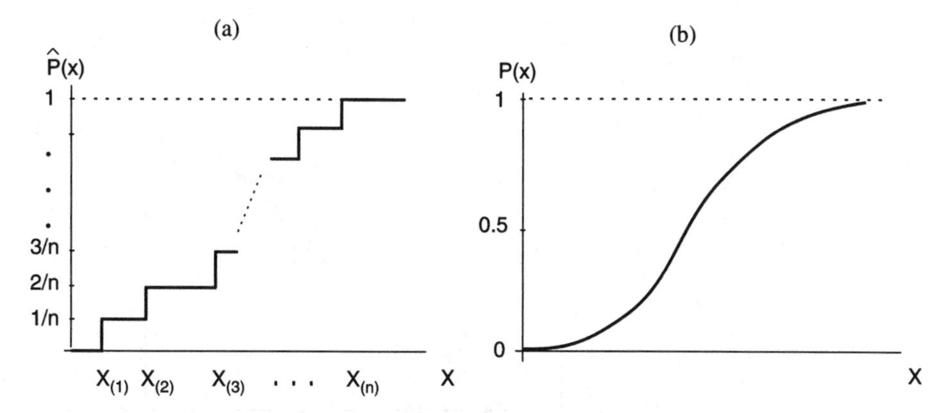

Figure 3.6. A "typical" empirical cumulative distribution function (ECDF) is shown in (*a*), a "typical" theoretical cumulative distribution function (CDF) in (*b*). $X_{(1)}$, $X_{(2)}, \ldots, X_{(n)}$ represent the data values ordered from smallest to largest. Note that the ordered data values are not, in general, equally spaced.

step occurs at an observation, and is of height $1/n$), while the CDF is typically smooth, making the comparison difficult.

The quantile comparison plot avoids this problem by never constructing the ECDF explicitly:

1. Order the data values from smallest to largest, denoted $X_{(1)}, X_{(2)}, \ldots, X_{(n)}$. The $X_{(i)}$ are called the *order statistics* of the sample.
2. By convention, the cumulative proportion of the data "below" $X_{(i)}$ is given by[9]

$$P_i = \frac{i - \frac{1}{2}}{n}$$

3. Use the inverse of the CDF to find the value z_i corresponding to the cumulative probability P_i; that is,[10]

$$z_i = P^{-1}\left(\frac{i - \frac{1}{2}}{n}\right)$$

4. Plot the z_i as horizontal coordinates against the $X_{(i)}$ as vertical coordinates. If X is sampled from the distribution P, then $X_{(i)} \simeq z_i$. That is, the plot should be approximately linear, with an intercept of 0 and slope of 1. This relationship is only approximate because of sampling error (see point 6). If the distributions are identical except for location, then the plot is approximately linear with a nonzero intercept, $X_{(i)} \simeq \mu + z_i$; if the distributions are identical except for scale, then the plot is approximately linear with a slope different from 1, $X_{(i)} \simeq \sigma z_i$; finally, if the distributions differ both in location and scale, then $X_{(i)} \simeq \mu + \sigma z_i$.

[9] This definition avoids cumulative proportions of 0 or 1, which would be an embarrassment in step 3 for distributions like the normal which never quite reach cumulative probabilities of 0 or 1. In effect, we count half of each observation below its exact value and half above.

[10] This operation assumes that the CDF has an inverse—that is, that P is a strictly increasing function (one that never quite levels off). The common continuous probability distributions in statistics—for example, the normal, t-, F-, and χ^2 distributions—all have this property.

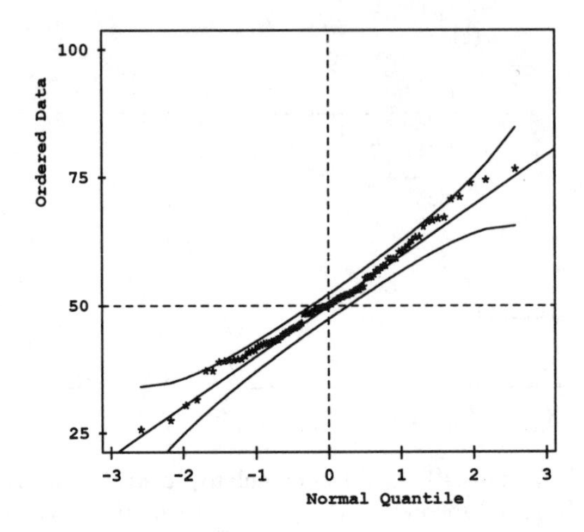

Figure 3.7. Normal quantile comparison plot for a sample of 100 observations drawn from a normal distribution with mean 50 and standard deviation 10.

5. It is often helpful to place a comparison line on the plot to facilitate the perception of departures from linearity. The line can be plotted by eye, attending to the central part of the data, or we can draw a line connecting the hinges (quartiles). For a normal quantile comparison plot—comparing the distribution of the data with the standard normal distribution—we can alternatively use the median as a robust estimator of μ and the hinge spread/1.349 as a robust estimator of σ. (The more conventional choices, $\hat{\mu} = \overline{X}$ and $\hat{\sigma} = S$ will not work well when the data are substantially nonnormal.)
6. We expect some departure from linearity because of sampling variation; it therefore assists interpretation to display the expected degree of sampling error in the plot. The estimated standard error of the order statistic $X_{(i)}$ is

$$\widehat{\mathrm{SE}}(X_{(i)}) = \frac{\hat{\sigma}}{p(z_i)}\sqrt{\frac{P_i(1-P_i)}{n}}$$

where $p(z)$ is the probability density function corresponding to the CDF $P(z)$. The values along the fitted line are given by $\hat{X}_{(i)} = \hat{\mu} + \hat{\sigma} z_i$. An approximate 95% confidence "envelope" around the fitted line is, therefore,[11]

$$\hat{X}_{(i)} \pm 2 \times \widehat{\mathrm{SE}}(X_{(i)})$$

Figures 3.7 to 3.10 display normal quantile comparison plots for several illustrative distributions:

- Figure 3.7 plots a sample of $n = 100$ observations from a normal distribution with mean $\mu = 50$ and standard deviation $\sigma = 10$. The plotted points are reasonably linear and stay within the rough 95% confidence envelope.

[11] By the method of construction, the 95% confidence level applies (pointwise) to each $\hat{X}_{(i)}$, not to the whole envelope: There is a greater probability that *at least one* point strays outside the envelope even if the data are sampled from the comparison distribution. Determining a true 95% confidence envelope would be a formidable task, since the order statistics are not independent.

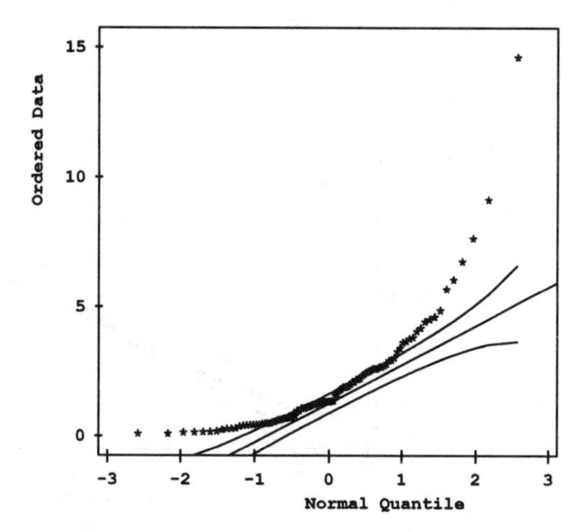

Figure 3.8. Normal quantile comparison plot for a sample of 100 observations from the positively skewed chi-square distribution with 2 degrees of freedom.

- Figure 3.8 plots a sample of $n = 100$ observations from the positively skewed chi-square distribution with 2 degrees of freedom. The positive skew of the data is reflected in points that lie above the comparison line in both tails of the distribution. (In contrast, negatively skewed data would lie below the comparison line.)
- Figure 3.9 plots a sample of $n = 100$ observations from the heavy-tailed t-distribution with 2 degrees of freedom. In this case, values in the upper tail lie above the corresponding normal quantiles, and values in the lower tail below the corresponding normal quantiles.
- Figure 3.10 shows a normal quantile comparison plot for the income data. The positive skew of the distribution is readily apparent. The bimodal character of the data, however, is not easily discerned in this display: Quantile comparison plots highlight the tails of distributions. This is important, because the behavior of the

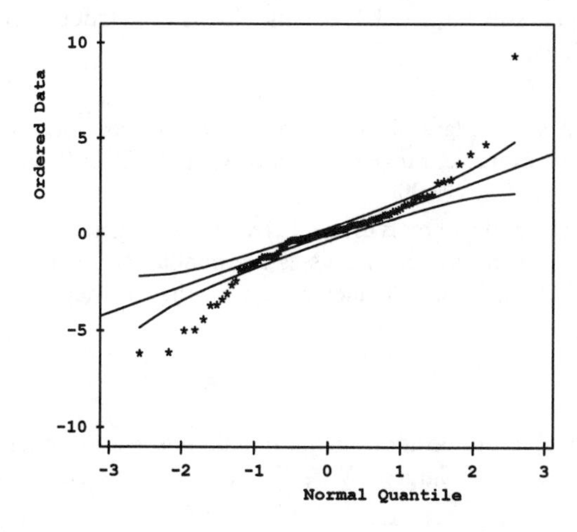

Figure 3.9. Normal quantile comparison plot for a sample of 100 observations from the heavy-tailed t-distribution with 2 degrees of freedom.

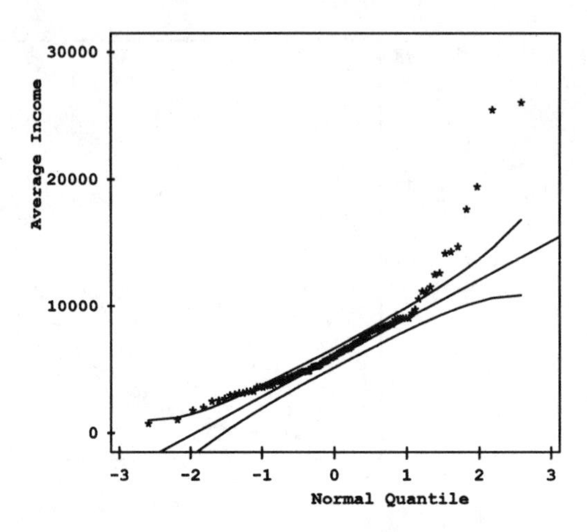

Figure 3.10. Normal quantile comparison plot for average income in the Canadian occupational prestige data. Note the positive skew.

tails is often problematic for standard estimation methods like least squares, but it is useful to supplement quantile comparison plots with other displays.

3.1.4 Boxplots

Unlike histograms, density plots, and quantile comparison plots, boxplots (due to Tukey, 1977) present only summary information on center, spread, and skewness, but they show individual, potentially outlying, observations in the tails of the distribution. Boxplots, therefore, are particularly useful when we require a compact representation of a distribution (as, for example, in the margins of a scatterplot), when we wish to compare the principal characteristics of several distributions,[12] or when we want to select a transformation that makes a distribution more symmetric.[13]

An illustrative boxplot, for income in the Canadian occupational prestige data, appears in Figure 3.11. This plot is constructed according to the following conventions:

1. A scale is laid off to accommodate the extremes of the data. The income data, for example, range between \$611 and \$25,879, so I have used a scale that runs from 0 to 30,000.
2. The central box is drawn between the hinges, which are simply defined quartiles, and therefore encompasses the middle half of the data. The line in the central box represents the median. Recall that the depth of the median is

$$\text{depth}(M) = \frac{n+1}{2}$$

giving the position of the middle observation after the data are ordered from smallest to largest: $X_{(1)}, X_{(2)}, \ldots, X_{(n)}$. When n is even, the depth of the me-

[12] See Section 3.2.
[13] See Section 4.2.

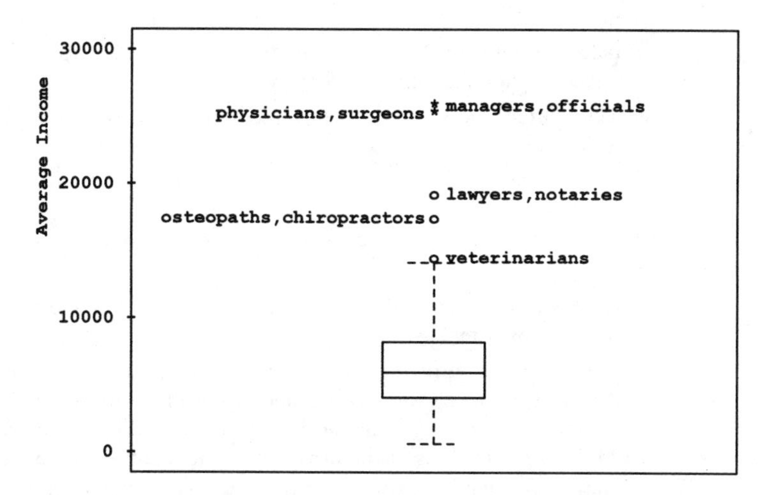

Figure 3.11. Boxplot for income in the Canadian occupational prestige data. The central box is drawn between the hinges; the position of the median is marked in the box; and outlying observations are displayed individually.

dian has a fractional part; using double brackets to represent truncation to an integer, we count in from either end to average the two observations at depth $[[(n+1)/2]]$. For example, for the 102 income scores, $\text{depth}(M) = (102+1)/2 = 51.5$; the two observations at depth $[[51.5]] = 51$ are 5902 and 5959, and so the median is $M = (5902 + 5959)/2 = 5930.5$.
Likewise, the depth of the hinges is

$$\text{depth}(H) = \frac{[[\text{depth}(M)]] + 1}{2}$$

If depth(H) has a fractional part, then, for each hinge, we average the two observations at the adjacent positions, that is, at $[[\text{depth}(H)]]$ and $[[\text{depth}(H) + 1]]$. For the income distribution, $\text{depth}(H) = ([[51.5]] + 1)/2 = 26$; the lower hinge is, therefore, $H_L = X_{(26)} = 4075$, and the upper hinge is $H_U = X_{(77)} = 8206$. (Note that counting down 26 observations from the top yields the subscript $102 - 26 + 1 = 77$.)

3. The following rules are used to identify possible outliers, which are shown individually in the boxplot:

   - The hinge spread (or interquartile range) is the difference between the hinges:

   $$H\text{-spread} = H_U - H_L$$

   - The "inner fences" are located 1.5 hinge spreads beyond the hinges:

   $$\text{IF}_L = H_L - 1.5 \times H\text{-spread}$$

   $$\text{IF}_U = H_U + 1.5 \times H\text{-spread}$$

   Observations beyond the inner fences (but within the outer fences, defined below) are termed "outside" and are represented by circles. The fences themselves are not shown in the display.

- The "outer fences" are located three hinge spreads beyond the hinges:

$$\text{OF}_L = H_L - 3 \times H\text{-spread}$$

$$\text{OF}_U = H_U + 3 \times H\text{-spread}$$

Observations beyond the outer fences are termed "far outside" and are represented by asterisks.

- The "whisker" growing from each end of the central box extends either to the extreme observation on its side of the distribution (as at the low end of the income data) or to the most extreme nonoutlying observation, called the "adjacent value" (as at the high end of the income distribution).[14]

Here is a rough justification for the fences: In a normal population, the hinge spread is 1.349 standard deviations, and so $1.5 \times H\text{-spread} = 1.5 \times 1.349 \times \sigma \simeq 2\sigma$. The hinges are located $1.349/2 \simeq 0.7$ standard deviations above and below the mean. The inner fences are, therefore, approximately at $\mu \pm 2.7\sigma$, and the outer fences at $\mu \pm 4.7\sigma$. From the standard normal table, $\Pr(Z > 2.7) \simeq .003$, so we expect slightly less than 1% of the observations beyond the inner fences ($2 \times .003 = .006$); likewise, because $\Pr(Z > 4.7) \simeq 1.3 \times 10^{-6}$, we expect less than one observation in 100,000 beyond the outer fences.

The boxplot of the income distribution in Figure 3.11 clearly reveals the skewness of the distribution: The lower whisker is shorter than the upper whisker; and there are several outside observations at the upper end of the income distribution, but not at the lower end. The central part of the distribution appears reasonably symmetric, however: The median is only slightly closer to the lower hinge than to the upper hinge. The apparent bimodality of the income data is not represented in the boxplot.

> There are many useful univariate displays, including the traditional histogram. The stem-and-leaf plot is a modern variant of the histogram constructed directly from numerical data. Nonparametric density estimation may be employed to smooth a histogram. Quantile comparison plots are useful for comparing data with a theoretical probability distribution. Boxplots summarize some of the most important characteristics of a distribution, including center, spread, skewness, and the presence of outliers.

[14] All of the folksy terminology—hinges, fences, whiskers, etc.—comes from Tukey (1977).

EXERCISE

3.1 Using the methods for displaying univariate distributions described in this section, examine the distributions of the following variables. In each case, examine the data for skewness, nonnormality, multiple modes, and unusual values.

(a) Moral integration, ethnic heterogeneity, and geographic mobility in Angell's data on 43 U.S. cities (in Table 2.3 and `angell.dat`).

(b) Prestige in Duncan's data on 45 U.S. occupations (in Table 3.2 and `duncan.dat`).

(c) Income and infant mortality in Leinhardt and Wasserman's data on 103 nations of the world (in `leinhard.dat`).

(d) Number of interlocking directorships and assets in Ornstein's data on 248 dominant Canadian corporations (in `ornstein.dat`).

TABLE 3.2 Prestige (Prs.), Income (Inc.), and Educational (Ed.) Levels of U.S. Occupations in 1950. "Prestige" is the percentage of ratings of good or better; "Income" is the percentage earning $3500 or more; "Education" is the percentage of high-school graduates

Occupation	*Prs.*	*Inc.*	*Ed.*	*Occupation*	*Prs.*	*Inc.*	*Ed.*
Professional and Managerial				Mail carrier	34	48	55
Physician	97	76	97	Store clerk	16	29	50
Professor	93	64	93	*Blue Collar*			
Banker	92	78	82	Railroad engineer	67	81	28
Architect	90	75	92	Machinist	57	36	32
Chemist	90	64	86	Electrician	53	47	39
Dentist	90	80	100	Policeman	41	34	47
Lawyer	89	76	98	Carpenter	33	21	23
Civil engineer	88	72	86	Plumber	29	44	25
Minister	87	21	84	Auto repairman	26	22	22
Pilot	83	72	76	Machine operator	24	21	20
Accountant	82	62	86	Barber	20	16	26
Factory owner	81	60	56	Streetcar motorman	19	42	26
Author	76	55	90	Cook	16	14	22
Contractor	76	53	45	Coal miner	15	7	7
Teacher	73	48	91	Truck driver	13	21	15
Welfare worker	59	41	84	Night watchman	11	17	25
Undertaker	57	42	74	Gas station attendant	10	15	29
Store manager	45	42	44	Taxi driver	10	9	19
White Collar				Waiter	10	8	32
Reporter	52	67	87	Janitor	8	7	20
Insurance agent	41	55	71	Bartender	7	16	28
Bookkeeper	39	29	72	Soda fountain clerk	6	12	30
Conductor	38	76	34	Shoe shiner	3	9	17

Source of Data: Duncan (1961).

3.2 Plotting Bivariate Data

The scatterplot—a direct geometric representation of observations on two quantitative variables (generically, Y and X)—is the most useful of all statistical graphs. The scatterplot is a natural representation of data partly because the media on which we draw plots—paper, computer screens—are intrinsically two dimensional. Scatterplots are as familiar and essentially simple as they are useful; I will therefore limit this presentation to a few points. There are many examples of bivariate scatterplots in this book, including in the previous chapter.

- In analyzing data, it is convenient to work in a computing environment that permits the interactive identification of observations in a scatterplot.
- Because relationships between variables in the social sciences are often weak, scatterplots can be dominated visually by "noise." It often helps, therefore, to plot a nonparametric regression of Y on X.
- Scatterplots in which one or both variables are highly skewed are difficult to examine because the bulk of the data congregate in a small part of the display. It often helps to "correct" substantial skews prior to examining the relationship between Y and X.[15]
- Scatterplots in which the variables are discrete are difficult to examine. An extreme instance of this phenomenon is shown in Figure 3.12, which plots scores on a 10-item vocabulary test included in NORC's General Social Survey against years of education. One solution—particularly useful when only X is discrete—is to focus on the conditional distribution of Y for each value of X. Boxplots, for example, can be employed to represent the conditional distributions (see Figure 3.14). Another solution is to separate overlapping points by adding a small random quantity to the discrete scores. In Figure 3.13, for example, I have added a uniform random variable on the interval $[-1/2, +1/2]$ to each of vocabulary and education. Paradoxically, the tendency for vocabulary to increase with education is much clearer in the randomly "jittered" display.[16]

> The bivariate scatterplot is a natural graphical display of the relationship between two quantitative variables. Interpretation of a scatterplot can often be assisted by graphing a nonparametric regression, which summarizes the relationship between the two variables. Scatterplots of the relationship between discrete variables can be enhanced by randomly jittering the data.

As mentioned, when the independent variable is discrete, parallel boxplots can be used to display the conditional distributions of Y. One common case occurs when the independent variable is a qualitative/categorical variable. An example is shown in Figure 3.14, using data collected by Michael Ornstein (1976)

---

[15] See Chapter 4.

[16] The idea of jittering a scatterplot, as well as the terminology, is due to Cleveland (1994).

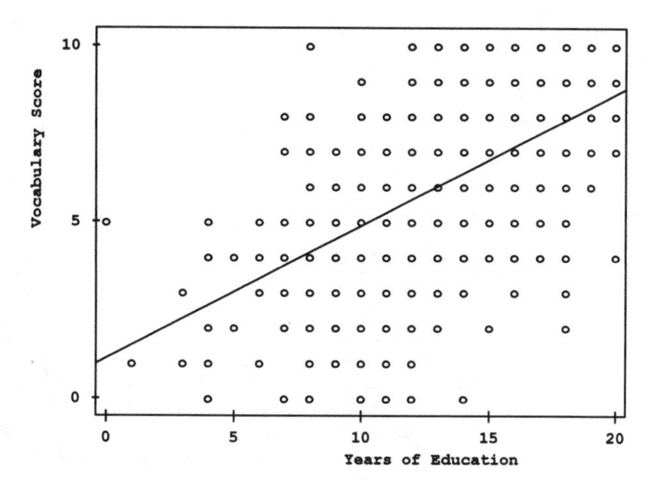

Figure 3.12. Scatterplot of scores on a 10-item vocabulary test versus years of education. Although there are $n = 968$ observations in the dataset, most of the plotted points fall on top of one another. The least-squares regression line is shown on the plot.

Source of Data: 1989 General Social Survey.

on interlocking directorates among the 248 largest Canadian firms. The dependent variable in this graph is the number of interlocking directorships and executive positions maintained by each firm with others in the group of 248. The independent variable is the nation in which the corporation is controlled, coded as Canada, United Kingdom, United States, and other foreign.

It is apparent from the graph that the average level of interlocking is greater among Canadian and other-foreign corporations than among corporations controlled in the United Kingdom and the United States. It is relatively difficult to discern detail in this display: first, because the conditional distributions of inter-

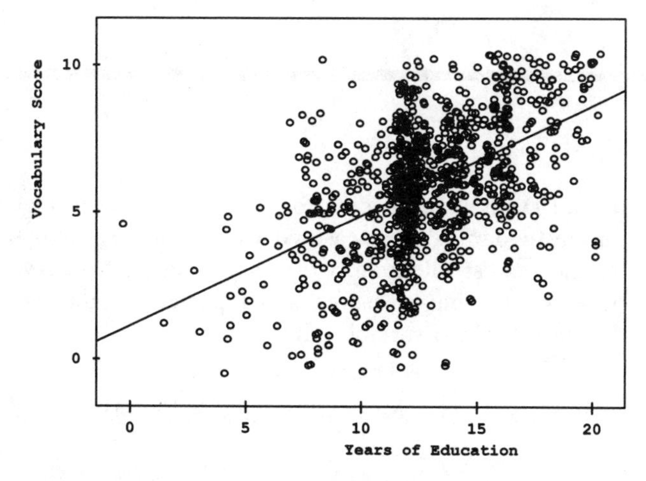

Figure 3.13. Jittered scatterplot for vocabulary score versus years of education. A uniformly distributed random quantity between $-1/2$ and $+1/2$ was added to each score for both variables. The original least-squares line is shown on the plot.

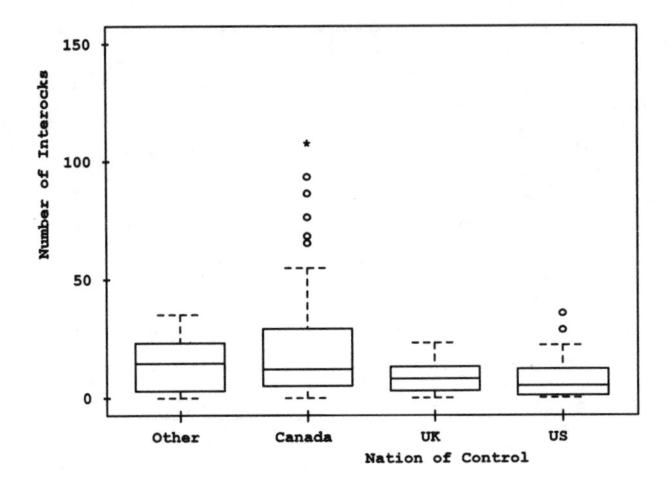

Figure 3.14. Number of interlocking directorate and executive positions by nation of control, for 248 dominant Canadian firms.

Source of Data: Personal communication from Michael Ornstein.

locks are positively skewed; and, second, because there is an association between level and spread—variation is also greater among Canadian and other-foreign firms than among U.K. and U.S. firms.[17]

> Parallel boxplots display the relationship between a quantitative dependent variable and a discrete independent variable.

EXERCISES

3.2 Recall Davis's weight data for women, plotted in Figure 2.3 (and included in `davis.dat`). Both measured weight and reported weight were recorded to the nearest kilogram. Correct the outlying observation, and investigate whether jittering measured and reported weight clarifies the relationship between the two variables. It might help to show the line

$$\widehat{\text{Measured}}\ \text{weight} = \text{Reported weight}$$

on the scatterplot.

[17] We shall revisit this example in Section 4.4.

3.3 Construct parallel comparison boxplots for each of the following pairs of variables. In each case, compare levels and spreads across the parallel boxplots, and look for unusual data values.

(a) Moral integration by region, for Angell's data on 43 U.S. cities (in Table 2.3 and `angell.dat`).

(b) Infant mortality by region, for Leinhardt and Wasserman's data on 103 nations of the world (in `leinhard.dat`).

3.3 Plotting Multivariate Data

Because paper and computer screens are two dimensional, graphical display of multivariate data is intrinsically difficult. Multivariate displays for quantitative data often project the higher-dimensional "point cloud" of the data onto a two-dimensional space. It is, of course, impossible to view a higher-dimensional scatterplot directly. The essential trick of effective multidimensional display is to select projections that reveal important characteristics of the data. In certain circumstances, projections can be selected on the basis of a statistical model fit to the data or on the basis of explicitly stated criteria.[18]

A simple approach to multivariate data, which does not require a statistical model, is to examine bivariate scatterplots for all pairs of variables. Arraying these plots in a "scatterplot matrix" produces a graphical analog to the correlation matrix.

An illustrative scatterplot matrix, for data on the prestige, education, and income levels of 45 U.S. occupations, appears in Figure 3.15. In this dataset, analyzed by Duncan (1961), "prestige" represents the percentage of respondents in a survey who rated an occupation as "good" or "excellent" in prestige; "education" represents the percentage of incumbents in the occupation in the 1950 U.S. census who were high-school graduates; and "income" represents the percentage of occupational incumbents who earned incomes in excess of $3500. Duncan's purpose was to use a regression analysis of prestige on income and education to predict the prestige levels of other occupations, for which data on income and education were available, but for which there were no direct prestige ratings.[19]

The variable names on the diagonal of the scatterplot matrix in Figure 3.15 label the rows and columns of the display: For example, the vertical axis for the two plots in the first row of the display is "prestige"; the horizontal axis for the two plots in the second column is "income."

It is important to understand an essential limitation of the scatterplot matrix as a device for analyzing multivariate data: By projecting the multidimensional point cloud onto pairs of axes, the plot focuses on the *marginal* relationships between the corresponding pairs of variables. The object of data analysis for several variables is typically to investigate *partial* relationships (between pairs of variables, "controlling" statistically for other variables), not marginal associations. For example, in the Duncan dataset, we are more interested in the

---

[18] We shall apply these powerful ideas in Chapters 11 and 12.

[19] We shall return to this regression problem in Chapter 5.

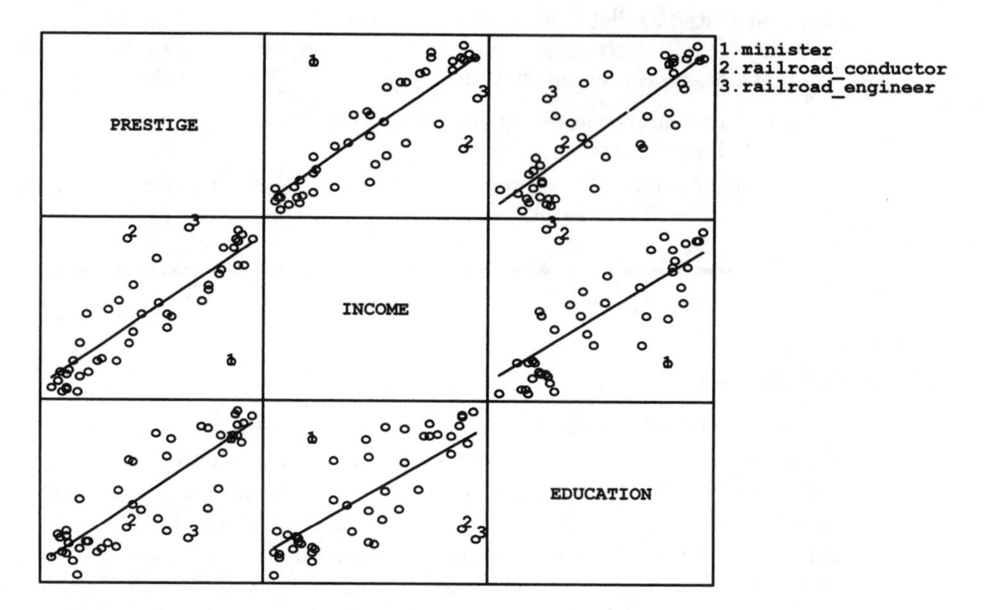

Figure 3.15. Scatterplot matrix for occupational prestige, level of education, and level of income, for 45 U.S. occupations. The least-squares regression line is shown on each plot. Three unusual observations were identified interactively using a "mouse."

Source of Data: Duncan (1961).

partial relationship of prestige to income holding education constant than in the marginal relationship between prestige and income ignoring education.

The dependent variable Y can be related marginally to a particular X, even when there is no partial relationship between the two variables controlling for other X's. It is also possible for there to be a partial association between Y and an X but no marginal association. Furthermore, if the X's themselves are nonlinearly related, then the marginal relationship between Y and a specific X can be nonlinear even when their partial relationship is linear.[20]

Despite this intrinsic limitation, scatterplot matrices often uncover interesting features of the data, and this is indeed the case in Figure 3.15, where the display reveals three unusual observations: *Ministers* have relatively low income for their relatively high level of education, and relatively high prestige for their relatively low income; *railroad conductors* and *railroad engineers* have relatively high incomes for their more-or-less average levels of education; *railroad conductors* also have relatively low prestige for their relatively high incomes.[21]

Information about a categorical third variable can be entered on a bivariate scatterplot by coding the plotting symbols. The most effective codes use different

---

[20] These ideas are explored in Chapter 5.

[21] This pattern bodes ill for the least-squares linear regression of prestige on income and education, as we shall see in Section 11.6.

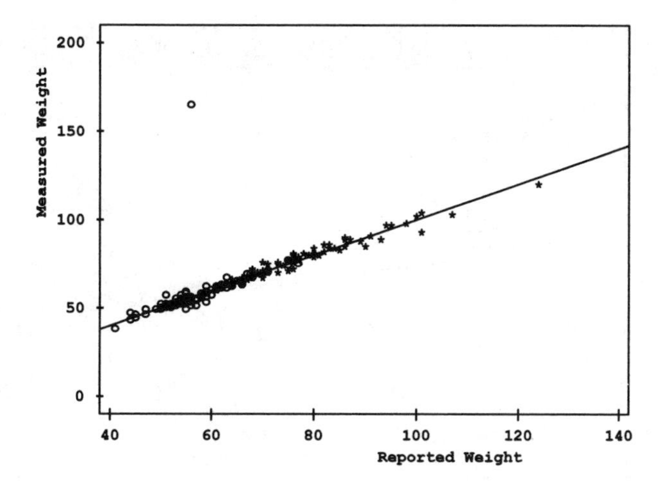

Figure 3.16. Davis's data on measured and reported weight, by gender. Data points for men are represented by asterisks, for women by circles. The line on the plot is $Y = X$.

colors to represent categories, but degrees of fill, distinguishable shapes, and distinguishable letters can also be effective.[22] Figure 3.16 shows a scatterplot of Davis's data on measured and reported weight.[23] Observations for men are displayed as asterisks, for women as circles. Note that except for the outlying point (which, recall, represents an error in the data), the points both for men and for women cluster near the line $Y = X$; it is also clear from the display that most men are heavier than most women, as one would expect.

Another useful multivariate display, directly applicable only to three variables at a time, is the three-dimensional scatterplot. This display is an illusion produced by modern statistical software: The graph really represents a projection of a three-dimensional scatterplot onto a two-dimensional computer screen. Nevertheless, motion (e.g., rotation) and the ability to interact with the display—sometimes combined with the effective use of perspective, color, depth cueing, and other visual devices—can produce a vivid impression of directly examining a three-dimensional space.

It is literally impossible to convey this impression adequately on the static, two-dimensional page of a book, but Figure 3.17 shows Duncan's prestige data rotated interactively into a particularly revealing orientation: Looking down the cigar-shaped scatter of most of the data, the three unusual observations stand out very clearly. Just as data can be projected onto a judiciously chosen plane in a two-dimensional plot, higher-dimensional data can be projected onto a three-dimensional space, which then can be examined and rotated interactively.

---

[22] See Spence and Lewandowsky (1990) for a fine review of the literature on graphical perception, including information on coded scatterplots.

[23] Davis's data were introduced in Chapter 2, where only the data for women were presented.

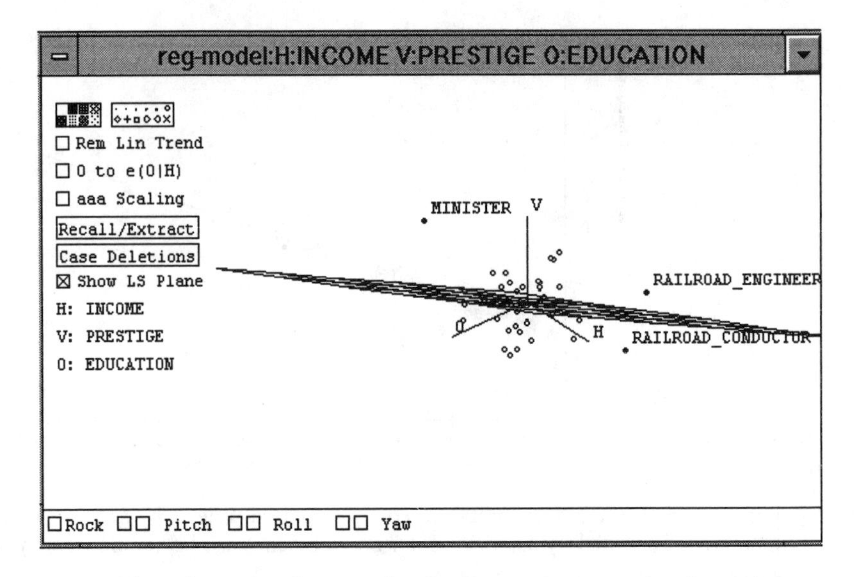

Figure 3.17. Three-dimensional scatterplot for Duncan's occupational prestige data, rotated into an orientation that reveals the three unusual observations. These observations were interactively identified with a mouse. From this orientation, the least-squares regression plane, also shown in the plot, is viewed nearly edge on. The plot was produced by a slightly modified version of the R-code/Lisp-Stat software described in Cook and Weisberg (1994). The "plot controls" at the bottom and to the left of the graph are used to manipulate the display.

> Visualizing multivariate data is intrinsically difficult because we cannot directly examine higher-dimensional scatterplots. Effective displays project the higher-dimensional point cloud onto two or three dimensions; these displays include the scatterplot matrix and the dynamic three-dimensional scatterplot.

EXERCISE

3.4 Using suitable statistical software, construct a scatterplot matrix and a three-dimensional scatterplot for each of the following datasets. If the software permits, link the scatterplot matrix to the three-dimensional plot. In each case, examine the pattern of association between and among the variables, and try to locate unusual data values.

(a) Moral integration, ethnic heterogeneity, and geographic mobility, for Angell's data on 43 U.S. cities (in Table 2.3 and `angell.dat`). Use a different plotting symbol or color for cities in different regions. In the three-dimensional plot, treat moral integration as the dependent variable (vertical axis).

(b) Prestige, income, education, and percentage of women for the Canadian occupational prestige data (in `prestige.dat`). Use the first three of these variables for the three-dimensional plot, treating prestige as the dependent variable (vertical axis).

3.4 Summary

- Statistical graphs are central to effective data analysis, both in the early stages of an investigation and in statistical modeling.
- There are many useful univariate displays, including the traditional histogram. The stem-and-leaf plot is a modern variant of the histogram constructed directly from numerical data. Nonparametric density estimation may be employed to smooth a histogram. Quantile comparison plots are useful for comparing data with a theoretical probability distribution. Boxplots summarize some of the most important characteristics of a distribution, including center, spread, skewness, and the presence of outliers.
- The bivariate scatterplot is a natural graphical display of the relationship between two quantitative variables. Interpretation of a scatterplot can often be assisted by graphing a nonparametric regression, which summarizes the relationship between the two variables. Scatterplots of the relationship between discrete variables can be enhanced by randomly jittering the data.
- Parallel boxplots display the relationship between a quantitative dependent variable and a discrete independent variable.
- Visualizing multivariate data is intrinsically difficult because we cannot directly examine higher-dimensional scatterplots. Effective displays project the higher-dimensional point cloud onto two or three dimensions; these displays include the scatterplot matrix and the dynamic three-dimensional scatterplot.

3.5 Recommended Reading

The literature—particularly the recent literature—on statistical graphics is truly voluminous. I shall furnish only the briefest of bibliographies:

- Fox (1992) presents a brief overview of statistical graphics, including information on the history of the subject. Tufte's (1983) influential book on graphical presentation is opinionated but well worth reading.
- Modern interest in statistical graphics is the direct result of John Tukey's work on exploratory data analysis; unfortunately, Tukey's idiosyncratic writing style makes his seminal book (Tukey, 1977) difficult to read. Velleman and Hoaglin (1981) provide a more digestible introduction to the topic. There is interesting information on the statistical theory underlying exploratory data analysis in two volumes edited by Hoaglin et al. (1983, 1985).

- Tukey's influence has made Bell Labs a center of current work on statistical graphics, much of which is described in two accessible and interesting books by William Cleveland (1993, 1994) and in Chambers et al. (1983). Cleveland (1994) is a good place to start.
- Modern statistical graphics is closely associated with advances in statistical computing: The New S statistical computing environment (Becker, et al., 1988; Chambers and Hastie, 1992), also a product of Bell Labs, is particularly strong in its graphical capabilities. An interesting—and free—competitor is Tierney's Lisp-Stat, based on the Lisp programming language, and described in Tierney (1990). Cook and Weisberg (1994) use Lisp-Stat to produce an impressive statistical package, called R-Code, which incorporates a variety of statistical graphics of particular relevance to regression analysis (including many of the methods described later in this text). Friendly (1991) describes how to construct modern statistical graphs using the SAS/Graph system. Brief presentations of these and other statistical computing environments appear in a book edited by Stine and Fox (1996).
- Atkinson (1985) presents a variety of innovative graphs in support of regression analysis, as do Cook and Weisberg (1994).

4

Transforming Data

"Classical" statistical models—linear least-squares regression, for example—make strong assumptions about the structure of data, assumptions which, more often than not, fail to hold in practice. One solution is to abandon classical methods in favor of more flexible alternatives, such as nonparametric regression analysis. These newer methods are valuable, and I expect that they will be used with increasing frequency, but they are more complex and have their own limitations, as we saw in Chapter 2.[1]

It is, alternatively, often feasible to transform the data so that they conform more closely to the restrictive assumptions of classical statistical models. In addition, and as we shall discover in this chapter, transformations can often assist in the examination of data, even in the absence of a statistical model. The chapter introduces two general families of transformations and shows how they can be used to make distributions symmetric, to make the relationship between two variables linear, and to equalize variation across groups.

> Transformations can often facilitate the examination and modeling of data.

[1] Also see Chapter 14.

4.1 The Family of Powers and Roots

There is literally an infinite variety of functions $f(x)$ that could be used to transform a quantitative variable X. In practice, of course, it helps to be more restrictive, and a particularly useful group of transformations is the "family" of powers and roots:

$$X \to X^p \qquad [4.1]$$

where the arrow indicates that we intend to replace X with the transformed variable X^p. If p is negative, then the transformation is an inverse power: For example, $X^{-1} = 1/X$, and $X^{-2} = 1/X^2$. If p is a fraction, then the transformation represents a root: For example, $X^{1/3} = \sqrt[3]{X}$ and $X^{-1/2} = 1/\sqrt{X}$.

We shall find it convenient to define the family of power transformations in a slightly more complex manner:

$$X \to X^{(p)} \equiv \frac{X^p - 1}{p} \qquad [4.2]$$

We use the parenthetical superscript (p) to distinguish this definition from the more straightforward one in Equation 4.1. Because $X^{(p)}$ is a linear function of X^p, the two transformations have the same essential effect on the data, but, as is apparent in Figure 4.1, the definition in Equation 4.2 reveals more transparently

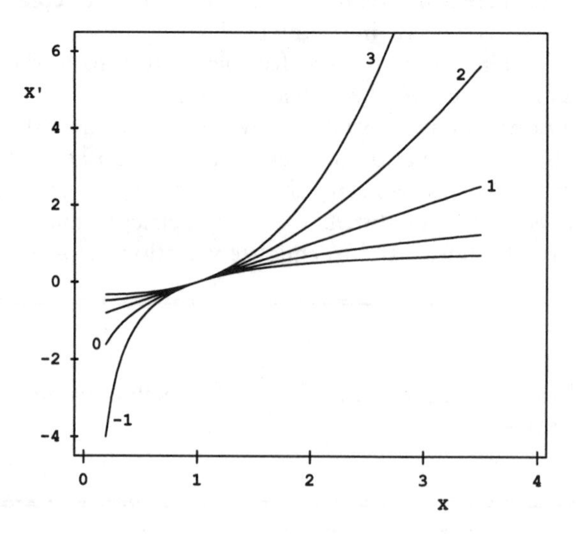

Figure 4.1. The family of power transformations X' of X. The curve labeled p is the transformation $X^{(p)}$, that is, $(X^p - 1)/p$; $X^{(0)}$ is $\log_e X$.

the essential unity of the family of powers and roots:

- Dividing by p preserves the direction of X, which otherwise would be reversed when p is negative, as illustrated in the following example:

X	X^{-1}	$\frac{X^{-1}}{-1}$
1	1	−1
2	1/2	−1/2
3	1/3	−1/3
4	1/4	−1/4

- The transformations $X^{(p)}$ are "matched" above $X = 1$ both in level and in slope: (1) $1^{(p)} = 0$, for all values of p; and (2) each transformation has a slope of 1 at $X = 1$.[2]
- The power transformation X^0 is useless, because it changes all values to 1, but we can think of the log transformation as a kind of "zeroth" power: As p gets very close to 0, the log function more and more closely approximates $X^{(p)}$.[3] Because the log transformation is so useful, we shall, by convention, take $X^{(0)} \equiv \log_e X$, where $e \simeq 2.718$ is the base of the natural logarithms.

In practice, it is generally more convenient to use logs to the base 10 or base 2, which are more easily interpreted than logs to the base e: For example, increasing $\log_{10} X$ by 1 is equivalent to multiplying X by 10; increasing $\log_2 X$ by 1 is equivalent to doubling X. Selection of a base for the log transformation is essentially arbitrary and inconsequential, however, because changing bases is equivalent to multiplying by a constant; for example,

$$\log_{10} X = \log_{10} e \times \log_e X \simeq 0.4343 \times \log_e X$$

Likewise, because of its relative simplicity, we usually use X^p in preference to $X^{(p)}$ when $p \neq 0$.

> The powers and roots are a particularly useful family of transformations: $X \rightarrow X^p$. We employ the log transformation in place of X^0.

Matching the transformations, as in Figure 4.1, facilitates comparisons among them and highlights their relative effects. In particular, descending the "ladder" of powers and roots toward $X^{(-1)}$ compresses the large values of X

---

[2] That is, the derivative of $X^{(p)}$ at $X = 1$ is 1; see Exercise 4.1.

[3] More formally,

$$\lim_{p \to 0} \frac{X^p - 1}{p} = \log_e X$$

and spreads out the small ones; ascending the ladder of powers and roots toward $X^{(2)}$ has the opposite effect.[4] As p moves from $p = 1$ (i.e., no transformation) in either direction, the transformation grows more powerful.

These effects are apparent in Figure 4.1 and in the following simple examples (in which the numbers by the braces give differences between adjacent values):

$-1/X$	$\log_2 X$	X	X^2	X^3
-1	0	1	1	1
$\frac{1}{2}$ {	1 {	} 1	} 3	} 7
$-1/2$	1	2	4	8
$\frac{1}{6}$ {	0.59 {	} 1	} 5	} 19
$-1/3$	1.59	3	9	27
$\frac{1}{12}$ {	0.41 {	} 1	} 7	} 37
$-1/4$	2	4	16	64

Power transformations are sensible only when all of the values of X are positive. First of all, some of the transformations, such as log and square root, are undefined for negative or zero values. Second, the power transformations are not monotone—that is, not order preserving—when there are both positive and negative values in the data; for example,

X	X^2
-2	4
-1	1
0	0
1	1
2	4

This is not, however, a practical limitation, because we can always add a positive constant (called a "start") to each data value to make all of the values positive, calculating the transformation $X \rightarrow (X + s)^p$; in the preceding example,

X	$(X+3)^2$
-2	1
-1	4
0	9
1	16
2	25

[4] The heuristic characterization of the family of powers and roots as a "ladder" follows Tukey (1977).

It is, finally, worth pointing out that power transformations are effective only when the ratio of the biggest data values to the smallest ones is sufficiently large; if, in contrast, this ratio is close to 1, then power transformations are nearly linear and, hence, are ineffective. Consider the following example, where the ratio of the largest to the smallest data value is only $1995/1991 = 1.002 \simeq 1$:

	X	$\log_{10} X$	
	1991	3.29907	
1 {			} 0.00022
	1992	3.29929	
1 {			} 0.00022
	1993	3.29951	
1 {			} 0.00022
	1994	3.29973	
1 {			} 0.00021
	1995	3.29994	

Using a negative start produces the desired effect:

	X	$\log_{10}(X - 1990)$	
	1991	0	
1 {			} 0.301
	1992	0.301	
1 {			} 0.176
	1993	0.477	
1 {			} 0.125
	1994	0.602	
1 {			} 0.097
	1995	0.699	

This strategy should be considered whenever the ratio of the largest to the smallest data value is less than about 5. When the ratio is sufficiently large—either initially or after subtracting a suitable start—an adequate power transformation can usually be found in the range $-2 \leq p \leq 3$. We usually select integer values of p, or simple fractions, such as $\frac{1}{2}$ or $\frac{1}{3}$.

> Power transformations preserve the order of the data only when all values are positive, and are effective only when the ratio of largest to smallest data values is itself large. When these conditions do not hold, we can impose them by adding a positive or negative start to all of the data values.

EXERCISE

4.1 *Show that the derivative of $f(X) = (X^p - 1)/p$ is equal to 1 at $X = 1$ regardless of the value of p.

4.2 Transforming Skewness

Power transformations can make a skewed distribution more symmetric. But why should we bother?

- Highly skewed distributions are difficult to examine because most of the observations are confined to a small part of the range of the data. Recall, for example, the distribution of average income in the Canadian occupational prestige data discussed in the previous chapter, and redisplayed in Figure 4.2.[5]
- Apparently outlying values in the direction of the skew are brought in toward the main body of the data when the distribution is made more symmetric. In contrast, unusual values in the direction opposite to the skew can be hidden prior to transforming the data.
- Some of the most common statistical methods summarize distributions using means. Least-squares regression, which traces the mean of Y conditional on X, comes immediately to mind.[6] The mean of a skewed distribution is not, however, a good summary of its center.

The following simple example illustrates how a power transformation can eliminate a positive skew:

	X	$\log_{10} X$	
	1	0	
9 {			} 1
	10	1	
90 {			} 1
	100	2	
900 {			} 1
	1000	3	

Descending the ladder of powers to $\log X$ makes the distribution more symmetric by pulling in the right tail. Ascending the ladder of powers (toward X^2 and X^3) can, similarly, "correct" a negative skew.

---

[5] Repeating Figure 2.7. If you did not read Section 3.1.2 on density estimation, simply think of the density estimates in this figure as smoothed histograms.

[6] See Chapter 5.

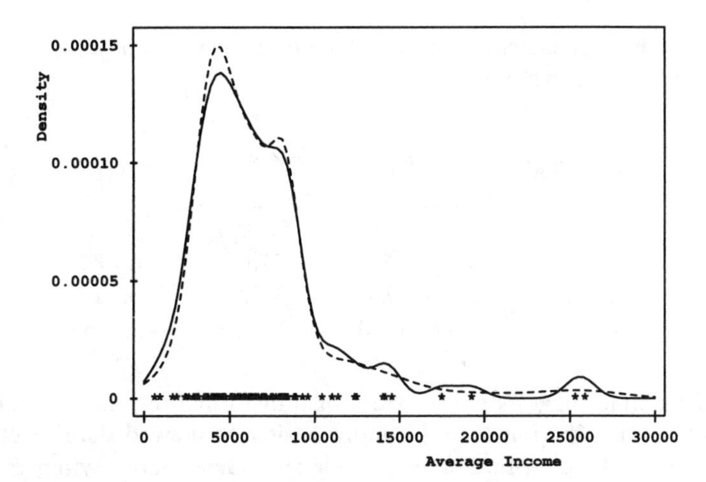

Figure 4.2. The distribution of income in the Canadian occupational prestige data. The solid line shows a kernel density estimate, the broken line an adaptive-kernel density estimate. The income values are displayed in the one-dimensional scatterplot at the bottom of the figure.

An effective transformation can be selected analytically or by trial and error.[7] Examining the median and the hinges, moreover, can provide some guidance to trial and error. A nice property of order statistics—including the median and hinges—is that they are preserved under nonlinear monotone transformations of the data, such as powers and roots;[8] that is, if $X' = X^{(p)}$, then $X'_{(i)} = [X_{(i)}]^{(p)}$, and thus median$(X')$ = [median$(X)]^{(p)}$. This is not the case for the mean and standard deviation.

In a symmetric distribution, the median is midway between the hinges, and, consequently, the ratio

$$\frac{\text{Upper hinge } - \text{Median}}{\text{Median} - \text{Lower hinge}}$$

is approximately 1. In contrast, a positive skew is reflected in a ratio that exceeds 1 and a negative skew in a ratio that is smaller than 1. Trial and error can begin, therefore, with a transformation that makes this ratio close to 1. Some statistical computing environments (e.g., Lisp-Stat) allow the transformation p to be selected interactively using a "slider," while a graph of the distribution—for example, a density plot—is updated every time the value of p changes. This is a particularly convenient and effective approach.

---

[7] See Section 12.5 for a discussion of analytic methods for selecting transformations.

[8] There is some slippage here since the median and hinges sometimes require *averaging* adjacent order statistics. Because the two averaged values are seldom very far apart, however, the distinction between the median of the transformed values and the transformation of the median is almost always trivial. The same is true for the hinges.

For the distribution of income in the Canadian occupational prestige data, for example, we have:

Transformation	H_U	Median	H_L	$\frac{H_U - \text{Median}}{\text{Median} - H_L}$
X	8206	5930.5	4075	1.23
$\sqrt{X}$	90.59	77.01	63.84	1.03
$\log_{10} X$	3.914	3.773	3.610	0.87
$-1/\sqrt{X}$	−0.01104	−0.01299	−0.01566	0.73

This table suggests the square-root transformation but, as it turns out, the tails of the distribution are a bit more positively skewed than its center; as a consequence, the distribution as a whole is rendered more symmetric by a log transformation. The result is shown in Figure 4.3. Notice that I have used a $\log_{10}$ axis for income in this plot, rather than (as would be equivalent) laying off the log-income values (i.e., 2, 3, 4, 5) on the horizontal axis.

The distribution is much more symmetric than before, but it is not unimodal: There are two clear modes, and the suggestion of a third, below $10,000, corresponding to groups of occupations with relatively similar levels of income; there is also the suggestion of another mode slightly to the right of $10,000. There are several outlying observations beyond the fences:

General managers	$25, 879
Physicians	$25, 308
Newsboys	$918
Babysitters	$611

The two outliers on the low end were not apparent prior to transforming income.

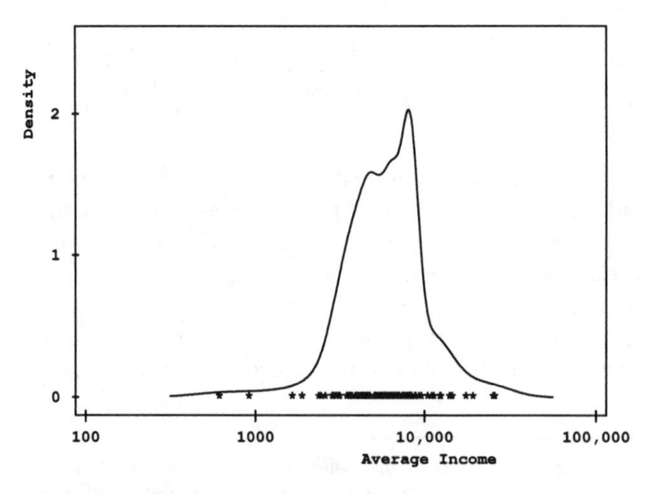

Figure 4.3. Adaptive-kernel density estimate for $\log_{10}$ average income in the Canadian occupational prestige data. The window width is 0.05 (on the log-income scale). A one-dimensional scatterplot of the data values appears at the bottom of the graph.

It is often the case, as here, that we have a choice between transformations that perform roughly equally well. Although we should try to avoid distorting the data, we may prefer one transformation to another because of interpretability. I have already mentioned that the log transformation has a convenient multiplicative interpretation. In certain contexts, other transformations may have specific substantive meanings. Here are a few common examples: The inverse of time taken to travel a fixed distance is speed; the inverse of response latency (as in a psychophysical experiment) is response speed; the square root of a measure of area is a linear measure of size; and the cube of a linear measure can be interpreted as a volume.

We particularly prefer interpretable transformations when variables are measured on familiar and meaningful scales. Conversely, because the rating "scales" that are ubiquitous in social research are not really measurements, there is typically no reason to prefer the original scores to a monotone transformation of them.[9]

> Descending the ladder of powers (e.g., to $\log X$) tends to correct a positive skew; ascending the ladder of powers (e.g., to X^2) tends to correct a negative skew.

EXERCISE

4.2 Attempt to make the distribution of each of the following variables more symmetric by employing transformations in the family of powers and roots:

(a) Geographic mobility and ethnic heterogeneity in Angell's data on 43 U.S. cities (in Table 2.3 and `angell.dat`).

(b) Income and infant mortality in Leinhardt and Wasserman's data on 103 nations of the world (in `leinhard.dat`).

---

[9] Rating scales are composed, for example, of items with response categories labeled "strongly agree," "agree," "disagree," "strongly disagree." A scale is constructed by assigning arbitrary numbers to the categories (e.g., 1–4) and adding or averaging the items. See Coombs et al. (1970, Chapters 2 and 3) for an elementary treatment of measurement issues in the social sciences, and Duncan (1984) for an interesting account of the history and practice of social measurement. I believe that social scientists should pay more attention to measurement issues, but that it is unproductive simply to discard rating scales and similar "measurements by fiat" (a felicitous term borrowed from Torgerson, 1958): There is a prima facie reasonableness to many rating scales, and to refuse to use them without adequate substitutes would be foolish.

4.3 Transforming Nonlinearity

Power transformations can also be used to make many nonlinear relationships more nearly linear. Again, we ask, why bother?

- Linear relationships—expressible in the form $\hat{Y} = A + BX$—are particularly simple. Recall that this equation specifies that the average value of the dependent variable Y is a linear function of the independent variable X, with intercept A and slope B. Linearity implies that a unit increase in X—regardless of the level of X—is associated, on average, with a change of B units in Y.[10] In particular, fitting a linear equation to data makes it relatively easy to answer certain questions about the data: For example, if B is positive, then Y tends to increase with X.
- Especially when there are several independent variables, the alternative of nonparametric regression may not be feasible because of the sparseness of the data. Even if we can fit a nonparametric regression with several X's, it may be difficult to visualize the multidimensional result.[11]
- There is a simple and elegant statistical theory for linear models, which we shall begin to explore in the next chapter. If these models are reasonable for the data, then their use is convenient.
- There are certain technical advantages to having linear relationships among the *independent* variables in a regression analysis.[12]

The following simple example suggests how a power transformation can serve to straighten a nonlinear relationship: Suppose that $Y = \frac{1}{5}X^2$ (with no residual), and that X takes on successive integer values between 1 and 5:

X	Y
1	0.2
2	0.8
3	1.8
4	3.2
5	5.0

These "data" are graphed in panel (*a*) of Figure 4.4, where the nonlinearity of the relationship between Y and X is apparent. Because of the manner in which the example was constructed, it is obvious that there are two simple ways to transform the data to achieve linearity:

1. We could replace Y by $Y' = \sqrt{Y}$, in which case $Y' = \sqrt{\frac{1}{5}}X$.
2. We could replace X by $X' = X^2$, in which case $Y = \frac{1}{5}X'$.

---

[10] Recall as well that we use the terms "increase" and "change" loosely here as a shorthand for static comparisons between average values of Y for X-values that differ by one unit: *Literal* change is not necessarily implied.

[11] See, however, the *additive regression models* discussed in Section 14.4.2, which overcome this deficiency.

[12] This point is developed in Section 12.6.

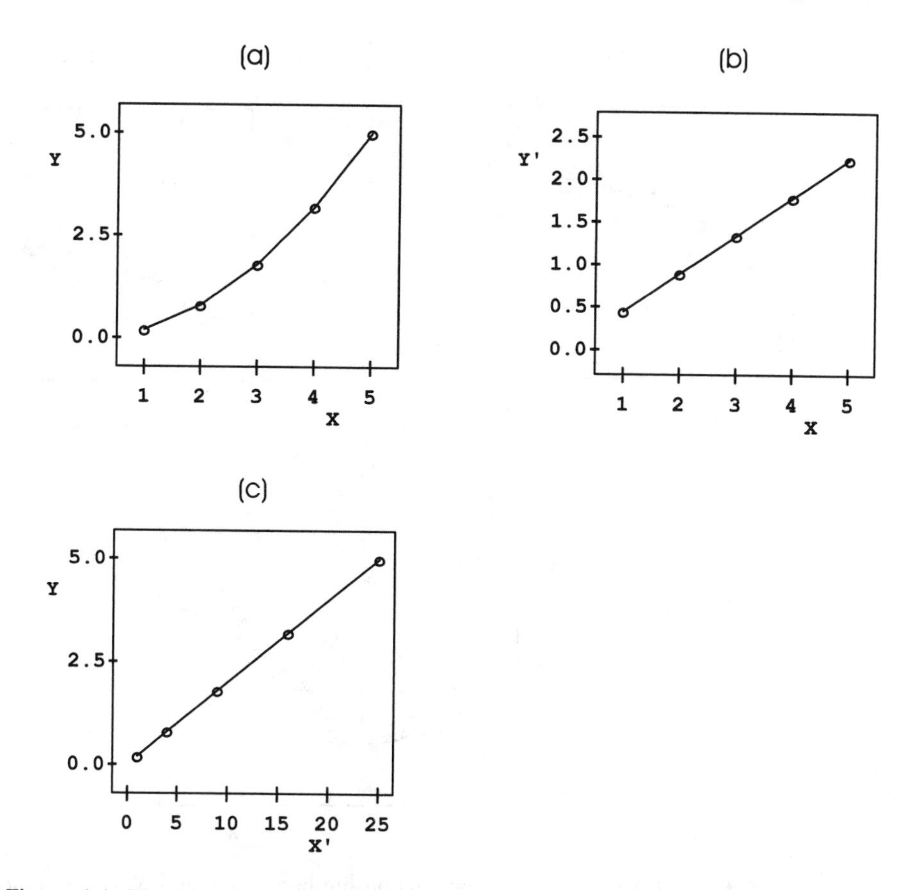

Figure 4.4. How a power transformation of Y or X can make a simple monotone nonlinear relationship linear. Panel (a) shows the relationship $Y = \frac{1}{5}X^2$. In panel (b), Y is replaced by the transformed value $Y' = Y^{1/2}$. In panel (c), X is replaced by the transformed value $X' = X^2$.

In either event, the relationship is rendered perfectly linear, as shown graphically in panels (b) and (c) of Figure 4.4. To achieve an intuitive understanding of this process, imagine that the original plot in panel (a) is drawn on a rubber sheet: Transforming Y "down" the ladder of powers to square root differentially stretches the rubber sheet vertically so that small values are spread out relative to large ones, stretching the curve in (a) into the straight line in (b). Likewise, transforming X "up" the ladder of powers spreads out the large values relative to the small ones, stretching the curve into the straight line in (c).

A power transformation works here because the relationship between Y and X is smooth, monotone (in this instance, strictly increasing), and simple. What I mean by "simple" in this context is that the direction of curvature of the function relating Y to X does not change. Figure 4.5 seeks to clarify these distinctions: The relationship in panel (a) is simple and monotone; the relationship in panel (b) is monotone but not simple; and the relationship in panel (c) is simple but not monotone. I like to use the term "curvilinear" for cases such as (c), to distinguish nonmonotone from monotone nonlinearity, but this is not standard terminology. In panel (c), no power transformation of Y or X can

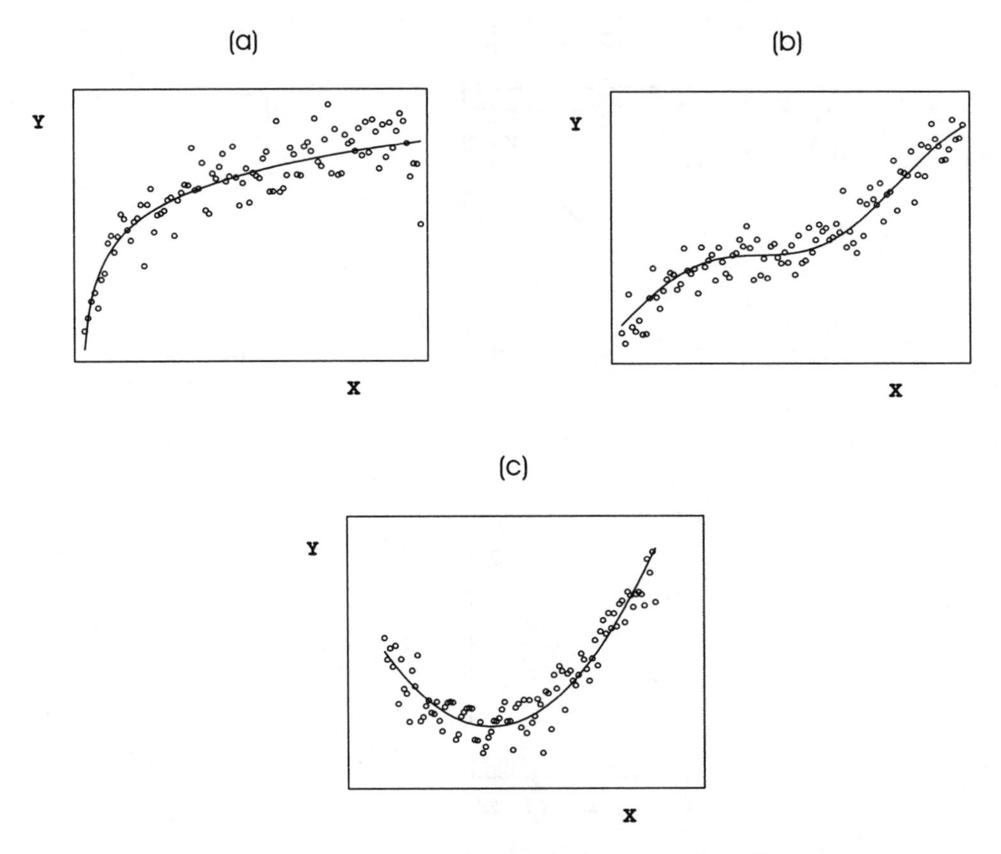

Figure 4.5. (*a*) A simple monotone relationship between Y and X; (*b*) a monotone relationship that is not simple; (*c*) a relationship that is simple but not monotone. A power transformation of Y or X can straighten (*a*), but not (*b*) or (*c*).

straighten the relationship between them, but we could capture this relationship with a quadratic model[13] of the form $\hat{Y} = A + B_1X + B_2X^2$.

As for transformations to reduce skewness, a transformation to correct nonlinearity can be selected analytically or by guided trial and error.[14] Figure 4.6 introduces Mosteller and Tukey's (1977) "bulging rule" for selecting a transformation: If the "bulge" points *down* and to the *right*, for example, we need to transform Y *down* the ladder of powers or X *up* (or both). This case corresponds to the example in Figure 4.4, and the general justification of the rule follows from the need to stretch an axis differentially to transform the curve into a straight line. Trial and error is simplest in a computing environment that provides "sliders" for the power transformations of X and Y, immediately displaying the effect of a change in either power on the scatterplot relating the two variables.

---

[13] Quadratic and other polynomial regression models are discussed in Section 14.2.1.

[14] See the discussion of "Box-Tidwell" regression models in Section 12.5.2 for an analytic method of selecting a linearizing transformation.

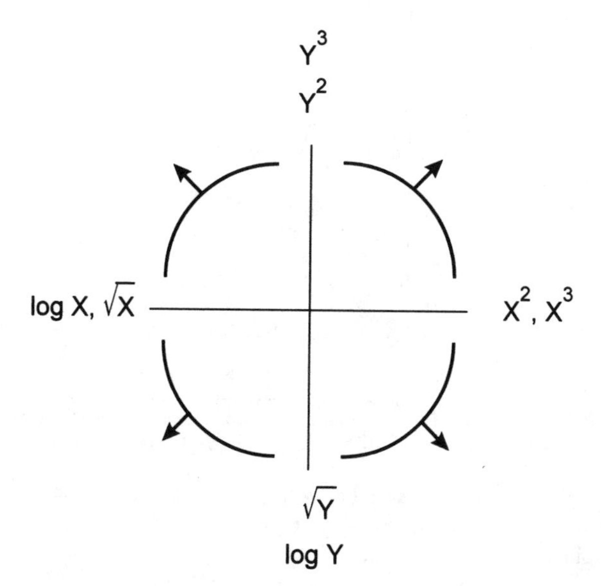

Figure 4.6. Tukey and Mosteller's "bulging rule": The direction of the "bulge" indicates the direction of the power transformation of Y and/or X to straighten the relationship between them.

> Simple monotone nonlinearity can often be corrected by a power transformation of X, of Y, or of both variables. Mosteller and Tukey's "bulging rule" assists in the selection of a transformation.

Now reexamine, in the light of this discussion, the relationship between prestige and income for the 102 Canadian occupations first encountered in Chapter 2, and shown in Figure 4.7.[15] The relationship between prestige and income is clearly monotone and nonlinear: Prestige rises with income, but the slope is steeper at the left of the plot, where income is low, than at the right, where it is high. The change in slope appears fairly abrupt rather than smooth, however, and we might do better to model the relationship with two straight lines (one for relatively small values of income, one for relatively large ones) than simply to transform prestige or income.[16]

Nevertheless, because the bulge points up and to the left, we can try transforming prestige up the ladder of powers or income down. Because the income distribution is positively skewed, I prefer to transform income rather than prestige, which is more symmetrically distributed. As shown in Figure 4.8, the cube-root transformation of income works reasonably well here: Some nonlinearity

[15] Repeating Figure 2.7.

[16] For an alternative interpretation of the relationship between prestige and income, see Exercise 12.11.

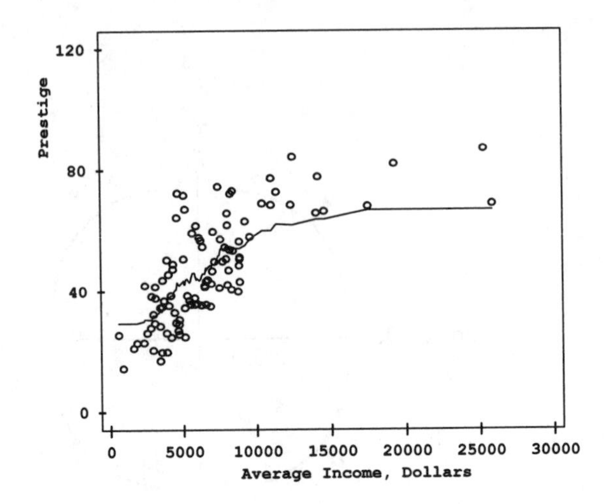

Figure 4.7. The relationship between prestige and income for the Canadian occupational prestige data. The nonparametric regression line on the plot is computed by local averaging.

remains, but it is not simple, and the linear regression of prestige on income no longer *grossly* distorts the relationship between the two variables. I would have preferred to use the log transformation, which (recall) I employed to make the income distribution more symmetric, but this transformation "overcorrected" the nonlinearity in the relationship between prestige and income.

A more extreme, and ultimately more successful, example appears in Figure 4.9, which shows the relationship between the infant mortality rate (number of infant deaths per 1000 live births) and per-capita income (in U.S. dollars) of 101 nations in 1970. The data for this example were reported by Leinhardt and Wasserman (1978). Notice that both variables are positively skewed, and that,

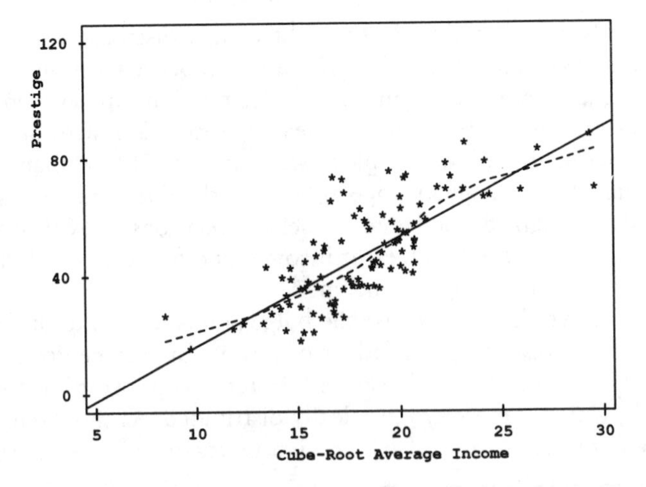

Figure 4.8. Scatterplot of prestige versus income$^{1/3}$ for 102 Canadian occupations in 1970. The solid line shows the least-squares linear regression, while the broken line shows a robust local regression.

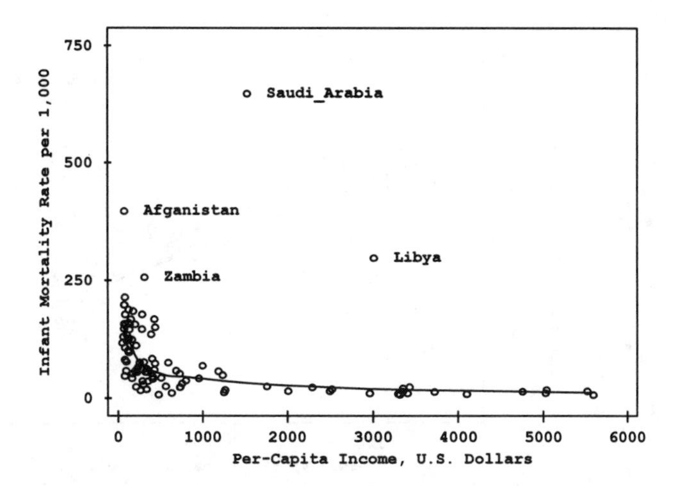

Figure 4.9. Scatterplot of infant mortality rate versus income in U.S. dollars, for 101 nations *circa* 1970. The nonparametric regression shown on the plot was calculated by robust local regression. Several outlying observations are flagged.

Source of Data: Leinhardt and Wasserman (1978).

consequently, most of the data are confined to a small region in the lower left of the plot. Several observations with very large infant mortality rates (in particular, Saudi Arabia, Afghanistan, Libya, and Zambia) are outliers: The infant mortality rates for these countries are so large that I question the accuracy of the data.

The skewness of infant mortality and income in Figure 4.9 makes the scatterplot difficult to interpret; nevertheless, the nonparametric regression shown on the plot reveals a nonlinear but monotone relationship between infant mortality and income.[17] The bulging rule suggests that infant mortality or income should be transformed down the ladder of powers and roots. In this case, transforming both variables by taking logs makes the relationship nearly perfectly linear (as shown in Figure 4.10), although Saudi Arabia and Libya continue to stand out from the rest of the data.

The least-squares regression line in Figure 4.10, calculated omitting the two outlying nations, has the equation

$$\log_{10} \widehat{\text{Infant mortality}} = 3.22 - 0.566 \times \log_{10} \text{Income}$$

Because both variables are expressed on log scales to the same base, the slope of this relationship has a simple interpretation: A 1% increase in per-capita income is associated, on average, with an approximate 0.57% decline in the infant mortality rate. Economists call this type of coefficient an "elasticity."[18]

---

[17] The nonparametric regression in Figure 4.9 is calculated by robust locally weighted regression, described in Section 14.4.1.

[18] Increasing X by 1% is equivalent to multiplying it by 1.01, which, in turn, implies that the log of X increases by $\log_{10} 1.01 = 0.00432$. The corresponding change in $\log Y$ is then $B \times 0.00432 = -0.566 \times 0.00432 = -0.00245$. Subtracting 0.00245 from $\log Y$ is equivalent to multiplying Y by $10^{-0.00245} = 0.99437$, that is, decreasing Y by $100 \times (1 - 0.99437) = 0.563 \simeq 0.566\%$. The approximation holds because the log function is nearly linear across the small domain of X-values between $\log 1$ and $\log 1.01$.

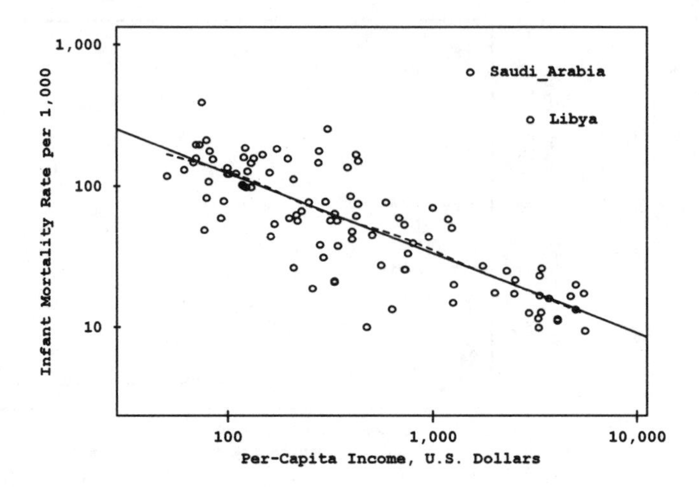

Figure 4.10. Scatterplot of $\log_{10}$ infant mortality rate versus $\log_{10}$ per-capita income for 101 nations. The solid line was calculated by least-squares linear regression, omitting Saudi Arabia and Libya; the broken line was calculated by robust local regression.

EXERCISES

4.3 Attempt to make each of the following relationships more nearly linear by transforming one or both variables:

(a) The relationship between Acres/Gardener and Consumers/Gardener in Sahlins's data on agricultural production in a primitive community (in Table 2.1 and `sahlins.dat`).

(b) The relationship between the total fertility rate and the percentage of women using contraception in Robey et al.'s data on 50 developing nations (in Table 2.2 and `robey.dat`).

(c) The relationship between moral integration and geographic mobility in Angell's data on 43 U.S. cities (in Table 2.3 and `angell.dat`).

(d) The relationship between number of interlocks and corporate assets in Ornstein's data on interlocking directorates among 248 dominant Canadian firms (in `ornstein.dat`).

4.4 Transforming Nonconstant Spread

When a variable has very different degrees of variation in different groups, it becomes difficult to examine the data and to compare differences in level across the groups. We encountered this problem in Section 3.2, where we compared the distribution of number of interlocking directorships by nation of control, employing Ornstein's data on 248 dominant Canadian corporations, shown in Figure 4.11.[19]

[19] Repeating Figure 3.14.

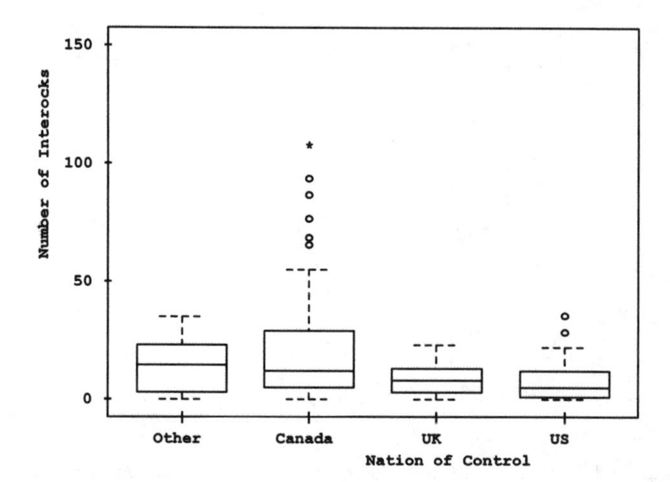

Figure 4.11. Number of interlocking directorate and executive positions by nation of control, for 248 dominant Canadian firms.

Differences in spread are often systematically related to differences in level: Groups with higher levels tend to have higher spreads. Using the median and hinge spread as indices of level and spread, respectively, the following table shows that there is indeed an association, if an imperfect one, between spread and level for Ornstein's data:

Nation of Control	*Lower Hinge*	*Median*	*Upper Hinge*	*Hinge Spread*
Other	3	14.5	23	20
Canada	5	12.0	29	24
United Kingdom	3	8.0	13	10
United States	1	5.0	12	11

Tukey (1977) suggests graphing the log hinge spread against the log median, as shown in Figure 4.12. Because some firms maintained 0 interlocks, I used a start of 1 to construct this graph—adding 1 to each median, but leaving the hinge spreads unchanged.

The slope of the linear "trend," if any, in the spread-level plot can be used to suggest a spread-stabilizing power transformation of the data: Express the linear fit as

$$\text{log spread} \simeq a + b \log \text{level}$$

Then the corresponding spread-stabilizing transformation uses the power $p = 1 - b$. When spread is positively related to level (i.e., $b > 0$), therefore, we select a transformation *down* the ladder of powers and roots.

Starting with this transformation, it is convenient to employ a computing environment that connects a "slider" for the power p to the spread-by-level plot and the parallel boxplots. Changing the value of p via the slider immediately

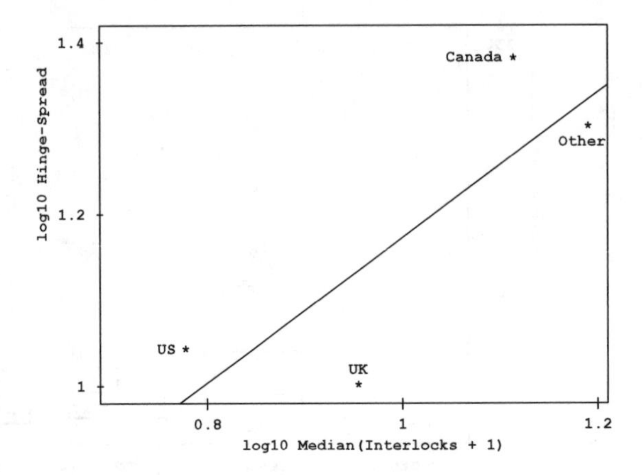

Figure 4.12. Spread ($\log_{10}$ hinge spread) versus level [$\log_{10}$(median + 1)]. The plot is for Ornstein's interlocking-directorate data, with groups defined by nation of control. The line on the plot was fit by least squares.

updates the plots, allowing us to assess the relative effects of different transformations.

> When there is a positive association between the level of a variable in different groups and its spread, the spreads can be made more constant by descending the ladder of powers. A negative association between level and spread is less common, but can be corrected by ascending the ladder of powers.

In Figure 4.12, a line was fit by least squares to the spread-level plot for the interlocking directorate data. The slope of this line, $b = 0.85$, suggests the power transformation, $p = 1 - 0.85 = 0.15 \simeq 0$. I decided, therefore, to try a log transformation. Figure 4.13 shows the result, employing logs to the base 2. The spreads of the several groups are now much more similar, and differences in level are easier to discern. As well, the within-group distributions are more symmetric.

The problems of unequal spread and skewness commonly occur together, because they often have a common origin: When, as here, the data represent frequency counts (*number* of interlocks), the impossibility of obtaining a negative count tends to produce positive skewness, together with a tendency for larger levels to be associated with larger spreads. The same is true of other types of variables that are bounded below (e.g., wage and salary income). Likewise, variables that are bounded above but not below (e.g., grades on a very simple exam) tend both to be negatively skewed and to show a negative association between

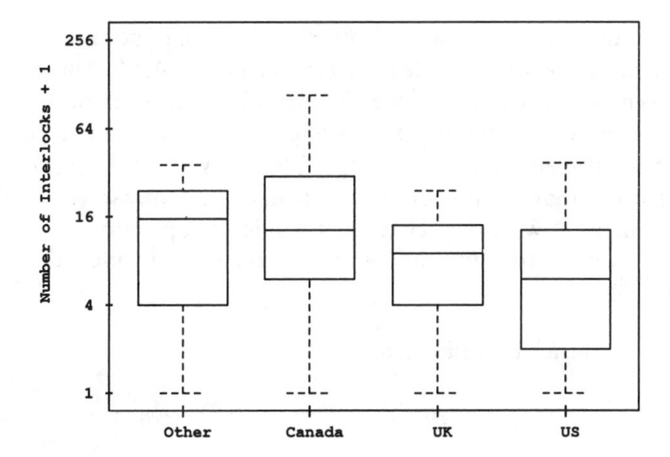

Figure 4.13. Parallel boxplots of number of interlocks by nation of control, plotting interlocks + 1 on the $\log_2$ scale. Compare this plot with Figure 4.11, where number of interlocks in not transformed.

spread and level. In the latter event, a transformation "up" the ladder of powers (e.g., to X^2) usually provides a remedy.[20]

EXERCISE

4.4 Construct parallel boxplots and a spread-by-level plot for infant mortality by region, using Leinhardt and Wasserman's data (in `leinhard.dat`). If you detect a relationship between spread and level, try to equalize the spreads by transforming infant mortality.

4.5 Transforming Proportions

Power transformations are often not helpful for proportions, because these quantities are bounded below by 0 and above by 1. Of course, if the data values do not approach these two boundaries, then proportions can be handled much like other sorts of data.

Percentages and many sorts of rates (e.g., the infant mortality rate per 1000 live births) are simply rescaled proportions, and therefore are similarly affected. It is, moreover, common to encounter "disguised" proportions, such as the number of questions correct on an exam of fixed length or the number of affirmative responses to a series of dichotomous attitude questions.

[20] Plotting log spread against log level to select a spread-stabilizing transformation is quite a general idea. In Section 12.2.1, for example, we shall use a version of the spread-level plot to find a variance-stabilizing transformation in regression analysis.

An example, drawn from the Canadian occupational prestige data, is shown in the stem-and-leaf display in Figure 4.14. The distribution is for the percentage of women among the incumbents of each of 102 occupations. There are many occupations with no women or a very small percentage of women, but the distribution is not simply positively skewed, because there are also occupations that are predominantly female. In contrast, relatively few occupations are balanced with respect to their gender composition.

Several transformations are commonly employed for proportions, including:

- The *logit* transformation:

$$P \to \text{logit}(P) = \log_e \frac{P}{1-P}$$

The logit transformation is the log of the "odds," $P/(1-P)$. The "trick" of the logit transformation is to remove the upper and lower boundaries of the scale, spreading out the tails of the distribution and making the resulting quantities symmetric about 0; for example,

P	$\frac{P}{1-P}$	logit
.05	1/19	−2.94
.1	1/9	−2.20
.3	3/7	−0.85
.5	1	0
.7	7/3	0.85
.9	9/1	2.20
.95	19/1	2.94

A graph of the logit transformation is shown in Figure 4.15. Note that the transformation is nearly linear in its center, between about $P = .2$ and $P = .8$.

```
Unit: 1      Lines/stem: 2
      1|2 <--> 12

depth
 32   0|00000000000000111111222233334444
 44   0|555566777899
  8)  1|01111333
 50   1|5557779
 43   2|1344
 39   2|57
 37   3|01334
 32   3|99
 30   4|
 30   4|678
 27   5|224
 24   5|67
 22   6|3
 21   6|789
 18   7|024
 15   7|5667
 11   8|233
  8   8|
  8   9|012
  5   9|56667
```

Figure 4.14. Stem-and-leaf display of percentage of women in each of 102 Canadian occupations in 1970. Notice how the data "stack up" against both boundaries.

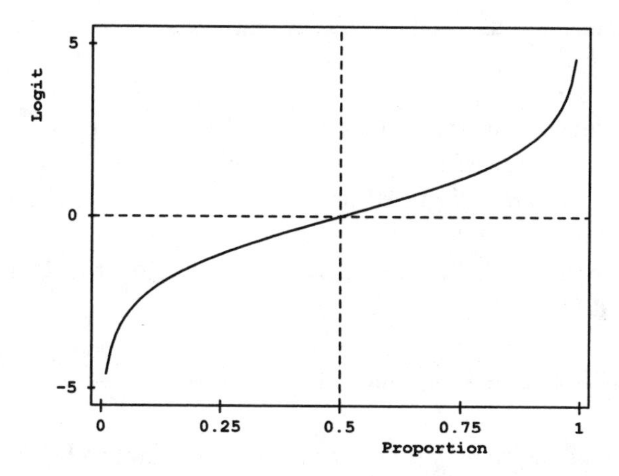

Figure 4.15. The logit transformation $\log[P/(1-P)]$ of a proportion P.

- The *probit* transformation:

$$P \rightarrow \text{probit}(P) = \Phi^{-1}(P)$$

where Φ^{-1} is the inverse distribution function for the standard normal distribution. Once their scales are equated, the logit and probit functions are, for practical purposes, indistinguishable: logit $\simeq (\pi/\sqrt{3})\times$ probit.[21]

- The *arcsine-square-root* transformation also has a similar shape:

$$P \rightarrow \sin^{-1}\sqrt{P}$$

Tukey has embedded these common transformations for proportions into the family of "folded" powers and roots, indexed by the power q, which takes on values between 0 and 1:

$$P \rightarrow P^q - (1-P)^q$$

When $q = 0$, we take the natural log, producing the logit transformation. Setting $q = 0.14$ yields (to a very close approximation) a multiple of the probit transformation. Setting $q = 0.41$ produces (again, to a close approximation) a multiple of the arcsine-square-root transformation. When $q = 1$, the transformation is just twice the "plurality" (i.e., the difference between P and $\frac{1}{2}$), leaving the shape of the distribution of P unaltered:

$$[P - (1-P)] = 2(P - \tfrac{1}{2})$$

---

[21] We shall encounter the logit and probit functions again in a different context when we take up the analysis of categorical data in Chapter 15.

> Power transformations are ineffective for proportions P that push the boundaries of 0 and 1, and for other variables (e.g., percentages, rates, disguised proportions) that are bounded both below and above. The folded powers $P \rightarrow P^q - (1 - P)^q$ are often effective in this context; for $q = 0$, we employ the logit transformation, $P \rightarrow \log[P/(1 - P)]$.

The logit and probit transformations cannot be applied to proportions of exactly 0 or 1. If, however, we have access to the original counts on which the proportions were based, then we can avoid this embarrassment by employing

$$P' = \frac{F + \frac{1}{2}}{N + 1}$$

in place of P. Here, F is the frequency count in the focal category (e.g., number of women) and N is the total count (total number of occupational incumbents, women plus men). If the original counts are not available, then we can use the expedient of remapping the proportions to an interval that excludes 0 and 1. For example, $P' = .005 + .99 \times P$ remaps proportions to the interval [.005, .995].

Employing this strategy for the Canadian occupational data produces the distribution for logit(P'_{women}) that appears in Figure 4.16. Spreading out the tails of the distribution has improved its behavior substantially, although there is still some stacking up of low and high values.

```
Unit: 0.1     Lines/stem: 2
        1|2 <--> 1.2

depth
  5   -4|77777
  8   -3|444
 16   -3|55667888
 21   -2|01124
 31   -2|5567888999
 39   -1|01112344
 48   -1|556779999
 10)  -0|0111333444
 44   -0|668889
 38    0|01233355889
 27    0|00122577889
 16    1|01111
 11    1|556
  8    2|23
  6    2|5
  5    3|00014
```

Figure 4.16. Stem-and-leaf display for the logit transformation of proportion of women in each of 102 Canadian occupations. Since there are some occupations with no women, the proportions were remapped to the interval .005 to .995 prior to calculating the logits.

EXERCISE

4.5 Using a histogram and a normal quantile comparison plot, examine the distribution of prestige (the percentage of raters classifying an occupation as good or better in prestige) in Duncan's data on 45 U.S. occupations (in Table 3.2 and `duncan.dat`). Transform the data using logits (and, possibly, other folded-power transformations) and reexamine the distribution. Does the transformation make the distribution of prestige more nearly normal?

4.6 Summary

- Transformations can often facilitate the examination and modeling of data.
- The powers and roots are a particularly useful family of transformations: $X \to X^p$. We employ the log transformation in place of X^0.
- Power transformations preserve the order of the data only when all values are positive, and are effective only when the ratio of largest to smallest data values is itself large. When these conditions do not hold, we can impose them by adding a positive or negative start to all of the data values.
- Descending the ladder of powers (e.g., to $\log X$) tends to correct a positive skew; ascending the ladder of powers (e.g., to X^2) tends to correct a negative skew.
- Simple monotone nonlinearity can often be corrected by a power transformation of X, of Y, or of both variables. Mosteller and Tukey's "bulging rule" assists in the selection of a transformation.
- When there is a positive association between the level of a variable in different groups and its spread, the spreads can be made more constant by descending the ladder of powers. A negative association between level and spread is less common, but can be corrected by ascending the ladder of powers.
- Power transformations are ineffective for proportions P that push the boundaries of 0 and 1, and for other variables (e.g., percentages, rates, disguised proportions) that are bounded both below and above. The folded powers $P \to P^q - (1 - P)^q$ are often effective in this context; for $q = 0$, we employ the logit transformation, $P \to \log[P/(1 - P)]$.

4.7 Recommended Reading

Because examination and transformation of data are closely related topics, most of the readings here were also listed at the end of the previous chapter.

- Tukey's important text on exploratory data analysis (Tukey, 1977), and the companion volume by Mosteller and Tukey (1977) on regression analysis, have a great deal of interesting information and many examples. As mentioned in the previous chapter, however, Tukey's writing style is opaque. Velleman and Hoaglin (1981) is easier to digest, but it is not as rich in material on transformations.

- Several papers in a volume edited by Hoaglin, et al. (1983) have valuable material on the family of power transformations, including a general paper by Emerson and Stoto; an extended discussion of the spread-versus-level plot in a paper on boxplots by Emerson and Strenio; and a more difficult paper by Emerson on the mathematics of transformations.
- The tools provided by the Lisp-Stat statistical computing environment (described in Tierney, 1990)—including the ability to associate a transformation with a slider and to link different plots—are especially helpful in selecting transformations. Cook and Weisberg (1994) have developed a system for data analysis and regression based on Lisp-Stat that includes these capabilities. Similar, if less flexible, facilities are built into some statistical packages.

PART II

Linear Models and Least Squares

5

Linear Least-Squares Regression

I have on several occasions emphasized the limitations of linear least-squares regression. Despite these limitations, linear least squares lies at the very heart of applied statistics:[1]

- Some data are adequately summarized by linear least-squares regression.
- The effective application of linear regression is considerably expanded through data transformations and techniques for diagnosing problems such as overly influential data.
- As we shall see, the general linear model—a direct extension of least-squares linear regression—is able to accommodate a very broad class of specifications, including, for example, qualitative independent variables and polynomial functions.
- Linear least-squares provides a computational basis for a variety of generalizations, including weighted regression, robust regression, nonparametric regression, and generalized linear models.

The current chapter describes the mechanics of linear least-squares regression. That is, I shall explain how the method of least squares can be employed to fit a line to a bivariate scatterplot, a plane to a three-dimensional scatterplot, and a general linear surface to multivariate data (which, of course, cannot be directly visualized).

[1] The extensions of linear least-squares regression mentioned here are the subject of subsequent chapters.

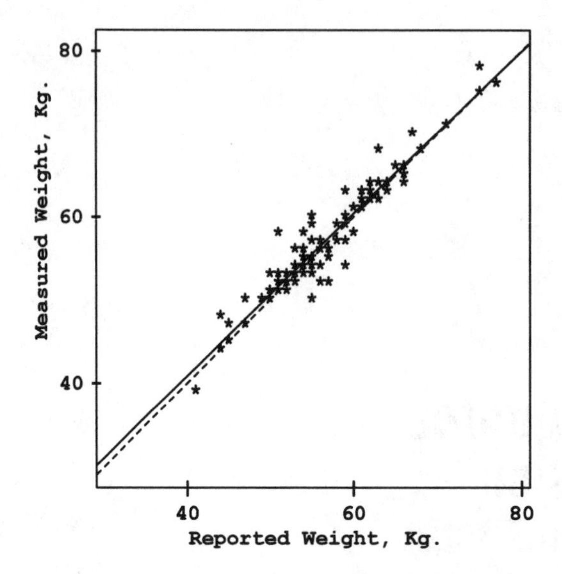

Figure 5.1. Scatterplot of Davis's data on the measured and reported weight of 101 women. The solid line gives the least-squares fit; the broken line is $Y = X$. Both variables are discrete, since weight is given to the nearest kilogram, but overplotting of points is not a serious problem here.

5.1 Simple Regression

5.1.1 Least-Squares Fit

Figure 5.1 shows Davis's data, introduced in Chapter 2, on the measured and reported weight in kilograms of 101 women who were engaged in regular exercise.[2] The relationship between measured and reported weight appears to be linear, so it is reasonable to fit a line to the plot. A line will help us to determine whether the subjects in Davis's study were accurate and unbiased reporters of their weights; and it can provide a basis for predicting the measured weight of similar women for whom only reported weight is available.

Denoting measured weight by Y and reported weight by X, a line relating the two variables has the equation $Y = A + BX$. It is obvious, however, that no line can pass perfectly through all of the data points, despite the strong linear relationship between these two variables. We introduce a residual, E , into the regression equation to reflect this fact; writing the regression equation for the ith of the $n = 101$ observations:

$$\begin{aligned} Y_i &= A + BX_i + E_i \\ &= \hat{Y}_i + E_i \end{aligned} \tag{5.1}$$

where $\hat{Y} = A + BX_i$ is the fitted value for observation i. The essential geometry is shown in Figure 5.2, which reveals that the residual

$$E_i = Y_i - \hat{Y}_i = Y_i - (A + BX_i)$$

[2] The misrecorded data value that produced an outlier in Figure 2.3 has been corrected.

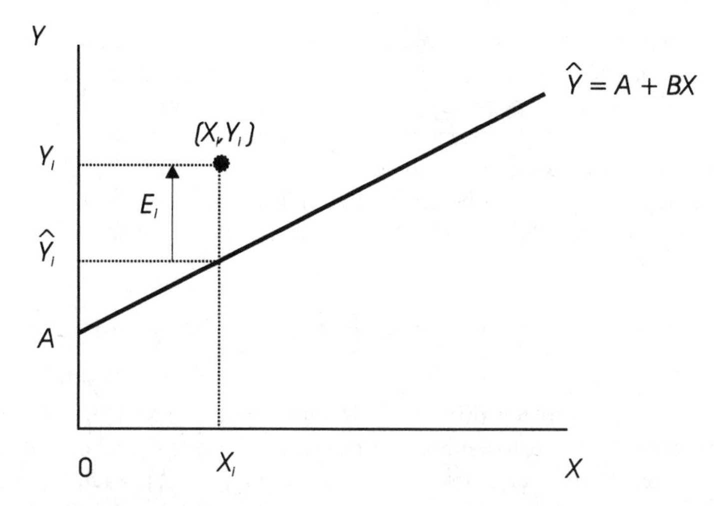

Figure 5.2. Linear regression of Y on X, showing the residual E_i for the ith observation.

is the signed vertical distance between the point and the line—that is, the residual is negative when the point lies below the line, and positive when the point is above the line.

A line that fits the data well therefore makes the residuals small, but in order to determine a line analytically we need to be more precise about what we mean by "small." First of all, we want residuals that are small in magnitude, because large negative residuals are as offensive as large positive ones. For example, simply requiring that the sum of residuals, $\sum_{i=1}^{n} E_i$, be small is futile, because large negative residuals can offset large positive ones.

Indeed, any line through the means of the variables—the point $(\overline{X}, \overline{Y})$—has $\sum E_i = 0$. Such a line satisfies the equation $\overline{Y} = A + B\overline{X}$. Subtracting this equation from Equation 5.1 produces

$$Y_i - \overline{Y} = B(X_i - \overline{X}) + E_i$$

Then, summing over all observations,

$$\sum_{i=1}^{n} E_i = \sum (Y_i - \overline{Y}) - B \sum (X_i - \overline{X}) = 0 - B \times 0 = 0 \qquad [5.2]$$

Two possibilities immediately present themselves: We can employ the unsigned vertical distances between the points and the line, that is, the absolute values of the residuals; or we can employ the squares of the residuals. The first possibility leads to least-absolute-value (LAV) regression:

Find A and B to minimize the sum of the absolute residuals, $\sum |E_i|$.

The second possibility leads to the least-squares criterion:

Find A and B to minimize the sum of squared residuals, $\sum E_i^2$.

Squares are more tractable mathematically than absolute values, so we shall focus on least squares here, but LAV regression should not be rejected out of hand, because it provides greater resistance to outlying observations.[3]

We need to consider the residuals in the aggregate, because it is no trick to produce a 0 residual for an individual point simply by placing the line directly through the point. The least-squares criterion therefore minimizes the *sum* of squared residuals over all observations; that is, we seek the values of A and B that minimize:

$$S(A, B) = \sum_{i=1}^{n} E_i^2 = \sum (Y_i - A - BX_i)^2$$

I have written this expression as a *function* $S(A, B)$ of the regression coefficients A and B to emphasize the dependence of the sum of squared residuals on the coefficients: For a fixed set of data $\{X_i, Y_i\}$, each possible choice of values for A and B corresponds to a specific sum of squares, $\sum E_i^2$; we want the pair of values for the regression coefficients that makes this sum of squares as small as possible.

*The most direct approach[4] to finding the least-squares coefficients is to take the partial derivatives of the sum-of-squares function with respect to the coefficients:

$$\frac{\partial S(A, B)}{\partial A} = \sum (-1)(2)(Y_i - A - BX_i)$$

$$\frac{\partial S(A, B)}{\partial B} = \sum (-X_i)(2)(Y_i - A - BX_i)$$

Setting the partial derivatives to 0 yields simultaneous linear equations for the least-squares coefficients, A and B.[5]

Simultaneous linear equations for the least-squares coefficients A and B, the so-called *normal equations*[6] for simple regression, are

$$An + B\sum X_i = \sum Y_i$$

$$A\sum X_i + B\sum X_i^2 = \sum X_i Y_i$$

Solving the normal equations produces the least-squares coefficients:

$$A = \overline{Y} - B\overline{X} \quad [5.3]$$

$$B = \frac{n\sum X_i Y_i - \sum X_i \sum Y_i}{n\sum X_i^2 - (\sum X_i)^2} = \frac{\sum (X_i - \overline{X})(Y_i - \overline{Y})}{\sum (X_i - \overline{X})^2}$$

The formula for A implies that the least-squares line passes through the point of means of the two variables. By Equation 5.2, therefore, the least-squares

---

[3] I shall return to LAV regression in Section 14.3, which discusses robust regression.

[4] In Chapter 10, I shall derive the least-squares solution by an alternative geometric approach.

[5] As a formal matter, it remains to be shown that the solution of the normal equations *minimizes* the least-squares function $S(A, B)$. See Section 9.2.

[6] The term "normal" here refers not to the normal distribution but to orthogonality (perpendicularity); see Chapter 10 on the vector geometry of regression.

residuals sum to 0. The second normal equation implies that $\sum X_i E_i = 0$, for

$$\sum X_i E_i = \sum X_i(Y_i - A - BX_i) = \sum X_i Y_i - A\sum X_i - B\sum X_i^2 = 0$$

Similarly,[7] $\sum \hat{Y}_i E_i = 0$. These properties, which will be useful to us below, imply that the least-squares residuals are uncorrelated with both the independent-variable values (the X's) and the fitted values (the $\hat{Y}$'s).[8]

It is clear from Equation 5.3 that the least-squares coefficients are uniquely defined as long as the independent-variable values are not all identical, for when there is no variation in X, the denominator of B vanishes. This result is intuitively plausible: Only if the independent-variable scores are spread out can we hope to fit a (unique) line to the X, Y scatter; if, alternatively, all of the X-values are the same (say, equal to x_0), then any line through the point $(x_0, \overline{Y})$ is a least-squares line.

I shall illustrate the least-squares calculations using Davis's data on measured weight (Y) and reported weight (X), for which

$$\begin{aligned}
n &= 101 \\
\overline{Y} &= \frac{5780}{101} = 57.23 \\
\overline{X} &= \frac{5731}{101} = 56.74 \\
\sum(X_i - \overline{X})(Y_i - \overline{Y}) &= 4435. \\
\sum(X_i - \overline{X})^2 &= 4539. \\
B &= \frac{4435}{4539} = 0.9771 \\
A &= 57.23 - 0.9771 \times 56.74 = 1.789
\end{aligned}$$

Thus, the least-squares regression equation is

$$\widehat{\text{Measured weight}} = 1.79 + 0.977 \times \text{Reported weight}$$

Interpretation of the least-squares slope coefficient is straightforward: $B = 0.977$ indicates that a 1-kg increase in reported weight is associated, on average, with just under a 1-kg increase in measured weight. Because the data are not longitudinal, the phrase "a unit increase" here implies not a literal change over time, but rather a static comparison between two individuals who differ by 1-kg in their reported weights.

Ordinarily, we may interpret the intercept A as the fitted value associated with $X = 0$, but it is, of course, impossible for an individual to have a reported weight equal to 0. The intercept A is usually of little direct interest, because the fitted value above $X = 0$ is rarely important. Here, however, if individuals' reports are unbiased predictions of their actual weights, then we should have

[7] See Exercise 5.1.

[8] See the next section for a definition of correlation.

the equation $\hat{Y} = X$—that is, an intercept of 0 and a slope of 1. The intercept $A = 1.79$ is indeed close to 0, and the slope $B = 0.977$ is close to 1.

> In simple linear regression, the least-squares coefficients are given by $A = \overline{Y} - B\overline{X}$ and $B = \sum(X_i - \overline{X})(Y_i - \overline{Y})/\sum(X_i - \overline{X})^2$. The slope coefficient B represents the average change in Y associated with a one-unit increase in X. The intercept A is the fitted value of Y when $X = 0$.

5.1.2 Simple Correlation

Having calculated the least-squares line, it is of interest to determine how closely the line fits the scatter of points. This is a vague question, which may be answered in a variety of ways. The standard deviation of the residuals, S_E, often called the *standard error of the regression*, provides one sort of answer. Because of estimation considerations,[9] the variance of the residuals is defined using *degrees of freedom* $n-2$, rather than the sample size n, in the denominator:

$$S_E^2 = \frac{\sum E_i^2}{n-2}$$

The standard error is, therefore,

$$S_E = \sqrt{\frac{\sum E_i^2}{n-2}}$$

Because it is measured in the units of the dependent variable, and represents a type of "average" residual, the standard error is simple to interpret. For example, for Davis's regression of measured on reported weight, the sum of squared residuals is $\sum E_i^2 = 418.9$, and thus the standard error of the regression is

$$S_E = \sqrt{\frac{418.9}{101-2}} = 2.06$$

On average, then, using the least-squares regression line to predict measured weight from reported weight results in an error of about 2 kg, which is small, but perhaps not negligible. Moreover, if the residuals are approximately normally distributed, then about 2/3 of the residuals are in the range ± 2, and about 95% are in the range ± 4. I believe that social scientists overemphasize correlation (described immediately below) and pay insufficient attention to the standard error of the regression as an index of fit.

In contrast to the standard error of the regression, the *correlation coefficient* provides a *relative* measure of fit: To what degree do our predictions of Y

[9] Estimation is discussed in the next chapter.

improve when we base these predictions on the linear relationship between Y and X? A relative index of fit requires a baseline—how well can Y be predicted if X is disregarded?

To disregard the independent variable is implicitly to fit the equation $\hat{Y}_i' = A'$, or, equivalently,

$$Y_i = A' + E_i'$$

By ignoring the independent variable, we lose our ability to differentiate among the observations; as a result, the fitted values are constant. The constant A' is generally different from the intercept A of the least-squares line, and the residuals E_i' are different from the least-squares residuals E_i.

How should we find the best constant A'? An obvious approach is to employ a least-squares fit—that is, to minimize

$$S(A') = \sum E_i'^2 = \sum (Y_i - A')^2$$

As you may be aware, the value of A' that minimizes this sum of squares is simply the dependent-variable mean, $\overline{Y}$.[10]

The residuals $E_i = Y_i - \hat{Y}_i$ from the linear regression of Y on X will generally be smaller than the residuals $E_i' = Y_i - \overline{Y}$, and it is necessarily the case that

$$\sum (Y_i - \hat{Y}_i)^2 \leq \sum (Y_i - \overline{Y})^2$$

This inequality holds because the "null model," $Y_i = A' + E_i'$, specifying no relationship between Y and X, is a special case of the more general linear regression "model," $Y_i = A + BX_i + E_i$: The two models are the same when $B = 0$. The null model therefore cannot have a smaller sum of squared residuals. After all, the least-squares coefficients A and B are selected precisely to minimize $\sum E_i^2$, so constraining $B = 0$ cannot improve the fit and will usually make it worse.

We call

$$\sum E_i'^2 = \sum (Y_i - \overline{Y})^2$$

the *total sum of squares* for Y, abbreviated TSS, while

$$\sum E_i^2 = \sum (Y_i - \hat{Y}_i)^2$$

is called the *residual sum of squares*, and is abbreviated RSS. The difference between the two, termed the regression sum of squares,

$$\text{RegSS} \equiv \text{TSS} - \text{RSS}$$

gives the reduction in squared error due to the linear regression. The ratio of RegSS to TSS, the proportional reduction in squared error, defines the square of the correlation coefficient:

$$r^2 \equiv \frac{\text{RegSS}}{\text{TSS}}$$

[10] See Exercise 5.3.

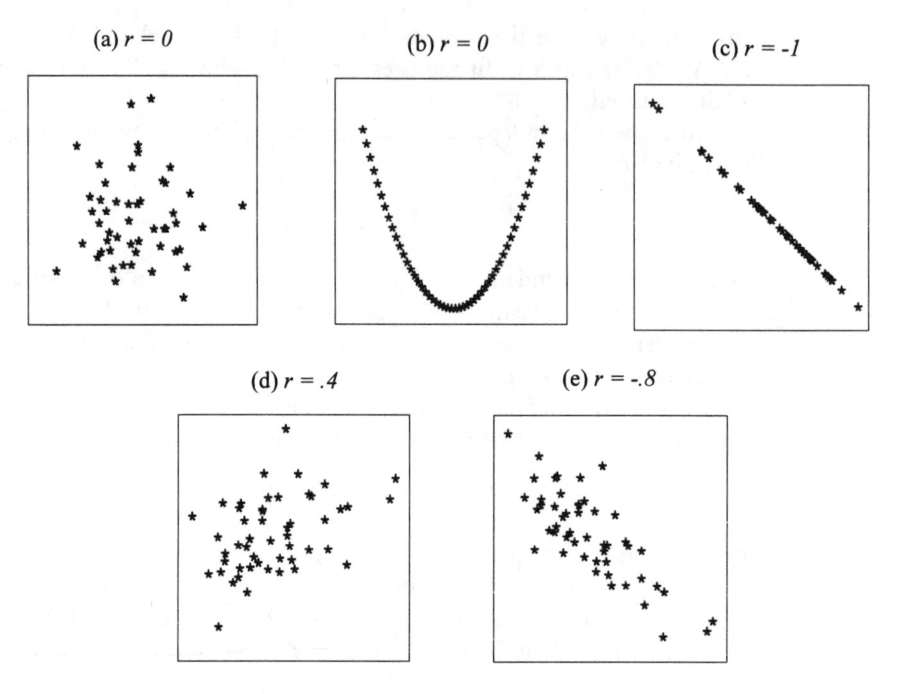

Figure 5.3. Scatterplots illustrating different levels of correlation: $r = 0$ in both (a) and (b); $r = -1$ in (c); $r = .4$ in (d); and $r = -.8$ in (e). All of the datasets have $n = 50$ observations.

To find the correlation coefficient r, we take the positive square root of r^2 when the simple-regression slope B is positive, and the negative square root when B is negative.

Thus, if there is a perfect positive linear relationship between Y and X (i.e., if all of the residuals are 0, and $B > 0$), then $r = 1$. A perfect negative linear relationship corresponds to $r = -1$. If there is no linear relationship between Y and X, then RSS = TSS, RegSS = 0, and $r = 0$. Between these extremes, r gives the direction of the linear relationship between the two variables, and r^2 can be interpreted as the proportion of the total variation of Y that is "captured" by its linear regression on X. Figure 5.3 illustrates several levels of correlation.

It is instructive to examine the three sums of squares more closely. Starting with an individual observation, we have the identity

$$Y_i - \overline{Y} = (Y_i - \hat{Y}_i) + (\hat{Y}_i - \overline{Y})$$

This equation is interpreted geometrically in Figure 5.4. Squaring both sides of the equation and summing over observations produces

$$\sum(Y_i - \overline{Y})^2 = \sum(Y_i - \hat{Y}_i)^2 + \sum(\hat{Y}_i - \overline{Y})^2 + 2\sum(Y_i - \hat{Y}_i)(\hat{Y}_i - \overline{Y})$$

The last term in this equation is 0,[11] and thus the regression sum of squares, which I previously defined as the difference TSS − RSS, may also be written

[11] See Exercise 5.1 and Section 10.1.

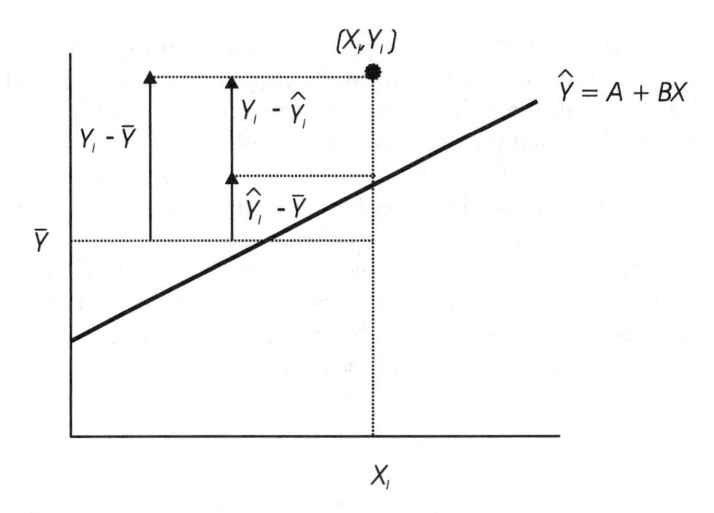

Figure 5.4. Decomposition of the total deviation $Y_i - \bar{Y}$ into components $Y_i - \hat{Y}_i$ and $\hat{Y}_i - \bar{Y}$.

directly as

$$\text{RegSS} = \sum(\hat{Y}_i - \bar{Y})^2$$

This decomposition of total variation into "explained" and "unexplained" components, paralleling the decomposition of each observation into a fitted value and a residual, is typical of linear models. The decomposition is called the *analysis of variance* for the regression: TSS = RegSS + RSS.

Although I have developed the correlation coefficient from the regression of Y on X, it is also possible to define r by analogy with the correlation $\rho = \sigma_{XY}/\sigma_X\sigma_Y$ between two random variables. First defining the *sample covariance* between X and Y,

$$S_{XY} \equiv \frac{\sum(X_i - \bar{X})(Y_i - \bar{Y})}{n-1}$$

we may then write

$$r = \frac{S_{XY}}{S_X S_Y} = \frac{\sum(X_i - \bar{X})(Y_i - \bar{Y})}{\sqrt{\sum(X_i - \bar{X})^2 \sum(Y_i - \bar{Y})^2}} \qquad [5.4]$$

where S_X and S_Y are, respectively, the sample standard deviations of X and Y.[12]

It is immediately apparent from the symmetry of Equation 5.4 that the correlation does not depend on which of the two variables is treated as the dependent variable. This property of r is surprising in light of the *asymmetry* of the regression equation used to define the sums of squares: Unless there is a perfect correlation between the two variables, the least-squares line for the regression of Y on X differs from the line for the regression of X on Y.[13]

[12] The equivalence of the two formulas for r is established in Section 10.1 on the geometry of simple regression analysis.

[13] See Exercise 5.2.

There is another central property, aside from symmetry, that distinguishes the correlation coefficient r from the regression slope B. The slope coefficient B is measured in the units of the dependent variable per unit of the independent variable. For example, if dollars of income are regressed on years of education, then the units of B are dollars/year. The correlation coefficient r, however, is unitless, as can be seen from either of its definitions. As a consequence, a change in scale of Y or X produces a compensating change in B, but does not affect r. If, for example, income is measured in thousands of dollars rather than in dollars, the units of the slope become \$1000's/year, and the value of the slope decreases by a factor of 1000, but r remains the same.

For Davis's regression of measured on reported weight,

$$\text{TSS} = 4753.8$$
$$\text{RSS} = 418.87$$
$$\text{RegSS} = 4334.9$$

Thus,

$$r^2 = \frac{4334.9}{4753.8} = .91188$$

and, because B is positive, $r = +\sqrt{.91188} = .9549$. The linear regression of measured on reported weight, therefore, captures 91% of the variation in measured weight. Equivalently,

$$S_{XY} = \frac{4435.9}{101 - 1} = 44.359$$
$$S_X^2 = \frac{4539.3}{101 - 1} = 45.393$$
$$S_Y^2 = \frac{4753.8}{101 - 1} = 47.538$$
$$r = \frac{44.359}{\sqrt{45.393 \times 47.538}} = .9549$$

EXERCISES

5.1 *Prove that the least-squares fit in simple-regression analysis has the following properties:

(a) $\sum \hat{Y}_i E_i = 0$.

(b) $\sum (Y_i - \hat{Y}_i)(\hat{Y}_i - \bar{Y}) = \sum E_i(\hat{Y}_i - \bar{Y}) = 0$.

5.2 *Suppose that the means and standard deviations of Y and X are the same: $\bar{Y} = \bar{X}$ and $S_Y = S_X$.

(a) Show that, under these circumstances,

$$B_{Y|X} = B_{X|Y} = r_{XY}$$

where $B_{Y|X}$ is the least-squares slope for the regression of Y on X; $B_{X|Y}$ is the least-squares slope for the simple regression of X on Y; and r_{XY} is the correlation between the two variables. Show that the intercepts are also the same, $A_{Y|X} = A_{X|Y}$.

(b) Why, if $A_{Y|X} = A_{X|Y}$ and $B_{Y|X} = B_{X|Y}$, is the least-squares line for the regression of Y on X different from the line for the regression of X on Y (as long as $r^2 < 1$)?

(c) "Regression toward the mean" (the original sense of the term "regression"): Imagine that X is father's height and Y is son's height for a sample of father-son pairs. Suppose that $S_Y = S_X$, that $\overline{Y} = \overline{X}$, and that the regression of sons' heights on fathers' heights is linear. Finally, suppose that $0 < r_{XY} < 1$ (i.e., fathers' and sons' heights are positively correlated, but not perfectly so). Show that the expected height of a son whose father is shorter than average is also less than average, but to a smaller extent; likewise, the expected height of a son whose father is taller than average is also greater than average, but to a smaller extent. Does this result imply a contradiction—that the standard deviation of son's height is in fact less than that of father's height?

(d) What is the expected height for a father whose son is shorter than average? Of a father whose son is taller than average?

(e) Regression effects in research design: Imagine that educational researchers wish to assess the efficacy of a new program to improve the reading performance of children. To test the program, they recruit a group of children who are reading substantially below grade level; after a year in the program, the researchers observe that the children, on average, have improved their reading performance. Why is this a weak research design? How could it be improved?

5.3 *Show that $A' = \overline{Y}$ minimizes the sum of squares

$$S(A') = \sum_{i=1}^{n}(Y_i - A')^2$$

5.4 #The following five observations were selected from among the 20 households in Sahlins's dataset on Mazulu village:

Household	*Consumers/ Gardener* X_i	*Acres/ Gardener* Y_i
1	1.00	1.71
5	1.20	2.21
10	1.46	2.09
15	1.65	2.41
20	2.30	2.36

(a) Construct a scatterplot for Y and X.

(b) Find A and B for the least-squares regression of Y on X, and draw the least-squares line on the scatterplot. Interpret A and B.

(c) Calculate the standard error of the regression, S_E, and the correlation coefficient, r. Interpret these statistics.

5.5 Linear transformation of X and Y:

(a) Suppose that the independent-variable values in the previous problem are transformed according to the equation $X' = X - 10$, and that Y is regressed on X'. Without redoing the regression calculations in detail, find A', B', S'_E, and r'. What happens to these quantities when $X' = 10X$? When $X' = 10(X - 1) = 10X - 10$?

(b) Now suppose that the dependent variable scores are transformed according to the formula $Y'' = Y + 10$, and that Y'' is regressed on X. Find A'', B'', S''_E, and r''. What happens to these quantities when $Y'' = 5Y$? When $Y'' = 5(Y + 2) = 5Y + 10$?

(c) In general, how are the results of a simple-regression analysis affected by linear transformations of X and Y? In what sense, then, is the scaling of these variable arbitrary? Why might one scaling be preferred to another?

5.6 Analyze Sahlins's data, given in Table 2.1 and `sahlins.dat`, by regressing Acres/Gardener on Consumers/Gardener. In a society characterized by "primitive communism," the social product of the village would be redistributed according to need, while each household would work in proportion to its capacity, implying a regression slope of 0. In contrast, in a society in which redistribution is purely through the market, each household should have to work in proportion to its consumption needs, suggesting a positive regression slope and an intercept of 0. Interpret the results of the regression in light of these observations. Examine and interpret the values of A, B, S_E, and r (or r^2). Do the results change if the fourth household is deleted? Plot the regression lines calculated with and without the fourth household on a scatterplot of the data. Does either regression do a good job of summarizing the relationship between Acres/Gardener and Consumers/Gardener? (Cf. Exercise 2.4.)

5.7 Recall Robey et al.'s data on fertility and contraception in developing countries; the data are in Table 2.2 and `robey.dat`. Perform a least-squares regression of fertility on contraception. Plot the data and the least-squares line. Does the line adequately summarize the relationship between these variables? Examine and interpret the values of A, B, S_E, and r (or r^2). (Cf. Exercise 2.5.)

5.8 Anscombe (1981) presents and analyzes the data on the U.S. states and Washington, D.C. given in Table 5.1 and `anscombe.dat`, for state per-capita public-school expenditures (in dollars), per-capita annual income (dollars), the proportion of residents under the age of 18 (per 1000), and the proportion of the population residing in urban areas (per 1000). Regress school expenditures on each of income, proportion under 18, and proportion urban. Plot the least-squares line for each regression on a scatterplot of the data. Does the line adequately capture the relationship between the variables shown in the plot? In each case, examine and interpret the values of A, B, S_E, and r (or r^2).

5.9 Recall Angell's data on the moral integration of U.S. cities, given in Table 2.3 and `angell.dat`. Regress moral integration on each of ethnic hetero-

TABLE 5.1 Data on the U.S. States and Washington, D.C. in 1970: State Per-Capita Public-School Expenditures (E, in Dollars); Per-Capita Annual Income (I, in Dollars); Proportion of Residents Under 18 (S, per 1000); and Proportion Residing in Urban Areas (U, per 1000)

State	*E*	*I*	*S*	*U*	*State*	*E*	*I*	*S*	*U*
ME	189	2824	350.7	508	NC	155	2664	354.1	450
NH	169	3259	345.9	564	SC	149	2380	376.7	476
VT	230	3072	348.5	322	GA	156	2781	370.6	603
MA	168	3835	335.3	846	FL	191	3191	336.0	805
RI	180	3549	327.1	871	KY	140	2645	349.3	523
CT	193	4256	341.0	774	TN	137	2579	342.8	588
NY	261	4151	326.2	856	AL	112	2337	362.2	584
NJ	214	3954	333.5	889	MS	130	2081	385.2	445
PA	201	3419	326.2	715	AR	134	2322	351.9	500
OH	172	3509	354.5	753	LA	162	2634	389.6	661
IN	194	3412	359.3	649	OK	135	2880	329.8	680
IL	189	3981	348.9	830	TX	155	3029	369.4	797
MI	233	3675	369.2	738	MT	238	2942	368.9	534
WI	209	3363	360.7	659	ID	170	2668	367.7	541
MN	262	3341	365.4	664	WY	238	3190	365.6	605
IA	234	3265	343.8	572	CO	192	3340	358.1	785
MO	177	3257	336.1	701	NM	227	2651	421.5	698
ND	177	2730	369.1	443	AZ	207	3027	387.5	796
SD	187	2876	368.7	446	UT	201	2790	412.4	804
NE	148	3239	349.9	615	NV	225	3957	385.1	809
KS	196	3303	339.9	661	WA	215	3688	341.3	726
DE	248	3795	375.9	722	OR	233	3317	332.7	671
MD	247	3742	364.1	766	CA	273	3968	348.4	909
DC	246	4425	352.1	1000	AK	372	4146	439.7	484
VA	180	3068	353.0	631	HI	212	3513	382.9	831
WV	149	2470	328.8	390					

Source of Data: Anscombe (1981).

geneity and geographic mobility. Plot the least-squares line for each regression on a scatterplot of the data, and comment on the results. Examine and interpret the values of A, B, S_E, and r (or r^2). (Cf. Exercise 2.6.)

5.2 Multiple Regression

5.2.1 Two Independent Variables

The linear multiple-regression equation

$$\hat{Y} = A + B_1X_1 + B_2X_2$$

for two independent variables, X_1 and X_2, describes a plane in the three-dimensional $\{X_1, X_2, Y\}$ space, as shown in Figure 5.5. As in the case of simple regression, it is unreasonable to expect that the regression plane will pass precisely through every point, so the fitted value for observation i in general differs

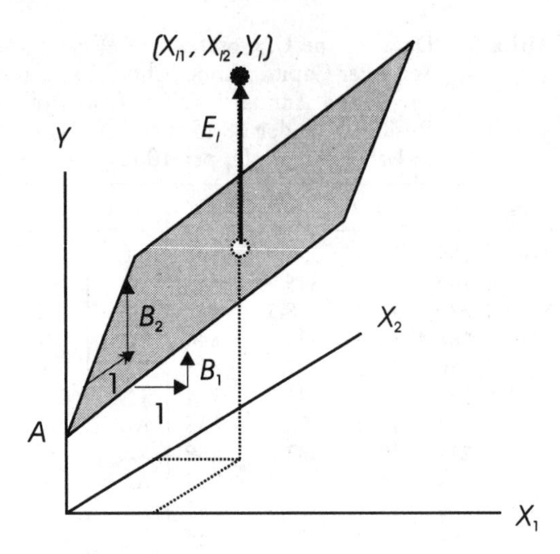

Figure 5.5. The multiple-regression plane, showing the partial slopes B_1 and B_2, and the residual E_i for the ith observation.

from the observed value. The residual is the signed vertical distance from the point to the plane:

$$E_i = Y_i - \hat{Y}_i = Y_i - (A + B_1 X_{i1} + B_2 X_{i2})$$

To make the plane come as close as possible to the points in the aggregate, we want the values of A, B_1, and B_2 that minimize the sum of squared residuals:

$$S(A, B_1, B_2) = \sum E_i^2 = \sum (Y_i - A - B_1 X_{i1} - B_2 X_{i2})^2$$

*As for simple regression, we can proceed by differentiating the sum-of-squares function with respect to the regression coefficients:

$$\begin{aligned}
\frac{\partial S(A, B_1, B_2)}{\partial A} &= \sum (-1)(2)(Y_i - A - B_1 X_{i1} - B_2 X_{i2}) \\
\frac{\partial S(A, B_1, B_2)}{\partial B_1} &= \sum (-X_{i1})(2)(Y_i - A - B_1 X_{i1} - B_2 X_{i2}) \\
\frac{\partial S(A, B_1, B_2)}{\partial B_2} &= \sum (-X_{i2})(2)(Y_i - A - B_1 X_{i1} - B_2 X_{i2})
\end{aligned}$$

Setting the partial derivatives to 0 and rearranging terms produces the normal equations for the regression coefficients A, B_1, and B_2.

The normal equations for the regression coefficients A, B_1, and B_2 are

$$\begin{aligned}
An + B_1 \sum X_{i1} + B_2 \sum X_{i2} &= \sum Y_i \\
A \sum X_{i1} + B_1 \sum X_{i1}^2 + B_2 \sum X_{i1} X_{i2} &= \sum X_{i1} Y_i \\
A \sum X_{i2} + B_1 \sum X_{i2} X_{i1} + B_2 \sum X_{i2}^2 &= \sum X_{i2} Y_i
\end{aligned} \qquad [5.5]$$

Because Equation 5.6 is a system of three linear equations in three unknowns, it usually provides a unique solution for the least-squares regression coefficients A, B_1, and B_2. We can write out the solution explicitly, if somewhat tediously: Dropping the subscript i for observations, and using asterisks to denote variables in mean-deviation form (e.g., $Y^* \equiv Y_i - \overline{Y}$),

$$A = \overline{Y} - B_1\overline{X}_1 - B_2\overline{X}_2 \qquad [5.6]$$

$$B_1 = \frac{\sum X_1^* Y^* \sum X_2^{*2} - \sum X_2^* Y^* \sum X_1^* X_2^*}{\sum X_1^{*2} \sum X_2^{*2} - (\sum X_1^* X_2^*)^2}$$

$$B_2 = \frac{\sum X_2^* Y^* \sum X_1^{*2} - \sum X_1^* Y^* \sum X_1^* X_2^*}{\sum X_1^{*2} \sum X_2^{*2} - (\sum X_1^* X_2^*)^2}$$

The denominator of B_1 and B_2 is nonzero—and, therefore, the least-squares coefficients are uniquely defined—as long as

$$\sum X_1^{*2} \sum X_2^{*2} \neq \left(\sum X_1^* X_2^*\right)^2$$

This condition is satisfied unless X_1 and X_2 are perfectly correlated or unless one of the independent variables is invariant.[14] If X_1 and X_2 are perfectly correlated, then they are said to be *collinear*.

To illustrate the computation of multiple-regression coefficients, I shall employ Duncan's occupational prestige data, which were introduced in Chapter 3. I shall, for the time-being, disregard the problems with these data that were revealed by graphical analysis. Recall that Duncan wished to predict the prestige of occupations (Y) from their educational and income levels (X_1 and X_2, respectively). I calculated the following quantities from Duncan's data:

$$n = 45$$

$$\overline{Y} = \frac{2146}{45} = 47.69$$

$$\overline{X}_1 = \frac{2365}{45} = 52.56$$

$$\overline{X}_2 = \frac{1884}{45} = 41.87$$

$$\sum X_1^{*2} = 38{,}971.$$

$$\sum X_2^{*2} = 26{,}271.$$

$$\sum X_1^* X_2^* = 23{,}182.$$

$$\sum X_1^* Y^* = 35{,}152.$$

$$\sum X_2^* Y^* = 28{,}383.$$

[14] The correlation between X_1 and X_2 is, in the current notation,

$$r_{12} = \frac{\sum X_1^* X_2^*}{\sqrt{\sum X_1^{*2} \sum X_2^{*2}}}$$

Substituting these results into Equation 5.6 produces $A = -6.070$, $B_1 = 0.5459$, and $B_2 = 0.5987$. The fitted least-squares regression equation is, therefore,

$$\widehat{\text{Prestige}} = -6.07 + 0.546 \times \text{Education} + 0.599 \times \text{Income}$$

Although the development of least-squares linear regression for two independent variables is very similar to the development for simple regression, there is this important difference in interpretation: The slope coefficients for the independent variables in multiple regression are *partial* coefficients, while the slope coefficient in simple regression gives the *marginal* relationship between the dependent variable and a single independent variable. That is, each slope in multiple regression represents the "effect" on the dependent variable of a one-unit increment in the corresponding independent variable *holding constant* the value of the other independent variable. The simple-regression slope effectively ignores the other independent variable.

This interpretation of the multiple-regression slope is apparent in Figure 5.5, which shows the multiple-regression plane. Because the regression plane is flat, its slope (B_1) in the direction of X_1, holding X_2 constant, does not depend on the specific value at which X_2 is fixed. Likewise, the slope in the direction of X_2 fixing the value of X_1 is always B_2.

Algebraically, let us fix X_2 to the specific value x_2 and see how $\hat{Y}$ changes as X_1 is increased by 1, from some specific value x_1 to $x_1 + 1$:

$$[A + B_1(x_1 + 1) + B_2x_2] - (A + B_1x_1 + B_2x_2) = B_1$$

Similarly, increasing X_2 by 1, fixing X_1 produces

$$[A + B_1x_1 + B_2(x_2 + 1)] - (A + B_1x_1 + B_2x_2) = B_2$$

*Because the regression surface

$$\hat{Y} = A + B_1X_1 + B_2X_2$$

is a plane, precisely the same results follow from differentiating the regression equation with respect to each of X_1 and X_2:

$$\frac{\partial \hat{Y}}{\partial X_1} = B_1$$

$$\frac{\partial \hat{Y}}{\partial X_2} = B_2$$

Nothing new is learned here, but differentiation is often a useful approach for understanding nonlinear statistical models, for which the regression surface is not flat.[15]

[15] See Section 14.2.

For Duncan's regression, then, a unit increase in education, holding income constant, is associated, on average, with an increase of 0.55 units in prestige (which, recall, is the percentage of respondents rating the prestige of the occupation as good or excellent). A unit increase in income, holding education constant, is associated, on average, with an increase of 0.60 units in prestige.

The regression intercept, $A = -6.1$, has the following literal interpretation: The fitted value of prestige is -6.1 for a hypothetical occupation with education and income levels both equal to 0. Literal interpretation of the intercept is problematic here, however. Although there are some observations in Duncan's dataset with small education and income levels, no occupations have levels of 0. Moreover, the dependent variable cannot take on negative values.

5.2.2 Several Independent Variables

The extension of least-squares regression to several independent variables is straightforward. For the general case of k independent variables, the multiple-regression equation is

$$\begin{aligned} Y_i &= A + B_1 X_{i1} + B_2 X_{i2} + \cdots + B_k X_{ik} + E_i \\ &= \hat{Y}_i + E_i \end{aligned}$$

It is, of course, not possible to visualize the point cloud of the data directly when $k > 2$, but it is a relatively simple matter to find the values of A and the B's that minimize the sum of squared residuals:

$$S(A, B_1, B_2, \ldots, B_k) = \sum_{i=1}^{n} [Y_i - (A + B_1 X_{i1} + B_2 X_{i2} + \cdots + B_k X_{ik})]^2$$

Minimization of the sum-of-squares function produces the normal equations for general multiple regression:[16]

$$\begin{aligned} An + B_1 \sum X_{i1} + B_2 \sum X_{i2} + \cdots + B_k \sum X_{ik} &= \sum Y_i \\ A \sum X_{i1} + B_1 \sum X_{i1}^2 + B_2 \sum X_{i1} X_{i2} + \cdots + B_k \sum X_{i1} X_{ik} &= \sum X_{i1} Y_i \\ A \sum X_{i2} + B_1 \sum X_{i2} X_{i1} + B_2 \sum X_{i2}^2 + \cdots + B_k \sum X_{i2} X_{ik} &= \sum X_{i2} Y_i \\ \vdots \qquad\qquad\qquad\qquad &\qquad \vdots \\ A \sum X_{ik} + B_1 \sum X_{ik} X_{i1} + B_2 \sum X_{ik} X_{i2} + \cdots + B_k \sum X_{ik}^2 &= \sum X_{ik} Y_i \end{aligned} \qquad [5.7]$$

We cannot write out a general solution to the normal equations without specifying the number of independent variables k, and even for k as small as 3, an explicit solution would be very complicated.[17] Nevertheless, because the normal equations are linear, and because there are as many equations as unknown

[16] See Exercise 5.10.

[17] As I shall show in Section 9.1, however, it is simple to write out a solution to the normal equations using matrices.

regression coefficients ($k + 1$), there is usually a unique solution for the coefficients $A, B_1, B_2, \ldots, B_k$. Only when one independent variable is a perfect linear function of others, or when one or more independent variables are invariant, will the normal equations not have a unique solution. Dividing the first normal equation through by n reveals that the least-squares surface passes through the point of means $(\overline{X}_1, \overline{X}_2, \ldots, \overline{X}_k, \overline{Y})$.

> The least-squares coefficients in multiple linear regression are found by solving the normal equations for the intercept A and the slope coefficients $B_1, B_2, \ldots, B_k$. The slope coefficient B_1 represents the average change in Y associated with a one-unit increase in X_1 when the other X's are held constant.

To illustrate the solution of the normal equations, let us return to the Canadian occupational prestige data, regressing the prestige of the occupations on average education, average income, and the percentage of women in each occupation. Recall that our graphical analysis of the data in Chapter 3 cast doubt on the appropriateness of the linear regression, but I shall disregard these problems for now.

The various sums, sums of squares, and sums of products that are required are given in Table 5.2. Notice that the sums of squares and products are very large, particularly for income, which is scaled in small units (dollars). Substituting these values into the four normal equations and solving for the regression coefficients produces

$$\begin{aligned} A &= -6.7943 \\ B_1 &= 4.1866 \\ B_2 &= 0.0013136 \\ B_3 &= -0.0089052 \end{aligned}$$

The fitted regression equation is, therefore,

$$\begin{aligned} \widehat{\text{Prestige}} = -6.794 + 4.187 \times \text{Education} + 0.001314 \times \text{Income} \\ -0.008905 \times \text{Percent women} \end{aligned}$$

In interpreting the regression coefficients, we need to keep in mind the units of each variable. Prestige scores are arbitrarily scaled, and range from a minimum of 14.8 to a maximum of 87.2 for these 102 occupations; the hinge spread of prestige is 24.4 points. Education is measured in years, and hence the impact of education on prestige is considerable—a little more than four points, on average, for each year of education, holding income and gender composition constant. Likewise, despite the small absolute size of its coefficient, the partial

TABLE 5.2 Sums of Squares (Diagonal), Sums of Products (Off Diagonal), and Sums (Last Row) for the Canadian Occupational Prestige Data

Variable	Prestige	Education	Income	Percentage of women
Prestige	253,618.	55,326.	37,748,108.	131,909.
Education	55,326.	12,513.	8,121,410.	32,281.
Income	37,748,108.	8,121,410.	6,534,383,460.	14,093,097.
Percentage of women	131,909.	32,281.	14,093,097.	187,312.
Sum	4,777.	1,095.	693,386.	2,956.

effect of income is also substantial—more than 0.001 points, on average, for an additional dollar of income, or more than 1 point for each $1000. In contrast, the impact of gender composition, holding education and income constant, is very small—an average decline of about 0.01 points for each 1% increase in the percentage of women in an occupation.

5.2.3 Multiple Correlation

As in simple regression, the standard error in multiple regression measures the "average" size of the residuals. As before, we divide by degrees of freedom, here $n - (k+1) = n - k - 1$, rather than by the sample size n, to calculate the variance of the residuals; thus, the standard error is

$$S_E = \sqrt{\frac{\sum E_i^2}{n - k - 1}}$$

Heuristically, we "lose" $k+1$ degrees of freedom by calculating the $k+1$ regression coefficients, $A, B_1, \ldots, B_k$.[18]

For Duncan's regression of occupational prestige on the income and educational levels of occupations, the standard error is

$$S_E = \sqrt{\frac{7506.7}{45 - 2 - 1}} = 13.37$$

Recall that the dependent variable here is the percentage of raters classifying the occupation as good or excellent in prestige; an average error of 13 is substantial given Duncan's purpose, which was to use the regression equation to calculate substitute prestige scores for occupations for which direct ratings were unavailable. For the Canadian occupational prestige data, regressing prestige scores on average education, average income, and gender composition, the standard error is

$$S_E = \sqrt{\frac{6033.6}{102 - 3 - 1}} = 7.846$$

which is also a substantial figure.

---

[18] A deeper understanding of the central concept of degrees of freedom is developed in Chapter 10.

The sums of squares in multiple regression are defined in the same manner as in simple regression:

$$\text{TSS} = \sum (Y_i - \overline{Y})^2$$
$$\text{RegSS} = \sum (\hat{Y}_i - \overline{Y})^2$$
$$\text{RSS} = \sum (Y_i - \hat{Y}_i)^2 = \sum E_i^2$$

Of course, the fitted values $\hat{Y}_i$ and residuals E_i now come from the multiple-regression equation. Moreover, we have a similar decomposition of variation:

$$\text{TSS} = \text{RegSS} + \text{RSS}$$

The least-squares residuals are uncorrelated with the fitted values and with each of the X's.[19]

> The linear regression decomposes the variation in Y into "explained" and "unexplained" components: TSS = RegSS + RSS. The least-squares residuals, E, are uncorrelated with the fitted values, $\hat{Y}$, and with the independent variables, $X_1, \ldots, X_k$.

The squared multiple correlation R^2, representing the proportion of variation in the dependent variable captured by the regression, is defined in terms of the sums of squares:

$$R^2 \equiv \frac{\text{RegSS}}{\text{TSS}}$$

Because there are now several slope coefficients, potentially with different signs, the multiple correlation coefficient is, by convention, the positive square root of R^2. The multiple correlation is also interpretable as the simple correlation between the fitted and observed Y-values.

> The standard error of the regression, $S_E = \sqrt{\sum E_i^2/(n-k-1)}$, gives the "average" size of the regression residuals; the squared multiple correlation, R^2 = RegSS/TSS, indicates the proportion of the variation in Y that is captured by its linear regression on the X's.

[19] These and other properties of the least-squares fit are derived in Chapters 9 and 10.

For Duncan's regression, we have the following sums of squares:

$$\text{TSS} = 43{,}688.$$
$$\text{RegSS} = 36{,}181.$$
$$\text{RSS} = 7506.7$$

The squared multiple correlation,

$$R^2 = \frac{36{,}181}{43{,}688} = .8282$$

indicates that more than 80% of the variation in prestige among the 45 occupations is accounted for by the linear regression of prestige on the income and educational levels of the occupations. For the Canadian prestige regression, the sums of squares and R^2 are as follows:

$$\text{TSS} = 29{,}895$$
$$\text{RegSS} = 23{,}862$$
$$\text{RSS} = 6033.6$$
$$R^2 = \frac{23{,}862}{29{,}895} = .7982.$$

Because the multiple correlation can only rise, never decline, when independent variables are added to the regression equation,[20] investigators sometimes penalize the value of R^2 by a "correction" for degrees of freedom. The corrected R^2 is defined as

$$\tilde{R}^2 \equiv 1 - \frac{S_E^2}{S_Y^2} = 1 - \frac{\dfrac{\text{RSS}}{n-k-1}}{\dfrac{\text{TSS}}{n-1}}$$

Unless the sample size is very small, however, $\tilde{R}^2$ will differ little from R^2. For Duncan's regression, for example,

$$\tilde{R}^2 = 1 - \frac{\dfrac{7506.7}{45-2-1}}{\dfrac{43{,}688}{45-1}} = .8200$$

5.2.4 Standardized Regression Coefficients

Social researchers often wish to compare the coefficients of different independent variables in a regression analysis. When the independent variables are commensurable (i.e., measured in the same units on the same scale), or when they can be reduced to a common standard, comparison is straightforward. In most instances, however, independent variables are not commensurable. Stan-

[20] See Exercise 5.11.

dardized regression coefficients permit a limited assessment of the relative effects of incommensurable independent variables.

To place standardized coefficients in perspective, let us first consider an example in which the independent variables are measured in the same units. Imagine that the annual dollar income of wage workers is regressed on their years of education, years of labor force experience, and some other independent variables, producing the fitted regression equation

$$\widehat{\text{Income}} = A + B_1 \times \text{Education} + B_2 \times \text{Experience} + \cdots$$

Because education and experience are each measured in years, the coefficients B_1 and B_2 are both expressed in dollars/year and, consequently, can be directly compared. If, for example, B_1 is larger than B_2, then (disregarding issues arising from sampling variation) a year's increment in education yields a greater average return in income than a year's increment in labor force experience, holding constant the other factors in the regression equation.

It is, as I have mentioned, much more common for independent variables to be measured in different units. In the Canadian occupational prestige regression, for example, the coefficient for education is expressed in points (of prestige) per year; the coefficient for income is expressed in points per dollar; and the coefficient of gender composition in points per percentage of women. I have already pointed out that the income coefficient (0.001314) is much smaller than the education coefficient (4.187) not because income is a much less important determinant of prestige, but because the unit of income (the dollar) is small, while the unit of education (the year) is relatively large. If we were to reexpress income in $1000's, then we would multiply the income coefficient by 1000.

By the very meaning of the term, incommensurable quantities cannot be directly compared. Still, in certain circumstances, incommensurables can be reduced to a common (e.g., monetary) standard. In most cases, however—as in the prestige regression—there is no obvious basis for this sort of reduction.

In the absence of a theoretically meaningful basis for comparison, an empirical comparison can be made by rescaling regression coefficients according to a measure of independent-variable spread. We can, for example, multiply each regression coefficient by the hinge spread of the corresponding independent variable. For the Canadian prestige data, the hinge spread of education is 4.28 years; of income, 4131 dollars; and of gender composition, 48.68% women. When each independent variable is manipulated over this range, holding the other independent variables constant, the corresponding average changes in prestige are

$$\begin{aligned} &\text{Education:} && 4.28 \times 4.187 &&= 17.92 \\ &\text{Income:} && 4131 \times 0.001314 &&= 5.428 \\ &\text{Gender:} && 48.68 \times -0.008905 &&= -0.4335 \end{aligned}$$

Thus, education has a larger effect than income over the central half of scores observed in the data, and the effect of gender is very small. Note that this conclusion is distinctly limited: For other data, where the variation in education and income may be different, the relative impact of the two variables may also differ, even if the regression coefficients are unchanged.

There is really no deep reason for equating the hinge spread of one independent variable to the hinge spread of another, as we have done here implicitly in calculating the relative "effect" of each. Indeed, the following observation should give you pause: If two independent variables are commensurable, and if their hinge spreads differ, then performing this calculation is, in effect, to adopt a rubber ruler. If expressing coefficients relative to a measure of spread potentially distorts their comparison when independent variables are commensurable, then why should the procedure magically allow us to compare coefficients that are measured in different units?

It is much more common to standardize regression coefficients using the standard deviations of the independent variables rather than their hinge spreads. Although I shall proceed to explain this procedure, keep in mind that the standard deviation is not a good measure of spread when the distributions of the independent variables are substantially nonnormal. The usual practice standardizes the dependent variable as well, but this is an inessential element of the computation of standardized coefficients, because the *relative* size of the slope coefficients does not change when Y is rescaled.

Beginning with the fitted multiple-regression equation

$$Y_i = A + B_1 X_{i1} + \cdots + B_k X_{ik} + E_i$$

let us express all of the variables in mean-deviation form by subtracting[21]

$$\overline{Y} = A + B_1 \overline{X}_1 + \cdots + B_k \overline{X}_k$$

which produces

$$Y_i - \overline{Y} = B_1(X_{i1} - \overline{X}_1) + \cdots + B_k(X_{ik} - \overline{X}_k) + E_i$$

Then divide both sides of the equation by the standard deviation of the dependent variable S_Y; simultaneously multiply and divide the jth term on the right-hand size of the equation by the standard deviation S_j of X_j. These operations serve to standardize each variable in the regression equation:

$$\frac{Y_i - \overline{Y}}{S_Y} = \left(B_1 \frac{S_1}{S_Y}\right) \frac{X_{i1} - \overline{X}_1}{S_1} + \cdots + \left(B_k \frac{S_k}{S_Y}\right) \frac{X_{ik} - \overline{X}_k}{S_k} + \frac{E_i}{S_Y}$$

$$Z_{iY} = B_1^* Z_{i1} + \cdots + B_k^* Z_{ik} + E_i^*$$

In this equation, $Z_{iY} \equiv (Y_i - \overline{Y})/S_Y$ is the standardized dependent variable, linearly transformed to a mean of 0 and a standard deviation of 1; $Z_{i1}, \ldots, Z_{ik}$ are the independent variables, similarly standardized; $E_i^* \equiv E_i/S_Y$ is the transformed residual which, note, *does not* have a standard deviation of 1; and $B_j^* \equiv B_j(S_j/S_Y)$ is the *standardized partial regression coefficient* for the jth independent variable. The standardized coefficient is interpretable as the average

---

[21] Recall that the least-squares regression surface passes through the point of means for the $k+1$ variables.

change in Y, in standard-deviation units, for a one standard-deviation increase in X_j, holding constant the other independent variables.

> By rescaling regression coefficients in relation to a measure of variation—such as the hinge spread or standard deviation—standardized regression coefficients permit a limited comparison of the relative impact of incommensurable independent variables.

For the Canadian prestige regression, we have the following calculations:

$$\begin{aligned} &\text{Education:} && 4.187 \times 2.728/17.20 &&= 0.6639 \\ &\text{Income:} && 0.001314 \times 4246/17.20 &&= 0.3242 \\ &\text{Gender:} && -0.008905 \times 31.72/17.20 &&= -0.01642 \end{aligned}$$

Because both income and gender composition have substantially nonnormal distributions, however, the use of standard deviations here is difficult to justify.

I have stressed the restricted extent to which standardization permits the comparison of coefficients for incommensurable independent variables. A common misuse of standardized coefficients is to employ them to make comparisons of the effects of the *same* independent variable in two or more samples drawn from different populations. If the independent variable in question has different spreads in these samples, then spurious differences between coefficients may result, even when unstandardized coefficients are similar; alternatively, differences in unstandardized coefficients can be masked by compensating differences in dispersion.

EXERCISES

5.10 *Derive the normal equations (see Equation 5.7) for the least-squares coefficients of the general multiple-regression model with k independent variables. [*Hint*: Differentiate the sum-of-squares function $S(A, B_1, \ldots, B_k)$ with respect to the regression coefficients, and set the partial derivatives to 0.]

5.11 Why is it the case that the multiple-correlation coefficient R^2 can never get smaller when an independent variable is added to the regression equation? (*Hint*: Recall that the model is fit by minimizing the residual sum of squares.)

5.12 #The following nine southeastern cities are selected from Angell's data on U.S. cities:

City	*Moral Integration* Y_i	*Heterogeneity* X_{i1}	*Mobility* X_{i2}
Richmond	10.4	65.3	24.9
Chattanooga	9.3	57.7	27.2
Nashville	8.6	57.4	25.4
Birmingham	8.2	83.1	25.9
Louisville	7.7	31.5	19.4
Jacksonville	6.0	73.7	27.7
Memphis	5.4	84.5	26.7
Miami	5.1	50.2	41.8
Atlanta	4.2	70.6	32.6

Employing the data for these nine cities:

(a) Find A, B_1, and B_2 for the linear least-squares regression of Y on X_1 and X_2.

(b) Calculate the standard error of the regression, S_E, and the multiple-correlation coefficient, R.

5.13 Using Angell's data on 43 U. S. cities, given in Table 2.3 and `angell.dat`, perform a multiple regression of moral integration on ethnic heterogeneity and geographic mobility. Interpret the least-squares coefficients from this regression, along with the standard error of the regression, S_E, and the multiple-correlation coefficient, R. Compare the results of the multiple regression with the two simple regressions performed in Exercise 5.9.

5.14 Using Anscombe's data on the U.S. states, given in Table 5.1 and `anscombe.dat`, perform a multiple regression of state education expenditures on income, proportion under 18, and proportion urban. Interpret the least-squares coefficients from this regression, along with the standard error of the regression, S_E, and the multiple-correlation coefficient, R. Compare the results of the multiple regression with the three simple regressions performed in Exercise 5.8.

5.15 Consider the general multiple-regression equation

$$Y = A + B_1X_1 + B_2X_2 + \cdots + B_kX_k + E$$

An alternative procedure for calculating the least-squares coefficient B_1 is as follows:

1. Regress Y on X_2 through X_k, obtaining residuals $E_{Y|2\ldots k}$.
2. Regress X_1 on X_2 through X_k, obtaining residuals $E_{1|2\ldots k}$.
3. Regress the residuals $E_{Y|2\ldots k}$ on the residuals $E_{1|2\ldots k}$. The slope for this simple regression is the multiple-regression slope for X_1, that is, B_1.

(a) Apply this procedure to Anscombe's multiple regression, computed in Exercise 5.14, confirming that the coefficient for income is properly recovered.

(b) Notice that the intercept for the simple regression in step 3 is 0. Why is this the case?

(c) In light of this procedure, is it reasonable to describe B_1 as the effect of X_1 on Y when the influence of $X_2, \ldots, X_k$ is removed from both X_1 and Y?

(d) The procedure in this problem reduces the multiple regression to a series of simple regressions (in step 3). Can you see any practical application for this procedure? (See the discussion of partial-regression plots in Section 11.6.)

5.16 Partial correlation: The *partial correlation* between X_1 and Y "controlling for" X_2 through X_k is defined as the simple correlation between the residuals $E_{Y|2\ldots k}$ and $E_{1|2\ldots k}$, given in the previous exercise. The partial correlation is denoted $r_{Y1|2\ldots k}$.

(a) Calculate the partial correlation between state education expenditures and income, using Anscombe's data (see the previous exercise).

(b) In light of the interpretation of a partial regression coefficient developed in the previous exercise, why is $r_{Y1|2\ldots k}$ 0 if and only if B_1 (from the multiple regression of Y on X_1 through X_k) is 0?

5.17 *Show that in simple-regression analysis, the standardized slope coefficient B^* is equal to the correlation coefficient r. (In general, however, standardized slope coefficients are not correlations, and can be outside of the range [0, 1].)

5.18 Calculate standardized regression coefficients for Angell's regression of moral integration of U.S. cities on ethnic heterogeneity and geographic mobility (Exercise 5.13). Calculate the standardized coefficients both in the usual manner, using the standard deviations of the variables, and using their hinge spreads (interquartile ranges).

5.19 Repeat Exercise 5.18 for Anscombe's regression of state education expenditures on income, proportion under 18, and proportion urban (Exercise 5.14).

5.3 Summary

- In simple linear regression, the least-squares coefficients are given by

$$B = \frac{\sum(X_i - \overline{X})(Y_i - \overline{Y})}{\sum(X_i - \overline{X})^2}$$

$$A = \overline{Y} - B\overline{X}$$

The slope coefficient B represents the average change in Y associated with a one-unit increase in X. The intercept A is the fitted value of Y when $X = 0$.

- The least-squares coefficients in multiple linear regression are found by solving the normal equations for the intercept A and the slope coefficients $B_1, B_2, \ldots, B_k$. The slope coefficient B_1 represents the average change in Y associated with a one-unit increase in X_1 when the other X's are held constant.

- The least-squares residuals, E, are uncorrelated with the fitted values, $\hat{Y}$, and with the independent variables, $X_1, \ldots, X_k$.
- The linear regression decomposes the variation in Y into "explained" and "unexplained" components: $\text{TSS} = \text{RegSS} + \text{RSS}$.
- The standard error of the regression, $S_E = \sqrt{\sum E_i^2/(n-k-1)}$, gives the "average" size of the regression residuals; the squared multiple correlation, $R^2 = \text{RegSS}/\text{TSS}$, indicates the proportion of the variation in Y that is captured by its linear regression on the X's.
- By rescaling regression coefficients in relation to a measure of variation—such as the hinge spread or standard deviation—standardized regression coefficients permit a limited comparison of the relative impact of incommensurable independent variables.

6

Statistical Inference for Regression

The previous chapter developed least-squares linear regression as a descriptive technique for fitting a linear surface to data. The subject of the present chapter, in contrast, is statistical inference. I shall discuss point estimation of regression coefficients, along with elementary but powerful procedures for constructing confidence intervals and performing hypothesis tests in simple and multiple regression.[1] I shall also develop two topics related to inference in regression: the distinction between empirical and structural relationships; and the consequences of random measurement error in regression.

6.1 Simple Regression

6.1.1 The Simple-Regression Model

Standard statistical inference in simple regression is based on a statistical model, assumed to be descriptive of the population or process that is sampled:

$$Y_i = \alpha + \beta X_i + \varepsilon_i$$

The coefficients α and β are the population regression parameters; the central object of simple-regression analysis is to estimate these coefficients. The *error* ε_i represents the aggregated, omitted causes of Y (i.e., the causes of Y beyond

[1] The focus here is on the procedures themselves: The statistical theory underlying these methods is developed in Chapters 9 and 10.

the independent variable X): other independent variables that could have been included in the regression model (at least in principle); measurement error in Y; and whatever component of Y is inherently random. The Greek letter epsilon is used for the errors because, without knowledge of the values of α and β, the errors are not directly observable.

The key assumptions of the simple-regression model concern the behavior of the errors—or, equivalently, of the distribution of Y conditional on X:

- *Linearity.* The expectation of the error—that is, the average value of ε given the value of X—is 0: $E(\varepsilon_i) \equiv E(\varepsilon|x_i) = 0$. Equivalently, the expected value of the dependent variable is a linear function of the independent variable:

$$\begin{aligned}\mu_i \equiv E(Y_i) \equiv E(Y|x_i) &= E(\alpha + \beta x_i + \varepsilon_i)\\ &= \alpha + \beta x_i + E(\varepsilon_i)\\ &= \alpha + \beta x_i + 0\\ &= \alpha + \beta x_i\end{aligned}$$

  We can remove $\alpha + \beta x_i$ from the expectation operator because α and β are fixed parameters, while the value of X is conditionally fixed[2] to x_i.
- *Constant variance.* The variance of the errors is the same regardless of the value of X: $V(\varepsilon|x_i) = \sigma_\varepsilon^2$. Because the distribution of the errors is the same as the distribution of the dependent variable around the population regression line, constant error variance implies constant conditional variance of Y given X:

$$V(Y|x_i) = E[(Y_i - \alpha - \beta x_i)^2] = E(\varepsilon_i^2) = \sigma_\varepsilon^2$$

  Notice that because the mean of ε_i is 0, its variance is simply $E(\varepsilon_i^2)$.
- *Normality.* The errors are normally distributed: $\varepsilon_i \sim N(0, \sigma_\varepsilon^2)$. Equivalently, the conditional distribution of the dependent variable is normal: $Y_i \sim N(\alpha + \beta x_i, \sigma_\varepsilon^2)$. The assumptions of linearity, constant variance, and normality are illustrated in Figure 6.1. It should be abundantly clear from this figure that the assumptions are very strong.
- *Independence.* The observations are sampled independently: Any pair of errors ε_i and ε_j (or, equivalently, of conditional dependent-variable values, Y_i and Y_j) are independent for $i \neq j$. The assumption of independence needs to be justified by the procedures of data collection. For example, if the data constitute a simple random sample drawn from a large population, then the assumption of independence will be met to a close approximation. In contrast, if the data comprise a time series, then the assumption of independence may be very wrong.[3]
- *Fixed X, or X independent of the error.* Depending on the design of a study, the values of the independent variable may be fixed in advance of data collection or they may be sampled along with the dependent variable. Fixed X corresponds almost exclusively to experimental research, in which the value of the independent variable is under the direct control of the researcher; if the experiment were

---

[2] I use a lowercase x here to stress that the value x_i is fixed—either literally, as in experimental research (see below), or by conditioning on the observed value x_i of X_i.

[3] Section 14.1 discusses regression with time series data.

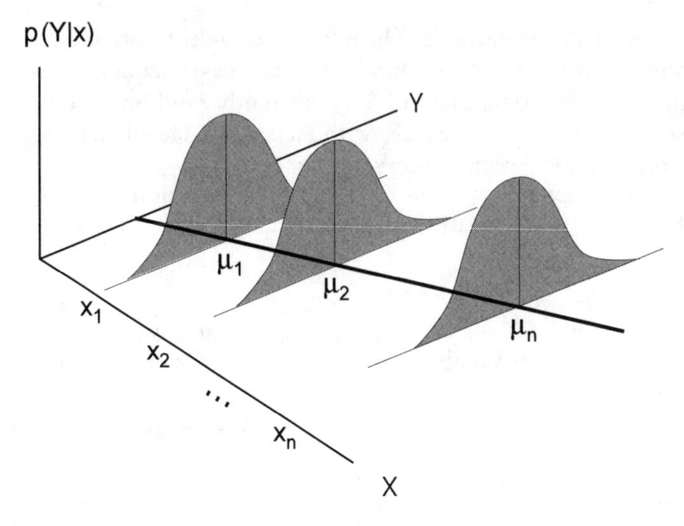

Figure 6.1. The assumptions of linearity, constant variance, and normality in simple regression. The figure shows the conditional population distributions of Y given X for several values of the independent variable, labeled $x_1, x_2, \ldots, x_n$. The conditional means of Y given X are denoted $\mu_1, \mu_2, \ldots, \mu_n$.

replicated, then—at least in principle—the values of X would remain the same. In most social research, X-values are sampled, not fixed by design. Under these circumstances, we assume that the independent variable and the error are independent in the population from which the sample is drawn: That is, the error has the same distribution $[N(0, \sigma_\varepsilon^2)]$ for every value of X in the population.

> Standard statistical inference for least-squares simple regression analysis is based on the statistical model $Y_i = \alpha + \beta x_i + \varepsilon_i$. The key assumptions of the model concern the behavior of the errors ε_i: (1) Linearity, $E(\varepsilon_i) = 0$; (2) constant variance, $V(\varepsilon_i) = \sigma_\varepsilon^2$; (3) normality, $\varepsilon_i \sim N(0, \sigma_\varepsilon^2)$; (4) independence, ε_i, ε_j are independent for $i \neq j$; and (5) the X-values are fixed or, if random, are independent of the errors.

6.1.2 Properties of the Least-Squares Estimator

Under the strong assumptions of the simple-regression model, the sample least-squares coefficients A and B have several desirable properties as estimators of the population regression coefficients α and β:[4]

---

[4] I shall simply state these properties here; derivations can be found in the exercises to this chapter and in Chapter 9.

- The least-squares intercept and slope are *linear estimators*, in the sense that they are linear functions of the observations Y_i. For example, for fixed independent-variable values x_i,

$$B = \sum_{i=1}^{n} m_i Y_i$$

where

$$m_i = \frac{x_i - \overline{x}}{\sum_{j=1}^{n}(x_j - \overline{x})^2}$$

While unimportant in itself, this property makes it simple to derive the sampling distributions of A and B.

- The sample least-squares coefficients are *unbiased estimators* of the population regression coefficients:

$$E(A) = \alpha$$
$$E(B) = \beta$$

Only the assumption of linearity is required to establish this result.[5]

- Both A and B have simple sampling variances:

$$V(A) = \frac{\sigma_\varepsilon^2 \sum x_i^2}{n \sum (x_i - \overline{x})^2}$$

$$V(B) = \frac{\sigma_\varepsilon^2}{\sum (x_i - \overline{x})^2}$$

The assumptions of linearity, constant variance, and independence are employed in the derivation of these simple formulas.[6] It is instructive to examine the formula for $V(B)$ more closely, in order to understand the conditions under which least-squares estimation is precise. Rewriting the formula,

$$V(B) = \frac{\sigma_\varepsilon^2}{(n-1)S_X^2}$$

Thus, the sampling variance of the slope estimate will be small when: (1) the error variance σ_ε^2 is small; (2) the sample size n is large; and (3) the independent-variable values are spread out. The estimate of the intercept has small sampling variance under similar circumstances but, in addition,[7] when the X-values are centered near 0 and, hence, $\sum x_i^2$ is not much larger than $\sum (x_i - \overline{x})^2$.

- Of all linear unbiased estimators, the least-squares estimators are most efficient—that is, have the smallest sampling variance and hence the smallest mean-squared error. This result, called the Gauss-Markov theorem, requires the assumptions of linearity, constant variance, and independence, but not the assumption of normality.[8] Under normality, moreover, the least-squares estimators are most

---

[5] See Exercise 6.1.

[6] See Exercise 6.2.

[7] See Exercise 6.3.

[8] The theorem is named after the 19th-century German mathematical genius Carl Friedrich Gauss and the 20th-century Russian mathematician A.A. Markov. Although Gauss worked in the context of measurement error in the physical sciences, much of the general statistical theory of linear models is due to him.

efficient among *all* unbiased estimators, not just among linear estimators. This is a much more compelling result, because the restriction to linear estimators is merely a matter of convenience. When the error distribution is heavier tailed than normal, for example, the least-squares estimators may be much less efficient than certain robust-regression estimators, which are not linear functions of the data.[9]

- Under the full suite of assumptions, the least-squares coefficients A and B are the maximum-likelihood estimators[10] of α and β.
- Under the assumption of normality, the least-squares coefficients are themselves normally distributed. Summing up,

$$A \sim N\left[\alpha, \frac{\sigma_\varepsilon^2 \sum x_i^2}{n\sum(x_i - \overline{x})^2}\right] \qquad [6.1]$$

$$B \sim N\left[\beta, \frac{\sigma_\varepsilon^2}{\sum(x_i - \overline{x})^2}\right]$$

Even if the errors are not normally distributed, the distributions of A and B are approximately normal, under very broad conditions, with the approximation improving as the sample size grows.[11]

> Under the assumptions of the regression model, the least-squares coefficients have certain desirable properties as estimators of the population regression coefficients. The least-squares coefficients are: linear functions of the data and therefore have simple sampling distributions; unbiased estimators of the population regression coefficients; the most efficient unbiased estimators of the population regression coefficients; maximum-likelihood estimators; and normally distributed.

6.1.3 Confidence Intervals and Hypothesis Tests

The distributions of A and B, given in Equation 6.1, cannot be directly employed for statistical inference because the error variance, σ_ε^2, is never known in practice. The variance of the residuals provides an unbiased estimator[12] of σ_ε^2:

$$S_E^2 = \frac{\sum E_i^2}{n-2}$$

---

[9] See Section 14.3.

[10] See Exercise 6.5.

[11] The asymptotic normality of A and B follows from the central-limit theorem, since the least-squares coefficients are linear functions of the Y_i's.

[12] See Section 10.3.

With the estimated error variance in hand, we can estimate the sampling variances of A and B:

$$\widehat{V(A)} = \frac{S_E^2 \sum x_i^2}{n \sum (x_i - \overline{x})^2}$$

$$\widehat{V(B)} = \frac{S_E^2}{\sum (x_i - \overline{x})^2}$$

As in statistical inference for the mean, the added uncertainty induced by estimating the error variance is reflected in the use of the t-distribution, in place of the normal distribution, for confidence intervals and hypothesis tests.

> The estimated standard error of the slope coefficient B in simple regression is $\widehat{\text{SE}(B)} = S_E / \sqrt{\sum (x_i - \overline{x})^2}$.

To construct a $100(1 - a)\%$ confidence interval for the slope, we take

$$\beta = B \pm t_{a/2} \widehat{\text{SE}(B)}$$

where $t_{a/2}$ is the critical value of t with $n-2$ degrees of freedom and a probability of $a/2$ to the right, and $\widehat{\text{SE}(B)}$ is the estimated standard error of B [i.e., the square root of $\widehat{V(B)}$]. For a 95% confidence interval, $t_{.025} \simeq 2$, unless n is very small. Similarly, to test the hypothesis H_0: $\beta = \beta_0$ that the population slope is equal to a specific value (most commonly, the null hypothesis H_0: $\beta = 0$), calculate the test statistic

$$t_0 = \frac{B - \beta_0}{\widehat{\text{SE}(B)}}$$

which is distributed as t with $n - 2$ degrees of freedom under the hypothesis H_0. Confidence intervals and hypothesis tests for α are usually of less interest, but they follow the same pattern.

For Davis's regression of measured on reported weight, for example, we have the following results:

$$S_E = \sqrt{\frac{418.87}{101 - 2}} = 2.0569$$

$$\widehat{\text{SE}(A)} = \frac{2.0569 \times \sqrt{329{,}731}}{\sqrt{101 \times 4539.3}} = 1.7444$$

$$\widehat{\text{SE}(B)} = \frac{2.0569}{\sqrt{4539.3}} = 0.030529$$

Thus, because $t_{.025}$ for $101-2 = 99$ degrees of freedom is 1.984, 95% confidence intervals for α and β are

$$\alpha = 1.778 \pm 1.984 \times 1.744 = 1.778 \pm 3.460$$

$$\beta = 0.9772 \pm 1.984 \times 0.03053 = 0.9772 \pm 0.06057$$

The estimates of α and β are therefore quite precise. Furthermore, the confidence intervals include the values $\alpha = 0$ and $\beta = 1$, which, recall, imply unbiased prediction of measured weight from reported weight.[13]

EXERCISES

6.1 *Demonstrate the unbias of the least-squares estimators A and B of α and β:

(a) Expressing the least-squares slope B as a linear function of the observations, $B = \sum m_i Y_i$ (as in the text), and using the assumption of linearity, $E(Y_i) = \alpha + \beta x_i$, show that $E(B) = \beta$. [*Hint*: $E(B) = \sum m_i E(Y_i)$.]

(b) Show that A can also be written as a linear function of the Y_i's. Then show that $E(A) = \alpha$.

6.2 *Using the assumptions of linearity, constant variance, and independence, along with the fact that A and B can each be expressed as a linear function of the Y_i's, derive the sampling variances of A and B. [*Hint*: $V(B) = \sum m_i^2 V(Y_i)$.]

6.3 Examining the formula for the sampling variance of A,

$$V(A) = \frac{\sigma_\varepsilon^2 \sum x_i^2}{n \sum (x_i - \overline{x})^2}$$

why is it intuitively sensible that the variance of A is large when the mean of the x's is far from 0? Illustrate your explanation with a graph.

6.4 The formula for the sampling variance of B,

$$V(B) = \frac{\sigma_\varepsilon^2}{\sum (x_i - \overline{x})^2}$$

shows that, to estimate β precisely, it helps to have spread-out x's. Explain why this result is intuitively sensible, illustrating your explanation with a graph. What happens to $V(B)$ when there is *no* variation in X?

---

[13] There is, however, a subtlety here: To construct separate confidence intervals for α and β is not quite the same as constructing a *joint confidence region* for both coefficients simultaneously. See Section 9.4.4 for a discussion of confidence regions in regression.

6.5 *Maximum-likelihood estimation: Deriving the maximum-likelihood estimators of α and β in simple regression is straightforward. Under the assumptions of the simple-regression model, the Y_i's are independently and normally distributed random variables with expectations $\alpha + \beta x_i$ and common variance σ_ε^2. Show that if these assumptions hold, then the least-squares coefficients A and B are the maximum-likelihood estimators of α and β, and that $\hat{\sigma}_\varepsilon^2 = \sum E_i^2/n$ is the maximum-likelihood estimator of σ_ε^2. Note that the MLE of the error variance is biased. [*Hints*: Because of the assumption of independence, the joint probability density for the Y_i's is the product of their marginal probability densities

$$p(y_i) = \frac{1}{\sqrt{2\pi\sigma_\varepsilon^2}} \exp\left[-\frac{(y_i - \alpha - \beta x_i)^2}{2\sigma_\varepsilon^2}\right]$$

Find the log likelihood function; take the partial derivatives of the log likelihood with respect to the parameters α, β, and σ_ε^2; set these partial derivatives to 0; and solve for the maximum-likelihood estimators.] A more general result is proven in Section 9.3.3.

6.6 #Recall Exercise 5.4, in which a simple regression of Acres/Gardener on Consumers/Gardener was calculated for five observations drawn from Sahlins's data on agricultural production in a primitive community. Assuming that the observations were independently sampled, find the estimated standard error $\widehat{SE(B)}$ of the slope coefficient, and calculate a 90% confidence interval for the population slope β.

6.7 In Exercise 5.6, you calculated a simple regression of Acres/Gardener on Consumers/Gardener for the 20 households of Mazulu village (the data are in Table 2.1 and `sahlins.dat`). Find the estimated standard errors of the least-squares intercept and slope. Can we conclude that the population slope is greater than 0? Can we conclude that the intercept is greater than 0? Repeat these computations omitting the fourth household.

6.8 Construct 95% confidence intervals for α and β for the least-squares simple-regression analyses performed in each of the following exercises:

(a) Exercise 5.7 (Robey, et al.'s data on fertility and contraception, in Table 2.2 and `robey.dat`).

(b) Exercise 5.8 (Anscombe's data on public-school expenditures in the U.S. states, in Table 5.1 and `anscombe.dat`).

(c) Exercise 5.9 (Angell's data on the moral integration of U.S. cities, in Table 2.3 and `angell.dat`).

6.9 Linear transformation of X and Y (continuation of Exercise 5.5):

(a) Suppose that the independent-variable values in Exercise 6.6 are transformed according to the equation $X' = 10(X-1)$, and that Y is regressed on X'. Without redoing the regression calculations in detail, find $\widehat{SE(B')}$ and $t_0' = B'/\widehat{SE(B')}$.

(b) Now suppose that the dependent-variable values are transformed according to the equation $Y'' = 5(Y+2)$, and that Y'' is regressed on X. Find $\widehat{SE(B'')}$ and $t_0'' = B''/\widehat{SE(B'')}$.

(c) In general, how are hypothesis tests and confidence intervals for β affected by linear transformations of X and Y?

6.2 Multiple Regression

Most of the results for multiple-regression analysis parallel those for simple regression.

6.2.1 The Multiple-Regression Model

The statistical model for multiple regression is

$$Y_i = \alpha + \beta_1 X_{i1} + \beta_2 X_{i2} + \cdots + \beta_k X_{ik} + \varepsilon_i$$

The assumptions underlying the model concern the errors, $\varepsilon_i \equiv \varepsilon | x_{i1}, \ldots, x_{ik}$, and are identical to the assumptions in simple regression:

- *Linearity.* $E(\varepsilon_i) = 0$.
- *Constant variance.* $V(\varepsilon_i) = \sigma_\varepsilon^2$.
- *Normality.* $\varepsilon_i \sim N(0, \sigma_\varepsilon^2)$.
- *Independence.* $\varepsilon_i, \varepsilon_j$ are independent for $i \neq j$.
- *Fixed X's or X's independent of* ε.

Under these assumptions (or particular subsets of them), the least-squares estimators $A, B_1, \ldots, B_k$ of $\alpha, \beta_1, \ldots, \beta_k$ are

- linear functions of the data, and hence relatively simple;
- unbiased;
- maximally efficient among unbiased estimators;
- maximum-likelihood estimators;
- normally distributed.

The slope coefficient B_j in multiple regression has sampling variance[14]

$$V(B_j) = \frac{1}{1 - R_j^2} \times \frac{\sigma_\varepsilon^2}{\sum_{i=1}^{n} (X_{ij} - \overline{X}_j)^2} \qquad [6.2]$$

where R_j^2 is the squared multiple correlation from the regression of X_j on all of the other X's. The second factor is essentially the sampling variance of the slope in simple regression, although the error variance σ_ε^2 is smaller than before, because some of the independent variables that were implicitly in the error in simple regression are now incorporated into the systematic part of the model.

---

[14] Although we are usually less interested in estimating α, it is also possible to find the sampling variance of the intercept A. See Section 9.3.1.

The first factor in Equation 6.2—called the *variance-inflation factor*—is new, however: The variance-inflation factor $1/(1 - R_j^2)$ is large when the independent variable X_j is strongly correlated with other independent variables.

We saw in Chapter 5 that when one independent variable is perfectly collinear with others, the least-squares regression coefficients are not uniquely determined; in this case, the variance-inflation factor is infinite. The variance-inflation factor tells us that strong, though less-than-perfect, collinearity presents a problem for estimation, for although we can calculate least-squares estimates under these circumstances, their sampling variances may be very large. Equation 6.2 reveals that the other sources of imprecision of estimation in multiple regression are the same as in simple regression: large error variance, a small sample, and independent variables with little variation.[15]

6.2.2 Confidence Intervals and Hypothesis Tests

Individual Slope Coefficients

Confidence intervals and hypothesis tests for individual coefficients closely follow the pattern of simple-regression analysis: To estimate the standard error of a slope coefficient, we need to substitute an estimate of the error variance for the unknown σ_ε^2 in Equation 6.2. The variance of the residuals provides an unbiased estimator:

$$S_E^2 = \frac{\sum E_i^2}{n - k - 1}$$

Then, the estimated standard error of B_j is

$$\widehat{\text{SE}(B_j)} = \frac{1}{\sqrt{1 - R_j^2}} \times \frac{S_E}{\sqrt{\sum(X_{ij} - \overline{X}_j)^2}}$$

Confidence intervals and tests, based on the t-distribution with $n - k - 1$ degrees of freedom, follow straightforwardly.

> The estimated standard error of the slope coefficient B_j in multiple regression is $\widehat{\text{SE}(B_j)} = S_E/\sqrt{(1 - R_j^2)\sum(X_{ij} - \overline{X}_j)^2}$. This estimated coefficient standard error can be used in t-intervals and hypothesis tests for β_j.

[15] Collinearity is discussed further in Chapter 13.

For example, for Duncan's regression of occupational prestige on education and income, we have

$$S_E^2 = \frac{7506.7}{45 - 2 - 1} = 178.73$$

$$r_{12} = .72451$$

$$\widehat{\text{SE}(B_1)} = \frac{1}{\sqrt{1 - .72451^2}} \times \frac{\sqrt{178.73}}{\sqrt{38{,}971}} = 0.098252$$

$$\widehat{\text{SE}(B_2)} = \frac{1}{\sqrt{1 - .72451^2}} \times \frac{\sqrt{178.73}}{\sqrt{26{,}271}} = 0.11967$$

Note that with only two independent variables, $R_1^2 = R_2^2 = r_{12}^2$; this simplicity and symmetry are peculiar to the two-independent-variable case. To construct 95% confidence intervals for the slope coefficients, we use $t_{.025} = 2.018$ from the t-distribution with $45 - 2 - 1 = 42$ degrees of freedom. Then,

$$\text{Education:} \quad \beta_1 = 0.5459 \pm 2.018 \times 0.09825 = 0.5459 \pm 0.1983$$
$$\text{Income:} \quad \beta_2 = 0.5987 \pm 2.018 \times 0.1197 = 0.5987 \pm 0.2415$$

Although they are far from 0, these confidence intervals are quite broad, indicating that the estimates of the education and income coefficients are imprecise—as is to be expected in a sample of only 45 occupations.

All Slopes

We can also test the null hypothesis that all of the regression slopes are 0:

$$H_0\colon \beta_1 = \beta_2 = \cdots = \beta_k = 0 \qquad [6.3]$$

To test this global or "omnibus" null hypothesis is not quite the same as testing the separate hypotheses

$$H_0^{(1)}\colon \beta_1 = 0;\ H_0^{(2)}\colon \beta_2 = 0; \ldots;\ H_0^{(k)}\colon \beta_k = 0$$

If the independent variables are very highly correlated, for example, we might be able to reject the omnibus hypothesis (Equation 6.3) without being able to reject *any* of the individual hypotheses.

An F-test for the null hypothesis (Equation 6.3) is given by

$$F_0 = \frac{\text{RegSS}/k}{\text{RSS}/(n - k - 1)}$$

$$= \frac{n - k - 1}{k} \times \frac{R^2}{1 - R^2}$$

Under the omnibus null hypothesis, this test statistic has an F-distribution with k and $n - k - 1$ degrees of freedom. The omnibus F-test follows from the analysis of variance for the regression, and the calculation of the test statistic can be organized in an *analysis-of-variance table*, which shows the partition of total variation into its components:

Source	*Sum of Squares*	*df*	*Mean Square*	*F*
Regression	RegSS	k	$\frac{\text{RegSS}}{k}$	$\frac{\text{RegMS}}{\text{RMS}}$
Residuals	RSS	$n-k-1$	$\frac{\text{RSS}}{n-k-1}$	
Total	TSS	$n-1$		

Notice that the degrees of freedom (*df*) add in the same manner as the sums of squares, and that the residual "mean square," RMS, is simply the estimated error variance, S_E^2.

It turns out that when the null hypothesis is true, the regression mean square, RegMS, provides an independent estimate of the error variance, so the ratio of the two mean squares should be close to 1. When, alternatively, the null hypothesis is false, the regression mean square estimates the error variance plus a positive quantity that depends on the β's, tending to make the numerator of F_0 larger than the denominator:

$$E(F_0) \simeq \frac{E(\text{RegMS})}{E(\text{RMS})} = \frac{\sigma_\varepsilon^2 + \text{Positive quantity}}{\sigma_\varepsilon^2} > 1$$

We consequently reject the omnibus null hypothesis for values of F_0 that are sufficiently larger than 1.[16]

> An omnibus F-test for the null hypothesis that all of the slopes are 0 can be calculated from the analysis of variance for the regression.

For Duncan's regression, we have the following analysis-of-variance table:

Source	*Sum of Squares*	*df*	*Mean Square*	*F*	*p*
Regression	36,181.	2	18,090.	101.2	<<.0001
Residuals	7,506.7	42	178.73		
Total	43,688.	44			

The p-value for the omnibus null hypothesis—that is, $\Pr(F > 101.2)$ for an F-distribution with 2 and 42 degrees of freedom—is very close to 0.

[16] The reasoning here is only approximate, since the expectation of the ratio of two independent random variables is not the ratio of their expectations. Nevertheless, when the sample size is large, the null distribution of the F-statistic has an expectation very close to 1 (see Appendix D, Section D.3.4).

A Subset of Slopes

It is, finally, possible to test a null hypothesis about a *subset* of the regression slopes

$$H_0: \beta_1 = \beta_2 = \cdots = \beta_q = 0 \qquad [6.4]$$

where $1 \leq q \leq k$. Purely for notational convenience, I have specified a hypothesis on the *first* q coefficients; we can, of course, equally easily test a hypothesis for *any* q slopes. The "full" regression model, including all of the independent variables, can be written as

$$Y_i = \alpha + \beta_1 X_{i1} + \cdots + \beta_q X_{iq} + \beta_{q+1} X_{i,q+1} + \cdots + \beta_k X_{ik} + \varepsilon_i$$

If the null hypothesis is correct, then the first q of the β's are 0, yielding the "null" model

$$\begin{aligned} Y_i &= \alpha + 0X_{i1} + \cdots + 0X_{iq} + \beta_{q+1} X_{i,q+1} + \cdots + \beta_k X_{ik} + \varepsilon_i \\ &= \alpha + \beta_{q+1} X_{i,q+1} + \cdots + \beta_k X_{ik} + \varepsilon_i \end{aligned}$$

In effect, then, the null model omits the first q independent variables, regressing Y on the remaining $k - q$ independent variables.

An F-test of the null hypothesis (Equation 6.4) is based on a comparison of these two models. Let RSS_1 and RegSS_1 represent, respectively, the residual and regression sums of squares for the full model; similarly, RSS_0 and RegSS_0 are the residual and regression sums of squares for the null model. Because the null model is a special case of the full model, in which the first q slopes are constrained to 0, $\text{RSS}_0 \geq \text{RSS}_1$. The residual and regression sums of squares in the two models add to the same total sum of squares; it follows that $\text{RegSS}_0 \leq \text{RegSS}_1$. If the null hypothesis is wrong and (some of) $\beta_1, \ldots, \beta_q$ are nonzero, then the *incremental* (or "*extra*") *sum of squares* due to fitting the additional independent variables

$$\text{RSS}_0 - \text{RSS}_1 = \text{RegSS}_1 - \text{RegSS}_0$$

should be large.

The F-statistic for testing the null hypothesis (Equation 6.4) is

$$\begin{aligned} F_0 &= \frac{(\text{RegSS}_1 - \text{RegSS}_0)/q}{\text{RSS}_1/(n-k-1)} \\ &= \frac{n-k-1}{q} \times \frac{R_1^2 - R_0^2}{1 - R_1^2} \end{aligned}$$

where R_1^2 and R_0^2 are the squared multiple correlations from the full and null models, respectively. Under the null hypothesis, this test statistic has an F-distribution with q and $n - k - 1$ degrees of freedom.

An F-test for the null hypothesis that a subset of slope coefficients is 0 is based on a comparison of the regression sums of squares for two models: the full regression model, and a null model that deletes the independent variables in the null hypothesis.

The motivation for testing a subset of coefficients will become clear in the next chapter, which takes up regression models that incorporate qualitative independent variables. I shall, for the present, illustrate the incremental F-test by applying it to the trivial case in which $q = 1$ (i.e., a single coefficient).

In Duncan's dataset, the regression of prestige on income alone produces $\text{RegSS}_0 = 30{,}665$, while the regression of prestige on both income and education produces $\text{RegSS}_1 = 36{,}181$ and $\text{RSS}_1 = 7506.7$. Consequently, the incremental sum of squares due to education is $36,181 - 30,665 = 5516$. The F-statistic for testing H_0: $\beta_{\text{Education}} = 0$ is, then,

$$F_0 = \frac{5516/1}{7506.7/(45-2-1)} = 30.86$$

with 1 and 42 degrees of freedom, for which $p < .0001$.

When, as here, $q = 1$, the incremental F-test is equivalent to the t-test obtained by dividing the regression coefficient by its estimated standard error: $F_0 = t_0^2$. For the current example,

$$t_0 = \frac{0.5459}{0.09825} = 5.556$$

$$t_0^2 = 5.556^2 = 30.87$$

(which is the same as F_0, within rounding error).

EXERCISES

6.10 Consider the regression model $Y = \alpha + \beta_1 X_1 + \beta_2 X_2 + \varepsilon$. How can the incremental sum-of-squares approach be used to test the hypothesis that the two population slopes are equal to each other, H_0: $\beta_1 = \beta_2$? [*Hint*: Under H_0, the model becomes $Y = \alpha + \beta X_1 + \beta X_2 + \varepsilon = Y = \alpha + \beta(X_1 + X_2) + \varepsilon$, where β is the common value of β_1 and β_2.] Under what circumstances would a hypothesis of this form be meaningful? (*Hint*: Consider the units of measurement of X_1 and X_2.)

6.11 #Recall Exercise 5.12, in which moral integration was regressed on ethnic heterogeneity and geographic mobility for nine cities in the Southeast, selected from among Angell's 43 U.S. cities.

(a) Assuming independently distributed errors, find the estimated standard errors of the slope coefficients for heterogeneity and mobility, $\widehat{SE}(B_1)$ and $\widehat{SE}(B_2)$. Separately test the null hypotheses H_0: $\beta_1 = 0$ and H_0: $\beta_2 = 0$.

(b) Construct an analysis-of-variance table for the regression, testing the omnibus null hypothesis, H_0: $\beta_1 = \beta_2 = 0$.

6.12 Compute coefficient standard errors and 95% confidence intervals for the coefficients in each of the following regressions. In each case, construct the analysis-of-variance table for the regression, and test the omnibus null hypothesis that all β's are 0. Compute an incremental F-test of the null hypothesis H_0: $\beta_1 = 0$, and verify that the F-statistic for this test is equal to the square of the t-value obtained when B_1 is divided by $\widehat{SE}(B_1)$.

(a) Angell's regression of moral integration on ethnic heterogeneity and geographic mobility for 43 U.S. cities (Exercise 5.13, Table 2.3, and `angell.dat`).

(b) Anscombe's regression of state public-school expenditures on income, proportion under 18, and proportion urban (Exercise 5.14, Table 5.1, and `anscombe.dat`).

6.3 Empirical Versus Structural Relations

There are two fundamentally different interpretations of regression coefficients, and failure to distinguish clearly between them is the source of much confusion. Borrowing Goldberger's (1973) terminology, we may interpret a regression descriptively, as an *empirical association* among variables, or causally, as a *structural relation* among variables.

I shall deal first with empirical associations, because the notion is simpler. Suppose that, in a population of interest, the relationship between two variables, Y and X_1, is well described by the simple-regression model:

$$Y = \alpha' + \beta_1' X_1 + \varepsilon'$$

That is to say, the conditional mean of Y is a linear function of X. We do not assume that X_1 necessarily causes Y or, if it does, that the omitted causes of Y, incorporated in ε', are independent of X_1. There is, quite simply, a linear empirical relationship between Y and X_1 in the population. If we proceed to draw a random sample from this population, then the least-squares sample slope B_1' is an unbiased estimator of β_1'.

Suppose, now, that we introduce a second independent variable, X_2, and that, in the same sense as before, the population relationship between Y and the two X's is linear:

$$Y = \alpha + \beta_1 X_1 + \beta_2 X_2 + \varepsilon$$

That is, the conditional mean of Y is a linear function of X_1 and X_2. The slope β_1 of the population regression plane can, and generally will, differ from β_1', the simple-regression slope (see below). The sample least-squares coefficients for the multiple regression, B_1 and B_2, are unbiased estimators of the corresponding population coefficients, β_1 and β_2.

That the simple-regression slope β'_1 differs from the multiple-regression slope β_1, and that therefore the sample *simple*-regression coefficient B'_1 is a *biased* estimator of the population *multiple*-regression slope β_1, is not problematic, for these are simply empirical relationships, and we do not, in this context, interpret a regression coefficient as the *effect* of an independent variable on the dependent variable. The issue of *specification error* —fitting a false model to the data—does not arise, as long as the linear regression model adequately describes the empirical relationship between the dependent variable and the independent variables in the population.

The situation is different, however, if we view the regression equation as representing a structural relation—that is, a model of how dependent-variable scores are determined.[17] Imagine now that dependent-variable scores are *constructed* according to the multiple-regression model

$$Y = \alpha + \beta_1 X_1 + \beta_2 X_2 + \varepsilon \quad [6.5]$$

where the error ε satisfies the usual regression assumptions; in particular, $E(\varepsilon) = 0$ and ε is independent of X_1 and X_2.

If we use least squares to fit this model to sample data, then we obtain unbiased estimators of β_1 and β_2. Suppose, however, that instead we fit the simple-regression model

$$Y = \alpha + \beta_1 X_1 + \varepsilon' \quad [6.6]$$

where, implicitly, the effect of X_2 on Y is absorbed by the error $\varepsilon' \equiv \varepsilon + \beta_2 X_2$, because X_2 is now among the omitted causes of Y. In the event that X_1 and X_2 are correlated, there is a correlation induced between X_1 and ε'. If we proceed to assume wrongly that X_1 and ε' are *uncorrelated*, as we do if we fit model [6.6] by least squares, then we make an error of specification. The consequence of this error is that our simple-regression estimator of β_1 is biased: Because X_1 and X_2 are correlated, and because X_2 is omitted from the model, part of the effect of X_2 is mistakenly attributed to X_1.

To make the nature of this specification error more precise, let us take the expectation of both sides of Equation 6.5, obtaining

$$\mu_Y = \alpha + \beta_1 \mu_1 + \beta_2 \mu_2 + 0 \quad [6.7]$$

where, for example, μ_Y is the population mean of Y; to obtain Equation 6.7, we use the fact that $E(\varepsilon)$ is 0. Subtracting this equation from Equation 6.5 has the effect of eliminating the constant α and expressing the variables as deviations from their means:

$$Y - \mu_Y = \beta_1(X_1 - \mu_1) + \beta_2(X_2 - \mu_2) + \varepsilon$$

[17] In the interest of clarity, I am making this distinction more categorically than I believe is justified. I argued in Chapter 1 that it is unreasonable to treat statistical models as literal representations of social processes. Nevertheless, it is useful to distinguish between purely empirical descriptions and descriptions from which we intend to infer causation.

Next, multiply this equation through by $X_1 - \mu_1$:

$$(X_1 - \mu_1)(Y - \mu_Y) = \beta_1(X_1 - \mu_1)^2 + \beta_2(X_1 - \mu_1)(X_2 - \mu_2) + (X_1 - \mu_1)\varepsilon$$

Taking the expectation of both sides of the equation produces

$$\sigma_{1Y} = \beta_1\sigma_1^2 + \beta_2\sigma_{12}$$

where σ_{1Y} is the covariance between X_1 and Y; σ_1^2 is the variance of X_1; and σ_{12} is the covariance of X_1 and X_2.[18] Solving for β_1, we get

$$\beta_1 = \frac{\sigma_{1Y}}{\sigma_1^2} - \beta_2\frac{\sigma_{12}}{\sigma_1^2} \qquad [6.8]$$

Recall that the least-squares coefficient for the simple regression of Y on X_1 is $B = S_{1Y}/S_1^2$. The simple regression therefore estimates not β_1 but rather $\sigma_{1Y}/\sigma_1^2 \equiv \beta_1'$. Solving Equation 6.8 for β_1' produces $\beta_1' = \beta_1 +$ *bias*, where the *bias* $= \beta_2\sigma_{12}/\sigma_1^2$.

It is instructive to take a closer look at the bias in the simple-regression estimator. For the bias to be nonzero, two conditions must be met: (1) X_2 must be a *relevant* independent variable—that is, $\beta_2 \neq 0$; and (2) X_1 and X_2 must be *correlated*—that is, $\sigma_{12} \neq 0$. Moreover, depending on the signs of β_2 and σ_{12}, the bias in the simple-regression estimator may be either positive or negative.

It is important to distinguish between interpreting a regression descriptively as an empirical association among variables and structurally as specifying causal relations among variables. In the latter event, but not in the former, it is sensible to speak of bias produced by omitting an independent variable that (1) is a cause of Y and (2) is correlated with an independent variable in the regression equation. Bias in least-squares estimation results from the correlation that is induced between the included independent variable and the error by incorporating the omitted independent variable in the error.

There is one final subtlety: The proper interpretation of the "bias" in the simple-regression estimator depends on the nature of the causal relationship between X_1 and X_2. Consider the situation depicted in Figure 6.2(a), where X_2 *intervenes* causally between X_1 and Y. Here, the "bias" term $\beta_2\sigma_{12}/\sigma_1^2$ is simply the *indirect effect* of X_1 on Y transmitted through X_2, because σ_{12}/σ_1^2 is the population slope for the regression of X_2 on X_1. If, however, as in Figure 6.2(b), X_2 is a *common prior cause* of both X_1 and Y, then the bias term represents a *spurious*—that is, noncausal—component of the empirical association between X_1 and Y. In the latter event, but not in the former, it is critical to control for X_2 in examining the relationship between Y and X_1.

---

[18] This result follows from the observation that the expectation of a mean-deviation product is a covariance, and the expectation of a mean-deviation square is a variance. Note that $E[(X_1 - \mu_1)\varepsilon] = \sigma_{1\varepsilon}$ is 0 because of the independence of X_1 and the error.

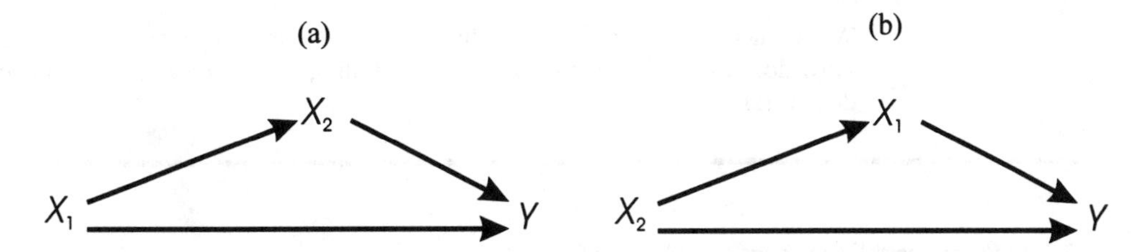

Figure 6.2. Two causal schemes relating a dependent variable to two independent variables: In (*a*) X_2 intervenes causally between X_1 and Y, while in (*b*) X_2 is a common prior cause of both X_1 and Y. In the second case, but not in the first, it is important to control for X_2 in examining the effect of X_1 on Y.

EXERCISES

6.13 Examples of specification error (also see the discussion in Section 1.2):

(a) Describe a nonexperimental research situation—real or contrived—in which failure to control statistically for an omitted variable induces a correlation between the error and an independent variable, producing erroneous conclusions. (For example: An educational researcher discovers that university students who study more get lower grades on average; the researcher concludes that studying has an adverse effect on students' grades.)

(b) Describe an experiment—real or contrived—in which faulty experimental practice induces an independent variable to become correlated with the error, compromising the validity of the results produced by the experiment. (For example: In an experimental study of a promising new therapy for depression, doctors administering the treatments tend to use the new therapy with patients for whom more traditional approaches have failed; it is discovered that subjects receiving the new treatment tend to do worse, on average, than those receiving older treatments or a placebo; the researcher concludes that the new treatment is not effective.)

(c) Is it fair to conclude that a researcher is *never* able absolutely to rule out the possibility that an independent variable of interest is correlated with the error? Is experimental research no better than observational research in this respect? Explain your answer.

6.14 Suppose that the "true" model generating a set of data is $Y = \alpha + \beta_1 X_1 + \varepsilon$, where the error ε conforms to the usual linear-regression assumptions. A researcher fits the model $Y = \alpha + \beta_1 X_1 + \beta_2 X_2 + \varepsilon$, which includes the irrelevant independent variable X_2—that is, the true value of β_2 is 0. Had the researcher fit the (correct) simple-regression model, the variance of B_1 would have been $V(B_1) = \sigma_\varepsilon^2 / \sum(X_{i1} - \overline{X}_1)^2$.

(a) Is the model $Y = \alpha + \beta_1 X_1 + \beta_2 X_2 + \varepsilon$ wrong? Is B_1 for this model a biased estimator of β_1?

(b) The variance of B_1 in the multiple-regression model is

$$V(B_1) = \frac{1}{1 - r_{12}^2} \times \frac{\sigma_\varepsilon^2}{\sum(X_{i1} - \overline{X}_1)^2}$$

What, then, is the cost of including the irrelevant independent variable X_2? How does this cost compare with that of failing to include a relevant independent variable?

6.4 Measurement Error in Independent Variables*

Variables are rarely—if ever—measured without error.[19] Even relatively straightforward characteristics, such as education, income, height, and weight, are imperfectly measured, especially when we rely on individuals' verbal reports. Measures of "subjective" characteristics, such as racial prejudice and conservatism, almost surely have substantial components of error. Measurement error affects not only characteristics of individuals: As you are likely aware, official statistics relating to crime, the economy, and so on, are subject to a variety of measurement errors.

The regression model accommodates measurement error in the *dependent* variable, because measurement error can be conceptualized as a component of the general error term ε, but the independent variables in regression analysis are assumed to be measured without error. In this section, I shall explain the consequences of violating this assumption. To do so, we shall examine the multiple-regression equation

$$Y = \beta_1 \tau + \beta_2 X_2 + \varepsilon \qquad [6.9]$$

To keep the notation as simple as possible, all of the variables in Equation 6.9 are expressed as deviations from their expectations, so the constant term disappears from the regression equation.[20] One of the independent variables, X_2, is measured without error, but the other, τ, is not directly observable. Instead, we have a fallible *indicator* X_1 of τ:

$$X_1 = \tau + \delta$$

where δ represents measurement error.

In addition to the usual assumptions about the regression errors ε, I shall assume that the measurement errors δ are "random" and "well behaved"; in particular:

- $E(\delta) = 0$, so there is no systematic tendency for measurements to be too large or too small.
- The measurement errors δ are uncorrelated with the "true-score" variable τ. This assumption could easily be wrong: If, for example, individuals who are lighter than average tend to overreport their weights, and individuals who are heavier than average tend to underreport their weights, then there will be a negative correlation between the measurement errors and true weight.

---

[19] Indeed, one of the historical sources of statistical theory in the 18th and 19th centuries was the investigation of measurement errors in the physical sciences by such great mathematicians as Gauss (mentioned previously) and Pierre Simon Laplace.

[20] There is no loss of generality here, since we can always subtract the mean from each variable. See the previous section.

- The measurement errors δ are uncorrelated with the regression errors ε and with the other independent variable X_2.

Because $\tau = X_1 - \delta$, we can rewrite Equation 6.9 as

$$\begin{aligned} Y &= \beta_1(X_1 - \delta) + \beta_2 X_2 + \varepsilon \\ &= \beta_1 X_1 + \beta_2 X_2 + (\varepsilon - \beta_1 \delta) \end{aligned} \qquad [6.10]$$

As in the previous section, we can proceed by multiplying Equation 6.10 through by X_1 and X_2 and taking expectations; because all variables are in mean-deviation form, expected products are covariances and expected squares are variances:[21]

$$\begin{aligned} \sigma_{Y1} &= \beta_1 \sigma_1^2 + \beta_2 \sigma_{12} - \beta_1 \sigma_\delta^2 \\ \sigma_{Y2} &= \beta_1 \sigma_{12} + \beta_2 \sigma_2^2 \end{aligned} \qquad [6.11]$$

Then, solving for the regression coefficients,

$$\begin{aligned} \beta_1 &= \frac{\sigma_{Y1}\sigma_2^2 - \sigma_{12}\sigma_{Y2}}{\sigma_1^2\sigma_2^2 - \sigma_{12}^2 - \sigma_\delta^2\sigma_2^2} \\ \beta_2 &= \frac{\sigma_{Y2}\sigma_1^2 - \sigma_{12}\sigma_{Y1}}{\sigma_1^2\sigma_2^2 - \sigma_{12}^2} - \frac{\beta_1\sigma_{12}\sigma_\delta^2}{\sigma_1^2\sigma_2^2 - \sigma_{12}^2} \end{aligned} \qquad [6.12]$$

Suppose, now, that we (temporarily) ignore the measurement error in X_1, and proceed by least-squares regression of Y on X_1 and X_2. The population analogs of the least-squares regression coefficients are:[22]

$$\begin{aligned} \beta_1' &= \frac{\sigma_{Y1}\sigma_2^2 - \sigma_{12}\sigma_{Y2}}{\sigma_1^2\sigma_2^2 - \sigma_{12}^2} \\ \beta_2' &= \frac{\sigma_{Y2}\sigma_1^2 - \sigma_{12}\sigma_{Y1}}{\sigma_1^2\sigma_2^2 - \sigma_{12}^2} \end{aligned} \qquad [6.13]$$

Comparing Equations 6.12 and 6.13 reveals the consequences of ignoring the measurement error in X_1. The denominator of β_1 in Equation 6.12 is necessarily positive, and its component $-\sigma_\delta^2\sigma_2^2$ is necessarily negative. Ignoring this component therefore inflates the denominator of β_1' in Equation 6.13, driving the coefficient β_1' toward 0. Put another way, ignoring measurement error in an independent variable tends to *attenuate* its coefficient, as makes intuitive sense.

The effect of measurement error in X_1 on the coefficient of X_2 is even more pernicious. Here, we can write $\beta_2' = \beta_2 + bias$, where

$$bias = \frac{\beta_1\sigma_{12}\sigma_\delta^2}{\sigma_1^2\sigma_2^2 - \sigma_{12}^2}$$

[21] See Exercise 6.15.

[22] See Exercise 6.16.

The bias term can be positive or negative, toward 0 or away from it. To get a better grasp on the bias in the least-squares estimand β_2', imagine that the measurement-error variance σ_δ^2 grows larger and larger. Because σ_δ^2 is a component of σ_1^2, this latter quantity also grows larger, but because the measurement errors δ are uncorrelated with variables other than X_1, other variances and covariances are unaffected.[23]

Using Equation 6.13,

$$\lim_{\sigma_\delta^2 \to \infty} \beta_2' = \frac{\sigma_{Y2}\sigma_1^2}{\sigma_1^2\sigma_2^2} = \frac{\sigma_{Y2}}{\sigma_2^2}$$

which is the population analog of the least-squares slope for the *simple* regression of Y on X_2 alone. Once more, the result is simple and intuitively reasonable: Substantial measurement error in X_1 renders it an ineffective statistical control, driving β_2' toward the marginal relationship between X_2 and Y, and away from the partial relationship between these two variables.[24]

Measurement error in an independent variable tends to attenuate its regression coefficient and to make the variable an imperfect statistical control.

Although there are statistical methods that attempt to estimate regression equations taking account of measurement errors, these methods are beyond the scope of the presentation in this book and, in any event, involve assumptions that are difficult to justify in practice.[25] Perhaps the most important lessons to be drawn from the results of this section are (1) that substantial measurement errors can invalidate a regression analysis; (2) that, therefore, where measurement errors are likely to be substantial, we should not view the results of a regression as definitive; and (3) that it is worthwhile to expend effort to improve the quality of social measurements.

---

[23] See Exercise 6.17.

[24] I am grateful to Georges Monette, of York University, for this insight. See Exercise 6.18 for an illustration.

[25] Measurement errors in independent variables are often discussed in the context of ***structural-equation models***, which are multiple-equation regression models in which the dependent variable in one equation can appear as an independent variable in others. Duncan (1975, Chapters 9 and 10) presents a fine elementary treatment of the topic, part of which I have adapted for the presentation in this section. A more advanced development may be found in Bollen (1989).

EXERCISES

6.15 Derive Equation 6.11 by multiplying Equation 6.10 through by each of X_1 and X_2. (*Hints*: Both X_1 and X_2 are uncorrelated with the regression error ε. Likewise, X_2 is uncorrelated with the measurement error δ. Show that the covariance of X_1 and δ is simply the measurement-error variance σ_δ^2 by multiplying $X_1 = \tau + \delta$ through by δ and taking expectations.)

6.16 Show that the population analogs of the regression coefficients can be written as in Equation 6.13. (*Hint*: Ignore the measurement errors, and derive the population analogs of the normal equations by multiplying the "model" $Y = \beta_1 X_1 + \beta_2 X_2 + \varepsilon$ through by each of X_1 and X_2, and taking expectations.)

6.17 Show that the variance of $X_1 = \tau + \delta$ can be written as the sum of "true-score variance," σ_τ^2, and error variance, σ_δ^2. (*Hint*: Square both sides of the equation for X_1 and take expectations.)

6.18 Recall Duncan's regression of occupational prestige on the educational and income levels of occupations. (The data are in Table 3.2 and `duncan.dat`.) Following Duncan, regress prestige on education and income. As well, perform a simple regression of prestige on income alone. Then add random measurement errors to education. Sample these measurement errors from a normal distribution with mean 0, repeating the exercise for each of the following measurement-error variances: $\sigma_\delta^2 = 10^2, 25^2, 50^2, 100^2$. In each case, recompute the regression of prestige on income and education. Then, treating the initial multiple regression as corresponding to $\sigma_\delta^2 = 0$, plot the coefficients of education and income as a function of σ_δ^2. What happens to the education coefficient as measurement error in education grows? What happens to the income coefficient?

6.5 Summary

- Standard statistical inference for least-squares regression analysis is based on the statistical model

$$Y_i = \alpha + \beta_1 X_{i1} + \cdots + \beta_k X_{ik} + \varepsilon_i$$

The key assumptions of the model concern the behavior of the errors ε_i:

1. *Linearity.* $E(\varepsilon_i) = 0$.
2. *Constant variance.* $V(\varepsilon_i) = \sigma_\varepsilon^2$.
3. *Normality.* $\varepsilon_i \sim N(0, \sigma_\varepsilon^2)$.
4. *Independence.* $\varepsilon_i, \varepsilon_j$ are independent for $i \neq j$.
5. The X-values are fixed or, if random, are independent of the errors.

- Under these assumptions, or particular subsets of them, the least-squares coefficients have certain desirable properties as estimators of the population regression coefficients. The least-squares coefficients are:
  1. linear functions of the data and therefore have simple sampling distributions;
  2. unbiased estimators of the population regression coefficients;
  3. the most efficient unbiased estimators of the population regression coefficients;
  4. maximum-likelihood estimators;
  5. normally distributed.
- The estimated standard error of the slope coefficient B in simple regression is

$$\widehat{\mathrm{SE}(B)} = \frac{S_E}{\sqrt{\sum(X_i - \overline{X})^2}}$$

The estimated standard error of the slope coefficient B_j in multiple regression is

$$\widehat{\mathrm{SE}(B_j)} = \frac{1}{\sqrt{1 - R_j^2}} \times \frac{S_E}{\sqrt{\sum(X_{ij} - \overline{X}_j)^2}}$$

In both cases, these estimated standard errors can be used in t-intervals and hypothesis tests for the corresponding population slope coefficients.
- An omnibus F-test for the null hypothesis that all of the slopes are 0 can be calculated from the analysis of variance for the regression:

$$F_0 = \frac{\mathrm{RegSS}/k}{\mathrm{RSS}/(n - k - 1)}$$

The omnibus F-statistic has k and $n - k - 1$ degrees of freedom.
- There is also an F-test for the null hypothesis that a subset of q slope coefficients is 0. This test is based on a comparison of the regression sums of squares for the full regression model (model 1) and for a null model (model 0) that deletes the independent variables in the null hypothesis:

$$F_0 = \frac{(\mathrm{RegSS}_1 - \mathrm{RegSS}_0)/q}{\mathrm{RSS}_1/(n - k - 1)}$$

This F-statistic has q and $n - k - 1$ degrees of freedom.
- It is important to distinguish between interpreting a regression descriptively as an empirical association among variables and structurally as specifying causal relations among variables. In the latter event, but not in the former, it is sensible to speak of bias produced by omitting an independent variable that (1) is a cause of Y and (2) is correlated with an independent variable in the regression equation. Bias in least-squares estimation results from the correlation that is induced between the included independent variable and the error by incorporating the omitted independent variable in the error.
- Measurement error in an independent variable tends to attenuate its regression coefficient and to make the variable an imperfect statistical control.

7

Dummy-Variable Regression

One of the serious limitations of multiple-regression analysis, as presented in Chapters 5 and 6, is that it accommodates only quantitative dependent and independent variables. In this chapter and the next, I shall explain how qualitative independent variables can be incorporated into a linear model.[1]

The current chapter begins with an explanation of how a *dummy-variable regressor* can be coded to represent a *dichotomous* (i.e., two-category) independent variable. I proceed to show how a set of dummy regressors can be employed to represent a *polytomous* (many-category) independent variable. Finally, I explain how interactions between quantitative and qualitative independent variables can be represented in dummy-regression models.

7.1 A Dichotomous Independent Variable

Let us consider the simplest case: one dichotomous and one quantitative independent variable. As in the two previous chapters, assume that relationships are *additive*—that is, that the partial effect of each independent variable is the same regardless of the specific value at which the other independent variable is held constant. As well, suppose that the other assumptions of the regression model hold: The errors are independent and normally distributed, with zero means and constant variance.

---

[1] Chapter 15 deals with qualitative *dependent* variables.

The general motivation for including a qualitative independent variable in a regression is essentially the same as for including an additional quantitative independent variable: (1) to account more fully for the dependent variable, by making the errors smaller; and (2) even more importantly, to avoid a biased assessment of the impact of an independent variable, as a consequence of omitting another independent variable that is related to it.

For concreteness, suppose that we are interested in investigating the relationship between education and income among women and men. Figure 7.1(*a*) and (*b*) represents two small (idealized) populations. In both cases, the within-gender regressions of income on education are parallel. Parallel regressions imply additive effects of education and gender on income: Holding education constant, the "effect" of gender is the difference between the two regression lines, which—for parallel lines—is everywhere the same. Likewise, holding gender constant, the "effect" of education is captured by the within-gender education slope, which—for parallel lines—is the same for men and women.[2]

In Figure 7.1(*a*), the independent variables gender and education are unrelated to each other: Women and men have identical distributions of education scores. In this circumstance, if we ignore gender and regress income on education alone, we obtain the same slope as is produced by the separate within-gender regressions. Because women have lower incomes than men of equal education, however, by ignoring gender we inflate the size of the errors.

The situation depicted in Figure 7.1(*b*) is importantly different. Here, gender and education are related, and therefore if we regress income on education alone, we arrive at a biased assessment of the effect of education on income: Because women have a higher average level of education than men, and because—for a given level of education—women's incomes are lower, on average, than men's, the overall regression of income on education has a *negative* slope even though the within-gender regressions have a positive slope.[3]

In light of these considerations, we might proceed to partition our sample by gender and perform separate regressions for women and men. This approach is reasonable, but it has its limitations: Fitting separate regressions makes it difficult to estimate and test for gender differences in income. Furthermore, if we can reasonably assume parallel regressions for women and men, then we can more efficiently estimate the common education slope by pooling sample data drawn from both groups. In particular, if the usual assumptions of the regression model hold, then it is desirable to fit the common-slope model by least squares.

One way of formulating the common-slope model is

$$Y_i = \alpha + \beta X_i + \gamma D_i + \varepsilon_i \qquad [7.1]$$

where D, called a *dummy-variable regressor* or an *indicator variable*, is coded 1 for men and 0 for women:

$$D_i = \begin{cases} 1 \text{ for men} \\ 0 \text{ for women} \end{cases}$$

---

[2] I shall consider nonparallel within-group regressions in Section 7.3.

[3] That marginal and partial relationships can differ in sign is called *Simpson's paradox*. Here, the marginal relationship between income and education is negative, while the partial relationship, controlling for gender, is positive.

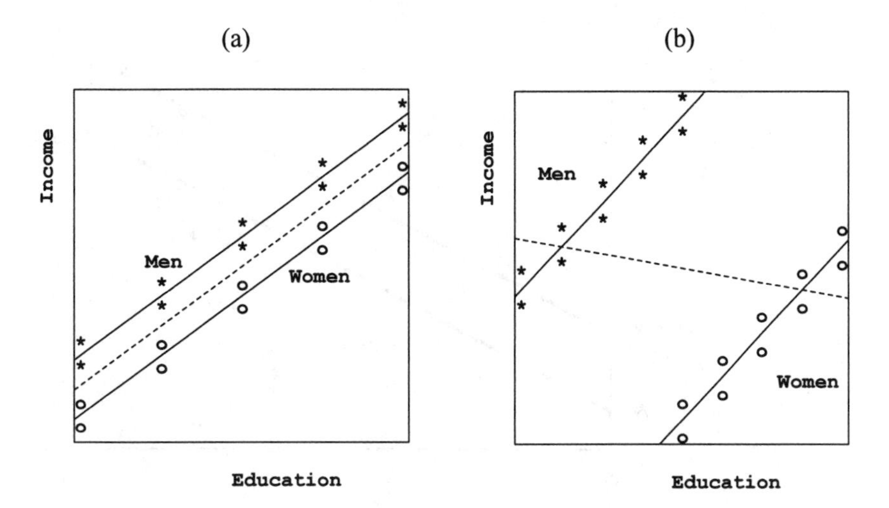

Figure 7.1. Idealized data representing the relationship between income and education for populations of men (asterisks) and women (circles). In (*a*), there is no relationship between education and gender; in (*b*), women have a higher average level of education than men. In both (*a*) and (*b*), the within-gender regressions (solid lines) are parallel. In each figure, the overall regression is given by the broken line.

Thus, for women the model becomes

$$Y_i = \alpha + \beta X_i + \gamma(0) + \varepsilon_i = \alpha + \beta X_i + \varepsilon_i$$

and for men

$$Y_i = \alpha + \beta X_i + \gamma(1) + \varepsilon_i = (\alpha + \gamma) + \beta X_i + \varepsilon_i$$

These regression equations are graphed in Figure 7.2.

This is our initial encounter with an idea that is fundamental to many linear models: the distinction between *independent variables* and *regressors*. Here, *gender* is a qualitative independent variable, with categories *male* and *female*. The dummy variable D is a regressor, representing the independent variable gender. In contrast, the quantitative independent variable *income* and the regressor X are one and the same. Were we to transform income, however, prior to entering it into the regression equation—say, by taking logs—then there would be a distinction between the independent variable (income) and the regressor (log income). In subsequent sections of this chapter, it will transpire that an independent variable can give rise to several regressors, and that some regressors are functions of more than one independent variable.

Returning to Equation 7.1 and Figure 7.2, the coefficient γ for the dummy regressor gives the difference in intercepts for the two regression lines. Because these regression lines are parallel, γ also represents the constant separation between the lines, and it may, therefore, be interpreted as the expected income advantage accruing to men when education is held constant. If men were *dis*advantaged relative to women, then γ would be *negative*. The coefficient α

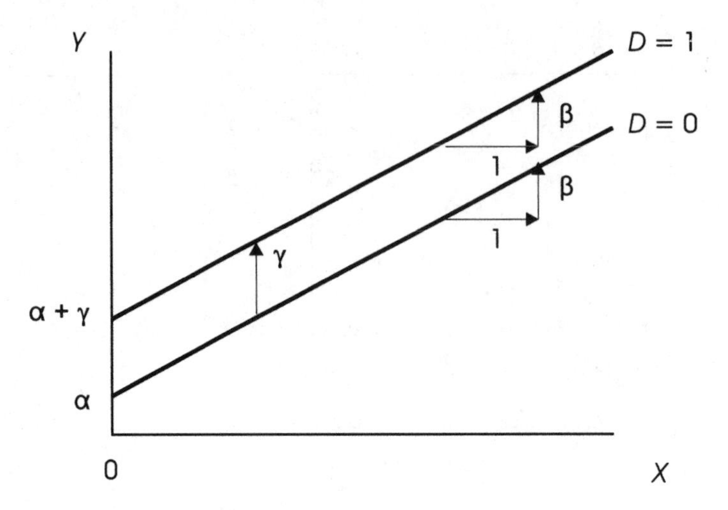

Figure 7.2. The additive dummy-variable regression model. The line labeled $D = 1$ is for men; the line labeled $D = 0$ is for women.

gives the intercept for women, for whom $D = 0$; and β is the common within-gender education slope.

Figure 7.3 reveals the fundamental geometric "trick" underlying the coding of a dummy regressor: We are, in fact, fitting a regression plane to the data, but the dummy regressor D is defined only at the values 0 and 1. The regression plane intersects the planes $\{X, Y|D = 0\}$ and $\{X, Y|D = 1\}$ in two lines, each with slope β. Because the difference between $D = 0$ and $D = 1$ is one unit, the difference in the Y-intercepts of these two lines is the slope of the plane in the D direction, that is γ. Indeed, Figure 7.2 is simply the projection of the two regression lines onto the $\{X, Y\}$ plane.

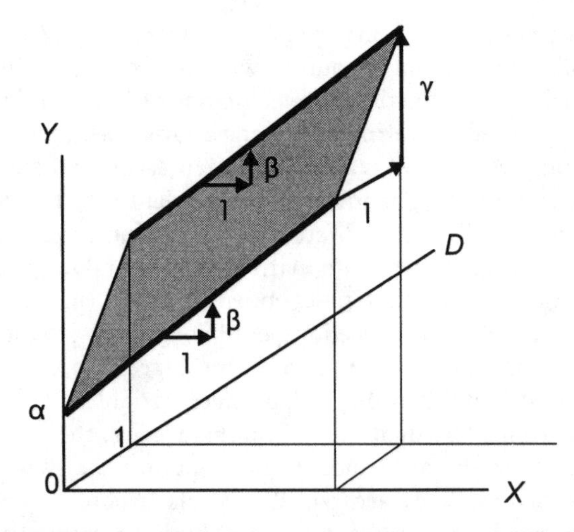

Figure 7.3. The geometric "trick" underlying dummy regression. The linear regression plane is defined only at $D = 0$ and $D = 1$, producing two regression lines with slope β and vertical separation γ.

Essentially similar results are obtained if we instead code D equal to 0 for men and 1 for women, making men the baseline category: The sign of γ is reversed, because it now represents the difference in intercepts between women and men (rather than vice versa), but its magnitude remains the same. The coefficient α now gives the income intercept for men. It is therefore immaterial which group is coded 1 and which is coded 0, as long as we are careful to interpret the coefficients of the model—for example, the sign of γ—in a manner consistent with the coding scheme that is employed.

To determine whether gender affects income, controlling for education, we can test H_0: $\gamma = 0$, either by a t-test, dividing the estimate of γ by its estimated standard error, or, equivalently, by dropping D from the regression model and formulating an incremental F-test. In either event, the procedures of the previous chapter apply.

Although I have developed dummy-variable regression for a single quantitative regressor, the method can be applied to any number of quantitative variables, as long as we are willing to assume that the slopes are the same in the two categories of the dichotomous independent variable—that is, that the regression surfaces are parallel in the two groups. In general, if we fit the model

$$Y_i = \alpha + \beta_1 X_{i1} + \cdots + \beta_k X_{ik} + \gamma D_i + \varepsilon_i$$

then, for $D = 0$, we have

$$Y_i = \alpha + \beta_1 X_{i1} + \cdots + \beta_k X_{ik} + \varepsilon_i$$

and, for $D = 1$,

$$Y_i = (\alpha + \gamma) + \beta_1 X_{i1} + \cdots + \beta_k X_{ik} + \varepsilon_i$$

> A dichotomous independent variable can be entered into a regression equation by formulating a dummy regressor, coded 1 for one category of the variable and 0 for the other category. A model incorporating a dummy regressor represents parallel regression surfaces, with the constant separation between the surfaces given by the coefficient of the dummy regressor.

EXERCISES

7.1 Using Davis's data on measured and reported height and weight (in `davis.dat`):

(a) Construct a dummy regressor for sex, and regress reported weight on measured weight and sex. Form the regression equations for the two groups and interpret each of the regression coefficients. Repeat this analysis, correcting the reported weight for subject 12 or omitting this subject.

(b) Repeat (a), regressing measured weight on reported weight and sex. Does the outlying observation have a different effect here than in part (a)? If so, why?

(c) Repeat (a) and (b) for reported and measured height.

7.2 Suppose that the values −1 and 1 are used for the dummy regressor D in Equation 7.1 instead of 0 and 1. Write out the regression equations for men and women, and explain how the parameters of the model are to be interpreted. Does this alternative coding of the dummy regressor adequately capture the effect of gender? Is it fair to conclude that the dummy-regression model will "work" properly as long as two distinct values of the dummy regressor are employed, one each for women and men? Is there a reason to prefer one coding to another?

7.2 Polytomous Independent Variables

The coding method of the previous section generalizes straightforwardly to polytomous independent variables. By way of illustration, recall (from the previous chapter) Duncan's regression of the rated prestige of 45 occupations on their education and income levels. I have classified Duncan's occupations into three rough categories: (1) professional and managerial occupations; (2) "white-collar" occupations; and (3) "blue-collar" occupations.[4] The *three*-category classification can be represented in the regression equation by introducing *two* dummy regressors, employing the following coding scheme:

Category	D_1	D_2
Professional and managerial	1	0
White collar	0	1
Blue collar	0	0

[7.2]

The regression model is then

$$Y_i = \alpha + \beta_1 X_{i1} + \beta_2 X_{i2} + \gamma_1 D_{i1} + \gamma_2 D_{i2} + \varepsilon_i$$

where X_1 is education and X_2 is income. This model describes three parallel regression planes, which can differ in their intercepts:

$$\begin{aligned}
\text{Professional:} \quad & Y_i = (\alpha + \gamma_1) + \beta_1 X_{i1} + \beta_2 X_{i2} + \varepsilon_i \\
\text{White collar:} \quad & Y_i = (\alpha + \gamma_2) + \beta_1 X_{i1} + \beta_2 X_{i2} + \varepsilon_i \\
\text{Blue collar:} \quad & Y_i = \quad \alpha \quad + \beta_1 X_{i1} + \beta_2 X_{i2} + \varepsilon_i
\end{aligned}$$

---

[4] The classification of some of the occupations—undertakers and reporters, for example—is ambiguous, but I believe my categorization to be reasonable, and the small number of occupations precludes a finer breakdown. The data are given in Table 3.2.

The coefficient α, therefore, gives the intercept for blue-collar occupations; γ_1 represents the constant vertical difference between the parallel regression planes for professional and blue-collar occupations (fixing the values of education and income); and γ_2 represents the constant vertical distance between the regression planes for white-collar and blue-collar occupations (again, fixing education and income).

Because blue-collar occupations are coded 0 for both dummy regressors, "blue collar" implicitly serves as a *baseline* category with which the other occupational categories are compared. The choice of a baseline category is essentially arbitrary, for we would fit precisely the same three regression planes regardless of which of the three occupational categories is selected for this role. The values (and meaning) of the individual dummy-variable coefficients γ_1 and γ_2 depend, however, on which category is chosen as the baseline.

It is sometimes natural to select a particular category as a basis for comparison—an experiment that includes a "control group" comes immediately to mind. In this instance, the individual dummy-variable coefficients are of interest, because they reflect differences between the "experimental" groups and the control group, holding other independent variables constant.

In most applications, however, the choice of a baseline category is entirely arbitrary, as it is for Duncan's occupational prestige regression. We are, therefore, interested in testing the null hypothesis of no effect of occupational type, controlling for education and income,

$$H_0\colon \gamma_1 = \gamma_2 = 0 \qquad [7.3]$$

but the individual hypotheses $H_0\colon \gamma_1 = 0$ and $H_0\colon \gamma_2 = 0$—which test, respectively, for differences between professional and blue-collar occupations, and between white-collar and blue-collar occupations—are of less intrinsic interest.[5] The null hypothesis in Equation 7.3 can be tested by the incremental-sum-of-squares approach.

I have demonstrated how to model the effects of a three-category qualitative independent variable by coding two dummy regressors. It may seem more natural to treat the three occupational categories symmetrically, coding *three* dummy regressors, rather than arbitrarily selecting one category as the baseline:

Category	D_1	D_2	D_3
Professional and Managerial	1	0	0
White collar	0	1	0
Blue collar	0	0	1

[7.4]

Then, for the jth occupational type, we would have

$$Y_i = (\alpha + \gamma_j) + \beta_1 X_{i1} + \beta_2 X_{i2} + \varepsilon_i$$

[5] The essential point here is not that the separate hypotheses are of *no* interest, but that they an arbitrary subset of the pairwise differences among the categories. In the present case, where there are three categories, the individual hypotheses represent two of the three pairwise group comparisons. The third comparison, between professional and white-collar occupations, is not *directly* represented in the model, although it is given indirectly by the difference $\gamma_1 - \gamma_2$.

The problem with this procedure is that there are too many parameters: We have used four parameters (α, γ_1, γ_2, γ_3) to represent only three group intercepts. As a consequence, we could not find unique values for these four parameters even if we knew the three population regression lines. Likewise, we cannot calculate unique least-squares estimates for the model, because the set of three dummy variables is perfectly collinear; for example, as is apparent from the table in Equation 7.4, $D_3 = 1 - D_1 - D_2$.

In general, then, for a polytomous independent variable with m categories, we need to code $m - 1$ dummy regressors. One simple scheme is to select the last category as the baseline, and to code $D_{ij} = 1$ when observation i falls in category j, and zero otherwise:

Category	D_1	D_2	$\cdots$	D_{m-1}
1	1	0	$\cdots$	0
2	0	1	$\cdots$	0
$\vdots$	$\vdots$	$\vdots$		$\vdots$
$m - 1$	0	0	$\cdots$	1
m	0	0	$\cdots$	0

> A polytomous independent variable can be entered into a regression by coding a set of 0/1 dummy regressors, one fewer than the number of categories of the variable. The "omitted" category, coded 0 for all dummy regressors in the set, serves as a baseline with which the other categories are compared. The model represents parallel regression surfaces, one for each category of the polytomy.

When there is more than one qualitative independent variable, and if we assume that the variables have additive effects, we can simply code a set of dummy regressors for each. To test the hypothesis that the effects of a qualitative independent variable are nil, we delete its dummy regressors from the model and compute an incremental F-test of the hypothesis that all of the associated coefficients are 0.

In the preceding chapter, I used Duncan's data to regress occupational prestige on education and income, obtaining the fitted equation

$$\begin{aligned} \widehat{Y} = -&6.065 + 0.5458X_1 + 0.5987X_2 \qquad R^2 = .8282 \\ &(4.272) \quad (0.0982) \quad (0.1197) \end{aligned}$$

As is common practice, I have shown the estimated standard error of each regression coefficient in parentheses beneath the coefficient. The three occupational categories differ substantially in their average levels of prestige:

Category	*Number of Cases*	*Mean Prestige*
Professional and managerial	18	80.44
White collar	6	36.67
Blue collar	21	22.76
All occupations	45	47.69

Inserting dummy variables for type of occupation into the regression equation, employing the coding scheme shown in Equation 7.2, produces the following results:

$$\begin{aligned}\widehat{Y} = & -0.1850 + 0.3453X_1 + 0.5976X_2 + 16.66D_1 - 14.66D_2 \\ & \quad (3.714) \qquad (0.1136) \qquad (0.0894) \qquad (6.99) \qquad (6.11)\end{aligned}$$

$$R^2 = .9131$$

The three fitted regression equations are, therefore,

$$\begin{aligned}\text{Professional:} \quad & \widehat{Y} = 16.48 + 0.3453X_1 + 0.5976X_2 \\ \text{White collar:} \quad & \widehat{Y} = -14.84 + 0.3453X_1 + 0.5976X_2 \\ \text{Blue collar:} \quad & \widehat{Y} = -0.1850 + 0.3453X_1 + 0.5976X_2\end{aligned}$$

Note that the coefficient for education (X_1)—but not the coefficient for income (X_2)—grows smaller when type of occupation is controlled. The dummy-variable coefficients (or, equivalently, the category intercepts) reveal that when education and income levels are controlled, the difference in average prestige between professional and blue-collar occupations declines, from $80.44 - 22.76 = 57.68$ points to 16.66 points. The difference between white-collar and blue-collar occupations is reversed when income and education are controlled, changing from $36.67 - 22.76 = +13.91$ points to -14.66 points. That is, the greater prestige of professional occupations compared with blue-collar occupations appears to be due mostly to differences in education and income between these two classes of occupations. While white-collar occupations have greater prestige, on average, than blue-collar occupations, they have lower prestige than blue-collar occupations of the same educational and income levels.

To test the null hypothesis of no partial effect of type of occupation,

$$H_0: \gamma_1 = \gamma_2 = 0$$

we can calculate the incremental F-statistic

$$\begin{aligned}F_0 &= \frac{n-k-1}{q} \times \frac{R_1^2 - R_0^2}{1 - R_1^2} \\ &= \frac{45-4-1}{2} \times \frac{.9131 - .8282}{1 - .9131} = 19.54\end{aligned}$$

with 2 and 40 degrees of freedom, for which $p < .0001$. The occupational type effect is therefore highly statistically significant, but (examining the coefficient standard errors) not very precisely estimated. The same is true of the education and income coefficients.

EXERCISES

7.3 Employing Angell's data on the moral integration of U.S. cities (in Table 2.3 and `angell.dat`), construct dummy regressors to represent the four regions East, Midwest, West, and South; regress moral integration on ethnic heterogeneity, geographic mobility, and region. Use an incremental F-test to test the null hypothesis of no region effects. Write out the fitted regression equation for each region.

7.4 Using Ornstein's data on interlocking directorates among major Canadian firms (in `ornstein.dat`), construct one set of dummy regressors to represent the 10 categories of industrial sector and another set to represent the four nations of control. Regress number of interlocks on assets, sector, and nation of control. Use incremental F-tests to test the null hypotheses of no sector and no nation-of-control effects.

7.5 Adjusted means: Let $\overline{Y}_1$ represent the ("unadjusted") mean moral-integration score of Angell's eastern cities, $\overline{Y}_2$ that of midwestern cities, $\overline{Y}_3$ that of western cities, and $\overline{Y}_4$ that of southern cities. Differences among the $\overline{Y}_j$ may partly reflect regional differences in heterogeneity and mobility. In the dummy-variable regression of Exercise 7.3, regional differences are "controlled" for heterogeneity and mobility, producing the fitted regression equation

$$\widehat{Y} = A + B_1X_1 + B_2X_2 + C_1D_1 + C_2D_2 + C_3D_3$$

Consequently, if we fix heterogeneity and mobility at particular values—say, $X_1 = x_1$ and $X_2 = x_2$—then the fitted moral-integration scores for the several regions are given by (treating "South" as the baseline region):

$$\begin{aligned}
\widehat{Y}_1 &= (A + C_1) + B_1x_1 + B_2x_2 \\
\widehat{Y}_2 &= (A + C_2) + B_1x_1 + B_2x_2 \\
\widehat{Y}_3 &= (A + C_3) + B_1x_1 + B_2x_2 \\
\widehat{Y}_4 &= \quad A \quad + B_1x_1 + B_2x_2
\end{aligned}$$

(a) Notice that the *differences* among the $\widehat{Y}_j$ depend only on the dummy-variable coefficients C_1, C_2, and C_3, and not on the values of x_1 and x_2. Why is this so?

(b) When $x_1 = \overline{X}_1$ and $x_2 = \overline{X}_2$, the $\widehat{Y}_j$ are called *adjusted means* and are denoted $\widetilde{Y}_j$. How can the adjusted means $\widetilde{Y}_j$ be interpreted? In what sense is $\widetilde{Y}_j$ an "adjusted" mean?

(c) Locate the "unadjusted" and adjusted means for women and men in each of Figures 7.1(*a*) and (*b*). Construct a similar figure in which the difference between adjusted means is *smaller* than the difference in unadjusted means.

(d) Using the results of Exercise 7.3, compute adjusted mean moral-integration scores for each of the four regions, controlling for heterogeneity and mobility. Find the unadjusted mean moral integration for each region and comment on the differences, if any, between the unadjusted and adjusted means.

7.6 An outlier test: Recall, from Exercise 5.6, Sahlins's regression of Acres/Gardener on Consumers/Gardener for the 20 households of Mazulu village. (Sahlins's data are in Table 2.1 and `sahlins.dat`.) To test whether the apparently outlying fourth household is discernibly different from the others, construct a dummy regressor coded 1 for this household and 0 for the others. Then refit the regression, using the dummy variable along with Consumers/Gardener. Is the coefficient of the dummy regressor statistically significant? How is this coefficient to be interpreted? What has happened to the residual of the fourth household? Why? Has the coefficient for Consumers/Gardener changed? How do the new intercept and slope coefficients compare with those obtained by simply deleting the fourth household? Is this outlier test valid if we decided to check the fourth household only after examining the data? (See Section 11.3 for further discussion of outliers in regression.)

7.3 Modeling Interactions

Two independent variables are said to *interact* in determining a dependent variable when the partial effect of one depends on the value of the other. The additive models that we have considered thus far therefore specify the absence of interactions. In this section, I shall explain how the dummy-variable regression model can be modified to accommodate interactions between qualitative and quantitative independent variables.[6]

The treatment of dummy-variable regression in the preceding two sections has assumed parallel regressions across the several categories of a qualitative independent variable. If these regressions are not parallel, then the qualitative independent variable interacts with one or more of the quantitative independent variables. The dummy-regression model can easily be modified to reflect these interactions.

For simplicity, I return to the contrived example of Section 7.1, examining the regression of income on gender and education. Consider the hypothetical data shown in Figure 7.4 (and contrast these examples with those shown in Figure 7.1, where the effects of gender and education are additive). In Figure 7.4(*a*) [as in Figure 7.1(a)], gender and education are independent, because women and men have identical education distributions; in Figure 7.4(*b*) [as in Figure 7.1(*b*)],

[6] Interactions between qualitative independent variables are taken up in the next chapter on analysis of variance; interactions between quantitative independent variables are discussed in Section 14.2.1 on polynomial regression.

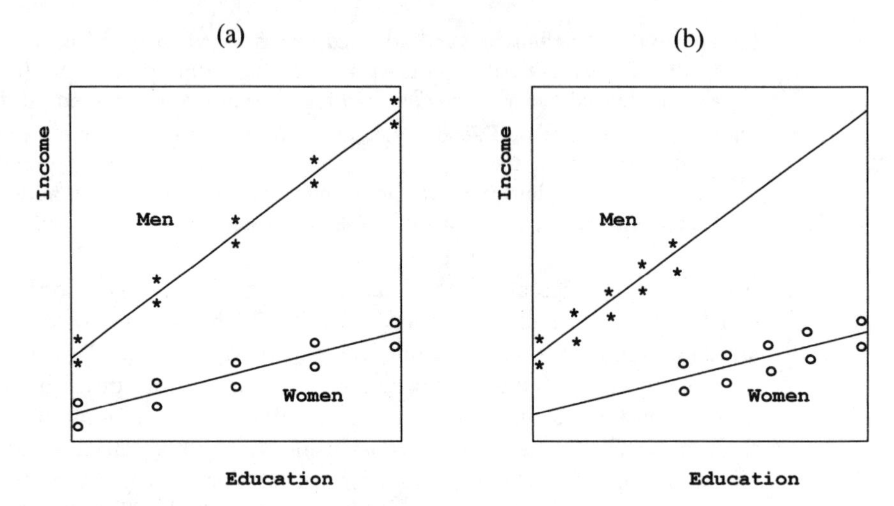

Figure 7.4. Idealized data representing the relationship between income and education for populations of men (asterisks) and women (circles). In (*a*), there is no relationship between education and gender; in (*b*), women have a higher average level of education than men. In both cases, the within-gender regressions (lines) are not parallel—the slope for men is greater than the slope for women—and, consequently, education and gender interact in affecting income.

gender and education are related, because women, on average, have higher levels of education than men.

It is apparent in both Figure 7.4(*a*) and Figure 7.4(*b*), however, that the within-gender regressions of income on education are not parallel: In both cases, the slope for men is larger than the slope for women. Because the effect of education varies by gender, education and gender interact in affecting income.

It is also the case, incidentally, that the effect of gender varies by education. Because the regressions are not parallel, the relative income advantage of men changes (indeed, grows) with education. Interaction, then, is a symmetric concept—the effect of education varies by gender, and the effect of gender varies by education.

The simple examples in Figures 7.1 and 7.4 illustrate an important and frequently misunderstood point: *Interaction* and *correlation* of independent variables are empirically and logically distinct phenomena. Two independent variables can interact *whether or not* they are related to one another statistically. Interaction refers to the manner in which independent variables combine to affect a dependent variable, not to the relationship between the independent variables themselves.

7.3.1 Constructing Interaction Regressors

We could model the data in Figure 7.4 by fitting separate regressions of income on education for women and men. As before, however, it is more convenient to fit a combined model, primarily because a combined model facilitates a test of the gender-by-education interaction. Moreover, a properly formulated unified model that permits different intercepts and slopes in the two groups produces the same fit as separate regressions: The full sample is composed of the two groups, and, consequently, the residual sum of squares for the full sample is minimized when the residual sum of squares is minimized in each group.[7]

The following model accommodates different intercepts and slopes for women and men:

$$Y_i = \alpha + \beta X_i + \gamma D_i + \delta(X_i D_i) + \varepsilon_i \qquad [7.5]$$

Along with the quantitative regressor X for education and the dummy regressor D for gender, I have introduced the *interaction regressor* XD into the regression equation. The interaction regressor is the *product* of the other two regressors; although XD is therefore a function of X and D, it is not a *linear* function, and perfect collinearity is avoided.[8]

For women, model (7.5) becomes

$$\begin{aligned} Y_i &= \alpha + \beta X_i + \gamma(0) + \delta(X_i \cdot 0) + \varepsilon_i \\ &= \alpha + \beta X_i + \varepsilon_i \end{aligned}$$

and for men

$$\begin{aligned} Y_i &= \alpha + \beta X_i + \gamma(1) + \delta(X_i \cdot 1) + \varepsilon_i \\ &= (\alpha + \gamma) + (\beta + \delta) X_i + \varepsilon_i \end{aligned}$$

These regression equations are graphed in Figure 7.5: The parameters α and β are, respectively, the intercept and slope for the regression of income on education among women; γ gives the difference in intercepts between the male and female groups; and δ gives the difference in slopes between the two groups. To test for interaction, therefore, we may simply test the hypothesis H_0: $\delta = 0$.

> Interactions can be incorporated by coding interaction regressors, taking products of dummy regressors with quantitative independent variables. The model permits different slopes in different groups—that is, regression surfaces that are not parallel.

In the additive, no-interaction model of Equation 7.1 and Figure 7.2, the dummy-regressor coefficient γ represents the unique partial effect of gender (i.e., the expected income difference between women and men of equal education,

[7] See Exercises 7.7 and 7.9.

[8] If this procedure seems illegitimate, then think of the interaction regressor as a new variable, say $Z \equiv XD$. The model is linear in X, D, and Z. The "trick' of introducing an interaction regressor is similar to the trick of formulating dummy regressors to capture the effects of a qualitative independent variable: In both cases, there is a distinction between independent variables and regressors. Unlike a dummy regressor, however, the interaction regressor is a function of *both* independent variables.

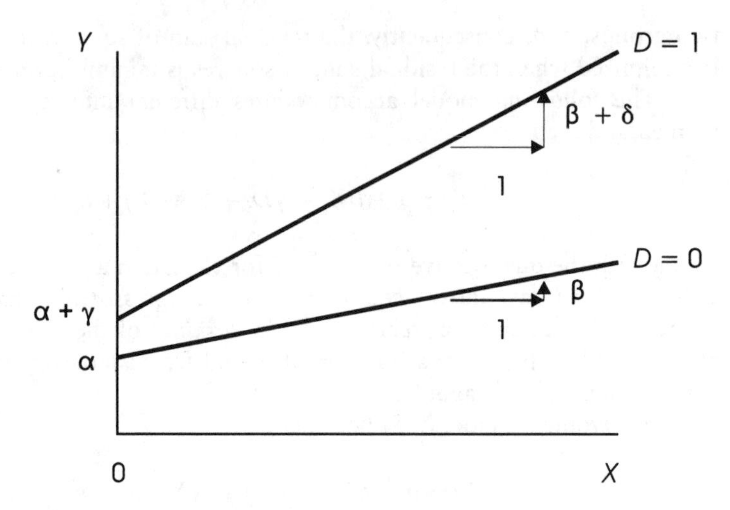

Figure 7.5. The dummy-variable regression model with an interaction regressor. The line labeled $D = 1$ is for men; the line labeled $D = 0$ is for women.

regardless of the value at which education is fixed), while the slope β represents the unique partial effect of education (i.e., the within-gender expected increment in income for a one-unit increase in education, for both women and men). In the interaction model of Equation 7.5 and Figure 7.5, in contrast, γ is no longer interpretable as the unqualified income difference between men and women of equal education.

Because the within-gender regressions are not parallel, the separation between the regression lines changes; here, γ is simply the separation at $X = 0$—that is, above the origin. It is generally no more important to assess the expected income difference between men and women of 0 education than at other educational levels, and therefore the difference-in-intercepts parameter γ is not of special interest in the interaction model.[9]

Likewise, in the interaction model, β is not the unqualified partial effect of education, but rather the effect of education among women. Although this coefficient is of interest, it is not necessarily more important than the effect of education among men ($\beta + \delta$), which does not appear directly in the model.

7.3.2 The Principle of Marginality

Following Nelder (1977), we say that the separate partial effects, or *main effects*, of education and gender are *marginal* to the education-by-gender interaction. In general, we neither test nor interpret the main effects of independent variables that interact. If, however, we can rule out interaction either on theoretical or on empirical grounds, then we can proceed to test, estimate, and interpret the main effects.

Furthermore, it does not generally make sense to specify and fit models that include interaction regressors but that delete main effects that are marginal

---

[9] Indeed, in many instances (although not here), the value $X = 0$ may not occur in the data or may be impossible (as, for example, if X is weight). In such cases, γ has no literal interpretation in the interaction model.

to them. This is not to say that such models—which violate the *principle of marginality*—are uninterpretable: They are, rather, not broadly applicable.

> The principle of marginality specifies that a model including a high-order term (such as an interaction) should normally also include the lower-order relatives of that term (the main effects that "compose" the interaction).

Suppose, for example, that we fit the model

$$Y_i = \alpha + \beta X_i + \delta(X_i D_i) + \varepsilon_i$$

which omits the dummy regressor D, but includes its "higher-order relative" XD. As shown in Figure 7.6(a), this model describes regression lines for women and men that have the same intercept but (potentially) different slopes, a specification that is peculiar and of no substantive interest. Similarly, the model

$$Y_i = \alpha + \gamma D_i + \delta(X_i D_i) + \varepsilon_i$$

graphed in Figure 7.6(b), constrains the slope for women to 0, which is needlessly restrictive.

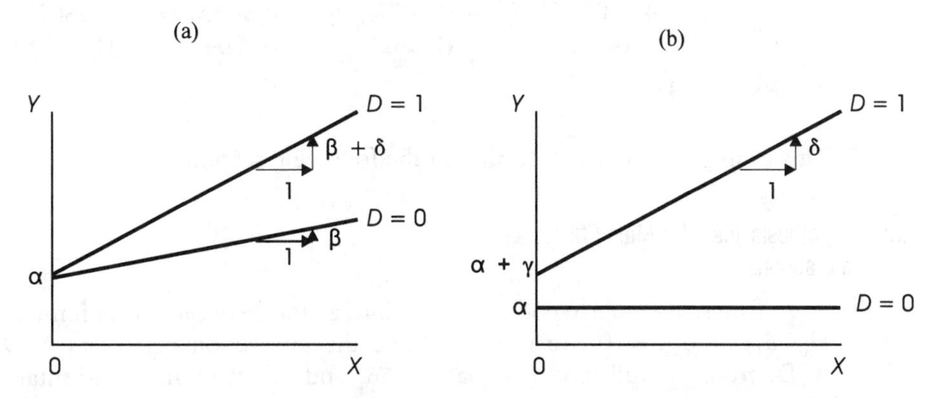

Figure 7.6. Two models that violate the principle of marginality: In (a), the dummy regressor D is omitted from the model $E(Y) = \alpha + \beta X + \delta(XD)$; in ($b$), the quantitative independent variable X is omitted from the model $E(Y) = \alpha + \gamma D + \delta(XD)$. These models violate the principle of marginality because they include the term XD, which is a higher-order relative of both X and D.

7.3.3 Interactions With Polytomous Independent Variables

The method for modeling interactions by forming product regressors is easily extended to polytomous independent variables, to several qualitative independent variables, and to several quantitative independent variables. I shall use Duncan's occupational prestige regression to illustrate the application of the method, entertaining the possibility that occupational type interacts both with education and with income:

$$Y_i = \alpha + \beta_1 X_{i1} + \beta_2 X_{i2} + \gamma_1 D_{i1} + \gamma_2 D_{i2} \\ + \delta_{11} X_{i1} D_{i1} + \delta_{12} X_{i1} D_{i2} + \delta_{21} X_{i2} D_{i1} + \delta_{22} X_{i2} D_{i2} + \varepsilon_i \quad [7.6]$$

Notice that we require one interaction regressor for each product of a dummy regressor with a quantitative independent variable. The regressors $X_1 D_1$ and $X_1 D_2$ capture the interaction between education and occupational type; $X_2 D_1$ and $X_2 D_2$ capture the interaction between income and occupational type. The model therefore permits different intercepts and slopes for the three types of occupations:

$$\begin{aligned} \text{Professional:} \quad & Y_i = (\alpha + \gamma_1) + (\beta_1 + \delta_{11}) X_{i1} + (\beta_2 + \delta_{21}) X_{i2} + \varepsilon_i \\ \text{White collar:} \quad & Y_i = (\alpha + \gamma_2) + (\beta_1 + \delta_{12}) X_{i1} + (\beta_2 + \delta_{22}) X_{i2} + \varepsilon_i \\ \text{Blue collar:} \quad & Y_i = \alpha + \beta_1 X_{i1} + \beta_2 X_{i2} + \varepsilon_i \end{aligned}$$

Blue-collar occupations, which are coded 0 for both dummy regressors, serve as the baseline for the intercepts and slopes of the other occupational types. As in the no-interaction model, the choice of baseline category is generally arbitrary, as it is here, and is inconsequential. Fitting model [7.6] to Duncan's data produces the following results:

$$\begin{aligned} \widehat{Y}_i = & -3.95 + 0.320X_1 + 0.783X_2 + 32.0D_1 - 7.04D_2 \\ & \ \ (6.79) \quad (0.280) \quad (0.131) \quad (14.1) \quad (20.6) \\ & + 0.0186X_1D_1 + 0.107X_1D_2 - 0.369X_2D_1 - 0.360X_2D_2 \\ & \quad (0.318) \qquad (0.362) \qquad (0.204) \qquad (0.260) \end{aligned} \quad [7.7]$$

$$R^2 = .9233$$

This example is discussed further in the following section.

7.3.4 Hypothesis Tests for Main Effects and Interactions

To test the null hypothesis of no interaction between education and type, H_0: $\delta_{11} = \delta_{12} = 0$, we need to delete the interaction regressors $X_1 D_1$ and $X_1 D_2$ from the full model (Equation 7.6) and calculate an incremental F-test; likewise, to test the null hypothesis of no interaction between income and type, H_0: $\delta_{21} = \delta_{22} = 0$, we delete the interaction regressors $X_2 D_1$ and $X_2 D_2$ from the full model. These tests, and tests for the main effects of occupational type, education, and income, are detailed in Tables 7.1 and 7.2: Table 7.1 gives the regression sums of squares for several models, which, along with the residual

TABLE 7.1 Regression Sums of Squares for Several Models Fit to Duncan's Occupational Prestige Data. These sums of squares are the building blocks of incremental *F*-tests for the main and interaction effects of the independent variables. The following code is used for "terms" in the model: *E*, Education; *I*, Income; *T*, Occupational type.

Model	*Terms*	*Parameters*	*Regression Sum of Squares*	*df*
1	$E, I, T, E \times T, I \times T$	$\alpha, \beta_1, \beta_2, \gamma_1, \gamma_2, \delta_{11}, \delta_{12}, \delta_{21}, \delta_{22}$	40,337.	8
2	$E, I, T, E \times T$	$\alpha, \beta_1, \beta_2, \gamma_1, \gamma_2, \delta_{11}, \delta_{12}$	39,965.	6
3	$E, I, T, I \times T$	$\alpha, \beta_1, \beta_2, \gamma_1, \gamma_2, \delta_{21}, \delta_{22}$	40,325.	6
4	E, I, T	$\alpha, \beta_1, \beta_2, \gamma_1, \gamma_2$	39,890.	4
5	E, I	α, β_1, β_2	36,181.	2
6	$E, T, E \times T$	$\alpha, \beta_1, \gamma_1, \gamma_2, \delta_{11}, \delta_{12}$	36,011.	5
7	$I, T, I \times T$	$\alpha, \beta_2, \gamma_1, \gamma_2, \delta_{21}, \delta_{22}$	39,434.	5

sum of squares for the full model, $\text{RSS}_1 = 3351$, are the building blocks of the incremental *F*-tests shown in Table 7.2. Table 7.3 shows the hypothesis tested by each of the incremental *F*-statistics in Table 7.2.

Although the analysis-of-variance table (Table 7.2) shows the tests for the main effects of education, income, and type before the education-by-type and income-by-type interactions, the logic of interpretation is to examine the interactions first: Conforming to the principle of marginality, the test for each main effect is computed assuming that the interactions that are higher-order relatives of that main effect are 0 (as shown in Table 7.3). Thus, for example, the test for the education main effect assumes that the education-by-type interaction is absent (i.e., that $\delta_{11} = \delta_{12} = 0$), but not that the income-by-type interaction is absent ($\delta_{21} = \delta_{22} = 0$).

TABLE 7.2 Analysis-of-Variance Table, Showing Incremental *F*-Tests for the Terms in Duncan's Occupational Prestige Regression

Source	*Models Contrasted*	*Sum of Squares*	*df*	*F*	*p*
Education	3 − 7	891.	1	9.6	.004
Income	2 − 6	3954.	1	42.5	<<.0001
Type	4 − 5	3709.	2	19.9	<.0001
Education × Type	1 − 3	12.	2	0.1	.91
Income × Type	1 − 2	372.	2	2.0	.15
Residuals		3351.	36		
Total		43,688.	44		

TABLE 7.3 Hypotheses Tested by the Incremental *F*-Tests in Table 7.2.

Source	*Models Contrasted*	*Null Hypothesis*
Education	$3-7$	$\beta_1 = 0 \mid \delta_{11} = \delta_{12} = 0$
Income	$2-6$	$\beta_2 = 0 \mid \delta_{21} = \delta_{22} = 0$
Type	$4-5$	$\gamma_1 = \gamma_2 = 0 \mid \delta_{11} = \delta_{12} = \delta_{21} = \delta_{22} = 0$
Education × Type	$1-3$	$\delta_{11} = \delta_{12} = 0$
Income × Type	$1-2$	$\delta_{21} = \delta_{22} = 0$

The principle of marginality serves as a guide to constructing incremental *F*-tests for the terms in a model that includes interactions.

There is no indication of an interaction between education and type, and weak evidence for an interaction between income and type, with the income slope steeper in the baseline blue-collar category than in the professional or white-collar categories.[10] Considering the small number of cases—especially for white-collar occupations—we are probably squeezing the data too hard. Assuming that their higher-order relatives are absent, the education, income, and occupational type main effects are all highly statistically significant.[11]

The degrees of freedom for the several sources of variation add to the total degrees of freedom, but—because the regressors in different sets are correlated—the sums of squares do not add to the total sum of squares. What is important here (and more generally) is that sensible hypotheses are tested, not that the sums of squares add to the total sum of squares.

EXERCISES

7.7 Repeat Exercise 7.1(*a*) (for Davis's regression of measured weight on sex and reported weight), including an interaction regressor for sex and reported weight. Write out the fitted regression equation for each sex. Show that precisely the same regression equations are obtained if measured weight

[10] See Equation 7.7 and Exercise 7.8.

[11] We tested the occupational type main effect in the previous section, but using an estimate of error variance based on model 4, which does not contain the interactions. In Table 7.2, the estimated error variance is based on the full model, model 1. A sound general practice is to use the largest model fit to the data to estimate the error variance even when, as is frequently the case, this model includes effects that are not statistically significant. The largest model necessarily has the smallest residual sum of squares, but it also has the fewest residual degrees of freedom. These two factors tend to offset one another, and it usually makes little difference whether the estimated error variance is based on the full model, or on a model that deletes nonsignificant terms. Nevertheless, using the full model ensures an unbiased estimate of the error variance.

is regressed on reported weight *separately* for male and female subjects. Show that the residual sum of squares for the combined regression is the sum of the residual sums of squares for the two separate regressions.

7.8 The fitted regression, using Duncan's data, of occupational prestige on education, income, and type of occupation is given in the text in Equation 7.7. Write out the fitted regression equation for each type of occupation, and verify (as stated in the text) that the income slope is steeper in the baseline blue-collar category than in the professional or white-collar categories.

7.9 Continuing (from Exercise 7.3) the analysis of Angell's data on the moral integration of U. S. cities, include interactions between region and each of the other two independent variables. Write out the fitted regression equation for each of the four regions. Test the statistical significance of each of these interactions with an incremental F-test. Then fit a separate regression of moral integration on heterogeneity and mobility for the 14 southern cities, and confirm that the fitted equation for the South is the same as that obtained from the combined regression.

7.10 Can the concept of an adjusted mean, introduced in Exercise 7.5, be extended to a model that includes interactions? If so, show how adjusted means can be found for the data in Figure 7.4(a) and (b).

7.4 A Caution Concerning Standardized Coefficients

In Chapter 5, I explained the use—and limitations—of standardized regression coefficients. It is appropriate to sound another cautionary note here: Inexperienced researchers sometimes report standardized coefficients for dummy regressors. As I have explained, an *unstandardized* coefficient for a dummy regressor is interpretable as the expected dependent-variable difference between a particular category and the baseline category for the dummy-regressor set (controlling, of course, for the other independent variables in the model).

If a dummy-regressor coefficient is standardized, then this straightforward interpretation is lost. Furthermore, because a 0/1 dummy regressor cannot be increased by one standard deviation, the usual interpretation of a standardized regression coefficient also does not apply. Standardization is a linear transformation, so many characteristics of the regression model—the value of R^2, for example—do not change, but the standardized coefficient itself is not directly interpretable. These difficulties can be avoided by standardizing only the dependent variable and quantitative independent variables in a regression, leaving dummy regressors in 0/1 form.

A similar point applies to interaction regressors. We may legitimately standardize a quantitative independent variable *prior* to taking its product with a dummy regressor, but to standardize the interaction regressor itself is not sensible: The interaction regressor cannot change independently of the main-effect regressors that compose it and are marginal to it.

It is not sensible to standardize dummy regressors or interaction regressors.

7.5 Summary

- A dichotomous independent variable can be entered into a regression equation by formulating a dummy regressor, coded 1 for one category of the variable and 0 for the other category. A model incorporating a dummy regressor represents parallel regression surfaces, with the constant separation between the surfaces given by the coefficient of the dummy regressor.
- A polytomous independent variable can be entered into a regression by coding a set of 0/1 dummy regressors, one fewer than the number of categories of the variable. The "omitted" category, coded 0 for all dummy regressors in the set, serves as a baseline with which the other categories are compared. The model represents parallel regression surfaces, one for each category of the polytomy.
- Interactions can be incorporated by coding interaction regressors, taking products of dummy regressors with quantitative independent variables. The model permits different slopes in different groups—that is, regression surfaces that are not parallel.
- The principle of marginality specifies that a model including a high-order term (such as an interaction) should normally also include the lower-order relatives of that term (the main effects that "compose" the interaction). The principle of marginality also serves as a guide to constructing incremental F-tests for the terms in a model that includes interactions.
- It is not sensible to standardize dummy regressors or interaction regressors.

8

Analysis of Variance

I introduced the term *analysis of variance* in Chapter 5 to describe the partition of the dependent-variable sum of squares into "explained" and "unexplained" components, noting that this decomposition applies generally to linear models. For historical reasons, *analysis of variance* (abbreviated ANOVA) also refers to procedures for fitting and testing linear models in which the independent variables are categorical.[1]

When there is a single categorical independent variable (also termed a *factor* or *classification*), then these procedures are called *one-way* analysis of variance, the subject of the first section of this chapter. Two factors produce *two-way* analysis of variance, three factors, *three-way* analysis of variance, and so on. Two-way ANOVA is taken up in Section 8.2, higher-way ANOVA in Section 8.3.

The dummy-variable regression model of the previous chapter incorporates both quantitative and categorical independent variables; in Section 8.4 we shall examine an alternative formulation of this model called *analysis of covariance* (ANCOVA).

Finally, I shall explain how *linear contrasts* can be used to "customize" hypothesis tests in analysis of variance.

---

[1] The methods and terminology of analysis of variance were introduced by the great British statistician R. A. Fisher (1925). Fisher's many other seminal contributions to statistics include the technique of randomization in experimental design and the method of maximum likelihood.

8.1 One-Way Analysis of Variance

In Chapter 7, we learned how to construct dummy regressors to represent the effects of a qualitative independent variable alongside those of quantitative independent variables. Suppose, however, that there are no quantitative independent variables—only a single, qualitative independent variable. For example, for a three-category classification we have the model

$$Y_i = \alpha + \gamma_1 D_{i1} + \gamma_2 D_{i2} + \varepsilon_i \qquad [8.1]$$

employing the following coding for the dummy regressors:

Group	D_1	D_2
1	1	0
2	0	1
3	0	0

The expectation of the dependent variable in each group is the population group mean, denoted μ_j for the jth group. Because the error ε has a mean of 0 under the usual linear-model assumptions, taking the expectation of both sides of the model (Equation 8.1) produces the following relationships between group means and model parameters:

$$\begin{aligned}
\text{Group 1:}\quad & \mu_1 = \alpha + \gamma_1 \times 1 + \gamma_2 \times 0 = \alpha + \gamma_1 \\
\text{Group 2:}\quad & \mu_2 = \alpha + \gamma_1 \times 0 + \gamma_2 \times 1 = \alpha + \gamma_2 \\
\text{Group 3:}\quad & \mu_3 = \alpha + \gamma_1 \times 0 + \gamma_2 \times 0 = \alpha
\end{aligned}$$

There are three parameters (α, γ_1, and γ_2) and three group means, so we can solve uniquely for the parameters in terms of the group means:

$$\begin{aligned}
\alpha &= \mu_3 \\
\gamma_1 &= \mu_1 - \mu_3 \\
\gamma_2 &= \mu_2 - \mu_3
\end{aligned}$$

It is not surprising that α represents the mean of the baseline category (group 3), and that γ_1 and γ_2 capture differences between the other group means and the mean of the baseline category.

One-way analysis of variance focuses on testing for differences among group means. The omnibus F-statistic for the model (Equation 8.1) tests H_0: $\gamma_1 = \gamma_2 = 0$, which corresponds to H_0: $\mu_1 = \mu_2 = \mu_3$, the null hypothesis of no differences among the population group means. Our consideration of one-way analysis of variance might well end here, but for a desire to develop methods that generalize easily to more complex situations in which there are several, potentially interacting, factors.

The first innovation is notational: Because observations are partitioned according to groups, it is convenient to let Y_{ij} denote the ith observation within the

jth of m groups. The number of observations in the jth group is n_j, and therefore the total number of observations is $n = \sum_{j=1}^{m} n_j$. As above, $\mu_j \equiv E(Y_{ij})$ represents the population mean in group j.

The one-way analysis-of-variance model is written in the following manner:

$$Y_{ij} = \mu + \alpha_j + \varepsilon_{ij} \tag{8.2}$$

where we would like μ to represent, in some reasonable sense, the general level of the dependent variable in the population; α_j should represent the effect on the dependent variable of membership in the jth group; and ε_{ij} is an error variable that follows the usual linear-model assumptions—that is, the ε_{ij} are independent and normally distributed with zero expectations and equal variances.

Upon taking expectations, Equation 8.2 becomes

$$\mu_j = \mu + \alpha_j$$

The parameters of the model are, therefore, underdetermined, for there are $m+1$ parameters (including μ) but only m population group means. For example, for $m = 3$, we have four parameters, but only three equations:

$$\mu_1 = \mu + \alpha_1$$
$$\mu_2 = \mu + \alpha_2$$
$$\mu_3 = \mu + \alpha_3$$

Even if we knew the three population group means, we could not solve uniquely for the parameters.

Because the parameters of the model (Equation 8.2) are themselves underdetermined, they cannot be uniquely estimated. To estimate the model, we would need to code one dummy regressor for each group-effect parameter α_j, and—as we discovered in the previous chapter—the resulting dummy regressors would be perfectly collinear.

One convenient way out of this dilemma is to place a linear restriction on the parameters of the model, of the form

$$w_0\mu + \sum_{j=1}^{m} w_j\alpha_j = 0$$

where the w's are prespecified constants, not all equal to 0. It turns out that any such restriction will do, in the sense that all linear restrictions yield the same F-test for the null hypothesis of no differences in population group means.[2] For example, if we employ the restriction $\alpha_m = 0$, we are, in effect, deleting the parameter for the last category, making it a baseline category. The result is the dummy-coding scheme of the previous chapter. Alternatively, we could use the restriction $\mu = 0$, which is equivalent to deleting the constant term from

[2] See Section 10.4 for an explanation of this surprising result.

the linear model, in which case the "effect" parameters and group means are identical: $\alpha_j = \mu_j$—a particularly simple solution.

There is, however, an advantage in selecting a restriction that produces easily interpretable parameters and estimates, and that generalizes usefully to more complex models. For these reasons, we shall impose the constraint

$$\sum_{j=1}^{m} \alpha_j = 0 \qquad [8.3]$$

Equation 8.3 is often called a *sigma constraint* or *sum-to-zero constraint*. Employing this restriction to solve for the parameters produces

$$\mu = \frac{\sum \mu_j}{m} \equiv \mu. \qquad [8.4]$$
$$\alpha_j = \mu_j - \mu.$$

The dot (in $\mu.$) indicates averaging over the range of a subscript, here over groups. The *grand* or *general mean* μ, then, is the average of the population group means, while α_j gives the difference between the mean of group j and the grand mean.[3] It is clear that, under the sigma constraint, the hypothesis of no differences in group means

$$H_0\colon \mu_1 = \mu_2 = \cdots = \mu_m$$

is equivalent to the hypothesis that all of the effect parameters are 0:

$$H_0\colon \alpha_1 = \alpha_2 = \cdots = \alpha_m = 0$$

All of this is well and good, but how can we estimate the one-way ANOVA model under the sigma constraint? One approach is to code *deviation regressors*, an alternative to the dummy-coding scheme, which (recall) implicitly imposes the constraint $\alpha_m = 0$. We require $m - 1$ deviation regressors, $D_1, D_2, \ldots, D_{m-1}$, the jth of which is coded according to the following rule:

$$D_j = \begin{cases} 1 & \text{for observations in group } j \\ -1 & \text{for observations in group } m \\ 0 & \text{for observations in all other groups} \end{cases}$$

For example, when $m = 3$,

Group	(α_1) D_1	(α_2) D_2
1	1	0
2	0	1
3	−1	−1

[3] There is a subtle distinction between $\mu.$, the mean of the group means, and the overall (i.e., unconditional) mean of Y in the population. In a real population, $\mu.$ and $E(Y)$ will generally differ if the groups have different numbers of observations. In an infinite or hypothetical population, we can speak of the grand mean but not of the overall (unconditional) mean $E(Y)$.

For ease of reference, I have shown in parentheses the parameter associated with each deviation regressor.

Writing out the equations for the group means in terms of the deviation regressors demonstrates how these regressors capture the sigma constraint on the parameters of the model:

$$\begin{aligned}
&\text{Group 1:} \quad \mu_1 = \mu + 1 \times \alpha_1 + 0 \times \alpha_2 = \mu + \alpha_1 \\
&\text{Group 2:} \quad \mu_2 = \mu + 0 \times \alpha_1 + 1 \times \alpha_2 = \mu + \alpha_2 \\
&\text{Group 3:} \quad \mu_3 = \mu - 1 \times \alpha_1 - 1 \times \alpha_2 = \mu - \alpha_1 - \alpha_2
\end{aligned}$$

The equation for the third group incorporates the sigma constraint, because $\alpha_3 = -\alpha_1 - \alpha_2$ is equivalent to $\alpha_1 + \alpha_2 + \alpha_3 = 0$.

The null hypothesis of no differences among population group means is tested by the omnibus F-statistic for the deviation-coded model: The omnibus F-statistic tests the hypothesis H_0: $\alpha_1 = \alpha_2 = 0$, which, under the sigma constraint, implies that α_3 is 0 as well.

One-way analysis of variance examines the relationship between a quantitative dependent variable and a categorical independent variable (or factor). The one-way ANOVA model $Y_{ij} = \mu + \alpha_j + \varepsilon_{ij}$ is underdetermined because it uses $m+1$ parameters to model m group means. The model can be solved, however, by placing a restriction on its parameters. Setting one of the α_j's to 0 leads to (0, 1) dummy-regressor coding. Constraining the α_j's to sum to 0 leads to (1, 0, −1) deviation-regressor coding. The two coding schemes are equivalent in that they provide the same fit to the data, producing the same regression and residual sums of squares.

Although it is often convenient to fit the one-way ANOVA model by least-squares regression, it is also possible to estimate the model and calculate sums of squares directly. The sample mean $\overline{Y}_j$ in group j is the least-squares estimator of the corresponding population mean μ_j. Estimates of μ and the α_j may therefore be written as follows (substituting estimates into Equation 8.4):

$$\begin{aligned}
M = \hat{\mu} &= \frac{\sum \overline{Y}_j}{m} = \bar{Y}. \\
A_j = \hat{\alpha}_j &= \overline{Y}_j - \bar{Y}.
\end{aligned}$$

Furthermore, the fitted Y-values are the group means:

$$\hat{Y}_{ij} = M + A_j = \bar{Y}. + (\overline{Y}_j - \bar{Y}.) = \bar{Y}_j$$

TABLE 8.1 General One-Way Analysis-of-Variance Table.

Source	*Sum of Squares*	*df*	*Mean Square*	*F*	H_0
Groups	$\sum n_j(\bar{Y}_j - \bar{Y})^2$	$m-1$	$\frac{\text{RegSS}}{m-1}$	$\frac{\text{RegMS}}{\text{RMS}}$	$\alpha_1 = \cdots = \alpha_m = 0$ $(\mu_1 = \cdots = \mu_m)$
Residuals	$\sum\sum(Y_{ij} - \bar{Y}_j)^2$	$n-m$	$\frac{\text{RSS}}{n-m}$		
Total	$\sum\sum(Y_{ij} - \bar{Y})^2$	$n-1$			

and the regression and residual sums of squares therefore take particularly simple forms in one-way analysis of variance:[4]

$$\text{RegSS} = \sum_{j=1}^{m}\sum_{i=1}^{n_j}(\hat{Y}_{ij} - \bar{Y})^2 = \sum_{j=1}^{m} n_j(\bar{Y}_j - \bar{Y})^2$$

$$\text{RSS} = \sum_{j=1}^{m}\sum_{i=1}^{n_j}(Y_{ij} - \hat{Y}_{ij})^2 = \sum\sum(Y_{ij} - \bar{Y}_j)^2$$

This information can be presented in an ANOVA table, as shown in Table 8.1.[5]

I shall use Duncan's data on the prestige of 45 U. S. occupations to illustrate one-way analysis of variance. Means, standard deviations, and frequencies for prestige within three types of occupations are as follows:

Type of Occupation	*Mean*	*Standard Deviation*	Frequency
Professional and managerial	80.44	14.11	18
White collar	36.67	11.79	6
Blue collar	22.76	18.06	21

Professional occupations therefore have the highest average level of prestige, followed by white-collar and blue-collar occupations, which are relatively closer to one another in their average levels. The standard deviation is greatest among the blue-collar occupations, and smallest among the white-collar occupations, but the differences are not very large, especially considering the small number of observations in the white-collar category.[6]

---

[4] If the n_j are unequal, as is usually the case in observational research, then the mean of the group means $\bar{Y}.$ generally differs from the overall sample mean $\bar{Y}$ of the dependent variable, for $\bar{Y} = (\sum\sum Y_{ij})/n = (\sum n_j\bar{Y}_j)/n$, while $\bar{Y}. = (\sum \bar{Y}_j)/m$. (Cf. footnote 3 for a similar point with respect to population means.)

[5] Although the notation may differ, this ANOVA table corresponds to the usual treatment in introductory statistics texts of one-way analysis of variance. It is common to call the regression sum of squares in one-way ANOVA "the between-group sum of squares," and the residual sum of squares "the within-group sum of squares."

[6] The assumption of constant error variance implies that the *population* variances should be the same in the several groups. See Section 12.4.2 for further discussion and a test for nonconstant variance in ANOVA.

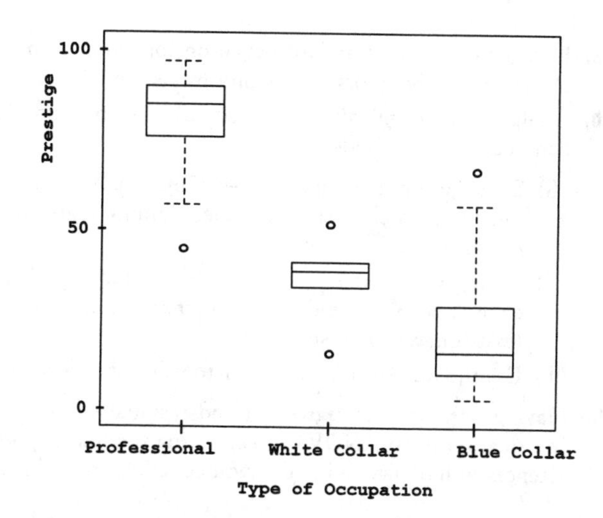

Figure 8.1. Parallel boxplots for occupational prestige by type of occupation.
Source of Data: Duncan (1961).

The parallel boxplots in Figure 8.1 reveal, however, that both the means and the standard deviations are a bit misleading: Because prestige is a percentage, bounded by 0 and 100, the prestige scores for the professional and managerial group, which are very high on average, are negatively skewed; similarly, the scores for the blue-collar group, which are relatively low, are positively skewed. The white-collar group has a much smaller hinge spread than the other two groups, but two outliers, one on the low end and one on the high end, inflate the standard deviation for this group.

The one-way ANOVA for the Duncan data is

Source	*Sum of Squares*	*df*	*Mean Square*	F	*p*
Groups	33,090.	2	16,545.	65.57	$\ll .0001$
Residuals	10,598.	42	252.33		
Total	43,688.	44			

We have, therefore, very strong evidence against the null hypothesis of no difference in average level of prestige across the occupational types. Occupational types account for three-quarters of the variation in prestige among these occupations ($R^2 = 33{,}090/43{,}688 = 0.757$), but the variation in prestige within occupational types is still substantial ($S_E = \sqrt{252.33} = 15.88$).

EXERCISES

8.1 Perform a one-way analysis of variance of geographic mobility by region, employing Angell's data on 43 U.S. cities, given in Table 2.3 and `angell.dat`.

(a) Find the mean and standard deviation for mobility in each of the four regions. Draw parallel boxplots of mobility by region. Comment on the results.

(b) Confirm that identical sums of squares are produced by the following three computational methods:

(i) Set $\alpha_4 = 0$ in the analysis-of-variance model; use the 0/1 coding scheme of the previous chapter to produce dummy regressors; and fit the model by least-squares regression.

(ii) Set $\alpha_4 = -(\alpha_1 + \alpha_2 + \alpha_3)$; use the $1, 0, -1$ coding scheme of the current chapter to produce deviation regressors; and fit the model to the data by least-squares regression.

(iii) Use the formulas in Table 8.1 to calculate the sums of squares directly.

(c) Having obtained the regression and residual sums of squares, summarize this information in an ANOVA table, testing the null hypothesis of no regional differences in mobility. What can you conclude from this test?

8.2 *The usual t-statistic for testing a difference between the means of two independently sampled groups, under the assumptions of normality and equal group variances, is

$$t_0 = \frac{\bar{Y}_1 - \bar{Y}_2}{\widehat{\text{SE}}(\bar{Y}_1 - \bar{Y}_2)}$$

where

$$\widehat{\text{SE}}(\bar{Y}_1 - \bar{Y}_2) = S\sqrt{\frac{1}{n_1} + \frac{1}{n_2}}$$

$$S^2 = \frac{\sum_{j=1}^{n_1}(Y_{i1} - \bar{Y}_1)^2 + \sum_{j=1}^{n_2}(Y_{i2} - \bar{Y}_2)^2}{n_1 + n_2 - 2}$$

Here, $\bar{Y}_1$ and $\bar{Y}_2$ are the means of the two groups; n_1 and n_2 are the numbers of observations in the groups; and Y_{i1} and Y_{i2} are the observations themselves. Let F_0 be the one-way ANOVA F-statistic for testing the null hypothesis H_0: $\mu_1 = \mu_2$. Prove that $t_0^2 = F_0$ and that, consequently, the two tests are equivalent.

8.2 Two-Way Analysis of Variance

The inclusion of a second factor permits us to model and test partial relationships, as well as to introduce interactions. Most issues pertaining to analysis of variance can be developed for the two-factor "design." Before immersing ourselves in the details of model specification and hypothesis testing for two-way analysis of variance, however, it is useful to step back and consider the patterns of relationship that can occur when a quantitative dependent variable is classified by two factors.

8.2.1 Patterns of Means in the Two-Way Classification

So as not to confuse ourselves with issues of estimation, we shall imagine at the outset that we have access to population means. Notation for the two-way classification is shown in the following table:

	C_1	C_2	$\cdots$	C_c	
R_1	μ_{11}	μ_{12}	$\cdots$	μ_{1c}	$\mu_{1\cdot}$
R_2	μ_{21}	μ_{22}	$\cdots$	μ_{2c}	$\mu_{2\cdot}$
$\vdots$	$\vdots$	$\vdots$		$\vdots$	$\vdots$
R_r	μ_{r1}	μ_{r2}	$\cdots$	μ_{rc}	$\mu_{r\cdot}$
	$\mu_{\cdot 1}$	$\mu_{\cdot 2}$	$\cdots$	$\mu_{\cdot c}$	$\mu_{\cdot\cdot}$

The two independent variables, R and C (for "rows" and "columns" of the table of means), have r and c categories, respectively. The factor categories are denoted R_j and C_k.

Within each *cell* of the design—that is, for each combination of categories $\{R_j, C_k\}$ of the two factors—there is a population cell mean μ_{jk} for the dependent variable. Extending the dot notation introduced in the previous section,

$$\mu_{j\cdot} \equiv \frac{\sum_{k=1}^{c} \mu_{jk}}{c}$$

is the *marginal mean* of the dependent variable in row j;

$$\mu_{\cdot k} \equiv \frac{\sum_{j=1}^{r} \mu_{jk}}{r}$$

is the marginal mean in column k; and

$$\mu_{\cdot\cdot} \equiv \frac{\sum_j \sum_k \mu_{jk}}{r \times c} = \frac{\sum_j \mu_{j\cdot}}{r} = \frac{\sum_k \mu_{\cdot k}}{c}$$

is the grand mean.

If R and C do not interact in determining the dependent variable, then the partial relationship between each factor and Y does not depend on the category at which the other factor is "held constant." The difference in cell means $\mu_{jk} - \mu_{j'k}$ across two categories of R (i.e., categories R_j and $R_{j'}$) is constant across all of the categories of C—that is, this difference is the same for all $k = 1, 2, \ldots, c$. Consequently, the difference in cell means across rows is equal to the corresponding difference in row marginal means:

$$\mu_{jk} - \mu_{j'k} = \mu_{jk'} - \mu_{j'k'} = \mu_{j\cdot} - \mu_{j'\cdot} \quad \text{for all } j, j' \text{ and } k, k'$$

This pattern is illustrated in Figure 8.2(a) for the simple case where $r = c = 2$. Interaction—where the row difference $\mu_{1k} - \mu_{2k}$ changes across columns $k = 1, 2$—is illustrated in Figure 8.2(b). Notice that no interaction implies parallel "profiles" of cell means. Parallel profiles also imply that the column difference $\mu_{j1} - \mu_{j2}$ for categories C_1 and C_2 is constant across rows $j = 1, 2$, and is equal to the difference in column marginal means $\mu_{\cdot 1} - \mu_{\cdot 2}$. As we discovered

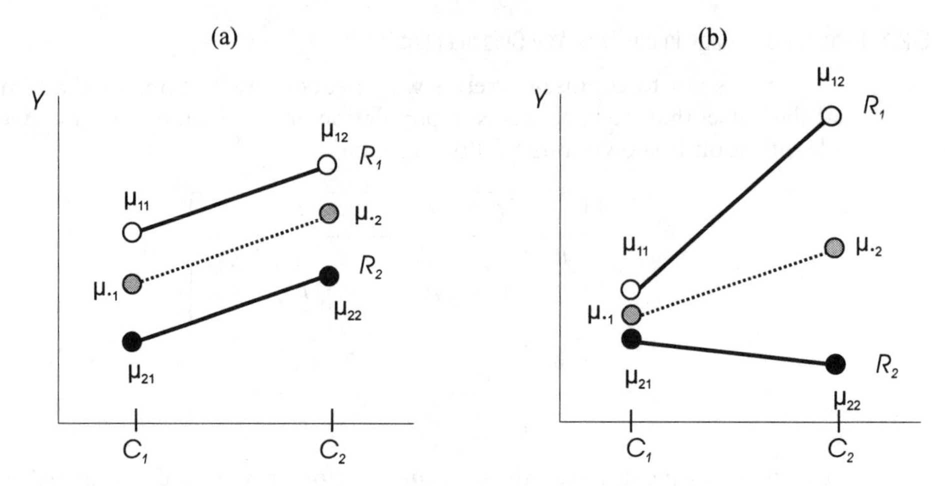

Figure 8.2. Interaction in the two-way classification. In (*a*), the parallel profiles of means (given by the white and black dots connected by solid lines) indicate that *R* and *C* do not interact in affecting *Y*. The *R*-effect—that is, the difference between the two profiles—is the same at both C_1 and C_2. Likewise, the *C*-effect—that is, the rise in the line from C_1 to C_2—is the same for both profiles. In (*b*), the *R*-effect differs at the two categories of *C*, and the *C*-effect differs at the two categories of *R*: That is, *R* and *C* interact in affecting *Y*. In both graphs, the column marginal means $\mu_{\cdot 1}$ and $\mu_{\cdot 2}$ are shown as averages of the cell means in each column (represented by the gray dots connected by broken lines).

in Chapter 7, interaction is a symmetric concept: If *R* interacts with *C*, then *C* interacts with *R*. When interactions are absent, the partial effect of each factor—the factor's *main effect*—is given by differences in the population marginal means.

Several patterns of relationship in the two-way classification, all showing no interaction, are graphed in Figure 8.3. Plots of means, incidentally, not only serve to clarify the ideas underlying analysis of variance, but are also a useful tool for summarizing and presenting data. Indeed, it is very difficult to inspect, understand, and interpret patterns of means in ANOVA *without* plotting the means. In the illustrations, factor *C* has three categories or *levels*, which are marked off along the horizontal axis. Because *C* is a qualitative variable, the order of its categories and the spacing between them are arbitrary.[7] Factor *R* has two categories. The six cell means are plotted as points, connected by lines (called *profiles*) according to the levels of factor *R*. The separation between the lines at level C_k (where *k* is 1, 2, or 3) represents the difference $\mu_{1k} - \mu_{2k}$. As noted above, when there is no interaction, therefore, the separation between the profiles is constant and the profiles themselves are parallel.

In Figure 8.3(*a*), both *R* and *C* have nonzero main effects. In Figure 8.3(b), the differences $\mu_{1k} - \mu_{2k} = \mu_{1\cdot} - \mu_{2\cdot}$ are 0, and, consequently, the *R* main effects are nil. In Figure 8.3(*c*), the *C* main effects are nil, because the differences $\mu_{jk} - \mu_{jk'} = \mu_{\cdot k} - \mu_{\cdot k'}$ are all 0. Finally, in Figure 8.3(*d*), both sets of main effects are nil.

[7] Analysis of variance is also useful when the levels of a factor are ordered ("low," "medium," "high," for example) or even discrete and quantitative (e.g., number of bedrooms for apartment dwellers—0, 1, 2, 3, 4), but, in general, I shall assume that factors are simply nominal.

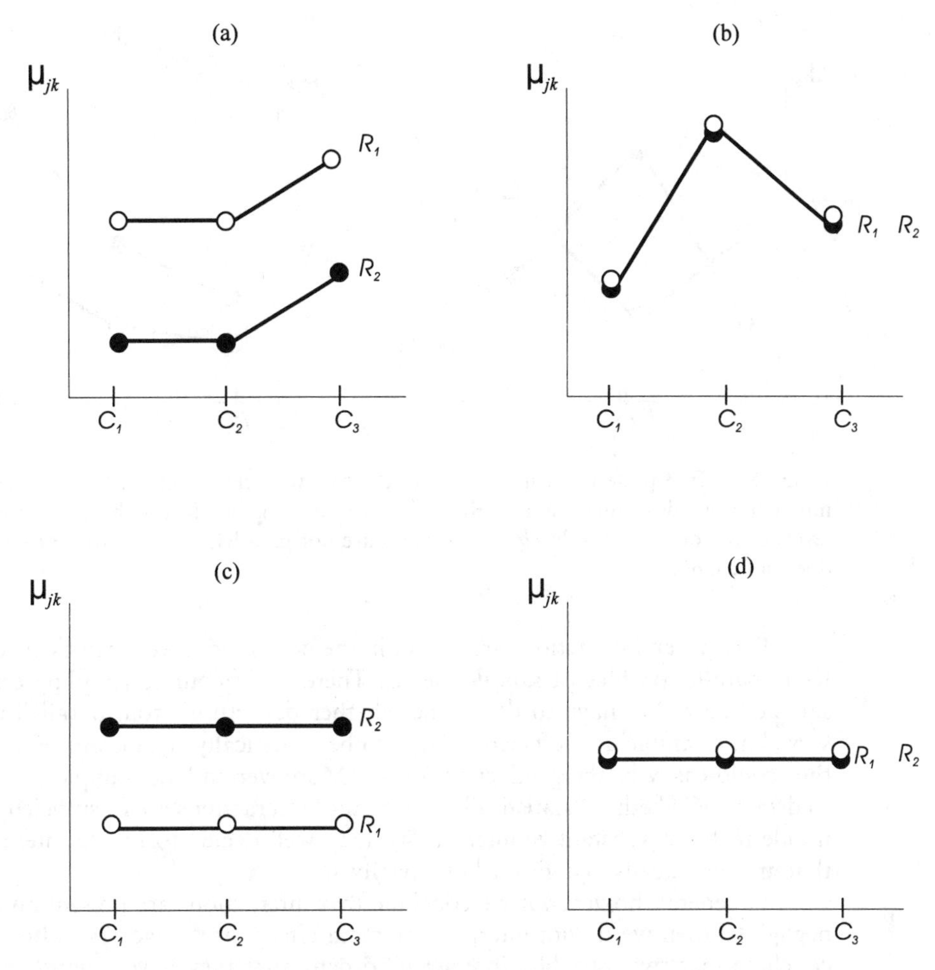

Figure 8.3. Several patterns of relationship in the two-way classification. In all of these cases, R and C do not interact. (a) Both R and C main effects. (b) C main effects (R main effects nil). (c) R main effects (C main effects nil). (d) No effects (both R and C main effects nil).

Figure 8.4 shows two different patterns of interactions. It is clear from the previous discussion that R and C interact when the profiles of means are not parallel—that is, when the row differences $\mu_{jk} - \mu_{j'k}$ change across the categories of the column factor, or, equivalently, when the column differences $\mu_{jk} - \mu_{jk'}$ change across the categories of the row factor. In Figure 8.4(a), the interaction is dramatic: The mean for level R_1 is above the mean for R_2 at levels C_1 and C_3, but at level C_2, the mean for R_2 is substantially above the mean for R_1. Likewise, the means for the three categories of C are ordered differently within R_1 and R_2. Interaction of this sort is sometimes called *disordinal.* In Figure 8.4 (b), in contrast, the profile for R_2 is above that for R_1 across all three categories of C, although the separation between the profiles of means changes. This less dramatic form of interaction can sometimes be transformed away (e.g., by taking logs).

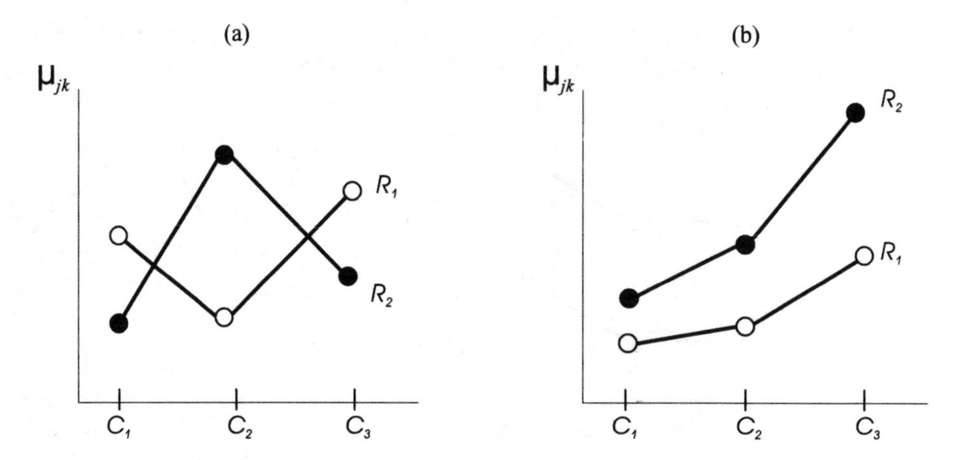

Figure 8.4. Two patterns of interaction in the two-way classification. In (*a*), the interaction is "disordinal" in that the order of means for one factor changes across the levels of the other factor. In (*b*), the profiles are not parallel, but the order of means does not change.

Even when interactions are absent in the population, we cannot expect perfectly parallel profiles of *sample* means: There is, of course, sampling error in sampled data. We have to determine whether departures from parallelism observed in a sample are sufficiently large to be statistically significant, or whether they could easily be the product of chance. Moreover, in large samples, we want to determine whether "statistically significant" interactions are of sufficient magnitude to be of substantive interest. We may well decide to ignore interactions that are statistically significant but trivially small.

In general, however, if we conclude that interactions are present and non-negligible, then we do not interpret the main effects of the factors—after all, to conclude that two variables interact is to deny that they have *separate* effects. This point is a reflection of the principle of marginality, introduced in Chapter 7 in the context of dummy-variable regression: Here, the R and C main effects are marginal to the RC interaction.[8]

> Two factors interact when the profiles of population means are not parallel; when the profiles of means are parallel, the effects of the two factors are additive.

Table 8.2 shows means, standard deviations, and cell frequencies for data from a social-psychological experiment reported by Moore and Krupat (1971);

[8] In cases of disordinal interaction, such as in Figure 8.4(a), interpreting main effects is clearly misleading, because it makes no sense to average over levels of one factor to examine the effect of the other. In cases such as Figure 8.4(b), however, there is some sense to examining the marginal means for one factor averaged over levels of the other, despite the interaction.

TABLE 8.2 Conformity by Authoritarianism and Partner's Status, for Moore and Krupat's (1971) Experiment. Each cell shows (from top to bottom) the Conformity mean, standard deviation, and cell frequency

Partner's Status		*Authoritarianism*		
		Low	*Medium*	*High*
Low	$\bar{Y}_{jk}$	8.900	7.250	12.63
	s_{jk}	2.644	3.948	7.347
	n_{jk}	10	4	8
High	$\bar{Y}_{jk}$	17.40	14.27	11.86
	s_{jk}	4.506	3.952	3.934
	n_{jk}	5	11	7

the data themselves appear in Table 8.3.[9] The experiment was designed to determine how the relationship between conformity and social status is influenced by "authoritarianism." The subjects in the experiment were asked to make perceptual judgments of stimuli that were intrinsically ambiguous. Upon forming an initial judgment, the subjects were presented with the judgment of another individual (their "partner") who was ostensibly participating in the experiment; the subjects were then asked for a final judgment. In fact, the partner's judgments were manipulated by the experimenters so that subjects were faced with nearly continuous disagreement.

The measure of conformity employed in the study was the number of times in 40 critical trials that subjects altered their judgments in response to disagreement. This measure is a disguised proportion. The 45 university student subjects in the study were randomly assigned to two experimental conditions: In one condition, the partner was described as of relatively high social status (a "physician"); in the other condition, the partner was described as of relatively low status (a "postal clerk").

A standard authoritarianism scale (the "*F*-scale") was administered to the subjects after the experiment was completed. This procedure was dictated by practical considerations, but it raises the possibility that authoritarianism scores were inadvertently influenced by the experimental manipulation of the partner's status. Note in Table 8.2 that the authors divided the authoritarianism scores into three categories.[10] A chi-square test of independence for the condition by authoritarianism frequency table (shown in Table 8.2) produces a p-value of .08, indicating that there is some ground for believing that the status manipulation affected the authoritarianism scores of the subjects.

[9] The data were generously made available by James Moore, Department of Sociology, York University.

[10] Moore and Krupat categorized authoritarianism *separately* within each condition. This approach is not strictly justified, but it serves to produce nearly equal cell frequencies—required by the method of computation employed by the authors—for the six combinations of partner's status and authoritarianism, and yields results similar to those reported here. It may have occurred to you that the dummy-regression procedures of the previous chapter are applicable here and do not require the arbitrary categorization of authoritiarianism. This analysis appears in Section 8.4. Moore and Krupat do report the difference between slopes for the within-condition regressions of conformity on authoritarianism.

TABLE 8.3 Data From Moore and Krupat's Conformity Experiment

	Partner's Status			
	Low		*High*	
Authoritarianism	*Conformity*	*F-Scale*	*Conformity*	*F-scale*
Low	8	37	21	16
	7	20	13	15
	10	36	15	30
	6	18	12	22
	13	31	16	35
	12	36		
	7	28		
	6	28		
	8	17		
	12	35		
Medium	12	51	12	41
	4	44	17	48
	4	42	14	42
	9	43	16	45
			20	44
			8	42
			14	42
			17	41
			7	50
			17	39
			15	44
High	4	57	19	68
	8	65	9	63
	13	56	7	54
	9	65	11	59
	24	57	14	52
	7	61	13	57
	23	57	10	52
	13	55		

Because of the conceptual-rigidity component of authoritarianism, Moore and Krupat expected that low-authoritarian subjects would be *more* responsive than high-authoritarian subjects to the social status of their partner. In other words, authoritarianism and partner's status are expected to interact—in a particular manner—in determining conformity. The cell means, graphed in Figure 8.5, appear to confirm the experimenters' expectations.

I note, for future reference, that the standard deviation of conformity in one cell (high-authoritarian, low-status partner) is appreciably larger than in the others. Upon graphical inspection of the data (Figure 8.6), it is clear that the relatively large dispersion in this cell is due to two subjects, numbers 16 and 19, who have atypically high conformity scores of 24 and 23.

8.2.2 The Two-Way ANOVA Model

Because interpretation of results in two-way analysis of variance depends crucially on the presence or absence of interaction, our first concern is to test

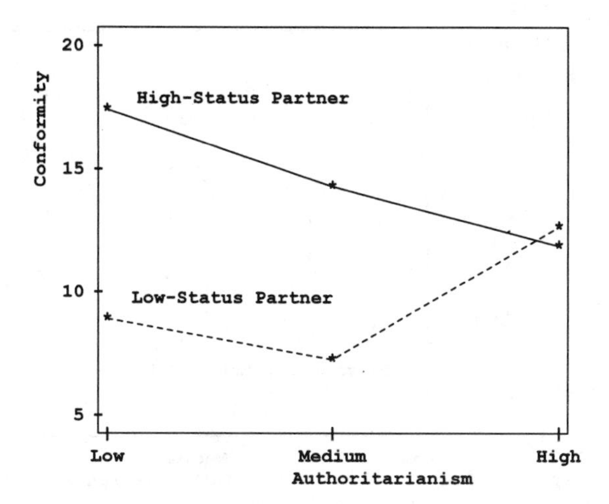

Figure 8.5. Cell means for the Moore and Krupat conformity experiment.

the null hypothesis of no interaction. Based on the discussion in the previous section, this hypothesis can be expressed in terms of the cell means:

$$H_0\colon\ \mu_{jk} - \mu_{j'k} = \mu_{jk'} - \mu_{j'k'} \quad \text{for all } j, j' \text{ and } k, k' \qquad [8.5]$$

In words: The row effects are the same within all levels of the column factor. By rearranging the terms in Equation 8.5, we can write the null hypothesis in the following manner:

$$H_0\colon\ \mu_{jk} - \mu_{jk'} = \mu_{j'k} - \mu_{j'k'} \quad \text{for all } j, j' \text{ and } k, k' \qquad [8.6]$$

That is, the column effects are invariant across rows. Once more, we see the symmetry of the concept of interaction.

It is convenient, following the presentation in the previous section, to express hypotheses concerning main effects in terms of the marginal means. Thus, for the row classification we have the null hypothesis

$$H_0\colon\ \mu_{1\cdot} = \mu_{2\cdot} = \cdots = \mu_{r\cdot} \qquad [8.7]$$

and for the column classification

$$H_0\colon\ \mu_{\cdot 1} = \mu_{\cdot 2} = \cdots = \mu_{\cdot c} \qquad [8.8]$$

The main-effect null hypotheses (Equations 8.7 and 8.8) are testable whether interactions are present or absent, but these hypotheses are generally of interest only when the interactions are nil.

The two-way ANOVA model, suitably defined, provides a convenient means for testing the hypotheses (Equations 8.5, 8.7, and 8.8). The model is

$$Y_{ijk} = \mu + \alpha_j + \beta_k + \gamma_{jk} + \varepsilon_{ijk} \qquad [8.9]$$

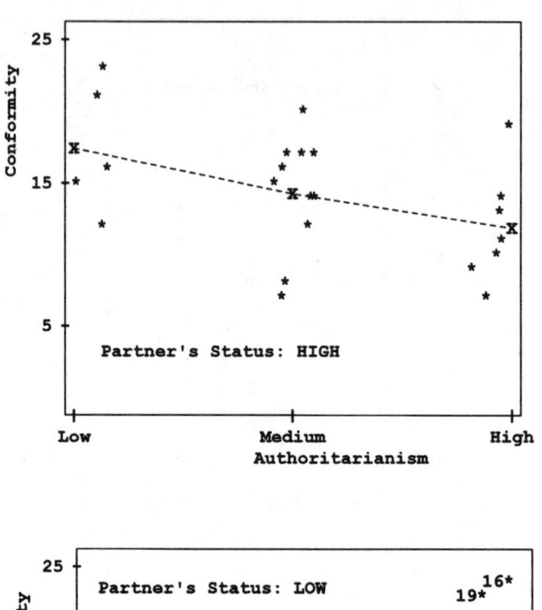

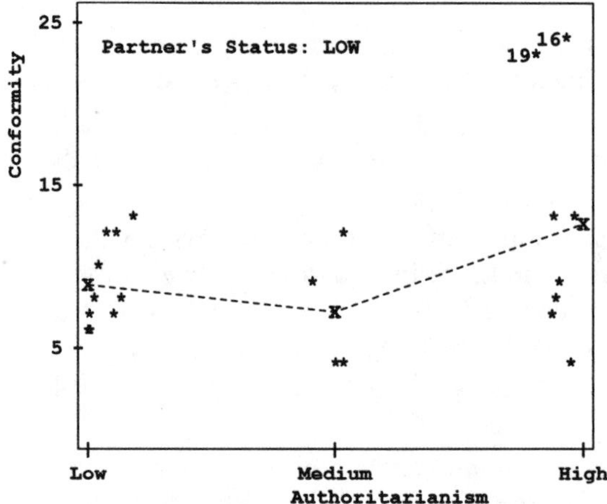

Figure 8.6. Data from Moore and Krupat's experiment on conformity and authoritarianism. The horizontal coordinates of the points have been jittered to separate overlapping points. The means are plotted as X's, with the profiles connected by broken lines. Note the two unusual points, for observations 16 and 19, in the high-authoritarian, low-status-partner group.

where Y_{ijk} is the ith observation in row j, column k of the RC table; μ is the general mean of Y; the α_j and β_k are main-effect parameters, for row effects and column effects, respectively; the γ_{jk} are interaction parameters; and the ε_{ijk} are errors satisfying the usual linear-model assumptions. Taking expectations, Equation 8.9 becomes

$$\mu_{jk} \equiv E(Y_{ijk}) = \mu + \alpha_j + \beta_k + \gamma_{jk} \qquad [8.10]$$

Because there are $r \times c$ population cell means and $1 + r + c + (r \times c)$ parameters in Equation 8.10, the parameters of the model are not uniquely determined by the cell means. By reasoning that is familiar from Section 8.1 on one-way ANOVA, the indeterminacy of Equation 8.10 can be overcome by imposing $1+r+c$ independent restrictions on its parameters. Although—from one point of view—any restrictions will do, it is convenient to select restrictions that make it simple to test the hypotheses of interest.[11]

With this purpose in mind, we specify the following sigma constraints on the model parameters:

$$\sum_{j=1}^{r} \alpha_j = 0 \qquad [8.11]$$

$$\sum_{k=1}^{c} \beta_k = 0$$

$$\sum_{j=1}^{r} \gamma_{jk} = 0 \quad \text{for all } k = 1, \ldots, c$$

$$\sum_{k=1}^{c} \gamma_{jk} = 0 \quad \text{for all } j = 1, \ldots, r$$

At first glance, it seems as if we have specified too many constraints, for Equation 8.11 defines $1+1+c+r$ restrictions. One of the restrictions on the interactions is redundant, however.[12] In shorthand form, the sigma constraints specify that each set of parameters sums to 0 over each of its coordinates.

The constraints produce the following solution for model parameters in terms of population cell and marginal means:

$$\begin{aligned} \mu &= \mu_{\cdot\cdot} \qquad [8.12] \\ \alpha_j &= \mu_{j\cdot} - \mu_{\cdot\cdot} \\ \beta_k &= \mu_{\cdot k} - \mu_{\cdot\cdot} \\ \gamma_{jk} &= \mu_{jk} - \mu - \alpha_j - \beta_k \\ &= \mu_{jk} - \mu_{j\cdot} - \mu_{\cdot k} + \mu_{\cdot\cdot} \end{aligned}$$

The hypothesis of no row main effects (Equation 8.7) is, therefore, equivalent to H_0: all $\alpha_j = 0$, for under this hypothesis

$$\mu_{1\cdot} = \mu_{2\cdot} = \cdots = \mu_{r\cdot} = \mu_{\cdot\cdot}$$

Likewise, the hypothesis of no column main effects (Equation 8.8) is equivalent to H_0: all $\beta_k = 0$, because then

$$\mu_{\cdot 1} = \mu_{\cdot 2} = \cdots = \mu_{\cdot c} = \mu_{\cdot\cdot}$$

---

[11] See Section 10.4 for an elaboration of this point.

[12] See Exercise 8.3.

Finally, it is not difficult to show that the hypothesis of no interactions (given in Equation 8.5 or 8.6) is equivalent to[13] H_0: all $\gamma_{jk} = 0$.

8.2.3 Fitting the Two-Way ANOVA Model to Data

Because the least-squares estimator of μ_{jk} is the sample cell mean

$$\bar{Y}_{jk} = \frac{\sum_{i=1}^{n_{jk}} Y_{ijk}}{n_{jk}}$$

least-squares estimators of the constrained model parameters follow immediately from Equation 8.12:

$$\begin{aligned}
M \equiv \hat{\mu} &= \bar{Y}.. = \frac{\sum\sum \bar{Y}_{jk}}{r \times c} \\
A_j \equiv \hat{\alpha}_j &= \bar{Y}_{j.} - \bar{Y}.. = \frac{\sum_k \bar{Y}_{jk}}{c} - \bar{Y}.. \\
B_k \equiv \hat{\beta}_k &= \bar{Y}_{\cdot k} - \bar{Y}.. = \frac{\sum_j \bar{Y}_{jk}}{r} - \bar{Y}.. \\
C_{jk} \equiv \hat{\gamma}_{jk} &= \bar{Y}_{jk} - \bar{Y}_{j.} - \bar{Y}_{\cdot k} + \bar{Y}..
\end{aligned}$$

The residuals are just the deviations of the observations from their cell means, because the fitted values are the cell means:

$$\begin{aligned}
E_{ijk} &= Y_{ijk} - (M + A_j + B_k + C_{jk}) \\
&= Y_{ijk} - \bar{Y}_{jk}
\end{aligned}$$

In testing hypotheses about sets of model parameters, however, we require incremental sums of squares for each set, and there is no way of calculating these sums of squares directly.[14] As in one-way analysis of variance, the restrictions on the two-way ANOVA model can be used to produce deviation-coded regressors. Incremental sums of squares can then be calculated in the usual manner. To illustrate this procedure, we shall first examine a two-row × three-column classification. The extension to the general $r \times c$ classification is straightforward, and is described below.

In light of the restriction $\alpha_1 + \alpha_2 = 0$ on the row effects of the 2×3 classification, α_2 can be deleted from the model, substituting $-\alpha_1$. Similarly, because $\beta_1 + \beta_2 + \beta_3 = 0$, the column main effect β_3 can be replaced by $-\beta_1 - \beta_2$. More generally, $-\sum_{j=1}^{r-1} \alpha_j$ replaces α_r, and $-\sum_{k=1}^{c-1} \beta_k$ replaces β_c. Because there are, then, $r - 1$ independent α_j parameters and $c - 1$ independent β_k parameters, the degrees of freedom for row and column main effects are, respectively, $r - 1$ and $c - 1$.

---

[13] See Exercise 8.4.

[14] An exception occurs when all of the cell frequencies are equal—see Section 8.2.5.

The interactions in the 2×3 classification satisfy the following constraints:[15]

$$\gamma_{11} + \gamma_{12} + \gamma_{13} = 0$$
$$\gamma_{21} + \gamma_{22} + \gamma_{23} = 0$$
$$\gamma_{11} + \gamma_{21} = 0$$
$$\gamma_{12} + \gamma_{22} = 0$$
$$\gamma_{13} + \gamma_{23} = 0$$

We can, as a consequence, delete all of the interaction parameters except γ_{11} and γ_{12}, substituting for the remaining four parameters in the following manner:

$$\gamma_{13} = -\gamma_{11} - \gamma_{12}$$
$$\gamma_{21} = -\gamma_{11}$$
$$\gamma_{22} = -\gamma_{12}$$
$$\gamma_{23} = -\gamma_{13} = \gamma_{11} + \gamma_{12}$$

More generally, we can write all $r \times c$ interaction parameters in terms of $(r-1)(c-1)$ of the γ_{jk}'s, and there are, therefore, $(r-1)(c-1)$ degrees of freedom for interaction.

These observations lead to the following coding of regressors for the 2×3 classification:

Cell		(α_1)	(β_1)	(β_2)	(γ_{11})	(γ_{12})
Row	*Column*	R_1	C_1	C_2	R_1C_1	R_1C_2
1	1	1	1	0	1	0
1	2	1	0	1	0	1
1	3	1	−1	−1	−1	−1
2	1	−1	1	0	−1	0
2	2	−1	0	1	0	−1
2	3	−1	−1	−1	1	1

Here, R_1 is the regressor for the row main effects; C_1 and C_2 are the regressors for the column main effects; and R_1C_1 and R_1C_2 are the interaction regressors. The notation for the interaction regressors is suggestive of multiplication, and, in fact, we can see that R_1C_1 is the product of R_1 and C_1, and that R_1C_2 is the product of R_1 and C_2.

[15] Recall that although there are five such constraints, the fifth follows from the first four—you may want to show this—and there are therefore only four *independent* constraints on the interaction parameters.

I have constructed these regressors to reflect the constraints on the model, but they can also be coded mechanically by applying these rules:

1. There are $r - 1$ regressors for the row main effects; the jth such regressor, R_j, is coded according to the deviation-coding scheme:

$$R_{ij} = \begin{cases} 1 & \text{if observation } i \text{ is in row } j \\ -1 & \text{if observation } i \text{ is in row } r \text{ (the last row)} \\ 0 & \text{if observation } i \text{ is in any other row} \end{cases}$$

2. There are $c - 1$ regressors for the column main effects; the kth such regressor, C_k, is coded according to the deviation-coding scheme:

$$C_{ik} = \begin{cases} 1 & \text{if observation } i \text{ is in column } k \\ -1 & \text{if observation } i \text{ is in column } c \text{ (the last column)} \\ 0 & \text{if observation } i \text{ is in any other column} \end{cases}$$

3. There are $(r - 1)(c - 1)$ regressors for the RC interactions. These interaction regressors consist of all pairwise products of the $r - 1$ main-effect regressors for rows and $c - 1$ main-effect regressors for columns.

The two-way analysis-of-variance model $Y_{ijk} = \mu + \alpha_j + \beta_k + \gamma_{jk} + \varepsilon_{ijk}$ incorporates the main effects and interactions of two factors. This model is overparameterized, but it may be fit to data by placing suitable restrictions on its parameters. A convenient set of restrictions is provided by sigma constraints, specifying that each set of parameters (α_j, β_k, and γ_{jk}) sums to 0 over each of its coordinates. As in one-way ANOVA, sigma constraints lead to deviation-coded regressors.

8.2.4 Testing Hypotheses in Two-Way ANOVA

I have specified constraints on the two-way ANOVA model so that testing hypotheses about the parameters of the constrained model is equivalent to testing hypotheses about interactions and main effects of the two factors. Tests for interactions and main effects can be constructed by the incremental-sum-of-squares approach.

For ease of reference, I shall write $SS(\alpha, \beta, \gamma)$ to denote the regression sum of squares for the full model, which includes both sets of main effects and the interactions. The regression sums of squares for other models are similarly represented. For example, for the no-interaction model, we have $SS(\alpha, \beta)$; and for the model that omits the column main-effect regressors, we have $SS(\alpha, \gamma)$. This last model violates the principle of marginality, because it includes the interaction regressors but omits the column main effects. Nevertheless, as I shall explain presently, the model plays a role in constructing the incremental sum of squares for testing the column main effects.

As usual, incremental sums of squares are given by differences between the regression sums of squares for alternative models. I shall use the following notation for incremental sums of squares in ANOVA:[16]

$$\begin{aligned} SS(\gamma|\alpha, \beta) &= SS(\alpha, \beta, \gamma) - SS(\alpha, \beta) \\ SS(\alpha|\beta, \gamma) &= SS(\alpha, \beta, \gamma) - SS(\beta, \gamma) \\ SS(\beta|\alpha, \gamma) &= SS(\alpha, \beta, \gamma) - SS(\alpha, \gamma) \\ SS(\alpha|\beta) &= SS(\alpha, \beta) - SS(\beta) \\ SS(\beta|\alpha) &= SS(\alpha, \beta) - SS(\alpha) \end{aligned}$$

We read $SS(\gamma|\alpha, \beta)$, for example, as "the sum of squares for interaction *after* the main effects," and $SS(\alpha|\beta)$ as "the sum of squares for the row main effects *after* the column main effects and *ignoring* the interactions." The residual sum of squares is

$$\begin{aligned} RSS &= \sum\sum\sum E_i^2 \\ &= \sum\sum\sum (Y_{ijk} - \bar{Y}_{jk})^2 \\ &= TSS - SS(\alpha, \beta, \gamma) \end{aligned}$$

The incremental sum of squares for interaction, $SS(\gamma|\alpha, \beta)$, is appropriate for testing the null hypothesis of no interaction, H_0: all $\gamma_{jk} = 0$. In the presence of interactions, we can use $SS(\alpha|\beta, \gamma)$ and $SS(\beta|\alpha, \gamma)$ to test hypotheses concerning main effects (i.e., differences among row and column marginal means), but—as I have explained—these hypotheses are usually not of interest when the interactions are important.

In the *absence* of interactions, $SS(\alpha|\beta)$ and $SS(\beta|\alpha)$ can be used to test for main effects, but the use of $SS(\alpha|\beta, \gamma)$ and $SS(\beta|\alpha, \gamma)$ is also appropriate. If, however, interactions are *present*, then F-tests based on $SS(\alpha|\beta)$ and $SS(\beta|\alpha)$ *do not* test the main-effect null hypotheses H_0: all $\alpha_j = 0$ and H_0: all $\beta_k = 0$; instead, the interaction parameters become implicated in these tests. These remarks are summarized in Tables 8.4 and 8.5.

Certain authors (e.g., Nelder, 1976, 1977) prefer main-effects tests based on $SS(\alpha|\beta)$ and $SS(\beta|\alpha)$ because, if interactions are absent, tests based on

---

[16] You may encounter variations of the SS notation. One common approach (used, for example, in Searle, 1971), is to include the grand mean μ in the arguments to the sum-of-squares function, and to let $R(\cdot)$ denote the "raw" (rather than mean-deviation) sum of squares. Thus, in this scheme, $R(\mu, \alpha, \beta) = \sum\sum\sum \hat{Y}_{ijk}^2$ is the raw sum of squares for the no-interaction model, while

$$\begin{aligned} R(\alpha, \beta|\mu) &= R(\mu, \alpha, \beta) - R(\mu) \\ &= \sum\sum\sum (\hat{Y}_{ijk} - \bar{Y})^2 \\ &= SS(\alpha, \beta) \end{aligned}$$

is the mean-deviation explained sum of squares for the same model. (The $\hat{Y}_{ijk}$ are the least-squares fitted values from the no-interaction model.)

TABLE 8.4 Two-Way Analysis of Variance, Showing Alternative Tests for Row and Column Main Effects. Each incremental F-test is formulated by dividing an effect mean-square by the residual mean-square

Source	*df*	*Sum of Squares*
R	$r-1$	$SS(\alpha \mid \beta, \gamma)$
		$SS(\alpha \mid \beta)$
C	$c-1$	$SS(\beta \mid \alpha, \gamma)$
		$SS(\beta \mid \alpha)$
RC	$(r-1)(c-1)$	$SS(\gamma \mid \alpha, \beta)$
Residuals	$n-rc$	$TSS - SS(\alpha, \beta, \gamma)$
Total	$n-1$	TSS

these sums of squares are more powerful than those based on $SS(\alpha \mid \beta, \gamma)$ and $SS(\beta \mid \alpha, \gamma)$. Other authors (e.g., Hocking and Speed, 1975) prefer $SS(\alpha \mid \beta, \gamma)$ and $SS(\beta \mid \alpha, \gamma)$ because, in the presence of interactions, tests based on these sums of squares have a straightforward (if usually uninteresting) interpretation. I believe that either approach is reasonable.[17]

It is important to understand, however, that while $SS(\alpha)$ and $SS(\beta)$ are useful as building blocks of $SS(\alpha \mid \beta)$ and $SS(\beta \mid \alpha)$, it is, in general, *inappropriate* to use $SS(\alpha)$ and $SS(\beta)$ to test hypotheses about the R and C main effects: Each of these sums of squares depends on the other set of main effects (and the interactions, if they are present). A main effect is a partial effect, so we need to control for rows in assessing the column main effects, and vice versa.

Testing hypotheses about the sigma-constrained parameters is equivalent to testing interaction-effect and main-effect hypotheses about cell and marginal means. There are two reasonable procedures for testing main-effect hypotheses in two-way ANOVA: Tests based on $SS(\alpha \mid \beta, \gamma)$ and $SS(\beta \mid \alpha, \gamma)$ employ models that violate the principle of marginality, but are valid whether or not interactions are present. Tests based on $SS(\alpha \mid \beta)$ and $SS(\beta \mid \alpha)$ conform to the principle of marginality, but are valid only if interactions are absent.

For the Moore and Krupat conformity data, factor R is partner's status and factor C is authoritarianism. Sums of squares for various models fit to the

[17] In the SAS statistical computer package, $SS(\alpha \mid \beta)$ and $SS(\beta \mid \alpha)$ are called "Type II" sums of squares, while $SS(\alpha \mid \beta, \gamma)$ and $SS(\beta \mid \alpha, \gamma)$ are called "Type III" sums of squares.

TABLE 8.5 Two-Way Analysis of Variance, Showing Alternative Tests for Row and Column Main Effects. The hypothesis tested by each such F-test Is expressed both in terms of constrained model parameters and in terms of cell or marginal means

Sum of Squares	H_0
$SS(\alpha\|\beta, \gamma)$	all $\alpha_j = 0$ $(\mu_{j\cdot} = \mu_{j'\cdot})$
$SS(\alpha\|\beta)$	all $\alpha_j = 0 \mid$ all $\gamma_{jk} = 0$ $(\mu_{j\cdot} = \mu_{j'\cdot} \mid$ no interaction$)$
$SS(\beta\|\alpha, \gamma)$	all $\beta_k = 0$ $(\mu_{\cdot k} = \mu_{\cdot k'})$
$SS(\beta\|\alpha)$	all $\beta_k = 0 \mid$ all $\gamma_{jk} = 0$ $(\mu_{\cdot k} = \mu_{\cdot k'} \mid$ no interaction$)$
$SS(\gamma\|\alpha, \beta)$	all $\gamma_{jk} = 0$ $(\mu_{jk} - \mu_{j'k} = \mu_{jk'} - \mu_{j'k'})$

data are as follows:

$$
\begin{aligned}
SS(\alpha, \beta, \gamma) &= 391.44 \\
SS(\alpha, \beta) &= 215.95 \\
SS(\alpha, \gamma) &= 355.42 \\
SS(\beta, \gamma) &= 151.87 \\
SS(\alpha) &= 204.33 \\
SS(\beta) &= 3.7333 \\
TSS &= 1209.2
\end{aligned}
$$

The ANOVA for the experiment is shown in Table 8.6. The predicted status × authoritarianism interaction proves to be statistically significant. A researcher would not normally report both sets of main-effect sums of squares; in this instance, where the interactions probably are not negligible, $SS(\alpha|\beta)$ and $SS(\beta|\alpha)$ do not test hypotheses about main effects, as I have explained.

8.2.5 Equal Cell Frequencies

Equal cell frequencies simplify—but do not change fundamentally—the procedures of the preceding section. When all of the cell frequencies are the same, the deviation regressors for *different* sets of effects are uncorrelated. Equal-cell-frequencies data are often termed *balanced* or *orthogonal*.[18]

---

[18] See Chapter 10, on the geometry of linear models, for an explanation of the term "orthogonal."

TABLE 8.6 Analysis-of-Variance Table for Moore and Krupat's Conformity Experiment. Alternative tests are shown for the Partner's status and Authoritarianism main effects

Source	*SS*	*df*	*MS*	*F*	*p*
Partner's status		1			
$\alpha\|\beta, \gamma$	239.57		239.57	11.43	.002
$\alpha\|\beta$	212.22		212.22	10.12	.003
Authoritarianism		2			
$\beta\|\alpha, \gamma$	36.02		18.01	0.86	.43
$\beta\|\alpha$	11.62		5.81	0.28	.76
Partner's status ×Authoritarianism	175.49	2	87.74	4.18	.02
Residuals	817.76	39	20.97		
Total	1209.2	44			

Uncorrelated main-effect and interaction regressors permit a unique decomposition of the regression sum of squares for the model, $SS(\alpha, \beta, \gamma)$, into components due to the three sets of effects. Indeed, for balanced data,

$$SS(\alpha|\beta, \gamma) = SS(\alpha|\beta) = SS(\alpha)$$
$$SS(\beta|\alpha, \gamma) = SS(\beta|\alpha) = SS(\beta)$$
$$SS(\gamma|\alpha, \beta) = SS(\gamma)$$

and hence

$$SS(\alpha, \beta, \gamma) = SS(\alpha) + SS(\beta) + SS(\gamma)$$

These results lead to particularly simple formulas for the several sums of squares:

$$SS(\alpha) = n'c \sum_{j=1}^{r} (\bar{Y}_{j\cdot} - \bar{Y}_{\cdot\cdot})^2$$

$$SS(\beta) = n'r \sum_{k=1}^{c} (\bar{Y}_{\cdot k} - \bar{Y}_{\cdot\cdot})^2$$

$$SS(\gamma) = n' \sum_{j=1}^{r} \sum_{k=1}^{c} (\bar{Y}_{jk} - \bar{Y}_{j\cdot} - \bar{Y}_{\cdot k} + \bar{Y}_{\cdot\cdot})^2$$

where $n' = n/rc$ is the number of observations in each cell of the RC table.

8.2.6 Some Cautionary Remarks

R. A. Fisher (1925) originally formulated analysis of variance for balanced data. Yet, as early as 1934, Fisher's colleague at the Rothamsted Experimental Station in England, F. Yates, indicated how the analysis of variance could be extended to unbalanced data. Apart from approximate methods motivated by the desire to reduce the effort of calculation, Yates (1934) suggested two approaches to the two-way classification, naming both for the computational techniques that he developed. The first approach, which he called "the method of weighted squares of means," calculates the main-effect sums of squares $SS(\alpha|\beta, \gamma)$ and $SS(\beta|\alpha, \gamma)$, and the interaction sum of squares $SS(\gamma|\alpha, \beta)$ (using our notation). Yates's second approach, which he called "the method of fitting constants," assumes that interactions are absent and calculates $SS(\alpha|\beta)$ and $SS(\beta|\alpha)$.

Considering the apparent simplicity of the two-way classification and the lucidity of Yates's treatment of it, it is ironic that the analysis of unbalanced data has become the subject of controversy and confusion. While it is not my purpose to present a complete account of the "debate" concerning the proper handling of unbalanced data—and while it is tempting to ignore this debate altogether—there are two reasons for addressing the topic briefly here: (1) You may encounter confused applications of ANOVA or may have occasion to consult other accounts of the method; and (2) computer programs for analysis of variance are occasionally misleading or vague in their documentation and output, or even incorrect in their calculations (see Francis, 1973).[19]

Much of the confusion about the analysis of unbalanced data has its source in the restrictions—or other techniques—that are used to solve the "over-parameterized" (i.e., unrestricted) two-way ANOVA model. Imagine, for example, that we use dummy coding rather than deviation coding to fit the model to the data.

Let $SS^*(\cdot)$ denote the regression sum of squares for a dummy-coded model. For the full model and the main-effects model, we obtain the same sums of squares as before; that is,

$$SS(\alpha, \beta, \gamma) = SS^*(\alpha, \beta, \gamma)$$
$$SS(\alpha, \beta) = SS^*(\alpha, \beta)$$

Likewise (because they are just the two one-way ANOVAs)

$$SS(\alpha) = SS^*(\alpha)$$
$$SS(\beta) = SS^*(\beta)$$

And because these regression sums of squares are the same, so are the incremental sums of squares that depend on them:

$$SS(\gamma|\alpha, \beta) = SS^*(\gamma|\alpha, \beta)$$
$$SS(\alpha|\beta) = SS^*(\alpha|\beta)$$
$$SS(\beta|\alpha) = SS^*(\beta|\alpha)$$

---

[19] With respect to the second point, it is good practice to test a computer program with known data before trusting it to analyze new data. This advice applies not just to ANOVA calculations, but generally.

In general, however,

$$\text{SS}(\alpha, \gamma) \neq \text{SS}^*(\alpha, \gamma)$$
$$\text{SS}(\beta, \gamma) \neq \text{SS}^*(\beta, \gamma)$$

and, consequently (also in general),

$$\text{SS}(\alpha|\beta, \gamma) \neq \text{SS}^*(\alpha|\beta, \gamma)$$
$$\text{SS}(\beta|\alpha, \gamma) \neq \text{SS}^*(\beta|\alpha, \gamma)$$

The general lesson to be drawn from these results is that tests that conform to the principle of marginality [here, those based on $\text{SS}(\gamma|\alpha, \beta)$, $\text{SS}(\alpha|\beta)$, and $\text{SS}(\beta|\alpha)$] do not depend on the specific restrictions that were employed to fit the model, while tests that "violate" the principle of marginality [those based on $\text{SS}(\alpha|\beta, \gamma)$ and $\text{SS}(\beta|\alpha, \gamma)$] do depend on the specific restrictions.

I showed that $\text{SS}(\alpha|\beta, \gamma)$ and $\text{SS}(\beta|\alpha, \gamma)$, based on the sigma constraints, are appropriate for testing hypotheses about main effects in the potential presence of interactions. It follows that $\text{SS}^*(\alpha|\beta, \gamma)$ and $\text{SS}^*(\beta|\alpha, \gamma)$ *do not* properly test these hypotheses. It is important, in this context, to select constraints that test reasonable hypotheses about cell and marginal means. The SS notation is frequently used carelessly, without attention to the constraints that are employed and to the hypotheses that follow from them.[20]

EXERCISES

8.3 *Show that one of the restrictions on the interaction parameters of the two-way analysis-of-variance model,

$$\sum_{j=1}^{r} \gamma_{jk} = 0 \quad \text{for } k = 1, \ldots, c$$

$$\sum_{k=1}^{c} \gamma_{jk} = 0 \quad \text{for } j = 1, \ldots, r$$

is redundant. (*Hint*: Construct a table of the interaction parameters, labeling the rows $1, 2, \ldots, r-1, r$ and the columns $1, 2, \ldots, c$. Insert a column for row sums after the last column, and a row for column sums after the last row. At the bottom right corner of the table is the overall sum of the interaction parameters, $\sum_{j=1}^{r} \sum_{k=1}^{c} \gamma_{jk}$. Then place a 0 in each entry of the column of row sums, corresponding to the r restrictions $\sum_{k=1}^{c} \gamma_{jk} = 0$. From

[20] Further discussion of the points raised in this section may be found in a variety of sources, including Hocking and Speed (1975). Speed and Hocking (1976), Speed, et al. (1978), Speed and Monlezun (1979), Searle, et al. (1981), and Steinhorst (1982). Also see Section 10.4.

these restrictions, show that $\sum\sum\gamma_{jk} = 0$, and place this 0 in the lower-right corner. Next, specify 0's for all but the last column sum, $\sum_{j=1}^{r}\gamma_{jk} = 0$, for $k = 1, \ldots, c-1$. Finally, show that the last column sum, $\sum_{j=1}^{r}\gamma_{jc}$, is necessarily 0.)

8.4 *Demonstrate that the hypothesis

$$H_0\text{: all } \gamma_{jk} = 0$$

for the sigma-constrained two-way ANOVA model is equivalent to the null hypothesis of no interaction stated in terms of the cell means:

$$\mu_{jk} - \mu_{j'k} = \mu_{jk'} - \mu_{j'k'} \quad \text{for all } j,\ j' \text{ and } k,\ k'$$

[*Hint*: Write out each of the interaction parameters γ_{jk}, $\gamma_{j'k}$, $\gamma_{jk'}$, and $\gamma_{j'k'}$ in terms of the cell and marginal means (e.g., $\gamma_{jk} = \mu_{jk} - \mu_{j.} - \mu_{.k} + \mu_{..}$). Then show that when the γ's are all 0, $\gamma_{jk} - \gamma_{j'k} = \gamma_{jk'} - \gamma_{j'k'}$ (i.e., $0-0=0-0$) implies that $\mu_{jk} - \mu_{j'k} = \mu_{jk'} - \mu_{j'k'}$.]

8.5 The following "balanced" (i.e., equal-cell-frequencies) data are from an experiment reported by Fox and Guyer (1978). Twenty four-person groups of subjects played 30 trials of a "prisoner's dilemma" game. On every trial of the experiment, each subject selected either a competitive or a cooperative choice. The value reported below for each group is the number of cooperative choices (of the 120 choices) made by subjects in that group. The data are also in `guyer.dat`. Ten of the groups recorded their choices anonymously, while the remaining 10 groups made public choices (i.e., subjects' choices were made known to other group members); half of the groups were composed of males and half of females. The experimenters expected to observe a higher level of cooperation in the public-choice condition, but did not make predictions about sex effects or sex-by-condition interaction.

	Sex	
Condition	Male	Female
Public choice	49	54
	64	61
	37	79
	52	64
	68	29
Anonymous	27	40
	58	39
	52	44
	41	34
	30	44

(a) Make a scatterplot of cooperation versus sex, using different colors or symbols to plot the points for the public-choice and anonymous conditions. Calculate the mean and standard deviation for cooperation in each cell, and graph the cell means. Comment on the descriptive results of the experiment: How do sex and condition appear to affect cooperation?

(b) Using the usual sigma constraints, form deviation regressors for condition, sex, and the condition × sex interaction. Show that the correlations among these regressors are all 0 (by virtue of the fact that there are equal numbers of observations in all of the cells of the design).

(c) Confirm that the sums of squares for the ANOVA can be obtained in the following four ways:

- **(i)** As $SS(\alpha|\beta)$, $SS(\beta|\alpha)$, $SS(\gamma|\alpha, \beta)$.
- **(ii)** As $SS(\alpha|\beta, \gamma)$, $SS(\beta|\alpha, \gamma)$, $SS(\gamma|\alpha, \beta)$.
- **(iii)** As $SS(\alpha)$, $SS(\beta)$, $SS(\gamma)$. *Note*: This last approach is valid only for balanced data.
- **(iv)** Using the special formulas given in Section 8.2.5.

(d) Form the ANOVA table for the experiment. What conclusions can you draw from the ANOVA about the effects of condition and sex on cooperation?

8.3 Higher-Way Analysis of Variance*

The methods of the previous section can be extended to any number of factors. I shall consider the three-way classification in some detail before commenting briefly on the general case.

8.3.1 The Three-Way Classification

It is convenient to label the factors in the three-way classification as A, B, and C, with a, b, and c levels, consecutively. A dependent-variable observation is represented by Y_{ijkm}, where the first subscript gives the index of the observation within its cell. The number of observations sampled in cell $\{j, k, m\}$ is n_{jkm}; and μ_{jkm} is the population mean in this cell. Quantities such as $\mu_{...}$, $\mu_{j..}$, and $\mu_{jk.}$ denote marginal means formed by averaging over the dotted subscripts.

The three-way ANOVA model is

$$
\begin{aligned}
Y_{ijkm} &= \mu_{jkm} + \varepsilon_{ijkm} \\
&= \mu + \alpha_{A(j)} + \alpha_{B(k)} + \alpha_{C(m)} + \alpha_{AB(jk)} \\
&\quad + \alpha_{AC(jm)} + \alpha_{BC(km)} + \alpha_{ABC(jkm)} + \varepsilon_{ijkm}
\end{aligned}
$$

Notice that, to avoid the proliferation of symbols, I have introduced a new and easily extended notation for model parameters: The first set of subscripts (e.g., AB) indicates the factors to which a parameter pertains, while the parenthetical subscripts [e.g., (j, k)] index factor categories.

We make the usual linear-model assumptions about the errors ε_{ijkm}, and constrain all sets of parameters to sum to 0 over every coordinate; for example,

$$\sum_{j=1}^{a} \alpha_{A(j)} = 0$$

$$\sum_{j=1}^{a} \alpha_{AB(jk)} = \sum_{k=1}^{b} \alpha_{AB(jk)} = 0 \quad \text{for all } j, k$$

$$\sum_{j=1}^{a} \alpha_{ABC(jkm)} = \sum_{k=1}^{b} \alpha_{ABC(jkm)} = \sum_{m=1}^{c} \alpha_{ABC(jkm)} = 0 \quad \text{for all } j, k, m$$

The sigma constraints for $\alpha_{B(k)}$, $\alpha_{C(m)}$, and $\alpha_{BC(km)}$ follow similar patterns.

The three-way ANOVA model includes parameters for main effects, for *two-way interactions* between each pair of factors, and for *three-way interactions* among all three factors. The two-way interactions have the same interpretation as in two-way ANOVA: If, for instance, A and B interact, then the effect of either factor on the dependent variable varies across the levels of the other factor. Similarly, if the ABC interaction is nonzero, then the joint effect of any pair of factors (say, A and B) varies across the categories of the remaining factor (C).

In formulating models and interpreting effects in three-way ANOVA, we may again appeal to the principle of marginality. Thus, main effects (e.g., of A) are generally not interpreted if they are marginal to nonnull interactions (AB, AC, or ABC). Likewise, a *lower-order* interaction (such as AB) is usually not interpreted if it has a nonnull *higher-order relative* (ABC): If the joint effects of A and B are different in different categories of C, then it is not generally sensible to speak of the unconditional AB effects, without reference to a specific category of C.

Deviation regressors for main effects in the three-way classification can be coded as before; regressors for interactions are formed by taking all possible products of the main effects that "compose" the interaction. Here, for example, is the coding for $a = 2, b = 2$, and $c = 3$:

Cell jkm	A	B	C_1	C_2	AB	AC_1	AC_2	BC_1	BC_2	ABC_1	ABC_2
111	1	1	1	0	1	1	0	1	0	1	0
112	1	1	0	1	1	0	1	0	1	0	1
113	1	1	−1	−1	1	−1	−1	−1	−1	−1	−1
121	1	−1	1	0	−1	1	0	−1	0	−1	0
122	1	−1	0	1	−1	0	1	0	−1	0	−1
123	1	−1	−1	−1	−1	−1	−1	1	1	1	1
211	−1	1	1	0	−1	−1	0	1	0	−1	0
212	−1	1	0	1	−1	0	−1	0	1	0	−1
213	−1	1	−1	−1	−1	1	1	−1	−1	1	1
221	−1	−1	1	0	1	−1	0	−1	0	1	0
222	−1	−1	0	1	1	0	−1	0	−1	0	1
223	−1	−1	−1	−1	1	1	1	1	1	−1	−1

The following points are noteworthy:

- The 12 cell means are expressed in terms of an equal number of independent parameters (including the general mean, μ), underscoring the point that three-way interactions may be required to account for the pattern of cell means. More generally in the three-way classification, there are *abc* cells and the same number of independent parameters:

$$\begin{aligned} 1 &+ (a-1) + (b-1) + (c-1) \\ &\quad +(a-1)(b-1) + (a-1)(c-1) + (b-1)(c-1) \\ &\quad +(a-1)(b-1)(c-1) \\ &= abc \end{aligned}$$

- The degrees of freedom for a set of effects correspond, as usual, to the number of independent parameters in the set. There are, for example, $a-1$ degrees of freedom for the *A* main effects; $(a-1)(b-1)$ degrees of freedom for the *AB* interactions; and $(a-1)(b-1)(c-1)$ degrees of freedom for the *ABC* interactions.

Solving for the constrained parameters in terms of populations means produces the following results:

$$\begin{aligned} \mu &= \mu_{\cdot\cdot\cdot} \\ \alpha_{A(j)} &= \mu_{j\cdot\cdot} - \mu_{\cdot\cdot\cdot} \\ \alpha_{AB(jk)} &= \mu_{jk\cdot} - \mu - \alpha_{A(j)} - \alpha_{B(k)} \\ &= \mu_{jk\cdot} - \mu_{j\cdot\cdot} - \mu_{\cdot k\cdot} + \mu_{\cdot\cdot\cdot} \\ \alpha_{ABC(jkm)} &= \mu_{jkm} - \mu - \alpha_{A(j)} - \alpha_{B(k)} - \alpha_{C(m)} \\ &\quad - \alpha_{AB(jk)} - \alpha_{AC(jm)} - \alpha_{BC(km)} \\ &= \mu_{jkm} - \mu_{jk\cdot} - \mu_{j\cdot m} - \mu_{\cdot km} + \mu_{j\cdot\cdot} + \mu_{\cdot k\cdot} + \mu_{\cdot\cdot m} - \mu_{\cdot\cdot\cdot} \end{aligned}$$

(The patterns for $\alpha_{B(k)}$, $\alpha_{C(m)}$, $\alpha_{AC(jm)}$, and $\alpha_{BC(km)}$ are similar, and are omitted for brevity.)

As in two-way analysis of variance, therefore, the null hypothesis

$$H_0\text{: all } \alpha_{A(j)} = 0$$

is equivalent to

$$H_0\text{: } \mu_{1\cdot\cdot} = \mu_{2\cdot\cdot} = \cdots = \mu_{a\cdot\cdot}$$

and the hypothesis

$$H_0\text{: all } \alpha_{AB(jk)} = 0$$

is equivalent to

$$H_0\text{: } \mu_{jk\cdot} - \mu_{j'k\cdot} = \mu_{jk'\cdot} - \mu_{j'k'\cdot} \quad \text{for all } i, i' \text{ and } j, j'$$

Likewise, some algebraic manipulation[21] shows that the null hypothesis

$$H_0\text{: all } \alpha_{ABC(jkm)} = 0$$

is equivalent to

$$\begin{aligned} H_0\text{: } &(\mu_{jkm} - \mu_{j'km}) - (\mu_{jk'm} - \mu_{j'k'm}) \\ &= (\mu_{jkm'} - \mu_{j'km'}) - (\mu_{jk'm'} - \mu_{j'k'm'}) \\ &\quad \text{for all } j, j'; k, k'; \text{ and } m, m' \end{aligned} \qquad [8.13]$$

The second-order differences in Equation 8.14 are equal when the pattern of AB interactions is invariant across categories of factor C—an intuitively reasonable extension of the notion of no interaction to three factors. Rearranging the terms in Equation 8.14 produces similar results for AC and BC, demonstrating that three-way interaction—like two-way interaction—is symmetric in the factors. As in two-way ANOVA, this simple relationship between model parameters and population means depends on the sigma constraints, which were imposed on the "over-parameterized" model.

Incremental F-tests can be constructed in the usual manner for the parameters of the three-way ANOVA model. A general ANOVA table, adapting the SS notation of Section 8.2.4 and showing alternative tests for main effects and lower-order interactions, is sketched in Table 8.7. Once more, for compactness, only tests involving factor A are shown. Note that a main-effect hypothesis such as H_0: all $\alpha_{A(j)} = 0$ is of interest even when the BC interactions are present, because A is not marginal to BC.[22]

8.3.2 Higher-Order Classifications

Extension of the analysis of variance to more than three factors is algebraically and computationally straightforward. The general p-way classification can be described by a model containing terms for every combination of factors; the highest-order term, therefore, is for the p-way interactions. If the p-way interactions are nonzero, then the joint effects of any $p - 1$ factors vary across the levels of the remaining factor. In general, we can be guided by the principle of marginality in interpreting effects.

Three-way interactions, however, are reasonably complex, and the even greater complexity of higher-order interactions can make their interpretation difficult. Yet, at times, we may expect to observe a high-order interaction of a particular sort, as when a specific *combination* of characteristics predisposes individuals to act in a certain manner.[23] On the other hand, it is common to find that high-order interactions are not statistically significant or that they are negligibly small relative to other effects.

---

[21] See Exercise 8.6.

[22] Data for an illustrative three-way ANOVA are given in Exercise 8.9.

[23] An alternative to specifying a high-order interaction would be simply to introduce a dummy regressor, coded 1 for the combination of categories in question and 0 elsewhere.

TABLE 8.7 General Three-Way ANOVA Table, Showing Incremental Sums of Squares for Terms Involving Factor A. Alternative tests are shown for the A main effects and AB interactions

Source	*df*	*Sum of Squares*	H_0
A	$a-1$	$SS(A \mid B, C, AB, AC, BC, ABC)$	$\alpha_A = 0$
		$SS(A \mid B, C, BC)$	$\alpha_A = 0 \mid \alpha_{AB} = \alpha_{AC} = \alpha_{ABC} = 0$
AB	$(a-1)(b-1)$	$SS(AB \mid A, B, C, AC, BC, ABC)$	$\alpha_{AB} = 0$
		$SS(AB \mid A, B, C, AC, BC)$	$\alpha_{AB} = 0 \mid \alpha_{ABC} = 0$
ABC	$(a-1)(b-1)(c-1)$	$SS(ABC \mid A, B, C, AB, AC, BC)$	$\alpha_{ABC} = 0$
Residuals	$n - abc$	$TSS - SS(A, B, C, AB, AC, BC, ABC)$	
Total	$n-1$	TSS	

There is, moreover, no rule of data analysis that requires us to fit and test all possible interactions. In working with higher-way classifications, we may limit consideration to effects that are of theoretical interest, or at least to effects that are substantively interpretable. It is fairly common, for example, for researchers to fit models containing only main effects:

$$Y_{ijk\ldots r} = \mu + \alpha_{A(j)} + \alpha_{B(k)} + \cdots + \alpha_{P(r)} + \varepsilon_{ijk\ldots r}$$

This approach, sometimes called *multiple-classification analysis* or *MCA*,[24] is analogous to an additive multiple regression. In a similar spirit, a researcher might entertain models that include only main effects and two-way interactions.

> The analysis-of-variance model and procedures for testing hypotheses about main effects and interactions extend straightforwardly to three-way and higher-way classifications. In each case, the highest-order interaction corresponds to the number of factors in the model. It is not necessary, however, to specify a model that includes all terms through the highest-order interaction.

[24] The term "multiple-classification analysis" is unfortunate because it is equally descriptive of any ANOVA model fit to the p-way classification.

8.3.3 Empty Cells in ANOVA

As the number of factors increases, the number of cells grows at a much faster rate: For p dichotomous factors, for example, the number of cells is 2^p. One consequence of this proliferation is that some combinations of factor categories may not be observed; that is, certain cells in the p-way classification may be empty.

Nevertheless, we can use our deviation-coding approach to estimation and testing in the presence of empty cells as long as the *marginal* frequency tables corresponding to the effects that we examine contain no empty cells. For example, in a two-way classification with an empty cell, we can safely fit the main-effects model (see below), because the one-way frequency counts for each factor separately contain no 0's. The full model with interactions is not covered by the rule, however, because the two-way table of counts contains a 0 frequency. By extension, the rule never covers the p-way interaction when there is a 0 cell in a p-way classification.[25]

To illustrate the difficulties produced by empty cells, I shall develop a very simple example for a 2 × 2 classification with cell frequencies:

	C_1	C_2	Row marginal
R_1	n_{11}	n_{12}	$n_{11} + n_{12}$
R_2	n_{21}	0	n_{21}
Column marginal	$n_{11} + n_{21}$	n_{12}	n

That is, the cell frequency n_{22} is 0. Because there are no observations in this cell, we cannot estimate the cell mean μ_{22}. Writing out the other cell means in terms of the restricted model parameters produces three equations:

$$\mu_{11} = \mu + \alpha_1 + \beta_1 + \gamma_{11}$$
$$\mu_{12} = \mu + \alpha_1 + \beta_2 + \gamma_{12} = \mu + \alpha_1 - \beta_1 - \gamma_{11}$$
$$\mu_{21} = \mu + \alpha_2 + \beta_1 + \gamma_{21} = \mu - \alpha_1 + \beta_1 - \gamma_{11}$$

There are, then, four independent parameters (μ, α_1, β_1, and γ_{11}) but only three population means in observed cells, so the parameters are not uniquely determined by the means.

Now imagine that we can reasonably specify the absence of two-way interactions for these data. Then, according to our general rule, we should be able to estimate and test the R and C main effects, because there are observations

---

[25] It may be possible to estimate and test effects not covered by this simple rule of thumb, but determining whether tests are possible and specifying meaningful hypotheses to be tested are substantially more complex in this instance. For details see, for example, Searle (1971, pp. 318–324), Hocking and Speed (1975, pp. 711–712), and Speed, et al. (1978, pp. 110–111). The advice given in Section 8.2.6 regarding care in the use of computer programs for analysis of variance of unbalanced data applies even more urgently when there are empty cells.

at each level of R and at each level of C. The equations relating cell means to independent parameters become

$$\mu_{11} = \mu + \alpha_1 + \beta_1$$
$$\mu_{12} = \mu + \alpha_1 + \beta_2 = \mu + \alpha_1 - \beta_1$$
$$\mu_{21} = \mu + \alpha_2 + \beta_1 = \mu - \alpha_1 + \beta_1$$

Solving for the parameters in terms of the cell means produces[26]

$$\mu = \frac{\mu_{12} + \mu_{21}}{2}$$
$$\alpha_1 = \frac{\mu_{11} - \mu_{21}}{2}$$
$$\beta_1 = \frac{\mu_{11} - \mu_{12}}{2}$$

These results make sense, for, in the *absence* of interaction:

- The cell means μ_{12} and μ_{21} are "balanced" with respect to both sets of main effects, and therefore their average serves as a suitable definition of the grand mean.
- The difference $\mu_{11} - \mu_{21}$ gives the effect of changing R while C is held constant (at level 1), which is a suitable definition of the main effect of R.
- The difference $\mu_{11} - \mu_{12}$ gives the effect of changing C while R is held constant (at level 1), which is a suitable definition of the main effect of C.

EXERCISES

8.6 *Show that in the sigma-constrained three-way ANOVA model, the null hypothesis

$$H_0\text{: all } \alpha_{ABC(jkm)} = 0$$

is equivalent to the hypothesis given in Equation 8.14. (*Hint*: See Exercise 8.4.)

---

[26] The 2×2 classification with one empty cell is particularly simple because the number of parameters in the main-effects model is equal to (i.e., no fewer than) the number of observed cell means. This is not generally the case, making a general analysis considerably more complex.

8.7 The geometry of effects in three-way ANOVA: Contrived parameter values for a three-way ANOVA model (each set satisfying the sigma constraints) are given in the following tables:

$\alpha_{A(j)}$	
A_1	A_2
2	−2

$\alpha_{B(k)}$	
B_1	B_2
−3	3

$\alpha_{C(m)}$		
C_1	C_2	C_3
1	−3	2

	$\alpha_{AB(jk)}$	
	B_1	B_2
A_1	−2	2
A_2	2	−2

	$\alpha_{AC(jm)}$		
	C_1	C_2	C_3
A_1	1	1	−2
A_2	−1	−1	2

	$\alpha_{BC(km)}$		
	C_1	C_2	C_3
B_1	0	3	−3
B_2	0	−3	3

	$\alpha_{ABC(jkm)}$		
	C_1	C_2	C_3
A_1B_1	1	−2	1
A_1B_2	−1	2	−1
A_2B_1	−1	2	−1
A_2B_2	1	−2	1

Use these parameter values to construct population cell means for each of the following models (simply sum the parameters that pertain to each of the 12 cells of the design):

(a) Main effects only

$$\mu_{jkm} = \mu + \alpha_{A(j)} + \alpha_{B(k)} + \alpha_{C(m)}$$

(b) One two-way interaction

$$\mu_{jkm} = \mu + \alpha_{A(j)} + \alpha_{B(k)} + \alpha_{C(m)} + \alpha_{AC(jm)}$$

(c) All three two-way interactions

$$\mu_{jkm} = \mu + \alpha_{A(j)} + \alpha_{B(k)} + \alpha_{C(m)} + \alpha_{AB(jk)} + \alpha_{AC(jm)} + \alpha_{BC(km)}$$

(d) The full model

$$\mu_{jkm} = \mu + \alpha_{A(j)} + \alpha_{B(k)} + \alpha_{C(m)} + \alpha_{AB(jk)} + \alpha_{AC(jm)} + \alpha_{BC(km)} + \alpha_{ABC(jkm)}$$

For each of these models:

(i) Draw a graph of the cell means, placing factor C on the horizontal axis. Use different lines (solid and broken) or line colors for the levels of factor A, and different symbols (e.g., circle and square) for the levels of factor B. Note that there will be four connected profiles of means on each of these plots, one profile for each combination of categories of A and B across the three levels of C. Attempt to interpret the graphs in terms of the effects that are included in each model.

(ii) Using the table of means generated from each of models (c) and (d), plot (for each model) the six differences across the levels of factor B, $\mu_{j1m} - \mu_{j2m}$, by the categories of factors A and C. Can you account for the different patterns of these two graphs in terms of the presence of three-way interactions in the second graph but not in the first?

8.8 Adjusted means (continued): The notion of an "adjusted" mean was introduced in Exercises 7.5 and 7.10. Consider the main-effects model for the p-way classification:

$$\mu_{jk...r} \equiv E(Y_{ijk...r}) = \mu + \alpha_{A(j)} + \alpha_{B(k)} + \cdots + \alpha_{P(r)}$$

(a) Show that if we constrain each set of effects to sum to 0, then the population marginal mean for category j of factor A is $\mu_{j....} = \mu + \alpha_{A(j)}$.

(b) Let us define the analogous sample quantity, $\tilde{Y}_{j....} \equiv M + A_{A(j)}$ to be the *adjusted mean* in category j of factor A. How is this quantity to be interpreted?

(c) Does the definition of the adjusted mean in part (b) depend fundamentally on the constraint that each set of effects sums to 0?

(d) Can the idea of an adjusted mean be extended to ANOVA models that include interactions?

8.9 ANOVA with equal cell frequencies: In higher-way ANOVA, as in two-way ANOVA, when cell frequencies are equal, the sum of squares for each set of effects can be calculated directly from the parameter estimates for the full model, or, equivalently, in terms of cell and marginal means. To get the sum of squares for a particular set of effects, we simply need to square the parameter estimate associated with each cell, sum over all cells, and multiply by the common cell frequency, n'. For example, for a balanced three-way ANOVA:

$$\begin{aligned} SS(\alpha_{AB}) &= n' \sum_{j=1}^{a} \sum_{k=1}^{b} \sum_{m=1}^{c} A^2_{AB(jk)} \\ &= n'c \sum_{j=1}^{a} \sum_{k=1}^{b} A^2_{AB(jk)} \\ &= n'c \sum_{j=1}^{a} \sum_{k=1}^{b} (\bar{Y}_{jk\cdot} - \bar{Y}_{j\cdot\cdot} - \bar{Y}_{\cdot k\cdot} + \bar{Y}_{\cdot\cdot\cdot})^2 \end{aligned}$$

(a) Write out similar expressions for SS(α_A) and SS(α_{ABC}) in three-way ANOVA. Show that

$$\text{RSS} = (n' - 1)\sum_{j=1}^{a}\sum_{k=1}^{b}\sum_{m=1}^{c} S_{jkm}^2$$

where S_{jkm}^2 is the variance in cell j, k, m of the design.

(b) The following table shows the results of an experiment on interpersonal attraction reported by Riordan, et al. (1982). Subjects in the study interacted with an experimenters' confederate whose attitudes were manipulated to be either similar or dissimilar to those of the subjects. In the course of the study, it was arranged for some subjects to request the confederate's help in completing a task, while other subjects did not ask for help. Finally, the confederate provided help to some subjects but not to others. These three factors combine to produce eight experimental conditions. Nine subjects were assigned to each condition. (Actually, two conditions contained eight subjects, but we shall disregard this slight complication.) At the beginning and again at the end of the study, subjects rated their attraction to the confederate on a two-item scale, with possible composite scores ranging from 2 through 14. The table reports means and standard deviations for changes in attraction over the course of the study.

			Attraction Change	
Attitude Similarity	*Help Requested*	*Help Provided*	*Mean*	*Standard Deviation*
Similar	Yes	Yes	0.22	1.2
		No	−4.62	1.3
	No	Yes	−0.12	1.2
		No	−2.56	1.4
Dissimilar	Yes	Yes	1.44	1.9
		No	−2.22	2.3
	No	Yes	2.10	2.1
		No	−1.98	1.3

Riordan et al. (1982, p. 364) make the following predictions with regard to these changes:

> *Changes in attraction should be associated with an interaction of all three factors. Subjects should show an increment in attraction when help is provided by a dissimilar other and the increment should be greater when the help was not requested than when it was requested. A decrement in attraction should occur when help is not given by a similar other and the decrement should be greater when help was requested. No change should occur when a dissimilar other does not provide help and it was not requested, and very little change in a negative direction when the dissimilar other does not provide help and it was requested. No change is also predicted when a similar other provides help and it was requested, and very little change in a positive direction when it was not requested.*

(i) Graph the cell means from the table. Do the results of the study appear to confirm the authors' predictions?

(ii) Using the results of part (a), compute an analysis-of-variance table for Riordan et al.'s data. What conclusions would you draw?

8.4 Analysis of Covariance

Analysis of covariance (ANCOVA) is a term used to describe linear models that contain both qualitative and quantitative independent variables. The method is, therefore, equivalent to dummy-variable regression, discussed in the previous chapter, although the ANCOVA model is parameterized differently from the dummy-regression model.[27] Traditional applications of analysis of covariance use an additive model (i.e., without interactions). The traditional additive ANCOVA model is a special case of the more general model that I present here.

In analysis of covariance, an ANOVA formulation is used for the main effects and interactions of the qualitative independent variables (or *factors*), and the quantitative independent variables (or *covariates*) are expressed as deviations from their means. Neither of these variations represents an essential change, however, for the ANCOVA model provides the same fit to the data as the dummy-regression model. Moreover, if tests are formulated following the principle of marginality, then precisely the same sums of squares are obtained for the two parameterizations. Nevertheless, the ANCOVA parameterization makes it simple to formulate meaningful (if ordinarily uninteresting) tests for lower-order terms in the presence of their higher-order relatives.

I shall use Moore and Krupat's study of conformity and authoritarianism to illustrate analysis of covariance. When we last encountered these data,[28] both independent variables—partner's status and authoritarianism—were treated as factors. Partner's status is dichotomous, but authoritarianism is a quantitative score (the "F-scale"), which was arbitrarily categorized for the two-way analysis of variance. Here, I shall treat authoritarianism more naturally as a covariate.

A dummy-regression formulation, representing authoritarianism by X and coding $D = 1$ in the low partner's status group and $D = 0$ in the high partner's status group, produces the following fit to the data (with estimated standard errors in parentheses below the coefficients):

$$\begin{aligned} \hat{Y} = \underset{(3.26)}{20.79} &- \underset{(0.0717)}{0.1511}X - \underset{(4.40)}{15.53}D + \underset{(0.0970)}{0.2611}(X \times D) \\ R^2 &= .2942 \end{aligned} \tag{8.14}$$

It makes sense, in this model, to test whether the interaction coefficient is statistically significant (clearly it is), but—as explained in the previous chapter—it is not sensible to construe the coefficients of X and D as "main effects" of authoritarianism and partner's status: The coefficient of X is the authoritarianism slope in the high-status group, while the coefficient of D is the difference in the regression lines for the two groups at an authoritarianism score of 0.

An ANCOVA model for the Moore and Krupat experiment is

$$Y_{ij} = \mu + \alpha_j + \beta(X_{ij} - \bar{X}) + \gamma_j(X_{ij} - \bar{X}) + \varepsilon_{ij} \tag{8.15}$$

---

[27] Usage here is not wholly standardized, and the terms "dummy regression" and "analysis of covariance" are often taken as synonymous.

[28] See Section 8.2.

where

- Y_{ij} is the conformity score for subject i in category j of partner's status;
- μ is the general level of conformity;
- α_j is the main effect of membership in group j of partner's status;
- β is the main-effect slope of authoritarianism, X;
- γ_j is the interaction between partner's status and authoritarianism for group j;
- ε_{ij} is the error; and
- the mean authoritarianism score $\bar{X}$ is computed over all of the data.

To achieve a concrete understanding of the model (Equation 8.15), let us—as is our usual practice—write out the model separately for each group:

$$\begin{aligned}
\text{Low status:}\quad Y_{i1} &= \mu + \alpha_1 + \beta(X_{i1} - \bar{X}) + \gamma_1(X_{i1} - \bar{X}) + \varepsilon_{i1} \\
&= \mu + \alpha_1 + (\beta + \gamma_1)(X_{i1} - \bar{X}) + \varepsilon_{i1} \\
\text{High status:}\quad Y_{i2} &= \mu + \alpha_2 + \beta(X_{i2} - \bar{X}) + \gamma_2(X_{i2} - \bar{X}) + \varepsilon_{i2} \\
&= \mu + \alpha_2 + (\beta + \gamma_2)(X_{i2} - \bar{X}) + \varepsilon_{i2}
\end{aligned}$$

It is immediately apparent that there are too many parameters: We are fitting one line in each of two groups, which requires four parameters, but there are six parameters in the model—μ, α_1, α_2, β, γ_1, and γ_2.

We require two restrictions, and to provide them we shall place sigma constraints on the α's and γ's:

$$\begin{aligned}
\alpha_1 + \alpha_2 = 0 \quad &\Rightarrow \quad \alpha_2 = -\alpha_1 \\
\gamma_1 + \gamma_2 = 0 \quad &\Rightarrow \quad \gamma_2 = -\gamma_1
\end{aligned}$$

Under these constraints, the two regression equations become

$$\begin{aligned}
\text{Low status:}\quad Y_{i1} &= \mu + \alpha_1 + (\beta + \gamma_1)(X_{i1} - \bar{X}) + \varepsilon_{i1} \\
\text{High status:}\quad Y_{i2} &= \mu - \alpha_1 + (\beta - \gamma_1)(X_{i2} - \bar{X}) + \varepsilon_{i2}
\end{aligned}$$

The parameters of the constrained model therefore have the following straightforward interpretations (see Figure 8.7):

- μ is midway between the two regression lines above the mean of the covariate, $\bar{X}$.
- α_1 is half the difference between the two regression lines, again above $\bar{X}$.
- β is the average of the slopes $\beta+\gamma_1$ and $\beta-\gamma_1$ for the two within-group regression lines.
- γ_1 is half the difference between the slopes of the two regression lines.

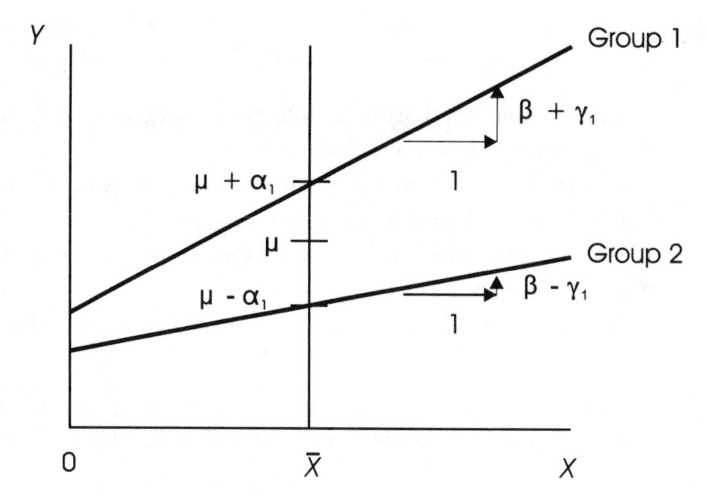

Figure 8.7. The analysis-of-covariance model for two groups, permitting different within-group slopes.

Notice, in particular, that the constrained parameters α_1 and β are reasonably interpreted as "main effects"—that is, the partial effect of one independent variable averaged over the other independent variable—even when interactions are present in the model.

To fit the model to the data, we need to code a deviation regressor S for partner's status:

Partner's Status	S
Low	1
High	−1

Then we can regress conformity on S, the mean-deviation scores for X, and the product of S and $X - \bar{X}$,

$$Y_{ij} = \mu + \alpha_1 S_{ij} + \beta(X_{ij} - \bar{X}) + \gamma_1[S_{ij}(X_{ij} - \bar{X})] + \varepsilon_{ij}$$

producing the following fit to the Moore and Krupat data:

$$\begin{aligned}\hat{Y} = \underset{(2.200)}{13.03} \; - \underset{(2.200)}{7.767S} - \underset{(0.0485)}{0.02055(X - \bar{X})} + \underset{(0.0485)}{0.1306[S(X - \bar{X})]} \quad [8.16]\\ R^2 = .2942\end{aligned}$$

Because each set of effects has 1 degree of freedom, incremental F-tests for the main effects and interactions are equivalent to the t-tests produced by dividing each coefficient by its standard error. It is apparent, then, that the partner's status × authoritarianism interaction is statistically significant, as is the status main effect, but the authoritarianism main effect is not. You can verify that the two regression lines derived from the fitted ANCOVA model (Equation 8.16) are the same as those derived from the dummy-regression model (Equation 8.14).[29]

> Analysis of covariance is an alternative parameterization of the dummy-regression model, employing deviation-coded regressors for factors and expressing covariates as deviations from their means. The ANCOVA model can incorporate interactions among factors and between factors and covariates.

EXERCISES

8.10 In Exercise 7.9, a dummy-regression model was fit to Angell's data on the moral integration of U.S. cities (in Table 2.3 and `angell.dat`), regressing moral integration on ethnic heterogeneity, geographic mobility, region, and the interactions between heterogeneity and region, and between mobility and region. Redo this analysis using the analysis-of-covariance formulation of the model.

8.11 Calculate the fitted regression equation for each group (low and high partner's status) in Moore and Krupat's conformity data using the dummy regression in Equation 8.14. Calculate the fitted regression equation for each group using the analysis of covariance in Equation 8.16. Why must the two sets of equations be the same (within rounding error)?

8.12 Adjusted means (concluded): The notion of an *adjusted mean* was discussed in Exercises 7.5, 7.10, and 8.8. Now consider the analysis-of-covariance model for two factors, R and C, and two covariates, X_1 and X_2:

$$Y_{ijk} = \mu + \alpha_j + \beta_k + \gamma_{jk} + \delta_1(X_{ijk1} - \bar{X}_1) + \delta_2(X_{ijk2} - \bar{X}) + \varepsilon_{ijk}$$

Note that this particular formulation of the model permits interactions between the factors but not between the factors and the covariates.

(a) How can the analysis-of-covariance model be used to compute adjusted cell means for the $r \times c$ combinations of levels of the factors R and C?

(b) In computing adjusted means, is anything gained by expressing the covariates as deviations from their respective means rather than as raw scores?

[29] See Exercise 8.11.

(c) If the interactions between the factors γ_{jk} are deleted from the model, how can we calculate adjusted means for the r categories of R and the c categories of C?

The calculation of adjusted means in additive ANCOVA models is a traditional use of the analysis of covariance. Further information on adjusted means can be found in Searle, et al. (1980).

8.5 Linear Contrasts of Means

I have explained how the overparameterized ANOVA model can be fit to data by placing a sufficient number of linear restrictions on the parameters of the model. Different restrictions produce different regressors and hence different parameter estimates, but identical sums of squares—at least for models that conform to the principle of marginality. We have examined in some detail two schemes for coding regressors for a factor: dummy (0, 1) coding and deviation (1, 0, −1) coding. The coefficients for a set of dummy-coded regressors compare each group of a factor with the baseline group, while the coefficients for a set of deviation-coded regressors compare each group (but the last) with the average of the groups.

We do not generally test hypotheses about individual coefficients for dummy-coded or deviation-coded regressors, but we can do so, if we wish. For dummy-coded regressors in one-way ANOVA, a t-test or F-test of H_0: $\alpha_1 = 0$, for example, is equivalent to testing for the difference in means between the first group and the baseline group, H_0: $\mu_1 = \mu_m$. For deviation-coded regressors, testing H_0: $\alpha_1 = 0$ is equivalent to testing for the difference between the mean of the first group and the average of all of the group means, H_0: $\mu_1 = \mu_{\cdot}$.

In this section, I shall explain a simple procedure for coding regressors that permits us to test specific hypotheses about *linear contrasts* (also called *linear comparisons*) among group means.[30] Although I shall develop this technique for one-way analysis of variance, contrast-coded regressors can also be employed for any factor in a two-way or higher-way ANOVA.[31]

For concreteness, let us examine the data in Table 8.8, which are drawn from an experimental study by Friendly and Franklin (1980) of the effects of presentation format on learning and memory.[32] Subjects participating in the experiment read a list of 40 words. Then, after performing a brief distracting task, the subjects were asked to recall as many of the words as possible. This procedure was repeated for five trials. Thirty subjects were randomly assigned to three conditions: In the control or "standard free recall" (*SFR*) condition, the order of presentation of the words on the list was randomized for each of the five trials of the experiment. In the two experimental conditions, recalled words were presented in the order in which they were listed by the subject on the previous trial. In one of these conditions (labeled *B*), the recalled words were presented

[30] A more general treatment of this topic may be found in Section 9.1.2.

[31] See Exercise 8.14.

[32] I am grateful to Michael Friendly of the Psychology Department, York University, for providing these data.

TABLE 8.8 Data From Friendly and Franklin's (1980) Experiment on the Effects of Presentation on Recall. The data in the table are the number of words correctly recalled by each subject on the final trial of the experiment.

Condition		
SFR	*B*	*M*
39	40	40
25	38	39
37	39	34
25	37	37
29	39	40
39	24	36
21	30	36
39	39	38
24	40	36
25	40	30

as a group *before* the forgotten ones, while in the other condition (labeled *M* for *meshed*), the recalled and forgotten words were interspersed. Friendly and Franklin expected that making the order of presentation contingent upon the subject's previous performance would enhance recall. The data recorded in the table are the number of words correctly recalled by each subject for the final trial of the experiment.

Means and standard deviations for Friendly and Franklin's memory data are as follows:

	Experimental Condition		
	SFR	*B*	*M*
Mean	30.30	36.60	36.60
Standard deviation	7.33	5.34	3.03

The mean number of words recalled is higher in the experimental conditions than in the control; the control group also has the largest standard deviation. A jittered scatterplot of the data, shown in Figure 8.8, reveals a potential problem: The data are disguised proportions (number correctly recalled of 40 words), and many subjects—particularly in the *B* and *M* conditions—are at or near the maximum. This "ceiling effect' produces negatively skewed distributions in the two experimental conditions.

A linear contrast in the group means tests the hypothesis that a particular linear combination of population group means is 0. For the Friendly and Franklin memory experiment, we might wish to test the null hypothesis that the mean for the control group is no different from the average of the means for the experimental groups:[33]

$$H_0\colon \mu_1 = \frac{\mu_2 + \mu_3}{2}$$

[33] It would also be reasonable to compare each experimental group with the control group. The comparison could be easily accomplished by using dummy coding, treating the control group as the baseline category.

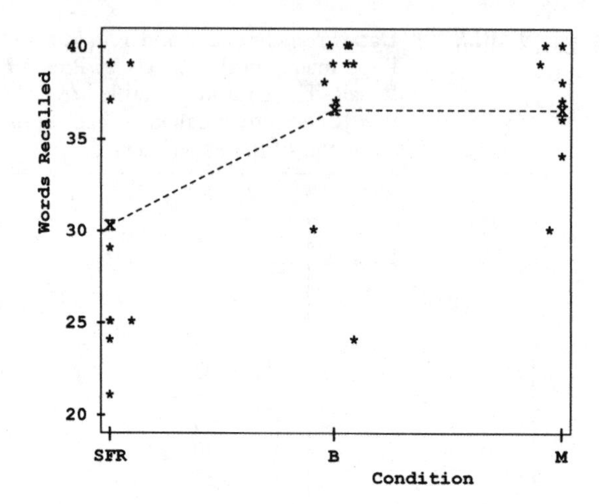

Figure 8.8. Number of words recalled (out of 40) by experimental condition, from Friendly and Franklin's (1980) memory experiment. The horizontal coordinates of the points are jittered to separate overlapping points. The mean of each group is plotted as an "X]' and the means are connected by a broken line.

and the hypothesis that the means for the two experimental groups are the same:

$$H_0\colon \mu_2 = \mu_3$$

The first hypothesis can be rewritten as

$$H_0\colon 1\mu_1 - \tfrac{1}{2}\mu_2 - \tfrac{1}{2}\mu_3 = 0$$

and the second hypothesis as

$$H_0\colon 0\mu_1 + 1\mu_2 - 1\mu_3 = 0$$

Then, the weights for the group means in these null hypotheses can be used to formulate two linear-contrast regressors, C_1 and C_2:

Group	C_1	C_2
(1) *SFR*	1	0
(2) *B*	$-\frac{1}{2}$	1
(3) *M*	$-\frac{1}{2}$	−1

This simple approach to coding linear contrasts will work as long as the following conditions are satisfied:[34]

1. We need one linear contrast for each degree of freedom. An m-category factor therefore requires $m - 1$ contrasts.

---

[34] See Section 9.1.2 for an explanation of these rules and for a more flexible and general approach to constructing contrasts.

2. Each column of the contrast-coding table must sum to 0.
3. The products of corresponding codes for *different* contrasts must also sum to 0. For the illustration:

$$(1 \times 0) + (-\tfrac{1}{2} \times 1) + (-\tfrac{1}{2} \times -1) = 0$$

When there are equal numbers of observations in the groups, these rules ensure that the contrast regressors are uncorrelated. As a consequence, the regression sum of squares for the ANOVA can be decomposed into components due to the contrasts. When the group frequencies are unequal, the regression sum of squares does not decompose in this simple manner, but properly formulated contrasts are still useful for testing hypotheses about the population group means. Because each contrast has 1 degree of freedom, we can test it by a t-test (dividing the estimated coefficient for the contrast by its standard error) or—equivalently—by an incremental F-test.

> Linear contrasts permit the researcher to test specific hypotheses about means within the framework of analysis of variance. A factor with m categories gives rise to $m-1$ contrasts, one for each degree of freedom. A simple procedure for constructing contrasts requires that the codes for each contrast sum to 0, and that the products of codes for each pair of contrasts also sum to 0.

For Friendly and Franklin's experiment, the fitted model is

$$\begin{aligned} Y = \underset{(1.008)}{34.5} &- \underset{(1.425)}{4.2C_1} + \underset{(1.234)}{0.0C_2} \\ R^2 &= .243 \end{aligned}$$

and the analysis-of-variance table is

Source	*SS*		*df*		MS	F	*p*
Groups	264.6		2		132.3	4.34	.02
C_1		264.6		1	264.6	8.68	.007
C_2		0.0		1	0.0	0.0	1.0
Residuals	822.9		27		30.48		
Total	1087.5		29				

Note that the coefficient and sum of squares for the second contrast are 0, because the sample group means are identical for the B and M groups. Because the sum of squares for C_2 is 0, the regression sum of squares is wholly captured by C_1. The first contrast—between the control group and the experimental groups—is highly statistically significant.

EXERCISES

8.13 #Testing contrasts using group means: Suppose that we wish to test a hypothesis concerning a contrast of group means:

$$H_0\colon\ c_1\mu_1 + c_2\mu_2 + \cdots + c_m\mu_m = 0$$

where $c_1 + c_2 + \cdots + c_k = 0$. Define the *sample value of the contrast* as

$$C \equiv c_1\bar{Y}_1 + c_2\bar{Y}_2 + \cdots + c_m\bar{Y}_m$$

and let

$$C'^2 \equiv \frac{C^2}{\dfrac{c_1^2}{n_1} + \dfrac{c_2^2}{n_2} + \cdots + \dfrac{c_m^2}{n_m}}$$

(a) *Show that under the null hypothesis:

(i) $E(C) = 0$.

(ii) $V(C) = \sigma_\varepsilon^2\left(\dfrac{c_1^2}{n_1} + \dfrac{c_2^2}{n_2} + \cdots + \dfrac{c_m^2}{n_m}\right)$.

(iii) $t_0 = C'/S_E$ follows a t-distribution with $n - m$ degrees of freedom. [*Hint*: The $\bar{Y}_j$ are independent, and each is distributed as $N(\mu_j, \sigma_\varepsilon^2/n_j)$.] C'^2 is the sum of squares for the contrast.

(b) Using Friendly and Franklin's memory data (in `friendly.dat`), verify that the test statistics obtained by the method of this exercise [i.e., in (a)(iii)] are the same as those produced by the incremental sum-of-squares approach, used in the text.

(c) Pecknold et al. (1982) describe an experiment in which psychiatric patients suffering from anxiety were assigned randomly to one of three groups: (i) a treatment group that received a standard antianxiety drug (diazepam); (ii) a treatment group that received a new antianxiety drug (fenobam); and (iii) a control group that received a placebo. At several points in the study, subjects were administered standard tests for anxiety, including the Hamilton Rating Scale; high scores on the scale are indicative of severe anxiety. After three weeks of treatment, the experiment produced the following results:

	Group		
	Diazepam	*Fenobam*	*Placebo*
Mean Hamilton score	16.00	15.11	23.54
Standard deviation	8.33	7.14	12.30
Number of subjects	9	9	11

Use the method derived in part (a) to test the null hypotheses that (i) the average anxiety score of patients receiving the drugs is no different from that of patients receiving the placebo; and (ii) the average anxiety score of patients receiving the new drug is the same as that of patients receiving diazepam. Note that the residual mean square can be calculated according to the formula

$$S_E^2 = \frac{\sum_{j=1}^{m} S_j^2(n_j - 1)}{n - m}$$

where S_j^2 is the variance within group j. *Why does this formula work?

8.14 *Contrasts in two-way ANOVA: A simple approach to formulating contrasts in two-way (and higher-way) ANOVA is first to specify contrasts separately for each set of main effects, obtaining interaction contrasts by forming all pairwise products of the main-effect contrasts. Then, as long as the main-effect contrasts meet the rules, the interaction contrasts will as well.

Imagine, for example, a 3×2 classification arising from an experiment in which the first factor consists of a control group (R_1) and two experimental groups (R_2 and R_3). The second factor is, say, gender, with categories male (C_1) and female (C_2). A possible set of main-effect contrasts for this experiment is

	Row contrast	
Group	A_1	A_2
R_1	2	0
R_2	−1	1
R_3	−1	−1

	Column contrast
Gender	B
C_1	1
C_2	−1

The following table shows the full set of main-effect and interaction contrasts for all six cells of the design (with the parameter for each contrast in parentheses):

Cell		(δ_1)	(δ_1)	(β)	(ζ_1)	(ζ_2)
Row	*Column*	A_1	A_2	B	A_1B	A_2B
1	1	2	0	1	2	0
1	2	2	0	−1	−2	0
2	1	−1	1	1	−1	1
2	2	−1	1	−1	1	−1
3	1	−1	−1	1	−1	−1
3	2	−1	−1	−1	1	1

Note that we use 2 degrees of freedom for condition main effects, 1 degree of freedom for the gender main effect, and 2 degrees of freedom for interaction.

Explain the meaning of the following null hypotheses:

(a) H_0: $\zeta_1 = 0$.
(b) H_0: $\zeta_2 = 0$.
(c) H_0: $\delta_1 = 0$.
(d) H_0: $\delta_2 = 0$.
(e) H_0: $\beta = 0$.

8.6 Summary

- One-way analysis of variance examines the relationship between a quantitative dependent variable and a categorical independent variable (or factor). The one-way ANOVA model

$$Y_{ij} = \mu + \alpha_j + \varepsilon_{ij}$$

is underdetermined because it uses $m + 1$ parameters to model m group means. The model can be solved, however, by placing a restriction on its parameters. Setting one of the α_j's to 0 leads to (0, 1) dummy-regressor coding. Constraining the α_j's to sum to 0 leads to (1, 0, -1) deviation-regressor coding. The two coding schemes are equivalent in that they provide the same fit to the data, producing the same regression and residual sums of squares.

- The two-way analysis-of-variance model

$$Y_{ijk} = \mu + \alpha_j + \beta_k + \gamma_{jk} + \varepsilon_{ijk}$$

incorporates the main effects and interactions of two factors. The factors interact when the profiles of population cell means are not parallel. The two-way ANOVA model is overparameterized, but it may be fit to data by placing suitable restrictions on its parameters. A convenient set of restrictions is provided by sigma constraints, specifying that each set of parameters (α_j, β_k, and γ_{jk}) sums to 0 over each of its coordinates. Testing hypotheses about the sigma-constrained parameters is equivalent to testing interaction-effect and main-effect hypotheses about cell and marginal means. There are two reasonable procedures for testing main-effect hypotheses in two-way ANOVA: Tests based on $SS(\alpha|\beta, \gamma)$ and $SS(\beta|\alpha, \gamma)$ employ models that violate the principle of marginality, but are valid whether or not interactions are present. Tests based on $SS(\alpha|\beta)$ and $SS(\beta|\alpha)$ conform to the principle of marginality, but are valid only if interactions are absent.

- The analysis-of-variance model and procedures for testing hypotheses about main effects and interactions extend straightforwardly to three-way and higher-way classifications. In each case, the highest-order interaction corresponds to the number of factors in the model. It is not necessary, however, to specify a model that includes all terms through the highest-order interaction. It is possible to fit an ANOVA model to a classification containing empty cells when the marginal frequency tables corresponding to the terms in the model have no empty cells.

- Analysis of covariance is an alternative parameterization of the dummy-regression model, employing deviation-coded regressors for factors and expressing covariates as deviations from their means. The ANCOVA model can incorporate interactions among factors and between factors and covariates.

- Linear contrasts permit the researcher to test specific hypotheses about means within the framework of analysis of variance. A factor with m categories gives rise to $m - 1$ contrasts, one for each degree of freedom. A simple procedure for constructing contrasts requires that the codes for each contrast sum to 0, and that the products of codes for each pair of contrasts also sum to 0.

9

Statistical Theory for Linear Models*

The purpose of this chapter is twofold: to deepen your knowledge of linear models and linear least-squares estimation; and to provide a basis for more advanced work in social statistics—in the remainder of this book and more generally. Relying on the mathematical tools of linear algebra and elementary calculus, we shall revisit with greater rigor many of the topics treated informally in Chapters 5 to 8, developing the statistical theory on which the methods described in those chapters depend. The next chapter, on the vector geometry of linear models, provides intuitive insight into this statistical theory.

9.1 Linear Models in Matrix Form

The general linear model is given by the equation

$$Y_i = \beta_0 + \beta_1 x_{i1} + \beta_2 x_{i2} + \cdots + \beta_k x_{ik} + \varepsilon_i$$

Notice that I have substituted the notationally more convenient β_0 for the constant α; I shall, for the time being, suppose that the X-values are fixed,[1] hence the lowercase x_{ij}.

Collecting the regressors into a row vector, appending a 1 for the constant, and placing the corresponding parameters in a column vector, permits us to

[1] See Section 9.5 for a discussion of random regressors.

rewrite the linear model as

$$Y_i = [1, x_{i1}, x_{i2}, \ldots, x_{ik}] \begin{bmatrix} \beta_0 \\ \beta_1 \\ \beta_2 \\ \vdots \\ \beta_k \end{bmatrix} + \varepsilon_i$$

$$= \underset{(1\times k+1)}{\mathbf{x}_i'} \underset{(k+1\times 1)}{\boldsymbol{\beta}} + \varepsilon_i$$

For a sample of n observations, we have n such equations, which can be combined into a single matrix equation:

$$\begin{bmatrix} Y_1 \\ Y_2 \\ \vdots \\ Y_n \end{bmatrix} = \begin{bmatrix} 1 & x_{11} & \cdots & x_{1k} \\ 1 & x_{21} & \cdots & x_{2k} \\ \vdots & \vdots & & \vdots \\ 1 & x_{n1} & \cdots & x_{nk} \end{bmatrix} \begin{bmatrix} \beta_0 \\ \beta_1 \\ \vdots \\ \beta_k \end{bmatrix} + \begin{bmatrix} \varepsilon_1 \\ \varepsilon_2 \\ \vdots \\ \varepsilon_n \end{bmatrix} \tag{9.1}$$

$$\underset{(n\times 1)}{\mathbf{y}} = \underset{(n\times k+1)}{\mathbf{X}} \underset{(k+1\times 1)}{\boldsymbol{\beta}} + \underset{(n\times 1)}{\boldsymbol{\varepsilon}}$$

As we shall see, with suitable specification of the contents of **X**, called the *model matrix*, Equation 9.1 serves not only for multiple regression, but for linear models generally.[2]

Because $\boldsymbol{\varepsilon}$ is a vector random variable, the assumptions of the linear model can be compactly restated in matrix form. The errors are assumed to be independent and normally distributed with zero expectation and common variance. Thus, $\boldsymbol{\varepsilon}$ follows a multivariate-normal distribution with expectation $E(\boldsymbol{\varepsilon}) = \underset{(n\times 1)}{\mathbf{0}}$ and covariance matrix $V(\boldsymbol{\varepsilon}) = E(\boldsymbol{\varepsilon}\boldsymbol{\varepsilon}') = \sigma_\varepsilon^2 \mathbf{I}_n$; in symbols, $\boldsymbol{\varepsilon} \sim N_n(\mathbf{0}, \sigma_\varepsilon^2 \mathbf{I}_n)$. The distribution of $\mathbf{y}$ follows immediately:

$$\begin{aligned} \boldsymbol{\mu} \equiv E(\mathbf{y}) &= E(\mathbf{X}\boldsymbol{\beta} + \boldsymbol{\varepsilon}) = \mathbf{X}\boldsymbol{\beta} + E(\boldsymbol{\varepsilon}) = \mathbf{X}\boldsymbol{\beta} \\ V(\mathbf{y}) &= E[(\mathbf{y} - \boldsymbol{\mu})(\mathbf{y} - \boldsymbol{\mu})'] = E[(\mathbf{y} - \mathbf{X}\boldsymbol{\beta})(\mathbf{y} - \mathbf{X}\boldsymbol{\beta})'] \\ &= E(\boldsymbol{\varepsilon}\boldsymbol{\varepsilon}') = \sigma_\varepsilon^2 \mathbf{I}_n \end{aligned} \tag{9.2}$$

Furthermore, because it is simply a translation of $\boldsymbol{\varepsilon}$ to a different expectation, $\mathbf{y}$ is also normally distributed: $\mathbf{y} \sim N_n(\mathbf{X}\boldsymbol{\beta}, \sigma_\varepsilon^2 \mathbf{I}_n)$.

---

[2] The model matrix is often called the *design matrix*, a term that is particularly appropriate in experimental applications where the independent variables, and hence the regressors that compose the **X** matrix, derive from the design of the experiment.

> The general linear model can be written in matrix form as $\mathbf{y} = \mathbf{X}\boldsymbol{\beta}+\boldsymbol{\varepsilon}$, where $\mathbf{y}$ is an $n \times 1$ vector of dependent-variable observations; $\mathbf{X}$ is an $n \times k+1$ matrix of regressors (called the model matrix), including an initial column of 1's for the constant regressor; $\boldsymbol{\beta}$ is a $k+1 \times 1$ vector of parameters to be estimated; and $\boldsymbol{\varepsilon}$ is an $n \times 1$ vector of errors. The assumptions of the linear model can be compactly written as $\boldsymbol{\varepsilon} \sim N_n(\mathbf{0}, \sigma_\varepsilon^2 \mathbf{I}_n)$.

9.1.1 Dummy Regression and Analysis of Variance

The model matrices for dummy-regression and analysis-of-variance models—particularly the latter—are strongly patterned. Consider the dummy-regression model

$$Y_i = \alpha + \beta x_i + \gamma d_i + \delta(x_i d_i) + \varepsilon_i$$

where Y is income, x is years of education, and the dummy regressor d is coded 1 for men and 0 for women.[3] In matrix form, this model becomes

$$\begin{bmatrix} Y_1 \\ \vdots \\ Y_{n_1} \\ \hline Y_{n_1+1} \\ \vdots \\ Y_n \end{bmatrix} = \begin{bmatrix} 1 & x_1 & 0 & 0 \\ \vdots & \vdots & \vdots & \vdots \\ 1 & x_{n_1} & 0 & 0 \\ \hline 1 & x_{n_1+1} & 1 & x_{n_1+1} \\ \vdots & \vdots & \vdots & \vdots \\ 1 & x_n & 1 & x_n \end{bmatrix} \begin{bmatrix} \alpha \\ \beta \\ \gamma \\ \delta \end{bmatrix} + \begin{bmatrix} \varepsilon_1 \\ \vdots \\ \varepsilon_{n_1} \\ \hline \varepsilon_{n_1+1} \\ \vdots \\ \varepsilon_n \end{bmatrix}$$

$$\mathbf{y} = \mathbf{X}\boldsymbol{\beta} + \boldsymbol{\varepsilon}$$

To emphasize the pattern of the model matrix, the n_1 observations for women (for whom d and hence xd are 0) precede the $n - n_1$ observations for men.

Now consider the overparameterized one-way analysis-of-variance model[4]

$$Y_{ij} = \mu + \alpha_j + \varepsilon_{ij} \quad \text{for groups } j = 1, \ldots, m$$

---

[3] This example was discussed in Chapter 7. Here, x and d are treated as fixed; random regressors are considered in Section 9.5.

[4] See Section 8.1.

The matrix form of the model is

$$
\left[\begin{array}{c}
Y_{11} \\ \vdots \\ Y_{n_1,1} \\ \hline Y_{12} \\ \vdots \\ Y_{n_2,2} \\ \hline \vdots \\ \hline Y_{1,m-1} \\ \vdots \\ Y_{n_{m-1},m-1} \\ \hline Y_{1m} \\ \vdots \\ Y_{n_m,m}
\end{array}\right]
=
\left[\begin{array}{cccccc}
1 & 1 & 0 & \cdots & 0 & 0 \\
\vdots & \vdots & \vdots & & \vdots & \vdots \\
1 & 1 & 0 & \cdots & 0 & 0 \\ \hline
1 & 0 & 1 & \cdots & 0 & 0 \\
\vdots & \vdots & \vdots & & \vdots & \vdots \\
1 & 0 & 1 & \cdots & 0 & 0 \\ \hline
\vdots & \vdots & \vdots & & \vdots & \vdots \\ \hline
1 & 0 & 0 & \cdots & 1 & 0 \\
\vdots & \vdots & \vdots & & \vdots & \vdots \\
1 & 0 & 0 & \cdots & 1 & 0 \\ \hline
1 & 0 & 0 & \cdots & 0 & 1 \\
\vdots & \vdots & \vdots & & \vdots & \vdots \\
1 & 0 & 0 & \cdots & 0 & 1
\end{array}\right]
\left[\begin{array}{c}
\mu \\ \alpha_1 \\ \alpha_2 \\ \vdots \\ \alpha_{m-1} \\ \alpha_m
\end{array}\right]
+
\left[\begin{array}{c}
\varepsilon_{11} \\ \vdots \\ \varepsilon_{n_1,1} \\ \hline \varepsilon_{12} \\ \vdots \\ \varepsilon_{n_2,2} \\ \hline \vdots \\ \hline \varepsilon_{1,m-1} \\ \vdots \\ \varepsilon_{n_{m-1},m-1} \\ \hline \varepsilon_{1m} \\ \vdots \\ \varepsilon_{n_m,m}
\end{array}\right]
$$

$$\mathbf{y} = \mathbf{X}\boldsymbol{\beta} + \boldsymbol{\varepsilon}$$

It is apparent that the model matrix is of rank m, one less than the number of columns, because the first column is the sum of the others. One solution is to delete a column, implicitly setting the corresponding parameter to 0. Deleting the last column of the model matrix, for example, sets $\alpha_m = 0$, establishing the last category as the baseline for a dummy-coding scheme.

Alternatively, imposing the sigma constraint $\sum_{j=1}^{m} \alpha_j = 0$ on the parameters leads to the following *full-rank* model matrix $\mathbf{X}_F$, composed of deviation-coded regressors; labeling each column of the matrix with the parameter to which it pertains,

$$
\underset{(n \times m)}{\mathbf{X}_F} =
\left[\begin{array}{ccccc}
(\mu) & (\alpha_1) & (\alpha_2) & \cdots & (\alpha_{m-1}) \\
1 & 1 & 0 & \cdots & 0 \\
\vdots & \vdots & \vdots & & \vdots \\
1 & 1 & 0 & \cdots & 0 \\ \hline
1 & 0 & 1 & \cdots & 0 \\
\vdots & \vdots & \vdots & & \vdots \\
1 & 0 & 1 & \cdots & 0 \\ \hline
\vdots & \vdots & \vdots & & \vdots \\ \hline
1 & 0 & 0 & \cdots & 1 \\
\vdots & \vdots & \vdots & & \vdots \\
1 & 0 & 0 & \cdots & 1 \\ \hline
1 & -1 & -1 & \cdots & -1 \\
\vdots & \vdots & \vdots & & \vdots \\
1 & -1 & -1 & \cdots & -1
\end{array}\right]
$$

There is, then, the following relationship between the group means $\boldsymbol{\mu} = \{\mu_j\}$ and the parameters of the constrained model:

$$\begin{bmatrix} \mu_1 \\ \mu_2 \\ \vdots \\ \mu_{m-1} \\ \mu_m \end{bmatrix} = \begin{bmatrix} 1 & 1 & 0 & 0 \\ 1 & 0 & 1 & 0 \\ \vdots & \vdots & \vdots & \vdots \\ 1 & 0 & 0 & 1 \\ 1 & -1 & -1 & -1 \end{bmatrix} \begin{bmatrix} \mu \\ \alpha_1 \\ \alpha_2 \\ \vdots \\ \alpha_{m-1} \end{bmatrix} \qquad [9.3]$$

$$\underset{(m\times 1)}{\boldsymbol{\mu}} = \underset{(m\times m)}{\mathbf{X}_B}\ \underset{(m\times 1)}{\boldsymbol{\beta}_F}$$

In this *parametric equation*, $\mathbf{X}_B$ is the *row basis* of the full-rank model matrix, consisting of the m unique rows of $\mathbf{X}_F$, one for each group; and $\boldsymbol{\beta}_F$ is the parameter vector associated with the full-rank model matrix. By construction, the $m \times m$ matrix $\mathbf{X}_B$ is of full column rank and hence nonsingular, allowing us to invert $\mathbf{X}_B$ and solve uniquely for the constrained parameters in terms of the cell means: $\boldsymbol{\beta}_F = \mathbf{X}_B^{-1}\boldsymbol{\mu}$. The solution forms a familiar pattern:[5]

$$\begin{bmatrix} \mu \\ \alpha_1 \\ \alpha_2 \\ \vdots \\ \alpha_{m-1} \end{bmatrix} = \begin{bmatrix} \mu. \\ \mu_1 - \mu. \\ \mu_2 - \mu. \\ \vdots \\ \mu_{m-1} - \mu. \end{bmatrix}$$

Let us next examine a two-way ANOVA model. To make the example manageable, suppose that there are two rows and three columns in the design. Imposing sigma constraints on the main effects and interactions produces the parametric equation:

$$\begin{bmatrix} \mu_{11} \\ \mu_{12} \\ \mu_{13} \\ \mu_{21} \\ \mu_{22} \\ \mu_{23} \end{bmatrix} = \begin{bmatrix} 1 & 1 & 1 & 0 & 1 & 0 \\ 1 & 1 & 0 & 1 & 0 & 1 \\ 1 & 1 & -1 & -1 & -1 & -1 \\ 1 & -1 & 1 & 0 & -1 & 0 \\ 1 & -1 & 0 & 1 & 0 & 1 \\ 1 & -1 & -1 & -1 & 1 & -1 \end{bmatrix} \begin{bmatrix} \mu \\ \alpha_1 \\ \beta_1 \\ \beta_2 \\ \gamma_{11} \\ \gamma_{12} \end{bmatrix} \qquad [9.4]$$

$$\underset{(6\times 1)}{\boldsymbol{\mu}} = \underset{(6\times 6)}{\mathbf{X}_B}\ \underset{(6\times 1)}{\boldsymbol{\beta}_F}$$

As in one-way ANOVA, the row basis of the full-rank model matrix is nonsingular by construction, yielding the following solution for the parameters in terms of the cell means:[6]

$$\begin{bmatrix} \mu \\ \alpha_1 \\ \beta_1 \\ \beta_2 \\ \gamma_{11} \\ \gamma_{12} \end{bmatrix} = \begin{bmatrix} \mu.. \\ \mu_{1}. - \mu.. \\ \mu._{1} - \mu.. \\ \mu._{2} - \mu.. \\ \mu_{11} - \mu_{1}. - \mu._{1} + \mu.. \\ \mu_{12} - \mu_{1}. - \mu._{2} + \mu.. \end{bmatrix}$$

---

[5] See Exercise 9.1(a) and Section 8.1.
[6] See Exercise 9.1(b) and Section 8.2.2.

> The model matrices for dummy-regression and analysis-of-variance models are strongly patterned. In analysis of variance, the relationship between group or cell means and the parameters of the linear model is expressed by the parametric equation $\boldsymbol{\mu} = \mathbf{X}_B \boldsymbol{\beta}_F$, where $\boldsymbol{\mu}$ is the vector of means; $\mathbf{X}_B$ is the row basis of the full-rank model matrix; and $\boldsymbol{\beta}_F$ is the parameter vector associated with the full-rank model matrix. Solving the parametric equation for the parameters yields $\boldsymbol{\beta}_F = \mathbf{X}_B^{-1} \boldsymbol{\mu}$.

9.1.2 Linear Contrasts

The relationship between group means and the parameters of the ANOVA model is given by the parametric equation $\boldsymbol{\mu} = \mathbf{X}_B \boldsymbol{\beta}_F$; thus, as I have explained, the parameters are linear functions of the group means, $\boldsymbol{\beta}_F = \mathbf{X}_B^{-1} \boldsymbol{\mu}$. The full-rank parameterizations of the one-way ANOVA model that we have considered—dummy coding and deviation coding—permit us to test the null hypothesis of no differences among group means, but the individual parameters are not usually of interest. In certain circumstances, however, we can formulate $\mathbf{X}_B$ so that the individual parameters of $\boldsymbol{\beta}_F$ incorporate interesting contrasts among group means.[7]

In Friendly and Franklin's (1980) memory experiment,[8] for example, subjects attempted to recall words under three experimental conditions:

1. the "standard free recall" (*SFR*) condition, in which words were presented in random order;
2. the "before" (*B*) condition, in which remembered words were presented before those forgotten on the previous trial; and
3. the "meshed" (*M*) condition, in which remembered words were interspersed with forgotten words, but were presented in the order in which they were recalled.

I defined linear contrasts to test two null hypotheses:

1. H_0: $\mu_1 = (\mu_2 + \mu_3)/2$, that the mean of the *SFR* condition does not differ from the average of the means of the other two conditions; and
2. H_0: $\mu_2 = \mu_3$, that the means of the *B* and *M* conditions do not differ.

These hypotheses can be written as linear functions of the group means: (1) H_0: $1\mu_1 - \frac{1}{2}\mu_2 - \frac{1}{2}\mu_3 = 0$; and (2) H_0: $0\mu_1 + 1\mu_2 - 1\mu_3 = 0$. Then each

[7] Linear contrasts were introduced in Section 8.5.

[8] See Section 8.5.

hypothesis can be coded in a parameter of the model, employing the following relationship between parameters and group means:[9]

$$\begin{bmatrix} \mu \\ \zeta_1 \\ \zeta_2 \end{bmatrix} = \begin{bmatrix} \frac{1}{3} & \frac{1}{3} & \frac{1}{3} \\ 1 & -\frac{1}{2} & -\frac{1}{2} \\ 0 & 1 & -1 \end{bmatrix} \begin{bmatrix} \mu_1 \\ \mu_2 \\ \mu_3 \end{bmatrix} \qquad [9.5]$$

$$\boldsymbol{\beta}_F = \mathbf{X}_B^{-1}\boldsymbol{\mu}$$

One parameter, μ, is used to code the average of the group means, leaving two parameters to represent differences among the three group means. The hypothesis H_0: $\zeta_1 = 0$ is equivalent to the first null hypothesis; H_0: $\zeta_2 = 0$ is equivalent to the second null hypothesis.

Because the *rows* of $\mathbf{X}_B^{-1}$ in Equation 9.5 are orthogonal, the *columns* of $\mathbf{X}_B$ are orthogonal as well: Each column of $\mathbf{X}_B$ is equal to the corresponding row of $\mathbf{X}_B^{-1}$ divided by the sum of squared entries in that row;[10] thus,

$$\begin{bmatrix} \mu_1 \\ \mu_2 \\ \mu_3 \end{bmatrix} = \begin{bmatrix} 1 & \frac{2}{3} & 0 \\ 1 & -\frac{1}{3} & \frac{1}{2} \\ 1 & -\frac{1}{3} & -\frac{1}{2} \end{bmatrix} \begin{bmatrix} \mu \\ \zeta_1 \\ \zeta_2 \end{bmatrix}$$

$$\boldsymbol{\mu} = \mathbf{X}_B\boldsymbol{\beta}_F$$

The one-to-one correspondence between rows of $\mathbf{X}_B^{-1}$ and columns of $\mathbf{X}_B$ makes it simple to specify the latter matrix directly. Moreover, we can rescale the columns of the row basis for more convenient coding, as shown in Equation 9.6, without altering the hypotheses incorporated in the contrast coefficients: If, for example, $\zeta_1 = 0$, then any multiple of ζ_1 is 0 as well.

$$\mathbf{X}_B = \begin{bmatrix} 1 & 2 & 0 \\ 1 & -1 & 1 \\ 1 & -1 & -1 \end{bmatrix} \qquad [9.6]$$

Although it is convenient to define contrasts that are orthogonal in the row basis, it is not necessary to do so. It is always possible to work backward from $\mathbf{X}_B^{-1}$ (which expresses the parameters of the model as linear functions of the population group means) to $\mathbf{X}_B$, as long as the comparisons specified by $\mathbf{X}_B^{-1}$ are linearly independent. Linear independence is required to ensure that $\mathbf{X}_B^{-1}$ is nonsingular.[11]

If there are equal numbers (say n') of observations in the several groups, then an orthogonal design-matrix basis $\mathbf{X}_B$ implies an orthogonal full-rank design matrix $\mathbf{X}_F$—because $\mathbf{X}_F$ is produced by repeating each of the rows of $\mathbf{X}_B$ an equal number (n') of times. The columns of an orthogonal design matrix

[9] Because we naturally form hypotheses as linear combinations of means, we specify $\mathbf{X}_B^{-1}$ directly, rather than $\mathbf{X}^{-1}$.

[10] See Exercise 9.2.

[11] See Exercise 9.3.

represent independent sources of variation in the dependent variable, and therefore a set of orthogonal contrasts partitions the regression sum of squares into 1-degree-of-freedom components, each testing a hypothesis of interest. When it is applicable, this is an elegant approach to linear-model analysis. Linear comparisons may well be of interest, however, even if group frequencies are unequal, causing contrasts that are orthogonal in $\mathbf{X}_B$ to be correlated in $\mathbf{X}_F$.[12]

EXERCISES

9.1 *Solving the parametric equations in one-way and two-way ANOVA:

(a) Show that the parametric equation (Equation 9.3) in one-way analysis of variance has the general solution

$$\begin{bmatrix} \mu \\ \alpha_1 \\ \alpha_2 \\ \vdots \\ \alpha_{m-1} \end{bmatrix} = \begin{bmatrix} \mu. \\ \mu_1 - \mu. \\ \mu_2 - \mu. \\ \vdots \\ \mu_{m-1} - \mu. \end{bmatrix}$$

(b) Show that the parametric equation (Equation 9.4) in two-way analysis of variance, with two rows and three columns, has the solution

$$\begin{bmatrix} \mu \\ \alpha_1 \\ \beta_1 \\ \beta_2 \\ \gamma_{11} \\ \gamma_{12} \end{bmatrix} = \begin{bmatrix} \mu.. \\ \mu_{1}. - \mu.. \\ \mu._1 - \mu.. \\ \mu._2 - \mu.. \\ \mu_{11} - \mu_{1}. - \mu._1 + \mu.. \\ \mu_{12} - \mu_{1}. - \mu._2 + \mu.. \end{bmatrix}$$

9.2 *Orthogonal contrasts: Consider the equation $\boldsymbol{\beta}_F = \mathbf{X}_B^{-1}\boldsymbol{\mu}$ relating the parameters $\boldsymbol{\beta}_F$ of the full-rank analysis-of-variance model to the cell means $\boldsymbol{\mu}$. Suppose that $\mathbf{X}_B^{-1}$ is constructed so that its rows are orthogonal. Show that the columns of the row basis $\mathbf{X}_B$ of the model matrix are also orthogonal, and further that each column of $\mathbf{X}_B$ is equal to the corresponding row of $\mathbf{X}_B^{-1}$ divided by the sum of squared entries in that row. (*Hint*: Multiply $\mathbf{X}_B^{-1}$ by its transpose.)

9.3 Nonorthogonal contrasts: Imagine that we want to compare each of three groups in a one-way ANOVA with a fourth (control) group. We know that coding three dummy regressors, treating group 4 as the baseline category, will accomplish this purpose. Starting with the equation

$$\begin{bmatrix} \mu \\ \gamma_1 \\ \gamma_2 \\ \gamma_3 \end{bmatrix} = \begin{bmatrix} 0 & 0 & 0 & 1 \\ 1 & 0 & 0 & -1 \\ 0 & 1 & 0 & -1 \\ 0 & 0 & 1 & -1 \end{bmatrix} \begin{bmatrix} \mu_1 \\ \mu_2 \\ \mu_3 \\ \mu_4 \end{bmatrix}$$

[12] These conclusions are supported by the vector geometry of linear models, described in Chapter 10.

show that the row basis $\mathbf{X}_B = (\mathbf{X}_B^{-1})^{-1}$ of the model matrix is equivalent to dummy coding.

9.2 Least-Squares Fit

To find the least-squares coefficients, we write the fitted linear model as

$$\mathbf{y} = \mathbf{X}\mathbf{b} + \mathbf{e}$$

where $\mathbf{b} = [B_0, B_1, \ldots, B_k]'$ is the vector of fitted coefficients, and $\mathbf{e} = [E_1, E_2, \ldots, E_n]'$ is the vector of residuals. We seek the coefficient vector $\mathbf{b}$ that minimizes the residual sum of squares expressed as a function of $\mathbf{b}$:

$$\begin{aligned} S(\mathbf{b}) &= \sum E_i^2 = \mathbf{e}'\mathbf{e} = (\mathbf{y} - \mathbf{X}\mathbf{b})'(\mathbf{y} - \mathbf{X}\mathbf{b}) \\ &= \mathbf{y}'\mathbf{y} - \mathbf{y}'\mathbf{X}\mathbf{b} - \mathbf{b}'\mathbf{X}'\mathbf{y} + \mathbf{b}'\mathbf{X}'\mathbf{X}\mathbf{b} \\ &= \mathbf{y}'\mathbf{y} - (2\mathbf{y}'\mathbf{X})\mathbf{b} + \mathbf{b}'(\mathbf{X}'\mathbf{X})\mathbf{b} \end{aligned} \qquad [9.7]$$

Although matrix multiplication is not generally commutative, each product in Equation 9.7 is (1×1); thus, $\mathbf{y}'\mathbf{X}\mathbf{b} = \mathbf{b}'\mathbf{X}\mathbf{y}$, justifying the transition to the last line of the equation.[13]

From the point of view of the coefficient vector $\mathbf{b}$, Equation 9.7 consists of a constant, a linear form in $\mathbf{b}$, and a quadratic form in $\mathbf{b}$. To minimize $S(\mathbf{b})$, we find its vector partial derivative with respect to $\mathbf{b}$:

$$\frac{\partial S(\mathbf{b})}{\partial \mathbf{b}} = 0 - 2\mathbf{X}'\mathbf{y} + 2\mathbf{X}'\mathbf{X}\mathbf{b}$$

Setting this derivative to $\mathbf{0}$ produces the matrix form of the normal equations for the linear model:

$$\mathbf{X}'\mathbf{X}\mathbf{b} = \mathbf{X}'\mathbf{y} \qquad [9.8]$$

There are $k + 1$ normal equations in the same number of unknown coefficients. If $\mathbf{X}'\mathbf{X}$ is nonsingular—that is, of rank $k + 1$—then we can uniquely solve for the least-squares coefficients:

$$\mathbf{b} = (\mathbf{X}'\mathbf{X})^{-1}\mathbf{X}'\mathbf{y}$$

---

[13] See Exercise 9.4.

The rank of $\mathbf{X}'\mathbf{X}$ is equal to the rank of $\mathbf{X}$:

- Because the rank of $\mathbf{X}$ can be no greater than the smaller of n and $k+1$, for the least-squares coefficients to be unique, we require at least as many observations (n) as there are coefficients in the model ($k+1$). This requirement is intuitively sensible: We cannot, for example, fit a unique line to a single data point, nor can we fit a unique plane to two data points. In most applications, n greatly exceeds $k+1$.
- The $k+1$ columns of $\mathbf{X}$ must be linearly independent. This requirement implies that no regressor can be a perfect linear functions of others, and that only the constant regressor can be invariant.[14]

In applications, these requirements are usually met: $\mathbf{X}'\mathbf{X}$, therefore, is generally nonsingular, and the least-squares coefficients are uniquely defined.[15]

The second partial derivative of the sum of squared residuals is

$$\frac{\partial^2 S(\mathbf{b})}{\partial \mathbf{b}\, \partial \mathbf{b}'} = 2\mathbf{X}'\mathbf{X}$$

Because $\mathbf{X}'\mathbf{X}$ is positive definite when $\mathbf{X}$ is of full rank, the solution $\mathbf{b} = (\mathbf{X}'\mathbf{X})^{-1}\mathbf{X}'\mathbf{y}$ represents a minimum of $S(\mathbf{b})$.

The least-squares coefficients are given by $\mathbf{b} = (\mathbf{X}'\mathbf{X})^{-1}\mathbf{X}'\mathbf{y}$.

The matrix $\mathbf{X}'\mathbf{X}$ contains sums of squares and products among the regressors (including the constant regressor, $X_0 = 1$); the vector $\mathbf{X}'\mathbf{y}$ contains sums of cross products between the regressors and the dependent variable. Forming these matrix products, and expressing the normal Equations 9.8 in scalar format, yields the familiar pattern:[16]

$$\begin{array}{llll}
B_0 n & +B_1 \sum x_{i1} & +\cdots + B_k \sum x_{ik} & = \sum Y_i \\
B_0 \sum x_{i1} & +B_1 \sum x_{i1}^2 & +\cdots + B_k \sum x_{i1}x_{ik} & = \sum x_{i1} Y_i \\
\vdots & & & \vdots \\
B_0 \sum x_{ik} & +B_1 \sum x_{ik}x_{i1} & +\cdots + B_k \sum x_{ik}^2 & = \sum x_{ik} Y_i
\end{array}$$

To write an explicit solution to the normal equations in scalar form would be impractical, even for small values of k.

---

[14] If another regressor is invariant, then it is a multiple of the constant regressor, $X_0 = 1$.

[15] We shall see in Section 13.1, however, that even when $\mathbf{X}$ is of rank $k+1$, *near*-collinearity of its columns can cause statistical difficulties.

[16] See Section 5.2.2.

For Duncan's regression of occupational prestige on the income and educational levels of 45 U. S. occupations, the sums of squares and products are as follows:[17]

$$\mathbf{X}'\mathbf{X} = \begin{bmatrix} 45 & 1884 & 2365 \\ 1884 & 105{,}148 & 122{,}197 \\ 2365 & 122{,}197 & 163{,}265 \end{bmatrix}$$

$$\mathbf{X}'\mathbf{y} = \begin{bmatrix} 2146 \\ 118{,}229 \\ 147{,}936 \end{bmatrix}$$

The inverse of $\mathbf{X}'\mathbf{X}$ is

$$(\mathbf{X}'\mathbf{X})^{-1} = \begin{bmatrix} 0.1021058996 & -0.0008495732 & -0.0008432006 \\ -0.0008495732 & 0.0000801220 & -0.0000476613 \\ -0.0008432006 & -0.0000476613 & 0.0000540118 \end{bmatrix}$$

and thus the least-squares regression coefficients are

$$\mathbf{b} = (\mathbf{X}'\mathbf{X})^{-1}\mathbf{X}'\mathbf{y} = \begin{bmatrix} -6.06466 \\ 0.59873 \\ 0.54583 \end{bmatrix}$$

EXERCISES

9.4 Verify that each of the terms in the sum-of-squares function

$$S(\mathbf{b}) = \mathbf{y}'\mathbf{y} - \mathbf{y}'\mathbf{X}\mathbf{b} - \mathbf{b}'\mathbf{X}'\mathbf{y} + \mathbf{b}'\mathbf{X}'\mathbf{X}\mathbf{b}$$

is (1×1), justifying writing

$$S(\mathbf{b}) = \mathbf{y}'\mathbf{y} - (2\mathbf{y}'\mathbf{X})\mathbf{b} + \mathbf{b}'(\mathbf{X}'\mathbf{X})\mathbf{b}$$

9.5 Using a calculator or a computer program that conveniently performs matrix computations, and working with the Canadian data for the regression of occupational prestige on income, education, and percentage of women (in `prestige.dat`):[18]

(a) Compute the least-squares regression coefficients, $\mathbf{b} = (\mathbf{X}'\mathbf{X})^{-1}\mathbf{X}'\mathbf{y}$.

[17] Cf. the scalar calculations for Duncan's regression, which appear in Section 5.2.1.

[18] Many computer programs (e.g., APL, Gauss, Lisp-Stat, Mathematica, S, and SAS/IML) include convenient facilities for matrix calculations.

TABLE 9.1 Correlations for Blau and Duncan's Stratification Data, $n \simeq 20{,}700$: X_1 = Father's Education; X_2 = Father's Occupational Status; X_3 = Respondent's Education; X_4 = Status of Respondent's First Job; Y = Respondent's Current Occupational Status.

	X_1	X_2	X_3	X_4	Y
X_1	1.000				
X_2	.516	1.000			
X_3	.453	.438	1.000		
X_4	.332	.417	.538	1.000	
Y	.322	.405	.596	.541	1.000

Source: Blau and Duncan (1967, p. 169).

(b) Verify that the least-squares slope coefficients $\mathbf{b}_1 = [B_1, B_2, B_3]'$ can be computed as $\mathbf{b}_1 = (\mathbf{X}^{*\prime}\mathbf{X}^*)^{-1}\mathbf{X}^{*\prime}\mathbf{y}^*$, where $\mathbf{X}^*$ and $\mathbf{y}^*$ contain mean deviations for the X's and Y, respectively.

9.6 Standardized regression coefficients: Standardized regression coefficients were introduced in Section 5.2.4.

(a) *Show that the standardized coefficients can be computed as $\mathbf{b}^* = \mathbf{R}_{XX}^{-1}\mathbf{r}_{Xy}$, where $\mathbf{R}_{XX}$ is the correlation matrix of the independent variables and $\mathbf{r}_{Xy}$ is the vector of correlations between the independent variables and the dependent variable. [*Hints*: Let $\underset{(n\times k)}{\mathbf{Z}_X} \equiv \{(X_{ij}-\overline{X}_j)/S_j\}$ contain the standardized independent variables, and let $\underset{(n\times 1)}{\mathbf{z}_y} \equiv \{(Y_i - \overline{Y})/S_Y\}$ contain the standardized dependent variable. The regression equation for the standardized variables in matrix form is $\mathbf{z}_y = \mathbf{Z}_X\mathbf{b}^* + \mathbf{e}^*$. Multiply both sides of this equation by $\mathbf{Z}_X'/(n-1)$.]

(b) The correlation matrix in Table 9.1 is taken from Blau and Duncan's (1967) work on social stratification. Using these correlations, along with the results in part (a), find the standardized coefficients for the regression of current occupational status on father's education, father's occupational status, respondent's education, and the status of the respondent's first job. Why is the slope for father's education so small? Is it reasonable to conclude that father's education is unimportant as a cause of respondent's occupational status (recall Section 6.3)?

(c) *Prove that the squared multiple correlation for the regression of Y on $X_1, \ldots, X_k$ can be written as

$$R^2 = B_1^* r_{r1} + \cdots + B_k^* r_{rk} = \mathbf{r}_{yX}'\mathbf{b}^*$$

[*Hint*: Multiply $\mathbf{z}_y = \mathbf{Z}_X\mathbf{b}^* + \mathbf{e}^*$ through by $\mathbf{z}_y'/(n-1)$.] Use this result to calculate the multiple correlation for Blau and Duncan's regression.

9.3 Properties of the Least-Squares Estimator

In this section, I derive a number of fundamental results concerning the least-squares estimator $\mathbf{b}$ of the linear-model parameter vector $\boldsymbol{\beta}$. These results serve several related purposes:

- They establish certain desirable properties of the least-squares estimator that hold under the assumptions of the linear model.
- They furnish a basis for using the least-squares coefficients to make statistical inferences about $\boldsymbol{\beta}$.[19]
- They provide a foundation for generalizing the linear model in several directions.[20]

9.3.1 The Distribution of the Least-Squares Estimator

With the model matrix $\mathbf{X}$ fixed, the least-squares coefficients $\mathbf{b}$ result from a linear transformation of the dependent-variable observations; that is, $\mathbf{b}$ is a *linear estimator*:

$$\mathbf{b} = (\mathbf{X}'\mathbf{X})^{-1}\mathbf{X}'\mathbf{y} = \mathbf{M}\mathbf{y}$$

defining $\mathbf{M} \equiv (\mathbf{X}'\mathbf{X})^{-1}\mathbf{X}'$. The expected value of $\mathbf{b}$ is easily established from the expectation of $\mathbf{y}$ (given previously in Equation 9.2):

$$E(\mathbf{b}) = E(\mathbf{M}\mathbf{y}) = \mathbf{M}E(\mathbf{y}) = (\mathbf{X}'\mathbf{X})^{-1}\mathbf{X}'(\mathbf{X}\boldsymbol{\beta}) = \boldsymbol{\beta}$$

The least-squares estimator $\mathbf{b}$ is therefore an unbiased estimator of $\boldsymbol{\beta}$.

The covariance matrix of the least-squares estimator is similarly derived:

$$V(\mathbf{b}) = \mathbf{M}V(\mathbf{y})\mathbf{M}' = [(\mathbf{X}'\mathbf{X})^{-1}\mathbf{X}']\sigma_\varepsilon^2\mathbf{I}_n[(\mathbf{X}'\mathbf{X})^{-1}\mathbf{X}']'$$

Moving the scalar error variance σ_ε^2 to the front of this expression, and noting that $(\mathbf{X}'\mathbf{X})^{-1}$ is the inverse of a symmetric matrix and is thus itself symmetric, we get

$$V(\mathbf{b}) = \sigma_\varepsilon^2(\mathbf{X}'\mathbf{X})^{-1}\mathbf{X}'\mathbf{X}(\mathbf{X}'\mathbf{X})^{-1} = \sigma_\varepsilon^2(\mathbf{X}'\mathbf{X})^{-1}$$

The sampling variances and covariances of the regression coefficients, therefore, depend only on the model matrix and the variance of the errors.

To derive $E(\mathbf{b})$ and $V(\mathbf{b})$, we do not require the assumption of normality—only the assumptions of linearity [i.e., $E(\mathbf{y}) = \mathbf{X}\boldsymbol{\beta}$], constant variance, and independence [$V(\mathbf{y}) = \sigma_\varepsilon^2\mathbf{I}_n$]. If $\mathbf{y}$ is normally distributed, however, then so is $\mathbf{b}$, for—as I have explained—$\mathbf{b}$ results from a linear transformation of $\mathbf{y}$:

$$\mathbf{b} \sim N_{k+1}[\boldsymbol{\beta},\ \sigma_\varepsilon^2(\mathbf{X}'\mathbf{X})^{-1}]$$

There is a striking parallel, noted by Wonnacott and Wonnacott (1979) and detailed in Table 9.2, between the scalar formulas for least-squares simple regression and the matrix formulas for the general linear model ("multiple regression"). This sort of structural parallel is common in statistical applications

---

[19] See Section 9.4.

[20] See, for example, Section 12.5 and Chapters 14 and 15.

TABLE 9.2 Comparison Between Simple Regression Using Scalars and Multiple Regression Using Matrices.

	Simple Regression	*Multiple Regression*
Model	$Y = \alpha + \beta x + \varepsilon$	$\mathbf{y} = \mathbf{X}\boldsymbol{\beta} + \boldsymbol{\varepsilon}$
Least-squares estimator	$B = \frac{\sum x^* Y^*}{\sum x^{*2}} = \left(\sum x^{*2}\right)^{-1} \sum x^* Y^*$	$\mathbf{b} = (\mathbf{X}'\mathbf{X})^{-1}\mathbf{X}'\mathbf{y}$
Sampling variance	$V(B) = \frac{\sigma_\varepsilon^2}{\sum x^{*2}} = \sigma_\varepsilon^2 \left(\sum x^{*2}\right)^{-1}$	$V(\mathbf{b}) = \sigma_\varepsilon^2(\mathbf{X}'\mathbf{X})^{-1}$
Distribution	$B \sim N[\beta, \sigma_\varepsilon^2(\sum x^{*2})^{-1}]$	$\mathbf{b} \sim N_{k+1}[\boldsymbol{\beta}, \sigma_\varepsilon^2(\mathbf{X}'\mathbf{X})^{-1}]$

Source: Adapted from Wonnacott and Wonnacott (1979, Table 12-1), *Econometrics, Second Edition*. Copyright © John Wiley & Sons, Inc. Reprinted by permission of John Wiley & Sons, Inc.

when matrix methods are used to generalize a scalar result: Matrix notation is productive because of the generality and simplicity that it achieves.

> Under the full set of assumptions for the linear model, $\mathbf{b} \sim N_{k+1}[\boldsymbol{\beta}, \ \sigma_\varepsilon^2(\mathbf{X}'\mathbf{X})^{-1}]$.

9.3.2 The Gauss-Markov Theorem

One of the primary theoretical justifications for least-squares estimation is the Gauss-Markov theorem, which states that if the errors are independently distributed with zero expectation and constant variance, then the least-squares estimator $\mathbf{b}$ is the most efficient linear unbiased estimator of $\boldsymbol{\beta}$. That is, of all unbiased estimators that are linear functions of the observations, the least-squares estimator has the smallest sampling variance and, hence, the smallest mean-squared error. For this reason, the least-squares estimator is sometimes termed *BLUE*, an acronym for best linear unbiased estimator.[21]

Let $\tilde{\mathbf{b}}$ represent the best linear unbiased estimator of $\boldsymbol{\beta}$. As we know, the least-squares estimator $\mathbf{b}$ is also a linear estimator, $\mathbf{b} = \mathbf{M}\mathbf{y}$. It is convenient to write $\tilde{\mathbf{b}} = (\mathbf{M} + \mathbf{A})\mathbf{y}$, where $\mathbf{A}$ gives the difference between the (as yet undetermined) transformation matrix for the BLUE and that for the least-squares estimator. To show that the BLUE and the least-squares estimator coincide—that is, to establish the Gauss-Markov theorem—we need to demonstrate that $\mathbf{A} = \mathbf{0}$.

---

[21] As I explained in Section 6.1.2, the comfort provided by the Gauss-Markov theorem is often an illusion, because the restriction to linear estimators is artificial. Under the additional assumption of normality, however, it is possible to show that the least-squares estimator is maximally efficient among *all* unbiased estimators (see, e.g., Rao, 1973, p. 319). The strategy of proof of the Gauss-Markov theorem employed in this section is borrowed from Wonnacott and Wonnacott (1979, pp. 428–430), where it is used in a slightly different context.

Because $\widetilde{\mathbf{b}}$ is *unbiased*,

$$\begin{aligned}\boldsymbol{\beta} = E(\widetilde{\mathbf{b}}) &= E[(\mathbf{M}+\mathbf{A})\mathbf{y}] = E(\mathbf{My}) + E(\mathbf{Ay}) \\ &= E(\mathbf{b}) + \mathbf{A}E(\mathbf{y}) = \boldsymbol{\beta} + \mathbf{AX}\boldsymbol{\beta}\end{aligned}$$

The matrix product $\mathbf{AX}\boldsymbol{\beta}$, then, is $\mathbf{0}$, regardless of the value of $\boldsymbol{\beta}$, and therefore $\mathbf{AX}$ must be $\mathbf{0}$.[22]

I have, to this point, made use of the linearity and unbias of $\widetilde{\mathbf{b}}$. Because $\widetilde{\mathbf{b}}$ is the *minimum-variance* linear unbiased estimator, the sampling variances of its elements—that is, the diagonal entries of $V(\widetilde{\mathbf{b}})$—are as small as possible.[23] The covariance matrix of $\widetilde{\mathbf{b}}$ is given by

$$\begin{aligned}V(\widetilde{\mathbf{b}}) &= (\mathbf{M}+\mathbf{A})V(\mathbf{y})(\mathbf{M}+\mathbf{A})' \\ &= (\mathbf{M}+\mathbf{A})\sigma_\varepsilon^2\mathbf{I}_n(\mathbf{M}+\mathbf{A})' \\ &= \sigma_\varepsilon^2(\mathbf{MM}' + \mathbf{MA}' + \mathbf{AM}' + \mathbf{AA}')\end{aligned} \qquad [9.9]$$

I have shown that $\mathbf{AX} = \mathbf{0}$; consequently, $\mathbf{AM}'$ and its transpose $\mathbf{MA}'$ are $\mathbf{0}$, for

$$\mathbf{AM}' = \mathbf{AX}(\mathbf{X}'\mathbf{X})^{-1} = \mathbf{0}(\mathbf{X}'\mathbf{X})^{-1} = \mathbf{0}$$

Equation 9.9 becomes

$$V(\widetilde{\mathbf{b}}) = \sigma_\varepsilon^2(\mathbf{MM}' + \mathbf{AA}')$$

The sampling variance of the coefficient $\widetilde{B}_j$ is the jth diagonal entry[24] of $V(\widetilde{\mathbf{b}})$:

$$V(\widetilde{B}_j) = \sigma_\varepsilon^2\left(\sum_{i=1}^{n} m_{ji}^2 + \sum_{i=1}^{n} a_{ji}^2\right)$$

Both sums in this equation are sums of squares and hence cannot be negative; because $V(\widetilde{B}_j)$ is as small as possible, all of the a_{ji} must be 0. This argument applies to each coefficient in $\widetilde{\mathbf{b}}$, and so every row of $\mathbf{A}$ must be $\mathbf{0}$, implying that $\mathbf{A} = 0$. Finally,

$$\widetilde{\mathbf{b}} = (\mathbf{M}+\mathbf{0})\mathbf{y} = \mathbf{My} = \mathbf{b}$$

demonstrating that the BLUE is the least-squares estimator.

---

[22] See Exercise 9.8.

[23] It is possible to prove a more general result: The best linear unbiased estimator of $\mathbf{a}'\beta$ (an arbitrary linear combination of regression coefficients) is $\mathbf{a}'\mathbf{b}$, where $\mathbf{b}$ is the least-squares estimator (see, e.g., Seber, 1977, p. 49).

[24] Actually, the variance of the constant $\widetilde{B}_0$ is the *first* diagonal entry of $V(\widetilde{\mathbf{b}})$; the variance of $\widetilde{B}_j$ is therefore the $(j+1)$th entry. To avoid this awkwardness, I shall index the covariance matrix of $\widetilde{\mathbf{b}}$ (and later, that of $\mathbf{b}$) from 0 rather than from 1.

9.3.3 Maximum-Likelihood Estimation

Under the assumptions of the linear model, the least-squares estimator **b** is also the maximum-likelihood estimator of $\boldsymbol{\beta}$. This result establishes an additional justification for least squares when the assumptions of the model are reasonable, but even more importantly, it provides a basis for generalizing the linear model.[25]

As I have explained, under the assumptions of the linear model, $\mathbf{y} \sim N_n(\mathbf{X}\boldsymbol{\beta}, \sigma_\varepsilon^2 \mathbf{I}_n)$. Thus, for the ith observation, $Y_i \sim N(\mathbf{x}_i'\boldsymbol{\beta}, \sigma_\varepsilon^2)$, where $\mathbf{x}_i'$ is the ith row of the model matrix $\mathbf{X}$. In equation form, the probability density for observation i is

$$p(y_i) = \frac{1}{\sigma_\varepsilon \sqrt{2\pi}} \exp\left[-\frac{(y_i - \mathbf{x}_i'\boldsymbol{\beta})^2}{2\sigma_\varepsilon^2}\right]$$

Because the n observations are independent, their joint probability density is the product of their marginal densities:

$$\begin{aligned} p(\mathbf{y}) &= \frac{1}{\left(\sigma_\varepsilon \sqrt{2\pi}\right)^n} \exp\left[-\frac{\sum (y_i - \mathbf{x}_i'\boldsymbol{\beta})^2}{2\sigma_\varepsilon^2}\right] \\ &= \frac{1}{(2\pi\sigma_\varepsilon^2)^{n/2}} \exp\left[-\frac{(\mathbf{y} - \mathbf{X}\boldsymbol{\beta})'(\mathbf{y} - \mathbf{X}\boldsymbol{\beta})}{2\sigma_\varepsilon^2}\right] \end{aligned} \quad [9.10]$$

Although this equation also follows directly from the multivariate-normal distribution of $\mathbf{y}$, the development from $p(y_i)$ to $p(\mathbf{y})$ will prove helpful when we consider random regressors.[26]

From Equation 9.10, the log likelihood is

$$\log_e L(\boldsymbol{\beta}, \sigma_\varepsilon^2) = -\frac{n}{2}\log_e 2\pi - \frac{n}{2}\log_e \sigma_\varepsilon^2 - \frac{1}{2\sigma_\varepsilon^2}(\mathbf{y} - \mathbf{X}\boldsymbol{\beta})'(\mathbf{y} - \mathbf{X}\boldsymbol{\beta}) \quad [9.11]$$

To maximize the likelihood, we require the partial derivatives of Equation 9.11 with respect to the parameters $\boldsymbol{\beta}$ and σ_ε^2. Differentiation is simplified when we notice that $(\mathbf{y} - \mathbf{X}\boldsymbol{\beta})'(\mathbf{y} - \mathbf{X}\boldsymbol{\beta})$ is the sum of squared errors:

$$\begin{aligned} \frac{\partial \log_e L(\boldsymbol{\beta}, \sigma_\varepsilon^2)}{\partial \boldsymbol{\beta}} &= -\frac{1}{2\sigma_\varepsilon^2}(2\mathbf{X}'\mathbf{X}\boldsymbol{\beta} - 2\mathbf{X}'\mathbf{y}) \\ \frac{\partial \log_e L(\boldsymbol{\beta}, \sigma_\varepsilon^2)}{\partial \sigma_\varepsilon^2} &= -\frac{n}{2}\left(\frac{1}{\sigma_\varepsilon^2}\right) + \frac{1}{\sigma_\varepsilon^4}(\mathbf{y} - \mathbf{X}\boldsymbol{\beta})'(\mathbf{y} - \mathbf{X}\boldsymbol{\beta}) \end{aligned}$$

Setting these partial derivatives to 0 and solving for the maximum-likelihood estimators $\hat{\boldsymbol{\beta}}$ and $\hat{\sigma}_\varepsilon^2$ produces

$$\begin{aligned} \hat{\boldsymbol{\beta}} &= (\mathbf{X}'\mathbf{X})^{-1}\mathbf{X}'\mathbf{y} \\ \hat{\sigma}_\varepsilon^2 &= \frac{(\mathbf{y} - \mathbf{X}\hat{\boldsymbol{\beta}})'(\mathbf{y} - \mathbf{X}\hat{\boldsymbol{\beta}})}{n} = \frac{\mathbf{e}'\mathbf{e}}{n} \end{aligned}$$

---

[25] See, for example, the discussions of transformations in Section 12.5 and of nonlinear least squares in Section 14.2.3.

[26] See Section 9.5.

The maximum-likelihood estimator $\hat{\boldsymbol{\beta}}$ is therefore the same as the least-squares estimator **b**. In fact, this identity is clear directly from Equation 9.10, without formal maximization of the likelihood: The likelihood is large when the negative exponent is small, and the numerator of the exponent contains the sum of squared errors; minimizing the sum of squared residuals, therefore, maximizes the likelihood.

The maximum-likelihood estimator $\hat{\sigma}_\varepsilon^2$ of the error variance is biased; consequently, we prefer the similar, unbiased estimator[27] $S_E^2 = \mathbf{e}'\mathbf{e}/(n-k-1)$ to $\hat{\sigma}_\varepsilon^2$. As n increases, however, the bias of $\hat{\sigma}_\varepsilon^2$ shrinks toward 0: As a maximum-likelihood estimator, $\hat{\sigma}_\varepsilon^2$ is consistent.

EXERCISES

9.7 Using the general result $V(\mathbf{b}) = \sigma_\varepsilon^2(\mathbf{X}'\mathbf{X})^{-1}$, show that the sampling variances of A and B in simple-regression analysis are

$$V(A) = \frac{\sigma_\varepsilon^2 \sum X_i^2}{n\sum(X_i - \overline{X})^2}$$

$$V(B) = \frac{\sigma_\varepsilon^2}{\sum(X_i - \overline{X})^2}$$

9.8 *A crucial step in the proof of the Gauss-Markov theorem uses the fact that the matrix product $\mathbf{AX}$ must be $\mathbf{0}$ because $\mathbf{AX}\boldsymbol{\beta} = \mathbf{0}$. Why is this the case? (*Hint*: The key here is that $\mathbf{AX}\boldsymbol{\beta} = \mathbf{0}$ regardless of the value of $\boldsymbol{\beta}$. Consider, for example, $\boldsymbol{\beta} = [1, 0, \ldots, 0]'$ (i.e., one possible value of $\boldsymbol{\beta}$). Show that this implies that the first row of $\mathbf{AX}$ is $\mathbf{0}$. Then consider $\boldsymbol{\beta} = [0, 1, \ldots, 0]'$, and so on.)

9.9 Using a calculator or a computer program that performs matrix computations, and working with the Canadian occupational prestige data (continuing Exercise 9.5):

(a) Calculate the estimated error variance, $S_E^2 = \mathbf{e}'\mathbf{e}/(n-4)$ (where $\mathbf{e} = \mathbf{y} - \mathbf{Xb}$), and the estimated covariance matrix of the coefficients, $\widehat{V(\mathbf{b})} = S_E^2(\mathbf{X}'\mathbf{X})^{-1}$.

(b) Verify that the estimated covariance matrix for the slope coefficients $\mathbf{b}_1 = [B_1, B_2, B_3]'$ in this regression can be calculated as $\widehat{V(\mathbf{b}_1)} = S_E^2(\mathbf{X}^{*\prime}\mathbf{X}^*)^{-1}$, where $\mathbf{X}^*$ is the mean-deviation matrix for the X's.

[27] See Section 10.3 for a derivation of the expectation of S_E^2.

9.4 Statistical Inference for Linear Models

The results of the previous section, along with some to be established in Chapter 10, provide a basis for statistical inference in linear models.[28] I have already shown that the least-squares coefficients **b** have certain desirable properties as point estimators of the parameters $\boldsymbol{\beta}$. In this section, I shall describe tests for individual coefficients, for several coefficients, and for general linear hypotheses.

9.4.1 Inference for Individual Coefficients

We saw[29] that the least-squares estimator **b** follows a normal distribution with expectation $\boldsymbol{\beta}$ and covariance matrix $\sigma_\varepsilon^2(\mathbf{X}'\mathbf{X})^{-1}$. Consequently, an individual coefficient B_j is normally distributed with expectation β_j and sampling variance $\sigma_\varepsilon^2 v_{jj}$, where v_{jj} is the jth diagonal entry[30] of $(\mathbf{X}'\mathbf{X})^{-1}$. The ratio $(B_j - \beta_j)/\sigma_\varepsilon\sqrt{v_{jj}}$, therefore, follows the unit-normal distribution $N(0, 1)$; and to test the hypothesis H_0: $\beta_j = \beta_j^{(0)}$, we can calculate the test statistic

$$Z_0 = \frac{B_j - \beta_j^{(0)}}{\sigma_\varepsilon\sqrt{v_{jj}}}$$

comparing the obtained value of the statistic to quantiles of the unit-normal distribution. This result is not of direct practical use, however, because in applications of linear models we do not know σ_ε^2.

Although the error variance is unknown, we have available the unbiased estimator $S_E^2 = \mathbf{e}'\mathbf{e}/(n - k - 1)$. Employing this estimator, we can estimate the covariance matrix of the least-squares coefficients:

$$\widehat{V(\mathbf{b})} = S_E^2(\mathbf{X}'\mathbf{X})^{-1} = \frac{\mathbf{e}'\mathbf{e}}{n - k - 1}(\mathbf{X}'\mathbf{X})^{-1}$$

An estimator of the standard error of the coefficient B_j is, therefore, given by $\widehat{\mathrm{SE}}(B_j) = S_E\sqrt{v_{jj}}$, the square root of the jth diagonal entry of $\widehat{V(\mathbf{b})}$.

It can be shown that $(n - k - 1)S_E^2/\sigma_\varepsilon^2 = \mathbf{e}'\mathbf{e}/\sigma_\varepsilon^2$ follows a chi-square distribution with $n - k - 1$ degrees of freedom.[31] We recently discovered that $(B_j - \beta_j)/\sigma_\varepsilon\sqrt{v_{jj}}$ is distributed as $N(0, 1)$. It can be further established that the estimators B_j and S_E^2 are independent,[32] and so the ratio

$$t = \frac{(B_j - \beta_j)/\sigma_\varepsilon\sqrt{v_{jj}}}{\sqrt{\dfrac{\mathbf{e}'\mathbf{e}/\sigma_\varepsilon^2}{n - k - 1}}} = \frac{B_j - \beta_j}{S_E\sqrt{v_{jj}}}$$

follows a t-distribution with $n - k - 1$ degrees of freedom. Heuristically, in

[28] The results of this section justify and extend the procedures for inference described in Chapter 6.

[29] See Section 9.3.1.

[30] Recall that we index the rows and columns of $(\mathbf{X}'\mathbf{X})^{-1}$ from 0 through k.

[31] See Section 10.3.

[32] See Exercise 9.10.

estimating σ_ε with S_E, we replace the normal distribution with the more spread-out t-distribution to reflect the additional source of variability.

To test the hypothesis H_0: $\beta_j = \beta_j^{(0)}$, therefore, we calculate the test statistic

$$t_0 = \frac{B_j - \beta_j^{(0)}}{\widehat{\text{SE}}(B_j)}$$

comparing the obtained value of t_0 with the quantiles of t_{n-k-1}. Likewise, a $100(1-a)\%$ confidence interval for β_j is given by

$$\beta_j = B_j \pm t_{a/2,\, n-k-1}\widehat{\text{SE}}(B_j)$$

where $t_{a/2,\, n-k-1}$ is the critical value of t_{n-k-1} with a probability of $a/2$ to the right.

For Duncan's occupational prestige regression, for example, the estimated error variance is $S_E^2 = 178.73$, and so the estimated covariance matrix of the regression coefficients is

$$\begin{aligned}\widehat{V(\mathbf{b})} &= 178.73(\mathbf{X}'\mathbf{X})^{-1}\\ &= \begin{bmatrix} 18.249387 & -0.151844 & -0.150705 \\ -0.151844 & 0.014320 & -0.008519 \\ -0.150705 & -0.008519 & 0.009653 \end{bmatrix}\end{aligned}$$

The estimated standard errors of the regression coefficients are

$$\begin{aligned}\widehat{\text{SE}}(B_0) &= \sqrt{18.249387} = 4.272\\ \widehat{\text{SE}}(B_1) &= \sqrt{0.014320} = 0.1197\\ \widehat{\text{SE}}(B_2) &= \sqrt{0.009653} = 0.09825\end{aligned}$$

> The estimated covariance matrix of the least-squares coefficients is $\widehat{V(\mathbf{b})} = S_E^2(\mathbf{X}'\mathbf{X})^{-1}$. The estimated standard errors of the regression coefficients are the diagonal entries of this matrix. Under the assumptions of the model, $(B_j - \beta_j)/\widehat{\text{SE}}(B_j) \sim t_{n-k-1}$, providing a basis for hypothesis tests and confidence intervals for individual coefficients.

9.4.2 Inference for Several Coefficients

Although we usually test regression coefficients individually, these tests may not be sufficient, for, in general, the least-squares estimators of different parameters are correlated: The off-diagonal entries of $V(\mathbf{b}) = \sigma_\varepsilon^2(\mathbf{X}'\mathbf{X})^{-1}$, giving the sampling covariances of the least-squares coefficients, are 0 only when the regressors themselves are uncorrelated.[33] Furthermore, in certain applications of linear models—such as dummy regression, analysis of variance, and polynomial regression—we are more interested in related sets of coefficients than in the individual members of these sets.

Simultaneous tests for sets of coefficients, taking their intercorrelations into account, can be constructed by the likelihood-ratio principle. Suppose that we fit the model

$$Y = \beta_0 + \beta_1 x_1 + \cdots + \beta_k x_k + \varepsilon \qquad [9.12]$$

obtaining the least-squares estimate $\mathbf{b} = [B_0, B_1, \ldots, B_k]'$, along with the maximum-likelihood estimate of the error variance, $\hat{\sigma}_\varepsilon^2 = \mathbf{e}'\mathbf{e}/n$. We wish to test the null hypothesis that a subset of regression parameters is 0; for convenience, let these coefficients be the first $q \leq k$, so that the null hypothesis is H_0: $\beta_1 = \cdots = \beta_q = 0$. The null hypothesis corresponds to the model

$$\begin{aligned} Y &= \beta_0 + 0x_1 + \cdots + 0x_q + \beta_{q+1}x_{q+1} + \cdots + \beta_k x_k + \varepsilon \qquad [9.13] \\ &= \beta_0 + \beta_{q+1}x_{q+1} + \cdots + \beta_k x_k + \varepsilon \end{aligned}$$

which is a specialization (or restriction) of the more general model (Equation 9.12). Fitting the restricted model (Equation 9.13) by least-squares regression of Y on x_{q+1} through x_k, we obtain $\mathbf{b}_0 = [B_0', 0, \ldots, 0, B_{q+1}', \ldots, B_k']'$, and $\hat{\sigma}_{\varepsilon_0}^2 = \mathbf{e}_0'\mathbf{e}_0/n$. Note that the coefficients in $\mathbf{b}_0$ generally differ from those in $\mathbf{b}$ (hence the primes), and that $\hat{\sigma}_\varepsilon^2 \leq \hat{\sigma}_{\varepsilon_0}^2$, because both models are fit by least squares.

The likelihood for the full model (Equation 9.12), evaluated at the maximum-likelihood estimates, can be obtained from Equation 9.10:[34]

$$L = \left(2\pi e \frac{\mathbf{e}'\mathbf{e}}{n}\right)^{-n/2}$$

Likewise, for the restricted model (Equation 9.13), the maximized likelihood is

$$L_0 = \left(2\pi e \frac{\mathbf{e}_0'\mathbf{e}_0}{n}\right)^{-n/2}$$

---

[33] This point pertains to sampling correlations among the *k slope* coefficients. The regression constant is correlated with the slope coefficients unless all of the regressors—save the constant regressor—have means of 0 (i.e., are in mean-deviation form). Expressing the regressors in mean-devation form, called *centering*, has certain computational advantages (it tends to reduce rounding errors in least-squares calculations), but it does not affect the slope coefficients or the sampling covariances among them.

[34] See Exercise 9.11. The notation here is potentially confusing: $e \simeq 2.718$ is the mathematical constant; $\mathbf{e}$ is the vector of residuals.

The likelihood ratio for testing H_0 is, therefore,

$$\frac{L_0}{L_1} = \left(\frac{\mathbf{e}_0'\mathbf{e}_0}{\mathbf{e}'\mathbf{e}}\right)^{-n/2} = \left(\frac{\mathbf{e}'\mathbf{e}}{\mathbf{e}_0'\mathbf{e}_0}\right)^{2/n}$$

Because $\mathbf{e}_0'\mathbf{e}_0 \geq \mathbf{e}'\mathbf{e}$, the likelihood ratio is small when the residual sum of squares for the restricted model is appreciably larger than for the general model—circumstances under which we should doubt the truth of the null hypothesis. A test of H_0 is provided by the generalized likelihood-ratio test statistic, $G_0^2 = -2\log_e(L_0/L_1)$, which is asymptotically distributed as χ_q^2 under the null hypothesis.

It is unnecessary to use this asymptotic result, however, for an exact test can be obtained:[35] As mentioned in the previous section, $\text{RSS}/\sigma_\varepsilon^2 = \mathbf{e}'\mathbf{e}/\sigma_\varepsilon^2$ is distributed as χ^2 with $n-k-1$ degrees of freedom. By a direct extension of this result, if the null hypothesis is true, then $\text{RSS}_0/\sigma_\varepsilon^2 = \mathbf{e}_0'\mathbf{e}_0/\sigma_\varepsilon^2$ is distributed as χ^2 with $n-(k-q)-1 = n-k+q-1$ degrees of freedom. Consequently, the difference $(\text{RSS}_0 - \text{RSS})/\sigma_\varepsilon^2$ has a χ^2 distribution with $(n-k+q-1)-(n-k-1) = q$ degrees of freedom, equal to the number of parameters set to 0 in the restricted model. It can be shown that $(\text{RSS}_0 - \text{RSS})/\sigma_\varepsilon^2$ and $\text{RSS}/\sigma_\varepsilon^2$ are independent, and so the ratio

$$F_0 = \frac{(\text{RSS}_0 - \text{RSS})/q}{\text{RSS}/(n-k-1)}$$

is distributed as F with q and $n-k-1$ degrees of freedom. This is, of course, the incremental F-statistic.[36]

Although it is sometimes convenient to find an incremental sum of squares by fitting alternative linear models to the data, it is also possible to calculate this quantity directly from the least-squares coefficient vector $\mathbf{b}$ and the matrix $(\mathbf{X}'\mathbf{X})^{-1}$ for the full model: Let $\mathbf{b}_1 = [B_1, \ldots, B_q]'$ represent the coefficients of interest selected from among the entries of $\mathbf{b}$; and let $\mathbf{V}_{11}$ represent the square submatrix consisting of the entries in the q rows and q columns of $(\mathbf{X}'\mathbf{X})^{-1}$ that pertain to the coefficients[37] in $\mathbf{b}_1$. Then it can be shown that the incremental sum of squares $\text{RSS}_0 - \text{RSS}$ is equal to $\mathbf{b}_1'\mathbf{V}_{11}^{-1}\mathbf{b}_1$, and thus the incremental F-statistic can be written $F_0 = \mathbf{b}_1'\mathbf{V}_{11}^{-1}\mathbf{b}_1/qS_E^2$. To test the more general hypothesis H_0: $\boldsymbol{\beta}_1 = \boldsymbol{\beta}_1^{(0)}$ (where $\boldsymbol{\beta}_1^{(0)}$ is not necessarily $\mathbf{0}$), we can compute

$$F_0 = \frac{(\mathbf{b}_1 - \boldsymbol{\beta}_1^{(0)})'\mathbf{V}_{11}^{-1}(\mathbf{b}_1 - \boldsymbol{\beta}_1^{(0)})}{qS_E^2} \qquad [9.14]$$

which is distributed as $F_{q,n-k-1}$ under H_0.

[35] The F-test that follows is exact when the assumptions of the model hold—including the assumption of normality. Of course, the asymptotically valid likelihood-ratio test also depends on these assumptions.

[36] See Section 6.2.2.

[37] Note the difference between the vector $\mathbf{b}_1$ (used here) and the vector $\mathbf{b}_0$ (used previously): $\mathbf{b}_1$ consists of coefficients extracted from $\mathbf{b}$, which, in turn, results from fitting the *full* model; in contrast, $\mathbf{b}_0$ consists of the coefficients—including those set to 0 in the hypothesis—that result from fitting the *restricted* model.

Recall that the omnibus F-statistic for the hypothesis H_0: $\beta_1 = \cdots = \beta_k = 0$ is

$$F_0 = \frac{\text{RegSS}/k}{\text{RSS}/(n-k-1)}$$

The denominator of this F-statistic estimates the error variance σ_ε^2, whether or not the null hypothesis is true.[38] The expectation of the regression sum of squares, it may be shown,[39] is

$$E(\text{RegSS}) = \boldsymbol{\beta}_1'(\mathbf{X}^{*\prime}\mathbf{X}^*)\boldsymbol{\beta}_1 + k\sigma_\varepsilon^2$$

where $\boldsymbol{\beta}_1 \equiv [\,\beta_1 = \cdots = \beta_k]'$ is the vector of regression coefficients, excluding the constant, and $\mathbf{X}^*_{(n\times k)} \equiv \{x_{ij} - \overline{x}_j\}$ is the matrix of mean-deviation regressors, omitting the constant regressor. When H_0 is true (and $\boldsymbol{\beta}_1 = \mathbf{0}$), the denominator of the F-statistic also estimates σ_ε^2; but, when H_0 is false, $E(\text{RegSS}/k) > \sigma_\varepsilon^2$, because $\mathbf{X}^{*\prime}\mathbf{X}^*$ is positive definite, and thus $\boldsymbol{\beta}_1'(\mathbf{X}^{*\prime}\mathbf{X}^*)\boldsymbol{\beta}_1 > 0$ for $\boldsymbol{\beta}_1 \neq \mathbf{0}$. Under these circumstances, we tend to observe numerators that are larger than denominators, and F-statistics that are bigger than 1.[40]

An incremental F-test for the hypothesis H_0: $\beta_1 = \cdots = \beta_q = 0$, where $1 \leq q \leq k$, is given by $F_0 = (n-k-1)(\text{RSS}_0 - \text{RSS})/q\,\text{RSS}$, where RSS is the residual sum of squares for the full model, and RSS_0 is the residual sum of squares for the model that deletes the q regressors in question. Under the null hypothesis, $F_0 \sim F_{q,n-k-1}$. The incremental F-statistic can also be computed directly as $F_0 = \mathbf{b}_1'\mathbf{V}_{11}^{-1}\mathbf{b}_1/qS_E^2$, where $\mathbf{b}_1 = [B_1, \ldots, B_q]'$ contains the coefficients of interest extracted from among the entries of $\mathbf{b}$; and $\mathbf{V}_{11}$ is the square submatrix of $(\mathbf{X}'\mathbf{X})^{-1}$ consisting of the q rows and columns pertaining to the coefficients in $\mathbf{b}_1$.

9.4.3 General Linear Hypotheses

Even more generally, we can test the linear hypothesis

$$H_0: \underset{(q\times k+1)}{\mathbf{L}}\;\underset{(k+1\times 1)}{\boldsymbol{\beta}} = \underset{(q\times 1)}{\mathbf{c}}$$

[38] See Section 10.3.

[39] See Seber (1977, Chapter 4).

[40] The expectation of F_0 is not precisely 1 when H_0 is true because the expectation of a ratio of random variables is not necessarily the ratio of their expectations. See Appendix D, Section D.3.4.

where $\mathbf{L}$ and $\mathbf{c}$ contain prespecified constants, and the *hypothesis matrix* $\mathbf{L}$ is of full row rank $q \leq k+1$. The resulting F-statistic,

$$F_0 = \frac{(\mathbf{Lb} - \mathbf{c})' [\mathbf{L}(\mathbf{X}'\mathbf{X})^{-1}\mathbf{L}']^{-1} (\mathbf{Lb} - \mathbf{c})}{qS_E^2} \qquad [9.15]$$

follows an F-distribution with q and $n-k-1$ degrees of freedom if H_0 is true.

To understand the structure of Equation 9.15, recall that $\mathbf{b} \sim N_{k+1}[\boldsymbol{\beta}, \sigma_\varepsilon^2(\mathbf{X}'\mathbf{X})^{-1}]$. As a consequence,

$$\mathbf{Lb} \sim N_q[\mathbf{L}\boldsymbol{\beta}, \sigma_\varepsilon^2\mathbf{L}(\mathbf{X}'\mathbf{X})^{-1}\mathbf{L}']$$

Under H_0, $\mathbf{L}\boldsymbol{\beta} = \mathbf{c}$, and thus

$$(\mathbf{Lb} - \mathbf{c})'[\mathbf{L}(\mathbf{X}'\mathbf{X})^{-1}\mathbf{L}']^{-1}(\mathbf{Lb} - \mathbf{c})/\sigma_\varepsilon^2 \sim \chi_q^2$$

Equation 9.15 is general enough to encompass all of the hypothesis tests that we have considered thus far, along with others. In Duncan's occupational prestige regression, for example, to test the omnibus null hypothesis H_0: $\beta_1 = \beta_2 = 0$, we can take

$$\mathbf{L} = \begin{bmatrix} 0 & 1 & 0 \\ 0 & 0 & 1 \end{bmatrix}$$

and $\mathbf{c} = [0, 0]'$. To test the hypothesis that the education and income coefficients are equal,[41] H_0: $\beta_1 = \beta_2$, which is equivalent to H_0: $\beta_1 - \beta_2 = 0$, we can take $\mathbf{L} = [0, 1, -1]$ and $\mathbf{c} = [0]$.

The F-statistic $F_0 = (\mathbf{Lb} - \mathbf{c})'[\mathbf{L}(\mathbf{X}'\mathbf{X})^{-1}\mathbf{L}']^{-1}(\mathbf{Lb} - \mathbf{c})/qS_E^2$ is used to test the general linear hypothesis H_0: $\underset{(q \times k+1)}{\mathbf{L}} \underset{(k+1 \times 1)}{\boldsymbol{\beta}} = \underset{(q \times 1)}{\mathbf{c}}$, where the rank-$q$ hypothesis matrix $\mathbf{L}$ and right-hand-side vector $\mathbf{c}$ contain pre-specified constants. Under the hypothesis, $F_0 \sim F_{q,n-k-1}$.

[41] Examples of these calculations appear in Exercise 9.12. The hypothesis that two regression coefficients are equal is sensible only if the corresponding independent variables are measured on the same scale. This is arguably the case for income and education in Duncan's regression, because both independent variables are percentages. Closer scrutiny suggests, however, that these independent variables are not commensurable: There is no reason to suppose that the percentage of occupational incumbents with at least high-school education is on the same scale as the percentage earning in excess of \$3500.

9.4.4 Joint Confidence Regions

The F-test of Equation 9.14 can be inverted to construct a *joint confidence region* for $\boldsymbol{\beta}_1$. If H_0: $\boldsymbol{\beta}_1 = \boldsymbol{\beta}_1^{(0)}$ is correct, then

$$\Pr\left[\frac{\left(\mathbf{b}_1 - \boldsymbol{\beta}_1^{(0)}\right)' \mathbf{V}_{11}^{-1} \left(\mathbf{b}_1 - \boldsymbol{\beta}_1^{(0)}\right)}{qS_E^2} \leq F_{a,\,q,\,n-k-1}\right] = 1 - a$$

where $F_{a,\,q,\,n-k-1}$ is the critical value of F with q and $n-k-1$ degrees of freedom, corresponding to a right-tail probability of a. The joint confidence region for $\boldsymbol{\beta}_1$ is thus

$$(\mathbf{b}_1 - \boldsymbol{\beta}_1)' \mathbf{V}_{11}^{-1} (\mathbf{b}_1 - \boldsymbol{\beta}_1) \leq qS_E^2 F_{a,\,q,\,n-k-1} \qquad [9.16]$$

That is, any parameter vector $\boldsymbol{\beta}_1$ that satisfies this inequality is within the confidence region and is acceptable as a hypothesis; any parameter vector that does not satisfy the inequality is unacceptable. The boundary of the joint confidence region (obtained when the left-hand side of Equation 9.16 equals the right-hand side) is an ellipsoid centered at the estimates $\mathbf{b}_1$ in the q-dimensional space of the parameters $\boldsymbol{\beta}_1$.

Like a confidence interval, a joint confidence region is a portion of the parameter space constructed so that, with repeated sampling, a preselected percentage of regions will contain the true parameter values. Unlike a confidence interval, however, which pertains to a single coefficient β_j, a joint confidence region encompasses all *combinations* of values for the parameters $\beta_1, \ldots, \beta_q$ that are *simultaneously* acceptable at the specified level of confidence. Indeed, the familiar confidence interval is just a one-dimensional confidence region; and there is a simple relationship between the confidence interval for a single coefficient and the confidence region for several coefficients (as I shall explain shortly).

The essential nature of joint confidence regions is clarified by considering the two-dimensional case, which can be directly visualized. To keep the mathematics as simple as possible, let us work with the slope coefficients β_1 and β_2 from the two-independent-variable model, $Y_i = \beta_0 + \beta_1 x_{i1} + \beta_2 x_{i2} + \varepsilon_i$. In this instance, the joint confidence region of Equation 9.16 becomes

$$[B_1 - \beta_1, B_2 - \beta_2] \begin{bmatrix} \sum x_{i1}^{*2} & \sum x_{i1}^* x_{i2}^* \\ \sum x_{i1}^* x_{i2}^* & \sum x_{i2}^{*2} \end{bmatrix} \begin{bmatrix} B_1 - \beta_1 \\ B_2 - \beta_2 \end{bmatrix} \leq 2S_E^2 F_{a,\,2,\,n-3} \qquad [9.17]$$

where the $x_{ij}^* \equiv x_{ij} - \overline{x}_j$ are deviations from the means of X_1 and X_2. The matrix $\mathbf{V}_{11}^{-1}$ contains mean-deviation sums of squares and products for the independent variables;[42] and the boundary of the confidence region, obtained when the equality holds, is an ellipse centered at (B_1, B_2) in the $\{\beta_1, \beta_2\}$ plane.

Illustrative joint confidence ellipses are shown in Figure 9.1. When the independent variables are uncorrelated, the sum of cross products $\sum x_{i1}^* x_{i2}^*$ vanishes,

[42] See Exercise 9.13.

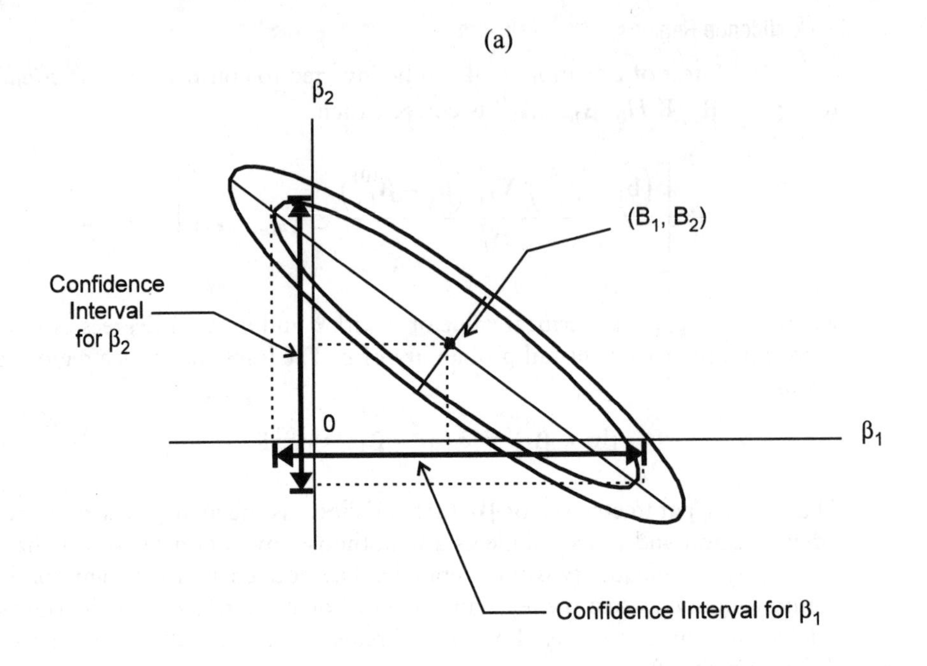

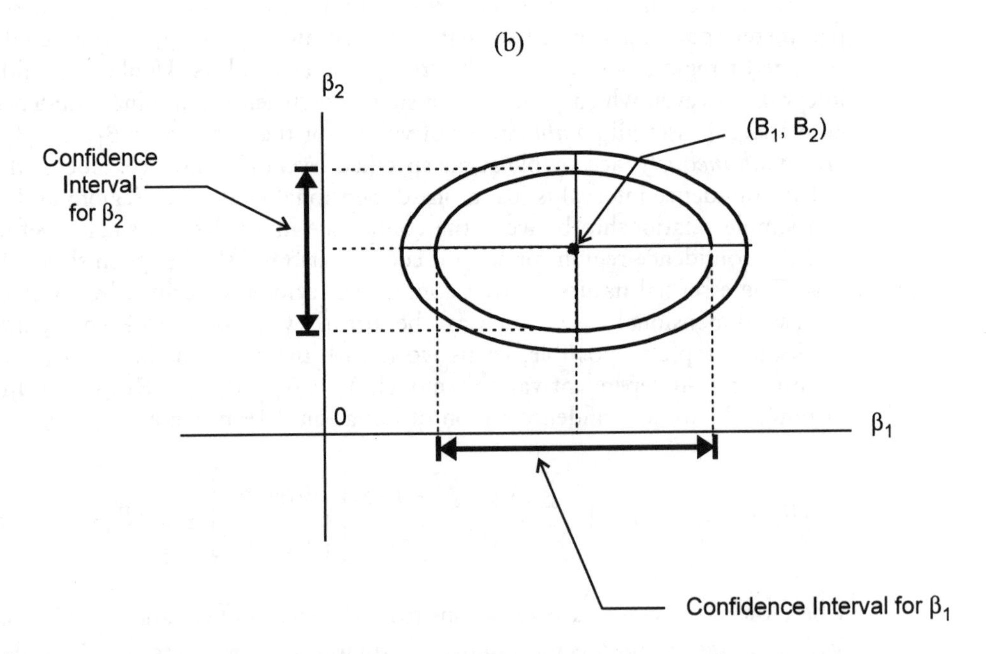

Figure 9.1. Illustrative joint confidence ellipses for the slope coefficients β_1 and β_2 in multiple-regression analysis. The outer ellipse is drawn at a level of confidence of 95%; the inner ellipse (the confidence-interval generating ellipse) is drawn so that its perpendicular shadows on the axes are 95% confidence intervals for the individual β's. In (*a*), the X's are positively correlated, producing a joint confidence ellipse that is negatively tilted. In (*b*), the X's are uncorrelated, producing a joint confidence ellipse with axes parallel to the axes of the parameter space.

and the axes of the confidence ellipse are parallel to the axes of the parameter space, as in Figure 9.1(*b*). When the independent variables are correlated, in contrast, the ellipse is "tilted," as in Figure 9.1(*a*).

Specializing (9.16) to a single coefficient produces the confidence *interval* for β_1:

$$(B_1 - \beta_1)^2 \frac{\sum x_{i2}^{*2}}{\sum x_{i1}^{*2} \sum x_{i2}^{*2} - (\sum x_{i1}^{*} x_{i2}^{*})^2} \leq S_E^2 F_{a,1,n-3} \qquad [9.18]$$

which is written, more conventionally, as[43]

$$B_1 - t_{a,n-3} \frac{S_E}{\sqrt{\dfrac{\sum x_{i1}^{*2}}{1 - r_{12}^2}}} \leq \beta_1 \leq B_1 + t_{a,n-3} \frac{S_E}{\sqrt{\dfrac{\sum x_{i1}^{*2}}{1 - r_{12}^2}}}$$

The individual confidence intervals for the regression coefficients are very nearly the perpendicular "shadows" (i.e., projections) of the joint confidence ellipse onto the β_1 and β_2 axes. The only slippage here is due to the right-hand-side constant: $2S_E^2 F_{a,2,n-3}$ for the joint confidence region, and $S_E^2 F_{a,1,n-3}$ for the confidence interval.

Consider a 95% region and interval, for example. If the residual degrees of freedom $n - 3$ are large, then $2F_{.05,2,n-3} \simeq \chi_{.05,2}^2 = 5.99$, while $F_{.05,1,n-3} \simeq \chi_{.05,1}^2 = 3.84$. Put another way, using $5.99S_E^2$ in place of $3.84S_E^2$ produces individual intervals at approximately the $1 - \Pr(\chi_1^2 > 5.99) = .986$ (rather than .95) level of confidence (but a *joint* 95% confidence region). Likewise, if we construct the joint confidence region using the multiplier 3.84, the resulting smaller ellipse produces shadows that give approximately 95% confidence intervals for *individual* coefficients [and a smaller *joint* level of confidence of $1 - \Pr(\chi_2^2 > 3.84) = .853$]. This *confidence-interval generating ellipse* is shown along with the joint confidence ellipse in Figure 9.1.[44]

Figure 9.1(*a*) illustrates how correlated regressors can lead to ambiguous inferences: Because the individual confidence intervals include 0, we cannot reject the separate hypotheses that *either* β_1 *or* β_2 is 0. Because the point (0, 0) is outside of the joint confidence region, however, we can reject the hypothesis that *both* β_1 *and* β_2 are 0. In contrast, in Figure 9.1(*b*), where the independent variables are uncorrelated, there is a close correspondence between inferences based on the separate confidence intervals and those based on the joint confidence region.

Still more generally, the confidence-interval generating ellipse can be projected onto *any* line through the origin of the $\{\beta_1, \beta_2\}$ plane. Each such line

---

[43] See Exercise 9.14.

[44] The individual intervals constructed from the larger joint confidence ellipse—called Scheffé intervals—can be thought of as incorporating a penalty for examining several coefficients simultaneously. The difference between the Scheffé interval—the shadow of the joint confidence region (for which the multiplier is $kS_E^2 F_{a,k,n-3}$)—and the individual confidence interval (for which the multiplier is $S_E^2 F_{a,1,n-3}$) grows larger as the number of coefficients k increases.

represents a specific linear combination of β_1 and β_2, and the shadow of the ellipse gives the corresponding confidence interval for that linear combination of the parameters.[45] This property is illustrated in Figure 9.2 for the linear combination $\beta_1 + \beta_2$; the line representing $\beta_1 + \beta_2$ is drawn through the origin and the point (1, 1), the coefficients of the parameters in the linear combination. Directions in which the ellipse is narrow, therefore, correspond to linear combinations of the parameters that are relatively precisely estimated.

It is illuminating to examine more closely the relationship between the joint confidence region for the regression coefficients and the joint distribution of the X-values. I have already remarked that the orientation of the confidence region reflects the correlation of the X's, but it is possible to be much more precise. Consider the quadratic form $(\mathbf{x} - \overline{\mathbf{x}})'\mathbf{S}_{XX}^{-1}(\mathbf{x} - \overline{\mathbf{x}})$, where $\mathbf{x}$ is a $k \times 1$ vector of independent-variable values, $\overline{\mathbf{x}}$ is the vector of means of the X's, and $\mathbf{S}_{XX}$ is the sample covariance matrix of the X's. Setting the quadratic form to 1 produces the equation of an ellipsoid—called the *standard data ellipsoid*—centered at the means of the independent variables.

For two independent variables, the standard data *ellipse* has the equation

$$\frac{n-1}{\sum x_{i1}^{*2} \sum x_{i2}^{*2} - (\sum x_{i1}^{*} x_{i2}^{*})^2}[x_1 - \overline{x}_1, x_2 - \overline{x}_2] \\ \times \begin{bmatrix} \sum x_{i2}^{*2} & -\sum x_{i1}^{*} x_{i2}^{*} \\ -\sum x_{i1}^{*} x_{i2}^{*} & \sum x_{i1}^{*2} \end{bmatrix} \begin{bmatrix} x_1 - \overline{x}_1 \\ x_2 - \overline{x}_2 \end{bmatrix} = 1 \quad [9.19]$$

representing an ellipse whose horizontal shadow is twice the standard deviation of X_1, and whose vertical shadow is twice the standard deviation of X_2. These properties are illustrated in Figure 9.3, which also shows the scatterplot of the X-values. The major axis of the data ellipse has a positive slope when the X's are positively correlated.

This representation of the data is most compelling when the independent variables are normally distributed. In this case, the means and covariance matrix of the X's are sufficient statistics for their joint distribution; and the standard data ellipsoid estimates a constant-density contour of the joint distribution. Even when—as is typical—the independent variables are not multivariate normal, however, the standard ellipsoid is informative because of the role of the means, variances, and covariance of the X's in the least-squares fit.

The joint confidence ellipse (Equation 9.17) for the slope coefficients and the standard data ellipse (Equation 9.19) of the X's are, except for a constant scale factor and their respective centers, inverses of each other—that is, the confidence ellipse is (apart from its size and location) the 90° rotation of the data ellipse. In particular, if the data ellipse is positively tilted, then the confidence ellipse is negatively tilted. Likewise, directions in which the data ellipse is relatively thick, reflecting a substantial amount of data, are directions in which the confidence ellipse is relatively thin, reflecting substantial

[45] See Monette (1990).

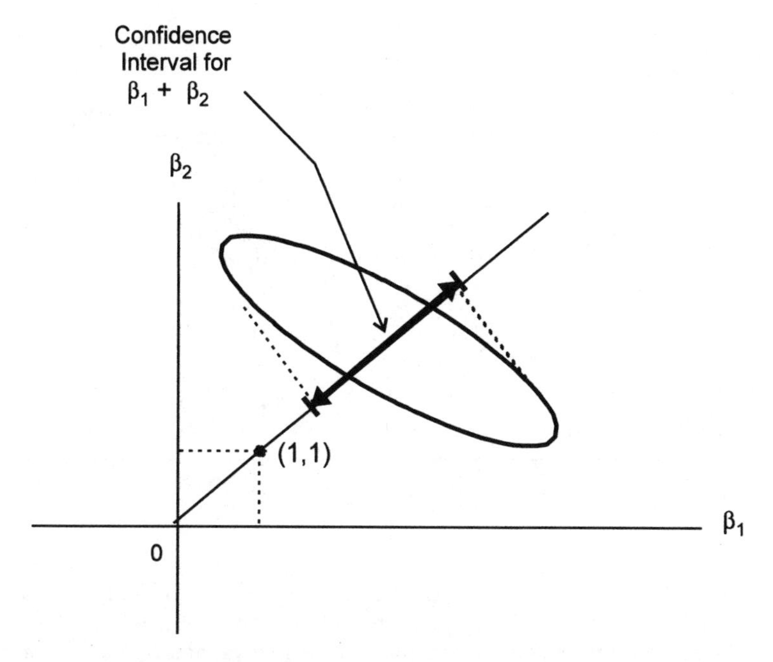

Figure 9.2. To find the 95% confidence interval for the linear combination of coefficients $\beta_1 + \beta_2$, find the perpendicular shadow of the confidence-interval generating ellipse on the line through the origin and the point (1, 1). Note that the regression coefficients (B_1, B_2) are not the same as in the previous figure.

Source: Adapted from Monette (1990, Figure 5.7*A*), in *Modern Methods of Data Analysis*, copyright © 1990 by Sage Publications, Inc. Reprinted with permission of Sage Publications, Inc.

information about the corresponding linear combination of regression coefficients. Thus, when the X's are strongly positively correlated (and assuming, for simplicity, that the standard deviations of X_1 and X_2 are similar), there is a great deal of information about $\beta_1 + \beta_2$ but little about $\beta_1 - \beta_2$ (as in Figure 9.2).[46]

> The joint confidence region for the q parameters $\boldsymbol{\beta}_1$, given by $(\mathbf{b}_1 - \boldsymbol{\beta}_1)' \mathbf{V}_{11}^{-1} (\mathbf{b}_1 - \boldsymbol{\beta}_1) \leq qS_E^2 F_{a,\, q,\, n-k-1}$, represents the combinations of values of these parameters that are jointly acceptable at the $1 - a$ level of confidence. The boundary of the joint confidence region is an ellipsoid in the q-dimensional parameter space, reflecting the correlational structure and dispersion of the X's.

[46] See Exercise 9.15.

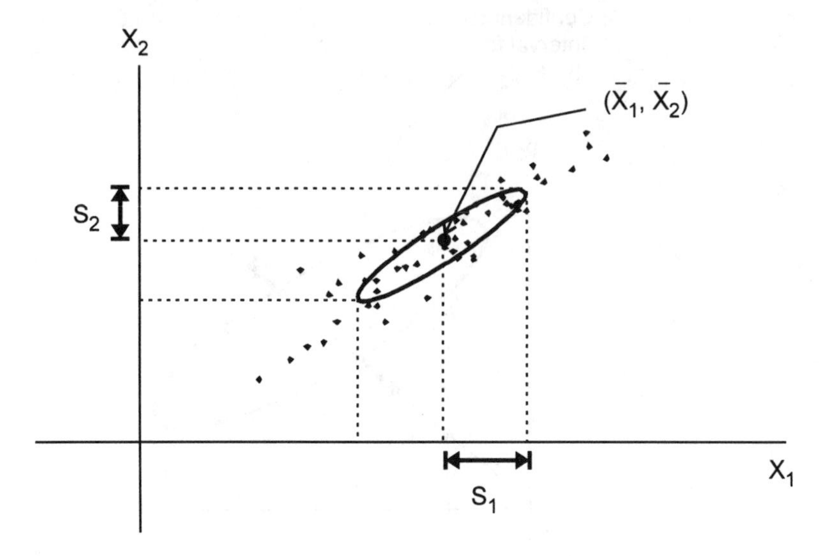

Figure 9.3. Scatterplot and standard data ellipse for two highly correlated regressors X_1 and X_2. The standard ellipse is centered at the point of means $(\overline{X}_1, \overline{X}_2)$; its shadows on the axes give the standard deviations of the two variables. (The standard deviations are the half-widths of the shadows.) The data in this figure (along with data on Y) gave rise to the joint confidence ellipse in Figure 9.2. The confidence ellipse is the rescaled and translated 90° rotation of the data ellipse.

Source: Adapted from Monette (1990, Figure 5.1*A*), in *Modern Methods of Data Analysis*, copyright © 1990 by Sage Publications, Inc. Reprinted with permission of Sage Publications, Inc.

EXERCISES

9.10 *For the statistic

$$t = \frac{B_j - \beta_j}{S_E\sqrt{v_{jj}}}$$

to have a t-distribution, the estimators B_j and S_E must be independent. [Here, v_{jj} is the jth diagonal entry of $(\mathbf{X}'\mathbf{X})^{-1}$.] The coefficient B_j is the jth element of $\mathbf{b}$, and $S_E = \sqrt{\mathbf{e}'\mathbf{e}/(n-k-1)}$ is a function of the residuals $\mathbf{e}$. Because both $\mathbf{b}$ and $\mathbf{e}$ are normally distributed, it suffices to prove that their covariance is $\mathbf{0}$. Demonstrate that this is the case. (*Hint*: Use $C(\mathbf{e}, \mathbf{b}) = E[\mathbf{e}(\mathbf{b} - \boldsymbol{\beta})']$, and begin by showing that $\mathbf{b} - \boldsymbol{\beta} = (\mathbf{X}'\mathbf{X})^{-1}\mathbf{X}'\boldsymbol{\varepsilon}$.]

9.11 *Using Equation 9.10, show that the maximized likelihood for the linear model can be written as

$$L = \left(2\pi e \frac{\mathbf{e}'\mathbf{e}}{n}\right)^{-n/2}$$

9.12 Using Duncan's regression of occupational prestige on income and education (the data for which are in Table 3.2 and duncan.dat), and performing the necessary calculations, show that the omnibus null hypothesis H_0: $\beta_1 = \beta_2 = 0$ can be tested as a general linear hypothesis, using the hypothesis matrix

$$\mathbf{L} = \begin{bmatrix} 0 & 1 & 0 \\ 0 & 0 & 1 \end{bmatrix}$$

and right-hand-side vector $\mathbf{c} = [0, 0]'$. Then show that the H_0: $\beta_1 = \beta_2$ can be tested using $\mathbf{L} = [0, 1, -1]$ and $\mathbf{c} = [0]$.

9.13 *Consider the model $Y_i = \beta_0 + \beta_1 x_{i1} + \beta_2 x_{i2} + \varepsilon_i$. Show that the matrix $\mathbf{V}_{11}^{-1}$ for the slope coefficients β_1 and β_2 contains mean-deviation sums of squares and products for the independent variables; that is,

$$\mathbf{V}_{11}^{-1} = \begin{bmatrix} \sum x_{i1}^{*2} & \sum x_{i1}^{*} x_{i2}^{*} \\ \sum x_{i1}^{*} x_{i2}^{*} & \sum x_{i2}^{*2} \end{bmatrix}$$

Now show, more generally, for the model $Y_i = \beta_0 + \beta_1 x_{i1} + \cdots + \beta_k x_{ik} + \varepsilon_i$, that the matrix $\mathbf{V}_{11}^{-1}$ for the slope coefficients $\beta_1, \ldots, \beta_k$ contains mean-deviation sums of squares and products for the independent variables.

9.14 *Show that Equation 9.18 for the confidence interval for β_1 can be written in the more conventional form

$$B_1 - t_{a,\,n-3} \frac{S_E}{\sqrt{\dfrac{\sum x_{i1}^{*2}}{1 - r_{12}^2}}} \leq \beta_1 \leq B_1 + t_{a,\,n-3} \frac{S_E}{\sqrt{\dfrac{\sum x_{i1}^{*2}}{1 - r_{12}^2}}}$$

9.15 Using Figure 9.2, show how the confidence-interval generating ellipse can be used to derive a confidence interval for the *difference* of the parameters $\beta_1 - \beta_2$. Compare the confidence interval for this linear combination with that for $\beta_1 + \beta_2$. Which combination of parameters is estimated more precisely? Why? What would happen if the regressors X_1 and X_2 were *negatively* correlated?

9.16 Prediction: One use of a fitted regression equation is to *predict* dependent-variable values for particular "future" combinations of independent-variable scores. Suppose, therefore, that we fit the model $\mathbf{y} = \mathbf{X}\boldsymbol{\beta} + \boldsymbol{\varepsilon}$, obtaining the least-squares estimate $\mathbf{b}$ of $\boldsymbol{\beta}$. Let $\mathbf{x}_0 = [1, x_{01}, \ldots, x_{0k}]$ represent a set of independent-variable scores for which a prediction is desired, and let Y_0 be the (generally unknown) corresponding value of Y. The independent-variable vector $\mathbf{x}_0$ does not necessarily correspond to an observation in the sample for which the model was fit.

(a) *If we use $\hat{Y}_0 = \mathbf{Xb}$ to estimate $E(Y_0)$, then the error in estimation is $\delta \equiv \hat{Y}_0 - E(Y_0)$. Show that if the model is correct, then $E(\delta) = 0$ [i.e., $\hat{Y}_0$ is an unbiased estimator of $E(Y_0)$], and that $V(\delta) = \sigma_\varepsilon^2 \mathbf{x}_0'(\mathbf{X}'\mathbf{X})^{-1}\mathbf{x}_0$.

(b) *We may be interested not in estimating the *expected* value of Y_0, but in predicting or forecasting the *actual* value $Y_0 = \mathbf{x}_0\boldsymbol{\beta} + \varepsilon_0$ that will be observed. The error in the forecast is then

$$D \equiv \hat{Y}_0 - Y_0 = \mathbf{x}_0'\mathbf{b} - (\mathbf{x}_0'\boldsymbol{\beta} + \varepsilon_0) = \mathbf{x}_0'(\mathbf{b} - \boldsymbol{\beta}) - \varepsilon_0$$

Show that $E(D) = 0$ and that $V(D) = \sigma_\varepsilon^2[1 + \mathbf{x}_0'(\mathbf{X}'\mathbf{X})^{-1}\mathbf{x}_0]$. Why is the variance of the forecast error D greater than the variance of δ found in part (a)?

(c) Use the results in parts (a) and (b), along with the Canadian occupational prestige data (Exercises 9.5 and 9.9 and `prestige.dat`), to predict the prestige score for an occupation with an average income of $12,000, an average education of 13 years, and 50% women. Place a 90% confidence interval around the prediction assuming (i) that you wish to estimate $E(Y_0)$, and (ii) that you wish to forecast an actual Y_0 score. (Because σ_ε^2 is not known, you will need to use S_E^2 and the t-distribution.)

(d) Suppose that the methods of this problem are used to forecast a value of Y for a combination of X's very different from the X-values in the data to which the model was fit. For example, calculate the estimated variance of the forecast error for an occupation with an average income of $50,000, an average education of 0 years, and 100% women. Is the estimated variance of the forecast error large or small? Does the variance of the forecast error adequately capture the uncertainty in using the regression equation to predict Y in this circumstance?

9.17 Suppose that the model matrix for the two-way ANOVA model

$$Y_{ijk} = \mu + \alpha_j + \beta_k + \gamma_{jk} + \varepsilon_{ijk}$$

is reduced to full rank by imposing the following constraints (for $r = 2$ rows and $c = 3$ columns):

$$\alpha_2 = 0$$
$$\beta_3 = 0$$
$$\gamma_{12} = \gamma_{22} = \gamma_{31} = \gamma_{32} = 0$$

These constraints imply dummy-variable (0/1) coding of the full-rank model matrix.

(a) Write out the row basis of the full-rank model matrix under these constraints.

(b) Solve for the parameters of the constrained model in terms of the cell means. What is the nature of the hypotheses H_0: all $\alpha_j = 0$ and H_0: all $\beta_k = 0$ for this parameterization of the model? Are these hypotheses generally sensible?

(c) Let $\mathrm{SS}^*(\alpha, \beta, \gamma)$ represent the regression sum of squares for the full model, calculated under the constraints defined above; let $\mathrm{SS}^*(\alpha, \beta)$ represent the regression sum of squares for the model that deletes the interaction regressors; and so on. Using the Moore and Krupat data (recorded in Table 8.3 and `moore.dat`, and discussed in Section 8.2), confirm that

$$\mathrm{SS}^*(\alpha|\beta) = \mathrm{SS}(\alpha|\beta)$$
$$\mathrm{SS}^*(\beta|\alpha) = \mathrm{SS}(\beta|\alpha)$$
$$\mathrm{SS}^*(\gamma|\alpha, \beta) = \mathrm{SS}(\gamma|\alpha, \beta)$$

but that

$$SS^*(\alpha|\beta, \gamma) \neq SS(\alpha|\beta, \gamma)$$
$$SS^*(\beta|\alpha, \gamma) \neq SS(\beta|\alpha, \gamma)$$

where $SS(\cdot)$ and $SS(\cdot|\cdot)$ give regression and incremental sums of squares under the usual sigma constraints and deviation-coded (1, 0, −1) regressors.

(d) Analyze the Moore and Krupat data using one or more computer programs available to you. How do the programs calculate sums of squares in two-way analysis of variance? Does the documentation accompanying the programs clearly explain how the sums of squares are computed?

9.5 Random Regressors

The theory of linear models developed in this chapter has proceeded from the premise that the model matrix $\mathbf{X}$ is fixed. If we repeat a study, we expect the dependent-variable observations $\mathbf{y}$ to change, but if $\mathbf{X}$ is fixed, then the independent-variable values are constant across replications of the study. This situation is realistically descriptive of an experiment, where the independent variables are manipulated by the researcher. Most research in the social sciences, however, is observational rather than experimental; and in an observational study (survey research, for example), we should typically obtain different independent-variable values upon replication of the study. In observational research, therefore, $\mathbf{X}$ is *random* rather than fixed.

It is remarkable that the statistical theory of linear models applies even when $\mathbf{X}$ is random, as long as certain assumptions are met. For fixed independent variables, the assumptions underlying the model take the form $\boldsymbol{\varepsilon} \sim N_n(\mathbf{0}, \sigma_\varepsilon^2\mathbf{I}_n)$. That is, the distribution of the error is the same for all combinations of independent-variable values represented by the rows of the model matrix. When $\mathbf{X}$ is random, we need to assume that this property holds for *all possible* combinations of independent-variable values in the population that is sampled: That is, $\mathbf{X}$ and $\boldsymbol{\varepsilon}$ are assumed to be independent, and thus the *conditional* distribution of the error for a sample of independent variable values $(\boldsymbol{\varepsilon}|\mathbf{X}_0)$ is $N_n(\mathbf{0}, \sigma_\varepsilon^2\mathbf{I}_n)$, regardless of the particular sample $\mathbf{X}_0 = \{x_{ij}\}$ that is chosen.

Because $\mathbf{X}$ is random, it has some (multivariate) probability distribution. It is not necessary to make assumptions about this distribution, however, beyond (1) requiring that $\mathbf{X}$ and $\boldsymbol{\varepsilon}$ are independent (as just explained), and (2) assuming that the distribution of $\mathbf{X}$ does not depend on the parameters $\boldsymbol{\beta}$ and σ_ε^2 of the linear model. In particular, we need *not* assume that the *regressors* (as opposed to the *errors*) are normally distributed. This is fortunate, for many regressors are highly nonnormal—dummy regressors and polynomial regressors come immediately to mind, not to mention many quantitative independent variables.

It would be unnecessarily tedious to recapitulate the entire argument of this chapter, but I shall show that some key results hold, under the new assumptions, when the independent variables are random. The other results of the chapter can be established for random regressors in a similar manner.

For a particular sample of X-values, $\mathbf{X}_0$, the conditional distribution of $\mathbf{y}$ is

$$E(\mathbf{y}|\mathbf{X}_0) = E[(\mathbf{X}\boldsymbol{\beta} + \boldsymbol{\varepsilon})|\mathbf{X}_0] = \mathbf{X}_0\boldsymbol{\beta} + E(\boldsymbol{\varepsilon}|\mathbf{X}_0)$$
$$= \mathbf{X}_0\boldsymbol{\beta}$$

Consequently, the conditional expectation of the least-squares estimator is

$$E(\mathbf{b}|\mathbf{X}_0) = E\left[(\mathbf{X}'\mathbf{X})^{-1}\mathbf{X}'\mathbf{y}|\mathbf{X}_0\right] = (\mathbf{X}_0'\mathbf{X}_0)^{-1}\mathbf{X}_0'E(\mathbf{y}|\mathbf{X}_0)$$
$$= (\mathbf{X}_0'\mathbf{X}_0)^{-1}\mathbf{X}_0'\mathbf{X}_0\boldsymbol{\beta} = \boldsymbol{\beta}$$

Because we can repeat this argument for any value of $\mathbf{X}$, the least-squares estimator $\mathbf{b}$ is conditionally unbiased for any and every such value; it is therefore *unconditionally* unbiased as well, $E(\mathbf{b}) = \boldsymbol{\beta}$.

Suppose now that we use the procedures of the previous section to perform statistical inference for $\boldsymbol{\beta}$. For concreteness, imagine that we calculate a p-value for the omnibus null hypothesis H_0: $\beta_1 = \cdots = \beta_k = 0$. Because $(\boldsymbol{\varepsilon}|\mathbf{X}_0) \sim N_n(\mathbf{0}, \sigma_\varepsilon^2\mathbf{I}_n)$, as was required when we treated $\mathbf{X}$ as fixed, the p-value obtained is correct for $\mathbf{X} = \mathbf{X}_0$ (i.e., for the sample at hand). There is, however, nothing special about a particular $\mathbf{X}_0$: The error vector $\boldsymbol{\varepsilon}$ is independent of $\mathbf{X}$, and so the distribution of $\boldsymbol{\varepsilon}$ is $N_n(\mathbf{0}, \sigma_\varepsilon^2\mathbf{I}_n)$ for any and every value of $\mathbf{X}$. The p-value, therefore, is *unconditionally* valid.

Finally, I shall show that the maximum-likelihood estimators of $\boldsymbol{\beta}$ and σ_ε^2 are unchanged when $\mathbf{X}$ is random, as long as the new assumptions are met: When $\mathbf{X}$ is random, sampled observations consist not just of dependent-variable values $(Y_1, \ldots, Y_n)$ but also of independent-variable values $(\mathbf{x}_1', \ldots, \mathbf{x}_n')$. The observations themselves are denoted $[Y_1, \mathbf{x}_1'], \ldots, [Y_n, \mathbf{x}_n']$. Because these observations are sampled independently, their joint probability density is the product of their marginal densities:

$$p(\mathbf{y}, \mathbf{X}) \equiv p\left([y_1, \mathbf{x}_1'], \ldots, [y_n, \mathbf{x}_n']\right) = p(y_1, \mathbf{x}_1') \times \cdots \times p(y_n, \mathbf{x}_n')$$

Now, the probability density $p(y_i, \mathbf{x}_i')$ for observation i can be written as $p(y_i|\mathbf{x}_i')p(\mathbf{x}_i')$. According to the linear model, the conditional distribution of Y_i given $\mathbf{x}_i'$ is normal:

$$p(y_i|\mathbf{x}_i') = \frac{1}{\sigma_\varepsilon\sqrt{2\pi}} \exp\left[-\frac{(y_i - \mathbf{x}_i'\boldsymbol{\beta})^2}{2\sigma_\varepsilon^2}\right]$$

Thus, the joint probability density for all observations becomes

$$p(\mathbf{y}, \mathbf{X}) = \prod_{i=1}^{n} p(\mathbf{x}_i') \frac{1}{\sigma_\varepsilon\sqrt{2\pi}} \exp\left[-\frac{(y_i - \mathbf{x}_i'\boldsymbol{\beta})^2}{2\sigma_\varepsilon^2}\right]$$
$$= \left[\prod_{i=1}^{n} p(\mathbf{x}_i')\right] \frac{1}{(2\pi\sigma_\varepsilon^2)^{n/2}} \exp\left[-\frac{(\mathbf{y} - \mathbf{X}\boldsymbol{\beta})'(\mathbf{y} - \mathbf{X}\boldsymbol{\beta})}{2\sigma_\varepsilon^2}\right]$$

As long as $p(\mathbf{x}_i')$ does not depend on the parameters $\boldsymbol{\beta}$ and σ_ε^2, we can ignore the joint density of the X's in maximizing $p(\mathbf{y}, \mathbf{X})$ with respect to the parameters. Consequently, the maximum-likelihood estimator of $\boldsymbol{\beta}$ is the least-squares estimator, as was the case for fixed $\mathbf{X}$.[47]

> The statistical theory of linear models, formulated under the supposition that the model matrix $\mathbf{X}$ is fixed with respect to repeated sampling, is also valid when $\mathbf{X}$ is random, as long as two additional requirements are satisfied: (1) the model matrix $\mathbf{X}$ and the errors $\boldsymbol{\varepsilon}$ are independent; and (2) the distribution of $\mathbf{X}$, which is otherwise unconstrained, does not depend on the parameters $\boldsymbol{\beta}$ and σ_ε^2 of the linear model.

9.6 Specification Error

To generalize our treatment of misspecified structural relationships,[48] it is convenient to work with probability limits. Suppose that the dependent variable Y is determined by the model

$$\mathbf{y}^* = \mathbf{X}^*\boldsymbol{\beta} + \boldsymbol{\varepsilon} = \mathbf{X}_1^*\boldsymbol{\beta}_1 + \mathbf{X}_2^*\boldsymbol{\beta}_2 + \boldsymbol{\varepsilon}$$

where the error $\boldsymbol{\varepsilon}$ behaves according to the usual assumptions. I have, for convenience, expressed each variable as deviations from its expectation [e.g., $\mathbf{y}^* \equiv \{Y_i - E(Y)\}$], and have partitioned the model matrix into two sets of regressors; the parameter vector is partitioned in the same manner.[49]

Imagine that we ignore $\mathbf{X}_2^*$, so that $\mathbf{y}^* = \mathbf{X}_1^*\boldsymbol{\beta}_1 + \widetilde{\boldsymbol{\varepsilon}}$, where $\widetilde{\boldsymbol{\varepsilon}} \equiv \mathbf{X}_2^*\boldsymbol{\beta}_2 + \boldsymbol{\varepsilon}$. The least-squares estimator for $\boldsymbol{\beta}_1$ in the model that omits $\mathbf{X}_2^*$ is

$$\begin{aligned}
\mathbf{b}_1 &= (\mathbf{X}_1^{*\prime}\mathbf{X}_1^*)^{-1}\mathbf{X}_1^{*\prime}\mathbf{y}^* \\
&= \left(\frac{1}{n}\mathbf{X}_1^{*\prime}\mathbf{X}_1^*\right)^{-1}\frac{1}{n}\mathbf{X}_1^{*\prime}\mathbf{y}^* \\
&= \left(\frac{1}{n}\mathbf{X}_1^{*\prime}\mathbf{X}_1^*\right)^{-1}\frac{1}{n}\mathbf{X}_1^{*\prime}(\mathbf{X}_1^*\boldsymbol{\beta}_1 + \mathbf{X}_2^*\boldsymbol{\beta}_2 + \boldsymbol{\varepsilon}) \\
&= \boldsymbol{\beta}_1 + \left(\frac{1}{n}\mathbf{X}_1^{*\prime}\mathbf{X}_1^*\right)^{-1}\frac{1}{n}\mathbf{X}_1^{*\prime}\mathbf{X}_2^*\boldsymbol{\beta}_2 + \left(\frac{1}{n}\mathbf{X}_1^{*\prime}\mathbf{X}_1^*\right)^{-1}\frac{1}{n}\mathbf{X}_1^{*\prime}\boldsymbol{\varepsilon}
\end{aligned}$$

---

[47] Cf. Section 9.3.3.

[48] See Section 6.3.

[49] Expressing the variables as deviations from their expectations eliminates the constant β_0.

Taking probability limits produces

$$\begin{aligned}\text{plim}\,\mathbf{b}_1 &= \boldsymbol{\beta}_1 + \boldsymbol{\Sigma}_{11}^{-1}\boldsymbol{\Sigma}_{12}\boldsymbol{\beta}_2 + \boldsymbol{\Sigma}_{11}^{-1}\boldsymbol{\sigma}_{1\varepsilon} \\ &= \boldsymbol{\beta}_1 + \boldsymbol{\Sigma}_{11}^{-1}\boldsymbol{\Sigma}_{12}\boldsymbol{\beta}_2\end{aligned}$$

where

- $\boldsymbol{\Sigma}_{11} \equiv \text{plim}(1/n)\mathbf{X}_1^{*\prime}\mathbf{X}_1^*$ is the population covariance matrix for $\mathbf{X}_1$;
- $\boldsymbol{\Sigma}_{12} \equiv \text{plim}(1/n)\mathbf{X}_1^{*\prime}\mathbf{X}_2^*$ is the matrix of population covariances between $\mathbf{X}_1$ and $\mathbf{X}_2$; and
- $\boldsymbol{\sigma}_{1\varepsilon} \equiv \text{plim}(1/n)\mathbf{X}_1^{*\prime}\boldsymbol{\varepsilon}$ is the vector of population covariances between $\mathbf{X}_1$ and $\boldsymbol{\varepsilon}$, and is $\mathbf{0}$ by the assumed independence of the error and the independent variables.

The asymptotic (or population) covariance of $\mathbf{X}_1$ and $\widetilde{\boldsymbol{\varepsilon}}$ is not generally $\mathbf{0}$, however, as is readily established:

$$\begin{aligned}\text{plim}\frac{1}{n}\mathbf{X}_1^{*\prime}\widetilde{\boldsymbol{\varepsilon}} &= \text{plim}\frac{1}{n}\mathbf{X}_1^{*\prime}\left(\mathbf{X}_2^*\boldsymbol{\beta}_2 + \boldsymbol{\varepsilon}\right) \\ &= \boldsymbol{\Sigma}_{12}\boldsymbol{\beta}_2 + \boldsymbol{\sigma}_{1\varepsilon} = \boldsymbol{\Sigma}_{12}\boldsymbol{\beta}_2\end{aligned}$$

The estimator $\mathbf{b}_1$, therefore, is consistent if $\boldsymbol{\Sigma}_{12}$ is $\mathbf{0}$—that is, if the excluded regressors in $\mathbf{X}_2$ are uncorrelated with the included regressors in $\mathbf{X}_1$. In this case, incorporating $\mathbf{X}_2^*$ in the error does not induce a correlation between $\mathbf{X}_1^*$ and the compound error $\widetilde{\boldsymbol{\varepsilon}}$. The estimated coefficients $\mathbf{b}_1$ are also consistent if $\boldsymbol{\beta}_2 = 0$: Excluding *irrelevant* regressors is unproblematic.[50]

> The omission of regressors from a linear model causes the coefficients of the included regressors to be inconsistent, unless (1) the omitted regressors are uncorrelated with the included regressors; or (2) the omitted regressors have coefficients of 0, and hence are irrelevant.

[50] *Including* irrelevant regressors also does not cause the least-squares estimator to become inconsistent; after all, if the assumptions of the model hold, then $\mathbf{b}$ is a consistent estimator of $\boldsymbol{\beta}$ even if some of the elements of $\boldsymbol{\beta}$ are 0. (Recall, however, Exercise 6.14.)

9.7 Summary

- The general linear model can be written in matrix form as $\mathbf{y} = \mathbf{X}\boldsymbol{\beta} + \boldsymbol{\varepsilon}$, where $\mathbf{y}$ is an $n \times 1$ vector of dependent-variable observations; $\mathbf{X}$ is an $n \times k+1$ matrix of regressors (called the model matrix), including an initial column of 1's for the constant regressor; $\boldsymbol{\beta}$ is a $k+1 \times 1$ vector of parameters to be estimated; and $\boldsymbol{\varepsilon}$ is an $n \times 1$ vector of errors. The assumptions of the linear model can be compactly written as $\boldsymbol{\varepsilon} \sim N_n(\mathbf{0}, \sigma_\varepsilon^2 \mathbf{I}_n)$.
- The model matrices for dummy-regression and analysis-of-variance models are strongly patterned. In analysis of variance, the relationship between group or cell means and the parameters of the linear model is expressed by the parametric equation $\boldsymbol{\mu} = \mathbf{X}_B \boldsymbol{\beta}_F$, where $\boldsymbol{\mu}$ is the vector of means; $\mathbf{X}_B$ is the row basis of the full-rank model matrix; and $\boldsymbol{\beta}_F$ is the parameter vector associated with the full-rank model matrix. Solving the parametric equation for the parameters yields $\boldsymbol{\beta}_F = \mathbf{X}_B^{-1}\boldsymbol{\mu}$. Linear contrasts are regressors that are coded to incorporate specific hypotheses about the group means in the parameters of the model.
- The least-squares coefficients are given by $\mathbf{b} = (\mathbf{X}'\mathbf{X})^{-1}\mathbf{X}'\mathbf{y}$. Under the full set of assumptions for the linear model, $\mathbf{b} \sim N_{k+1}[\boldsymbol{\beta}, \sigma_\varepsilon^2(\mathbf{X}'\mathbf{X})^{-1}]$. The least-squares estimator is also the most efficient unbiased estimator of $\boldsymbol{\beta}$ and the maximum-likelihood estimator of $\boldsymbol{\beta}$.
- The estimated covariance matrix of the least-squares coefficients is $\widehat{V(\mathbf{b})} = S_E^2(\mathbf{X}'\mathbf{X})^{-1}$. The estimated standard errors of the regression coefficients are the diagonal entries of this matrix. Under the assumptions of the model, $(B_j - \beta_j)/\widehat{\mathrm{SE}}(B_j) \sim t_{n-k-1}$, providing a basis for hypothesis tests and confidence intervals for individual coefficients.
- An incremental F-test for the hypothesis H_0: $\beta_1 = \cdots = \beta_q = 0$, where $1 \le q \le k$, is given by

$$F_0 = \frac{(\mathrm{RSS}_0 - \mathrm{RSS})/q}{\mathrm{RSS}/(n-k-1)}$$

  where RSS is the residual sum of squares for the full model, and RSS_0 is the residual sum of squares for the model that deletes the q regressors in question. Under the null hypothesis, $F_0 \sim F_{q,\, n-k-1}$. The incremental F-statistic can also be computed directly as $F_0 = \mathbf{b}_1'\mathbf{V}_{11}^{-1}\mathbf{b}_1/qS_E^2$, where $\mathbf{b}_1 = [B_1, \ldots, B_q]'$ contains the coefficients of interest extracted from among the entries of $\mathbf{b}$, and $\mathbf{V}_{11}$ is the square submatrix of $(\mathbf{X}'\mathbf{X})^{-1}$ consisting of the q rows and columns pertaining to the coefficients in $\mathbf{b}_1$.
- The F-statistic

$$F_0 = \frac{(\mathbf{L}\mathbf{b} - \mathbf{c})'[\mathbf{L}(\mathbf{X}'\mathbf{X})^{-1}\mathbf{L}']^{-1}(\mathbf{L}\mathbf{b} - \mathbf{c})}{qS_E^2}$$

  is used to test the general linear hypothesis H_0: $\underset{(q\times k+1)}{\mathbf{L}}\underset{(k+1\times 1)}{\boldsymbol{\beta}} = \underset{(q\times 1)}{\mathbf{c}}$, where the rank-$q$ hypothesis matrix $\mathbf{L}$ and right-hand-side vector $\mathbf{c}$ contain prespecified constants. Under the hypothesis, $F_0 \sim F_{q,\, n-k-1}$.
- The joint confidence region for the q parameters $\boldsymbol{\beta}_1$, given by

$$(\mathbf{b}_1 - \boldsymbol{\beta}_1)'\mathbf{V}_{11}^{-1}(\mathbf{b}_1 - \boldsymbol{\beta}_1) \le qS_E^2 F_{a,\, q,\, n-k-1}$$

  represents the combinations of values of these parameters that are jointly acceptable at the $1 - a$ level of confidence. The boundary of the joint confidence region is an ellipsoid in the q-dimensional parameter space, reflecting the correlational structure and dispersion of the X's.

- The statistical theory of linear models, formulated under the supposition that the model matrix **X** is fixed with respect to repeated sampling, is also valid when **X** is random, as long as two additional requirements are satisfied:

   1. the model matrix **X** and the errors $\boldsymbol{\varepsilon}$ are independent; and
   2. the distribution of **X**, which is otherwise unconstrained, does not depend on the parameters $\boldsymbol{\beta}$ and σ_ε^2 of the linear model.

- The omission of regressors from a linear model causes the coefficients of the included regressors to be inconsistent, unless

   1. the omitted regressors are uncorrelated with the included regressors; or
   2. the omitted regressors have coefficients of 0, and hence are irrelevant.

9.8 Recommended Reading

There are many texts that treat the theory of linear models more abstractly, more formally, and with greater generality than I have in this chapter.

- Seber (1977) is a reasonably accessible text that develops in a statistically more sophisticated manner most of the topics discussed in the last five chapters. Seber also pays more attention to issues of computation, and develops some topics that I do not.
- Searle (1971) presents a very general treatment of linear models, including a much broader selection of ANOVA models, stressing the analysis of unbalanced data. Searle directly analyzes model matrices of less than full rank, an approach that—in my opinion—makes the subject more complex than it needs to be. Despite its relative difficulty, however, the presentation is of exceptionally high quality.
- Hocking (1985) and Searle (1987) cover much the same ground as Searle (1971), but stress the use of "cell-means" models, avoiding some of the complications of over-parameterized models for analysis of variance. These books also contain a very general presentation of the theory of linear statistical models.
- A fine paper by Monette (1990) develops in more detail the geometric representation of regression analysis using ellipses (a topic that is usually treated only in difficult sources).

10

The Vector Geometry of Linear Models*

As is clear from the previous chapter, linear algebra is the algebra of linear models. Vector geometry provides a spatial representation of linear algebra, and therefore furnishes a powerful tool for understanding linear models. The geometric understanding of linear models is venerable. Fisher's development of the central notion of degrees of freedom in linear models was closely tied to vector geometry, for example.

Few points in this book are developed exclusively in geometric terms. The reader who takes the time to master the geometry of linear models, however, will find the effort worthwhile: Certain ideas—including degrees of freedom—are most simply developed or understood from the geometric perspective.[1]

The chapter begins by developing the geometric vector representation of simple and multiple regression. Then, the vector representation is employed to explain the connection between degrees of freedom and unbiased estimation of the error variance in linear models. Finally, vector geometry is used to illuminate the essential nature of overparameterized analysis-of-variance models.

10.1 Simple Regression

We can write the simple-regression model in vector form in the following manner:

$$\mathbf{y} = \alpha \mathbf{1}_n + \beta \mathbf{x} + \boldsymbol{\varepsilon} \tag{10.1}$$

[1] The basic vector geometry on which this chapter depends is developed in Appendix B.

where $\mathbf{y} \equiv [Y_1, Y_2, \ldots, Y_n]'$, $\mathbf{x} \equiv [x_1, x_2, \ldots, x_n]'$, $\boldsymbol{\varepsilon} \equiv [\varepsilon_1, \varepsilon_2, \ldots, \varepsilon_n]'$, and $\mathbf{1}_n \equiv [1, 1, \ldots, 1]'$; α and β are the population regression coefficients.[2] The fitted regression equation is, similarly,

$$\mathbf{y} = A\mathbf{1}_n + B\mathbf{x} + \mathbf{e} \qquad [10.2]$$

where $\mathbf{e} \equiv [E_1, E_2, \ldots, E_n]'$ is the vector of residuals, and A and B are the least-squares regression coefficients. From Equation 10.1, we have

$$E(\mathbf{y}) = \alpha\mathbf{1}_n + \beta\mathbf{x}$$

Analogously, from Equation 10.2,

$$\hat{\mathbf{y}} = A\mathbf{1}_n + B\mathbf{x}$$

We are familiar with a seemingly natural geometric representation of $\{X, Y\}$ data—the scatterplot—in which the axes of a two-dimensional coordinate space are defined by the variables X and Y, and where the observations are represented as points in the space according to their $\{x_i, Y_i\}$ coordinates. The scatterplot is a valuable data-analytic tool as well as a device for thinking about regression analysis.

I shall now exchange the familiar roles of variables and observations, defining an n-dimensional coordinate space for which the *axes* are given by the *observations* and in which the *variables* are plotted as *vectors*. Of course, because there are generally many more than three observations, it is not possible to visualize the full vector space of the observations.[3] Our interest, however, often inheres in two- and three-dimensional subspaces of this larger n-dimensional vector space. In these instances, as we shall see presently, graphical representation is both possible and illuminating. Moreover, the geometry of higher-dimensional subspaces can be grasped by analogy to the two- and three-dimensional case.

The two-dimensional *variable space* (i.e., in which the variables define the axes) and the n-dimensional *observation space* (in which the observations define the axes) each contains a complete representation of the $(n \times 2)$ data matrix $[\mathbf{x}, \mathbf{y}]$. The formal duality of these spaces means that properties of the data, or of models meant to describe them, have equivalent representations in both spaces. Sometimes, however, the geometric representation of a property will be easier to understand in one space or the other.

The simple-regression model of Equation 10.1 is shown geometrically in Figure 10.1. The subspace depicted in this figure is of dimension 3, and is spanned by the vectors $\mathbf{x}$, $\mathbf{y}$, and $\mathbf{1}_n$. Because $\mathbf{y}$ is a vector random variable which varies from sample to sample, the vector diagram necessarily represents a particular sample. The other vectors shown in the diagram clearly lie in the subspace spanned by $\mathbf{x}$, $\mathbf{y}$, and $\mathbf{1}_n$: $E(\mathbf{y})$ is a linear combination of $\mathbf{x}$ and $\mathbf{1}_n$ (and thus lies in the $\{\mathbf{1}_n, \mathbf{x}\}$ plane); and the error vector $\boldsymbol{\varepsilon}$ is $\mathbf{y} - \alpha\mathbf{1}_n - \beta\mathbf{x}$. Although $\boldsymbol{\varepsilon}$ is nonzero in this sample, on average, over many samples, $E(\boldsymbol{\varepsilon}) = \mathbf{0}$.

---

[2] Notice that the X-values are treated as fixed. As in the previous chapter, the development of the vector geometry of linear models is simpler for fixed X, but the results apply as well when X is random.

[3] See Exercise 10.1 for a scaled-down example, however.

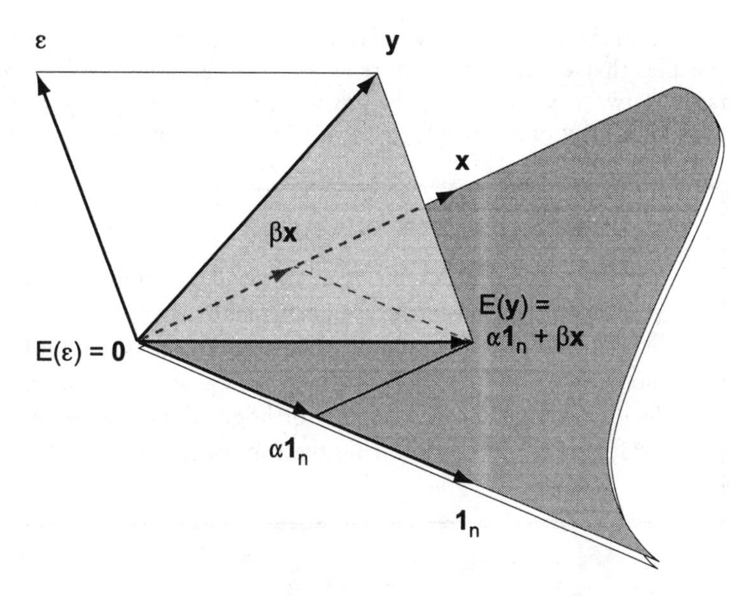

Figure 10.1. The vector geometry of the simple-regression model, showing the three-dimensional subspace spanned by the vectors **x**, **y**, and $\mathbf{1}_n$. Because the expected error is 0, the expected-Y vector, $E(\mathbf{y})$, lies in the plane spanned by $\mathbf{1}_n$ and **x**.

Figure 10.2 represents the least-squares simple regression of Y on X, for the same data as shown in Figure 10.1. The peculiar geometry of Figure 10.2 requires some explanation: We know that the fitted values are a linear combination of $\mathbf{1}_n$ and **x**, and hence lie in the $\{\mathbf{1}_n, \mathbf{x}\}$ plane. The residual vector $\mathbf{e} = \mathbf{y} - \hat{\mathbf{y}}$ has length $||\mathbf{e}|| = \sqrt{\sum E_i^2}$—that is, the square root of the residual

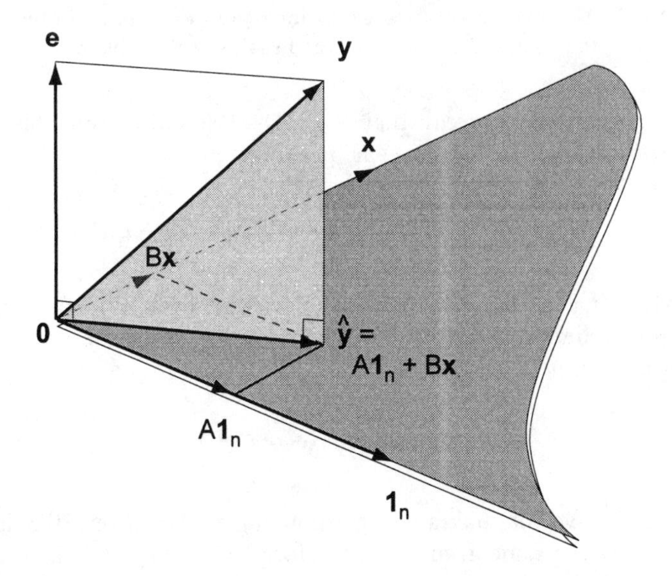

Figure 10.2. The vector geometry of least-squares fit in simple regression. Minimizing the residual sum of squares is equivalent to making the **e** vector as short as possible. The $\hat{\mathbf{y}}$ vector is, therefore, the orthogonal projection of **y** onto the $\{\mathbf{1}_n, \mathbf{x}\}$ plane.

sum of squares. The least-squares criterion interpreted geometrically, therefore, specifies that $\mathbf{e}$ must be as short as possible. Because the length of $\mathbf{e}$ is the distance between $\mathbf{y}$ and $\hat{\mathbf{y}}$, this length is minimized by taking $\hat{\mathbf{y}}$ as the orthogonal projection of $\mathbf{y}$ onto the $\{\mathbf{1}_n, \mathbf{x}\}$ plane, as shown in the diagram.

> Variables, such as X and Y in simple regression, can be treated as vectors—$\mathbf{x}$ and $\mathbf{y}$—in the n-dimensional space whose axes are given by the observations. Written in vector form, the simple-regression model is $\mathbf{y} = \alpha\mathbf{1}_n + \beta\mathbf{x} + \boldsymbol{\varepsilon}$. The least-squares regression, $\mathbf{y} = A\mathbf{1}_n + B\mathbf{x} + \mathbf{e}$, is found by projecting $\mathbf{y}$ orthogonally onto the plane spanned by $\mathbf{1}_n$ and $\mathbf{x}$, thus minimizing the sum of squared residuals, $||\mathbf{e}||^2$.

10.1.1 Variables in Mean-Deviation Form

We can simplify the vector representation for simple regression by eliminating the constant regressor $\mathbf{1}_n$ and, with it, the intercept coefficient A. This simplification is worthwhile for two reasons:

1. Our diagram is reduced from three to two dimensions. When we turn to multiple regression—introducing a second independent variable—eliminating the constant leaves us to contend with a three-dimensional rather than a four-dimensional subspace.
2. The analysis of variance for the regression appears in the vector diagram when the constant is eliminated, as I shall shortly explain.

To delete A, recall that $\overline{Y} = A + B\overline{x}$; subtracting this equation from the fitted model $Y_i = A + Bx_i + E_i$ produces

$$Y_i - \overline{Y} = B(x_i - \overline{x}) + E_i$$

Expressing the variables in mean-deviation form eliminates the regression constant. Defining $\mathbf{y}^* \equiv \{Y_i - \overline{Y}\}$ and $\mathbf{x}^* \equiv \{x_i - \overline{x}\}$, the vector form of the fitted regression model becomes

$$\mathbf{y}^* = B\mathbf{x}^* + \mathbf{e} \qquad [10.3]$$

The vector diagram corresponding to Equation 10.3 is shown in Figure 10.3. By the same argument as before,[4] $\hat{\mathbf{y}}^* \equiv \{\hat{Y}_i - \overline{Y}\}$ is a multiple of $\mathbf{x}^*$, and

[4] The mean deviations for the fitted values are $\{\hat{Y}_i - \overline{Y}\}$ because the mean of the fitted values is the same as the mean of Y. See Exercise 10.2.

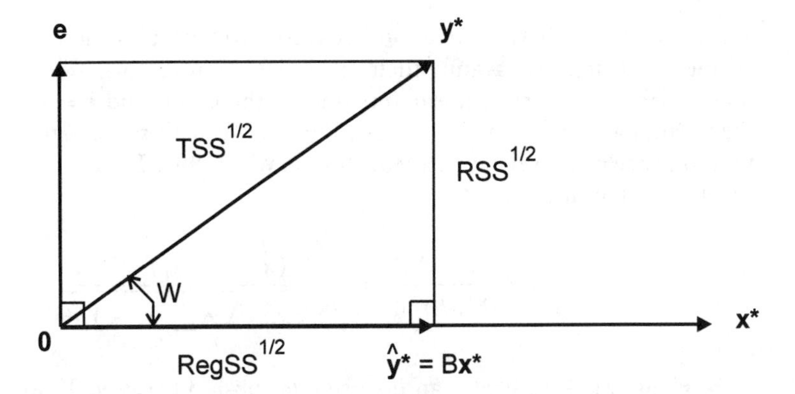

Figure 10.3. The vector geometry of least-squares fit in simple regression for variables in mean-deviation form. The analysis of variance for the regression follows from the Pythagorean theorem. The correlation between X and Y is the cosine of the angle W separating the $\mathbf{x}^*$ and $\mathbf{y}^*$ vectors.

the length of $\mathbf{e}$ is minimized by taking $\hat{\mathbf{y}}^*$ as the orthogonal projection of $\mathbf{y}^*$ onto $\mathbf{x}^*$. Thus,

$$B = \frac{\mathbf{x}^* \cdot \mathbf{y}^*}{||\mathbf{x}^*||^2} = \frac{\sum(x_i - \overline{x})(Y_i - \overline{Y})}{\sum(x_i - \overline{x})^2}$$

which is the familiar formula for the least-squares slope in simple regression.[5]

Sums of squares appear on the vector diagram as the squared lengths of vectors. I have already remarked that

$$\text{RSS} = \sum E_i^2 = ||\mathbf{e}||^2$$

Similarly,

$$\text{TSS} = \sum(Y_i - \overline{Y})^2 = ||\mathbf{y}^*||^2$$

and

$$\text{RegSS} = \sum(\hat{Y}_i - \overline{Y})^2 = ||\hat{\mathbf{y}}^*||^2$$

The analysis of variance for the regression, TSS = RegSS + RSS, follows from the Pythagorean theorem.

The correlation coefficient is

$$r = \sqrt{\frac{\text{RegSS}}{\text{TSS}}} = \frac{||\hat{\mathbf{y}}^*||}{||\mathbf{y}^*||}$$

The vectors $\hat{\mathbf{y}}^*$ and $\mathbf{y}^*$ are, respectively, the adjacent side and hypotenuse for the angle W in the right triangle whose vertices are given by the tips of $\mathbf{0}$, $\mathbf{y}^*$, and $\hat{\mathbf{y}}^*$.

[5] See Section 5.1.

Thus, $r = \cos W$: The correlation between two variables (here, X and Y) is the cosine of the angle separating their mean-deviation vectors. When this angle is 0, one variable is a perfect linear function of the other, and $r = \cos 0 = 1$. When the vectors are orthogonal, $r = \cos 90° = 0$. We shall see shortly that when two variables are negatively correlated,[6] $90° < W \leq 180°$. The correlation $r = \cos W$ can be written directly as[7]

$$r = \frac{\mathbf{x}^* \cdot \mathbf{y}^*}{||\mathbf{x}^*|| \, ||\mathbf{y}^*||} = \frac{\sum(x_i - \overline{x})(Y_i - \overline{Y})}{\sqrt{\sum(x_i - \overline{x})^2 \sum(Y_i - \overline{Y})^2}} \quad [10.4]$$

Figure 10.4 illustrates an inverse relationship between X and Y. All of the conclusions that we based on Figure 10.3 still hold. Because B is now negative, $\hat{\mathbf{y}}^* = B\mathbf{x}^*$ is a negative multiple of the $\mathbf{x}^*$ vector, pointing in the opposite direction from $\mathbf{x}^*$. The correlation is still the cosine of W, but now we need to take the negative root of $\sqrt{||\hat{\mathbf{y}}^*||^2/||\mathbf{y}^*||^2}$ if we wish to define r in terms of vector lengths; Equation 10.4 produces the proper sign, because $\mathbf{x}^* \cdot \mathbf{y}^*$ is negative.

> Writing X and Y in mean-deviation form, as the vectors $\mathbf{x}^*$ and $\mathbf{y}^*$, eliminates the constant term and thus permits representation of the fitted regression in two (rather than three) dimensions: $\mathbf{y}^* = B\mathbf{x}^* + \mathbf{e}$. The analysis of variance for the regression, TSS = RegSS + RSS, is represented geometrically as $||\mathbf{y}^*||^2 = ||\hat{\mathbf{y}}^*||^2 + ||\mathbf{e}||^2$. The correlation between X and Y is the cosine of the angle separating the vectors $\mathbf{x}^*$ and $\mathbf{y}^*$.

10.1.2 Degrees of Freedom

The vector representation of simple regression helps to clarify the concept of degrees of freedom. In general, sums of squares for linear models are the squared lengths of variable vectors. The degrees of freedom associated with a sum of squares represent the dimension of the subspace to which the corresponding vector is confined.

- Consider, first off, the vector $\mathbf{y}$ in Figure 10.2: This vector can be located anywhere in the n-dimensional observation space. The *uncorrected* sum of squares $\sum Y_i^2 = ||\mathbf{y}||$, therefore, has n degrees of freedom.

---

[6] We need only consider angles between 0° and 180° for we can always examine the smaller of the two angles separating $\mathbf{x}^*$ and $\mathbf{y}^*$. Because $\cos W = \cos(360° - W)$, this convention is of no consequence.

[7] This is the alternative formula for the correlation coefficient presented in Section 5.1. The vector representation of simple regression, therefore, demonstrates the equivalence of the two formulas for r—the direct formula and the definition in terms of sums of squares.

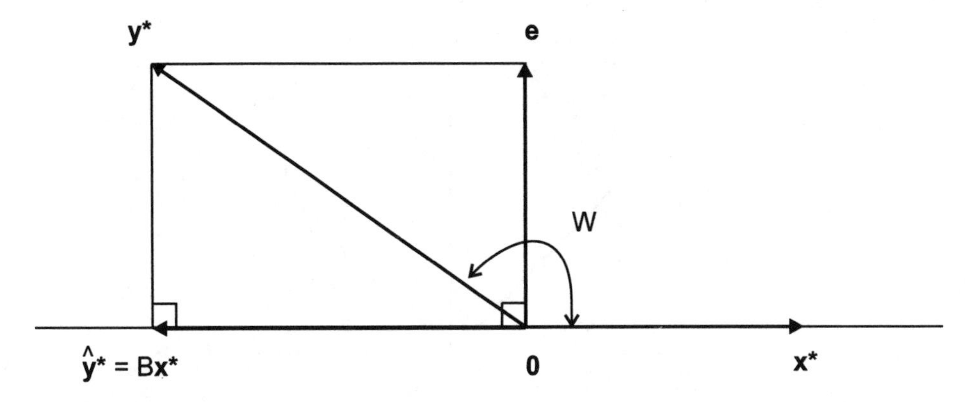

Figure 10.4. The vector geometry of least-squares fit for a negative relationship between X and Y.

- When we convert Y to mean-deviation form (as in Figure 10.3), we confine the $\mathbf{y}^*$ vector to an $(n-1)$-dimensional subspace, "losing" 1 degree of freedom in the process. This is easily seen for vectors in two-dimensional space: Let $\mathbf{y} = [Y_1, Y_2]'$, and $\mathbf{y}^* = [Y_1 - \overline{Y}, Y_2 - \overline{Y}]'$. Then, because $\overline{Y} = (Y_1 + Y_2)/2$, we can write

$$\mathbf{y}^* = \left[Y_1 - \frac{Y_1 + Y_2}{2}, Y_2 - \frac{Y_1 + Y_2}{2}\right]' = \left[\frac{Y_1 - Y_2}{2}, \frac{Y_2 - Y_1}{2}\right]' = [Y_1^*, -Y_1^*]'$$

  Thus, all vectors $\mathbf{y}^*$ lie on a line through the origin, as shown in Figure 10.5: The subspace of all vectors $\mathbf{y}^*$ is one dimensional. Algebraically, by subtracting the mean from each of its coordinates, we have imposed a linear restriction on $\mathbf{y}^*$, ensuring that its entries sum to zero, $\sum(Y_i - \overline{Y}) = 0$; among the n values of $Y_i - \overline{Y}$, only $n-1$ are linearly independent. The total sum of squares $\text{TSS} = \sum(Y_i - \overline{Y})^2$, therefore, has $n-1$ degrees of freedom.

We can extend this reasoning to the residual and regression sums of squares:

- In Figure 10.3, $\hat{\mathbf{y}}^*$ is a multiple of $\mathbf{x}^*$. The vector $\mathbf{x}^*$, in turn, is fixed, and spans a one-dimensional subspace. Because $\hat{\mathbf{y}}^*$ necessarily lies somewhere in this one-dimensional subspace, $\text{RegSS} = ||\hat{\mathbf{y}}^*||^2$ has 1 degree of freedom.
- The degrees of freedom for the residual sum of squares can be determined from either Figure 10.2 or Figure 10.3. In Figure 10.2, $\mathbf{y}$ lies somewhere in the n-dimensional observation space. The vectors $\mathbf{x}$ and $\mathbf{1}_n$ are fixed and together span a subspace of dimension 2 within the larger observation space. The location of the residual vector $\mathbf{e}$ depends on $\mathbf{y}$, but, in any event, $\mathbf{e}$ is orthogonal to the plane spanned by $\mathbf{x}$ and $\mathbf{1}_n$. Consequently, $\mathbf{e}$ lies in a subspace of dimension $n-2$, and $\text{RSS} = ||\mathbf{e}||^2$ has $n-2$ degrees of freedom. Algebraically, the least-squares residuals $\mathbf{e}$ satisfy two independent linear restrictions—$\sum E_i = 0$ (i.e., $\mathbf{e} \cdot \mathbf{1}_n = 0$) and $\sum E_i x_i = 0$ (i.e., $\mathbf{e} \cdot \mathbf{x} = 0$)—accounting for the "loss" of 2 degrees of freedom.[8]

[8] It is also the case that $\sum E_i \hat{Y}_i = \mathbf{e} \cdot \hat{\mathbf{y}} = 0$, but this constraint follows from the other two. See Exercise 10.3.

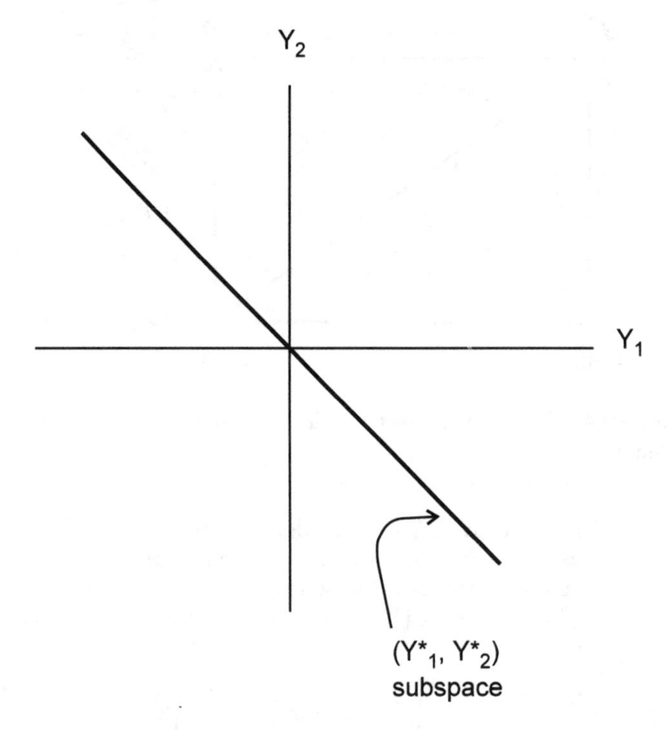

Figure 10.5. When $n = 2$, the mean-deviation vector $\mathbf{y}^* = [Y_1 - \overline{Y}, Y_2 - \overline{Y}]'$ is confined to a one-dimensional subspace (i.e., a line) of the two-dimensional observation space.

- Alternatively, referring to Figure 10.3, $\mathbf{y}^*$ lies in the $(n-1)$-dimensional subspace of mean deviations; the residual vector $\mathbf{e}$ is orthogonal to $\mathbf{x}^*$, which also lies in the $(n-1)$-dimensional mean-deviation subspace; hence, RSS has $(n-1)-1 = n-2$ degrees of freedom.

Degrees of freedom in simple regression correspond to the dimensions of subspaces to which variable vectors associated with sums of squares are confined: (1) The $\mathbf{y}^*$ vector lies in the $(n-1)$-dimensional subspace of mean deviations, but is otherwise unconstrained; TSS, therefore, has $n - 1$ degrees of freedom. (2) The $\hat{\mathbf{y}}^*$ vector lies somewhere along the one-dimensional subspace spanned by $\mathbf{x}^*$; RegSS, therefore, has 1 degree of freedom. (3) The $\mathbf{e}$ vector lies in the $(n-1)$-dimensional subspace of mean deviations, and is constrained to be orthogonal to $\mathbf{x}^*$; RSS, therefore, has $(n-1)-1 = n-2$ degrees of freedom.

EXERCISES

10.1 Here is a very small (contrived) dataset with two variables and two observations:

	Variables	
Observation	X	Y
1	1	2
2	3	5

Construct a scatterplot for the two observations in the $\{X, Y\}$ variable space; then construct a vector diagram showing **x** and **y** in the observation space.

10.2 Show that the average fitted value, $\overline{\hat{Y}}$, is the same as the average dependent-variable value, $\overline{Y}$. [*Hint*: Form the sum, $\sum Y_i = \sum(\hat{Y}_i + E_i)$.]

10.3 *Show that the constraints $\mathbf{e} \cdot \mathbf{x} = 0$ and $\mathbf{e} \cdot \mathbf{1}_n = 0$ imply that $\mathbf{e} \cdot \hat{\mathbf{y}} = 0$. (*Hint*: $\hat{\mathbf{y}}$ lies in the plane spanned by $\mathbf{x}$ and $\mathbf{1}_n$.)

10.4 Using Duncan's occupational prestige data (located in Table 3.2 and `duncan.dat`, and discussed, e.g., in Chapter 5), construct the geometric vector representation for the regression of prestige on education, showing the $\mathbf{x}^*$, $\mathbf{y}^*$, $\hat{\mathbf{y}}^*$, and $\mathbf{e}$ vectors drawn to scale. Find the angle between $\mathbf{x}^*$ and $\mathbf{y}^*$.

10.2 Multiple Regression

To develop the vector geometry of multiple regression, I shall work primarily with the two-independent-variable model: Virtually all important points can be developed for this case; and by expressing the variables in mean-deviation form (and, consequently, eliminating the constant regressor), the subspace of interest is confined to three dimensions.

Consider, then, the fitted model

$$\mathbf{y} = A\mathbf{1}_n + B_1\mathbf{x}_1 + B_2\mathbf{x}_2 + \mathbf{e} \quad [10.5]$$

where $\mathbf{y}$ is, as before, the vector of dependent-variable observations; $\mathbf{x}_1$ and $\mathbf{x}_2$ are independent-variable vectors; $\mathbf{e}$ is the vector of residuals; and $\mathbf{1}_n$ is a vector of 1's. The least-squares regression coefficients are A, B_1, and B_2. From each observation of Equation 10.5, let us subtract $\overline{Y} = A + B_1\overline{x}_1 + B_2\overline{x}_2$, obtaining

$$\mathbf{y}^* = B_1\mathbf{x}_1^* + B_2\mathbf{x}_2^* + \mathbf{e} \quad [10.6]$$

In Equation 10.6, $\mathbf{y}^*$, $\mathbf{x}_1^*$, and $\mathbf{x}_2^*$ are vectors of mean deviations.

Figure 10.6(a) shows the three-dimensional vector diagram for the fitted model of Equation 10.6, while Figure 10.6(b) depicts the independent-variable plane. The fitted values $\hat{\mathbf{y}}^* = B_1\mathbf{x}_1^* + B_2\mathbf{x}_2^*$ are a linear combination of the regressors, and the vector $\hat{\mathbf{y}}^*$, therefore, lies in the $\{\mathbf{x}_1^*, \mathbf{x}_2^*\}$ plane. By familiar reasoning, the least-squares criterion implies that the residual vector $\mathbf{e}$ is orthogonal to the independent-variable plane and, consequently, that $\hat{\mathbf{y}}^*$ is the orthogonal projection of $\mathbf{y}^*$ onto this plane.

The regression coefficients B_1 and B_2 are uniquely defined as long as $\mathbf{x}_1^*$ and $\mathbf{x}_2^*$ are not collinear. This is the geometric version of the requirement that the independent variables may not be perfectly correlated. If the regressors are collinear, then they span a line rather than a plane; although we can still find the fitted values by orthogonally projecting $\mathbf{y}^*$ onto this line, as shown in Figure 10.7, we cannot express $\hat{\mathbf{y}}^*$ *uniquely* as a linear combination of $\mathbf{x}_1^*$ and $\mathbf{x}_2^*$.

The analysis of variance for the multiple-regression model appears in the plane spanned by $\mathbf{y}^*$ and $\hat{\mathbf{y}}^*$, as illustrated in Figure 10.8. The residual vector also lies in this plane (because $\mathbf{e} = \mathbf{y}^* - \hat{\mathbf{y}}^*$), while the regressor plane $\{\mathbf{x}_1^*, \mathbf{x}_2^*\}$ is perpendicular to it. As in simple-regression analysis, TSS $= ||\mathbf{y}^*||^2$, RegSS $= ||\hat{\mathbf{y}}^*||^2$, and RSS $= ||\mathbf{e}||^2$. The identity TSS $=$ RegSS $+$ RSS follows from the Pythagorean theorem.

It is also clear from Figure 10.8 that $R = \sqrt{\text{RegSS}/\text{TSS}} = \cos W$. Thus, the multiple correlation is the simple correlation between the observed and fitted dependent-variable values, Y and $\hat{Y}$. If there is a perfect linear relationship between Y and the independent variables, then $\mathbf{y}^*$ lies in the regressor plane,

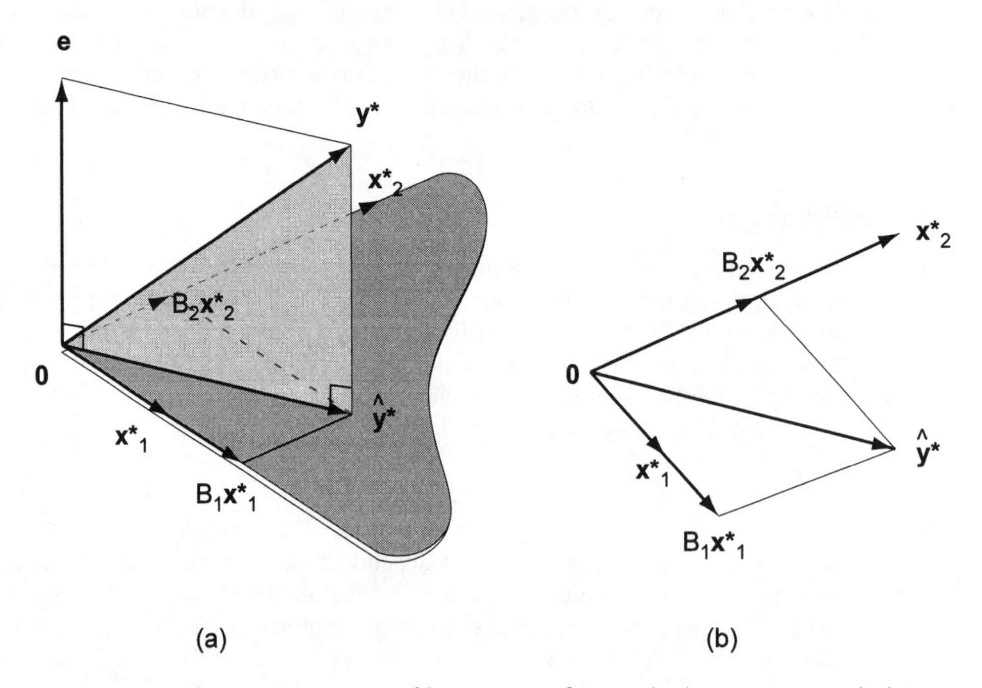

Figure 10.6. The vector geometry of least-squares fit in multiple regression, with the variables in mean-deviation form. The vectors $\mathbf{y}^*$, $\mathbf{x}_1^*$, and $\mathbf{x}_2^*$ span a three-dimensional subspace, shown in (a). The fitted Y vector, $\hat{\mathbf{y}}^*$, is the orthogonal projection of $\mathbf{y}^*$ onto the plane spanned by $\mathbf{x}_1^*$ and $\mathbf{x}_2^*$. The $\{\mathbf{x}_1^*, \mathbf{x}_2^*\}$ plane is shown in (b).

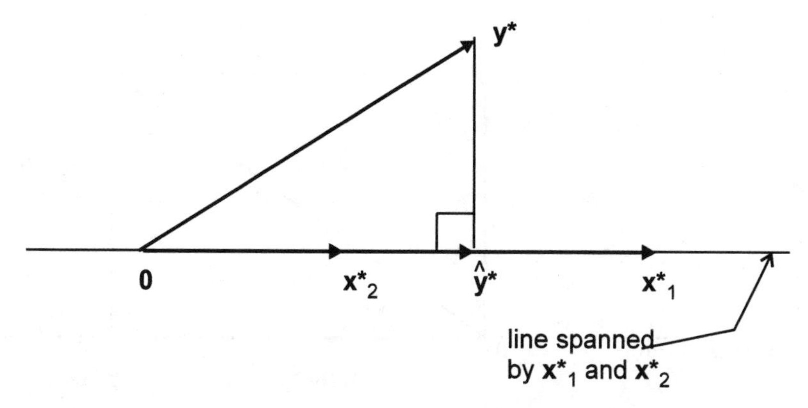

Figure 10.7. When the independent variables are perfectly collinear, $\mathbf{x}_1^*$ and $\mathbf{x}_2^*$ span a line rather than a plane. The $\hat{\mathbf{y}}^*$ vector can still be found by projecting $\mathbf{y}^*$ orthogonally onto this line, but the regression coefficients B_1 and B_2, expressing $\mathbf{y}^*$ as a linear combination of $\mathbf{x}_1^*$ and $\mathbf{x}_2^*$, are not unique.

$\mathbf{y}^* = \hat{\mathbf{y}}^*$, $\mathbf{e} = \mathbf{0}$, $W = 0°$, and $R = 1$; if, at the other extreme, there is no linear relationship between Y and the independent variables, then $\mathbf{y}^* = \mathbf{e}$, $\hat{\mathbf{y}}^* = \mathbf{0}$, $W = 90°$, and $R = 0$.

> The fitted multiple-regression model for two independent variables is written in vector form as $\mathbf{y} = A\mathbf{1}_n + B_1\mathbf{x}_1 + B_2\mathbf{x}_2 + \mathbf{e}$. Writing Y and the X's in mean-deviation form eliminates the constant, $\mathbf{y}^* = B_1\mathbf{x}_1^* + B_2\mathbf{x}_2^* + \mathbf{e}$, and permits a representation in three (rather than four) dimensions. The fitted values, $\hat{\mathbf{y}}^* = B_1\mathbf{x}_1^* + B_2\mathbf{x}_2^*$, are found by projecting $\mathbf{y}^*$ orthogonally onto the plane spanned by $\mathbf{x}_1^*$ and $\mathbf{x}_2^*$. The analysis of variance for the regression, which is essentially the same as in simple regression, appears in the plane spanned by $\mathbf{y}^*$ and $\hat{\mathbf{y}}^*$. The multiple correlation R is the cosine of the angle separating $\mathbf{y}^*$ and $\hat{\mathbf{y}}^*$, and, consequently, is the simple correlation between the observed and fitted Y-values.

Figure 10.9 shows the vector geometry of the incremental F-test for the hypothesis H_0: $\beta_1 = 0$. RegSS—the regression sum of squares from the full model, where Y is regressed on all of the X's—is decomposed into two orthogonal components: RegSS_0 (for the regression of Y on $X_2, \ldots, X_k$) and the incremental sum of squares $\text{RegSS} - \text{RegSS}_0$.

The vector representation of regression analysis helps clarify the relationship between simple and multiple regression. Figure 10.10(a) is drawn for two

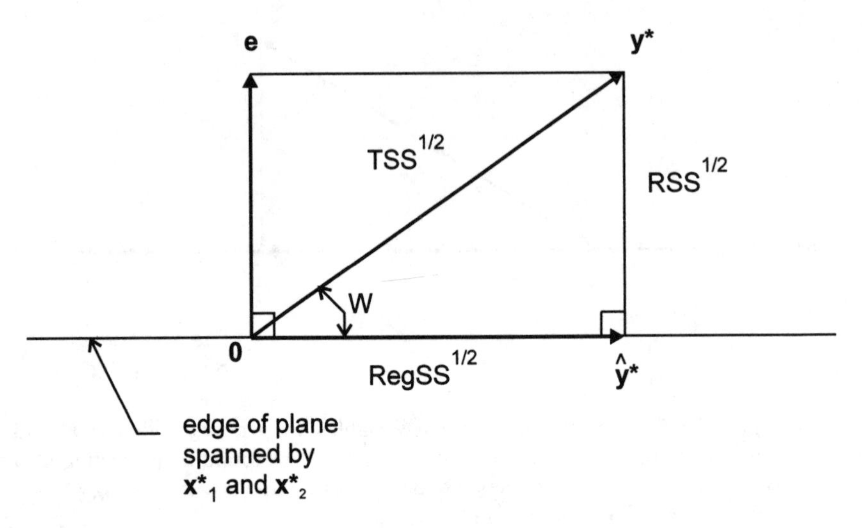

Figure 10.8. The analysis of variance for multiple regression appears in the plane spanned by $\mathbf{y}^*$ and $\hat{\mathbf{y}}^*$. The multiple correlation is the cosine of the angle W separating $\mathbf{y}^*$ and $\hat{\mathbf{y}}^*$.

positively correlated regressors. The fitted dependent-variable vector is, from our previous work, the orthogonal projection of $\mathbf{y}^*$ onto the $\{\mathbf{x}_1^*, \mathbf{x}_2^*\}$ plane. To find the multiple-regression coefficient B_1, we project $\hat{\mathbf{y}}^*$ parallel to $\mathbf{x}_2^*$, locating $B_1\mathbf{x}_1^*$, as shown in Figure 10.10(b), which depicts the regressor plane. The coefficient B_2 is located similarly.

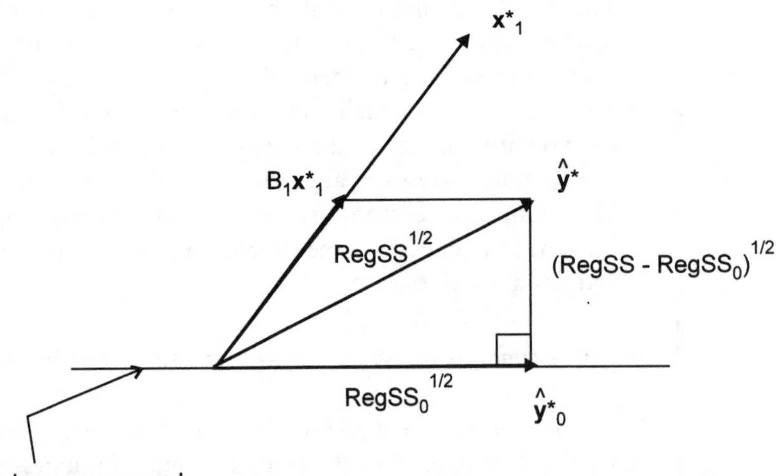

Figure 10.9. The incremental sum of squares for the hypothesis H_0: $\beta_1 = 0$. The vector $\hat{\mathbf{y}}_0^*$ is for the regression of Y on $X_2, \ldots, X_k$, while the vector $\hat{\mathbf{y}}^*$ is for the regression of Y on all of the X's, including X_1.

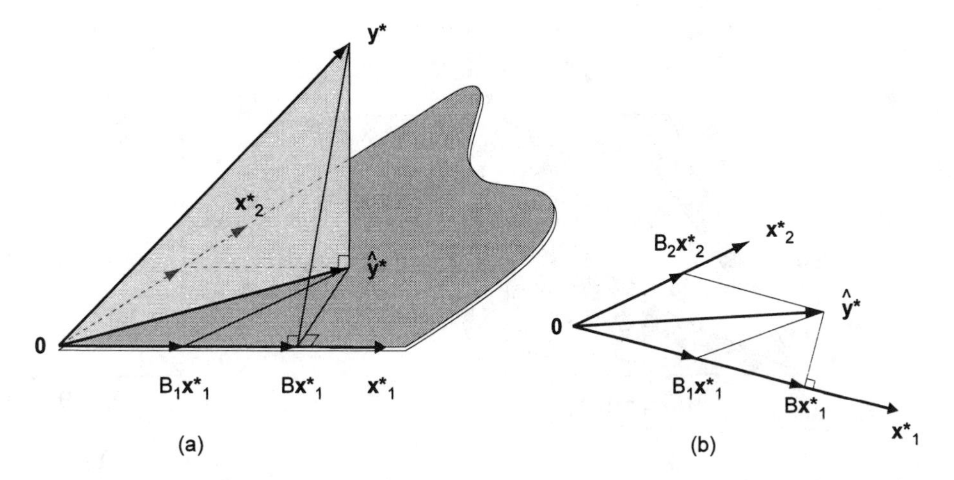

Figure 10.10. When the X's are correlated (here positively), the slope B for the simple regression of Y on X_1 differs from the slope B_1 in the multiple regression of Y on both X_1 and X_2. The least-squares fit is in (a), the regressor plane in (b).

To find the slope coefficient B for the *simple* regression of Y on X_1, we need to project $\mathbf{y}^*$ onto $\mathbf{x}_1^*$ alone, obtaining $B\mathbf{x}_1^*$; this result also appears in Figure 10.10(a). Because $\mathbf{x}_1^* \cdot \mathbf{y}^* = \mathbf{x}_1^* \cdot \hat{\mathbf{y}}^*$,[9] the vector $B_1\mathbf{x}_1^*$ is also the orthogonal projection of $\hat{\mathbf{y}}^*$ onto $\mathbf{x}_1^*$, as shown in Figure 10.10(a) and (b). In this instance, projecting $\hat{\mathbf{y}}^*$ perpendicular to $\mathbf{x}_1^*$ (simple regression) rather than parallel to $\mathbf{x}_2^*$ (multiple regression) causes the simple-regression slope B to exceed the multiple-regression slope B_1.

The situation changes fundamentally if the independent variables X_1 and X_2 are uncorrelated, as illustrated in Figure 10.11(a) and (b). Here, $B = B_1$. Another advantage of orthogonal regressors is revealed in Figure 10.11(b): There is a unique partition of the regression sum of squares into components due to each of the two regressors. We have[10]

$$\begin{aligned} \text{RegSS} &= \hat{\mathbf{y}}^* \cdot \hat{\mathbf{y}}^* \\ &= B_1^2 \mathbf{x}_1^* \cdot \mathbf{x}_1^* + B_2^2 \mathbf{x}_2^* \cdot \mathbf{x}_2^* \end{aligned}$$

In contrast, when the regressors are correlated, as in Figure 10.10(b), no such partition is possible, for then

$$\begin{aligned} \text{RegSS} &= \hat{\mathbf{y}}^* \cdot \hat{\mathbf{y}}^* \\ &= B_1^2 \mathbf{x}_1^* \cdot \mathbf{x}_1^* + B_2^2 \mathbf{x}_2^* \cdot \mathbf{x}_2^* + 2B_1B_2\mathbf{x}_1^* \cdot \mathbf{x}_2^* \end{aligned} \qquad [10.7]$$

[9] See Exercise 10.5.

[10] See Exercise 10.6.

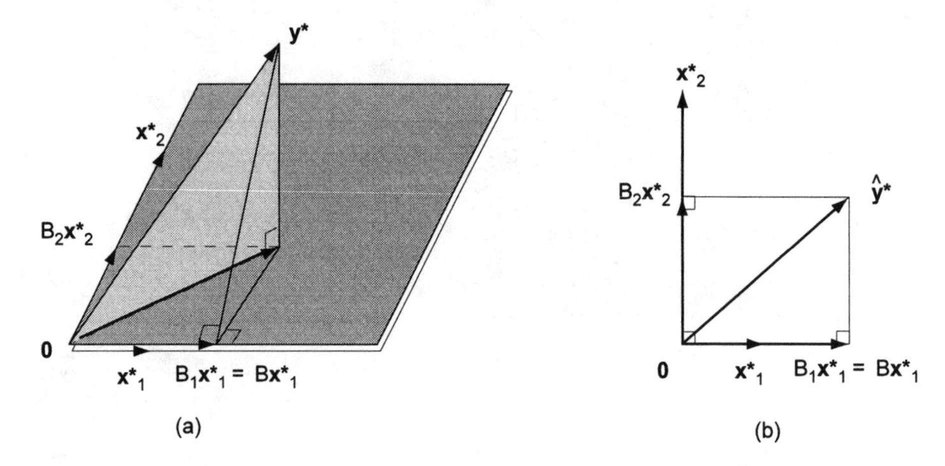

Figure 10.11. When the X's are uncorrelated, the simple-regression slope B and the multiple-regression slope B_1 are the same. The least-squares fit is in (a), the regressor plane in (b).

The last term[11] in Equation 10.7 can be positive or negative, depending on the signs of the regression coefficients and of the correlation between X_1 and X_2.

> When the independent variables in multiple regression are orthogonal (uncorrelated), the regression sum of squares can be partitioned into components due to each independent variable: $||\hat{\mathbf{y}}^*||^2 = B_1^2||\mathbf{x}_1^*||^2 + B_2^2||\mathbf{x}_2^*||^2$. When the independent variables are correlated, however, no such partition is possible.

As in simple regression, degrees of freedom in multiple regression correspond to the dimension of subspaces of the observation space. Because the $\mathbf{y}^*$ vector, as a vector of mean deviations, is confined to a subspace of dimension $n - 1$, there are $n - 1$ degrees of freedom for TSS. The fitted-value vector $\hat{\mathbf{y}}^*$ necessarily lies in the fixed $\{\mathbf{x}_1^*, \mathbf{x}_2^*\}$ plane, which is a subspace of dimension 2; thus, RegSS has 2 degrees of freedom. Finally, the residual vector $\mathbf{e}$ is orthogonal to the independent-variable plane, and, therefore, RSS has $(n - 1) - 2 = n - 3$ degrees of freedom.

More generally, k noncollinear regressors in mean-deviation form generate a subspace of dimension k. The fitted dependent-variable vector $\hat{\mathbf{y}}^*$ is the orthog-

[11] Some researchers seek to interpret $2B_1B_2\mathbf{x}_1^* \cdot \mathbf{x}_2^*$ as the variation in Y due to the "overlap" between the correlated independent variables X_1 and X_2. That this interpretation is nonsense follows from the observation that the overlap can be negative.

onal projection of $\mathbf{y}^*$ onto this subspace, and, therefore, RegSS has k degrees of freedom. Likewise, because $\mathbf{e}$ is orthogonal to the k-dimensional regressor subspace, RSS has $(n-1)-k=n-k-1$ degrees of freedom.

> As in simple regression, degrees of freedom in multiple regression follow from the dimensionality of the subspaces to which the various vectors are confined.

EXERCISES

10.5 Prove that $\mathbf{x}_1^* \cdot \mathbf{y}^* = \mathbf{x}_1^* \cdot \hat{\mathbf{y}}^*$. (*Hint*: $\mathbf{y}^* = \hat{\mathbf{y}}^* + \mathbf{e}$, and $\mathbf{e}$ is orthogonal to $\mathbf{x}_1^*$.)

10.6 Show that when X_1 and X_2 are uncorrelated, the regression sum of squares can be written as

$$\text{RegSS} = \hat{\mathbf{y}}^* \cdot \hat{\mathbf{y}}^* = B_1^2 \mathbf{x}_1^* \cdot \mathbf{x}_1^* + B_2^2 \mathbf{x}_2^* \cdot \mathbf{x}_2^*$$

(*Hint*: Use $\hat{\mathbf{y}}^* = B_1\mathbf{x}_1^* + B_2\mathbf{x}_2^*$.)

10.7 Exercise 10.4 (continued): Using Duncan's occupational prestige data, construct the geometric representation for the regression of prestige Y on income X_1 and education X_2. Draw separate graphs for (a) the $\{\mathbf{x}_1^*, \mathbf{x}_2^*\}$ plane, showing the $\hat{\mathbf{y}}^*$ vector, B_1, and B_2; and (b) the $\{\mathbf{y}^*, \hat{\mathbf{y}}^*\}$ plane, showing $\mathbf{e}$. Draw all vectors to scale. (*Hint*: Calculate the correlation between X_1 and X_2 to find the angle between $\mathbf{x}_1^*$ and $\mathbf{x}_2^*$.)

10.8 Nearly collinear regressors: Construct the geometric vector representation of a regression with two independent variables in mean-deviation form, $\hat{\mathbf{y}}^* = B_1\mathbf{x}_1^* + B_2\mathbf{x}_2^*$, distinguishing between two cases: (a) X_1 and X_2 are highly correlated, so that the angle separating the $\mathbf{x}_1^*$ and $\mathbf{x}_2^*$ vectors is small; and (b) X_1 and X_2 are uncorrelated, so that the $\mathbf{x}_1^*$ and $\mathbf{x}_2^*$ vectors are orthogonal. By examining the regressor plane, show that slight changes in the position of the $\hat{\mathbf{y}}^*$ vector (due, e.g., to sampling fluctuations) can cause dramatic changes in the regression coefficients B_1 and B_2 in case (a) but not in case (b). The problem of collinearity is discussed further in Chapter 13.

10.9 Partial correlation (see Exercise 5.16):

(a) Illustrate how the partial correlation $r_{Y1|2}$ can be represented using geometric vectors. Draw the vectors $\mathbf{y}^*, \mathbf{x}_1^*$, and $\mathbf{x}_2^*$, and define $\mathbf{e}_1 \equiv \{E_{i1|2}\}$ and $\mathbf{e}_Y \equiv \{E_{iY|2}\}$ (where i is the subscript for observations).

(b) *Use the vector diagram in part (a) to show that the incremental F-test for the hypothesis H_0: $\beta_1 = 0$ can be written as

$$F_0 = \frac{(n-k-1)r_{Y1|2}^2}{1-r_{Y1|2}^2}$$

Recalling part (b) of Exercise 5.16, why is this result intuitively plausible?

10.3 Estimating the Error Variance

The connection between degrees of freedom and unbiased variance estimation is subtle, but yields *relatively* simply to the geometric point of view. This section uses the vector geometry of regression to show that $S_E^2 = \sum E_i^2/(n-k-1)$ is an unbiased estimator of the error variance, σ_ε^2.

Even when the errors in a linear model are independent and normally distributed with zero means and constant variance, $\boldsymbol{\varepsilon} \sim N_n(\mathbf{0}, \sigma_\varepsilon^2\mathbf{I}_n)$, the least-squares residuals are correlated and generally have different variances, $\mathbf{e} \sim N_n(\mathbf{0}, \sigma_\varepsilon^2\mathbf{Q})$. The matrix $\mathbf{Q} \equiv \mathbf{I}_n - \mathbf{X}(\mathbf{X}'\mathbf{X})^{-1}\mathbf{X}'$ is nondiagonal, singular, and of rank $n-k-1$.[12]

Following Putter (1967), we can transform the least-squares residuals into an independent and identically distributed set by selecting an orthonormal basis for the error subspace, defining transformed residuals in the following manner:

$$\underset{(n-k-1\times 1)}{\mathbf{z}} \equiv \underset{(n-k-1\times n)}{\mathbf{G}} \underset{(n-1)}{\mathbf{e}}$$

The transformation matrix $\mathbf{G}$ is selected so that it is orthonormal and orthogonal to $\mathbf{X}$:

$$\mathbf{G}\mathbf{G}' = \mathbf{I}_{n-k-1}$$
$$\mathbf{G}\mathbf{X} = \underset{(n-k-1\times k+1)}{\mathbf{0}}$$

The transformed residuals then have the following properties:[13]

$$\mathbf{z} = \mathbf{G}\mathbf{y}$$
$$E(\mathbf{z}) = \mathbf{0}$$
$$V(\mathbf{z}) = \sigma_\varepsilon^2\mathbf{I}_{n-k-1}$$

If the elements of $\boldsymbol{\varepsilon}$ are independent and normally distributed with constant variance, then so are the elements of $\mathbf{z}$. There are, however, n of the former and $n-k-1$ of the latter. Furthermore, the transformation matrix $\mathbf{G}$ (and hence $\mathbf{z}$)

---

[12] See Exercise 10.10.
[13] See Exercise 10.11.

is not unique—there are infinitely many ways of selecting an orthonormal basis for the error subspace.[14]

Transforming $\mathbf{e}$ to $\mathbf{z}$ suggests a simple method for deriving an estimator of the error variance σ_ε^2. The entries of $\mathbf{z}$ have zero expectations and common variance σ_ε^2, so

$$E(\mathbf{z}'\mathbf{z}) = \sum_{i=1}^{n-k-1} E(Z_i^2) = (n-k-1)\sigma_\varepsilon^2$$

Thus, an unbiased estimator of the error variance is given by

$$S_E^2 \equiv \frac{\mathbf{z}'\mathbf{z}}{n-k-1}$$

Moreover, because the Z_i are independent and normally distributed,

$$\frac{\mathbf{z}'\mathbf{z}}{\sigma_\varepsilon^2} = \frac{(n-k-1)S_E^2}{\sigma_\varepsilon^2} \sim \chi_{n-k-1}^2$$

The estimator S_E^2 can be computed without finding transformed residuals, for the length of the least-squares residual vector $\mathbf{e}$ is the same as the length of the vector of transformed residuals $\mathbf{z}$; that is, $\sqrt{\mathbf{e}'\mathbf{e}} = \sqrt{\mathbf{z}'\mathbf{z}}$. This result follows from the observation that $\mathbf{e}$ and $\mathbf{z}$ are the *same* vector represented according to alternative bases: (1) $\mathbf{e}$ gives the coordinates of the residuals relative to the natural basis of the n-dimensional observation space; (2) $\mathbf{z}$ gives the coordinates of the residuals relative to an arbitrary orthonormal basis for the $(n-k-1)$-dimensional error subspace. A vector does not change its length when the basis changes, and, therefore,

$$S_E^2 = \frac{\mathbf{z}'\mathbf{z}}{n-k-1} = \frac{\mathbf{e}'\mathbf{e}}{n-k-1}$$

which is our usual estimator of the error variance.

Heuristically, although $\mathbf{e}$ contains n elements, there are, as I have explained, $k+1$ linear dependencies among them. In calculating an unbiased estimator of the error variance, we need to divide by the residual degrees of freedom rather than by the number of observations.

---

[14] The transformed residuals are useful not only for exploring properties of least-squares estimation, but also in diagnosing certain linear-model problems (see, e.g., Theil, 1971, Chapter 5; Putter, 1967).

An unbiased estimator of the error variance σ_ε^2 can be derived by transforming the n correlated residuals $\mathbf{e}$ to $n - k - 1$ independently and identically distributed residuals $\mathbf{z}$, employing an orthonormal basis $\mathbf{G}$ for the $(n - k - 1)$-dimensional error subspace: $\mathbf{z} = \mathbf{Ge}$. If the errors are independent and normally distributed, with zero means and common variance σ_ε^2, then so are the elements of $\mathbf{z}$. Thus, $\mathbf{z}'\mathbf{z}/(n - k - 1)$ is an unbiased estimator of the error variance; and because $\mathbf{z}$ and $\mathbf{e}$ are the same vector represented according to alternative bases, $\mathbf{z}'\mathbf{z}/(n - k - 1) = \mathbf{e}'\mathbf{e}/(n - k - 1)$, which is our usual estimator of error variance, S_E^2.

EXERCISES

10.10 *Show that the matrix $\mathbf{Q} = \mathbf{I}_n - \mathbf{X}(\mathbf{X}'\mathbf{X})^{-1}\mathbf{X}'$ is nondiagonal, singular, and of rank $n - k - 1$. (*Hints*: Verify that the rows of $\mathbf{Q}$ satisfy the $k + 1$ constraints implied by $\mathbf{QX} = \mathbf{0}$. If $\mathbf{Q}$ is singular and diagonal, then some of its diagonal entries must be 0; show that this is not generally the case.)

10.11 *Prove that when the least-squares residuals are transformed according to the equation $\mathbf{z} = \mathbf{Ge}$, where the $n - k - 1 \times n$ transformation matrix $\mathbf{G}$ is orthonormal and orthogonal to $\mathbf{X}$, the transformed residuals $\mathbf{z}$ have the following properties: $\mathbf{z} = \mathbf{Gy}$, $E(\mathbf{z}) = \mathbf{0}$, and $V(\mathbf{z}) = \sigma_\varepsilon^2\mathbf{I}_{n-k-1}$.

10.4 Analysis-of-Variance Models

Recall the overparameterized one-way ANOVA model[15]

$$Y_{ij} = \mu + \alpha_j + \varepsilon_{ij} \quad \text{for } i = 1, \ldots, n_j; j = 1, \ldots, m$$

[15] See Section 8.1.

The $\mathbf{X}$ matrix for this model (with parameters labeling the columns) is

$$\underset{(n\times m+1)}{\mathbf{X}} = \left[\begin{array}{cccccc}
(\mu) & (\alpha_1) & (\alpha_2) & \cdots & (\alpha_{m-1}) & (\alpha_m) \\
1 & 1 & 0 & \cdots & 0 & 0 \\
\vdots & \vdots & \vdots & & \vdots & \vdots \\
1 & 1 & 0 & \cdots & 0 & 0 \\
\hline
1 & 0 & 1 & \cdots & 0 & 0 \\
\vdots & \vdots & \vdots & & \vdots & \vdots \\
1 & 0 & 1 & \cdots & 0 & 0 \\
\hline
\vdots & \vdots & \vdots & & \vdots & \vdots \\
\hline
1 & 0 & 0 & \cdots & 1 & 0 \\
\vdots & \vdots & \vdots & & \vdots & \vdots \\
1 & 0 & 0 & \cdots & 1 & 0 \\
\hline
1 & 0 & 0 & \cdots & 0 & 1 \\
\vdots & \vdots & \vdots & & \vdots & \vdots \\
1 & 0 & 0 & \cdots & 0 & 1
\end{array}\right]$$

The $m+1$ columns of the model matrix span a subspace of dimension m. We can project the dependent-variable vector $\mathbf{y}$ onto this subspace, locating the fitted-value vector $\hat{\mathbf{y}}$. The columns of $\mathbf{X}$ do not provide a basis for the subspace that they span, however, and, consequently, the individual parameter estimates are not uniquely determined. This situation is illustrated in Figure 10.12 for $m = 2$. Even in the absence of uniquely determined parameters, we have no trouble calculating the regression sum of squares for the model, because we can find $\hat{\mathbf{y}}$ by picking an arbitrary basis for the column space of $\mathbf{X}$. The dummy-coding and deviation-coding schemes of Chapter 8 select alternative bases for the column space of the model matrix: Dummy coding simply deletes the last column to provide a basis for the column space of $\mathbf{X}$; deviation coding constructs a new basis for the column space of $\mathbf{X}$.

> In the overparameterized one-way ANOVA model, $Y_{ij} = \mu + \alpha_j + \varepsilon_{ij}$, the $m + 1$ columns of the model matrix $\mathbf{X}$ are collinear and span a subspace of dimension m. We can, however, still find $\hat{\mathbf{y}}$ for the model by projecting $\mathbf{y}$ orthogonally onto this subspace, most simply by selecting an arbitrary basis for the column space of the model matrix. Conceived in this light, dummy coding and deviation coding are two techniques for constructing a basis for the column space of $\mathbf{X}$.

Let us turn next to the overparameterized two-way ANOVA model:

$$Y_{ijk} = \mu + \alpha_j + \beta_k + \gamma_{jk} + \varepsilon_{ijk} \quad \text{for } i = 1, \ldots, n_{jk}; j = 1, \ldots, r; k = 1, \ldots, c$$

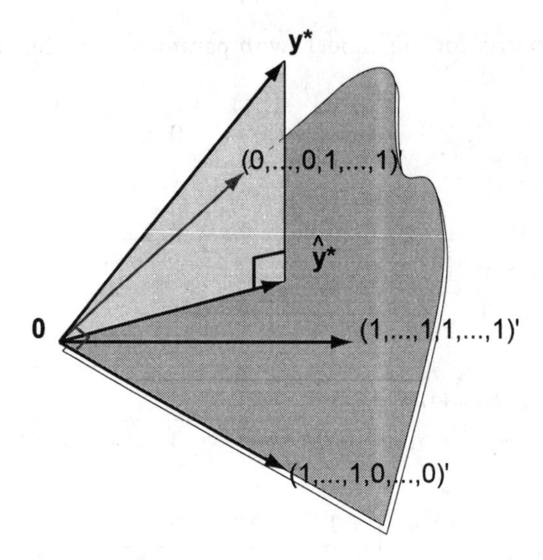

Figure 10.12. The vector geometry of least-squares fit for the overparameterized one-way ANOVA model when there are two groups. The $m+1=3$ columns of the model matrix are collinear and span a subspace of dimension $m=2$.

We shall consider the simplest case, where $j=k=2$. It suffices to examine the parametric equation for the model, relating the four cell means μ_{jk} to the nine parameters of the model:

$$
\begin{bmatrix} \mu_{11} \\ \mu_{12} \\ \mu_{21} \\ \mu_{22} \end{bmatrix} =
\left[\begin{array}{c|cc|cc|cccc}
1 & 1 & 0 & 1 & 0 & 1 & 0 & 0 & 0 \\
1 & 1 & 0 & 0 & 1 & 0 & 1 & 0 & 0 \\
1 & 0 & 1 & 1 & 0 & 0 & 0 & 1 & 0 \\
1 & 0 & 1 & 0 & 1 & 0 & 0 & 0 & 1
\end{array}\right]
\left[\begin{array}{c} \mu \\ \hline \alpha_1 \\ \alpha_2 \\ \hline \beta_1 \\ \beta_2 \\ \hline \gamma_{11} \\ \gamma_{12} \\ \gamma_{21} \\ \gamma_{22} \end{array}\right]
$$

$$\boldsymbol{\mu} = \mathbf{X}_B \boldsymbol{\beta}$$

Notice that the four columns in $\mathbf{X}_B$ representing the interactions are linearly independent, and hence the corresponding columns of $\mathbf{X}$ span the full column space of the model matrix. The subspaces spanned by the main effects, each consisting of two linearly independent columns, lie in the space spanned by the interaction regressors—the main-effect subspaces are literally marginal to (i.e., contained in) the interaction subspace. Finally, the constant regressor is marginal both to the interaction subspace and to each of the main-effect subspaces: The constant regressor is simply the sum of the interaction regressors or of either set of main-effect regressors. Understood in this light, the deviation-coding method of Chapter 8 selects a convenient full-rank basis for the model matrix.

In the overparameterized two-way ANOVA model, $Y_{ijk} = \mu + \alpha_j + \beta_k + \gamma_{jk} + \varepsilon_{ijk}$, the subspace spanned by the interaction regressors provides a basis for the full column space of the model matrix. The model matrix columns for the two sets of main effects are therefore marginal to (i.e., subspaces of) the interaction space. The column for the constant regressor is marginal to the main-effect subspaces as well as to the interactions.

EXERCISES

10.12 Exercise 9.17 (continued): Let SS(·) give sums of squares for the two-way ANOVA model

$$Y_{ijk} = \mu + \alpha_j + \beta_k + \gamma_{jk} + \varepsilon_{ijk}$$

using deviation-coded regressors (i.e., employing sigma constraints to reduce the model matrix to full rank); and let SS*(·) give sums of squares for the same model using dummy-coded regressors. Working with the model for $r = 2$ rows and $c = 3$ columns, use the geometric vector representation of the model to explain why

$$\begin{aligned} SS^*(\alpha|\beta) &= SS(\alpha|\beta) \\ SS^*(\beta|\alpha) &= SS(\beta|\alpha) \\ SS^*(\gamma|\alpha, \beta) &= SS(\gamma|\alpha, \beta) \end{aligned}$$

but that, in general,

$$\begin{aligned} SS^*(\alpha|\beta, \gamma) &\neq SS(\alpha|\beta, \gamma) \\ SS^*(\beta|\alpha, \gamma) &\neq SS(\beta|\alpha, \gamma) \end{aligned}$$

[*Hints*: Show that (i) the subspaces spanned by the deviation and dummy regressors for each of the two sets of main effects are the same; (ii) the subspaces spanned by the deviation and dummy regressors for the full set of effects (main effects and interactions) are the same; (iii) the subspaces spanned by the deviation and dummy interaction regressors are *different*.]

10.5 Summary

- Variables, such as X and Y in simple regression, can be treated as vectors—$\mathbf{x}$ and $\mathbf{y}$—in the n-dimensional space whose axes are given by the observations. Written in vector form, the simple-regression model is $\mathbf{y} = \alpha\mathbf{1}_n + \beta\mathbf{x} + \boldsymbol{\varepsilon}$. The least-squares regression, $\mathbf{y} = A\mathbf{1}_n + B\mathbf{x} + \mathbf{e}$, is found by projecting $\mathbf{y}$ orthogonally onto the plane spanned by $\mathbf{1}_n$ and $\mathbf{x}$, thus minimizing the sum of squared residuals $||\mathbf{e}||^2$.
- Writing X and Y in mean-deviation form, as the vectors $\mathbf{x}^*$ and $\mathbf{y}^*$, eliminates the constant term and thus permits representation of the fitted regression in two (rather than three) dimensions: $\mathbf{y}^* = B\mathbf{x}^* + \mathbf{e}$. The analysis of variance for the regression, $\text{TSS} = \text{RegSS} + \text{RSS}$, is represented geometrically as $||\mathbf{y}^*||^2 = ||\hat{\mathbf{y}}^*||^2 + ||\mathbf{e}||^2$. The correlation between X and Y is the cosine of the angle separating the vectors $\mathbf{x}^*$ and $\mathbf{y}^*$.
- Degrees of freedom in simple regression correspond to the dimensions of subspaces to which variable vectors associated with sums of squares are confined:

  - The $\mathbf{y}^*$ vector lies in the $(n-1)$-dimensional subspace of mean deviations, but is otherwise unconstrained; TSS, therefore, has $n-1$ degrees of freedom.
  - The $\hat{\mathbf{y}}^*$ vector lies somewhere along the one-dimensional subspace spanned by $\mathbf{x}^*$; RegSS, therefore, has 1 degree of freedom.
  - The $\mathbf{e}$ vector lies in the $(n-1)$-dimensional subspace of mean deviations, and is constrained to be orthogonal to $\mathbf{x}^*$; RSS, therefore, has $(n-1) - 1 = n-2$ degrees of freedom.

- The fitted multiple-regression model for two independent variables is written in vector form as $\mathbf{y} = A\mathbf{1}_n + B_1\mathbf{x}_1 + B_2\mathbf{x}_2 + \mathbf{e}$. Writing Y and the X's in mean-deviation form eliminates the constant, $\mathbf{y}^* = B_1\mathbf{x}_1^* + B_2\mathbf{x}_2^* + \mathbf{e}$, and permits a representation in three (rather than four) dimensions. The fitted values, $\hat{\mathbf{y}}^* = B_1\mathbf{x}_1^* + B_2\mathbf{x}_2^*$, are found by projecting $\mathbf{y}^*$ orthogonally onto the plane spanned by $\mathbf{x}_1^*$ and $\mathbf{x}_2^*$. The analysis of variance for the regression, which is essentially the same as in simple regression, appears in the plane spanned by $\mathbf{y}^*$ and $\hat{\mathbf{y}}^*$. The multiple correlation R is the cosine of the angle separating $\mathbf{y}^*$ and $\hat{\mathbf{y}}^*$, and, consequently, is the simple correlation between the observed and fitted Y-values.
- When the independent variables in multiple regression are orthogonal (uncorrelated), the regression sum of squares can be partitioned into components due to each independent variable: $||\hat{\mathbf{y}}^*||^2 = B_1^2||\mathbf{x}_1^*||^2 + B_2^2||\mathbf{x}_2^*||^2$. When the independent variables are correlated, however, no such partition is possible.
- As in simple regression, degrees of freedom in multiple regression follow from the dimensionality of the subspaces to which the various vectors are confined:

  - The $\mathbf{y}^*$ vector lies in the $(n-1)$-dimensional subspace of mean deviations; TSS, therefore, has $n-1$ degrees of freedom.
  - The $\hat{\mathbf{y}}^*$ vector lies somewhere in the plane spanned by $\mathbf{x}_1^*$ and $\mathbf{x}_2^*$; RegSS, therefore, has 2 degrees of freedom. More generally, k independent variables $\mathbf{x}_1^*, \mathbf{x}_2^*, \ldots, \mathbf{x}_k^*$ span a subspace of dimension k, and $\hat{\mathbf{y}}^*$ is the orthogonal projection of $\mathbf{y}^*$ onto this subspace; thus, RegSS has k degrees of freedom.
  - The $\mathbf{e}$ vector is constrained to be orthogonal to the two-dimensional subspace spanned by $\mathbf{x}_1^*$ and $\mathbf{x}_2^*$; RSS, therefore, has $(n-1) - 1 = n-2$

degrees of freedom. More generally, e is orthogonal to the k-dimensional subspace spanned by $\mathbf{x}_1^*, \mathbf{x}_2^*, \ldots, \mathbf{x}_k^*$, and so RSS has $(n-1)-k = n-k-1$ degrees of freedom.

- An unbiased estimator of the error variance σ_ε^2 can be derived by transforming the n correlated residuals e to $n-k-1$ independently and identically distributed residuals z, employing an orthonormal basis G for the $(n-k-1)$-dimensional error subspace: $\mathbf{z} = \mathbf{G}\mathbf{e}$. If the errors are independent and normally distributed, with zero means and common variance σ_ε^2, then so are the elements of z. Thus, $\mathbf{z}'\mathbf{z}/(n-k-1)$ is an unbiased estimator of the error variance; and because z and e are the same vector represented according to alternative bases, $\mathbf{z}'\mathbf{z}/(n-k-1) = \mathbf{e}'\mathbf{e}/(n-k-1)$, which is our usual estimator of error variance, S_E^2.
- In the overparameterized one-way ANOVA model, $Y_{ij} = \mu + \alpha_j + \varepsilon_{ij}$, the $m+1$ columns of the model matrix **X** are collinear and span a subspace of dimension m. We can, however, still find $\hat{\mathbf{y}}$ for the model by projecting y orthogonally onto this subspace, most simply by selecting an arbitrary basis for the column space of the model matrix. Conceived in this light, dummy coding and deviation coding are two techniques for constructing a basis for the column space of **X**.
- In the overparameterized two-way ANOVA model, $Y_{ijk} = \mu + \alpha_j + \beta_k + \gamma_{jk} + \varepsilon_{ijk}$, the subspace spanned by the interaction regressors provides a basis for the full column space of the model matrix. The model matrix columns for the two sets of main effects are therefore marginal to (i.e., subspaces of) the interaction space. The column for the constant regressor is marginal to the main-effect subspaces as well as to the interactions.

10.6 Recommended Reading

- There are several advanced texts that treat linear models from a strongly geometric perspective, including Dempster (1969) and Stone (1987). Both of these books describe multivariate (i.e., multiple dependent-variable) generalizations of linear models, and both demand substantial mathematical sophistication.
- In a text on matrix algebra, vector geometry, and associated mathematical topics, Green and Carroll (1978) focus on the geometric properties of linear models and related multivariate methods. The pace of the presentation is relatively leisurely, and the strongly geometric orientation provides insight into both the mathematics and the statistics.
- Wonnacott and Wonnacott (1979) invoke vector geometry to explain a variety of statistical topics, including some not covered in the present text—such as instrumental-variables estimation and structural-equation models.

PART III

Linear-Model Diagnostics

11

Unusual and Influential Data

As we have seen, linear statistical models—particularly linear regression analysis—make strong assumptions about the structure of data, assumptions that often do not hold in applications. The method of least squares, which is typically used to fit linear models to data, is very sensitive to the structure of the data, and can be markedly influenced by one or a few unusual observations.

We could abandon linear models and least-squares estimation in favor of nonparametric regression and robust estimation.[1] A less drastic response is also possible, however: We can adapt and extend the methods for examining and transforming data described in Chapters 3 and 4 to diagnose problems with a linear model that has been fit to data, and—often—to suggest solutions.

I shall pursue this strategy in this and the next two chapters. The current chapter deals with unusual and influential data. Chapter 12 takes up a variety of problems, including nonlinearity, nonconstant error variance, and nonnormality. Collinearity is the subject of Chapter 13.

Taken together, the diagnostic and corrective methods described in these chapters substantially extend the practical application of linear models. These methods are often the difference between a crude, mechanical data analysis, and a careful, nuanced analysis that accurately describes the data and therefore supports meaningful interpretation of them.

---

[1] Methods for nonparametric and robust regression were introduced informally in Chapter 2 and will be described in more detail in Chapter 14.

11.1 Outliers, Leverage, and Influence

Unusual data are problematic in linear models fit by least squares because they can unduly influence the results of the analysis, and because their presence may be a signal that the model fails to capture important characteristics of the data. Some central distinctions are illustrated in Figure 11.1 for the simple regression model $Y = \alpha + \beta X + \varepsilon$.

In simple regression, an *outlier* is an observation whose dependent-variable value is conditionally unusual given the value of the independent variable. In contrast, a univariate outlier is a value of Y or X that is unconditionally unusual; such a value may or may not be a regression outlier.

Regression outliers appear in Figure 11.1(a) and (b). In Figure 11.1(a), the outlying observation has an X-value that is at the center of the X-distribution; as a consequence, deleting the outlier has little impact on the least-squares fit,

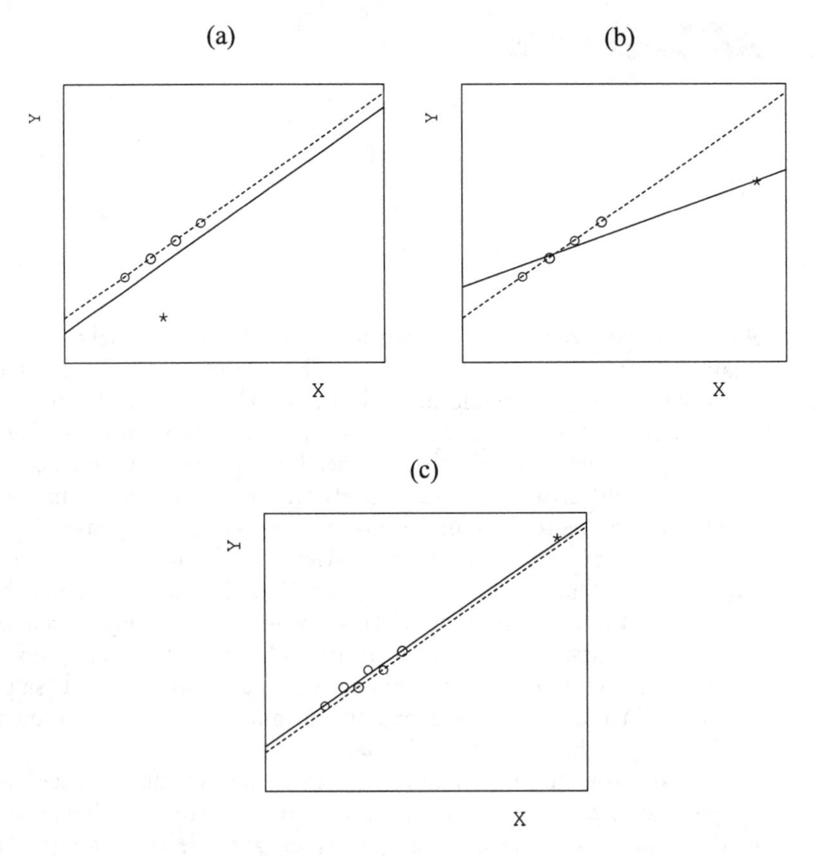

Figure 11.1. Leverage and influence in simple regression. In each graph, the solid line gives the least-squares regression for all of the data, while the broken line gives the least-squares regression with the unusual data point (the asterisk) omitted. (a) An outlier near the mean of X has low leverage and little influence on the regression coefficients. (b) An outlier far from the mean of X has high leverage and substantial influence on the regression coefficients. (c) A high-leverage observation in line with the rest of the data does not influence the regression coefficients. In panel (c), the two regression lines are separated slightly for visual effect, but are, in fact, coincident.

leaving the slope B unchanged, and affecting the intercept A only slightly. In Figure 11.1(b), however, the outlier has an unusual X-value, and thus its deletion markedly affects both the slope and the intercept. Because of its unusual X-value, the outlying last observation in Figure 11.1(b) exerts strong *leverage* on the regression coefficients, while the outlying middle observation in Figure 11.1(a) is at a low-leverage point. The combination of high leverage with a regression outlier therefore produces substantial *influence* on the regression coefficients. In Figure 11.1(c), the last observation has no influence on the regression coefficients even though it is a high-leverage point, because this observation is in line with the rest of the data—it is not a regression outlier.

The following heuristic formula helps to distinguish among the three concepts of influence, leverage, and discrepancy ("outlyingness"):

$$\text{Influence on coefficients} = \text{Leverage} \times \text{Discrepancy}$$

A simple and transparent example, with real data from Davis (1990), appears in Figure 11.2. These data record the measured and reported weight of 183 male and female subjects who engage in programs of regular physical exercise.[2] Davis's data can be treated in two ways:

- We could regress reported weight (RW) on measured weight (MW), a dummy variable for sex (F, coded 1 for women and 0 for men), and an interaction regressor (formed as the product $MW \times F$). This specification follows from the reasonable assumption that measured weight, and possibly sex, can affect reported weight. The results are as follows (with coefficient standard errors in parentheses):

$$\begin{array}{l} \widehat{RW} = \underset{(3.28)}{1.36} + \underset{(0.043)}{0.990MW} + \underset{(3.9)}{40.0F} - \underset{(0.056)}{0.725(MW \times F)} \\ R^2 = 0.89 \qquad S_E = 4.66 \end{array}$$

Were these results taken seriously, we would conclude that men are unbiased reporters of their weights (because $A = 1.36 \simeq 0$ and $B_1 = 0.990 \simeq 1$), while women tend to overreport their weights if they are relatively light and underreport if they are relatively heavy (the intercept for women is $1.36 + 40.0 = 41.4$ and the slope is $0.990 - 0.725 = 0.265$). Figure 11.2, however, makes it clear that the differential results for women and men are due to one female subject whose reported weight is about average (for women), but whose measured weight is extremely large. Recall that this subject's measured weight in kilograms and height in centimeters were erroneously switched. Correcting the data produces the regression

$$\begin{array}{l} \widehat{RW} = \underset{(1.58)}{1.36} + \underset{(0.021)}{0.990MW} + \underset{(2.45)}{1.98F} - \underset{(0.0385)}{0.0567(MW \times F)} \\ R^2 = 0.97 \qquad S_E = 2.24 \end{array}$$

which suggests that both women and men are unbiased reporters of their weight.

---

[2] Davis's data were introduced in Chapter 2.

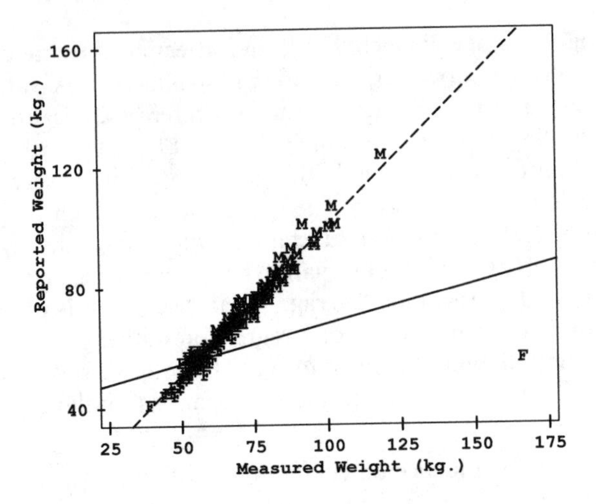

Figure 11.2. Davis's data on reported and measured weight for women (F) and men (M), showing the least-squares linear regression line for each group (the broken line for men, the solid line for women). The outlying observation has a substantial effect on the fitted line for women.

- We could (as in our previous analysis of Davis's data) regress measured weight on reported weight, sex, and their interaction, reflecting a desire to use reported weight as a predictor of measured weight. For the *uncorrected* data:

$$\widehat{MW} = \underset{(5.92)}{1.79} + \underset{(0.076)}{0.969RW} + \underset{(9.30)}{2.07F} - \underset{(0.147)}{0.00953(RW \times F)}$$

$$R^2 = 0.70 \qquad S_E = 8.45$$

The outlier does not have much impact on the coefficients for this regression (both the dummy-variable coefficient and the interaction coefficient are small) precisely because the value of RW for the outlying observation is near $\overline{RW}$ for women. There is, however, a marked effect on the multiple correlation and standard error: For the corrected data, $R^2 = 0.97$ and $S_E = 2.25$.

> Unusual data are problematic in linear models fit by least squares because they can substantially influence the results of the analysis, and because they may indicate that the model fails to capture important features of the data.

11.2 Assessing Leverage: Hat Values

The so-called *hat value* h_i is a common measure of leverage in regression.[3] These values are so named because it is possible to express the fitted values $\widehat{Y}_j$ ("Y-hat") in terms of the observed values Y_i:

$$\widehat{Y}_j = h_{1j}Y_1 + h_{2j}Y_2 + \cdots + h_{jj}Y_j + \cdots + h_{nj}Y_n = \sum_{i=1}^{n} h_{ij}Y_i$$

Thus, the weight h_{ij} captures the contribution of observation Y_i to the fitted value $\widehat{Y}_j$: If h_{ij} is large, then the ith observation can have a substantial impact on the jth fitted value. It can be shown that $h_{ii} = \sum_{j=1}^{n} h_{ij}^2$, and so the hat value $h_i \equiv h_{ii}$ summarizes the potential influence (the leverage) of Y_i on *all* of the fitted values. The hat values are bounded between $1/n$ and 1 (i.e., $1/n \le h_i \le 1$), and the average hat value is $\overline{h} = (k+1)/n$ (where k is the number of regressors in the model, excluding the constant).

In simple-regression analysis,[4] the hat values measure distance from the mean of X:

$$h_i = \frac{1}{n} + \frac{(X_i - \overline{X})^2}{\sum_{j=1}^{n}(X_j - \overline{X})^2}$$

In multiple regression, h_i measures distance from the centroid (point of means) of the X's, taking into account the correlational and variational structure of the X's, as illustrated for $k = 2$ in Figure 11.3. Multivariate outliers in the X-space are thus high-leverage observations. The dependent-variable values are not at all involved in determining leverage.

For Davis's regression of reported weight on measured weight, the largest hat value by far belongs to the 12th subject, whose measured weight was wrongly recorded as 166 kg: $h_{12} = 0.714$. This quantity is many times the average hat value, $\overline{h} = (3+1)/183 = 0.0219$.

> Observations with unusual combinations of independent-variable values have high *leverage* in a least-squares regression. The hat values h_i provide a measure of leverage. The average hat value is $\overline{h} = (k+1)/n$.

[3] For derivations of this and other properties of leverage, outlier, and influence diagnostics, see Section 11.8.

[4] See Exercise 11.3. Note that the sum in the denominator is over the subscript j because the subscript i is already in use.

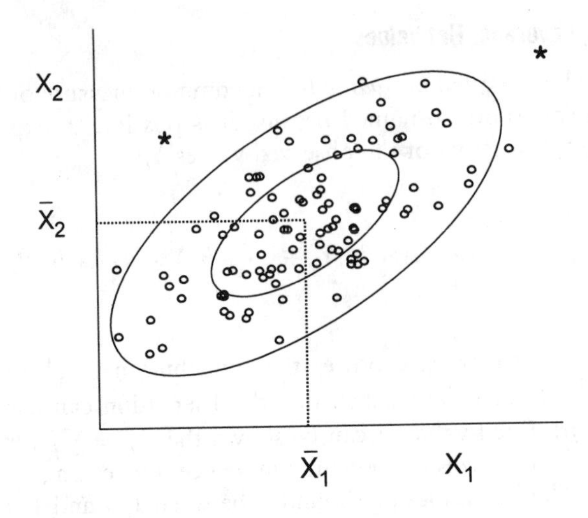

Figure 11.3. Elliptical contours of constant leverage (constant hat values h_i) for $k = 2$ independent variables. Two high-leverage points appear, both represented by asterisks. One point has unusually large values for each of X_1 and X_2, but the other is unusual only in combining a moderately large value of X_2 with a moderately small value of X_1. (These contours of constant leverage are proportional to the standard data ellipse, introduced in Chapter 9.)

11.3 Detecting Outliers: Studentized Residuals

To identify an outlying observation, we need an index of the unusualness of Y given the X's. Discrepant observations usually have large residuals, but it turns out that even if the errors ε_i have equal variances (as assumed in the general linear model), the residuals E_i do not: $V(E_i) = \sigma_\varepsilon^2(1 - h_i)$. High-leverage observations, therefore, tend to have small residuals—an intuitively sensible result, because these observations can coerce the regression surface to be close to them.

Although we can form a *standardized residual* by calculating

$$E'_i = \frac{E_i}{S_E\sqrt{1 - h_i}}$$

this measure is slightly inconvenient because its numerator and denominator are not independent, preventing E'_i from following a t-distribution: When $|E_i|$ is large, $S_E = \sqrt{\sum E_i^2/(n - k - 1)}$, which contains E_i^2, tends to be large as well. Suppose, however, that we refit the model deleting the ith observation, obtaining an estimate $S_{E(-i)}$ of σ_ε that is based on the remaining $n - 1$ observations. Then the *studentized residual*

$$E^*_i = \frac{E_i}{S_{E(-i)}\sqrt{1 - h_i}} \qquad [11.1]$$

has independent numerator and denominator, and follows a t-distribution with $n - k - 2$ degrees of freedom.

An alternative, but equivalent, procedure for finding the studentized residuals employs a "mean-shift" outlier model:

$$Y_j = \alpha + \beta_1 X_{j1} + \cdots + \beta_k X_{jk} + \gamma D_j + \varepsilon_j \qquad [11.2]$$

where D is a dummy regressor set to 1 for observation i and 0 for all other observations:

$$D_j = \begin{cases} 1 & \text{for } j = i \\ 0 & \text{otherwise} \end{cases}$$

Thus,

$$E(Y_i) = \alpha + \beta_1 X_{i1} + \cdots + \beta_k X_{ik} + \gamma$$
$$E(Y_j) = \alpha + \beta_1 X_{j1} + \cdots + \beta_k X_{jk} \quad \text{for } j \neq i$$

It would be natural to specify the model in Equation 11.2 if, before examining the data, we suspected that observation i differed from the others. Then, to test H_0: $\gamma = 0$ (i.e., the null hypothesis that the ith observation is *not* an outlier), we can calculate $t_0 = \hat{\gamma}/\widehat{\mathrm{SE}}(\hat{\gamma})$. This test statistic is distributed as t_{n-k-2} under H_0, and (it turns out) is the studentized residual E_i^* of Equation 11.1.

Hoaglin and Welsch (1978) arrive at the studentized residuals by successively omitting each observation, calculating its residual based on the regression coefficients obtained for the remaining sample, and dividing the resulting residual by its standard error. Finally, Beckman and Trussell (1974) demonstrate the following simple relationship between studentized and standardized residuals:

$$E_i^* = E_i' \sqrt{\frac{n-k-2}{n-k-1-E_i'^2}} \qquad [11.3]$$

If n is large, then the factor under the square root in Equation 11.3 is close to 1, and the distinction between standardized and studentized residuals essentially disappears.[5] Moreover, for large n, the hat values are generally small, and thus it is usually the case that

$$E_i^* \simeq E_i' \simeq \frac{E_i}{S_E}$$

[5] Here, as elsewhere in statistics, terminology is not wholly standard: E_i^* is sometimes called a *deleted studentized residual*, an *externally studentized residual*, or even a standardized residual; likewise, E_i' is sometimes called an *internally studentized residual*, or simply a studentized residual. It is therefore important, especially in small samples, to determine exactly what is being calculated by a computer program before using these quantities.

11.3.1 Testing for Outliers in Linear Models

Because in most applications we do not suspect a particular observation in advance, but rather want to look for *any* outliers that may occur in the data, we can, in effect, refit the mean-shift model n times,[6] once for each observation, producing studentized residuals $E_1^*, E_2^*, \ldots, E_n^*$. Usually, our interest then focuses on the largest absolute E_i^*, denoted $E_{\max}^*$. Because we have picked the biggest of n test statistics, however, it is not legitimate simply to use t_{n-k-2} to find a p-value for $E_{\max}^*$: For example, even if our model is wholly adequate, and disregarding for the moment the dependence among the E_i^*'s, we would expect to obtain about 5% of E_i^*'s beyond $t_{.025} \simeq \pm 2$, about 1% beyond $t_{.005} \simeq \pm 2.6$, and so forth.

One solution[7] to the problem of simultaneous inference is to perform a Bonferroni adjustment to the p-value for the largest absolute E_i^*. The Bonferroni test requires either a special t-table or, even more conveniently, a computer program that returns accurate p-values for values of t far into the tail of the t-distribution. In the latter event, suppose that $p' = \Pr(\, t_{n-k-2} > E_{\max}^*)$. Then the Bonferroni p-value for testing the statistical significance of $E_{\max}^*$ is $p = 2np'$. The factor 2 reflects the two-tail character of the test: We want to detect large negative as well as large positive outliers.

Beckman and Cook (1983) have shown that the Bonferroni adjustment is usually exact in testing the largest studentized residual. Note that a much larger $E_{\max}^*$ is required for a statistically significant result than would be the case for an ordinary individual t-test.

In Davis's regression of reported weight on measured weight, the largest studentized residual by far belongs to the incorrectly coded 12th observation, with $E_{12}^* = -24.3$. Here, $n - k - 2 = 183 - 3 - 2 = 178$, and $\Pr(t_{178} > 24.3) \simeq 10^{-58}$. The Bonferroni p-value for the outlier test is thus $p \simeq 2 \times 183 \times 10^{-58} \simeq 4 \times 10^{-56}$, an unambiguous result.

Put alternatively, the 5% critical value for $E_{\max}^*$ in this regression is the value of t_{178} with probability $.025/183 = 0.0001366$ to the right. That is, $E_{\max}^* = t_{178,\,.0001366} = 3.714$; this critical value contrasts with $t_{178.,\,.025} = 1.973$, which would be appropriate for testing an individual studentized residual identified in advance of inspecting the data.

11.3.2 Anscombe's Insurance Analogy

Thus far, I have treated the identification (and, implicitly, the potential correction, removal, or accommodation) of outliers as a hypothesis-testing problem. Although this is by far the most common procedure in practice, a more reasonable (if subtle) general approach is to assess the potential costs and benefits for estimation of rejecting an unusual observation.

---

[6] It is not necessary literally to perform n auxiliary regressions. Equation 11.3, for example, permits the computation of studentized residuals with little effort.

[7] A graphical alternative is to construct a quantile-comparison plot for the studentized residuals, comparing the sample distribution of these quantities with the t-distribution for $n-k-2$ degrees of freedom. See the discussion of nonnormality in Section 12.1.

Imagine, for the moment, that the observation with the largest E_i^* is simply an unusual data point, but one generated by the assumed statistical model:

$$Y_i = \alpha + \beta_1 X_{i1} + \cdots + \beta_k X_{ik} + \varepsilon_i$$

with independent errors ε_i that are each distributed as $N(0, \sigma_\varepsilon^2)$. To discard an observation under these circumstances would decrease the efficiency of estimation, because when the model—including the assumption of normality—is correct, the least-squares estimators are maximally efficient among all unbiased estimators of the regression coefficients.

If, however, the observation in question does not belong with the rest (e.g., because the mean-shift model applies), then to eliminate it may make estimation more efficient. Anscombe (1960) developed this insight by drawing an analogy to insurance: To obtain protection against "bad" data, one purchases a policy of outlier rejection, a policy paid for by a small premium in efficiency when the policy inadvertently rejects "good" data.[8]

Let q denote the desired premium, say 0.05—that is, a 5% increase in estimator mean-squared error if the model holds for all of the data. Let z represent the unit-normal deviate corresponding to a tail probability of $q(n - k - 1)/n$. Following the procedure derived by Anscombe and Tukey (1963), compute $m = 1.4 + 0.85z$, and then find

$$E'_q = m\left(1 - \frac{m^2 - 2}{4(n - k - 1)}\right)\sqrt{\frac{n - k - 1}{n}} \qquad [11.4]$$

The largest absolute *standardized* residual can be compared with E'_q to determine whether the corresponding observation should be rejected as an outlier. This cutoff can be translated to the studentized-residual scale using Equation 11.3:

$$E_q^* = E'_q\sqrt{\frac{n - k - 2}{n - k - 1 - E'^2_q}} \qquad [11.5]$$

In a real application, of course, we should inquire about discrepant observations rather than simply throwing them away.[9]

For example, for Davis's regression of reported on measured weight, $n = 183$ and $k = 3$; so, for the premium $q = 0.05$, we have

$$\frac{q(n - k - 1)}{n} = \frac{0.05(183 - 3 - 1)}{183} = 0.0489$$

From the unit-normal table, $z = 1.66$, from which $m = 1.4 + 0.85 \times 1.66 = 2.81$. Then, using Equation 11.4, $E'_q = 2.76$, and using Equation 11.5, $E_q^* = 2.81$. Because $E_{\max}^* = |E_{12}^*| = 24.3$ is much larger than E_q^*, the 12th observation is identified as an outlier.

---

[8] An alternative is to employ a robust estimator, which is a bit less efficient than least squares when the model is correct, but much more efficient when outliers are present. See Section 14.3.

[9] See the discussion in Section 11.7.

A regression *outlier* is an observation with an unusual dependent-variable value given its combination of independent-variable values. The studentized residuals E_i^* can be used to identify outliers, through graphical examination, a Bonferroni test for the largest absolute E_i^*, or Anscombe's insurance analogy. If the model is correct (and there are no true outliers), then each studentized residual follows a t-distribution with $n - k - 2$ degrees of freedom.

11.4 Measuring Influence

As noted previously, influence on the regression coefficients combines leverage and discrepancy. The most direct measure of influence simply expresses the impact on each coefficient of deleting each observation in turn:

$$D_{ij} = B_j - B_{j(-i)} \quad \text{for } i = 1, \ldots, n \text{ and } j = 0, 1, \ldots, k$$

where the B_j are the least-squares coefficients calculated for all of the data, and the $B_{j(-i)}$ are the least-squares coefficients calculated with the ith observation omitted. (So as not to complicate the notation here, I denote the least-squares intercept A as B_0.) To assist in interpretation, it is useful to scale the D_{ij} by (deleted) estimates of the coefficient standard errors:

$$D_{ij}^* = \frac{D_{ij}}{\widehat{\text{SE}}_{(-i)}(B_j)}$$

Following Belsley, et al. (1980), the D_{ij} are often termed DFBETA_{ij}, and the D_{ij}^* are called DFBETAS_{ij}.

One problem associated with using the D_{ij} or the D_{ij}^* is their large number—$n(k + 1)$ of each. Of course, these values can be more quickly and effectively examined graphically than in numerical tables. We can, for example, construct an *index plot* of the D_{ij}^*'s for each coefficient, $j = 0, 1, \ldots, k$—a simple scatterplot with D_{ij}^* on the vertical axis versus the observation index i on the horizontal axis. A more informative, if more complex, alternative is to construct a scatterplot matrix of the D_{ij}^* with index plots (or some other univariate display) on the diagonal.[10] Nevertheless, it is useful to have a single summary index of the influence of each observation on the least-squares fit.

Cook (1977) has proposed measuring the "distance" between the B_j and the corresponding $B_{j(-i)}$ by calculating the F-statistic for the "hypothesis" that $\beta_j = B_{j(-i)}$, for $j = 0, 1, \ldots, k$. This statistic is recalculated for each observation

[10] This interesting display was suggested to me by Michael Friendly of the Psychology Department, York University.

$i = 1, \ldots, n$. The resulting values should not literally be interpreted as F-tests—Cook's approach merely exploits an analogy to testing to produce a measure of distance that is independent of the scales of the X variables. Cook's statistic can be written (and simply calculated) as

$$D_i = \frac{E_i'^2}{k+1} \times \frac{h_i}{1-h_i}$$

In effect, the first term in the formula for Cook's D is a measure of discrepancy, and the second is a measure of leverage. We look for values of D_i that are substantially larger than the rest.

> Observations that combine high leverage with a large studentized residual exert substantial *influence* on the regression coefficients. Cook's D-statistic provides a summary index of influence on the coefficients.

Belsley et al. (1980) have suggested the very similar measure[11]

$$\text{DFFITS}_i = E_i^* \sqrt{\frac{h_i}{1-h_i}}$$

Except for unusual data configurations, $D_i \simeq \text{DFFITS}_i^2/(k+1)$.

Because all of the deletion statistics depend on the hat values and residuals, a graphical alternative to either of these general influence measures is to plot the E_i^* against the h_i and to look for observations for which both are big. A slightly more sophisticated (and more informative) version of this plot displays circles of area proportional to Cook's D instead of points (see Figure 11.6 on page 285). We can follow up by examining the D_{ij} or D_{ij}^* for the observations with the largest few D_i, $|\text{DFFITS}_i|$, or combination of large h_i and $|E_i^*|$.

For Davis's regression of reported weight on measured weight, all of the indices of influence point to the obviously discrepant 12th observation:

$$\text{Cook's } D_{12} = 85.9 \text{ (next largest, } D_{21} = 0.065)$$

$$\text{DFFITS}_{12} = -38.4 \text{ (next largest, DFFITS}_{50} = 0.512)$$

$$\text{DFBETAS}_{0,12} = \text{DFBETAS}_{1,12} = 0$$

$$\text{DFBETAS}_{2,12} = 20.0, \text{DFBETAS}_{3,12} = -24.8$$

Notice that the outlying observation 12, which is for a female subject, has no impact on the male intercept B_0 (i.e., A) and slope B_1.

[11] Other global measures of influence are available (see Chatterjee and Hadi, 1988, Chapter 4, for a comparative treatment).

11.4.1 Influence on Standard Errors

In developing the concept of influence in regression, I have focused on changes in regression coefficients. Other regression outputs are also subject to influence, however. One important regression output is the set of coefficient sampling variances and covariances, which capture the precision of estimation in regression.

Recall, for example, Figure 11.1(*c*), in which a high-leverage observation exerts no influence on the regression coefficients because it is in line with the rest of the data. Recall, as well, that the estimated standard error of the least-squares slope in simple regression is

$$\widehat{\mathrm{SE}}(B) = \frac{S_E}{\sqrt{\sum(X_i - \overline{X})^2}}$$

By increasing the variance of X, therefore, a high-leverage in-line observation serves to decrease $\widehat{\mathrm{SE}}(B)$ even though it does not influence the regression coefficients A and B. Depending on the context, such an observation may be considered beneficial—because it increases the precision of estimation—or it may cause us to exaggerate our confidence in the estimate B.

In multiple regression, we can examine the impact of deleting each observation in turn on the size of the joint confidence region for the regression coefficients.[12] The size of the joint confidence region is analogous to the length of a confidence interval for an individual regression coefficient, which, in turn, is proportional to the standard error of the coefficient. The squared length of a confidence interval is, therefore, proportional to the sampling variance of the coefficient, and, analogously, the squared size of a joint confidence region is proportional to the "generalized variance" of a set of coefficients.

An influence measure proposed by Belsley et al. (1980) closely approximates the squared ratio of volumes of the deleted and full-data confidence regions for the regression coefficients:[13]

$$\mathrm{COVRATIO}_i = \frac{1}{(1 - h_i)\left(\dfrac{n - k - 2 + E_i^{*2}}{n - k - 1}\right)^{k+1}}$$

Observations that increase the precision of estimation have values of COVRATIO that are larger than 1; those that decrease the precision of estimation have values smaller than 1. Look for values of COVRATIO, therefore, that differ substantially from 1.

As was true of measures of influence on the regression coefficients, both the hat value and the (studentized) residual figure in COVRATIO. A large hat

---

[12] See Section 9.4.4 for a discussion of joint confidence regions.

[13] Alternative, similar measures have been suggested by several authors. Chatterjee and Hadi (1988, Chapter 4) provide a comparative discussion.

value produces a large COVRATIO, however, even when—indeed, especially when—the studentized residual is small, because a high-leverage in-line observation improves the precision of estimation. In contrast, a discrepant, low-leverage observation might not change the coefficients much, but it decreases the precision of estimation by increasing the estimated error variance; such an observation, with small h_i and large E_i^*, produces a COVRATIO_i substantially below 1.

For Davis's regression of reported weight on measured weight, sex, and their interaction, by far the most extreme value is $\text{COVRATIO}_{12} = 0.0103$. The 12th observation, therefore, *decreases* the precision of estimation by a factor of $1/0.0103 \simeq 100$. In this instance, a very large leverage, $h_{12} = 0.714$, is more than offset by a massive residual, $E_{12}^* = -24.3$.

11.4.2 Influence on Collinearity

Other characteristics of a regression analysis can also be influenced by individual observations, including the degree of collinearity among the independent variables.[14] I shall not address this issue in any detail, but the following points may prove helpful:[15]

- Influence on collinearity is one of the factors reflected in influence on coefficient standard errors. Measures such as COVRATIO, however, also reflect influence on the error variance and on the variation of the X's. Moreover, COVRATIO and similar measures examine the sampling variances and covariances of all of the regression coefficients, including the regression constant, while a consideration of collinearity generally excludes the constant. Nevertheless, our concern for collinearity reflects its impact on the precision of estimation, which is precisely what is addressed by COVRATIO.
- Collinearity-influential points are those that either induce or weaken correlations among the X's. Such points usually—but not always—have large hat values. Conversely, points with large hat values often influence collinearity.
- Individual points that induce collinearity are obviously problematic. More subtly, points that substantially weaken collinearity also merit examination, because they may cause us to be overly confident in our results.
- It is frequently possible to detect collinearity-influential points by plotting independent variables against each other, as in a scatterplot matrix or a three-dimensional rotating plot. This approach may fail, however, if the collinear relations in question involve more than two or three independent variables at a time.

11.5 Numerical Cutoffs for Diagnostic Statistics

I have deliberately refrained from suggesting specific numerical criteria for identifying noteworthy observations on the basis of measures of leverage and influence: I believe that it is generally more effective to examine the distributions of these quantities directly to locate unusual values. For studentized residuals, the hypothesis-testing and insurance approaches provide numerical cutoffs, but even these criteria are no substitute for graphical examination of the residuals.

---

[14] See Chapter 13 for a general treatment of collinearity.

[15] See Chatterjee and Hadi (1988, Chapter 4 and 5) for more information about influence on collinearity.

Nevertheless, numerical cutoffs can be of some use, as long as they are not given too much weight, and especially when they are employed to enhance graphical displays. A line can be drawn on a graph at the value of a numerical cutoff, and observations that exceed the cutoff can be identified individually.[16]

Cutoffs for a diagnostic statistic may be derived from statistical theory, or they may result from examination of the sample distribution of the statistic. Cutoffs may be absolute, or they may be adjusted for sample size.[17] For some diagnostic statistics, such as measures of influence, absolute cutoffs are unlikely to identify noteworthy observations in large samples. In part, this characteristic reflects the ability of large samples to absorb discrepant data without changing the results substantially, but it is still often of interest to identify *relatively* influential points, even if no observation has strong *absolute* influence.

The cutoffs presented below are, as explained briefly here, derived from statistical theory. An alternative, very simple, and universally applicable data-based criterion is to examine the most extreme (e.g., 5% of) values of a diagnostic statistic.

11.5.1 Hat Values

Belsley et al. (1980) suggest that hat values exceeding about twice the average $\bar{h} = (k + 1)/n$ are noteworthy. This size-adjusted cutoff was derived as an approximation identifying the most extreme 5% of cases when the X's are multivariate normal, and the number of regressors k and degrees of freedom for error $n - k - 1$ are relatively large. The cutoff is nevertheless recommended by these authors as a rough general guide even when the regressors are not normally distributed. In small samples, using $2 \times \bar{h}$ tends to nominate too many points for examination, and $3 \times \bar{h}$ can be used instead.[18]

11.5.2 Studentized Residuals

Beyond the issues of "statistical significance" and estimator robustness and efficiency discussed above, it sometimes helps to call attention to residuals that are relatively large. Recall that, under ideal conditions, about 5% of studentized residuals are outside the range $|E_i^*| \leq 2$. It is, therefore, reasonable, for example, to draw lines at ± 2 on a display of studentized residuals to draw attention to observations outside this range.

11.5.3 Measures of Influence

Many cutoffs have been suggested for different measures of influence. A few are presented here:

- *Standardized change in regression coefficients*. The D_{ij}^* are scaled by standard errors, and, consequently, $|D_{ij}^*| > 1$ or 2 suggests itself as an absolute cutoff. As explained above, however, this criterion is unlikely to nominate observations

---

16 An example appears in Figure 11.6 on page 285.

17 See Belsley et al. (1980, Chapter 2) for further discussion of these distinctions.

18 See Chatterjee and Hadi (1988, Chapter 4) for a discussion of alternative cutoffs for hat values.

in large samples. Belsley et al. (1980) propose the size-adjusted cutoff $2/\sqrt{n}$ for identifying noteworthy D_{ij}^*'s.

- *Cook's D and DFFITS*. Several numerical cutoffs have been recommended for Cook's D and for DFFITS—exploiting the analogy between D and an F-statistic, for example. Chatterjee and Hadi (1988) suggest the size-adjusted cutoff[19]

$$|\text{DFFITS}_i| > 2\sqrt{\frac{k+1}{n-k-1}}$$

Because of the approximate relationship between DFFITS and Cook's D, it is simple to translate this criterion into

$$D_i > \frac{4}{n-k-1}$$

Absolute cutoffs for D, such as $D_i > 1$, risk missing relatively influential data.

- *COVRATIO*. Belsley et al. (1980) suggest the size-adjusted cutoff

$$|\text{COVRATIO}_i - 1| > \frac{3(k+1)}{n}$$

11.6 Joint Influence and Partial-Regression Plots

As illustrated in Figure 11.4, subsets of observations can be *jointly influential* or can offset each other's influence. Influential subsets or multiple outliers can often be identified by applying single-observation diagnostics, such as Cook's D and studentized residuals, sequentially. It can be important, however, to refit the model after deleting each point, because the presence of a single influential value can dramatically affect the fit at other points. Still, the sequential approach is not always successful.

Although it is possible to generalize deletion statistics to subsets of several points, the very large number of subsets usually renders this approach impractical.[20] An attractive alternative is to employ graphical methods, and a particularly useful influence graph is the *partial-regression plot* (also called a *partial-regression leverage plot* or an *added-variable plot*).

Let $Y_i^{(1)}$ represent the residuals from the least-squares regression of Y on all of the X's with the exception of X_1—that is, the residuals from the fitted regression equation

$$Y_i = A^{(1)} + B_2^{(1)}X_{i2} + \cdots + B_k^{(1)}X_{ik} + Y_i^{(1)}$$

---

[19] Also see Cook (1977), Belsley et al. (1980), and Velleman and Welsch (1981).

[20] Cook and Weisberg (1980), for example, extend the D-statistic to a subset of p observations indexed by the vector subscript $\mathbf{i} = (i_1, i_2, \ldots, i_p)'$:

$$D_{\mathbf{i}} = \frac{\mathbf{d}_{\mathbf{i}}'(\mathbf{X}'\mathbf{X})\mathbf{d}_{\mathbf{i}}}{(k+1)S_E^2}$$

where $\mathbf{d}_{\mathbf{i}} = \mathbf{b} - \mathbf{b}_{(-\mathbf{i})}$ gives the impact on the regression coefficients of deleting the subset $\mathbf{i}$. See Belsley et al. (1980, Chapter 2) and Chatterjee and Hadi (1988) for further discussion of deletion diagnostics based on subsets of observations. Note that there are $n!/[p!(n-p)!]$ subsets of size p—typically a prohibitively large number, even for modest values of p.

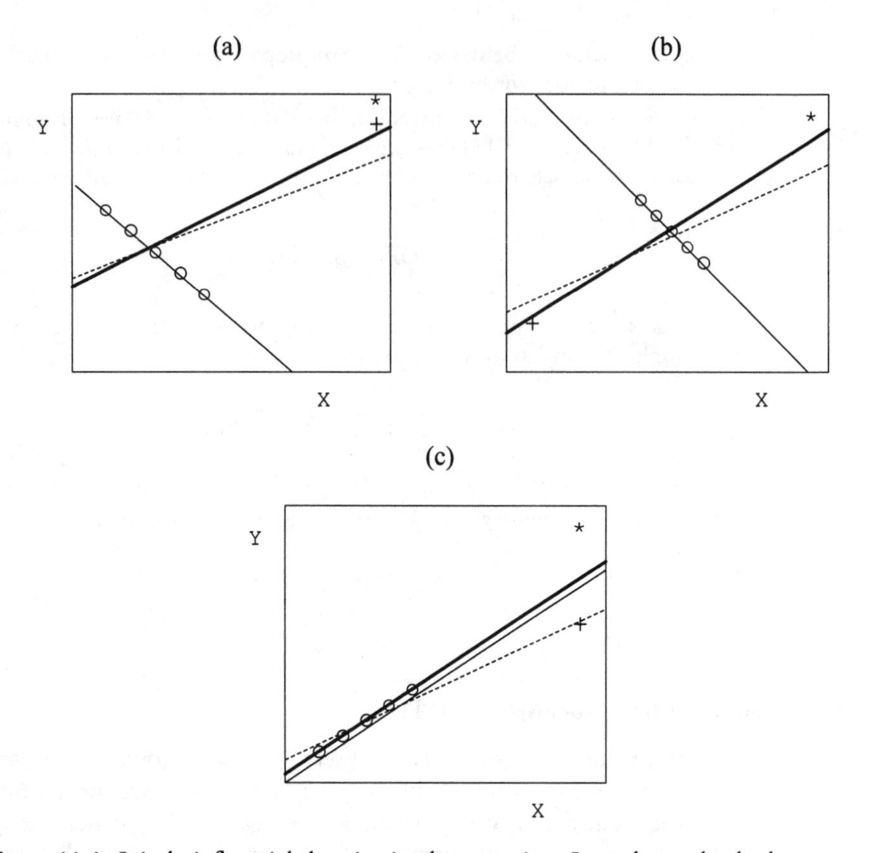

Figure 11.4. Jointly influential data in simple regression. In each graph, the heavy solid line gives the least-squares regression for all of the data; the broken line gives the regression with the asterisk deleted; and the light solid line gives the regression with both the asterisk and the plus deleted. (*a*) Jointly influential observations located close to one another: Deletion of both observations has a much greater impact than deletion of only one. (*b*) Jointly influential observations located on opposite sides of the data. (*c*) Observations that offset one another: The regression with both observations deleted is the same as for the whole dataset (the two lines are separated slightly for visual effect).

The parenthetical superscript (1) indicates the omission of X_1 from the right-hand side of the regression equation. Likewise, $X_i^{(1)}$ is the residual from the least-squares regression of X_1 on all the other X's:

$$X_{i1} = C^{(1)} + D_2^{(1)} X_{i2} + \cdots + D_k^{(1)} X_{ik} + X_i^{(1)}$$

The notation emphasizes the interpretation of the residuals $Y^{(1)}$ and $X^{(1)}$ as the parts of Y and X_1 that remain when the effects of $X_2, \ldots, X_k$ are "removed." The residuals $Y^{(1)}$ and $X^{(1)}$ have the following interesting properties:

1. The slope from the least-squares regression of $Y^{(1)}$ on $X^{(1)}$ is simply the least-squares slope B_1 from the full multiple regression.

2. The residuals from the simple regression of $Y^{(1)}$ on $X^{(1)}$ are the same as those from the full regression; that is,

$$Y_i^{(1)} = B_1 X_i^{(1)} + E_i \qquad [11.6]$$

No constant is required here, because both $Y^{(1)}$ and $X^{(1)}$ are least-squares residuals and therefore have means of 0.

3. The variation of $X^{(1)}$ is the conditional variation of X_1 holding the other X's constant and, as a consequence, the standard error of B_1 in the auxiliary simple regression (Equation 11.6),

$$\widehat{\text{SE}(B_1)} = \frac{S_E}{\sqrt{\sum X_i^{(1)2}}}$$

is the same[21] as the multiple-regression standard error of B_1. Unless X_1 is uncorrelated with the other X's, its conditional variation is smaller than its marginal variation—much smaller, if X_1 is strongly collinear with the other X's.

Plotting $Y^{(1)}$ against $X^{(1)}$ permits us to examine leverage and influence on B_1. Because of properties 1–3, this plot also provides a visual impression of the precision of the estimate B_1. Similar partial-regression plots can be constructed for the other regressors:[22]

$$\text{Plot } Y^{(j)} \text{ versus } X^{(j)} \text{ for each } j = 1, \ldots, k$$

> Subsets of observations can be jointly influential. Partial-regression plots are useful for detecting joint influence on the regression coefficients. The partial-regression plot for the regressor X_j is formed using the residuals from the least-squares regressions of X_j and Y on all of the other X's.

Illustrative partial-regression plots are shown in Figure 11.5, using data from Duncan's regression of occupational prestige on the income and educational levels of 45 U.S. occupations. Recall (from Chapter 5) that Duncan's regression yields the following least squares fit:

$$\begin{aligned}\widehat{\text{Prestige}} = \underset{(4.27)}{-6.06} + \underset{(0.120)}{0.599} \times \text{ Income} + \underset{(0.098)}{0.546} \times \text{ Education}\end{aligned}$$

$$R^2 = 0.83 \qquad S_E = 13.4$$

The partial-regression plot for income [Figure 11.5(*a*)] reveals three observations that exert substantial leverage on the income coefficient. Two of these observations serve to decrease the income slope: *ministers*, whose income is un-

[21] There is slight slippage here with respect to the degrees of freedom for error: S_E is from the multiple regression, with $n - k - 1$ degrees of freedom for error. We need not subtract the mean of $X_i^{(1)}$ to calculate the standard error of the slope since the mean is already 0.

[22] We can also construct a partial-regression plot for the intercept A, by regressing the "constant regressor" $X_0 = 1$ and Y on X_1 through X_k, with no constant in these regression equations.

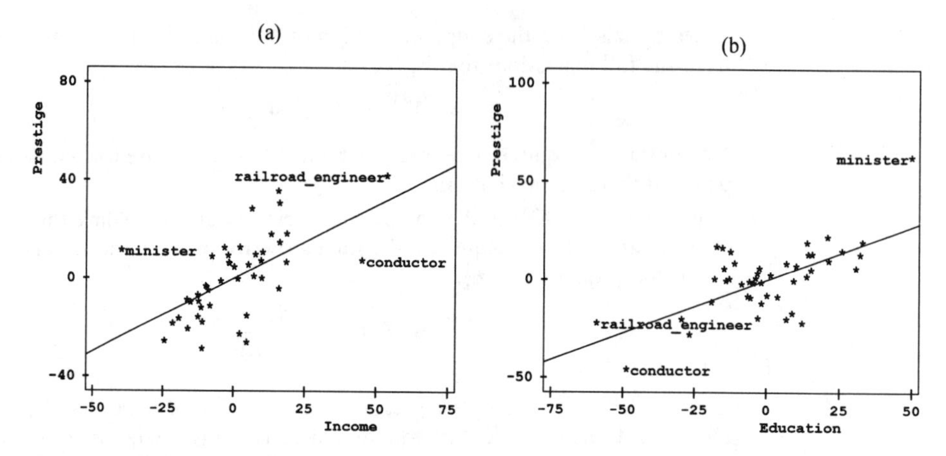

Figure 11.5. Partial-regression plots for Duncan's regression of occupational prestige on the income (*a*) and educational levels (*b*) of 45 U.S. occupations in 1950. Three potentially influential observations (*ministers, railroad conductors*, and *railroad engineers*) are identified on the plots. The partial-regression plot for the intercept A is not shown.

usually low given the educational level of the occupation; and *railroad conductors*, whose income is unusually high given education. The third occupation, *railroad engineers*, is above the fitted regression, but is not as discrepant; it, too, has relatively high income given education. Remember that the horizontal variable in this partial-regression plot is the residual from the regression of income on education, and thus values far from 0 in this direction are for occupations with incomes that are unusually high or low given their levels of education.

The partial-regression plot for education [Figure 11.5(*b*)] shows that the same three observations have relatively high leverage on the education coefficient: *Ministers* and *railroad conductors* tend to increase the education slope, while *railroad engineers* appear to be closer in line with the rest of the data.

Examining the single-observation deletion statistics for Duncan's regression reveals that *ministers* have the largest Cook's D ($D_6 = 0.566$) and the largest studentized residual ($E_6^* = 3.14$). This studentized residual is not especially big, however: The Bonferroni p-value for the outlier test is $\Pr(t_{41} > 3.14) \times 2 \times 45 = 0.14$. Figure 11.6 displays a plot of studentized residuals versus hat values, with the areas of the plotted circles proportional to values of Cook's D. The lines on the plot are at $E^* = \pm 2$ (on the vertical axis) and at $h = 2\bar{h}$ and $3\bar{h}$ (on the horizontal axis). Four observations that exceed these cutoffs are identified on the plot. *Reporters* have a relatively large residual but are at a low-leverage point, while *railroad engineers* have high leverage but a small studentized residual.

Deleting *ministers* and *conductors* produces the fitted regression

$$\widehat{\text{Prestige}} = \underset{(3.65)}{-6.41} + \underset{(0.122)}{0.867} \times \text{Income} + \underset{(0.099)}{0.332} \times \text{Education}$$

$$R^2 = 0.88 \qquad S_E = 11.4$$

which, as expected from the partial-regression plots, has a larger income slope and smaller education slope than the original regression. The estimated standard

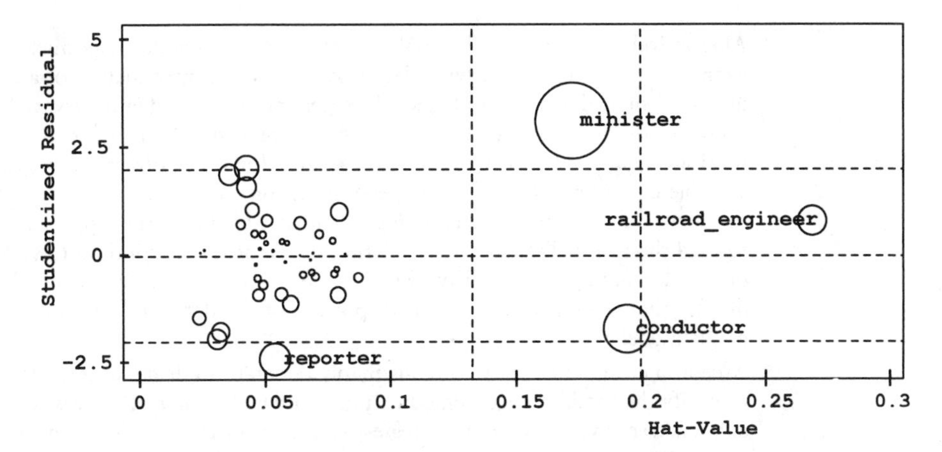

Figure 11.6. "Bubble plot" of Cook's D's, studentized residuals, and hat values, for Duncan's regression of occupational prestige on income and education. Each point is plotted as a circle with area proportional to D. Horizontal reference lines are drawn at studentized residuals of 0 and ±2; vertical reference lines are drawn at hat values of $2\bar{h}$ and $3\bar{h}$. Several observations are identified on the plot: *Ministers* and *conductors* have large hat values and relatively large residuals; *reporters* have a relatively large residual, but a small hat value; *railroad engineers* have a large hat value, but a small residual.

errors are likely optimistic, however, because relative outliers have been trimmed away. Deleting *railroad engineers*, along with *ministers* and *conductors*, further increases the income slope and decreases the education slope, but the change is not dramatic: $B_{\text{Income}} = 0.931$, $B_{\text{Education}} = 0.285$.

Partial-regression plots can be straightforwardly extended to pairs of regressors. We can, for example, regress each of X_1, X_2, and Y on the remaining regressors, $X_3, \ldots, X_k$, obtaining residuals $X_{i1}^{(12)}$, $X_{i2}^{(12)}$, and $Y_i^{(12)}$; $Y^{(12)}$ is then plotted against $X_1^{(12)}$ and $X_2^{(12)}$ to produce a dynamic three-dimensional scatterplot on which the partial-regression plane can be displayed.[23]

11.7 Should Unusual Data Be Discarded?

The discussion thus far in this chapter has implicitly assumed that outlying and influential data are simply discarded. Although problematic data should not be ignored, they also should not be deleted automatically and without reflection:

- It is important to investigate why an observation is unusual. Truly bad data (e.g., an error in data entry as in Davis's regression) can often be corrected or, if correction is not possible, thrown away. When a discrepant data point is correct, we may be able to understand why the observation is unusual. For Duncan's regression, for example, it makes sense that ministers enjoy prestige not accounted for by the income and educational levels of the occupation. In a case like this, we may choose to deal separately with an outlying observation.

[23] See Cook and Weisberg (1989) for a discussion of three-dimensional partial-regression plots. An alternative, two-dimensional extension of partial-regression plots to subsets of coefficients is described in Section 11.8.4.

- Alternatively, outliers or influential data may motivate model respecification. For example, the pattern of outlying data may suggest the introduction of additional independent variables. If, in Duncan's regression, we can identify a variable that produces the unusually high prestige of ministers (net of their income and education), and if we can measure that variable for other observations, then the variable could be added to the regression. In some instances, transformation of the dependent variable or of an independent variable may draw apparent outliers toward the rest of the data, by rendering the error distribution more symmetric or by eliminating nonlinearity. We must, however, be careful to avoid "overfitting" the data—permitting a small portion of the data to determine the form of the model.[24]
- Except in clear-cut cases, we are justifiably reluctant to delete observations or to respecify the model to accommodate unusual data. Some researchers reasonably adopt alternative estimation strategies, such as robust regression, which continuously downweights outlying data rather than simply discarding them. Because these methods assign zero or very small weight to highly discrepant data, however, the result is generally not very different from careful application of least squares, and, indeed, robust-regression weights can be used to identify outliers.[25]

> Outlying and influential data should not be ignored, but they also should not simply be deleted without investigation. "Bad" data can often be corrected. "Good" observations that are unusual may provide insight into the structure of the data, and may motivate respecification of the statistical model used to summarize the data.

EXERCISES

11.1 Employ the methods of this chapter to look for unusual data in each of the following linear-model analyses. In each case, consider the impact of any unusual data that you discover on the results of the analysis, and—within the limits of your knowledge of the datasets—suggest how unusual data should be treated.

(a) Sahlins's regression of acres tended per gardener on consumers per gardener for the households of Mazulu village. (See Exercise 5.6; the data are in Table 2.1 and `sahlins.dat`.)

(b) Angell's dummy-variable regression of moral integration of U.S. cities on ethnic heterogeneity, geographic mobility, and region. (See Exercise 7.3; the data are in Table 2.3 and `angell.dat`.) Are partial-regression plots for dummy regressors interpretable? If so, how?

---

24 See Chapter 12.

25 See Sections 2.3 and 14.3.

(c) Moore and Krupat's two-way analysis of variance of conformity by authoritarianism and partner's status. (See Section 8.2; the data are in Table 8.3 and `moore.dat`.) Repeat your analysis for the analysis-of-covariance model fit to Moore and Krupat's data, treating authoritarianism as a covariate. (See Section 8.4.) Are partial-regression plots for deviation-coded regressors interpretable? If so, how?

(d) Anscombe's regression of state education expenditures on income, proportion under 18, and proportion urban. (See Exercise 5.14; the data are in Table 5.1 and `anscombe.dat`.)

11.2 In order to test two theories of peasant revolt, Chirot and Ragin (1975) gathered data (in Table 11.1 and `chirot.dat`) on a 1907 rebellion in 32 counties of Romania.[26] The dependent variable in their analysis was the intensity of the rebellion (I), an index constructed from the reported level of violence and the degree to which the rebellion spread within a county. According to the "transitional society" theory of peasant rebellion, intensity should be high when *both* the level of commercialization of agriculture (C) and the level of traditionalism (T) are high. Commercialization—the penetration of market forces—was measured by the percentage of land in the county devoted to cultivation of wheat, the major cash crop raised in the region. Traditionalism was measured by the percentage of illiterates. The "structural" theory of peasant revolt implies that the rebellion should be intense where middle peasants (M) are relatively strong and where the inequality of land tenure (G) is high. The strength of the middle peasantry was assessed by the percentage of rural households owning between 7 and 50 hectares of land; inequality of land tenure was measured by a Gini coefficient. Chirot and Ragin tested the two theories by regressing I on C, T, the product of C and T (i.e., $C \times T$), M, and G. The first theory predicts a positive coefficient for $C \times T$, while the second predicts positive coefficients for M and G. Redo Chirot and Ragin's linear-model analysis, using the methods of this chapter to look for unusual data. Consider the impact of any unusual observations that you discover on the results of the study.

11.8 Some Statistical Details*

11.8.1 Hat Values and the Hat Matrix

Recall, from Chapter 9, the matrix form of the general linear model, $\mathbf{y} = \mathbf{X}\boldsymbol{\beta} + \boldsymbol{\varepsilon}$. Recall, as well, that the fitted model is given by $\mathbf{y} = \mathbf{Xb} + \mathbf{e}$, in which the vector of least-squares estimates is $\mathbf{b} = (\mathbf{X}'\mathbf{X})^{-1}\mathbf{X}'\mathbf{y}$.

The least-squares fitted values are therefore a linear function of the observed dependent-variable values:

$$\hat{\mathbf{y}} = \mathbf{Xb} = \mathbf{X}(\mathbf{X}'\mathbf{X})^{-1}\mathbf{X}'\mathbf{y} = \mathbf{Hy}$$

---

[26] I am grateful to Michael Gillespie of the Sociology Department, University of Alberta, for bringing this dataset to my attention.

TABLE 11.1 Data on the 1907 Romanian Peasant Rebellion: I, Intensity of the Rebellion (Corrected from the original); C, Commercialization of Agriculture; T, Traditionalism; M, Market Forces; and G, Inequality of Land Tenure.

County	*I*	C	T	M	G
1	−1.39	13.8	86.2	6.2	0.60
2	0.65	20.4	86.7	2.9	0.72
3	1.89	27.6	79.3	16.9	0.66
4	−0.15	18.6	90.1	3.4	0.74
5	−0.86	17.2	84.5	9.0	0.70
6	0.11	21.5	81.5	5.2	0.60
7	−0.51	11.6	82.6	5.1	0.52
8	−0.86	20.4	82.4	6.3	0.64
9	−0.24	19.5	87.5	4.8	0.68
10	−0.77	8.9	85.6	9.5	0.58
11	−0.24	25.8	82.2	10.9	0.68
12	−1.57	24.1	83.5	8.4	0.74
13	−0.51	2.0	88.3	6.2	0.70
14	−1.57	24.2	84.9	6.1	0.62
15	−0.51	30.6	76.1	1.3	0.76
16	−1.13	33.9	85.5	5.8	0.70
17	−1.22	28.6	84.2	2.9	0.58
18	−1.22	36.5	78.1	4.3	0.72
19	−0.86	40.9	84.4	2.3	0.64
20	−1.39	6.8	76.3	3.6	0.58
21	2.81	41.9	89.7	6.6	0.66
22	−1.04	25.4	83.2	2.5	0.68
23	1.57	30.5	80.2	4.1	0.76
24	4.32	48.2	91.0	4.2	0.70
25	3.79	46.0	90.5	3.7	0.68
26	3.79	45.1	85.5	5.1	0.64
27	−1.75	12.5	83.8	7.2	0.50
28	0.82	39.3	85.6	4.9	0.60
29	2.59	47.7	87.6	5.2	0.58
30	−0.86	15.2	87.3	10.8	0.42
31	−1.84	11.7	82.3	81.7	0.42
32	−1.84	25.6	80.1	68.4	0.26

Source of Data: Chirot and Ragin (1975).

Here, $\mathbf{H} = \mathbf{X}(\mathbf{X}'\mathbf{X})^{-1}\mathbf{X}'$ is the *hat matrix*, so named because it transforms $\mathbf{y}$ into $\hat{\mathbf{y}}$. The hat matrix is symmetric ($\mathbf{H} = \mathbf{H}'$) and idempotent ($\mathbf{H}^2 = \mathbf{H}$), as can easily be verified.[27] Consequently, the diagonal entries of the hat matrix $h_i \equiv h_{ii}$, which we called the *hat values*, are

$$h_i \equiv \mathbf{h}_i'\mathbf{h}_i = \sum_{j=1}^{n} h_{ij}^2 = h_i^2 + \sum_{j \neq i} h_{ij}^2 \qquad [11.7]$$

where (because of symmetry) the elements of $\mathbf{h}_i$ comprise both the ith row and the ith column of $\mathbf{H}$.

Equation 11.7 implies that $0 \leq h_i \leq 1$. If the model matrix $\mathbf{X}$ includes the constant regressor, then $1/n \leq h_i$. Because $\mathbf{H}$ is a projection matrix,[28] projecting

[27] See Exercise 11.4.

[28] See Chapter 10, on the vector geometry of linear models.

$\mathbf{y}$ orthogonally onto the $(k+1)$-dimensional subspace spanned by the columns of $\mathbf{X}$, it follows that $\sum h_i = k+1$, and thus $\bar{h} = (k+1)/n$ (as stated in Section 11.2).

I mentioned as well that when there are several independent variables in the model, the leverage h_i of the ith observation is directly related to the distance of this observation from the center of the independent-variable scatter. To demonstrate this property of the hat-values, it is convenient to rewrite the fitted model with all variables in mean-deviation form: $\mathbf{y}^* = \mathbf{X}^*\mathbf{b}_1 + \mathbf{e}$, where $\mathbf{y}^* \equiv \{Y_i - \overline{Y}\}$ is the "centered" dependent-variable vector; $\mathbf{X}^* \equiv \{X_{ij} - \overline{X}_j\}$ contains the centered independent variables, but no constant regressor, which is no longer required; and $\mathbf{b}_1$ is the vector of least-squares slopes (suppressing the regression intercept). Then the hat value for the ith observation is

$$h_i^* = \mathbf{h}_i^{*\prime}\mathbf{h}_i^* = \mathbf{x}_i^{*\prime}(\mathbf{X}^{*\prime}\mathbf{X}^*)^{-1}\mathbf{x}_i^* = h_i - \frac{1}{n}$$

where $\mathbf{x}_i^{*\prime} = [X_{i1} - \overline{X}_1, \ldots, X_{ik} - \overline{X}_k]$ is the ith row of $\mathbf{X}^*$ (and $\mathbf{x}_i^*$ is the ith row of $\mathbf{X}^*$ written as a column vector).

As Weisberg (1985, p. 112) has pointed out, $(n-1)h_i^*$ is the *generalized* or *Mahalanobis distance* between $\mathbf{x}_i'$ and $\bar{\mathbf{x}}'$, where $\bar{\mathbf{x}}' = [\overline{X}_1, \ldots, \overline{X}_k]$ is the mean vector or *centroid* of the independent variables. The Mahalanobis distances, and hence the hat values, do not change if the independent variables are rescaled. Indeed, the Mahalanobis distances and hat values are invariant with respect to any nonsingular linear transformation of $\mathbf{X}$.

11.8.2 The Distribution of the Least-Squares Residuals

The least-squares residuals are given by

$$\begin{aligned}
\mathbf{e} &= \mathbf{y} - \hat{\mathbf{y}} \\
&= (\mathbf{X}\boldsymbol{\beta} + \boldsymbol{\varepsilon}) - \mathbf{X}(\mathbf{X}'\mathbf{X})^{-1}\mathbf{X}'(\mathbf{X}\boldsymbol{\beta} + \boldsymbol{\varepsilon}) \\
&= (\mathbf{I} - \mathbf{H})\boldsymbol{\varepsilon}
\end{aligned}$$

Thus,

$$E(\mathbf{e}) = (\mathbf{I} - \mathbf{H})E(\boldsymbol{\varepsilon}) = (\mathbf{I} - \mathbf{H})\mathbf{0} = \mathbf{0}$$

and

$$V(\mathbf{e}) = (\mathbf{I} - \mathbf{H})V(\mathbf{e})(\mathbf{I} - \mathbf{H})' = \sigma_\varepsilon^2(\mathbf{I} - \mathbf{H})$$

because $\mathbf{I} - \mathbf{H}$, like $\mathbf{H}$ itself, is symmetric and idempotent. The matrix $\mathbf{I} - \mathbf{H}$ is not diagonal, and therefore the residuals are generally correlated, even when the errors are (as assumed here) independent. The diagonal entries of $\mathbf{I} - \mathbf{H}$ generally differ from one another, and so the residuals generally have different variances (as stated in Section 11.3):[29] $V(E_i) = \sigma_\varepsilon^2(1 - h_i)$.

[29] Balanced ANOVA models are an exception: Here, all the hat values are equal. (Why?)

11.8.3 Deletion Diagnostics

Let $\mathbf{b}_{(-i)}$ denote the vector of least-squares regression coefficients calculated with the ith observation omitted. Then $\mathbf{d}_i \equiv \mathbf{b} - \mathbf{b}_{(-i)}$ represents the influence of observation i on the regression coefficients. The influence vector $\mathbf{d}_i$ can be calculated efficiently as[30]

$$\mathbf{d}_i = (\mathbf{X}'\mathbf{X})^{-1}\mathbf{x}_i \frac{E_i}{1 - h_i} \qquad [11.8]$$

where $\mathbf{x}_i'$ is the ith row of the model matrix $\mathbf{X}$ (and $\mathbf{x}_i$ is the ith row written as a column vector).

Cook's D_i is the F-statistic for testing the "hypothesis" that $\boldsymbol{\beta} = \mathbf{b}_{(-i)}$:

$$\begin{aligned} D_i &= \frac{(\mathbf{b} - \mathbf{b}_{(-i)})'\mathbf{X}'\mathbf{X}(\mathbf{b} - \mathbf{b}_{(-i)})}{(k+1)S_E^2} \\ &= \frac{(\hat{\mathbf{y}} - \hat{\mathbf{y}}_{(-i)})'(\hat{\mathbf{y}} - \hat{\mathbf{y}}_{(-i)})}{(k+1)S_E^2} \end{aligned}$$

An alternative interpretation of D_i, therefore, is that it measures the aggregate influence of observation i on the fitted values $\hat{\mathbf{y}}$. This is why Belsley et al. (1980) call their similar statistic "DFFITS." Using Equation 11.8,

$$\begin{aligned} D_i &= \frac{E_i^2}{S_E^2(k+1)} \times \frac{h_i}{(1-h_i)^2} \\ &= \frac{E_i'^2}{k+1} \times \frac{h_i}{1-h_i} \end{aligned}$$

which is the formula for Cook's D given in Section 11.4.

11.8.4 Partial-Regression Plots

In vector form, the fitted multiple-regression model is

$$\begin{aligned} \mathbf{y} &= A\mathbf{1}_n + B_1\mathbf{x}_1 + B_2\mathbf{x}_2 + \cdots + B_k\mathbf{x}_k + \mathbf{e} \\ &= \hat{\mathbf{y}} + \mathbf{e} \end{aligned} \qquad [11.9]$$

where the fitted-value vector $\hat{\mathbf{y}}$ is the orthogonal projection of $\mathbf{y}$ onto the subspace spanned by the regressors[31] $\mathbf{1}_n$, $\mathbf{x}_1$, $\mathbf{x}_2, \ldots, \mathbf{x}_k$. Let $\mathbf{y}^{(1)}$ and $\mathbf{x}^{(1)}$ be the projections of $\mathbf{y}$ and $\mathbf{x}_1$, respectively, onto the orthogonal complement of the subspace spanned by $\mathbf{1}_n$ and $\mathbf{x}_2, \ldots, \mathbf{x}_k$ (i.e., the residual vectors from the least-squares regressions of Y and X_1 on the other X's). Then, by the geometry of projections, the orthogonal projection of $\mathbf{y}^{(1)}$ onto $\mathbf{x}^{(1)}$ is $B_1\mathbf{x}^{(1)}$, and

[30] See Exercise 11.5.

[31] See Chapter 10.

$\mathbf{y}^{(1)} - B_1\mathbf{x}^{(1)} = \mathbf{e}$, the residual vector from the overall least-squares regression, given in Equation 11.9.[32]

Sall (1990) suggests the following generalization of partial-regression plots, which he terms *leverage plots*: Consider the general linear hypothesis[33]

$$H_0\colon \underset{(q\times k+1)}{\mathbf{L}}\ \underset{(k+1\times 1)}{\boldsymbol{\beta}} = \underset{(q\times 1)}{\mathbf{0}} \qquad [11.10]$$

For example, in the regression of occupational prestige (Y) on education (X_1), income (X_2), and type of occupation (represented by the dummy regressors D_1 and D_2),

$$Y = \alpha + \beta_1 X_1 + \beta_2 X_2 + \gamma_1 D_1 + \gamma_2 D_2 + \varepsilon$$

the hypothesis matrix

$$\mathbf{L} = \begin{bmatrix} 0 & 0 & 0 & 1 & 0 \\ 0 & 0 & 0 & 0 & 1 \end{bmatrix}$$

is used to test the hypothesis $H_0\colon \gamma_1 = \gamma_2 = 0$ that there is no effect of type of occupation.[34]

The residuals for the full model, unconstrained by the hypothesis (Equation 11.10), are the usual least-squares residuals, $\mathbf{e} = \mathbf{y} - \mathbf{X}\mathbf{b}$. The estimated regression coefficients under the hypothesis are[35]

$$\mathbf{b}_0 = \mathbf{b} - (\mathbf{X}'\mathbf{X})^{-1}\mathbf{L}'\mathbf{u}$$

and the residuals constrained by the hypothesis are given by

$$\mathbf{e}_0 = \mathbf{e} + \mathbf{X}(\mathbf{X}'\mathbf{X})^{-1}\mathbf{L}'\mathbf{u}$$

where

$$\mathbf{u} \equiv [\mathbf{L}(\mathbf{X}'\mathbf{X})^{-1}\mathbf{L}']^{-1}\mathbf{L}\mathbf{b}$$

Thus, the incremental sum of squares for H_0 is[36]

$$||\mathbf{e}_0 - \mathbf{e}||^2 = \mathbf{b}'\mathbf{L}'[\mathbf{L}(\mathbf{X}'\mathbf{X})^{-1}\mathbf{L}']^{-1}\mathbf{L}\mathbf{b}$$

The leverage plot is a scatterplot with

$$\mathbf{v}_x \equiv \mathbf{X}(\mathbf{X}'\mathbf{X})^{-1}\mathbf{L}'\mathbf{u}$$

on the horizontal axis, and

$$\mathbf{v}_y \equiv \mathbf{v}_x + \mathbf{e}$$

---

32 See Exercises 11.6 and 11.7.

33 See Section 9.4.3.

34 See Exercise 11.9.

35 For this and other results pertaining to leverage plots, see Sall (1990).

36 See Exercise 11.8.

on the vertical axis. The leverage plot, so defined, has the following properties:

- The residuals around the horizontal line at $V_y = 0$ are the constrained least-squares residuals E_{0i} under the hypothesis H_0.
- The least-squares line fit to the leverage plot has an intercept of 0 and a slope of 1; the residuals about this line are the unconstrained least-squares residuals, E_i. The incremental sum of squares for H_0 is thus the regression sum of squares for the line.
- When the hypothesis matrix $\mathbf{L}$ is formulated with a single row to test the coefficient of an individual regressor, the leverage plot specializes to the usual partial-regression plot, with the horizontal axis rescaled so that the least-squares intercept is 0 and the slope 1.

EXERCISES

11.3 Show that, in simple-regression analysis, the hat value is

$$h_i = \frac{1}{n} + \frac{(X_i - \overline{X})^2}{\sum_{j=1}^{n}(X_j - \overline{X})^2}$$

[*Hint*: Evaluate $\mathbf{x}_i'(\mathbf{X}'\mathbf{X})^{-1}\mathbf{x}_i$ for $\mathbf{x}_i' = (1, X_i)$.]

11.4 Show that the hat matrix $\mathbf{H} = \mathbf{X}(\mathbf{X}'\mathbf{X})^{-1}\mathbf{X}'$ is symmetric ($\mathbf{H} = \mathbf{H}'$) and idempotent ($\mathbf{H}^2 = \mathbf{H}$).

11.5 Using Duncan's regression of occupational prestige on the educational and income levels of occupations (the data are in Table 3.2 and `duncan.dat`), verify that the influence vector for the deletion of *ministers* on the regression coefficients, $\mathbf{d}_i = \mathbf{b} - \mathbf{b}_{(-i)}$, can be written as

$$\mathbf{d}_i = (\mathbf{X}'\mathbf{X})^{-1}\mathbf{x}_i \frac{E_i}{1 - h_i}$$

where $\mathbf{x}_i$ is the ith row of the model matrix $\mathbf{X}$ (i.e., the row for *ministers*) written as a column. [A much more difficult problem is to show that this formula works in general; see, e.g., Belsley, et al. (1980, pp. 69–83) or Velleman and Welsch (1981).]

11.6 *Consider the two-independent-variable linear-regression model, with variables written as vectors in mean-deviation form (as in Section 10.2): $\mathbf{y}^* = B_1\mathbf{x}_1^* + B_2\mathbf{x}_2^* + \mathbf{e}$. Let $\mathbf{x}^{(1)}$ and $\mathbf{y}^{(1)}$ represent the residual vectors from the regression (i.e., orthogonal projection) of $\mathbf{x}_1^*$ and $\mathbf{y}^*$, respectively, on $\mathbf{x}_2^*$. Drawing the three-dimensional diagram of the subspace spanned by $\mathbf{x}_1^*$, $\mathbf{x}_2^*$, and $\mathbf{y}^*$, prove geometrically that the coefficient for the orthogonal projection of $\mathbf{y}^{(1)}$ onto $\mathbf{x}^{(1)}$ is B_1.

11.7 *Now consider the more general model $\mathbf{y}^* = B_1\mathbf{x}_1^* + B_2\mathbf{x}_2^* + \cdots + B_k\mathbf{x}_k^* + \mathbf{e}$. Let $\mathbf{x}^{(1)}$ and $\mathbf{y}^{(1)}$ represent the residual vectors from the projections of $\mathbf{x}_1^*$ and $\mathbf{y}^*$, respectively, onto the subspace spanned by $\mathbf{x}_2^*, \ldots, \mathbf{x}_k^*$. Prove that the coefficient for the orthogonal projection of $\mathbf{y}^{(1)}$ onto $\mathbf{x}^{(1)}$ is B_1.

11.8 *Show that the incremental sum of squares for the general linear hypothesis H_0: $\mathbf{L}\boldsymbol{\beta} = \mathbf{0}$ can be written as

$$||\mathbf{e}_0 - \mathbf{e}||^2 = \mathbf{b}'\mathbf{L}'[\mathbf{L}(\mathbf{X}'\mathbf{X})^{-1}\mathbf{L}']^{-1}\mathbf{L}\mathbf{b}$$

[*Hint*: $||\mathbf{e}_0 - \mathbf{e}||^2 = (\mathbf{e}_0 - \mathbf{e})'(\mathbf{e}_0 - \mathbf{e})$.]

11.9 Using Duncan's data on the prestige of 45 U. S. occupations, regress prestige on education, income, and two dummy variables to represent the effects of three occupational types. (See Section 7.2; Duncan's data are in Table 3.2 and `duncan.dat`.)

(a) Construct partial-regression plots for education, income, and the two dummy regressors for occupational type.

(b) Construct leverage plots for education, income, and occupational type. Confirm that the leverage plots for education and income are identical to the partial-regression plots in part (a), except for the scaling of the horizontal axis. Compare the information obtained from the leverage plot for occupational type with the two partial-regression plots for the occupational type coefficients in part (a).

11.9 Summary

- Unusual data are problematic in linear models fit by least squares because they can substantially influence the results of the analysis, and because they may indicate that the model fails to capture important features of the data.
- Observations with unusual combinations of independent-variable values have high *leverage* in a least-squares regression. The hat values h_i provide a measure of leverage. A rough cutoff for noteworthy hat values is $h_i > 2\bar{h} = 2(k+1)/n$.
- A regression *outlier* is an observation with an unusual dependent-variable value given its combination of independent-variable values. The studentized residuals E_i^* can be used to identify outliers, through graphical examination, a Bonferroni test for the largest absolute E_i^*, or Anscombe's insurance analogy. If the model is correct (and there are no true outliers), then each studentized residual follows a t-distribution with $n - k - 2$ degrees of freedom.
- Observations that combine high leverage with a large studentized residual exert substantial *influence* on the regression coefficients. Cook's D-statistic provides a summary index of influence on the coefficients. A rough cutoff for noteworthy values of D is $D_i > 4/(n - k - 1)$.
- It is also possible to investigate the influence of individual observations on other regression "outputs," such as coefficient standard errors and collinearity.

- Subsets of observations can be jointly influential. Partial-regression plots are useful for detecting joint influence on the regression coefficients. The partial-regression plot for the regressor X_j is formed using the residuals from the least-squares regressions of X_j and Y on all of the other X's.
- Outlying and influential data should not be ignored, but they also should not simply be deleted without investigation. "Bad" data can often be corrected. "Good" observations that are unusual may provide insight into the structure of the data, and may motivate respecification of the statistical model used to summarize the data.

11.10 Recommended Reading

There is a large journal literature on methods for identifying unusual and influential data. Fortunately, there are several texts that present this literature in a more digestible form:[37]

- Although it is now more than a decade old, Cook and Weisberg (1982) is, in my opinion, still the best book-length presentation of methods for assessing leverage, outliers, and influence. There are also good discussions of others problems, such as nonlinearity and transformations of the dependent and independent variables.
- Chatterjee and Hadi (1988) is a thorough and reasonably up-to-date text dealing primarily with influential data and collinearity; other problems—such as nonlinearity and nonconstant error variance—are treated briefly.
- Belsley, et al. (1980) is a seminal text that discusses influential data and the detection of collinearity.[38]
- Barnett and Lewis (1994) present an encyclopedic survey of methods for outlier detection, including methods for detecting outliers in linear models.

[37] Also see the recommended readings given at the end of the following chapter.

[38] I believe that Belsley et al.'s (1980) approach to diagnosing collinearity is fundamentally flawed—see the discussion of collinearity in Chapter 13.

12

Diagnosing Nonlinearity, Nonconstant Error Variance, and Nonnormality

Chapters 11, 12, and 13 show how to detect and correct problems with linear models that have been fit to data. The previous chapter focused on problems with specific observations. The current chapter and the next deal with more general problems with the specification of the model.

The first three sections of this chapter take up the problems of nonnormally distributed errors, nonconstant error variance, and nonlinearity. The treatment here stresses simple graphical methods for detecting these problems, along with transformations of the data to correct problems that are detected.

Subsequent sections describe tests of nonconstant error variance and nonlinearity for discrete independent variables; diagnostic methods based on embedding the usual linear model in a more general nonlinear model that incorporates transformations as parameters; and diagnostics that seek to detect the underlying dimensionality of the regression.

12.1 Nonnormally Distributed Errors

The assumption of normally distributed errors is almost always arbitrary. Nevertheless, the central-limit theorem assures that, under very broad conditions, inference based on the least-squares estimator is approximately valid in all but small samples. Why, then, should we be concerned about nonnormal errors?

- Although the *validity* of least-squares estimation is robust—the levels of tests and confidence intervals are approximately correct in large samples even when the assumption of normality is violated—the *efficiency* of least squares is not robust: Statistical theory assures us that the least-squares estimator is the most efficient unbiased estimator only when the errors are normal. For some types of error distributions, however, particularly those with heavy tails, the efficiency of least-squares estimation decreases markedly. In these cases, the least-squares estimator becomes much less efficient than robust estimators (or least-squares augmented by diagnostics).[1] To a substantial extent, heavy-tailed error distributions are problematic because they give rise to outliers, a problem that I addressed in the previous chapter.

A commonly quoted justification of least-squares estimation—called the Gauss-Markov theorem—states that the least-squares coefficients are the most efficient unbiased estimators that are *linear* functions of the observations Y_i. This result depends on the assumptions of linearity, constant error variance, and independence, but does not require the assumption of normality.[2] Although the restriction to linear estimators produces simple formulas for coefficient standard errors, it is not compelling in the light of the vulnerability of least squares to heavy-tailed error distributions.

- Highly skewed error distributions, aside from their propensity to generate outliers in the direction of the skew, compromise the interpretation of the least-squares fit. This fit is a conditional mean (of Y given the X's), and the mean is not a good measure of the center of a highly skewed distribution. Consequently, we may prefer to transform the data to produce a symmetric error distribution.
- A multimodal error distribution suggests the omission of one or more discrete independent variables that divide the data naturally into groups. An examination of the distribution of the residuals may, therefore, motivate respecification of the model.

Although there are tests for nonnormal errors, I shall instead describe graphical methods for examining the distribution of the residuals, employing univariate displays introduced in Chapter 3.[3] These methods are more useful for pinpointing the character of a problem and for suggesting solutions.

One such graphical display is the quantile comparison plot. We typically compare the sample distribution of the studentized residuals, E_i^*, with the quantiles of the unit-normal distribution, $N(0, 1)$, or with those of the t-distribution for $n - k - 2$ degrees of freedom. Unless n is small, of course, the normal and t-distributions are nearly identical. We choose to plot studentized residuals because they have equal variances and are t-distributed, but, in larger samples, standardized or raw residuals will convey much the same impression.

Even if the model is correct, however, the studentized residuals are not an independent random sample from t_{n-k-2}: Different residuals are correlated with one another.[4] These correlations depend on the configuration of the X-values, but they are generally negligible unless the sample size is small. Furthermore, at the cost of some computation, it is possible to adjust for the dependencies among the residuals in interpreting a quantile comparison plot.[5]

---

[1] Robust estimation is discussed in Section 14.3.

[2] A proof of the Gauss-Markov theorem appears in Section 9.3.2.

[3] See the discussion of Box-Cox transformations in Section 12.5.1, however.

[4] Different residuals are correlated because the off-diagonal entries of the hat-matrix (i.e., h_{ij} for $i \neq j$) are generally nonzero; see Section 11.8.

[5] See Section 12.1.1.

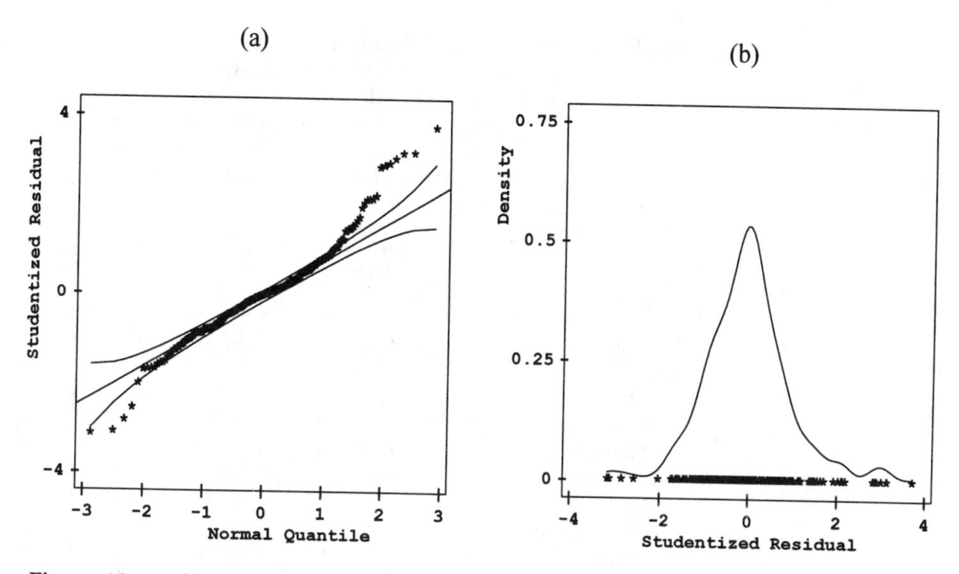

Figure 12.1. The distribution of the studentized residuals from Ornstein's interlocking-directorate regression. A normal quantile comparison plot is shown in (*a*). The 95% confidence envelope is based on the standard errors of the order statistics for an independent normal sample. A nonparametric density estimate is shown in (b).

The quantile comparison plot is particularly effective in displaying the tail behavior of the residuals: Outliers, skewness, heavy tails, or light tails all show up clearly. Other univariate graphical displays effectively supplement the quantile comparison plot. In large samples, a histogram with many bars conveys a good impression of the shape of the residual distribution, and generally reveals multiple modes more clearly than the quantile comparison plot does. In smaller samples, a more stable impression is formed by smoothing the histogram of the residuals with a nonparametric density estimator.

Figure 12.1 shows plots of the studentized residuals for a regression model fit to Ornstein's interlocking-directorate data, first discussed in Chapter 3. The dependent variable in the regression is the number of executive and director interlocks maintained by each of 248 dominant Canadian firms with other companies in this group. The independent variables are the assets of the firm (in millions of dollars), the industrial sector in which the firm operates (10 categories), and the nation in which the firm is controlled (four categories). The results of the regression are shown on the left of Table 12.1.[6] Because the residual degrees of freedom are relatively large (234), the studentized residuals are plotted against the normal distribution in the quantile comparison plot of Figure 12.1(*a*). The quantile comparison plot suggests that the distribution of the residuals has heavy tails—particularly the upper tail. The density estimate [in Figure 12.1(*b*)] suggests that there may be two groups of observations somewhat separated from the others, one group at the low end of the residual distribution, another at the high end.

---

[6] The square root of assets is used in place of assets to make the regression more nearly linear. See Section 12.3 for a discussion of nonlinearity.

TABLE 12.1 Regression of Number of Interlocking Directorate and Executive Positions Maintained by 248 Dominant Canadian Corporations on Corporate Assets, Sector, and Nation of Control. The baseline category for Sector is *Heavy Manufacturing*; for Nation of Control, *Canada*

	Interlocks		$\sqrt{Interlocks + 1}$	
Regressor	*B*	$\widehat{\mathrm{SE}}$	*B*	$\widehat{\mathrm{SE}}$
Constant	4.19	1.85	2.33	0.23
$\sqrt{\text{Assets}}$	0.252	0.019	0.0260	0.0023
Sector				
Wood, paper	5.15	2.68	0.786	0.335
Mining, metals	0.342	2.01	0.356	0.252
Transport	−0.381	2.82	0.354	0.353
Merchandizing	−0.867	2.63	0.148	0.329
Agriculture, food, Light industry	−1.20	2.04	−0.0567	0.255
Holding companies	−2.43	4.01	−0.245	0.502
Construction	−5.13	4.70	−0.740	0.588
Other financials	−5.70	2.93	−0.0880	0.366
Banking	−14.4	5.58	−2.25	0.697
Nation of Control				
Other	−1.16	2.66	−0.114	0.333
Britain	−4.44	2.65	−0.527	0.331
United States	−8.09	1.48	−1.11	0.185
R^2	.655		.580	

A positive skew in the residuals can usually be corrected by moving the dependent variable down the ladder of powers and roots. In the present case, *both* tails of the residual distribution are heavy, but I decided to try power transformations because (1) the upper tail appears heavier than the lower tail; (2) the distribution of number of interlocks (i.e., the dependent variable) is positively skewed; and (3) transformations down the ladder of powers, particularly square root and log, often are effective when the dependent variable is (as here) a count. Because some of the firms maintained 0 interlocks, I used a start of 1 for the transformations. Trial and error suggests that the square-root transformation of *number of interlocks*+1 renders the distribution of the residuals close to normal, as shown in Figure 12.2.

The results of Ornstein's regression using $\sqrt{\text{interlocks} + 1}$ as the dependent variable is shown on the right of Table 12.1. Although we cannot compare the coefficients directly across these two models—the scale of the dependent variable is different in the two cases—the general character of the results does not change much: In both models, assets has a substantial impact on interlocks; the rankings of the nation-of-control categories are identical in the two models; and the rankings of the sectors are nearly the same.[7]

[7] Exercise 12.1 suggests a more precise comparison of the two sets of results using "adjusted" means.

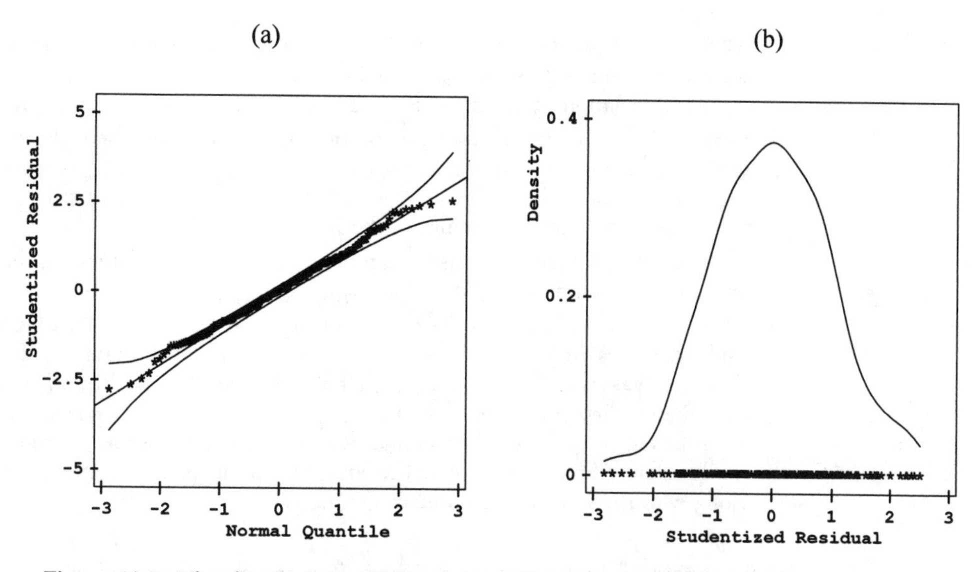

Figure 12.2. The distribution of the studentized residuals from Ornstein's interlocking-directorate regression, after transforming the dependent variable. A normal quantile comparison plot is shown in (*a*), a nonparametric density estimate in (*b*).

> Heavy-tailed errors threaten the efficiency of least-squares estimation; skewed and multimodal errors compromise the interpretation of the least-squares fit. Nonnormality can often be detected by examining the distribution of the least-squares residuals, and frequently can be corrected by transforming the data.

12.1.1 Confidence Envelopes by Simulated Sampling*

Atkinson (1985) has suggested the following procedure for constructing an approximate confidence "envelope" in a quantile comparison plot, taking into account the correlational structure of the independent variables. Atkinson's procedure employs simulated sampling, and uses the assumption of normally distributed errors.[8]

1. Fit the regression model as usual, obtaining fitted values $\hat{Y}_i$ and the estimated standard error S_E.
2. Construct m samples, each consisting of n simulated Y-values; for the jth such sample, the simulated value for observation i is

$$Y_{ij}^s = \hat{Y}_i + S_E Z_{ij}$$

[8] The notion of simulated sampling from a population constructed from the observed data is the basis of "bootstrapping," discussed in Section 16.1. Atkinson's procedure described here is an example of the parametric bootstrap.

where Z_{ij} is a random draw from the unit-normal distribution. In other words, we sample from a "population" in which the expectation of Y_i is $\hat{Y}_i$; the true standard deviation of the errors is S_E; and the errors are normally distributed.

3. Regress the n simulated observations for sample j on the X's in the original sample, obtaining simulated studentized residuals, $E^*_{1j}, E^*_{2j}, \ldots, E^*_{nj}$. Because this regression employs the original X-values, the simulated studentized residuals reflect the correlational structure of the X's.
4. Order the studentized residuals for sample j from smallest to largest, as required by a quantile comparison plot: $E^*_{(1)j}, E^*_{(2)j}, \ldots, E^*_{(n)j}$.
5. To construct an estimated $(100-a)$% confidence interval for $E^*_{(i)}$ (the ith ordered studentized residual), find the $a/2$ and $1 - a/2$ empirical quantiles of the m simulated values $E^*_{(i)1}, E^*_{(i)2}, \ldots, E^*_{(i)m}$. For example, if $m = 20$ and $a = .05$, then the smallest and largest[9] of the $E^*_{(1)j}$ provide a 95% confidence interval for $E^*_{(1)}$: $(E^*_{(1)(1)}, E^*_{(1)(20)})$. The confidence limits for the n ordered studentized residuals are graphed as a confidence envelope on the quantile comparison plot, along with the studentized residuals themselves.

A weakness of Atkinson's procedure is that the probability of *some* studentized residual straying outside of the confidence limits by chance is greater than a, which is the probability that an *individual* studentized residual falls outside of its confidence interval. Because the joint distribution of the studentized residuals is complicated, however, to construct a correct joint-confidence envelope would require even more calculation. As well, in small samples, where there are few residual degrees of freedom, even radical departures from normally distributed errors can give rise to apparently normally distributed residuals; Andrews (1979) presents an example of this phenomenon, which is sometimes termed "supernormality."

EXERCISE

12.1 Use adjusted means (see Exercises 7.5, 7.10, 8.8, and 8.12) to compare the two regressions for Ornstein's interlocking-directorate data summarized in Table 12.1. For the first regression, in which interlocks is the dependent variable, calculate the adjusted mean number of interlocks for each sector and nation of control. For the second regression, in which $\sqrt{\text{interlocks} + 1}$ is the dependent variable, first calculate the adjusted dependent-variable mean for each sector and nation of control, and then translate back to the interlocks scale by squaring and subtracting 1 from each of these quantities. The partial relationship between interlocks and assets for each model

[9] Selecting the smallest and largest of the 20 simulated values corresponds to our simple convention that the proportion of the data below the jth of m order statistics is $(j - 1/2)/m$. Here, $(1 - 1/2)/20 = .025$ and $(20 - 1/2)/20 = .975$, defining 95% confidence limits. Atkinson uses a slightly different convention. To estimate the confidence limits more accurately, it would help to make m larger, and perhaps to use a more sophisticated version of the bootstrap (see Section 16.1), but the approximate nature of the entire enterprise makes it difficult to justify the additional computation that would be required.

can be displayed graphically by setting each dummy variable to its mean (i.e., the proportion in the corresponding category of sector or nation of control) and substituting these values into the regression equation. Then, letting assets run over its range of values (roughly \$50 million to \$150,000 million), substitute $\sqrt{\text{assets}}$ into the regression equation to calculate the corresponding fitted values for the dependent variable, connecting the fitted values by a smooth curve (as in Figure 12.7 on page 314). For the second regression, remember to translate back to the interlocks scale to facilitate the comparison of the two results. (Ornstein's data are in `ornstein.dat`.)

12.2 Nonconstant Error Variance

As we know, one of the assumptions of the regression model is that the variation of the dependent variable around the regression surface—the error variance—is everywhere the same:

$$V(\varepsilon) = V(Y|x_1, \ldots, x_k) = \sigma_\varepsilon^2$$

Nonconstant error variance is sometimes termed "heteroscedasticity." Although the least-squares estimator is unbiased and consistent even when the error variance is not constant, the efficiency of the least-squares estimator is impaired, and the usual formulas for coefficient standard errors are inaccurate—the degree of the problem depending on the degree to which error variances differ. In this section, I shall describe graphical methods for detecting nonconstant error variances, and methods for dealing with the problem when it is detected.[10]

12.2.1 Residual Plots

Because the regression surface is k-dimensional and embedded in a space of $k+1$ dimensions, it is generally impractical to assess the assumption of constant error variance by direct graphical examination of the data when k is larger than 1 or 2. Nevertheless, it is common for error variance to increase as the expectation of Y grows larger, or there may be a systematic relationship between error variance and a particular X. The former situation can often be detected by plotting residuals against fitted values, and the latter by plotting residuals against each X.[11]

Plotting residuals against Y (as opposed to $\hat{Y}$) is generally unsatisfactory, because the plot is "tilted": Because $Y = \hat{Y} + E$, the linear correlation[12] between Y and E is $\sqrt{1 - R^2}$. In contrast, the least-squares fit ensures that the correlation between $\hat{Y}$ and E is precisely 0, producing a plot that is much easier to examine for evidence of nonconstant spread.

---

[10] Tests for heteroscedasticity are discussed in Section 12.4 on discrete data, and in Section 12.5 on maximum-likelihood methods.

[11] These displays are not infallible, however: See Cook (1994), and the discussion in Section 12.6.

[12] See Exercise 12.2.

Because the least-squares *residuals* have unequal variances even when the assumption of constant *error* variance is correct, it is preferable to plot studentized residuals against fitted values. A pattern of changing spread is often more easily discerned in a plot of absolute studentized residuals, $|E_i^*|$, or squared studentized residuals, E_i^{*2}, against $\hat{Y}$. Finally, if the values of $\hat{Y}$ are all positive, then we can plot $\log |E_i^*|$ (log spread) against $\log \hat{Y}$ (log level). A line, with slope b fit to this plot, suggests the variance-stabilizing transformation[13] $Y^{(p)}$, with $p = 1 - b$.

Recall Ornstein's interlocking-directorate regression, described in the previous section. Figure 12.3(a) shows the plot of studentized residuals against fitted values for this regression. Although the substantial positive skew in the fitted values makes the plot difficult to examine, there appears to be a tendency for the residual scatter to get wider at larger values[14] of $\hat{Y}$. The log-spread versus log-level plot for the regression, in Figure 12.3(b), is easier to examine. Because there are negative fitted values, I used $\log(\hat{Y} + 2)$ to construct the plot.[15] The least-squares line fit to the plot has slope $b = 0.497$, suggesting the variance-stabilizing transformation $p = 1 - 0.497 = 0.503$. The positive trend in the spread-versus-level plot translates into a transformation *down* the ladder of powers and roots.

In the previous section, the transformation $\sqrt{\text{interlocks} + 1}$—that is, $p = 0.5$—made the distribution of the studentized residuals more nearly normal. The same transformation nearly stabilizes the residual variance, as illustrated in the spread-versus-level plot shown in Figure 12.4.[16] This outcome is not surprising, because the heavy right tail of the residual distribution and nonconstant spread are both common consequences of the lower bound of 0 for the dependent variable.

Transforming Y changes the shape of the error distribution, but it also alters the shape of the regression of Y on the X's. At times, eliminating nonconstant spread also makes the relationship of Y to the X's more nearly linear, but this is not a necessary consequence of stabilizing the error variance, and it is important to check for nonlinearity following transformation of the dependent variable. Of course, because there is generally no reason to suppose that the regression is linear prior to transforming Y, we should check for nonlinearity in any event.[17]

Nonconstant residual spread sometimes is symptomatic of the omission of important effects from the model. Suppose, for example, that there is an omitted

---

[13] This is an application of Tukey's rule, presented in Section 4.4. Other analytic methods for choosing a variance-stabilizing transformation are discussed in Section 12.5.

[14] Part of the tendency for the residual spread to increase with $\hat{Y}$ is due to the lower bound of 0 for Y: Because $E = Y - \hat{Y}$, the smallest possible residual corresponding to a particular $\hat{Y}$ is $E = 0 - \hat{Y} = -\hat{Y}$; the boundary $E = -\hat{Y}$ is a line with slope -1 at the lower left of the residual versus fitted-value plot. When there are many observations with 0 values, it may be more appropriate to use a Poisson regression model, as described in Section 15.4.

[15] Several observations have negative fitted values, the smallest of which is -1.57.

[16] Figure 12.4 still shows some relationship between spread and level, but the log transformation substantially overcorrects the original problem, inducing a negative association between spread and level. The start of 1 is not really required here, because the square-root transformation is defined for $Y = 0$. In this dataset, using $\sqrt{\text{interlocks}}$, which is a slightly more powerful transformation than $\sqrt{\text{interlocks} + 1}$, nearly perfectly stablilizes the variance of the residuals.

[17] See Section 12.3.

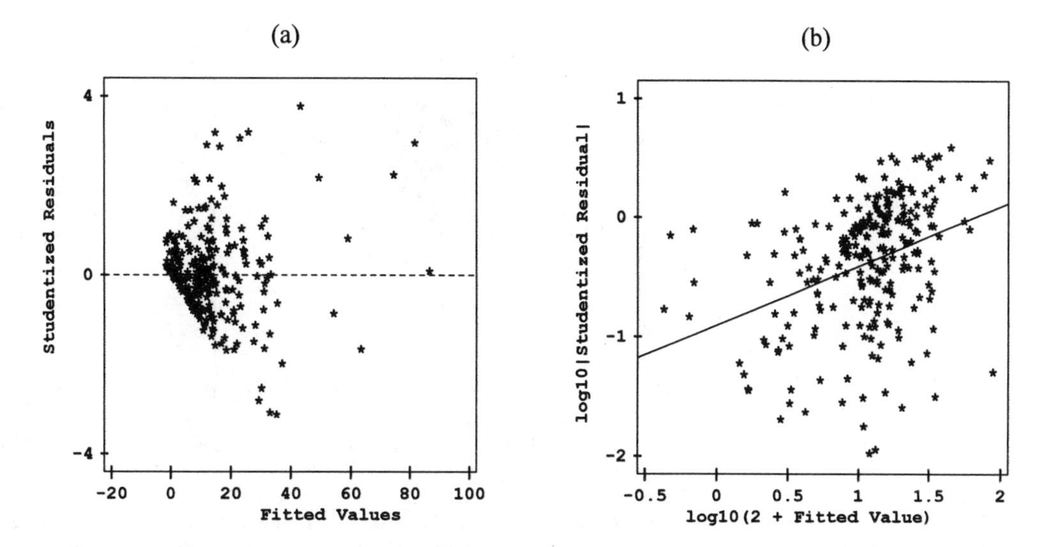

Figure 12.3. Detecting nonconstant spread in Ornstein's interlocking-directorate regression. (*a*) A plot of studentized residuals versus fitted values. (*b*) A plot of log spread (log absolute studentized residuals) versus log level (log fitted values). The least-squares line is shown on the plot.

categorical independent variable, such as regional location, that interacts with assets in affecting interlocks; in particular, suppose that the assets slope, although positive in every region, is steeper in some regions than in others. Then the omission of region and its interaction with assets could produce a fan-shaped residual plot even if the errors from the correct model have constant variance.[18] The detection of this type of specification error requires insight into the process generating the data and cannot rely on diagnostics alone.

12.2.2 Weighted-Least-Squares Estimation*

Weighted-least-squares regression provides an alternative approach to estimation in the presence of nonconstant error variance. Suppose that the errors from the linear regression model $\mathbf{y} = \mathbf{X\beta} + \boldsymbol{\varepsilon}$ are independent and normally distributed, with zero means but *different* variances: $\varepsilon_i \sim N(0, \sigma_i^2)$. Suppose further that the variances of the errors are known up to a constant of proportionality σ_ε^2, so that $V(\varepsilon_i) = \sigma_i^2 = \sigma_\varepsilon^2/w_i^2$. Then, the likelihood for the model is[19]

$$L(\boldsymbol{\beta}, \sigma_\varepsilon^2) = \frac{1}{(2\pi)^{n/2}|\boldsymbol{\Sigma}|^{1/2}} \exp\left[-\frac{1}{2}(\mathbf{y} - \mathbf{X\beta})'\boldsymbol{\Sigma}(\mathbf{y} - \mathbf{X\beta})\right]$$

where $\boldsymbol{\Sigma}$ is the covariance matrix of the errors,

$$\boldsymbol{\Sigma} = \sigma_\varepsilon^2 \times \text{diag}\{1/w_1^2, \ldots, 1/w_n^2\} \equiv \sigma_\varepsilon^2 \times \mathbf{W}^{-1}$$

---

[18] See Exercise 12.3 for an illustration of this phenomenon.

[19] See Exercise 12.4 for this and other results pertaining to weighted-least-squares estimation.

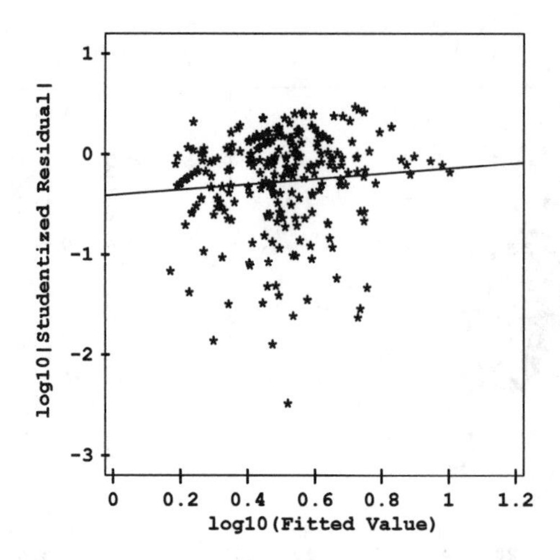

Figure 12.4. Plot of log spread versus log level for Ornstein's interlocking-directorate regression, after transforming the dependent variable. The least-squares line is shown on the plot.

The maximum-likelihood estimators of $\boldsymbol{\beta}$ and σ_ε^2 are

$$\hat{\boldsymbol{\beta}} = (\mathbf{X}'\mathbf{W}\mathbf{X})^{-1}\mathbf{X}'\mathbf{W}\mathbf{y}$$

$$\hat{\sigma}_\varepsilon^2 = \frac{\sum(E_i/w_i)^2}{n}$$

where the residuals E_i are defined in the usual manner. This procedure is equivalent to minimizing the weighted sum of squares $\sum w_i^2 E_i^2$, according greater weight to observations with smaller variance—hence the term *weighted least squares (WLS)*. The estimated asymptotic covariance matrix of $\hat{\boldsymbol{\beta}}$ is given by

$$\hat{\mathscr{V}}(\hat{\beta}) = \hat{\sigma}_\varepsilon^2(\mathbf{X}'\mathbf{W}\mathbf{X})^{-1}$$

In practice, we would need to estimate the weights w_i or know that the error variance is systematically related to some observable variable. In the first instance, for example, we could use the residuals from a preliminary *ordinary-least-squares (OLS)* regression to obtain estimates of the error variance within different categories of the data, partitioned by one or more categorical variables. Basing the weights on a preliminary estimate of error variances can, however, seriously bias the estimated covariance matrix $\hat{\mathscr{V}}(\hat{\beta})$, because the sampling error in the estimates should reflect the additional source of uncertainty.[20]

In the second instance, suppose that inspection of a residual plot for the preliminary OLS fit suggests that the magnitude of the errors is proportional to the first independent variable, X_1. We can then use $1/X_{i1}$ as the weights w_i. Dividing both sides of the regression equation by X_{i1} produces

$$\frac{Y_i}{X_{i1}} = \alpha\frac{1}{X_{i1}} + \beta_1 + \beta_2\frac{X_{i2}}{X_{i1}} + \cdots + \beta_k\frac{X_{ik}}{X_{i1}} + \frac{\varepsilon_i}{X_{i1}} \qquad [12.1]$$

[20] In this case, it is probably better to obtain an honest estimate of the coefficient covariance matrix from the bootstrap, described in Section 16.1.

Because the standard deviations of the errors are proportional to X_1, the "new" errors $\varepsilon_i' \equiv \varepsilon_i/X_{i1}$ have constant variance, and Equation 12.1 can be estimated by OLS regression of Y/X_1 on $1/X_1$, $X_2/X_1, \ldots, X_k/X_1$. Notice that the constant from this regression estimates β_1, while the coefficient of $1/X_1$ estimates α; the remaining coefficients are straightforward.[21]

> It is common for the variance of the errors to increase with the level of the dependent variable. This pattern of nonconstant error variance ("heteroscedasticity") can often be detected in a plot of residuals against fitted values. Strategies for dealing with nonconstant error variance include transformation of the dependent variable to stabilize the variance; the substitution of weighted-least-squares estimation for ordinary least squares; and the correction of coefficient standard errors for heteroscedasticity. A rough rule is that nonconstant error variance seriously degrades the least-squares estimator only when the ratio of the largest to smallest variance is about 10 or more.

12.2.3 Correcting OLS Standard Errors for Nonconstant Variance*

The covariance matrix of the ordinary-least-squares estimator is

$$V(\mathbf{b}) = (\mathbf{X}'\mathbf{X})^{-1}\mathbf{X}'V(\mathbf{y})\mathbf{X}(\mathbf{X}'\mathbf{X})^{-1} \qquad [12.2]$$

Under the standard assumptions, including the assumption of constant error variance, $V(\mathbf{y}) = \sigma_\varepsilon^2\mathbf{I}_n$, Equation 12.2 simplifies to the usual formula, $V(\mathbf{b}) = \sigma_\varepsilon^2(\mathbf{X}'\mathbf{X})^{-1}$. If, however, the errors are heteroscedastic but independent, then $\boldsymbol{\Sigma} \equiv V(\mathbf{y}) = \text{diag}\{\sigma_1^2, \ldots, \sigma_n^2\}$, and

$$V(\mathbf{b}) = (\mathbf{X}'\mathbf{X})^{-1}\mathbf{X}'\boldsymbol{\Sigma}\mathbf{X}(\mathbf{X}'\mathbf{X})^{-1}$$

Because $E(\varepsilon_i) = 0$, the variance of the ith error is $\sigma_i^2 = E(\varepsilon_i^2)$, which suggests the possibility of estimating $V(\mathbf{b})$ by

$$\tilde{V}(\mathbf{b}) = (\mathbf{X}'\mathbf{X})^{-1}\mathbf{X}'\hat{\boldsymbol{\Sigma}}\mathbf{X}(\mathbf{X}'\mathbf{X})^{-1} \qquad [12.3]$$

with $\hat{\boldsymbol{\Sigma}} = \text{diag}\{E_1^2, \ldots, E_n^2\}$, and where E_i is the OLS residual for observation i. White (1980) shows that Equation 12.3 provides a consistent estimator[22] of $V(\mathbf{b})$.

---

[21] An application of WLS regression to Ornstein"s interlocking-directorate data is given in Exercise 12.6.

[22] See Exercise 12.7 for an application of White's heteroscedasticity correction to Ornstein's interlocking-directorate regression.

An advantage of White's approach is that knowledge of the *pattern* of nonconstant error variance (e.g., increased variance with the level of Y or with an X) is not required. If, however, the heteroscedasticity problem is severe, and the corrected coefficient standard errors therefore are substantially larger than those produced by the usual formula, then discovering the pattern of nonconstant variance and correcting for it—by a transformation or WLS estimation—offers the possibility of more efficient estimation. In any event, as the next section shows, unequal error variance is worth correcting only when the problem is severe.

12.2.4 How Nonconstant Error Variance Affects the OLS Estimator*

The impact of nonconstant error variance on the efficiency of the ordinary least-squares estimator and on the validity of least-squares inference depends on several factors, including the sample size, the degree of variation in the σ_i^2, the configuration of the X-values, and the relationship between the error variance and the X's. It is therefore not possible to develop wholly general conclusions concerning the harm produced by heteroscedasticity, but the following simple case is nevertheless instructive.

Suppose that $Y_i = \alpha + \beta X_i + \varepsilon_i$, where the errors are independent and normally distributed, with zero means but with different standard deviations proportional to X, so that $\sigma_i = \sigma_\varepsilon X_i$. Then the OLS estimator B is less efficient than the WLS estimator $\hat{\beta}$, which, under these circumstances, is the most efficient unbiased estimator[23] of β.

Formulas for the sampling variances of B and $\hat{\beta}$ are easily derived.[24] The efficiency of the OLS estimator relative to the optimal WLS estimator is given by $V(\hat{\beta})/V(B)$, and the relative precision of the OLS estimator is the square root of this ratio, that is, $\text{SE}(\hat{\beta})/\text{SE}(B)$.

Now suppose that X is uniformly distributed over the interval $[x_0, ax_0]$, where both x_0 and a are positive numbers, so that a is the ratio of the largest to the smallest value of X (and, consequently, of the largest to the smallest σ_i). The relative precision of the OLS estimator stabilizes quickly as the sample size grows, and exceeds 90% when $a = 2$, and 85% when $a = 3$, even when n is as small as 20. For $a = 10$, the penalty for using OLS is greater, but even here the relative precision of OLS exceeds 65% for $n \geq 20$.

The validity of statistical inferences based on OLS estimation is even less sensitive to common patterns of nonconstant error variance. Here, we need to compare the expectation of the usual estimator of $V(B)$, which is typically biased when the error variance is not constant, with the true sampling variance of B. The square root of $E[\widehat{V(B)}]/V(B)$ expresses the result in relative standard-error terms. For the illustration, where the standard deviation of the errors is proportional to X, and where X is uniformly distributed, this ratio is 98% when $a = 2$; 97% when $a = 3$; and 93% when $a = 10$; all for $n \geq 20$.

The results in this section suggest that nonconstant error variance is a serious problem only when the magnitude (i.e., the standard deviation) of the

---

[23] This property of the WLS estimator requires the assumption of normality. Without normal errors, the WLS estimator is still the most efficient *linear* unbiased estimator—an extension of the Gauss-Markov theorem. See Exercise 12.5.

[24] See Exercise 12.8 for this and other results described in this section.

errors varies by more than a factor of about 3—that is, when the largest error variance is more than about 10 times the smallest.

EXERCISES

12.2 *Show that the correlation between the least-squares residuals E_i and the dependent-variable values Y_i is $\sqrt{1-R^2}$. (*Hint*: Use the geometric vector representation of multiple regression, examining the plane in which the $\mathbf{e}$, $\mathbf{y}^*$, and $\hat{\mathbf{y}}^*$ vectors lie.)

12.3 Nonconstant variance and specification error: Generate 100 observations according to the following model:

$$Y = 10 + (1 \times X) + (1 \times D) + (2 \times X \times D) + \varepsilon$$

where $\varepsilon \sim N(0, 10^2)$; the values of X are $1, 2, \ldots, 50, 1, 2, \ldots, 50$; the first 50 values of D are 0; and the last 50 values of D are 1. Then regress Y on X alone (i.e., omitting D and XD), $Y = A + BX + E$. Plot the residuals E from this regression against the fitted values $\hat{Y}$. Is the variance of the residuals constant? How do you account for the pattern in the plot?

12.4 *Weighted-least-squares estimation: Suppose that the errors from the linear regression model $\mathbf{y} = \mathbf{X}\boldsymbol{\beta} + \boldsymbol{\varepsilon}$ are independent and normally distributed, but with different variances, $\varepsilon_i \sim N(0, \sigma_i^2)$, and that $\sigma_i^2 = \sigma_\varepsilon^2 / w_i^2$. Show that:

(a) The likelihood for the model is

$$L(\boldsymbol{\beta}, \sigma_\varepsilon^2) = \frac{1}{(2\pi)^{n/2}|\boldsymbol{\Sigma}|^{1/2}} \exp\left[-\frac{1}{2}(\mathbf{y} - \mathbf{X}\boldsymbol{\beta})'\boldsymbol{\Sigma}(\mathbf{y} - \mathbf{X}\boldsymbol{\beta})\right]$$

where

$$\boldsymbol{\Sigma} = \sigma_\varepsilon^2 \times \text{diag}\{1/w_1^2, \ldots, 1/w_n^2\} \equiv \sigma_\varepsilon^2 \times \mathbf{W}^{-1}$$

(b) The maximum-likelihood estimators of $\boldsymbol{\beta}$ and σ_ε^2 are

$$\hat{\boldsymbol{\beta}} = (\mathbf{X}'\mathbf{W}\mathbf{X})^{-1}\mathbf{X}'\mathbf{W}\mathbf{y}$$

$$\hat{\sigma}_\varepsilon^2 = \frac{\sum(E_i/w_i)^2}{n}$$

where $\mathbf{e} = \{E_i\} = \mathbf{y} - \mathbf{X}\hat{\boldsymbol{\beta}}$.

(c) The MLE is equivalent to minimizing the weighted sum of squares $\sum w_i^2 E_i^2$.

(d) The estimated asymptotic covariance matrix of $\hat{\boldsymbol{\beta}}$ is given by

$$\hat{\mathscr{V}}(\hat{\beta}) = \hat{\sigma}_\varepsilon^2(\mathbf{X}'\mathbf{W}\mathbf{X})^{-1}$$

12.5 *Show that when the covariance matrix of the errors is

$$\mathbf{\Sigma} = \sigma_\varepsilon^2 \times \text{diag}\{1/w_1^2, \ldots, 1/w_n^2\} \equiv \sigma_\varepsilon^2 \times \mathbf{W}^{-1}$$

the weighted-least-squares estimator

$$\begin{aligned}\hat{\boldsymbol{\beta}} &= (\mathbf{X}'\mathbf{W}\mathbf{X})^{-1}\mathbf{X}'\mathbf{W}\mathbf{y} \\ &= \mathbf{M}\mathbf{y}\end{aligned}$$

is the minimum-variance linear unbiased estimator of $\boldsymbol{\beta}$. (*Hint*: Adapt the proof of the Gauss-Markov theorem for OLS estimation given in Section 9.3.2.)

12.6 *Apply weighted-least-squares estimation to Ornstein's regression of number of interlocking directorates on square-root assets, sector, and nation of control, supposing that the standard deviation of the errors is proportional to the square root of assets. (The OLS regression is reported in Table 12.1; the data are in `ornstein.dat`.) How do the results of WLS estimation compare with those of OLS estimation? With OLS following the square-root transformation of number of interlocks (also shown in Table 12.1)?

12.7 *Using White's correction for nonconstant variance, recalculate coefficient standard errors for Ornstein's OLS regression of number of interlocking directorates on square-root assets, sector, and nation of control (given in Table 12.1). How do the corrected standard-error estimates compare with those computed by the usual approach?

12.8 *The impact of nonconstant error variance on OLS estimation: Suppose that $Y_i = \alpha + \beta x_i + \varepsilon_i$, with independent errors, $\varepsilon_i \sim N(0, \sigma_i^2)$, and $\sigma_i = \sigma_\varepsilon x_i$. Let B represent the OLS estimator and $\hat{\beta}$ the WLS estimator of β.

(a) Show that the sampling variance of the OLS estimator is

$$V(B) = \frac{\sum(X_i - \overline{X})^2\sigma_i^2}{\left[\sum(X_i - \overline{X})^2\right]^2}$$

and that the sampling variance of the WLS estimator is

$$V(\hat{\beta}) = \frac{\sigma_\varepsilon^2}{\sum w_i^2(X_i - \tilde{X})^2}$$

where $\tilde{X} \equiv (\sum w_i^2 X_i)/(\sum w_i^2)$. (*Hint*: Write each slope estimator as a linear function of the Y_i.)

(b) Now suppose that x is uniformly distributed over the interval $[x_0, ax_0]$, where $x_0 > 0$ and $a > 0$, so that a is the ratio of the largest to the smallest σ_i. The efficiency of the OLS estimator relative to the optimal WLS estimator is $V(\hat{\beta})/V(B)$, and the relative precision of the OLS estimator is the square root of this ratio, that is, $\text{SE}(\hat{\beta})/\text{SE}(B)$. Calculate the relative precision of the OLS estimator for all combinations of a = 2, 3, 5, 10, and n = 5, 10, 20, 50, 100. For example, when $a = 3$ and $n = 10$, you can take the x-values as 1, 1.222, 1.444, ..., 2.778, 3. Under what circumstances is the OLS estimator substantially less precise than the WLS estimator?

(c) The usual variance estimate for the OLS slope (assuming constant error variance) is

$$\widehat{V(B)} = \frac{S_E^2}{\sum(X_i - \overline{X})^2}$$

where $S_E^2 = \sum E_i^2/(n-2)$. Kmenta (1986, Section 8.2) shows that the expectation of this variance estimator (under nonconstant error variance σ_i^2) is

$$E[\widehat{V(B)}] = \frac{\overline{\sigma}^2}{\sum(X_i - \overline{X})^2} - \frac{\sum(X_i - \overline{X})^2(\sigma_i^2 - \overline{\sigma}^2)}{(n-2)[\sum(X_i - \overline{X})^2]^2}$$

where $\overline{\sigma}^2 \equiv \sum \sigma_i^2/n$. (*Prove this result.) Kmenta also shows that the true variance of the OLS slope estimator, $V(B)$ [derived in part (a)], is generally different from $E[\widehat{V(B)}]$. If $\sqrt{E[\widehat{V(B)}]/V(B)}$ is substantially below 1, then the usual formula for the standard error of B will lead us to believe that the OLS estimator is more precise than it really is. Calculate $\sqrt{E[\widehat{V(B)}]/V(B)}$ under the conditions of part (b), for $a = 5, 10, 20, 50$, and $n = 5, 10, 20, 50, 100$. What do you conclude about the robustness of validity of OLS inference with respect to nonconstant error variance?

12.3 Nonlinearity

The assumption that the average error, $E(\varepsilon)$, is everywhere 0 implies that the specified regression surface accurately reflects the dependency of the conditional average value of Y on the X's. Conversely, violating the assumption of linearity implies that the model fails to capture the systematic pattern of relationship between the dependent and independent variables. The term "nonlinearity," therefore, is not used in the narrow sense here, although it includes the possibility that a partial relationship assumed to be linear is, in fact, nonlinear: If, for example, two independent variables specified to have additive effects instead interact, then the average error is not 0 for all combinations of X-values.

If nonlinearity, in the broad sense, is slight, then the fitted model can be a useful approximation even though the regression surface $E(Y|X_1, \ldots X_k)$ is not captured precisely. In other instances, however, the model can be seriously misleading.

The regression surface is generally high dimensional, even after accounting for regressors (such as dummy variables, interactions, and polynomial terms) that are functions of a smaller number of fundamental independent variables.[25] As in the case of nonconstant error variance, therefore, it is necessary to focus on particular patterns of departure from linearity. The graphical diagnostics discussed in this section are two-dimensional (and three-dimensional) projections of the $(k+1)$-dimensional point cloud of observations $\{Y_i, X_{i1}, \ldots, X_{ik}\}$.

[25] Polynomial regression—for example, the model $Y = \alpha + \beta_1 X + \beta_2 X^2 + \varepsilon$—is discussed in Sections 12.4 and 14.2.1. In this simple quadratic model, there are two regressors, but only one independent variable.

12.3.1 Partial-Residual Plots

Although it is useful in multiple regression to plot Y against each X (e.g., in one row of a scatterplot matrix), these plots often do not tell the whole story—and can be misleading—because our interest centers on the *partial* relationship between Y and each X (controlling for the other X's), not on the *marginal* relationship between Y and an individual X (ignoring the other X's). Residual-based plots are consequently more promising in the specific context of multiple regression.

Plotting residuals or studentized residuals against each X, perhaps augmented by a nonparametric-regression smooth, is frequently helpful for detecting departures from linearity. As Figure 12.5 illustrates, however, simple residual plots cannot distinguish between monotone and nonmonotone nonlinearity. This distinction is lost in the residual plots because the least-squares fit ensures that the residuals are linearly uncorrelated with each X. The distinction is important because monotone nonlinearity frequently can be "corrected" by simple transformations.[26] In Figure 12.5, for example, case (*b*) might be modeled by $Y = \alpha + \beta\sqrt{X} + \varepsilon$, while case (*a*) cannot be linearized by a power transformation of X, and might instead be dealt with by the quadratic regression,[27] $Y = \alpha + \beta_1 X + \beta_2 X^2 + \varepsilon$.

In contrast to simple residual plots, partial-regression plots, introduced in the previous chapter for detecting influential data, can reveal nonlinearity and suggest whether a relationship is monotone. These plots are not always useful for locating a transformation, however: The partial-regression plot adjusts X_j for the other X's, but it is the unadjusted X_j that is transformed in respecifying the model. The similarly named *partial-residual plots*, also called *component-plus-residual plots*, are often an effective alternative. Partial-residual plots are not as suitable as partial-regression plots for revealing leverage and influence.[28]

Define the partial residual for the jth independent variable as

$$E_i^{(j)} = E_i + B_j X_{ij}$$

In words, add back the linear component of the partial relationship between Y and X_j to the least-squares residuals, which may include an unmodeled nonlinear component. Then plot $E^{(j)}$ versus X_j. By construction, the multiple-regression coefficient B_j is the slope of the simple linear regression of $E^{(j)}$ on X_j, but nonlinearity may be apparent in the plot as well. Again, a nonparametric regression may help in interpreting the plot.

The partial-residual plots in Figure 12.6 are for a regression of the rated prestige (P) of 102 Canadian occupations on the average education (S—"schooling") in years, average income (I) in dollars, and percentage of women

---

[26] Recall the material in Section 4.3 on linearizing transformations.

[27] Case (*b*) could, however, be accommodated by a more complex transformation of X, of the form $Y = \alpha + \beta(X - \gamma)^\lambda + \varepsilon$. In the illustration, γ could be taken as $\overline{X}$, and λ as 2. More generally, γ and λ could be estimated from the data, for example, by nonlinear least squares (as described in Section 14.2.3). I shall not pursue this approach here.

[28] The argument that partial-residual plots are more suitable than partial-regression plots for diagnosing nonlinearity reflects common experience and advice, but it does not hold in every instance. See Cook (1996).

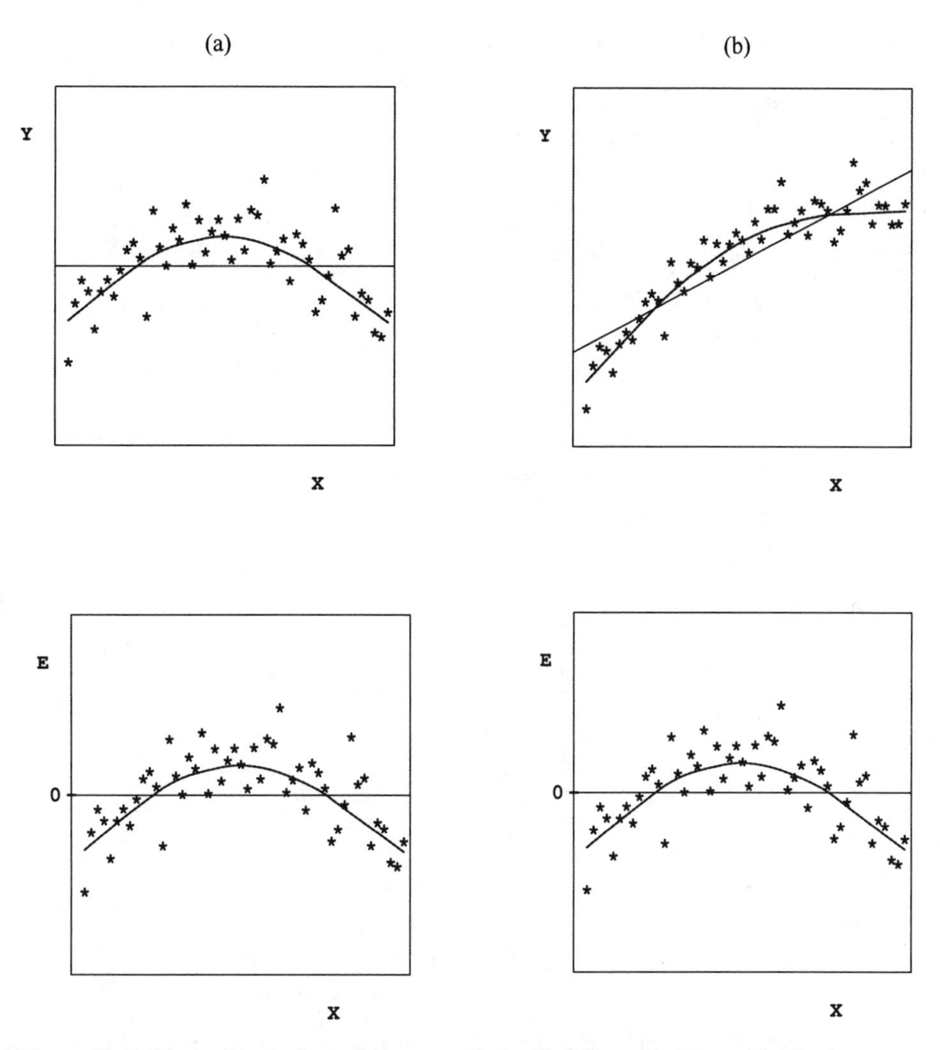

Figure 12.5. The residual plots of E versus X (in the lower panels) are identical, even though the regression of Y on X in (a) is nonmonotone while that in (b) is monotone.

(W) in the occupations in 1971.[29] A nonparametric-regression smooth is shown on each of the plots. The results of the regression are as follows:

$$\underset{}{\hat{P}} = \underset{(3.24)}{-6.79} + \underset{(0.39)}{4.19S} + \underset{(0.00028)}{0.00131I} - \underset{(0.0304)}{0.00891W}$$

$$R^2 = 0.80 \qquad S_E = 7.85$$

There is apparent monotone nonlinearity in the partial-residual plots for education [Figure 12.6(a)] and, much more strongly, income [Figure 12.6(b)]; there is also a small apparent tendency [in Figure 12.6(c)] for occupations with

[29] This regression was first fit in Section 5.2.2.

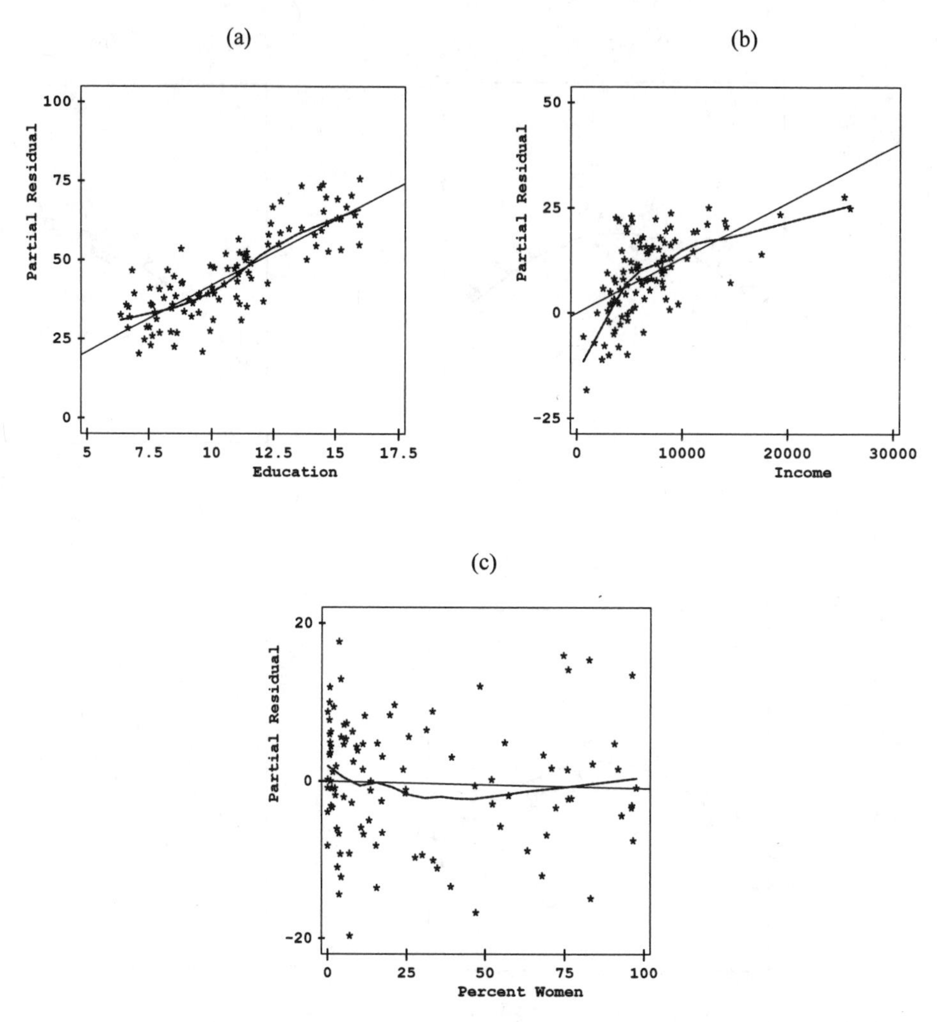

Figure 12.6. Partial-residual plots for the regression of occupational prestige on (*a*) education, (*b*) income, and (*c*) percentage of women. The data are for 102 Canadian occupations in 1971. The least-squares line and a nonparametric-regression smooth are shown on each plot.

intermediate percentages of women to have lower prestige, controlling for education and income, as if occupations with a gender mix pay a small penalty in prestige. To my eye, the patterns in the partial-residual plots for education and percentage of women are not easily discernible without the nonparametric-regression smooth: The departures from linearity in these plots are not great.

The nonlinear pattern for income is simple as well as monotone, suggesting a power transformation; because the bulge points upward and toward the left, we can try to transform prestige *up* the ladder of powers and roots or income *down*. In multiple regression, we are generally loath to transform Y (as opposed to an X), because the relationship between Y and every X would be affected—unless, of course, there is a similar nonlinear pattern to the relationship between

Y and all of the X's. Some experimentation indicates that a log transformation of income straightens its partial relationship to prestige.

The nonlinear pattern for education, in contrast, is monotone (or nearly monotone) but not simple, making a power transformation of education unpromising.[30] The S-shaped pattern in Figure 12.6(a) can be captured, however, by a cubic regression in education. The nonlinear pattern for percentage of women in Figure 12.6(c) is simple but not monotone, suggesting a quadratic regression rather than a power transformation of percentage of women.

The revised fit is as follows:

$$\begin{aligned}\hat{P} = \underset{(56.9)}{20.8} + \underset{(1.27)}{8.78}\log_2 I - \underset{(0.085)}{0.179}W + \underset{(0.00092)}{0.00250}W^2 \\ -\underset{(15.3)}{29.9}S + \underset{(1.41)}{2.91}S^2 - \underset{(0.042)}{0.0807}S^3\end{aligned}$$

$$R^2 = 0.86 \qquad S_E = 6.72$$

- The quadratic term for percentage of women is "statistically significant," but the partial effect of this independent variable is relatively small, ranging from a minimum of −3.2 prestige points, for a hypothetical occupation with 32% women, to a maximum of 7.1 points, for a hypothetical occupation consisting entirely of women.[31]
- The partial effect of education is substantial, in contrast, but the departure from linearity is not great, except at very low levels of education (and the coefficient for S^3 is not quite statistically significant at the 5% level, two tailed). Figure 12.7 traces the partial effect of education on prestige, setting the other two independent variables to their average levels, illustrating how a nonlinear relationship can be presented graphically. The curve plotted in the figure is

$$\begin{aligned}\hat{P} = 20.8 + 8.78 \log_2 6798 - 0.179 \times 29.0 + 0.00250 \times 29.0^2 \\ - 29.9S + 2.91S^2 - 0.0807S^3\end{aligned} \qquad [12.4]$$

  where 6798 and 29.0 are, respectively, the means of income and percentage of women. The points in the plot are partial residuals from the cubic education fit, obtained by adding the least-squares residuals to the fitted values determined by Equation 12.4.
- Income also has a large partial effect: Doubling income is associated, on average, with about a 9-point increment in prestige.

In summary, although the small quadratic effect of percentage of women is substantively interesting, not much is gained by including percentage of women in the model. Likewise, little is gained by modeling the effect of education as a third-degree polynomial, even though the leveling off of prestige at low education is potentially interesting. The transformation of income, however, is compelling.

---

[30] Because, for most of the data, the "bulge" points down and to the right, the transformation $S \to S^2$ does help to straighten the regression.

[31] These numbers are determined by finding the minimum and maximum of $f(W) = -0.179W + 0.00250W^2$ for $0 \leq W \leq 100$.

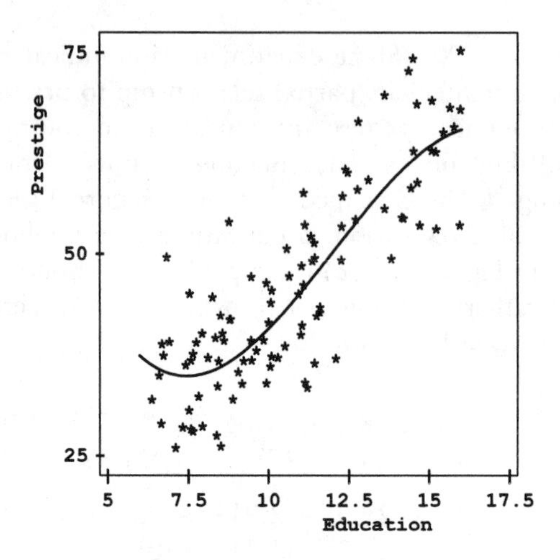

Figure 12.7. The partial relationship between prestige and education, holding income and percentage of women at their average levels. The curve shows the cubic fit for education. The points are partial residuals, obtained by adding the least-squares residuals to the education fit.

12.3.2 When Do Partial-Residual Plots Work?

Circumstances under which regression plots, including partial-residual plots, are informative about the structure of data are an active area of statistical research.[32] It is unreasonable to expect that lower-dimensional displays can always uncover structure in a higher-dimensional problem. We may, for example, discern an interaction between two independent variables in a three-dimensional scatterplot, but could not in two separate two-dimensional plots, one for each independent variable.

It is important, therefore, to understand when graphical displays work and why they sometimes fail: First, understanding the circumstances under which a plot is effective may help us to produce those circumstances. Second, understanding why plots succeed and why they fail may help us to construct more effective displays. Both of these aspects will be developed below.

To provide a point of departure for this discussion, imagine that the following model accurately describes the data:

$$Y_i = \alpha + f(X_{i1}) + \beta_2 X_{i2} + \cdots + \beta_k X_{ik} + \varepsilon_i \qquad [12.5]$$

That is, the partial relationship between Y and X_1 is (potentially) nonlinear, characterized by the function $f(X_1)$, while the other independent variables, $X_2, \ldots, X_k$, enter the model linearly.

We do not know in advance the shape of the function $f(X_1)$, and indeed do not know that the partial relationship between Y and X_1 is nonlinear. Instead

---

[32] Much of this work is due to Cook and his colleagues; see, in particular, Cook (1993), on which the current section is based, and Cook (1994). Cook and Weisberg (1994) provide an accessible summary.

of fitting the true model (Equation 12.5) to the data, therefore, we fit the "working model":

$$Y_i = \alpha' + \beta'_1 X_{i1} + \beta'_2 X_{i2} + \cdots + \beta'_k X_{ik} + \varepsilon'_i$$

The primes indicate that the estimated coefficients for this model do not, in general, estimate the parameters of the true model (Equation 12.5), nor is the "error" of the working model the same as the error of the true model.

Suppose, now, that we construct a partial-residual plot for the working model. The partial residuals estimate

$$\varepsilon_i^{(1)} = \beta'_1 X_{i1} + \varepsilon'_i \tag{12.6}$$

What we would really like to estimate, however, is $f(X_{i1})+\varepsilon_i$, which, apart from random error, will tell us the partial relationship between Y and X_1. Cook (1993) shows that $\varepsilon_i^{(1)} = f(X_{i1}) + \varepsilon_i$, as desired, under either of two circumstances:

1. The function $f(X_1)$ is linear after all, in which case the population analogs of the partial residuals in Equation 12.6 are appropriately linearly related to X_1.
2. The *other* independent variables $X_2, \ldots, X_k$ are each linearly related to X_1. That is,

$$E(X_{ij}) = \alpha_{j1} + \beta_{j1} X_{i1} \quad \text{for } j = 2, \ldots, k \tag{12.7}$$

   If, in contrast, there are *nonlinear* relationships between the other X's and X_1, then the partial-residual plot for X_1 may not reflect the true partial regression $f(X_1)$.[33]

The second result suggests a practical procedure for improving the chances that partial-residual plots will provide accurate evidence of nonlinearity: If possible, transform the independent variables to linearize the relationships among them. Evidence suggests that weak nonlinearity is not especially problematic, but strong nonlinear relationships among the independent variables can invalidate the partial-residual plot as a useful diagnostic display.[34]

> Simple forms of nonlinearity can often be detected in partial-residual plots. Once detected, nonlinearity can frequently be accommodated by variable transformations or by altering the form of the model (to include a quadratic term in an independent variable, for example). Partial-residual plots adequately reflect nonlinearity when the independent variables are themselves linearly related.

[33] Notice that each of the other X's is regressed on X_1, not vice versa.

[34] See Exercise 12.12.

Mallows (1986) has suggested a variation on the partial-residual plot that sometimes reveals nonlinearity more clearly. I shall focus on X_1, but the spirit of Mallows's suggestion is to construct a plot for each X in turn. First, construct a working model with a quadratic term in X_1 along with the usual linear term:

$$Y_i = \alpha' + \beta_1' X_{i1} + \gamma_1 X_{i1}^2 + \beta_2' X_{i2} + \cdots + \beta_k' X_{ik} + \varepsilon_i'$$

Then, after fitting the model, form the "augmented" partial residual

$$E_i'^{(1)} = E_i' + B_1' X_{i1} + C_1 X_{i1}^2$$

Note that B_1' generally differs from the regression coefficient for X_1 in the original model, which does not include the squared term. Finally, plot $E'^{(1)}$ versus X_1.

The circumstances under which the augmented partial residuals accurately capture the true partial-regression function $f(X_1)$ are closely analogous to the linear case (see Cook, 1993); either

1. the function $f(X_1)$ is a quadratic in X_1, or
2. the regressions of the other independent variables on X_1 are quadratic:

$$E(X_{ij}) = \alpha_{j1} + \beta_{j1} X_{i1} + \gamma_{j1} X_{i1}^2 \quad \text{for } j = 2, \ldots, k \qquad [12.8]$$

This is a potentially useful result if we cannot transform away nonlinearity among the independent variables—as is the case, for example, when the relationships among the independent variables are not monotone.

The premise of this discussion, expressed in Equation 12.5, is that Y is a nonlinear function of X_1, but linearly related to the other X's. In real applications of partial-residual plots, however, it is quite possible that there is more than one nonlinear partial relationship, and we typically wish to examine each independent variable in turn. Suppose, for example, that the relationship between Y and X_1 is linear; that the relationship between Y and X_2 is nonlinear; and that X_1 and X_2 are correlated. The partial-residual plot for X_1 can, in this situation, show apparent nonlinearity—sometimes termed a "leakage" effect. If more than one partial-residual plot shows evidence of nonlinearity, it may, therefore, be advisable to refit the model and reconstruct the partial-residual plots after correcting the most dramatic instance of nonlinearity.[35]

[35] Exercise 12.9 applies this procedure to the Canadian occupational prestige regression. An iterative formalization of the procedure provides the basis for nonparametric additive regression models, discussed in Section 14.4.2.

*CERES Plots**

Cook (1993) provides a still more general procedure, which he calls *CERES* (for "Combining conditional Expectations and RESiduals"): Let

$$\hat{X}_{ij} = \hat{g}_{j1}(X_{i1})$$

represent the estimated regression of X_j on X_1, for $j = 2, \ldots, k$. These regressions may be linear (as in Equation 12.7), quadratic, (as in Equation 12.8), or they may be nonparametric. Of course, the function $\hat{g}_{j1}(X_1)$ will generally be different for different X_j's. Once the regression functions for the other independent variables are found, form the working model

$$Y_i = \alpha'' + \beta_2''X_{i2} + \cdots + \beta_k''X_{ik} + \gamma_{12}\hat{X}_{i2} + \cdots + \gamma_{1k}\hat{X}_{ik} + \varepsilon_i''$$

The residuals from this model are then combined with the estimates of the γ's:

$$E_i''^{(1)} = E_i'' + C_{12}\hat{X}_{i2} + \cdots + C_{1k}\hat{X}_{ik}$$

and plotted against X_1.

EXERCISES

12.9 The partial-residual plot for the Canadian occupational prestige data showing the most severe nonlinearity is the plot for income (see Figure 12.6). Reconstruct the three partial-residual plots *after* transforming income. Do the resulting plots for education and percentage of women differ substantially from those shown in Figure 12.6(a) and (c)? (The data are in `prestige.dat`.)

12.10 Apply Mallows's procedure to construct augmented partial-residual plots for the Canadian occupational prestige regression. (The data are in `prestige.dat`.) *Then apply Cook's CERES procedure to this regression. Compare the results of these two procedures with each other and with the ordinary partial-residual plots shown in Figure 12.6. Do the more complex procedures give clearer indications of nonlinearity in this case?

12.11 Consider the following alternative analysis of the Canadian occupational prestige data: Regress prestige on income, education, percentage of women, and on dummy regressors for type of occupation (professional and managerial, white collar, blue collar); include interactions between type of occupation and each of income, education, and percentage of women. Why is it that the interaction between income and type of occupation can induce a nonlinear relationship between prestige and income when the interaction is ignored? (*Hint*: Construct a scatterplot of prestige versus income, labeling the points in the plot by occupational type, and plotting the separate regression line for each occupational type.)

12.12 Experimenting with partial-residual plots: Generate random samples of 100 observations according to each of the following schemes. In each case, construct the partial-residual plots for X_1 and X_2. Do these plots accurately capture the partial relationships between Y and each of X_1 and X_2? Whenever they appear, E and U are $N(0, 1)$ and independent of each other and of the other variables.

(a) Independent X's and a linear regression: X_1 and X_2 independent and uniformly distributed on the interval $[0, 1]$; $Y = X_1 + X_2 + 0.1E$.

(b) Linearly related X's and a linear regression: X_1 uniformly distributed on the interval $[0, 1]$; $X_2 = X_1 + 0.1U$; $Y = X_1 + X_2 + 0.1E$.

(c) Independent X's and a nonlinear regression on one X: X_1 and X_2 independent and uniformly distributed on the interval $[0, 1]$; $Y = 2(X_1 - 0.5)^2 + X_2 + 0.1E$.

(d) Linearly related X's and a nonlinear regression on one X: X_1 uniformly distributed on the interval $[0, 1]$; $X_2 = X_1 + 0.1U$; $Y = 2(X_1 - 0.5)^2 + X_2 + 0.1E$. (Note the "leakage" here from X_1 to X_2.)

(e) Nonlinearly related X's and a linear regression: X_1 uniformly distributed on the interval $[0, 1]$; $X_2 = |X_1 - 0.5|$; $Y = X_1 + X_2 + 0.02E$.

(f) Nonlinearly related X's and a linear regression on one X: X_1 uniformly distributed on the interval $[0, 1]$; $X_2 = |X_1 - 0.5|$; $Y = 2(X_1 - 0.5)^2 + X_2 + 0.02E$. (Note how strong a nonlinear relationship between the X's, and how small an error variance in the regression, are required for the effects in this example to be noticeable.)

12.4 Discrete Data

As explained in Chapter 3, discrete independent and dependent variables often lead to plots that are difficult to interpret, a problem that can be partially rectified by "jittering" the plotted points.[36] A discrete *dependent* variable also violates the assumption that the errors in a linear model are normally distributed. This problem, like that of a limited dependent variable (i.e., one that is bounded below or above), is only serious in extreme cases—for example, when there are very few response categories, or where a large proportion of the data is in a small number of categories, conditional on the values of the independent variables. In these cases, it is best to use statistical models for categorical dependent variables.[37]

Discrete *independent* variables, in contrast, are perfectly consistent with the general linear model, which makes no distributional assumptions about the X's, other than independence between the X's and the errors. Indeed, because it partitions the data into groups, a discrete X (or combination of X's) facilitates straightforward tests of nonlinearity and nonconstant error variance.

12.4.1 Testing for Nonlinearity ("Lack of Fit")

Recall the data on vocabulary and education collected in the 1989 General Social Survey.[38] Years of education in this dataset range between 0 and 20.

[36] See Section 3.2.
[37] See Chapter 15.
[38] This dataset is described in Section 3.2.

TABLE 12.2 Analysis of Variance for Vocabulary-Test Scores, Showing the Incremental *F*-Test for Nonlinearity of the Relationship Between Vocabulary and Education

Source	*df*	*SS*	*F*	*p*
Education (*Model 12.10*)	19	1261.7	18.1	<<.0001
Linear (*Model 12.9*)	1	1175.1	320.0	<<.0001
Nonlinear (*"lack of fit"*)	18	86.58	1.31	.17
Error (*"pure error"*)	948	3472.8		
Total	967	4734.5		

Suppose that we model the relationship between vocabulary score and education in two ways:

1. Fit a linear regression of vocabulary on education:

$$Y_i = \alpha + \beta X_i + \varepsilon_i \qquad [12.9]$$

2. Model education with a set of dummy regressors. Although there are 21 conceivable education scores, none of the individuals in the sample has 2 years of education, yielding 20 categories and 19 dummy regressors (treating 0 years of education as the baseline category):

$$Y_i = \alpha' + \gamma_1 D_{i1} + \gamma_3 D_{i3} + \cdots + \gamma_{20} D_{i,20} + \varepsilon_i' \qquad [12.10]$$

Contrasting the two models produces a test for nonlinearity, because the model in Equation 12.9, specifying a linear relationship between vocabulary and education, is a special case of the model given in Equation 12.10, which can capture *any* pattern of relationship between $E(Y)$ and X. The resulting incremental F-test for nonlinearity appears in Table 12.2. There is, therefore, very strong evidence of a *linear* relationship between vocabulary and education, but little evidence of nonlinearity.

The incremental F-test for nonlinearity can easily be extended to a discrete independent variable—say X_1—in a multiple-regression model. Here, we need to contrast the general model:

$$Y_i = \alpha + \gamma_1 D_{i1} + \cdots + \gamma_{m-1} D_{i,m-1} + \beta_2 X_{i2} + \cdots + \beta_k X_{ik} + \varepsilon_i$$

with the model specifying a linear effect of X_1:

$$Y_i = \alpha \ + \beta_1 X_{i1} + \beta_2 X_{i2} + \cdots + \beta_k X_{ik} + \varepsilon_i$$

where $D_1, \ldots, D_{m-1}$ are dummy regressors constructed to represent the m distinct values of X_1.

Another approach to testing for nonlinearity exploits the fact that a polynomial of degree $m - 1$ can perfectly capture the relationship between Y and a discrete X with m categories, regardless of the specific form of this relationship. We remove one term at a time from the model

$$Y_i = \alpha + \beta_1 X_i + \beta_2 X_i^2 + \cdots + \beta_{m-1} X_i^{m-1} + \varepsilon_i$$

beginning with X^{m-1}. If the decrement in the regression sum of squares is nonsignificant (by an incremental F-test on 1 degree of freedom), then we proceed to remove X^{m-2}, and so on.[39] This approach has the potential advantage of parsimony, because we may well require more than one term (i.e., a linear relationship) but fewer than $m-1$ (i.e., a relationship of arbitrary form). High-degree polynomials, however, are usually difficult to interpret.[40]

12.4.2 Testing for Nonconstant Error Variance

A discrete X (or combination of X's) partitions the data into m groups (as in analysis of variance). Let Y_{ij} denote the ith of n_j dependent-variable scores in group j. If the error variance is constant across groups, then the within-group sample variances

$$S_j^2 = \frac{\sum_{i=1}^{n_j}(Y_{ij} - \overline{Y}_j)^2}{n_j - 1}$$

should be similar. Tests that examine the S_j^2 directly, such as Bartlett's (1937) classic (and commonly employed) test, do not maintain their validity well when the distribution of the errors is nonnormal.

Many alternative tests have been proposed. In a large-scale simulation study, Conover et al. (1981) found that the following simple F-test (called "Levine's test") is both robust and powerful: Calculate the values

$$Z_{ij} \equiv |Y_{ij} - \tilde{Y}_j|$$

where $\tilde{Y}_j$ is the median dependent-variable value in group j. Then perform a one-way analysis of variance of the Z_{ij} over the m groups. If the error variance is not constant across the groups, then the group means $\overline{Z}_j$ will tend to differ, producing a large value of the F-test statistic.[41]

For the vocabulary data, for example, where education partitions the 968 observations into $m = 20$ groups, this test gives $F_0 = 1.48$, with 19 and 948

[39] As usual, the estimate of error variance in the denominator of these F-tests is taken from the full model with all $m - 1$ terms.

[40] *There is a further, technical difficulty with this procedure: The several powers of X are usually highly correlated, sometimes to the point that least-squares calculations break down. A solution is to orthogonalize the power regressors prior to fitting the model: Let X^{2*} represent the residual from the regression (i.e., orthogonal projection) of X^2 on X; let X^{3*} represent the residual from the regression of X^3 on X and X^{2*}; and so on. The set of regressors $X, X^{2*}, \ldots, X^{m-1*}$ is orthogonal and spans the same subspace as the original set of powers. Because the new regressors are orthogonal, it is no longer necessary to fit successively smaller models sequentially; t-tests for the individual terms in the full model provide the same results as sequential incremental F-tests.

[41] This test ironically exploits the robustness of the F-test in one-way ANOVA. (The irony lies in the common use of tests of constant variance as a preliminary to tests of differences in means.)

degrees of freedom, for which $p = .08$. There is, therefore, weak evidence of nonconstant spread in vocabulary across the categories of education.

> Discrete independent variables divide the data into groups. A simple incremental F-test for nonlinearity compares the sum of squares accounted for by the linear regression of Y on X with the sum of squares accounted for by differences in the group means. Likewise, tests of nonconstant variance can be based on comparisons of spread in the different groups.

EXERCISE

12.13 Recall (from Section 8.2) Moore and Krupat's analysis of variance of conformity by authoritarianism and partner's status. The data are in Table 8.3 and `moore.dat`.

(a) Treating the three categories of authoritarianism as evenly spaced, fit a model to the data that incorporates the linear effect of this factor (e.g., coding the categories as 1, 2, and 3). Include the interaction between authoritarianism and partner's status in the model. Compare this model with the standard two-way ANOVA model to determine whether there is a significant departure from linearity.

(b) Test for nonconstant variance across the six cells of Moore and Krupat's design.

12.5 Maximum-Likelihood Methods*

A statistically sophisticated approach to selecting a transformation of Y or an X is to embed the usual linear model in a more general nonlinear model that contains a parameter for the transformation. If several variables are potentially to be transformed, or if the transformation is complex, then there may be several such parameters.[42]

Suppose that the transformation is indexed by a single parameter λ (e.g., $Y \rightarrow Y^{\lambda}$), and that we can write down the likelihood for the model as a function of the transformation parameter and the usual regression parameters: $L(\lambda, \alpha, \beta_1, \ldots, \beta_k, \sigma_\varepsilon^2)$. Maximizing the likelihood yields the maximum-likelihood estimate of λ along with the MLEs of the other parameters. Now

[42] Models of this type are fundamentally nonlinear, and can be treated by the general methods of Section 14.2.3, as well as by the methods described in the present section.

suppose that $\lambda = \lambda_0$ represents *no* transformation (e.g., $\lambda_0 = 1$ for the power transformation Y^λ). A likelihood-ratio test, Wald test, or score test of H_0: $\lambda = \lambda_0$ assesses the evidence that a transformation is required.

A disadvantage of the likelihood-ratio and Wald tests in this context is that they require finding the MLE, which usually necessitates iteration (i.e., a repetitive process of successively closer approximations). In contrast, the slope of the log likelihood at λ_0—on which the score test depends—generally can be assessed or approximated without iteration, and therefore is faster to compute.

Often, the score test can be formulated as the t-statistic for a new regressor, called a *constructed variable*, to be added to the linear model. A partial-regression plot for the constructed variable then can reveal whether one or a small group of observations is unduly influential in determining the transformation, or, alternatively, whether evidence for the transformation is spread throughout the data.

12.5.1 Box-Cox Transformation of *Y*

Box and Cox (1964) have suggested a power transformation of Y with the object of normalizing the error distribution, stabilizing the error variance, and straightening the relationship of Y to the X's.[43] The general Box-Cox model is

$$Y_i^{(\lambda)} = \alpha + \beta_1 X_{i1} + \cdots + \beta_k X_{ik} + \varepsilon_i$$

where the errors ε_i are independently $N(0, \sigma_\varepsilon^2)$, and

$$Y_i^{(\lambda)} = \begin{cases} \dfrac{Y_i^\lambda - 1}{\lambda} & \text{for } \lambda \neq 0 \\ \log_e Y_i & \text{for } \lambda = 0 \end{cases}$$

Note that all of the Y_i must be positive.[44]

For a particular choice of λ, the conditional maximized log likelihood is[45]

$$\log_e L(\alpha, \beta_1, \ldots, \beta_k, \sigma_\varepsilon^2 | \lambda) = -\frac{n}{2}(1 + \log_e 2\pi) \\ - \frac{n}{2} \log_e \hat{\sigma}_\varepsilon^2(\lambda) + (\lambda - 1) \sum_{i=1}^{n} \log_e Y_i$$

where $\hat{\sigma}_\varepsilon^2(\lambda) = \sum E_i^2(\lambda)/n$, and where the $E_i(\lambda)$ are the residuals from the least-squares regression of $Y^{(\lambda)}$ on the X's. The least-squares coefficients from this regression are the maximum-likelihood estimates of α and the β's, conditional on the value of λ.

---

[43] Subsequent work (Hernandez and Johnson, 1980) suggests that Box and Cox's method principally serves to normalize the error distribution.

[44] Strictly speaking, the requirement that the Y_i are positive precludes the possibility that they are normally distributed (because the normal distribution is unbounded), but this is not a serious practical difficulty unless many Y-values stack up near 0.

[45] See Exercise 12.14.

A simple procedure for finding the maximum-likelihood estimator $\hat{\lambda}$, then, is to evaluate the maximized $\log_e L$ (called the "profile log likelihood") for a range of values of λ, say between -2 and $+2$. If this range turns out not to contain the maximum of the log likelihood, then the range can be expanded. To test H_0: $\lambda = 1$, calculate the likelihood-ratio statistic

$$G_0^2 = -2[\log_e L(\lambda = 1) - \log_e L(\lambda = \hat{\lambda})]$$

which is asymptotically distributed as χ^2 with 1 degree of freedom under H_0. Alternatively (but equivalently), a 95% confidence interval for λ includes those values for which

$$\log_e L(\lambda) > \log_e L(\lambda = \hat{\lambda}) - 1.92$$

The figure 1.92 comes from $1/2 \times \chi^2_{1,.05} = 1/2 \times 1.96^2$.

Figure 12.8 shows a plot of the maximized log likelihood against λ for Ornstein's interlocking-directorate regression. The maximum-likelihood estimate of λ is $\hat{\lambda} = 0.31$, and a 95% confidence interval, marked out by the intersection of the line near the top of the graph with the log likelihood, runs from 0.20 to 0.41.[46]

Atkinson (1985) has proposed an approximate score test for the Box-Cox model, based on the constructed variable

$$G_i = Y_i\left[\log_e\left(\frac{Y_i}{\tilde{Y}}\right) - 1\right]$$

where $\tilde{Y}$ is the *geometric mean* of Y:[47]

$$\tilde{Y} \equiv (Y_1 \times Y_2 \times \cdots \times Y_n)^{1/n}$$

This constructed variable is obtained by a linear approximation to the Box-Cox transformation $Y^{(\lambda)}$ evaluated at $\lambda = 1$. The augmented regression, including the constructed variable, is then

$$Y_i = \alpha' + \beta_1' X_{i1} + \cdots + \beta_k' X_{ik} + \phi G_i + \varepsilon_i'$$

The t-test of H_0: $\phi = 0$, that is, $t_0 = \hat{\phi}/\widehat{\text{SE}}(\hat{\phi})$, assesses the need for a transformation. The quantities $\hat{\phi}$ and $\widehat{\text{SE}}(\hat{\phi})$ are obtained from the least-squares regression of Y on $X_1, \ldots, X_k$ and G. An estimate of λ (though not the MLE) is given by $\tilde{\lambda} = 1 - \hat{\phi}$; and the partial-regression plot for the constructed variable G shows influence and leverage on $\hat{\phi}$, and hence on the choice of λ.

Atkinson's constructed-variable plot for the interlocking-directorate regression is shown in Figure 12.9. Although the trend in the plot is not altogether

---

[46] Recall that we previously employed a square-root transformation for these data to make the residual distribution more nearly normal and to stabilize the error variance.

[47] It is more practical to compute the geometric mean as $\tilde{Y} = \exp[(\sum \log_e Y_i)/n]$.

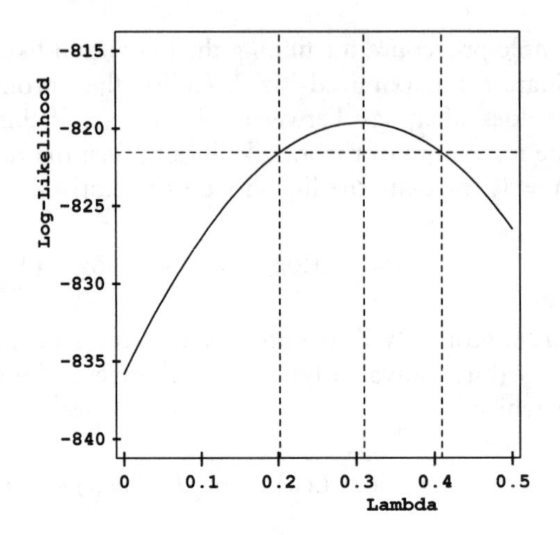

Figure 12.8. Box-Cox transformations for Ornstein's interlocking-directorate regression. The maximized log likelihood is plotted against the transformation parameter λ. The intersection of the line near the top of the graph with the log likelihood curve marks off a 95% confidence interval for λ. The maximum of the log likelihood corresponds to the MLE of λ.

linear, it appears that evidence for the transformation of Y is spread throughout the data and does not depend unduly on a small number of observations. The coefficient of the constructed variable in the regression is $\hat{\phi} = 0.585$, with $\widehat{\text{SE}}(\hat{\phi}) = 0.031$, providing very strong evidence of the need to transform Y. The suggested transformation, $\tilde{\lambda} = 1 - 0.585 = 0.415$, is close to the MLE (but just at the boundary of the narrow 95% confidence interval constructed around the MLE).

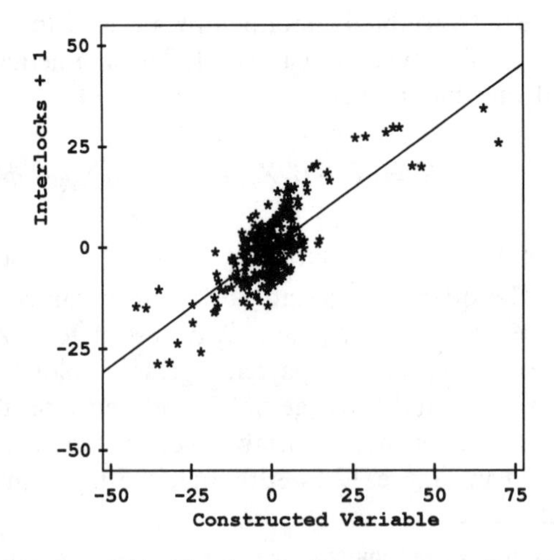

Figure 12.9. Constructed-variable plot for the Box-Cox transformation of Ornstein's interlocking-directorate regression. The least-squares line is shown on the plot.

12.5.2 Box-Tidwell Transformation of the X's

Now, consider the model

$$Y_i = \alpha + \beta_1 X_{i1}^{\gamma_1} + \cdots + \beta_k X_{ik}^{\gamma_k} + \varepsilon_i$$

where the errors are independently distributed as $\varepsilon_i \sim N(0, \sigma_\varepsilon^2)$, and all of the X_{ij} are positive. The parameters of this model—α, $\beta_1, \ldots, \beta_k$, $\gamma_1, \ldots, \gamma_k$, and σ_ε^2—could be estimated by general nonlinear least squares, but Box and Tidwell (1962) suggest instead a computationally more efficient procedure that also yields a constructed-variable diagnostic:[48]

1. Regress Y on $X_1, \ldots, X_k$, obtaining A, $B_1, \ldots, B_k$.
2. Regress Y on $X_1, \ldots, X_k$ and the constructed variables $X_1 \log_e X_1$, $\ldots$, $X_k \log_e X_k$, obtaining A', $B'_1, \ldots, B'_k$ and $D_1, \ldots, D_k$. Because of the presence of the constructed variables in this second regression, in general $A \neq A'$ and $B_j \neq B'_j$. As in the Box-Cox model, the constructed variables result from a linear approximation[49] to $X_j^{\gamma_j}$ evaluated at $\gamma_j = 1$.
3. The constructed variable $X_j \log_e X_j$ can be used to assess the need for a transformation of X_j by testing the null hypothesis H_0: $\delta_j = 0$, where δ_j is the population coefficient of $X_j \log_e X_j$ in step 2. Partial-regression plots for the constructed variables are useful for assessing leverage and influence on the decision to transform the X's.
4. A preliminary estimate of the transformation parameter γ_j (not the MLE) is given by

$$\tilde{\gamma}_j = 1 + \frac{D_j}{B_j}$$

Recall that B_j is from the *initial* (i.e., step 1) regression (not from step 2).

This procedure can be iterated through steps 1, 2, and 4 until the estimates of the transformation parameters stabilize, yielding the MLEs $\hat{\gamma}_j$.

For the Canadian occupational prestige data, leaving the regressors for percentage of women (W and W^2) untransformed, the coefficients of $S \log_e S$ (education) and $I \log_e I$ (income) are, respectively, $D_S = 5.30$ with $\widehat{\text{SE}}(D_S) = 2.20$, and $D_I = -0.00243$ with $\widehat{\text{SE}}(D_I) = 0.00046$. There is, consequently, much stronger evidence of the need to transform income than education.

The first-step estimates of the transformation parameters are

$$\tilde{\gamma}_S = 1 + \frac{D_S}{B_S} = 1 + \frac{5.30}{4.26} = 2.2$$

$$\tilde{\gamma}_I = 1 + \frac{D_I}{B_I} = 1 + \frac{-0.00243}{0.00127} = -0.91$$

The fully iterated MLEs of the transformation parameters are $\hat{\gamma}_S = 2.2$ and $\hat{\gamma}_I = -0.038$. Compare these values with the square and log transformations

[48] Nonlinear least-squares regression is described in Section 14.2.3.
[49] See Exercise 12.15.

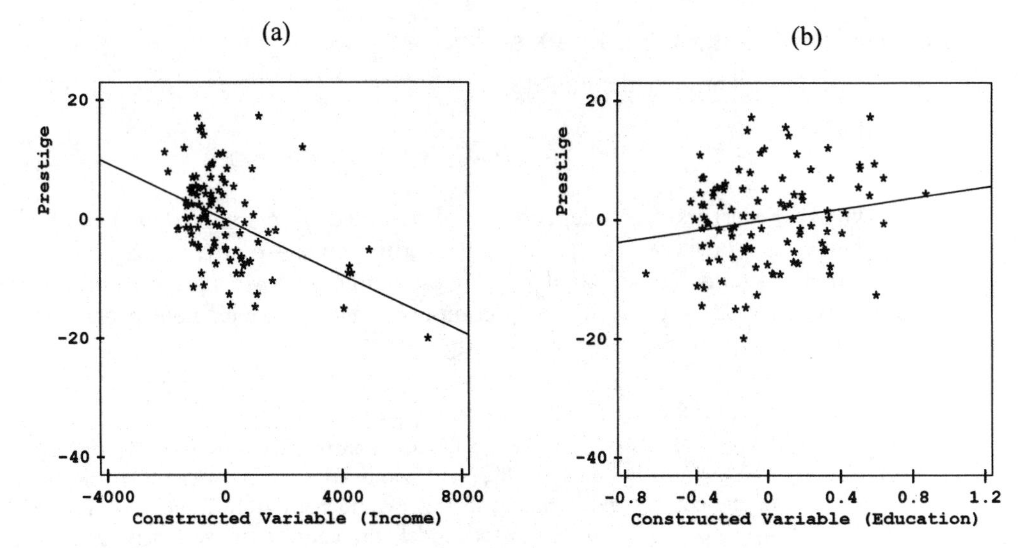

Figure 12.10. Constructed-variable plots for the Box-Tidwell transformation of (a) income and (b) education in the regression of occupational prestige on income, education, and percentage of women.

discovered following trial and error in Section 12.3.1.[50] Constructed-variable plots for the transformation of education and income, shown in Figure 12.10, suggest that there is general evidence for these transformations, although there are some high-leverage in-line observations in the income plot.

> A statistically sophisticated general approach to selecting a transformation of Y or an X is to embed the linear-regression model in a more general model that contains a parameter for the transformation. The Box-Cox procedure selects a power transformation of Y to normalize the errors. The Box-Tidwell procedure selects power transformations of the X's to linearize the regression of Y on the X's. In both cases, "constructed-variable" plots help us to decide whether individual observations are unduly influential in determining the transformation parameters.

12.5.3 Nonconstant Error Variance Revisited

Breusch and Pagan (1979) develop a score test for heteroscedasticity based on the specification:

$$\sigma_i^2 \equiv V(\varepsilon_i) = g(\gamma_0 + \gamma_1 Z_{i1} + \cdots + \gamma_p Z_{ip})$$

[50] Recall from Section 12.3.1 that the power transformation of education is not wholly appropriate because the partial relationship between prestige and education did not appear to be simple.

where $Z_1, \ldots, Z_p$ are known variables, and where the function $g(\cdot)$ is quite general (and need not be explicitly specified). The same test was independently derived by Cook and Weisberg (1983). The score statistic for the hypothesis that the σ_i^2 are all the same, which is equivalent to H_0: $\gamma_1 = \cdots = \gamma_p = 0$, can be formulated as an auxiliary-regression problem.

Let $U_i \equiv E_i^2/\hat{\sigma}_\varepsilon^2$, where $\hat{\sigma}_\varepsilon^2 = \sum E_i^2/n$ is the MLE of the error variance.[51] The U_i are a type of standardized squared residual. Regress U on the Z's:

$$U_i = \eta_0 + \eta_1 Z_{i1} + \cdots + \eta_p Z_{ip} + \omega_i \qquad [12.11]$$

Breusch and Pagan (1979) show that the score statistic

$$S_0^2 = \frac{\sum(\hat{U}_i - \overline{U})^2}{2}$$

is asymptotically distributed as χ^2 with p degrees of freedom under the null hypothesis of constant error variance. Here, the $\hat{U}_i$ are fitted values from the regression of U on the Z's, and thus S_0^2 is half the regression sum of squares from fitting Equation 12.11.

To apply this result, it is, of course, necessary to select Z's, the choice of which depends on the suspected pattern of nonconstant error variance. If several patterns are suspected, then several score tests can be performed. Employing $X_1, \ldots, X_k$ in the auxiliary regression (Equation 12.11), for example, permits detection of a tendency of the error variance to increase with the values of one or more of the independent variables in the main regression.

Likewise, Cook and Weisberg (1983) suggest regressing U on the fitted values from the main regression (i.e., fitting the auxiliary regression $U_i = \eta_0 + \eta_1 \hat{Y}_i + \omega_i$), producing a 1-degree-of-freedom score test to detect the common tendency of the error variance to increase with the level of the dependent variable. When the error variance follows this pattern, the auxiliary regression of U on $\hat{Y}$ provides a more powerful test than the more general regression of U on the X's. A similar, but more complex, procedure is described by Anscombe (1961), who suggests correcting detected heteroscedasticity by transforming Y to $Y^{(\tilde{\lambda})}$ with $\tilde{\lambda} = 1 - \frac{1}{2}\hat{\eta}_1 \overline{Y}$.

Finally, White (1980) proposes a score test based on a comparison of his heteroscedasticity-corrected estimator of coefficient sampling variance with the usual estimator of coefficient variance.[52] If the two estimators are sufficiently different, then doubt is cast on the assumption of constant error variance. White's test can be implemented as an auxiliary regression of the squared residuals from the main regression, E_i^2, on all of the X's together with all of the squares and pairwise products of the X's. Thus, for $k = 2$ independent variables in the main regression, we would fit the model

$$E_i^2 = \delta_0 + \delta_1 X_{i1} + \delta_2 X_{i2} + \delta_{11} X_{i1}^2 + \delta_{22} X_{i2}^2 + \delta_{12} X_{i1} X_{i2} + v_i$$

[51] Note the division by n rather than by $n - 1$ in $\hat{\sigma}_\varepsilon^2$. See Section 9.3.3.

[52] White's coefficient-variance estimator is described in Section 12.2.3.

In general, there will be $p = k(k+3)/2$ terms in the auxiliary regression, plus the constant.

The score statistic for testing the null hypothesis of constant error variance is $S_0^2 = nR^2$, where R^2 is the squared multiple correlation from the auxiliary regression. Under the null hypothesis, S_0^2 follows an asymptotic χ^2 distribution with p degrees of freedom.

Because all of these score tests are potentially sensitive to violations of model assumptions other than constant error variance, it is important, in practice, to supplement the tests with graphical diagnostics, as suggested by Cook and Weisberg (1983). When there are several Z's, a simple diagnostic is to plot U_i against $\hat{U}_i$, the fitted values from the auxiliary regression. We can also construct partial-regression plots for the Z's in the auxiliary regression. When U_i is regressed on $\hat{Y}_i$, these plots convey essentially the same information as the plot of studentized residuals against fitted values proposed in Section 12.2.

> Simple score tests are available to determine the need for a transformation and to test for nonconstant error variance.

Applied to Ornstein's interlocking-directorate data, an auxiliary regression of U on $\hat{Y}$ yields $\hat{U} = 0.134 + 0.0594\hat{Y}$, and $S_0^2 = 147.6/2 = 73.8$ on 1 degree of freedom. There is, consequently, very strong evidence that the error variance increases with the level of the dependent variable. The suggested variance-stabilizing transformation using Anscombe's rule is $\tilde{\lambda} = 1 - \frac{1}{2}(0.0594)(14.58) = 0.57$. Compare this value with those produced by the Box-Cox model ($\hat{\lambda} = 0.3$, in Section 12.5.1) and by trial and error ($\lambda = 0.5$, in Section 12.2).

An auxiliary regression of U on the independent variables in the main regression yields $S_0^2 = 172.6/2 = 86.3$ on $k = 13$ degrees of freedom, and thus also provides strong evidence against constant error variance. Examination of the coefficients from the auxiliary regression (not shown here) indicates, in particular, a tendency of the error variance to increase with assets. The score statistic for the more general test is not much larger than that for the regression of U on $\hat{Y}$, however, suggesting that the pattern of nonconstant error variance is indeed for the spread of the errors to increase with the level of Y. Assets are, of course, an important component of $\hat{Y}$. Because White's test requires 104 regressors for this problem, it was not performed.

EXERCISES

12.14 *Box-Cox transformations of Y: In matrix form, the Box-Cox regression model given in Section 12.5.1 can be written as

$$\mathbf{y}^{(\lambda)} = \mathbf{X}\boldsymbol{\beta} + \boldsymbol{\varepsilon}$$

(a) Show that the probability density for the observations is given by

$$p(y) = \frac{1}{(2\pi\sigma_\varepsilon^2)^{n/2}} \exp\left[-\frac{\sum_{i=1}^{n}(Y_i^{(\lambda)} - \mathbf{x}_i'\boldsymbol{\beta})^2}{2\sigma_\varepsilon^2} \right] \prod_{i=1}^{n} Y_i^{\lambda-1}$$

where $\mathbf{x}_i'$ is the ith row of $\mathbf{X}$. (*Hint*: $Y_i^{\lambda-1}$ is the Jacobian of the transformation from Y_i to ε_i.)

(b) For a given value of λ, the *conditional* maximum-likelihood estimator of $\boldsymbol{\beta}$ is the least-squares estimator

$$\mathbf{b}_\lambda = (\mathbf{X}'\mathbf{X})^{-1}\mathbf{X}'\mathbf{y}^{(\lambda)}$$

(Why?) Show that the maximized log likelihood can be written as

$$\log_e L(\alpha, \beta_1, \ldots, \beta_k, \sigma_\varepsilon^2 | \lambda)$$
$$= -\frac{n}{2}(1 + \log_e 2\pi) - \frac{n}{2}\log_e \hat{\sigma}_\varepsilon^2(\lambda) + (\lambda - 1)\sum_{i=1}^{n} \log_e Y_i$$

as stated in the text.

12.15 *Box-Tidwell transformations of the X's: Recall the Box-Tidwell model

$$Y = \alpha + \beta_1 X_1^{\gamma_1} + \cdots + \beta_k X_k^{\gamma_k} + \varepsilon$$

and focus on the first regressor, X_1. Show that the first-order Taylor series approximation for $X_1^{\gamma_1}$ at $\gamma_1 = 1$ is

$$X_1^{\gamma_1} \simeq X_1 + (\gamma_1 - 1)X_1 \log_e X_1$$

providing the basis for the constructed variable $X_1 \log_e X_1$.

12.6 Structural Dimension*

In discussing the use and potential failure of partial-residual plots as a diagnostic for nonlinearity, I explained that it is unreasonable to expect that a collection of two- or three-dimensional graphs can, in every instance, adequately capture the dependence of Y on the X's: The surface representing this dependence lies, after all, in a space of $k + 1$ dimensions. Relying primarily on Cook (1994), I shall now briefly consider the geometric notion of dimension in regression analysis, along with the implications of this notion for diagnosing problems with regression models that have been fit to data.[53] The *structural dimension* of a regression problem corresponds to the dimensionality of the smallest subspace of the X's required to represent the dependency of Y on the X's.

[53] An extended discussion of structural dimension, at a much simpler level than Cook (1994), may be found in Cook and Weisberg (1994).

Let us initially suppose that the distribution of Y is completely independent of the independent variables $X_1, \ldots, X_k$. Then, in Cook and Weisberg's (1994) terminology, an "ideal summary" of the data is simply the univariate, unconditional distribution of Y—represented, say, by the density $p(y)$. In a sample, we could compute a density estimate, a histogram, or some other univariate display. In this case, the *structural dimension* of the data is 0.

Now suppose that Y depends on the X's only through the regression equation

$$Y = \alpha + \beta_1 X_1 + \cdots + \beta_k X_k + \varepsilon$$

where $E(\varepsilon) = 0$ and the distribution of the error is independent of the X's. Then the expectation of Y conditional on the X's is a linear function of the X's:

$$E(Y|x_1, \ldots, x_k) = \alpha + \beta_1 x_1 + \cdots + \beta_k x_k$$

A plot of Y against $\alpha+\beta_1 X_1+\cdots+\beta_k X_k$, therefore, constitutes an ideal summary of the data. This two-dimensional plot shows the systematic component of Y in an edge-on view of the regression hyperplane, and also shows the conditional variation of Y around the hyperplane (i.e., the variation of the errors).

Because the subspace spanned by the linear combination $\alpha + \beta_1 X_1 + \cdots + \beta_k X_k$ is one dimensional, the structural dimension of the data is 1. In a sample, the ideal summary is a two-dimensional scatterplot of Y_i against $\hat{Y}_i = A + B_1 X_{i1} + \cdots + B_k X_{ik}$; the regression line in this plot is an edge-on view of the fitted least-squares surface.

The structural dimension of the data can be 1 even if the regression is nonlinear or if the errors are not identically distributed, as long as the expectation of Y and the distribution of the errors depend only on a single linear combination of the X's—that is, a subspace of dimension 1. The structural dimension is 1, for example, if

$$E(Y|x_1, \ldots, x_k) = f(\alpha + \beta_1 x_1 + \cdots + \beta_k x_k) \qquad [12.12]$$

and

$$V(Y|x_1, \ldots, x_k) = g(\alpha + \beta_1 x_1 + \cdots + \beta_k x_k) \qquad [12.13]$$

where the mean function $f(\cdot)$ and the variance function $g(\cdot)$, though generally different, depend on the *same* linear function of the X's. In this case, a plot of Y against $\alpha + \beta_1 X_1 + \cdots + \beta_k X_k$ is still an ideal summary of the data, showing the nonlinear dependency of the expectation of Y on the X's, along with the pattern of nonconstant error variance.

Similarly, we hope to see these features of the data in a sample plot of Y against $\hat{Y}$ from the *linear* regression of Y on the X's (even though the linear regression does not itself capture the dependency of Y on the X's). It turns out, however, that the plot of Y against $\hat{Y}$ can fail to reflect the mean and variance functions accurately if the X's themselves are not linearly related—even when the true structural dimension is 1 (i.e., when Equations 12.12 and 12.13 hold).[54] This, then, is another context in which linearly related independent variables are desirable.[55] Linearly related independent variables are not required here if the true regression is linear—something that, however, we are typically not in a position to know prior to examining the data.

> The structural dimension of a regression is the dimensionality of the smallest subspace of the independent variables required, along with the dependent variable, to represent the dependence of Y on the X's. When Y is completely independent of the X's, the structural dimension is 0, and an ideal summary of the data is simply the unconditional distribution of Y. When the linear-regression model holds—or when the conditional expectation and variance of Y are a function of a single linear combination of the X's—the structural dimension is 1.

The structural dimension of the data exceeds 1 if Equations 12.12 and 12.13 do not both hold. If, for example, the mean function depends on one linear combination of the X's:

$$E(Y|x_1, \ldots, x_k) = f(\alpha + \beta_1 x_1 + \cdots + \beta_k x_k)$$

and the variance function on a different linear combination

$$V(Y|x_1, \ldots, x_k) = g(\gamma + \delta_1 x_1 + \cdots + \delta_k x_k)$$

then the structural dimension is 2.

Correspondingly, if the mean function depends on two different linear combinations of the X's, implying interaction among the X's,

$$E(Y|x_1, \ldots, x_k) = f(\alpha + \beta_1 x_1 + \cdots + \beta_k x_k, \gamma + \delta_1 x_1 + \cdots + \delta_k x_k)$$

while the errors are independent of the X's, then the structural dimension is also 2. When the structural dimension is 2, a plot of Y against $\hat{Y}$ (from the linear regression of Y on the X's) is necessarily incomplete.

---

[54] See Exercise 12.16.

[55] The requirement of linearity here is, in fact, stronger than pairwise linear relationships among the X's: The regression of any linear function of the X's on any set of linear functions of the X's must be linear. If the X's are multivariate normal, then this condition is necessarily satisfied (although it may be satisfied even if the X's are not normal). It is not possible to check for linearity in this strict sense when there are more than two or three X's, but there is some evidence that checking pairs—and perhaps triples—of X's is usually sufficient. See Cook and Weisberg (1994). Cf. Section 12.3.2 for the conditions under which partial-residual plots are informative.

These observations are interesting, but their practical import—beyond the advantage of linearly related regressors—is unclear: Short of modeling the regression of Y on the X's nonparametrically, we can never be sure that we have captured all of the structure of the data in a lower-dimensional subspace of the independent variables.

There is, however, a further result that does have direct practical application: Suppose that the independent variables are linearly related and that there is one-dimensional structure. Then the *inverse regressions* of each of the independent variables on the dependent variable have the following character:[56]

$$E(X_j|y) = \mu_j + \eta_j m(y) \qquad [12.14]$$
$$V(X_j|y) \simeq \sigma_j^2 + \eta_j^2 v(y)$$

Equation 12.14 has two special features that are useful in checking whether one-dimensional structure is reasonable for a set of data:[57]

1. Most important, the functions $m(\cdot)$ and $v(\cdot)$, through which the means and variances of the X's depend on Y, are the same for all of the X's. Consequently, if the scatterplot of X_1 against Y shows a linear relationship, for example, then the scatterplots of each of $X_2, \ldots, X_k$ against Y must also show linear relationships. If one of these relationships is quadratic, in contrast, then the others must be quadratic. Likewise, if the variance of X_1 increases linearly with the level of Y, then the variances of the other X's must also be linearly related to Y. There is only one exception: The constant η_j can be 0, in which case the mean and variance of the corresponding X_j are *unrelated* to Y.
2. The constant η_j appears in the formula for the conditional mean of X_j, and η_j^2 in the formula for its conditional variance, placing constraints on the patterns of these relationships. If, for example, the mean of X_1 is unrelated to Y, then so should the variance.

The sample inverse regressions of the X's on Y can be conveniently examined in the first column of the scatterplot matrix for $\{Y, X_1, \ldots, X_k\}$.[58]

> If the structural dimension is 1, and if the independent variables are linearly related to one another, then the inverse regressions of the independent variables on the dependent variable all have the same general form.

[56] See Exercise 12.17 for illustrative applications.

[57] Equation 12.14 is the basis for formal dimension-testing methods, such as *sliced inverse regression* (Duan and Li, 1991) and related techniques. See Cook and Weisberg (1994) for an introductory treatment of dimension testing and for additional references.

[58] See, for example, Figure 3.15.

EXERCISES

12.16 Experimenting with structural dimension: Generate random samples of 100 observations according to each of the following schemes. In each case, fit the linear regression of Y on X_1 and X_2, and plot the values of Y against the resulting fitted values $\hat{Y}$. Do these plots accurately capture the dependence of Y on X_1 and X_2 ? To decide this question in each case, it may help (1) to draw graphs of $E(Y|x_1, x_2) = f(\alpha + \beta_1 x_1 + \beta_2 x_2)$ and $V(Y|x_1, x_2) = g(\alpha + \beta_1 x_1 + \beta_2 x_2)$ over the observed range of values for $\alpha + \beta_1 X_1 + \beta_2 X_2$; and (2) to plot a nonparametric-regression smooth in the plot of Y against $\hat{Y}$. Whenever they appear, E and U are $N(0, 1)$ and independent of each other and of the other variables.

(a) Independent X's, a linear regression, and constant error variance: X_1 and X_2 independent and uniformly distributed on the interval $[0, 1]$; $E(Y|x_1, x_2) = x_1 + x_2$; $V(Y|x_1, x_2) = 0.1E$.

(b) Independent X's, mean and variance of Y dependent on the same linear function of the X's: X_1 and X_2 independent and uniformly distributed on the interval $[0, 1]$; $E(Y|x_1, x_2) = (x_1 + x_2 - 1)^2$; $V(Y|x_1, x_2) = 0.1 \times |x_1 + x_2 - 1| \times E$.

(c) Linearly related X's, mean and variance of Y dependent on the same linear function of the X's: X_1 uniformly distributed on the interval $[0, 1]$; $X_2 = X_1 + 0.1U$; $E(Y|x_1, x_2) = (x_1 + x_2 - 1)^2$; $V(Y|x_1, x_2) = 0.1 \times |x_1 + x_2 - 1| \times E$.

(d) Nonlinearly related X's, mean and variance of Y dependent on the same linear function of the X's: X_1 uniformly distributed on the interval $[0, 1]$; $X_2 = |X_1 - 0.5|$; $E(Y|x_1, x_2) = (x_1 + x_2 - 1)^2$; $V(Y|x_1, x_2) = 0.1 \times |x_1 + x_2 - 1| \times E$.

12.17 Dimension checking: Apply the dimension-checking conditions

$$E(X_j|y) = \mu_j + \eta_j m(y)$$

$$V(X_j|y) \simeq \sigma_j^2 + \eta_j^2 v(y)$$

to each of the following regression analyses. In each case, construct the scatterplot matrix for the independent variables and the dependent variable. If the independent variables do not appear to be linearly related, attempt to make their relationship more nearly linear by transforming one or more of the X's. Then examine the column of the scatterplot matrix that shows the relationship of each X to Y. Are these relationships qualitatively similar, as required for one-dimensional structure?

(a) Duncan's regression of prestige on income and education, for 45 U.S. occupations (Table 3.2 and `duncan.dat`).

(b) The regression of prestige on income, education, and percentage of women, for the Canadian occupational prestige data (`prestige.dat`).

(c) Angell's regression of moral integration on ethnic heterogeneity and geographic mobility, for 43 U.S. cities (Table 2.3 and `angell.dat`).

(d) Anscombe's regression of state education expenditures on per-capita income, proportion under 18 years of age, and proportion urban (Table 5.1 and `anscombe.dat`).

12.7 Summary

- Heavy-tailed errors threaten the efficiency of least-squares estimation; skewed and multimodal errors compromise the interpretation of the least-squares fit. Nonnormality can often be detected by examining the distribution of the least-squares residuals, and frequently can be corrected by transforming the data.
- It is common for the variance of the errors to increase with the level of the dependent variable. This pattern of nonconstant error variance ("heteroscedasticity") can often be detected in a plot of residuals against fitted values. Strategies for dealing with nonconstant error variance include transformation of the dependent variable to stabilize the variance; the substitution of weighted-least-squares estimation for ordinary least squares; and the correction of coefficient standard errors for heteroscedasticity. A rough rule is that nonconstant error variance seriously degrades the least-squares estimator only when the ratio of the largest to smallest variance is about 10 or more.
- Simple forms of nonlinearity can often be detected in partial-residual plots. Once detected, nonlinearity can frequently be accommodated by variable transformations or by altering the form of the model (to include a quadratic term in an independent variable, for example). Partial-residual plots adequately reflect nonlinearity when the independent variables are themselves linearly related. More complex versions of these displays, such as augmented partial-residual plots and CERES plots, are more robust.
- Discrete independent variables divide the data into groups. A simple incremental F-test for nonlinearity compares the sum of squares accounted for by the linear regression of Y on X with the sum of squares accounted for by differences in the group means. Likewise, tests of nonconstant variance can be based on comparisons of spread in the different groups.
- A statistically sophisticated general approach to selecting a transformation of Y or an X is to embed the linear-regression model in a more general model that contains a parameter for the transformation. The Box-Cox procedure selects a power transformation of Y to normalize the errors. The Box-Tidwell procedure selects power transformations of the X's to linearize the regression of Y on the X's. In both cases, "constructed-variable" plots help us to decide whether individual observations are unduly influential in determining the transformation parameters.
- Simple score tests are available to determine the need for a transformation and to test for nonconstant error variance.
- The structural dimension of a regression is the dimensionality of the smallest subspace of the independent variables required, along with the dependent variable, to represent the dependence of Y on the X's. When Y is completely independent of the X's, the structural dimension is 0, and an ideal summary of the data is simply the unconditional distribution of Y. When the linear-regression model

holds—or when the conditional expectation and variance of Y are a function of a single linear combination of the X's—the structural dimension is 1. If the structural dimension is 1, and if the independent variables are linearly related to one another, then the inverse regressions of the independent variables on the dependent variable all have the same general form.

EXERCISES

12.18 Use the methods of this chapter to check for nonnormality, nonconstant error variance, and nonlinearity in each of the following regressions. In each case, attempt to correct any problems that are detected. Because many different methods are discussed in this chapter, you might find the following strategy useful: Use relatively simple diagnostics to check for problems and more sophisticated methods to follow up. To check for nonnormality, construct a quantile comparison plot and a histogram of the studentized residuals; to check for nonconstant error variance, plot studentized residuals against fitted values; to check for nonlinearity, examine partial-residual plots.

(a) Angell's regression of moral integration of U.S. cities on ethnic heterogeneity and geographic mobility (Table 2.3 and `angell.dat`).

(b) Anscombe's regression of state education expenditures on income, proportion under 18, and proportion urban (Table 5.1 and `anscombe.dat`).

12.19 Using Leinhardt and Wasserman's data on national infant mortality rates (given in `leinhard.dat`), regress infant mortality on income and dummy regressors for region. Using the methods of this chapter and the previous one, check the adequacy of the model and attempt to correct any problems that you find.

12.8 Recommended Reading

Methods for diagnosing problems in regression analysis and for visualizing regression data are an active area of research in statistics. The following texts summarize the current state of the art and include extensive references to the journal literature.

- Cook and Weisberg (1994) present a lucid and accessible treatment of many of the topics discussed in this chapter. They also describe a computer program, written in Lisp-Stat, that implements the graphical methods presented in their book (and much more). A copy of the program, called R-Code, and many programmed demonstrations, are included with the book.
- Cleveland (1993) describes novel graphical methods for regression data, including two-dimensional, three-dimensional, and higher-dimensional displays.

- Atkinson (1985) has written an interesting, if somewhat idiosyncratic, book which stresses the author's important contributions to regression diagnostics. There is, therefore, an emphasis on diagnostics that yield constructed-variable plots. This text includes a strong treatment of transformations, and a discussion of the extension of least-squares diagnostics to generalized linear models (e.g., logistic regression, as described in Chapter 15).

13

Collinearity and Its Purported Remedies

As I have explained,[1] when there is a perfect linear relationship among the regressors in a linear model, the least-squares coefficients are not uniquely defined. A strong, but less-than-perfect, linear relationship among the X's causes the least-squares coefficients to be unstable: Coefficient standard errors are large, reflecting the imprecision of estimation of the β's; consequently, confidence intervals for the β's are broad. Small changes in the data—even, in extreme cases, due to rounding errors—can substantially alter the least-squares coefficients; and relatively large changes in the coefficients from the least-squares values hardly increase the sum of squared residuals from its minimum (i.e., the least-squares coefficients are not sharply defined).

This chapter describes methods for detecting collinearity and techniques that are often employed for dealing with collinearity when it is present. I need to make three important points at the outset, however:

1. Except in certain specific contexts—such as time series regression[2]—collinearity is a comparatively rare problem in social science applications of linear models. Insufficient variation in independent variables, small samples, and large error variance (i.e., weak relationships) are much more frequently the source of imprecision in estimation.
2. Methods that are commonly employed as cures for collinearity—in particular, biased estimation and variable selection—can easily be worse than the disease.

[1] See Sections 5.2 and 9.2.
[2] See Section 14.1.

A principal goal of this chapter is to explain the substantial limitations of this statistical snake oil.

3. It is not at all obvious that the detection of collinearity in data has practical implications. There are, as mentioned in point 1, several sources of imprecision in estimation, which can augment or partially offset each other. The standard errors of the regression estimates are the bottom line: If these estimates are sufficiently precise, then the degree of collinearity is irrelevant; if the estimates are insufficiently precise, then knowing that the culprit is collinearity is of use only if the study can be redesigned to decrease the correlations among the X's. In observational studies, where the X's are sampled along with Y, it is usually impossible to influence their correlational structure, but it may very well be possible to increase the precision of estimation by increasing the sample size or by decreasing the error variance.[3]

13.1 Detecting Collinearity

We have encountered the notion of collinearity at several points, and it is therefore useful to summarize what we know:

- When there is a perfect linear relationship among the X's,

$$c_1 X_{i1} + c_2 X_{i2} + \cdots + c_k X_{ik} = c_0$$

  1. the least-squares normal equations do not have a unique solution; and
  2. the sampling variances of the regression coefficients are infinite.

  *Points 1 and 2 follow from the observation that the matrix $\mathbf{X}'\mathbf{X}$ of sums of squares and products is singular. Moreover, because the columns of $\mathbf{X}$ are perfectly collinear, the regressor subspace is of deficient dimension.

  Perfect collinearity is usually the product of some error in formulating the linear model, such as failing to employ a baseline category in dummy regression.

- When collinearity is less than perfect:

  1. The sampling variance of the least-squares slope coefficient B_j is

$$V(B_j) = \frac{1}{1 - R_j^2} \times \frac{\sigma_\varepsilon^2}{(n-1)S_j^2}$$

  where R_j^2 is the squared multiple correlation for the regression of X_j on the other X's, and $S_j^2 = \sum(X_{ij} - \overline{X}_j)^2/(n-1)$ is the variance of X_j. The term $1/(1 - R_j^2)$, called the *variance inflation factor* (VIF), directly and straightforwardly indicates the impact of collinearity on the precision of the estimate B_j. Because the precision of estimation of β_j is most naturally expressed as the width of the confidence interval for this parameter, and because the width of the confidence interval is proportional to the standard error of B_j (not its variance), I recommend examining the square root of the VIF in preference to the VIF itself.

[3] The error variance can sometimes be decreased by improving the procedures of the study or by introducing additional independent variables. The latter remedy may, however, increase collinearity, and may change the nature of the research. It may be possible, in some contexts, to increase precision by increasing the variation of the X's, but only if their values are under the control of the researcher, in which case collinearity could also be reduced. Sometimes, however, researchers may be able to exert indirect control over the variational and correlational structure of the X's by selecting a research setting judiciously or by designing an advantageous sampling procedure.

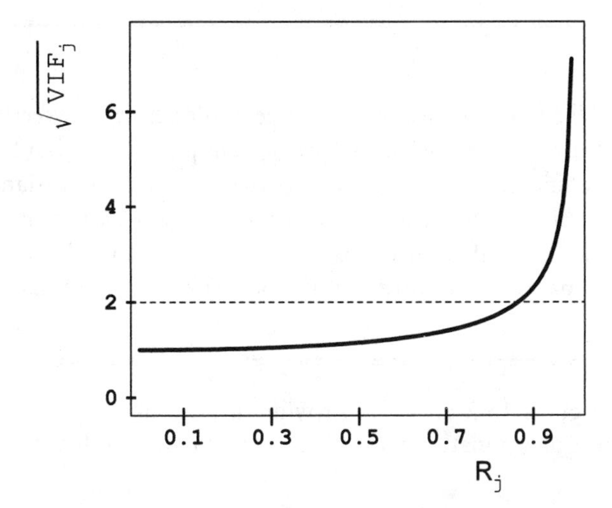

Figure 13.1. Precision of estimation (square root of the variance inflation factor) of β_j as a function of the multiple correlation between X_j and the other independent variables. It is not until the multiple correlation gets very large that the precision of estimation is seriously degraded.

Figure 13.1 reveals that the linear relationship among the X's must be very strong before collinearity seriously impairs the precision of estimation: It is not until R_j approaches .9 that the precision of estimation is halved.

Because of its simplicity and direct interpretation, the VIF (or its square root) is the principal diagnostic for collinearity. It is not, however, applicable to sets of related regressors, such as sets of dummy-variable coefficients, or coefficients for polynomial regressors.[4]

2. When X_1 is strongly collinear with the other regressors, the residuals $X^{(1)}$ from the regression of X_1 on $X_2, \ldots, X_k$ show little variation—most of the variation in X_1 is accounted for by its regression on the other X's. The partial-regression plot graphs the residuals from the regression of Y on $X_2, \ldots, X_k$ against $X^{(1)}$, converting the multiple regression into a simple regression.[5] Because the independent variable in this plot, $X^{(1)}$, is nearly invariant, the slope B_1 is subject to substantial sampling variation.[6]
3. *Confidence intervals for individual regression coefficients are projections of the confidence interval generating ellipse. Because this ellipse is the inverse—that is, the rescaled, 90° rotation—of the data ellipse for the independent variables, the individual confidence intervals for the coefficients are wide. If the correlations among the X's are positive, however, then there is substantial information in the data about the *sum* of the regression coefficients, if not about individual coefficients.[7]

---

[4] Section 13.1.2 describes a generalization of variance inflation to sets of related regressors.

[5] More precisely, the multiple regression is converted into a sequence of simple regressions, for each X in turn. Partial-regression plots are discussed in Section 11.6.

[6] See Stine (1995) for a nice graphical interpretation of this point.

[7] See Section 9.4.

When the regressors in a linear model are perfectly collinear, the least-squares coefficients are not unique. Strong, but less-than-perfect, collinearity substantially increases the sampling variances of the least-squares coefficients and can render them useless as estimators. The variance inflation factor $\text{VIF}_j = 1/(1 - R_j^2)$ indicates the deleterious impact of collinearity on the precision of the estimate B_j.

Figures 13.2 and 13.3 provide further insight into collinearity, illustrating its effect on estimation when there are two independent variables in a regression.

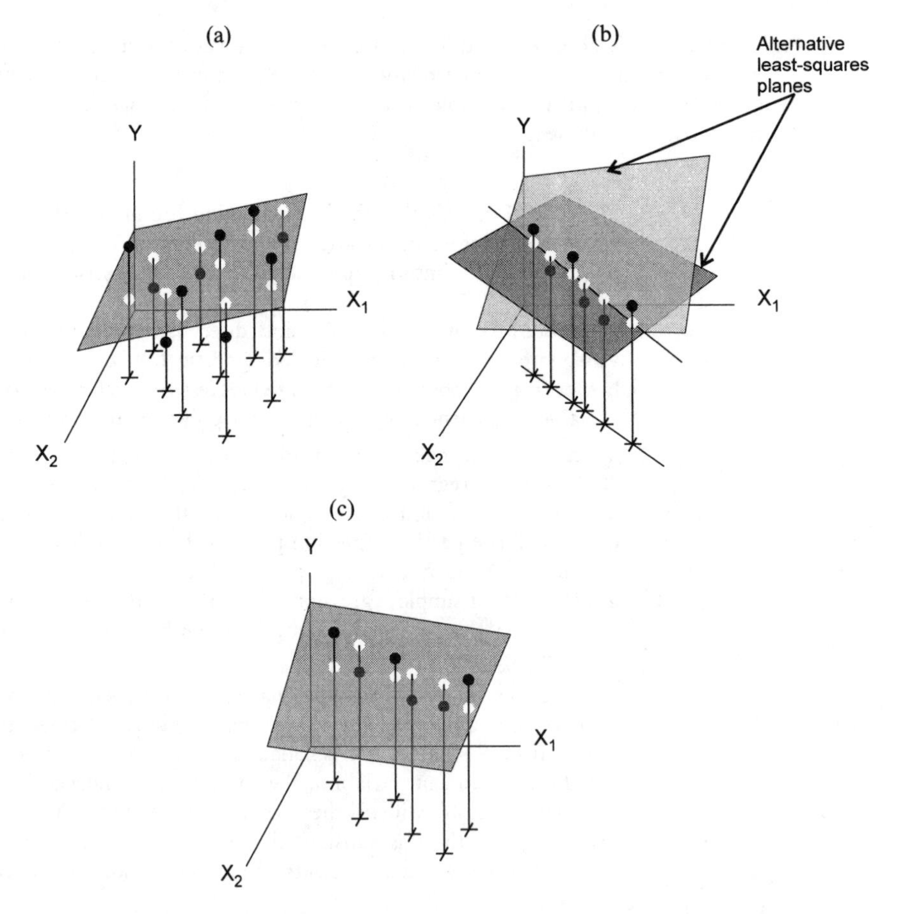

Figure 13.2. The impact of collinearity on the stability of the least-squares regression plane. In (*a*), the correlation between X_1 and X_2 is small, and the regression plane therefore has a broad base of support. In (*b*), X_1 and X_2 are perfectly correlated; the least-squares plane is not uniquely defined. In (*c*), there is a strong, but less-than-perfect, linear relationship between X_1 and X_2; the least-squares plane is uniquely defined, but it is not well supported by the data.

The black and gray dots in Figure 13.2 represent the data points (the gray dots are below the regression plane), while the white dots represent fitted values lying in the regression plane; the +'s show the projection of the data points onto the X_1, X_2 plane. Figure 13.3 shows the sum of squared residuals as a function of the slope coefficients B_1 and B_2. The residual sum of squares is at a minimum,

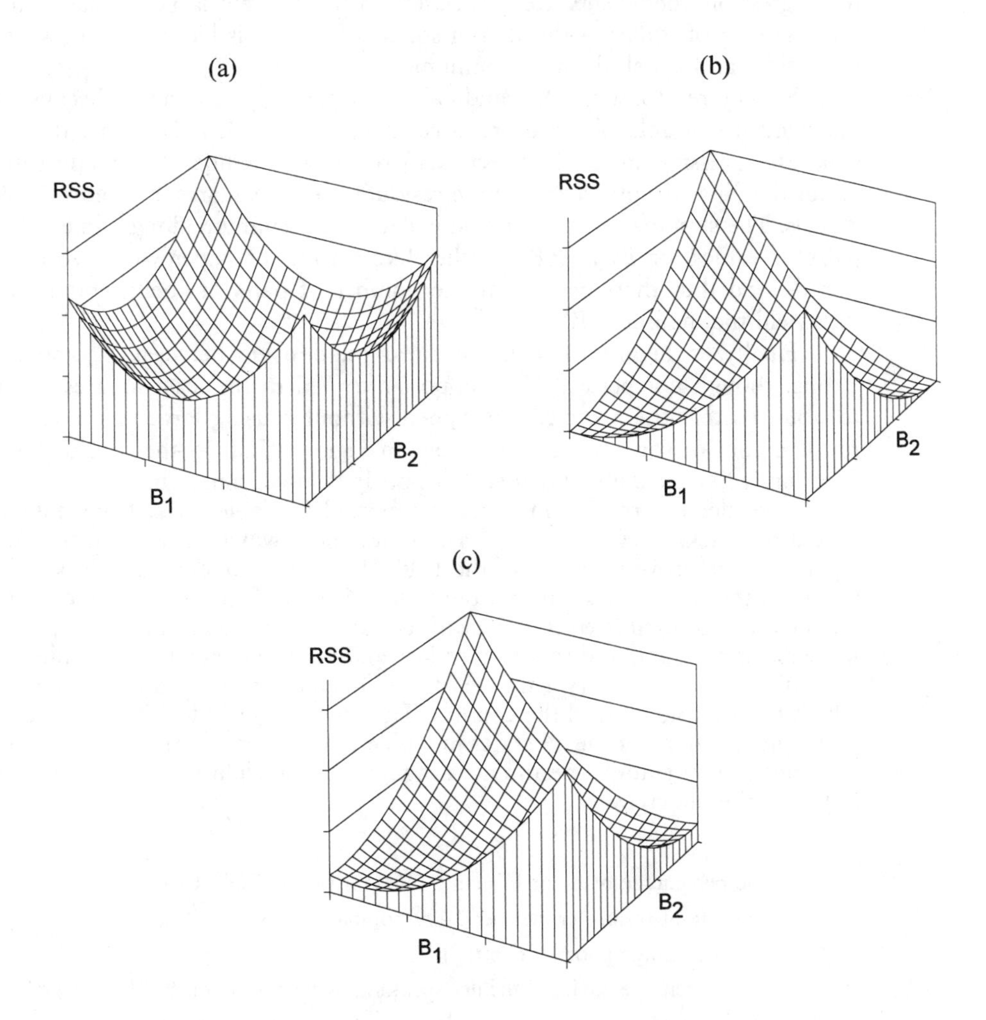

Figure 13.3. The residual sum of squares as a function of the slope coefficients B_1 and B_2. In each graph, the vertical axis is scaled so that the least-squares value of RSS is at the bottom of the axis. When, as in (*a*), the correlation between the independent variables X_1 and X_2 is small, the residual sum of squares has a well-defined minimum, much like a deep bowl. When there is a perfect linear relationship between X_1 and X_2, as in (*b*), the residual sum of squares is flat at its minimum, above a line in the B_1, B_2 plane: The least-squares values of B_1 and B_2 are not unique. When, as in (*c*), there is a strong, but less-than-perfect, linear relationship between X_1 and X_2, the residual sum of squares is nearly flat at its minimum, so values of B_1 and B_2 quite different from the least-squares values are associated with residual sums of squares near the minimum.

of course, when the B's are equal to the least-squares estimates; the vertical axis is scaled so that the minimum is at the "floor" of the figures.[8]

In Figure 13.2(a), the correlation between the independent variables X_1 and X_2 is slight, as indicated by the broad scatter of points in the X_1, X_2 plane. The least-squares regression plane, also shown in this figure, therefore has a firm base of support. Correspondingly, Figure 13.3(a) shows that small changes in the regression coefficients are associated with relatively large increases in the residual sum of squares—the sum-of-squares function is like a deep bowl, with steep sides and a well-defined minimum.

In Figure 13.2(b), X_1 and X_2 are perfectly collinear. Because the independent-variable observations form a line in the X_1, X_2 plane, the least-squares regression plane, in effect, also reduces to a line. The plane can tip about this line without changing the residual sum of squares, as Figure 13.3(b) reveals: The sum-of-squares function is flat at its minimum along a line defining pairs of values for B_1 and B_2—rather like a sheet of paper with two corners raised—and thus there are an infinite number of pairs of coefficients (B_1, B_2) that yield the minimum RSS.

Finally, in Figure 13.2(c), the linear relationship between X_1 and X_2 is strong, though not perfect. The support afforded to the least-squares plane is tenuous, so that the plane can be tipped without causing large increases in the residual sum of squares, as is apparent in Figure 13.3(c)—the sum-of-squares function is like a shallow bowl with a poorly defined minimum.

Consider the regression analysis reported in Table 13.1, from data presented by Ericksen et al. (1989).[9] The object here was to develop a prediction equation to improve estimates of the 1980 U.S. Census undercount. It is well established that the census fails to count all residents of the country, and that the likelihood of being missed is greater for certain categories of individuals, such as nonwhites, the poor, and residents of large cities. The dependent variable in the regression is a preliminary estimate of the undercount for each of 66 areas into which the authors divided the country. The 66 areas include 16 large cities, the remaining portions of the 16 states in which the cities are located, and the other 34 states. The preliminary estimates are regressed on eight predictors thought to influence the undercount:

1. the percentage black or Hispanic ("Minority" in Table 13.1);
2. the rate of serious crimes per 1000 population ("Crime");
3. the percentage poor ("Poverty");
4. the percentage having difficulty speaking or writing English ("Language");
5. the percentage aged 25 or older who have *not* finished high school ("High School");
6. the percentage of housing in small, multiunit buildings ("Housing");
7. a dummy variable coded 1 for cities, 0 for states or state remainders ("City"); and

---

[8] For each pair of slopes B_1 and B_2, the intercept A is chosen to make the residual sum of squares as small as possible.

[9] The authors employed a weighted-least-squares regression (see Section 12.2.2) to take account of differences in precision of initial estimates of the undercount in the 66 areas. The results reported here, in contrast, are for an ordinary-least-squares regression.

TABLE 13.1 Regression of Estimated 1980 U.S. Census Undercount on Area Characteristics, for 66 Central Cities, State Remainders, and States

Predictor	*Coefficient*	*Standard Error*	$\sqrt{VIF}$
Constant	−1.77	1.38	—
Minority	0.0798	0.0226	2.24
Crime	0.0301	0.0130	1.83
Poverty	−0.178	0.0849	2.15
Language	0.215	0.0922	1.28
High school	0.0613	0.0448	2.15
Housing	−0.0350	0.0246	1.37
City	1.16	0.77	1.88
Conventional	0.0370	0.0093	1.30
R^2	.708		

Source of Data: Ericksen et al. (1989).

8. the percentage of households counted by "conventional" personal enumeration, as opposed to mail-back questionnaire with follow-ups ("Conventional").

Correlations among the eight predictors appear in Table 13.2. Although some of the pairwise correlations are fairly large—the biggest are about .75—none is close to 1. It is apparent from the square-root VIFs shown in Table 13.1, however, that the precision of several of the regression estimates—in particular, the coefficients for Minority, Poverty, and High School—suffer from moderate collinearity. This result illustrates that collinearity in multiple regression is not restricted to pairwise relationships between regressors; sometimes the term *multicollinearity* is employed to emphasize this point.

13.1.1 Principal Components*

The method of principal components, developed in the early part of the 20th century by K. Pearson and H. Hotelling, provides a useful representation of the correlational structure of a set of variables. I shall develop the method briefly here, with particular reference to its application to collinearity in regression; more complete accounts can be obtained from texts on multivariate

TABLE 13.2 Correlations Among Eight Predictors of the 1980 U.S. Census Undercount

Predictor	*Minority*	*Crime*	*Poverty*	*Language*	*High School*	*Housing*	*City*
Crime	.655						
Poverty	.738	.369					
Language	.395	.512	.152				
High school	.535	.067	.751	−.116			
Housing	.357	.532	.335	.340	.235		
City	.758	.729	.538	.480	.315	.566	
Conventional	−.334	−.233	−.157	−.108	−.414	−.086	−.269

statistics (e.g., Morrison, 1976, Chapter 8). Because the material in this section is relatively complex, the section includes a summary; you may, on first reading, wish to pass lightly over most of the section and refer primarily to the summary, and to the two-variable case, which is treated immediately prior to the summary.

We begin with the vectors of standardized regressors, $\mathbf{z}_1, \mathbf{z}_2, \ldots, \mathbf{z}_k$. Because vectors have length equal to the square root of their sum of squared elements, each $\mathbf{z}_j$ has length $\sqrt{n-1}$. As we shall see, the *principal components* $\mathbf{w}_1, \mathbf{w}_2, \ldots, \mathbf{w}_p$ provide an orthogonal basis for the regressor subspace.[10] The first principal component, $\mathbf{w}_1$, is oriented so as to account for maximum collective variation in the $\mathbf{z}_j$; the second principal component, $\mathbf{w}_2$, is orthogonal to $\mathbf{w}_1$, and—under this restriction of orthogonality—is oriented to account for maximum remaining variation in the $\mathbf{z}_j$; the third component, $\mathbf{w}_3$, is orthogonal to $\mathbf{w}_1$ and $\mathbf{w}_2$; and so on. Each principal component is scaled so that its variance is equal to the combined regressor variance for which it accounts.

There are as many principal components as there are linearly independent regressors: $p \equiv \text{rank}(\mathbf{z}_X)$, where $\mathbf{z}_X \equiv [\mathbf{z}_1, \mathbf{z}_2, \ldots, \mathbf{z}_k]$. Although the method of principal components is more general, I shall assume throughout most of this discussion that the regressors are not perfectly collinear and, consequently, that $p = k$.

Because the principal components lie in the regressor subspace, each is a linear combination of the regressors. Thus, the first principal component can be written as

$$\underset{(n\times 1)}{\mathbf{w}_1} = A_{11}\mathbf{z}_1 + A_{21}\mathbf{z}_2 + \cdots + A_{k1}\mathbf{z}_k$$
$$= \underset{(n\times k)}{\mathbf{Z}_X}\underset{(k\times 1)}{\mathbf{a}_1}$$

The variance of the first component is

$$S^2_{W_1} = \frac{1}{n-1}\mathbf{w}_1'\mathbf{x}_1 = \frac{1}{n-1}\mathbf{a}_1'\mathbf{Z}_X'\mathbf{Z}_X\mathbf{a}_1 = \mathbf{a}_1'\mathbf{R}_{XX}\mathbf{a}_1$$

where $\mathbf{R}_{XX} \equiv [1/(n-1)]\mathbf{Z}_X'\mathbf{Z}_X$ is the correlation matrix of the regressors.

We want to maximize $S^2_{W_1}$, but, to make maximization meaningful, it is necessary to constrain the coefficients $\mathbf{a}_1$. In the absence of a constraint, $S^2_{W_1}$ can be made arbitrarily large simply by picking large coefficients. The normalizing constraint

$$\mathbf{a}_1'\mathbf{a}_1 = 1 \qquad [13.1]$$

proves convenient, but any constraint of this general form would do.[11]

---

[10] It is also possible to find principal components of the *unstandardized* regressors $\mathbf{x}_1, \mathbf{x}_2, \ldots, \mathbf{x}_k$, but these are not generally interpretable unless all of the X's are measured on the same scale.

[11] Normalizing the coefficients so that $\mathbf{a}_1'\mathbf{a}_1 = 1$ causes the variance of the first principal component to be equal to the combined variance of the standardized regressors accounted for by this component, as will become clear presently.

We can maximize $S^2_{W_1}$ subject to the restriction of Equation 13.1 by employing a Lagrange multiplier L_1, defining[12]

$$F_1 \equiv \mathbf{a}_1'\mathbf{R}_{XX}\mathbf{a}_1 - L_1(\mathbf{a}_1'\mathbf{a}_1 - 1)$$

Then, differentiating this equation with respect to $\mathbf{a}_1$ and L_1 ,

$$\frac{\partial F_1}{\partial \mathbf{a}_1} = 2\mathbf{R}_{XX}\mathbf{a}_1 - 2L_1\mathbf{a}_1$$
$$\frac{\partial F_1}{\partial L_1} = -(\mathbf{a}_1'\mathbf{a}_1 - 1)$$

Setting the partial derivatives to 0 produces the equations

$$\begin{aligned}(\mathbf{R}_{XX} - L_1\mathbf{I}_k)\mathbf{a}_1 &= \mathbf{0} \\ \mathbf{a}_1'\mathbf{a}_1 &= 1\end{aligned} \qquad [13.2]$$

The first formula in Equation 13.2 has nontrivial solutions for $\mathbf{a}_1$ only when $(\mathbf{R}_{XX} - L_1\mathbf{I}_k)$ is singular—that is, when $|\mathbf{R}_{XX} - L_1\mathbf{I}_k| = 0$. The multiplier L_1, therefore, is an eigenvalue of $\mathbf{R}_{XX}$, and $\mathbf{a}_1$ is the corresponding eigenvector, scaled so that $\mathbf{a}_1'\mathbf{a}_1 = 1$.

There are, however, k solutions to Equation 13.2, corresponding to the k eigenvalue-eigenvector pairs of $\mathbf{R}_{XX}$, so we must decide which solution to choose. From the first formula in Equation 13.2, we have $\mathbf{R}_{XX}\mathbf{a}_1 = L_1\mathbf{a}_1$. Consequently,

$$S^2_{W_1} = \mathbf{a}_1'\mathbf{R}_{XX}\mathbf{a}_1 = L_1\mathbf{a}_1'\mathbf{a}_1 = L_1$$

Because our purpose is to maximize $S^2_{W_1}$ (subject to the constraint on $\mathbf{a}_1$), we must select the *largest* eigenvalue of $\mathbf{R}_{XX}$ to define the first principal component.

The second principal component is derived similarly, under the further restriction that it is orthogonal to the first; the third that it is orthogonal to the first two; and so on.[13] It turns out that the second principal component corresponds to the second-largest eigenvalue of $\mathbf{R}_{XX}$, the third to the third-largest eigenvalue, and so forth. We order the eigenvalues of $\mathbf{R}_{XX}$ so that[14]

$$L_1 \geq L_2 \geq \cdots \geq L_k > 0$$

---

[12] See Appendix C, Section C.2, for an explanation of the method of Lagrange multipliers for constrained optimization.

[13] See Exercise 13.1.

[14] Recall that we are assuming that $\mathbf{R}_{XX}$ is of full rank, and hence none of its eigenvalues is 0. It is possible, but unlikely, that two or more eigenvalues of $\mathbf{R}_{XX}$ are equal. In this event, the orientation of the principal components corresponding to the equal eigenvalues is not unique, although the subspace spanned by these components—and for which they constitute a basis—is unique.

The matrix of principal-component coefficients

$$\underset{(k\times k)}{\mathbf{A}} \equiv [\mathbf{a}_1, \mathbf{a}_2, \ldots, \mathbf{a}_k]$$

contains normalized eigenvectors of $\mathbf{R}_{XX}$. This matrix is, therefore, orthonormal: $\mathbf{A}'\mathbf{A} = \mathbf{A}\mathbf{A}' = \mathbf{I}_k$.

The principal components

$$\underset{(n\times k)}{\mathbf{W}} = \underset{(n\times k)}{\mathbf{Z}_X}\underset{(k\times k)}{\mathbf{A}} \qquad [13.3]$$

have covariance matrix

$$\frac{1}{n-1}\mathbf{W}'\mathbf{W} = \frac{1}{n-1}\mathbf{A}'\mathbf{Z}_X'\mathbf{Z}_X\mathbf{A}$$
$$= \mathbf{A}'\mathbf{R}_{XX}\mathbf{A} = \mathbf{A}'\mathbf{A}\mathbf{L} = \mathbf{L}$$

where $\mathbf{L} \equiv \mathrm{diag}[L_1, L_2, \ldots, L_k]$ is the matrix of eigenvalues of $\mathbf{R}_{XX}$; the covariance matrix $\mathbf{W}$ of the principal components is, therefore, orthogonal, as required. Furthermore,

$$\mathrm{trace}(\mathbf{L}) = \sum_{j=1}^{k} L_j = k = \mathrm{trace}(\mathbf{R}_{XX})$$

and thus the principal components partition the combined variance of the standardized variables $Z_1, Z_2, \ldots, Z_k$.

Solving Equation 13.3 for $\mathbf{Z}_X$ produces

$$\mathbf{Z}_X = \mathbf{W}\mathbf{A}^{-1} = \mathbf{W}\mathbf{A}'$$

and, consequently,

$$\mathbf{R}_{XX} = \frac{1}{n-1}\mathbf{Z}_X'\mathbf{Z}_X = \frac{1}{n-1}\mathbf{A}\mathbf{W}'\mathbf{W}\mathbf{A}' = \mathbf{A}\mathbf{L}\mathbf{A}'$$

Finally,

$$\mathbf{R}_{XX}^{-1} = (\mathbf{A}')^{-1}\mathbf{L}^{-1}\mathbf{A}^{-1} = \mathbf{A}\mathbf{L}^{-1}\mathbf{A}' \qquad [13.4]$$

We shall use this result presently in our investigation of collinearity.

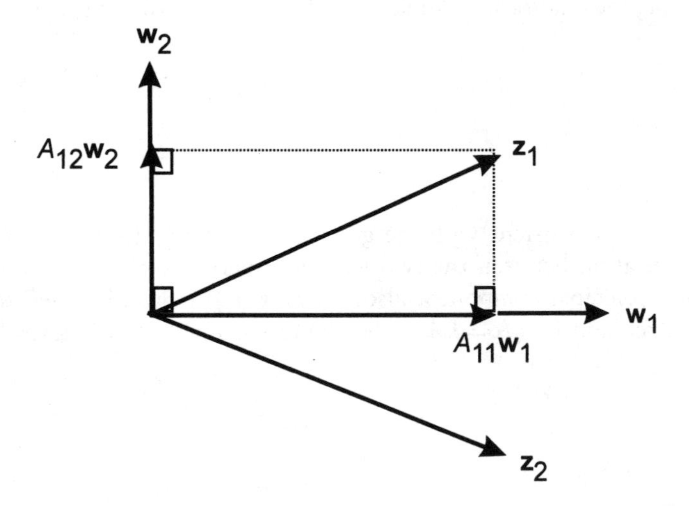

Figure 13.4. Vector geometry of principal components for two, positively correlated, standardized variables $\mathbf{z}_1$ and $\mathbf{z}_2$.

Two Variables

The vector geometry of principal components is illustrated for two variables in Figure 13.4. The symmetry of this figure is peculiar to the two-dimensional case. The length of each principal-component vector is the square root of the sum of squared orthogonal projections of $\mathbf{z}_1$ and $\mathbf{z}_2$ on the component. The direction of $\mathbf{w}_1$ is chosen to maximize the combined length of these projections, and hence to maximize the length of $\mathbf{w}_1$. Because the subspace spanned by $\mathbf{z}_1$ and $\mathbf{z}_2$ is two dimensional, $\mathbf{w}_2$ is simply chosen to be orthogonal to $\mathbf{w}_1$. Note that[15] $\|\mathbf{w}_j\|^2 = L_j(n-1)$.

It is clear from the figure that as the correlation between Z_1 and Z_2 increases, the first principal component grows at the expense of the second; thus, L_1 gets larger and L_2 smaller. If, alternatively, $\mathbf{z}_1$ and $\mathbf{z}_2$ are orthogonal, then $\|\mathbf{w}_1\| = \|\mathbf{w}_2\| = \sqrt{n-1}$, and $L_1 = L_2 = 1$.

The algebra of the two-variable case is also particularly simple. The eigenvalues of $\mathbf{R}_{XX}$ are the solutions of the characteristic equation

$$\begin{vmatrix} 1-L & r_{12} \\ r_{12} & 1-L \end{vmatrix} = 0$$

that is,

$$(1-L)^2 - r_{12}^2 = L^2 - 2L + 1 - r_{12}^2 = 0$$

[15] There is a small subtlety here: The subspace spanned by each component is one dimensional, and the length of each component is fixed by the corresponding eigenvalue, but these factors determine the orientation of the component only up to a rotation of 180°—that is, a change in sign.

Using the quadratic formula to find the roots of the characteristic equation yields

$$L_1 = 1 + \sqrt{r_{12}^2} \qquad [13.5]$$
$$L_2 = 1 - \sqrt{r_{12}^2}$$

And so, consistent with the geometry of Figure 13.4, as the magnitude of the correlation between the two variables increases, the variation attributed to the first principal component also grows. If r_{12} is positive, then solving for $\mathbf{A}$ from the relation $\mathbf{R}_{XX}\mathbf{A} = \mathbf{L}\mathbf{A}$ under the restriction $\mathbf{A}'\mathbf{A} = \mathbf{I}_2$ gives[16]

$$\mathbf{A} = \begin{bmatrix} \frac{\sqrt{2}}{2} & \frac{\sqrt{2}}{2} \\ \frac{\sqrt{2}}{2} & -\frac{\sqrt{2}}{2} \end{bmatrix}$$

The generalization to k standardized regressors is straightforward: If the variables are orthogonal, then all $L_j = 1$ and all $\|\mathbf{w}_j\| = \sqrt{n-1}$. As collinearities among the variables increase, some eigenvalues become large while others grow small. Small eigenvalues and the corresponding short principal components represent dimensions along which the regressor subspace has (nearly) collapsed. Perfect collinearities are associated with eigenvalues of 0.

The Data Ellipsoid

The principal components have an interesting interpretation in terms of the standard data ellipsoid for the Z's.[17] The data ellipsoid is given by the equation

$$\mathbf{z}'\mathbf{R}_{XX}^{-1}\mathbf{z} = 1$$

where $\mathbf{z} \equiv (Z_1, \ldots, Z_k)'$ is a vector of values for the k standardized regressors. Because the variables are standardized, the data ellipsoid is centered at the origin, and the shadow of the ellipsoid on each axis is of length 2 (i.e., 2 standard deviations). It can be shown that the principal components correspond to the principal axes of the data ellipsoid, and, further, that the half-length of each axis is equal to the square root of the corresponding eigenvalue[18] L_j of $\mathbf{R}_{XX}$.

These properties are depicted in Figure 13.5 for $k = 2$. When the variables are uncorrelated, the data ellipse becomes circular, and each axis has a half-length of 1.

---

[16] Exercise 13.2 derives the solution for $r_{12} < 0$.

[17] The standard data ellipsoid was introduced in Section 9.4.

[18] See Exercise 13.3. These relations also hold for *unstandardized* variables. That is, the principal components calculated from the covariance matrix $\mathbf{S}_{XX}$ give the principal axes of the standard data ellipsoid $(\mathbf{x} - \bar{\mathbf{x}})'\mathbf{S}_{XX}^{-1}(\mathbf{x} - \bar{\mathbf{x}})$; and the half-length of the jth principal axis of this ellipsoid is equal to the square root of the jth eigenvalue of $\mathbf{S}_{XX}$.

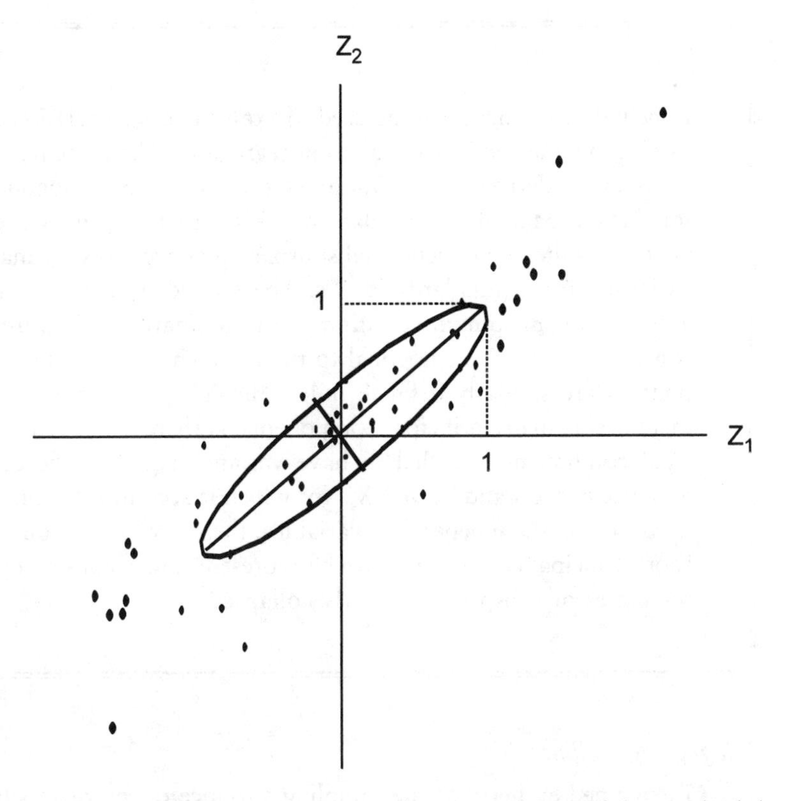

Figure 13.5. The principal components are the principal axes of the standard data ellipse $\mathbf{z}'\mathbf{R}_{XX}^{-1}\mathbf{z} = 1$. The first eigenvalue L_1 of $\mathbf{R}_{XX}$ gives the half-length of the major axis of the ellipse; the second eigenvalue L_2 gives the half-length of the minor axis. In this illustration, the two variables are highly correlated, so L_1 is large and L_2 is small.

Summary

- The principal components of the k standardized regressors $\mathbf{Z}_X$ are a new set of k variables derived from $\mathbf{Z}_X$ by a linear transformation: $\mathbf{W} = \mathbf{Z}_X\mathbf{A}$, where $\mathbf{A}$ is the $(k \times k)$ transformation matrix.
- The transformation $\mathbf{A}$ is selected so that the columns of $\mathbf{W}$ are orthogonal—that is, the principal components are uncorrelated. In addition, $\mathbf{A}$ is constructed so that the first component accounts for maximum variance in the Z's; the second for maximum variance under the constraint that it is orthogonal to the first; and so on. Each principal component is scaled so that its variance is equal to the variance in the Z's for which it accounts. The principal components therefore partition the variance of the Z's.
- The transformation matrix $\mathbf{A}$ contains (by columns) normalized eigenvectors of $\mathbf{R}_{XX}$, the correlation matrix of the regressors. The columns of $\mathbf{A}$ are ordered by their corresponding eigenvalues: The first column corresponds to the largest eigenvalue, and the last column to the smallest. The eigenvalue L_j associated with the jth component represents the variance attributable to that component.
- If there are perfect collinearities in $\mathbf{Z}_X$, then some eigenvalues of $\mathbf{R}_{XX}$ will be 0, and there will be fewer than k principal components, the number of components corresponding to $\text{rank}(\mathbf{Z}_X) = \text{rank}(\mathbf{R}_{XX})$. Near collinearities are associated with small eigenvalues and correspondingly short principal components.

Principal components can be used to explicate the correlational structure of the independent variables in regression. The principal components are a derived set of variables that form an orthogonal basis for the subspace of the standardized X's. The first principal component spans the one-dimensional subspace that accounts for maximum variation in the standardized X's. The second principal component accounts for maximum variation in the standardized X's, under the constraint that it is orthogonal to the first. The other principal components are similarly defined; unless the X's are perfectly collinear, there are as many principal components as there are X's. Each principal component is scaled to have variance equal to the collective variance in the standardized X's for which it accounts. Collinear relations among the independent variables, therefore, correspond to very short principal components, which represent dimensions along which the regressor subspace has nearly collapsed.

Diagnosing Collinearity

I explained earlier that the sampling variance of the regression coefficient B_j is

$$V(B_j) = \frac{\sigma_\varepsilon^2}{(n-1)S_j^2} \times \frac{1}{1-R_j^2}$$

It can be shown that $\text{VIF}_j = 1/(1 - R_j^2)$ is the jth diagonal entry of $\mathbf{R}_{XX}^{-1}$ (see Theil, 1971, p. 166). Using Equation 13.4, the variance inflation factors can be expressed as functions of the eigenvalues of $\mathbf{R}_{XX}$ and the principal components; specifically,

$$\text{VIF}_j = \sum_{l=1}^{k} \frac{A_{jl}^2}{L_l}$$

Thus, it is only the small eigenvalues that contribute to large sampling variance, but only for those regressors that have large coefficients associated with the corresponding short principal components. This result is sensible, for small eigenvalues and their short components correspond to collinear relations among the regressors; regressors with large coefficients for these components are the regressors implicated in the collinearities (see below).

The relative size of the eigenvalues serves as an indicator of the degree of collinearity present in the data. The square root of the ratio of the largest to smallest eigenvalue, $K \equiv \sqrt{L_1/L_k}$, called the *condition number*, is a commonly employed standardized index of the global instability of the least-squares regression coefficients: A large condition number (say, 10 or more) indicates that relatively small changes in the data tend to produce large changes in the least-squares solution. In this event, $\mathbf{R}_{XX}$ is said to be *ill conditioned.*

It is instructive to examine the condition number in the simplified context of the two-regressor model. From Equation 13.5,

$$K = \sqrt{\frac{L_1}{L_2}} = \sqrt{\frac{1 + \sqrt{r_{12}^2}}{1 - \sqrt{r_{12}^2}}}$$

and thus $K = 10$ corresponds to $r_{12}^2 = .9608$, for which VIF $= 26$.

Belsley et al. (1980, Chapter 3) define a *condition index* $K_j \equiv \sqrt{L_1/L_j}$ for each principal component[19] of $\mathbf{R}_{XX}$. Then, the number of large condition indices points to the number of different collinear relations among the regressors.

Chatterjee and Price (1991, Chapter 7) employ the principal-component coefficients to estimate these near collinearities: A component $\mathbf{w}_l$ associated with a very small eigenvalue $L_l \simeq 0$ is itself approximately equal to the zero vector; consequently,

$$A_{1l}\mathbf{z}_1 + A_{2l}\mathbf{z}_2 + \cdots + A_{kl}0_k \simeq \mathbf{0}$$

and we can use the large A_{jl}'s to specify a linear combination of the Z's that is approximately equal to 0.

13.1.2 Generalized Variance Inflation*

The methods for detecting collinearity described thus far are not fully applicable to models that include related sets of regressors, such as dummy regressors constructed from a polytomous categorical variable or polynomial regressors. The reasoning underlying this qualification is subtle, but can be illuminated by appealing to the vector representation of linear models.

The correlations among a set of dummy regressors are affected by the choice of baseline category. Similarly, the correlations among a set of polynomial regressors in an independent variable X are affected by adding a constant

---

[19] Primarily for computational accuracy, Belsley et al. (1980, Chapter 3) develop diagnostic methods for collinearity in terms of the *singular-value decomposition* of the regressor matrix, scaled so that each variable has a sum of squares of 1. I employ an equivalent eigenvalue-eigenvector approach because of its conceptual simplicity and broader familiarity. The eigenvectors of $\mathbf{R}_{XX}$, it turns out, are the squares of the singular values of $(1/\sqrt{n-1})\mathbf{Z}_X$. Indeed, the condition number K defined here is actually the condition number of $(1/\sqrt{n-1})\mathbf{Z}_X$ (and hence of $\mathbf{Z}_X$). Information on the singular-value decomposition and its role in linear-model analysis can be found in Belsley et al. (1980, Chapter 3) and in Mandel (1982).

A more substantial difference between my approach and that of Belsley et al. is that they base their analysis not on the correlation matrix of the X's, but rather on $\tilde{\mathbf{X}}'\tilde{\mathbf{X}}$, where $\tilde{\mathbf{X}}$ is the regressor matrix, including the constant regressor, with columns normed to unit length. Consider an independent variable that is uncorrelated with the others, but which has scores that are far from 0. Belsley et al. would say that this independent variable is "collinear with the constant regressor." This seems to me a corruption of the notion of collinearity, which deals fundamentally with the inability to separate the effects of highly correlated independent variables, and should not change with linear transformations of individual independent variables. See Belsley (1984), and the associated commentary, for various points of view on this issue.

to the X-values. Neither of these changes alters the fit of the model to the data, however, so neither is fundamental. It is, indeed, always possible to select an orthogonal basis for the dummy-regressor or polynomial-regressor subspace (although such a basis does not employ dummy variables or simple powers of X). What is at issue is the subspace itself, and not the arbitrarily chosen basis for it.[20]

We are not concerned, therefore, with the "artificial" collinearity among dummy regressors or polynomial regressors in the same set. We are instead interested in the relationships between the subspaces generated to represent the effects of *different* independent variables. As a consequence, we can legitimately employ variance inflation factors to examine the impact of collinearity on the coefficients of numerical regressors, or on any single-degree-of-freedom effects, even when sets of dummy regressors or polynomial regressors are present in the model.

Fox and Monette (1992) generalize the notion of variance inflation to sets of related regressors. Rewrite the linear model as

$$\mathbf{y} = \alpha\mathbf{1} + \mathbf{X}_1\boldsymbol{\beta}_1 + \mathbf{X}_2\boldsymbol{\beta}_2 + \boldsymbol{\varepsilon}$$

where the p regressors of interest (e.g., a set of dummy regressors) are in $\mathbf{X}_1$, while the remaining $k - p$ regressors (with the exception of the constant) are in $\mathbf{X}_2$. Fox and Monette (1992) show that the squared ratio of the size of the joint confidence region for $\boldsymbol{\beta}_1$ to the size of the same region for orthogonal but otherwise similar data is

$$\text{GVIF}_1 = \frac{\det \mathbf{R}_{11} \det \mathbf{R}_{22}}{\det \mathbf{R}}$$

Here, $\mathbf{R}_{11}$ is the correlation matrix for $\mathbf{X}_1$; $\mathbf{R}_{22}$ is the correlation matrix for $\mathbf{X}_2$; and $\mathbf{R}$ is the matrix of correlations among all of the variables. The generalized variance inflation factor (GVIF) is independent of the bases selected for the subspaces spanned by the columns of $\mathbf{X}_1$ and $\mathbf{X}_2$. If $\mathbf{X}_1$ contains only one column, then the GVIF reduces to the familiar variance inflation factor.

> The notion of variance inflation can be extended to sets of related regressors, such as dummy regressors and polynomial regressors, by considering the size of the joint confidence region for the related coefficients.

[20] A particular basis may be a poor computational choice, however, if it produces numerically unstable results. Consequently, researchers are sometimes advised to pick a category with many cases to serve as the baseline for a set of dummy regressors, or to subtract the mean from X prior to constructing polynomial regressors; the latter procedure is called "centering." Neither of these practices fundamentally alters the model, but may lead to more accurate computations.

EXERCISES

13.1 *The second principal component is

$$\underset{(n\times 1)}{\mathbf{w}_2} = A_{12}\mathbf{z}_1 + A_{22}\mathbf{z}_2 + \cdots + A_{k2}\mathbf{z}_k$$
$$= \underset{(n\times k)}{\mathbf{Z}_X}\underset{(k\times 1)}{\mathbf{a}_2}$$

with variance

$$S^2_{W_2} = \mathbf{a}_2'\mathbf{R}_{XX}\mathbf{a}_2$$

We need to maximize this variance subject to the *normalizing constraint* $\mathbf{a}_2'\mathbf{a}_2 = 1$ and the *orthogonality constraint* $\mathbf{w}_1'\mathbf{w}_2 = 0$. Show that the orthogonality constraint is equivalent to $\mathbf{a}_1'\mathbf{a}_2 = 0$. Then, using *two* Lagrange multipliers, one for the normalizing constraint and the other for the orthogonality constraint, show that $\mathbf{a}_2$ is an eigenvector corresponding to the second-largest eigenvalue of $\mathbf{R}_{XX}$. Explain how this procedure can be extended to derive the remaining $k - 2$ principal components.

13.2 *Find the matrix **A** of principal-component coefficients when $k = 2$ and r_{12} is negative.

13.3 *Show that when $k = 2$, the principal components of $\mathbf{R}_{XX}$ correspond to the principal axes of the data ellipse for the standardized regressors Z_1 and Z_2; show that the half-length of each axis is equal to the square root of the corresponding eigenvalue of $\mathbf{R}_{XX}$. Now extend this reasoning to the principal axes of the data ellipsoid for the standardized regressors when $k > 2$.

13.4 *The data that follow were constructed by Mandel (1982) to illustrate the problem of collinearity:

X_1	X_2	Y
16.85	1.46	41.38
24.81	−4.61	31.01
18.85	−0.21	37.41
12.63	4.93	50.05
21.38	−1.36	39.17
18.78	−0.08	38.86
15.58	2.98	46.14
16.30	1.73	44.47

(a) Compute the mean and standard deviation of each variable. Find the correlations among X_1, X_2, and Y, and use these correlations to calculate the standardized coefficients B_1^* and B_2^* for the regression of Y on X_1 and X_2. Find the unstandardized coefficients A, B_1, and B_2.

(b) Perform a principal-components analysis for X_1 and X_2. Draw the geometric vector representation of the principal-components analysis. Find the variance inflation factors for the coefficients B_1 and B_2, and calculate the condition number K for the regression.

(c) Use the second principal component to approximate the near-collinear relation between the standardized regressors Z_1 and Z_2. Express this relation as a linear relationship between the unstandardized regressors X_1 and X_2.

(d) Now regress X_1 on X_2. How does the fitted regression equation compare with the linear relationship found in part (c)?

(e) Draw the data ellipse for X_1 and X_2, and the 95% joint confidence ellipse for B_1 and B_2.

13.5 Time series data on Canadian women's labor-force participation in the first three decades of the postwar period are given in Table 13.3 and `bfox.dat`. B. Fox (1980) was interested in determining how women's labor-force participation rate (L, measured as the percentage of adult women in the work force) responds to a variety of factors indicative of the supply of and demand for women's labor. The independent variables in the analysis include:

- the total fertility rate (F), the expected number of births to a hypothetical cohort of 1000 women proceeding through their child-bearing years at current age-specific fertility rates;
- men's (M) and women's (W) average weekly earnings, in constant 1935 dollars and adjusted for current tax rates;
- per-capita consumer debt (D), in constant dollars; and
- the availability of part-time work (P), measured as the percentage of the active labor force working 34 hours a week or less.

Women's earnings, consumer debt, and the availability of part-time work were expected to affect women's labor-force participation positively. Fertility and men's earnings were expected to have negative effects. Because all of the series, including that for the dependent variable, manifest strong linear trends over the 20-year period of the study, year (T), coded from 1946 to 1975, was also included as an independent variable in the regression.

(a) Regress L on T, F, M, W, D, and P. What do you find: Are the researcher's expectations borne out? Are the estimates sufficiently precise?

(b) Employ the methods of this section to diagnose collinearity in B. Fox's data.

13.2 Coping With Collinearity: No Quick Fix

When X_1 and X_2 are strongly collinear, the data contain little information about the impact of X_1 on Y holding X_2 constant statistically, because there is little variation in X_1 when X_2 is fixed. Of course, the same is true for X_2 fixing X_1. Because B_1 estimates the partial effect of X_1 controlling for X_2, this estimate is imprecise.

Although there are several strategies for dealing with collinear data, none magically extracts nonexistent information from the data. Rather, the research problem is redefined, often subtly and implicitly. Sometimes the redefinition is reasonable; usually it is not. The ideal solution to the problem of collinearity is

TABLE 13.3 B. Fox's Canadian Women's Labor-Force Participation Data. T is year; L is women's labor-force participation rate, in percent; F is the total fertility rate, per 1000; M is men's average weekly wages in 1935 dollars; W is women's average weekly wages; D is per-capita consumer debt; and P is the percentage of part-time workers.

T	L	F	M	W	D	P
1946	25.3	3748	25.35	14.05	18.18	10.28
1947	24.4	3996	26.14	14.61	28.33	9.28
1948	24.2	3725	25.11	14.23	30.55	9.51
1949	24.2	3750	24.45	14.61	35.81	8.87
1950	23.7	3669	26.79	15.26	38.39	8.54
1951	24.2	3682	26.33	14.58	26.52	8.84
1952	24.1	3845	27.89	15.66	45.65	8.60
1953	23.8	3905	29.15	16.30	52.99	5.49
1954	23.6	4047	29.52	16.57	54.84	6.67
1955	24.3	4043	32.05	17.99	65.53	6.25
1956	25.1	4092	32.98	18.33	72.56	6.32
1957	26.2	4168	32.25	17.64	69.49	7.30
1958	26.6	4073	32.52	18.16	71.71	8.65
1959	26.9	4100	33.95	18.58	78.89	8.80
1960	27.9	4119	34.63	18.95	84.99	9.39
1961	29.1	4159	35.14	18.78	87.71	10.23
1962	29.9	4134	34.49	18.74	95.31	10.77
1963	29.8	4017	35.99	19.71	104.40	10.84
1964	30.9	3886	36.68	20.06	116.80	11.70
1965	32.1	3467	37.96	20.94	130.99	12.33
1966	33.2	3150	38.68	21.20	135.25	12.18
1967	34.5	2879	39.65	21.95	142.93	13.67
1968	35.1	2681	41.20	22.68	155.47	13.82
1969	36.1	2563	42.44	23.75	165.04	14.91
1970	36.9	2571	42.02	25.63	164.53	15.52
1971	37.0	2503	45.32	26.79	169.63	15.47
1972	37.9	2302	45.61	27.51	190.62	15.85
1973	40.1	2931	45.59	27.35	209.60	15.40
1974	40.6	1875	48.06	29.64	216.66	16.23
1975	42.2	1866	46.12	29.33	224.34	16.71

to collect new data in such a manner that the problem is avoided—for example, by experimental manipulation of the X's, or by a research setting (or sampling procedure) in which the independent variables of interest are not strongly related. Unfortunately, these solutions are rarely practical.

Several less adequate strategies for coping with collinear data are briefly described in this section. I have devoted most space to variable selection, because selection techniques are commonly abused by social scientists, because the rationale for variable selection is straightforward, and because variable selection is a reasonable approach in certain (limited) circumstances. Variable selection also has applications outside of the context of collinearity.

13.2.1 Model Respecification

Although collinearity is a data problem, not (necessarily) a deficiency of the model, one approach to the problem is to respecify the model. Perhaps, after further thought, several regressors in the model can be conceptualized as alternative indicators of the same underlying construct. Then these measures can be combined in some manner, or one can be chosen to represent the others. In this context, high correlations among the X's in question indicate high reliability—a fact to be celebrated, not lamented. Imagine, for example, an international analysis of factors influencing infant mortality, in which gross national product per capita, energy use per capita, and televisions per capita are among the independent variables and are highly correlated. A researcher may choose to treat these variables as indicators of the general level of economic development.

Alternatively, we can reconsider whether we really need to control for X_2 (for example) in examining the relationship of Y to X_1. Generally, though, respecification of this variety is possible only where the original model was poorly thought out, or where the researcher is willing to abandon some of the goals of the research. For example, suppose that in a time series regression examining determinants of married women's labor-force participation, collinearity makes it impossible to separate the effects of men's and women's wage levels.[21] There may be good theoretical reason to want to know the effect of women's wage level on their labor-force participation, holding men's wage level constant, but the data are simply uninformative about this question. It may still be of interest, however, to determine the partial relationship between general wage level and women's labor-force participation, controlling for other independent variables in the analysis.

13.2.2 Variable Selection

A common, but usually misguided, approach to collinearity is variable selection, where some procedure is employed to reduce the regressors in the model to a less highly correlated set. Forward selection methods add independent variables to the model one at a time. At each step, the variable that yields the largest increment in R^2 is selected. The procedure stops, for example, when the increment is smaller than a preset criterion.[22] Backward elimination methods are similar, except that the procedure starts with the full model and deletes variables one at a time. Forward/backward—or *stepwise*—methods combine the two approaches.

These methods frequently are abused by naive researchers who seek to interpret the order of entry of variables into the regression equation as an index of their "importance." This practice is potentially misleading: For example, suppose that there are two highly correlated independent variables that have nearly identical large correlations with Y; only one of these independent variables will enter the regression equation, because the other can contribute little additional information. A small modification to the data, or a new sample, could easily reverse the result.

A technical objection to stepwise methods is that they can fail to turn up the optimal subset of regressors of a given size (i.e., the subset that maximizes R^2). Advances in computer power and in computing procedures make

---

[21] See Exercises 13.5 and 13.9.

[22] More commonly, the stopping criterion is calibrated by the incremental F for adding a variable to the model.

it feasible to examine all subsets of regressors even when k is quite large.[23] Aside from optimizing the selection criterion, subset techniques also have the advantage of revealing alternative, nearly equivalent models, and thus avoid the misleading appearance of producing a uniquely "correct" result.[24]

One popular approach to subset selection is based on the total (normed) mean-squared error of estimating $E(Y)$ from $\hat{Y}$—that is, estimating the population regression surface over the observed X's from the fitted regression surface:

$$\begin{aligned} \gamma_p &\equiv \frac{1}{\sigma_\varepsilon^2}\sum_{i=1}^{n}\text{MSE}(\hat{Y}_i) \\ &= \frac{1}{\sigma_\varepsilon^2}\sum_{i=1}^{n}\{V(\hat{Y}_i) + [E(\hat{Y}_i) - E(Y_i)]^2\} \end{aligned} \qquad [13.6]$$

where the fitted values $\hat{Y}_i$ are based on a model containing $p \leq k+1$ regressors (counting the constant, which is always included in the model). Using the error in estimating $E(Y)$ as a criterion for model quality is reasonable if the goal is literally to predict Y from the X's, and if new observations on the X's for which predictions are required will be similar to those included in the data.

The term $[E(\hat{Y}_i) - E(Y_i)]^2$ in Equation 13.6 represents the squared bias of $\hat{Y}_i$ as an estimator of the population regression surface $E(Y_i)$. When collinear regressors are deleted from the model, generally $V(\hat{Y}_i)$ will decrease, but—depending on the configuration of data points and the true β's for the deleted regressors—bias may be introduced into the fitted values. Because the prediction MSE is the sum of variance and squared bias, the essential question is whether the decrease in variance offsets any increase in bias.

Mallows's (1973) C_p-statistic estimates γ_p as

$$\begin{aligned} C_p &= \frac{\sum E_i^2}{\hat{\sigma}_\varepsilon^2} + 2p - n \\ &= (k+1-p)(F_p - 1) + p \end{aligned}$$

where the residuals are from the subset model in question; the error variance estimate $\hat{\sigma}_\varepsilon^2$ is S_E^2 for the *full* model containing all k independent variables; and F_p is the incremental F-statistic for testing the hypothesis that the regressors omitted from the current subset have population coefficients of 0.[25] If this hypothesis is true, then $E(F_p) \simeq 1$, and thus $E(C_p) \simeq p$. A good model, therefore, has C_p close to or below p. As well, minimizing C_p minimizes the sum of squared residuals, and thus maximizes R^2. For the full model, C_{k+1} necessarily equals $k+1$.

Because a good model has C_p close to p, we can identify good models by plotting C_p against p, labeling each point in the plot with a mnemonic representing the independent variables included in the model, and superimposing the

[23] For k independent variables, the number of subsets, excluding the null subset with no predictors, is $2^k - 1$. See Exercise 13.6.

[24] There are algorithms available to find the optimal subset of a given size without examining all possible subsets (see, e.g., Furnival and Wilson, 1974). When the data are highly collinear, however, the optimal subset of a given size may be only trivially "better" than many of its competitors.

[25] See Exercise 13.7.

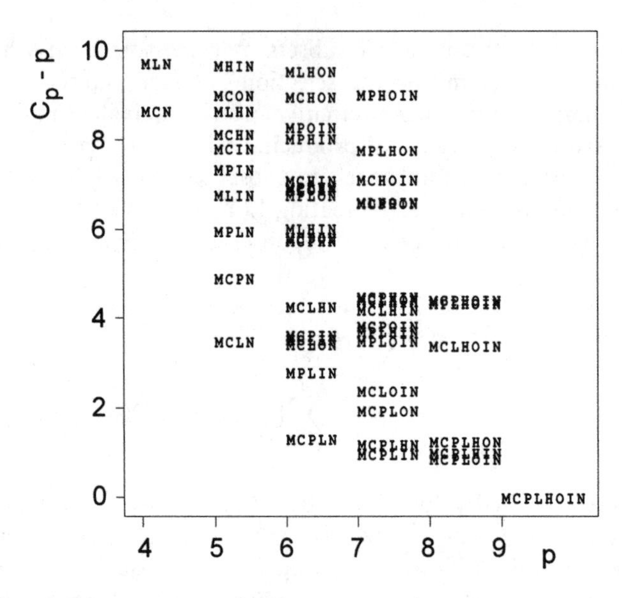

Figure 13.6. Plot of $C_p - p$ against p for the census-undercount regression. Only subsets for which $C_p - p < 10$ are shown. The following capitalized letters are employed to label the predictors in each subset: Minority, Crime, Poverty, Language, High school, hOusing, cIty, and coNventional. Ericksen et al. (1989) selected the predictor subset MCN (i.e., Minority, Crime, and coNventional).

line $C_p = p$ on the plot: Good models are close to or below the reference line. I find that the graph is easier to inspect if it is "detrended" by plotting $C_p - p$ against p (i.e., subtracting the reference line from each point). Now we can look for models with values of $C_p - p$ near or below 0.

An illustrative detrended C_p plot for the census-undercount data is given in Figure 13.6. Only models for which $C_p - p \leq 10$ are shown, including 52 of the $2^8 - 1 = 255$ predictor subsets. Ericksen et al. (1989) employed the subset labeled MCN on the plot (with predictors Minority, Crime, and Conventional).[26] For this subset, $p = 4$ and $C_p = 12.7$, suggesting that there is room for improvement by including more predictors. The regression equation for this subset, and equations for the "best" subsets of four predictors (MCLN, adding Language: $p = 5$ and $C_p = 8.5$) and five predictors (MCPLN, adding Poverty: $p = 6$ and $C_p = 7.3$) appear in Table 13.4. For this dataset, backward and forward/backward stepwise procedures identify the "best" subsets of three, four, and five predictors, but the forward method does not.[27]

In applying variable selection, it is essential to keep the following caveats in mind:

- Most important, variable selection results in a respecified model that usually does not address the research questions that were originally posed. In particular, if the original model is correctly specified, and if the included and omitted variables

[26] Recall, however, that Ericksen et al. (1989) adopted a more complex estimation strategy than ordinary-least-squares regression.

[27] See Exercise 13.8.

TABLE 13.4 "Best" Subset Regression Models for Ericksen et al.'s Census-Undercount Data. Coefficient standard errors are in parentheses.

Predictor	*Coefficients*		
	$p = 4$	$p = 5$	$p = 6$
Constant	−2.22	−1.98	−0.793
	(0.56)	(0.55)	(0.860)
Minority	0.0786	0.0752	0.101
	(0.0147)	(0.0143)	(0.020)
Crime	0.0363	0.0272	0.0243
	(0.0100)	(0.0104)	(0.0103)
Conventional	0.0280	0.0273	0.0293
	(0.0081)	(0.0078)	(0.0077)
Language		0.209	0.184
		(0.087)	(0.086)
Poverty			−0.110
			(0.062)
R^2	.638	.669	.686
C_p	12.7	8.51	7.32

are correlated, then coefficient estimates following variable selection are biased.[28] Consequently, these methods are most useful for pure prediction problems, in which the values of the regressors for the data to be predicted will be within the configuration of X-values for which selection was employed—as in the census-undercount example. In this case, it is possible to get good estimates of $E(Y)$ even though the regression coefficients themselves are biased. If, however, the X-values for a new observation differ substantially from those used to obtain the estimates, then the predicted Y can be badly biased.

- When regressors occur in sets (e.g., of dummy variables), then these sets should generally be kept together during selection. Likewise, when there are hierarchical relations among regressors, these relations should be respected: For example, an interaction regressor should not appear in a model that does not contain the main effects marginal to the interaction.
- Because variable selection optimizes the fit of the model to the sample data, coefficient standard errors calculated following independent-variable selection—and hence confidence intervals and hypothesis tests—almost surely overstate the precision of results. There is, therefore, a very substantial risk of capitalizing on chance characteristics of the sample.[29]
- Variable selection has applications to statistical modeling even when collinearity is not an issue. It is generally not problematic to eliminate regressors that have small, precisely estimated coefficients, thus producing a more parsimonious model. Indeed, in a very large sample, we may feel justified in deleting regressors with trivially small but "statistically significant" coefficients.

13.2.3 Biased Estimation

Still another general approach to collinear data is biased estimation. The essential idea here is to trade a small amount of bias in the coefficient estimates

[28] See Sections 6.3, 9.6, and 13.2.5.

[29] See Exercise 13.10 and the discussion of cross-validation in Section 16.2.

for a substantial reduction in coefficient sampling variance. The hoped-for result is a smaller mean-squared error of estimation of the β's than is provided by the least-squares estimates. By far the most common biased estimation method is *ridge regression* (due to Hoerl and Kennard, 1970a, 1970b).

Like variable selection, biased estimation is not a magical panacea for collinearity. Ridge regression involves the arbitrary selection of a "ridge constant," which controls the extent to which ridge estimates differ from the least-squares estimates: The larger the ridge constant, the greater the bias and the smaller the variance of the ridge estimator. Unfortunately, but as one might expect, to pick an optimal ridge constant—or even a good one—generally requires knowledge about the unknown β's that we are trying to estimate. My principal reason for mentioning biased estimation here is to caution against its routine use.

*Ridge Regression**

The ridge-regression estimator for the *standardized* regression coefficients is given by

$$\mathbf{b}_d^* \equiv (\mathbf{R}_{XX} + d\mathbf{I}_k)^{-1}\mathbf{r}_{Xy} \qquad [13.7]$$

where $\mathbf{R}_{XX}$ is the correlation matrix for the predictors; $\mathbf{r}_{Xy}$ is the vector of correlations between the predictors and the dependent variable; and $d \geq 0$ is a scalar constant. When $d = 0$, the ridge and least-squares estimators coincide: $\mathbf{b}_0^* = \mathbf{b}^* = \mathbf{R}_{XX}^{-1}\mathbf{r}_{Xy}$. When the data are collinear, some off-diagonal entries of $\mathbf{R}_{XX}$ are generally large, making this matrix ill conditioned. Heuristically, the ridge-regression method improves the conditioning of $\mathbf{R}_{XX}$ by inflating its diagonal entries.

Although the least-squares estimator $\mathbf{b}^*$ is unbiased, its entries tend to be too large in absolute value, a tendency that is magnified as collinearity increases. In practice, researchers working with collinear data often compute wildly large regression coefficients. The ridge estimator may be thought of as a "shrunken" version of the least-squares estimator, correcting the tendency of the latter to produce coefficients that are too far from 0.

The ridge estimator of Equation 13.7 can be rewritten as[30]

$$\mathbf{b}_d^* = \mathbf{U}\mathbf{b}^* \qquad [13.8]$$

where $\mathbf{U} \equiv (\mathbf{I}_k + d\mathbf{R}_{XX}^{-1})^{-1}$. As d increases, the entries of $\mathbf{U}$ tend to grow smaller, and, therefore, $\mathbf{b}_d^*$ is driven toward $\mathbf{0}$. Hoerl and Kennard (1970a) show that for any value of $d > 0$, the squared length of the ridge estimator is less than that of the least-squares estimator: $\mathbf{b}_d^{*\prime}\mathbf{b}_d^* < \mathbf{b}^{*\prime}\mathbf{b}^*$.

The expected value of the ridge estimator can be determined from its relation to the least-squares estimator, given in Equation 13.8; treating the X-values, and hence $\mathbf{R}_{XX}$ and $\mathbf{U}$, as fixed,

$$E(\mathbf{b}_d^*) = \mathbf{U}E(\mathbf{b}^*) = \mathbf{U}\boldsymbol{\beta}^*$$

[30] See Exercise 13.11.

The bias of $\mathbf{b}_d^*$ is, therefore,

$$\text{bias}(\mathbf{b}_d^*) \equiv E(\mathbf{b}_d^*) - \boldsymbol{\beta}^* = (\mathbf{U} - \mathbf{I}_k)\boldsymbol{\beta}^*$$

and because the departure of $\mathbf{U}$ from $\mathbf{I}_k$ increases with d, the bias of the ridge estimator is an increasing function of d.

The variance of the ridge estimator is also simply derived:[31]

$$V(\mathbf{b}_d^*) = \frac{\sigma_\varepsilon^{*2}}{n-1}(\mathbf{R}_{XX} + d\mathbf{I}_k)^{-1}\mathbf{R}_{XX}(\mathbf{R}_{XX} + d\mathbf{I}_k)^{-1} \qquad [13.9]$$

where σ_ε^{*2} is the error variance for the standardized regression. As d increases, the inverted term $(\mathbf{R}_{XX} + d\mathbf{I}_k)^{-1}$ is increasingly dominated by $d\mathbf{I}_k$. The sampling variance of the ridge estimator, therefore, is a decreasing function of d. This result is intuitively reasonable, because the estimator itself is driven toward 0.

The mean-squared error of the ridge estimator is the sum of its squared bias and sampling variance. Hoerl and Kennard (1970a) prove that it is always possible to choose a positive value of the ridge constant d so that the mean-squared error of the ridge estimator is less than the mean-squared error of the least-squares estimator. As mentioned, however, the optimal value of d depends on the unknown population regression coefficients.

The central problem in applying ridge regression is to find a value of d for which the trade-off of bias against variance is favorable. In deriving the properties of the ridge estimator, I treated d as fixed. If d is determined from the data, however, it becomes a random variable, casting doubt upon the conceptual basis for the ridge estimator. A number of methods have been proposed for selecting d. Some of these are rough and qualitative, while others incorporate specific formulas or procedures for estimating the optimal value of d. All of these methods, however, have only ad hoc justifications.[32]

There have been many random-sampling simulation experiments exploring the properties of ridge estimation along with other methods meant to cope with collinear data. While these studies are by no means unanimous in their conclusions, the ridge estimator often performs well in comparison with least-squares estimation and in comparison with other biased estimation methods. On the basis of evidence from simulation experiments, it would, however, be misleading to recommend a particular procedure for selecting the ridge constant d, and, indeed, the dependence of the optimal value of d on the unknown regression parameters makes it unlikely that there is a generally best way of finding d. Several authors critical of ridge regression (e.g., Draper and Smith, 1981, p. 324) have noted that simulations supporting the method generally incorporate restrictions on parameter values particularly suited to ridge regression.[33]

---

[31] See Exercise 13.12.

[32] Exercise 13.13 describes a qualitative method proposed by Hoerl and Kennard in their 1970 papers.

[33] See Section 13.2.5. Simulation studies of ridge regression and other biased estimation methods are too numerous to cite individually here. References to and comments on this literature can be found in many sources, including Draper and Van Nostrand (1979), Vinod (1978), and Hocking (1976). Vinod and Ullah (1981) present an extensive treatment of ridge regression and related methods.

Because the ridge estimator is biased, standard errors based on Equation 13.9 cannot be used in the normal manner for statistical inferences concerning the population regression coefficients. Indeed, as Obenchain (1977) has pointed out, under the assumptions of the linear model, confidence intervals centered at the least-squares estimates paradoxically retain their optimal properties regardless of the degree of collinearity: In particular, they are the *shortest* possible intervals at the stated level of confidence (Scheffé, 1959, Chapter 2). An interval centered at the ridge estimate of a regression coefficient is, therefore, *wider* than the corresponding least-squares interval, even if the ridge estimator has smaller mean-squared error than the least-squares estimator.

13.2.4 Prior Information about the Regression Coefficients

A final approach to estimation with collinear data is to introduce additional prior information (i.e., relevant information external to the data at hand) that reduces the ambiguity produced by collinearity. There are several different ways that prior information can be brought to bear on a regression, including Bayesian analysis, but I shall present a particularly simple case to illustrate the general point. More complex methods are beyond the scope of this discussion and are, in any event, difficult to apply in practice.[34]

Suppose that we wish to estimate the model

$$Y = \alpha + \beta_1 X_1 + \beta_2 X_2 + \beta_3 X_3 + \varepsilon$$

where Y is savings, X_1 is income from wages and salaries, X_2 is dividend income from stocks, and X_3 is interest income. Imagine that we have trouble estimating β_2 and β_3 because X_2 and X_3 are highly correlated in our data. Suppose further that we have reason to believe that $\beta_2 = \beta_3$, and denote the common quantity β_*. If X_2 and X_3 were not so highly correlated, then we could reasonably test this belief as a hypothesis. In the current situation, we can fit the model

$$Y = \alpha + \beta_1 X_1 + \beta_*(X_2 + X_3) + \varepsilon$$

incorporating our belief in the equality of β_2 and β_3 in the specification of the model, and thus eliminating the collinearity problem (along with the possibility of testing the belief).[35]

13.2.5 Some Comparisons

Although I have presented them separately, the several approaches to collinear data have much in common:

- Model respecification can involve variable selection, and variable selection, in effect, respecifies the model.

---

[34] See, for example, Belsley et al. (1980, pp. 193–204) and Theil (1971, pp. 346–352).

[35] To test H_0: $\beta_2 = \beta_3$ simply entails contrasting the two models (see Exercise 6.10). In the present context, however, where X_2 and X_3 are very highly correlated, this test has virtually no power: If the second model is wrong, then we cannot, as a practical matter, detect it. We need either to accept the second model on theoretical grounds or to admit that we cannot estimate β_2 and β_3.

- Variable selection implicitly constrains the coefficients of deleted regressors to 0.
- Variable selection produces biased coefficient estimates if the deleted variables have nonzero β's and are correlated with the included variables (as they will be for collinear data).[36] As in ridge regression and similar biased estimation methods, we might hope that the trade-off of bias against variance is favorable, and that, therefore, the mean-squared error of the regression estimates is smaller following variable selection than before. Because the bias, and hence the mean-squared error, depend on the unknown regression coefficients, however, we have no assurance that this will be the case. Even if the coefficients obtained following selection have smaller mean-squared error, their superiority can easily be due to the very large variance of the least-squares estimates when collinearity is high than to acceptably small bias.
- Certain types of prior information (as in the hypothetical example presented in the previous section) result in a respecified model.
- It can be demonstrated that biased-estimation methods like ridge regression place prior constraints on the values of the β's. Ridge regression imposes the restriction $\sum_{j=1}^{k} B_j^{*2} \leq c$, where c is a decreasing function of the ridge constant d; the ridge estimator finds least-squares coefficients subject to this constraint (Draper and Smith, 1981, pp. 320–321). In effect, large absolute standardized coefficients are ruled out a priori, but the specific constraint is imposed implicitly.

The primary lesson to be drawn from these remarks is that mechanical model selection and modification procedures disguise the substantive implications of modeling decisions. Consequently, these methods generally cannot compensate for weaknesses in the data and are no substitute for judgment and thought.

Several methods have been proposed for dealing with collinear data. Although these methods are sometimes useful, none can be recommended generally: When the X's are highly collinear, the data contain little information about the partial relationship between each X and Y, controlling for the other X's. To resolve the intrinsic ambiguity of collinear data it is necessary either to introduce information external to the data or to redefine the research question asked of the data. Neither of these general approaches should be undertaken mechanically. Methods that are commonly (and, more often than not, unjustifiably) employed with collinear data include: model respecification; variable selection (stepwise and subset methods); biased estimation (e.g., ridge regression); and the introduction of additional prior information. Comparison of the several methods shows that they have more in common than it appears at first sight.

[36] Bias due to the omission of independent variables is discussed in a general context in Sections 6.3 and 9.6.

EXERCISES

13.6 Why are there $2^k - 1$ distinct subsets of k predictors? Evaluate this quantity for $k = 2, 3, \ldots, 15$.

13.7 *Prove that Mallows's C_p-statistic for a subset of p predictors

$$C_p = \frac{\sum E_i^2}{\hat{\sigma}_\varepsilon^2} + 2p - n$$

can also be written as

$$C_p = (k + 1 - p)(F_p - 1) + p$$

Recall that $\hat{\sigma}_\varepsilon^2$ is the estimated error variance based on the model including all k predictors, and that F_p is the incremental F-statistic for testing the hypothesis that the $k-p$ omitted predictors all have zero coefficients. Why is C_p a reasonable estimator of the total normed mean-squared error of prediction?

$$\gamma_p = \frac{1}{\sigma_\varepsilon^2} \sum_{i=1}^{n} \text{MSE}(\hat{Y}_i)$$

[*Hint*: See Weisberg (1985, Appendix 8A.1).]

13.8 Apply the backward, forward, and forward/backward stepwise-regression methods to Ericksen et al.'s census-undercount data (in `ericksen.dat`). Compare the results of these procedures with those shown in Figure 13.6, based on the application of Mallows's C_p-statistic to all subsets of predictors.

13.9 Apply the variable-selection methods of this section to B. Fox's women's labor-force participation regression (described in Exercise 13.5).

13.10 Cross-validation and variable selection (cf. Hurvich and Tsai, 1990): Perform the following computer-simulation experiment: Independently sample 51 variables, with 200 observations each, from the unit-normal distribution. Call the first variable Y, and the remaining ones $X_1, X_2, \ldots, X_{50}$.

(a) Using all 200 observations, regress Y on $X_1, X_2, \ldots, X_{50}$. Calculate the omnibus F-statistic for the regression, along with the individual t-statistic for each regressor. Construct a quantile comparison plot for the 50 t-statistics, comparing the distribution of the t-values with the normal distribution (or with t for $n - k - 1 = 149$ degrees of freedom). Is the omnibus F statistically significant (say at the 5% level)? How many of the individual B's are statistically significant?

(b) Randomly divide the data in half—placing $n/2 = 100$ observations in each subsample. Employing one or another method of variable selection, and using the data from the first half-sample, find the "best" regression equation that includes $p = 5$ of the $k = 50$ independent variables. Calculate the omnibus F-test and the five individual t-tests for this regression equation. What do you find?

(c) Now recalculate the F- and t-tests in part (b) using the *same* five independent variables but employing the second half-sample. How do these tests compare with those in part (b)?

13.11 *Show that the ridge-regression estimator of the standardized regression coefficients,

$$\mathbf{b}_d^* = (\mathbf{R}_{XX} + d\mathbf{I}_k)^{-1}\mathbf{r}_{Xy}$$

can be written as a linear transformation $\mathbf{b}_d^* = \mathbf{U}\mathbf{b}^*$ of the usual least-squares estimator $\mathbf{b}^* = \mathbf{R}_{XX}^{-1}\mathbf{r}_{Xy}$, where the transformation matrix is $\mathbf{U} \equiv (\mathbf{I}_k + d\mathbf{R}_{XX}^{-1})^{-1}$.

13.12 *Show that the variance of the ridge estimator is

$$V(\mathbf{b}_d^*) = \frac{\sigma_\varepsilon^{*2}}{n-1}(\mathbf{R}_{XX} + d\mathbf{I}_k)^{-1}\mathbf{R}_{XX}(\mathbf{R}_{XX} + d\mathbf{I}_k)^{-1}$$

[*Hint*: Express the ridge estimator as a linear transformation of the standardized dependent-variable values, $\mathbf{b}_d^* = (\mathbf{R}_{XX} + d\mathbf{I}_k)^{-1}[1/(n-1)]\mathbf{Z}_X'\mathbf{z}_y$.]

13.13 *Finding the ridge constant d: Hoerl and Kennard suggest plotting the entries in $\mathbf{b}_d^*$ against values of d ranging between 0 and 1. The resulting graph, called a *ridge trace*, both furnishes a visual representation of the instability due to collinearity and (ostensibly) provides a basis for selecting a value of d. When the data are collinear, we generally observe dramatic changes in regression coefficients as d is gradually increased from 0. As d is increased further, the coefficients eventually stabilize, and then are driven slowly toward 0. The estimated error variance, S_E^{*2}, which is minimized at the least-squares solution ($d = 0$), rises slowly with increasing d. Hoerl and Kennard recommend choosing d so that the regression coefficients are stabilized and the error variance is not unreasonably inflated from its minimum value. (A number of other methods have been suggested for selecting d, but none avoids the fundamental difficulty of ridge regression—that good values of d depend on the unknown β's.)

(a) Construct a ridge trace, including the standard error S_E^*, for Ericksen et al.'s census-undercount regression (in `ericksen.dat`). Use this information to select a value of the ridge constant d, and compare the resulting ridge estimates of the regression parameters with the least-squares estimates. Make this comparison for both standardized and unstandardized coefficients.

(b) Repeat part (a) for B. Fox's women's labor-force participation data (in Table 13.3 and `bfox.dat`; see Exercises 13.5 and 13.9). In applying ridge regression to these data, B. Fox selected $d = 0.05$.

13.3 Summary

- When the regressors in a linear model are perfectly collinear, the least-squares coefficients are not unique. Strong, but less-than-perfect, collinearity substantially increases the sampling variances of the least-squares coefficients, and can render them useless as estimators.
- The sampling variance of the least-squares slope coefficient B_j is

$$V(B_j) = \frac{1}{1 - R_j^2} \times \frac{\sigma_\varepsilon^2}{(n-1)S_j^2}$$

  where R_j^2 is the squared multiple correlation for the regression of X_j on the other X's, and $S_j^2 = \sum(X_{ij} - \overline{X}_j)^2/(n-1)$ is the variance of X_j. The variance inflation factor $\text{VIF}_j = 1/(1 - R_j^2)$ indicates the deleterious impact of collinearity on the precision of the estimate B_j. The notion of variance inflation can be extended to sets of related regressors, such as dummy regressors and polynomial regressors, by considering the size of the joint confidence region for the related coefficients.
- Principal components can be used to explicate the correlational structure of the independent variables in regression. The principal components are a derived set of variables that form an orthogonal basis for the subspace of the standardized X's. The first principal component spans the one-dimensional subspace that accounts for maximum variation in the standardized X's. The second principal component accounts for maximum variation in the standardized X's, under the constraint that it is orthogonal to the first. The other principal components are similarly defined; unless the X's are perfectly collinear, there are as many principal components as are there are X's. Each principal component is scaled to have variance equal to the collective variance in the standardized X's for which it accounts. Collinear relations among the independent variables, therefore, correspond to very short principal components, which represent dimensions along which the regressor subspace has nearly collapsed.
- Several methods have been proposed for dealing with collinear data. Although these methods are sometimes useful, none can be recommended generally: When the X's are highly collinear, the data contain little information about the partial relationship between each X and Y, controlling for the other X's. To resolve the intrinsic ambiguity of collinear data, it is necessary either to introduce information external to the data or to redefine the research question asked of the data. Neither of these general approaches should be undertaken mechanically. Methods that are commonly (and, more often than not, unjustifiably) employed with collinear data include: model respecification; variable selection (stepwise and subset methods); biased estimation (e.g., ridge regression); and the introduction of additional prior information. Comparison of the several methods shows that they have more in common than it appears at first sight.

PART IV

Beyond Linear Least Squares

14

Extending Linear Least Squares: Time Series, Nonlinear, Robust, and Nonparametric Regression*

This chapter introduces four important generalizations of linear least-squares regression:

- *Generalized least squares* can be used to fit a linear regression model to time series data in which the errors are correlated over time rather than independent.
- *Nonlinear regression* fits a specific nonlinear function of the independent variables by least squares.
- *Robust regression*, introduced informally in Chapter 2, employs criteria for fitting a linear model that are not as sensitive as least squares to heavy-tailed error distributions.
- *Nonparametric regression*, also introduced in Chapter 2, does not assume a specific functional relationship relating the dependent variable to the independent variables.

Taken together, the methods of this chapter substantially expand the range of application of regression analysis.

14.1 Time Series Regression and Generalized Least Squares

The standard linear model of Chapters 5 through 10 assumes independently distributed errors. The assumption of independence is rarely (if ever) quite right, but it is often a reasonable approximation. When the observations comprise a *time series*, however, dependencies among the errors can be very strong.

In time series data, a single individual (person, nation, etc.) is tracked over many time periods or points of time.[1] These time periods or time points are usually evenly spaced, at least approximately, and I shall assume that this is the case. Economic statistics for Canada, for example, are reported on a daily, monthly, quarterly, and yearly basis. Crime statistics, likewise, are reported on a yearly basis. Later in this section, we shall use yearly time series for the period 1935 to 1968 to examine the relationship between Canadian women's crime rates and certain other factors, including fertility, women's labor-force participation, women's participation in higher education, and men's crime rates.

It is not generally reasonable to suppose that the errors in a time series regression are independent: After all, time periods that are close to one another are more likely to be similar than time periods that are relatively remote. This similarity may well extend to the errors, which represent (most importantly) the omitted causes of the dependent variable. Although the time dependence among the errors may turn out to be negligible, it is unwise to assume a priori that this is the case.

> In time series data, a single individual is tracked over many time periods or points of time. It is not generally reasonable to suppose that the errors in a time series regression are independent.

14.1.1 Generalized Least-Squares Estimation

I shall first address dependencies among the errors in a very general context. Consider the usual linear model

$$\underset{(n\times 1)}{\mathbf{y}} = \underset{(n\times k+1)}{\mathbf{X}}\;\underset{(k+1\times 1)}{\boldsymbol{\beta}} + \underset{(n\times 1)}{\boldsymbol{\varepsilon}}$$

but, rather than assuming that the errors are independently distributed, let us instead assume that

$$\boldsymbol{\varepsilon} \sim N_n(\mathbf{0}, \boldsymbol{\Sigma}_{\varepsilon\varepsilon})$$

where the order-n matrix $\boldsymbol{\Sigma}_{\varepsilon\varepsilon}$ is symmetric and positive definite. Nonzero off-diagonal entries in the covariance matrix $\boldsymbol{\Sigma}_{\varepsilon\varepsilon}$ correspond to correlated errors.[2]

---

[1] Temperature, for example, may be recorded at evenly spaced time points. Gross national product is cumulated over the period of a year. Most social data are collected in time periods rather than at time points.

[2] Because of the assumption of normality, dependence implies correlation. As for the standard linear model with independent errors, however, most of the results of this section do not require the assumption of normality. Notice that unequal *diagonal* entries of $\boldsymbol{\Sigma}_{\varepsilon\varepsilon}$ correspond to unequal error variances, a problem discussed in Section 12.2.

> To capture serial dependence among the errors in the regression model $\mathbf{y} = \mathbf{X}\boldsymbol{\beta} + \boldsymbol{\varepsilon}$, we drop the assumption that the errors are independent of one another; instead, we assume $\boldsymbol{\varepsilon} \sim N_n(\mathbf{0}, \boldsymbol{\Sigma}_{\varepsilon\varepsilon})$, where nonzero off-diagonal entries in the error covariance matrix $\boldsymbol{\Sigma}_{\varepsilon\varepsilon}$ correspond to correlated errors.

Let us assume unrealistically (and only for the moment) that we know $\boldsymbol{\Sigma}_{\varepsilon\varepsilon}$. Then the log likelihood for the model is[3]

$$\log_e L(\boldsymbol{\beta}) = -\frac{n}{2}\log_e 2\pi - \frac{1}{2}\log_e(\det \boldsymbol{\Sigma}_{\varepsilon\varepsilon}) - \frac{1}{2}(\mathbf{y} - \mathbf{X}\boldsymbol{\beta})'\boldsymbol{\Sigma}_{\varepsilon\varepsilon}^{-1}(\mathbf{y} - \mathbf{X}\boldsymbol{\beta}) \quad [14.1]$$

It is clear that the log likelihood is maximized when the *generalized sum of squares* $(\mathbf{y} - \mathbf{X}\boldsymbol{\beta})'\boldsymbol{\Sigma}_{\varepsilon\varepsilon}^{-1}(\mathbf{y} - \mathbf{X}\boldsymbol{\beta})$ is minimized.[4] Differentiating the generalized sum of squares with respect to $\boldsymbol{\beta}$, setting the partial derivatives to 0, and solving for $\boldsymbol{\beta}$ produces the *generalized least-squares (GLS) estimator*

$$\mathbf{b}_{\text{GLS}} = (\mathbf{X}'\boldsymbol{\Sigma}_{\varepsilon\varepsilon}^{-1}\mathbf{X})^{-1}\mathbf{X}'\boldsymbol{\Sigma}_{\varepsilon\varepsilon}^{-1}\mathbf{y} \quad [14.2]$$

It is simple to show that the GLS estimator is unbiased, $E(\mathbf{b}_{\text{GLS}}) = \boldsymbol{\beta}$; that its sampling variance is

$$V(\mathbf{b}_{\text{GLS}}) = (\mathbf{X}'\boldsymbol{\Sigma}_{\varepsilon\varepsilon}^{-1}\mathbf{X})^{-1}$$

and that, by an extension of the Gauss-Markov theorem,[5] $\mathbf{b}_{\text{GLS}}$ is the minimum-variance linear unbiased estimator of $\boldsymbol{\beta}$. None of these results (with the exception of the one establishing the GLS estimator as the ML estimator) requires the assumption of normality.

Here is another way of thinking about the GLS estimator: Let $\boldsymbol{\Gamma}_{(n\times n)}$ be a "square-root" of $\boldsymbol{\Sigma}_{\varepsilon\varepsilon}^{-1}$, in the sense that $\boldsymbol{\Gamma}'\boldsymbol{\Gamma} = \boldsymbol{\Sigma}_{\varepsilon\varepsilon}^{-1}$. From Equation 14.2,

$$\begin{aligned}\mathbf{b}_{\text{GLS}} &= (\mathbf{X}'\boldsymbol{\Gamma}'\boldsymbol{\Gamma}\mathbf{X})^{-1}\mathbf{X}'\boldsymbol{\Gamma}'\boldsymbol{\Gamma}\mathbf{y} \\ &= (\mathbf{X}^{*\prime}\mathbf{X}^{*})^{-1}\mathbf{X}^{*\prime}\mathbf{y}^{*}\end{aligned}$$

where $\mathbf{X}^* \equiv \boldsymbol{\Gamma}X$ and $\mathbf{y}^* \equiv \boldsymbol{\Gamma}\mathbf{y}$. Thus, the GLS estimator is the ordinary-least-squares (OLS) estimator for the regression of $\mathbf{y}^*$ on $\mathbf{X}^*$—that is, following the linear transformation of $\mathbf{y}$ and $\mathbf{X}$ using the transformation matrix $\boldsymbol{\Gamma}$.

---

[3] See Exercise 14.1 for this and other results described in this section.

[4] Recall that $\boldsymbol{\Sigma}_{\varepsilon\varepsilon}^{-1}$ is assumed to be known.

[5] The Gauss-Markov theorem is discussed in Section 9.3 in the context of ordinary-least-squares regression.

If the error covariance matrix $\mathbf{\Sigma}_{\varepsilon\varepsilon}$ is known, then the maximum-likelihood estimator of $\boldsymbol{\beta}$ is the generalized least-squares estimator $\mathbf{b}_{\text{GLS}} = (\mathbf{X}'\mathbf{\Sigma}_{\varepsilon\varepsilon}^{-1}\mathbf{X})^{-1}\mathbf{X}'\mathbf{\Sigma}_{\varepsilon\varepsilon}^{-1}\mathbf{y}$. The sampling variance-covariance matrix of $\mathbf{b}_{\text{GLS}}$ is $V(\mathbf{b}_{\text{GLS}}) = (\mathbf{X}'\mathbf{\Sigma}_{\varepsilon\varepsilon}^{-1}\mathbf{X})^{-1}$. The generalized least-squares estimator can also be expressed as the OLS estimator $(\mathbf{X}^{*\prime}\mathbf{X}^{*})^{-1}\mathbf{X}^{*\prime}\mathbf{y}^{*}$ for the transformed variables $\mathbf{X}^{*} \equiv \mathbf{\Gamma}\mathbf{X}$ and $\mathbf{y}^{*} \equiv \mathbf{\Gamma}\mathbf{y}$, where the transformation matrix $\mathbf{\Gamma}$ is a square root of $\mathbf{\Sigma}_{\varepsilon\varepsilon}^{-1}$.

14.1.2 Serially Correlated Errors

I have, thus far, left the covariance matrix of the errors $\mathbf{\Sigma}_{\varepsilon\varepsilon}$ very general: Because of its symmetry, there are $n(n+1)/2$ potentially different elements in $\mathbf{\Sigma}_{\varepsilon\varepsilon}$. Without further assumptions concerning the structure of this matrix, we cannot hope to estimate its elements from only n observations if—as is always the case in real applications of time series regression—$\mathbf{\Sigma}_{\varepsilon\varepsilon}$ is not known.

Suppose, however, that the process generating the errors is *stationary*. Stationarity means that the errors all have the same expectation (which, indeed, we have already assumed to be 0); that the errors have a common variance (σ_ε^2); and that the covariance of two errors depends only on their separation in time. Let ε_t denote the error for time period t, and ε_{t+s} the error for time period $t+s$. Stationarity implies that, for any t, the covariance between ε_t and ε_{t+s} is

$$C(\varepsilon_t, \varepsilon_{t+s}) = E(\varepsilon_t \varepsilon_{t+s}) = \sigma_\varepsilon^2 \rho_s = C(\varepsilon_t, \varepsilon_{t-s})$$

where ρ_s, called the *autocorrelation* (or *serial correlation*) at lag s, is the correlation between two errors separated by s time periods.

The error covariance matrix, then, has the following pattern:

$$\mathbf{\Sigma}_{\varepsilon\varepsilon} = \sigma_\varepsilon^2 \begin{bmatrix} 1 & \rho_1 & \rho_2 & \cdots & \rho_{n-1} \\ \rho_1 & 1 & \rho_1 & \cdots & \rho_{n-2} \\ \rho_2 & \rho_1 & 1 & \cdots & \rho_{n-3} \\ \vdots & \vdots & \vdots & \ddots & \vdots \\ \rho_{n-1} & \rho_{n-2} & \rho_{n-3} & \cdots & 1 \end{bmatrix} = \sigma_\varepsilon^2 \mathbf{P} \tag{14.3}$$

The situation is much improved, but it is not good enough: There are now n distinct parameters to estimate in $\mathbf{\Sigma}_{\varepsilon\varepsilon}$—that is, σ_ε^2 and $\rho_1, \ldots, \rho_{n-1}$.

When, more realistically, the error covariance matrix $\mathbf{\Sigma}_{\varepsilon\varepsilon}$ is unknown, we need to estimate its contents along with the regression coefficients $\boldsymbol{\beta}$. Without restricting its form, however, $\mathbf{\Sigma}_{\varepsilon\varepsilon}$ contains too many unique elements to estimate directly. Supposing that the errors are generated by a stationary time series process reduces the number of independent parameters in $\mathbf{\Sigma}_{\varepsilon\varepsilon}$ to n, including the error variance σ_ε^2 and the autocorrelations at various lags, $\rho_1, \ldots, \rho_{n-1}$.

To proceed, we need to specify a stationary process for the errors that depends on fewer parameters. The process that is by far most commonly used in practice is the *first-order autoregressive process*[6]

$$\varepsilon_t = \rho\varepsilon_{t-1} + \nu_t \qquad [14.4]$$

where the error in time period t depends directly only on the error in the previous time period, ε_{t-1}, and on a random contemporaneous "shock" ν_t. Unlike the regression errors ε_t, we shall assume that the random shocks ν_t are independent of each other (and of ε's from earlier time periods), and that $\nu_t \sim N(0, \sigma_\nu^2)$. Serial correlation in the regression errors, therefore, is wholly generated by the partial dependence of each error on the error of the previous time period.

For Equation 14.4 to specify a stationary process, it is necessary that $|\rho| < 1$. Otherwise, the errors will tend to grow without bound. If the process is stationary, and if all errors have zero expectations, then

$$\begin{aligned}\sigma_\varepsilon^2 &\equiv V(\varepsilon_t) = E(\varepsilon_t^2) \\ &= V(\varepsilon_{t-1}) = E(\varepsilon_{t-1}^2)\end{aligned}$$

[6] The first-order autoregressive process is a simple member of the family of autoregressive, moving-average processes. In the pth-order autoregressive process [abbreviated AR(p)], ε_t depends on the previous p errors:

$$\varepsilon_t = \phi_1\varepsilon_{t-1} + \phi_2\varepsilon_{t-2} + \cdots + \phi_p\varepsilon_{t-p} + \nu_t$$

(where we use ϕ rather than ρ because the autoregressive coefficients are no longer correlations). In the order-q moving-average process [MA(q)], the error at time t depends on the random shock at time t and on the shocks in the previous q time periods:

$$\varepsilon_t = \nu_t + \theta_1\nu_{t-1} + \theta_2\nu_{t-2} + \cdots + \theta_q\nu_{t-q}$$

The autoregressive, moving-average process ARMA(p, q) combines these elements:

$$\varepsilon_t = \phi_1\varepsilon_{t-1} + \phi_2\varepsilon_{t-2} + \cdots + \phi_p\varepsilon_{t-p} + \nu_t + \theta_1\nu_{t-1} + \theta_2\nu_{t-2} + \cdots + \theta_q\nu_{t-q}$$

These more general ARMA processes are capable of modeling a wider variety of patterns of autocorrelation. See, for example, Judge et al. (1985, Chapters 7 and 8) for details.

Squaring both sides of Equation 14.4 and taking expectations,

$$\begin{aligned} E(\varepsilon_t^2) &= \rho^2 E(\varepsilon_{t-1}^2) + E(\nu_t^2) + 2\rho E(\varepsilon_{t-1}\nu_t) \\ \sigma_\varepsilon^2 &= \rho^2 \sigma_\varepsilon^2 + \sigma_\nu^2 \end{aligned}$$

because $E(\varepsilon_{t-1}\nu_t) = C(\varepsilon_{t-1}, \nu_t) = 0$. Solving for the variance of the regression errors yields

$$\sigma_\varepsilon^2 = \frac{\sigma_\nu^2}{1-\rho^2}$$

It is also a simple matter to find the autocorrelation at lag s. For example, at lag 1, we have the covariance

$$\begin{aligned} C(\varepsilon_t, \varepsilon_{t-1}) &= E(\varepsilon_t \varepsilon_{t-1}) \\ &= E[(\rho\varepsilon_{t-1} + \nu_t)\varepsilon_{t-1}] \\ &= \rho\sigma_\varepsilon^2 \end{aligned}$$

So the correlation at lag 1 is just

$$\begin{aligned} \rho_1 &= \frac{C(\varepsilon_t, \varepsilon_{t-1})}{\sqrt{V(\varepsilon_t) \times V(\varepsilon_{t-1})}} \\ &= \frac{\rho\sigma_\varepsilon^2}{\sigma_\varepsilon^2} \\ &= \rho \end{aligned}$$

Likewise, at lag 2,

$$\begin{aligned} C(\varepsilon_t, \varepsilon_{t-2}) &= E(\varepsilon_t \varepsilon_{t-2}) \\ &= E\{[\rho(\rho\varepsilon_{t-2} + \nu_{t-1}) + \nu_t]\varepsilon_{t-2}\} \\ &= \rho^2\sigma_\varepsilon^2 \end{aligned}$$

and, therefore, $\rho_2 = \rho^2$. More generally, for the first-order autoregressive process, $\rho_s = \rho^s$, and because $|\rho| < 1$, the autocorrelations of the errors decay exponentially toward 0.

To reduce the number of parameters in $\mathbf{\Sigma}_{\varepsilon\varepsilon}$ further, we can adopt a particular time series model for the errors. The most commonly employed such model is the first-order autoregressive process $\varepsilon_t = \rho\varepsilon_{t-1} + \nu_t$, where $|\rho| < 1$ and the random shocks ν_t are independently distributed as $N(0, \sigma_\nu^2)$. Under this specification, two errors, ε_t and ε_{t+s}, separated by s time periods have covariance $\rho^s\sigma_\varepsilon^2$. The variance of the regression errors is $\sigma_\varepsilon^2 = \sigma_\nu^2/(1-\rho^2)$.

Some "realizations" of time series generated by Equation 14.4, with $\nu_t \sim N(0, 1)$, are shown in Figure 14.1. In Figure 14.1(a), $\rho = 0$, and, consequently, the ε_t are uncorrelated, a time series process sometimes termed *white noise*. In Figure 14.1(b), $\rho = .9$; notice how values of the series close to one another tend to be similar. In Figure 14.1(c), $\rho = -.7$; notice here how the series tends to bounce from negative to positive values. Negatively autocorrelated series are not common in the social sciences. Finally, Figure 14.1(d) illustrates a non-stationary process, with $\rho = 1.01$. The *autocorrelation functions* for $\rho = .9$ and $\rho = -.7$—showing the value of ρ_s as a function of the lag s—appear in Figure 14.2.

Figure 14.1(b) also provides some intuitive insight into the problems for estimation posed by autocorrelated errors:

- Because errors that are close in time are likely to be similar, there is much less information in a highly autocorrelated time series sample than in an independent sample of the same size. It is, for example, often unproductive to proliferate observations by using more closely spaced time periods (e.g., monthly or quarterly rather than yearly data[7]). To do so will likely increase the autocorrelation of the errors.[8]
- Over a relatively short period of time, a highly autocorrelated series is likely to rise or to fall—that is, show a positive or negative trend. This is true even though the series is stationary and, therefore, will eventually return to its expectation of 0. If our sample consisted of the first 25 observations in Figure 14.1(b), then there would be a negative trend in the errors; if our sample consisted of the last 25 observations, then there would be a positive trend.

  Because independent variables in a time series regression also often manifest directional trends, a rise or fall in the errors of a short time series can induce a correlation between an independent variable and the errors *for this specific sample*. It is important, however, to understand that there is no implication that the OLS estimates are biased because of correlation between the independent variables and the errors: Over many samples, there will sometimes be negative correlations between the errors and the independent variables, sometimes positive correlations, and sometimes no correlation. The correlations—sometimes negative, sometimes positive—that occur in specific samples can substantially increase the *variance* of the OLS estimator, however.[9]
- Finally, because the OLS estimator forces zero *sample* correlations between the independent variables and the residuals (as opposed to the unobserved errors),

---

[7] Monthly or quarterly data also raise the possibility of "seasonal" effects. One simple approach to seasonal effects is to include dummy regressors for months or quarters. Likewise, dummy regressors for days of the week might be appropriate for some daily time series. More sophisticated approaches to seasonal effects are described in the references at the end of the chapter.

[8] This point is nicely illustrated by considering the sampling variance of the sample mean $\overline{Y}$. From elementary statistics, we know that the variance of $\overline{Y}$ in an independent random sample of size n is σ^2/n, where σ^2 is the population variance. If instead we sample observations from a first-order autoregressive process with parameter ρ, the variance of $\overline{Y}$ is

$$\frac{\sigma^2}{n} \times \frac{1+\rho}{1-\rho}$$

The sampling variance of $\overline{Y}$ is, therefore, substantially larger than σ^2/n when the autocorrelation ρ is close to 1. Put another way, the "effective" number of observations is $n(1-\rho)/(1+\rho)$ rather than n. I am grateful to Robert Stine for suggesting this illustration.

[9] The effect of autocorrelated errors on OLS estimation is explored in Exercise 14.2.

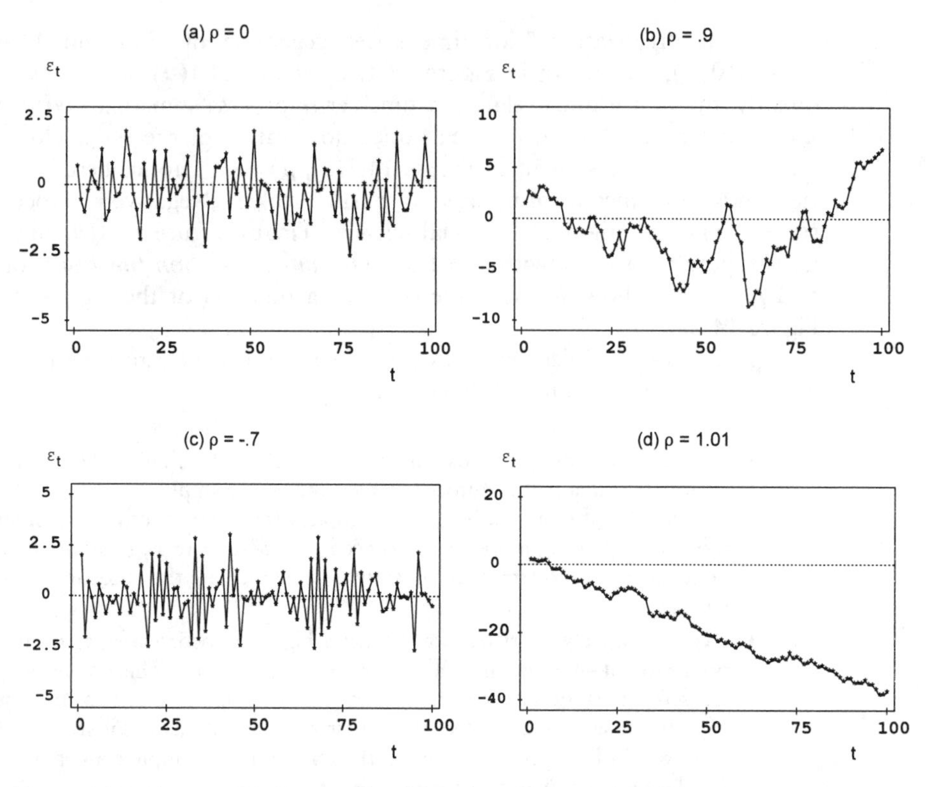

Figure 14.1. Four realizations, each of sample size $n = 100$, of the autoregressive process $\varepsilon_t = \rho\varepsilon_{t-1} + \nu_t$. (*a*) $\rho = 0$ ("white noise"). (*b*) $\rho = .9$. (*c*) $\rho = -.7$. (*d*) $\rho = 1.01$ (a nonstationary process).

the sampling variances of the OLS coefficients may be grossly underestimated by $S_E^2(\mathbf{X}'\mathbf{X})^{-1}$. Recall that in a short series, the highly autocorrelated errors will often manifest a trend in a *particular* sample.

GLS Estimation With Autoregressive Errors

If the errors follow a first-order autoregressive process, then the covariance matrix of the regression errors, given in general form in Equation 14.3, takes the relatively simple form

$$\boldsymbol{\Sigma}_{\varepsilon\varepsilon}(\rho, \sigma_\nu^2) = \frac{\sigma_\nu^2}{1-\rho^2}\begin{bmatrix} 1 & \rho & \rho^2 & \cdots & \rho^{n-1} \\ \rho & 1 & \rho & \cdots & \rho^{n-2} \\ \rho^2 & \rho & 1 & \cdots & \rho^{n-3} \\ \vdots & \vdots & \vdots & \ddots & \vdots \\ \rho^{n-1} & \rho^{n-2} & \rho^{n-3} & \cdots & 1 \end{bmatrix} \qquad [14.5]$$

As the notation implies, the error covariance matrix depends on only two parameters: σ_ν^2 and ρ—or, alternatively, on ρ and $\sigma_\varepsilon^2 = \sigma_\nu^2/(1-\rho^2)$. If we knew the values of these parameters, then we could form $\boldsymbol{\Sigma}_{\varepsilon\varepsilon}$ and proceed directly to GLS estimation.

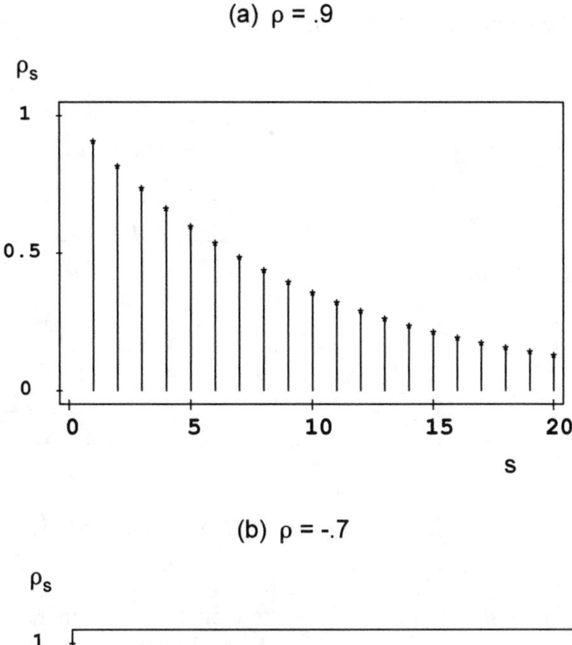

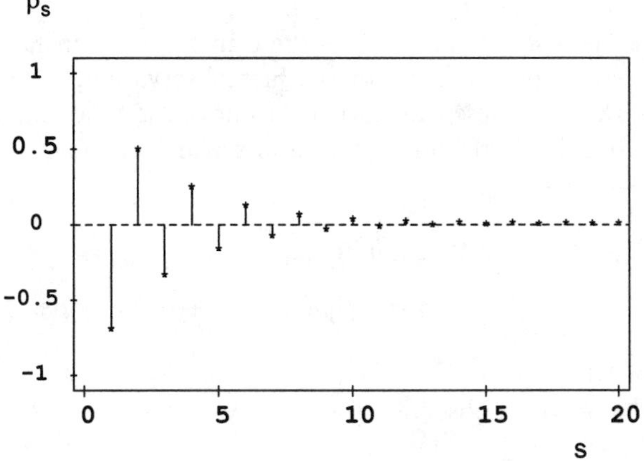

Figure 14.2. Theoretical autocorrelations ρ_s for the first-order autoregressive process $\varepsilon_t = \rho\varepsilon_{t-1} + \nu_t$, with (a) $\rho = .9$ and (b) $\rho = -.7$.

Recall that GLS estimation can be realized as OLS following a transformation of $\mathbf{y}$ and $\mathbf{X}$. In the present case (ignoring a constant factor), the transformation matrix is[10]

$$\mathbf{\Gamma} = \begin{bmatrix} \sqrt{1-\rho^2} & 0 & 0 & \cdots & 0 & 0 \\ -\rho & 1 & 0 & \cdots & 0 & 0 \\ 0 & -\rho & 1 & \cdots & 0 & 0 \\ \vdots & \vdots & \vdots & \ddots & \vdots & \vdots \\ 0 & 0 & 0 & \cdots & -\rho & 1 \end{bmatrix}$$

[10] See Exercise 14.3.

Then the transformed variables are

$$\mathbf{y}^* = \mathbf{\Gamma y} = \begin{bmatrix} \sqrt{1-\rho^2}Y_1 \\ Y_2 - \rho Y_1 \\ \vdots \\ Y_n - \rho Y_{n-1} \end{bmatrix} \qquad [14.6]$$

and

$$\mathbf{X}^* = \mathbf{\Gamma X} = \begin{bmatrix} \sqrt{1-\rho^2} & \sqrt{1-\rho^2}X_{11} & \cdots & \sqrt{1-\rho^2}X_{1k} \\ 1-\rho & X_{21} - \rho X_{11} & \cdots & X_{2k} - \rho X_{1k} \\ \vdots & \vdots & & \vdots \\ 1-\rho & X_{n1} - \rho X_{n-1,\,1} & \cdots & X_{nk} - \rho X_{n-1,\,k} \end{bmatrix} \qquad [14.7]$$

where the first column of $\mathbf{X}^*$ is the transformed constant regressor.

Notice that, except for the first observation, $Y_t^* = Y_t - \rho Y_{t-1}$ and $X_{tj}^* = X_{tj} - \rho X_{t-1,\,j}$. These transformations have the following intuitive interpretation: Write out the regression equation in scalar form as

$$\begin{aligned} Y_t &= \alpha + \beta_1 X_{t1} + \cdots + \beta_k X_{tk} + \varepsilon_t \\ &= \alpha + \beta_1 X_{t1} + \cdots + \beta_k X_{tk} + \rho\varepsilon_{t-1} + \nu_t \end{aligned} \qquad [14.8]$$

For the previous observation $(t-1)$, we have

$$Y_{t-1} = \alpha + \beta_1 X_{t-1,\,1} + \cdots + \beta_k X_{t-1,\,k} + \varepsilon_{t-1} \qquad [14.9]$$

Multiplying Equation 14.9 through by ρ and subtracting the result from Equation 14.8 produces

$$\begin{aligned} Y_t - \rho Y_{t-1} &= \alpha(1-\rho) + \beta_1(X_{t1} - \rho X_{t-1,\,1}) \\ &\quad + \cdots + \beta_k(X_{tk} - \rho X_{t-1,\,k}) + \nu_t \\ Y_t^* &= \alpha 1^* + \beta_1 X_{t1}^* + \cdots + \beta_k X_{tk}^* + \nu_t \quad \text{for } t = 2, \ldots, n \end{aligned} \qquad [14.10]$$

Because the errors in Equation 14.10 are the ν_t, which are independent, the transformed equation can legitimately be fit by OLS regression. The only slippage here is that the first observation is lost, for there are no data at $t-1 = 1-1 = 0$. Applying OLS to Equation 14.10 is, therefore, not quite the same as GLS.

Empirical GLS Estimation

All of this presupposes that we know the value of the error autocorrelation ρ. In practice, of course, we need to estimate ρ along with the regression parameters $\alpha, \beta_1, \ldots, \beta_k$ and the variance of the random shocks σ_ν^2 (or, alternatively, the variance of the regression errors σ_ε^2). One approach to this problem is first to estimate ρ. Then, using the estimate (say $\hat{\rho}$) as if ρ were known, we can calculate GLS estimates and their standard errors—either directly or, equivalently, by OLS following transformation of $\mathbf{y}$ and $\mathbf{X}$. This approach is called *empirical generalized least squares (EGLS)*.[11]

A particularly simple option is to base the estimate of ρ on the lag-one autocorrelation of the residuals from the OLS regression[12] of $\mathbf{y}$ on $\mathbf{X}$:

$$r_1 = \frac{\sum_{t=2}^{n} E_t E_{t-1}}{\sum_{t=1}^{n} E_t^2} \qquad [14.11]$$

where the E_t are the OLS residuals. Notice that the sum in the numerator of Equation 14.11 is over observations $t = 2, \ldots, n$ (because E_{t-1}—i.e., E_0—is unavailable for $t = 1$). Using $\hat{\rho} = r_1$ in Equations 14.6 and 14.7 produces transformed variables from which to calculate the EGLS estimates by OLS regression. The variance of the residuals from this OLS regression estimates σ_ν^2.

> To apply GLS estimation for first-order autoregressive errors, we can first estimate the autocorrelation of the errors from the sample autocorrelation of the OLS residuals: $\hat{\rho} = r_1 = (\sum_{t=2}^{n} E_t E_{t-1}) / (\sum_{t=1}^{n} E_t^2)$. We can then use $\hat{\rho}$ to form an estimate of the correlation matrix of the errors or to transform $\mathbf{y}$ and $\mathbf{X}$. Except for the first observation, these transformations take a particularly simple form: $Y_t^* = Y_t - \hat{\rho} Y_{t-1}$ and $X_{tj}^* = X_{tj} - \hat{\rho} X_{t-1,j}$. This procedure is called empirical GLS estimation.

Figure 14.3, based on data from Fox and Hartnagel (1979), shows a yearly time series plot of the theft-conviction rate per 100,000 Canadian women aged 15 and above, for the period 1935 to 1968.[13] The theft-conviction rate rose during the early 1940s, then declined until the mid-1950s, and subsequently rose dramatically.

---

[11] We could alternatively try to estimate all of the parameters—ρ, σ_ν^2, and β—directly and simultaneously, for example, by maximum likelihood. This approach is described in Exercise 14.4.

[12] Although there are other methods to obtain a preliminary estimate of ρ, none holds a particular advantage. For details, see the references on time series regression given at the end of the chapter.

[13] Because the basis for reporting convictions for theft changed in 1949, the data for the period 1950–1968 have been adjusted. The adjustment used here is very slightly different from the one employed by Fox and Hartnagel (1979).

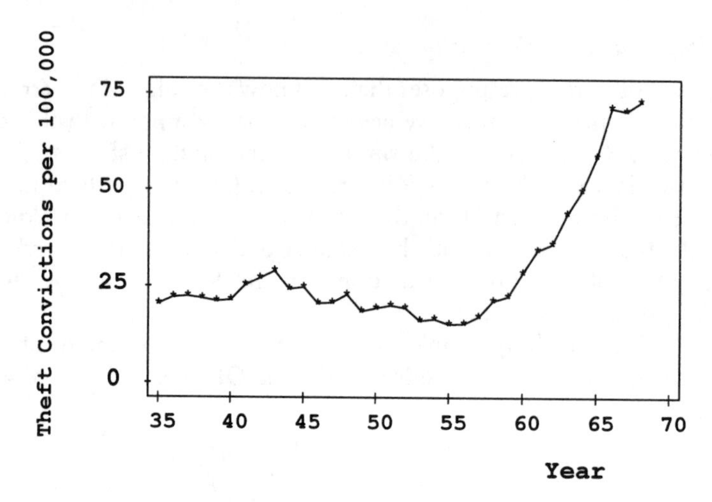

Figure 14.3. Canadian women's theft-conviction rate per 100,000 population, for the period 1935–1968.

Fox and Hartnagel were interested in relating variations in women's crime rates to changes in their position within Canadian society. To this end, they regressed the women's theft-conviction rate on the following independent variables:

- The *total fertility rate* (*TFR*)—the number of births to an imaginary cohort of 1000 women who live through their child-bearing years at current age-specific fertility rates.
- *Women's labor-force participation rate* (*LFPR*) per 1000 population.
- *Women's postsecondary-degree rate* (*PSDR*) per 10,000 population.
- *Men's theft-conviction rate* (*MTCR*) per 100,000 population. This independent variable was meant to represent factors affecting women's theft-conviction rate that are not specifically included in the model.

Ordinary-least-squares estimates for the regression of women's theft-conviction rate on the four independent variables are shown in Table 14.1,

TABLE 14.1 Regressions of Canadian Women's Theft-Conviction Rate on Several Independent Variables, for the Period 1935 to 1968: Total Fertility Rate (TFR); Labor-Force Participation Rate (LFPR); Postsecondary Degree Rate (PSDR); and Men's Theft-Conviction Rate (MTCR). The first set of EGLS estimates uses all 34 observations; the second set of EGLS estimates drops the first observation

	OLS		*EGLS*(1)		*EGLS*(2)	
Coefficient	*B*	$\widehat{SE}(B)$	*B*	$\widehat{SE}(B)$	*B*	$\widehat{SE}(B)$
Constant	−7.337	9.438	−6.645	11.319	−5.520	11.739
TFR	−0.006090	0.0014	−0.005879	0.0018	−0.006079	0.0019
LFPR	0.1199	0.023	0.1156	0.028	0.1136	0.028
PSDR	0.5516	0.043	0.5359	0.051	0.5342	0.051
MTCR	0.03933	0.018	0.03994	0.022	0.04071	0.022
S_E	3.812		3.730		3.722	

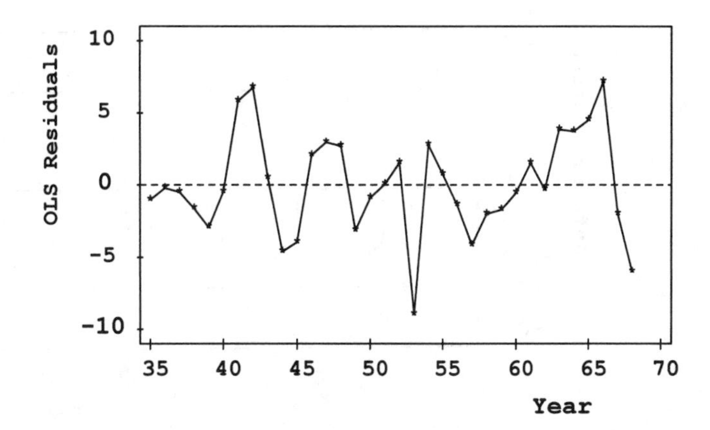

Figure 14.4. Residuals from the OLS regression of women's theft-conviction rate on several independent variables.

along with two EGLS regressions. The OLS regression fits the data very well, with a small standard error and $R^2 = .953$. The residuals from the OLS regression, graphed in Figure 14.4, are slightly autocorrelated, with $r_1 = .2442$.

Using the sample autocorrelation to transform the variables in the model produces the EGLS estimates shown in Table 14.1. The estimates labeled EGLS(1) are based on all of the data, including the first observation, which is transformed using $\sqrt{1 - r_1^2}$. The estimates labeled EGLS(2) omit the first observation. In this instance, where the autocorrelation of the residuals is small, all three sets of estimates are similar (although the estimated standard errors of the EGLS estimates are a bit larger than the standard errors of the OLS estimates).

14.1.3 Diagnosing Serially Correlated Errors

We need to ask whether the data support the hypothesis that the errors are serially correlated, because, in the absence of serially correlated errors, we can legitimately employ OLS estimation. As usual, our key to the behavior of the unobservable errors is the least-squares residuals.[14]

A useful first step is to plot the residuals against time, as in Figure 14.4. Then we can calculate sample autocorrelations for the residuals:

$$r_s = \frac{\sum_{t=s+1}^{n} E_t E_{t-s}}{\sum_{t=1}^{n} E_t^2}$$

for lags $s = 1, 2, \ldots, m$, where the maximum lag m should be no larger than about $n/4$. If the errors follows a first-order autoregressive process, then the residual correlations should (roughly) decay exponentially toward 0.

A graph of the residual autocorrelations, called a *correlogram*, for Fox and Hartnagel's regression is shown in Figure 14.5. The residual autocorrelations are all small and do not appear to decay exponentially: While r_1 is positive, r_2 and

[14] Cf. the discussion of diagnostics in Chapters 11 and 12.

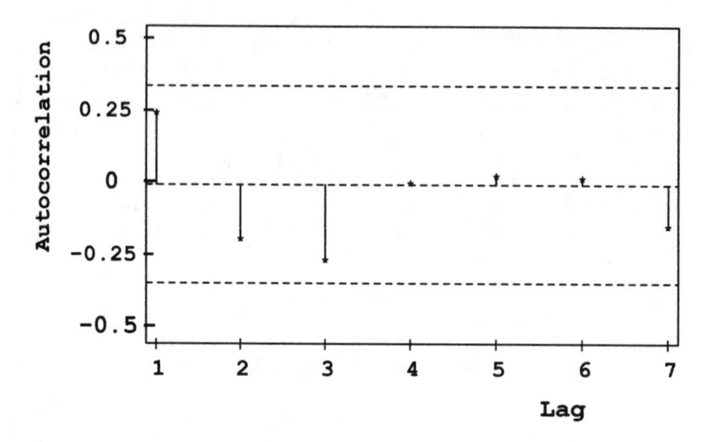

Figure 14.5. Residual autocorrelations from the OLS regression of women's theft-conviction rates on several independent variables. The horizontal lines are at 0 and ±2 approximate standard errors.

r_3 are both negative. In interpreting these correlations, however, it would help to have an indication of their sampling variability.

If the residuals were independently distributed (which, recall they are not—even if the errors are independent[15]), then the standard error of each r_s would be approximately $1/\sqrt{n}$, and the autocorrelations would be asymptotically normally distributed.[16] A rough guide to the "statistical significance" of the residual autocorrelations is therefore to place reference lines in the correlogram at $\pm 2/\sqrt{n}$. The correlogram in Figure 14.5, for example, suggests that none of the residual autocorrelations is statistically significant.[17]

Because a simple test for autocorrelation based on the r_s is, at best, rough, more precise methods have been proposed. The most popular method of testing for autocorrelated errors is due to Durbin and Watson (1950, 1951). Durbin and Watson's test statistic is based on the assumption that the errors follow a first-order autoregressive process, and tests the null hypothesis that the autoregressive parameter ρ is 0:

$$D \equiv \frac{\sum_{t=2}^{n}(E_t - E_{t-1})^2}{\sum_{t=1}^{n} E_t^2}$$

When n is large,[18] $D \simeq 2(1 - r_1)$. If the null hypothesis is correct, therefore, we expect to observe values of D close to 2; if the null hypothesis is wrong, and the errors are positively autocorrelated (i.e., $\rho > 0$), then we expect to observe values of D that are substantially smaller than 2. The range of D values is 0 to 2.

The sampling distribution of the Durbin-Watson statistic is complex and, unfortunately, depends on the configuration of the independent variables. Durbin

---

[15] See Section 11.8.

[16] See, for example, Chatfield (1989, Section 4.1).

[17] A further reason for caution in interpreting the correlogram is that there are many sample autocorrelations which are themselves correlated, creating a problem of simultaneous inference.

[18] See Exercise 14.5.

and Watson initially calculated critical values of D for two extreme scenarios: the X-configuration producing the smallest critical value, and the X-configuration producing the largest critical value. This approach leads to an extensive set of tables (because different critical values are required for different combinations of sample size n and number of independent variables k), and to ambiguous results if—as is common—the observed value of D falls between the two extreme critical values.[19]

Modern statistical software for time series regression, however, typically calculates a p-value for D based on the X-values in the sample at hand. For Fox and Hartnagel's regression, for example, $D = 1.42$, for which $p = .006$ (obtained from the AUTOREG procedure in SAS).

> To diagnose serial correlation in the errors, we can examine the correlogram of the OLS residuals, calculating the serial correlations $r_s = \left(\sum_{t=s+1}^{n} E_t E_{t-s}\right) / \left(\sum_{t=1}^{n} E_t^2\right)$ for a number of lags $s = 1, 2, \ldots, m$. If the serial dependence of the residuals is well described by a first-order autoregressive process, then these serial correlations should decay exponentially toward 0. The Durbin-Watson statistic $D \equiv \left[\sum_{t=2}^{n}(E_t - E_{t-1})^2\right] / \left(\sum_{t=1}^{n} E_t^2\right) \simeq 2(1 - r_1)$ can be used to test for serial correlation of the errors.

14.1.4 Concluding Remarks

It is tempting to conclude that the theoretical advantage of GLS regression should mandate its use, particularly if the residuals are significantly autocorrelated, but there are several factors that suggest caution:[20]

- The extent to which GLS estimation is more efficient that OLS estimation, and the extent to which the usual formula for OLS standard errors produces misleading results, depend on a number of complex factors, including the process generating the errors, the degree of autocorrelation of the errors, and the distribution of the independent-variable values.[21]
- When the errors are highly autocorrelated, and when independent-variable values manifest a linear trend, the advantage of GLS estimation can be strongly dependent on retaining the first observation (i.e., using the transformation $\sqrt{1-\rho^2}$ of the first observation).[22] Precisely in these circumstances, the first observation can become influential in the transformed regression, however. The comfort that

---

[19] Tables of critical values of D were published by Durbin and Watson (1951), and are reproduced in many texts.

[20] For an elaboration of some of these points, see, for example, the discussion in Judge et al. (1985, Chapter 8).

[21] See Exercise 14.2.

[22] See Exercise 14.2.

we derive from GLS may, therefore, be tenuous. It is, consequently, useful to examine influential-data diagnostics for the *transformed* equation.[23]

- Many of the properties of GLS and EGLS estimators depend on asymptotic results, but time series datasets are usually quite small.[24] Indeed, long time series raise the possibility that regression relationships—for example, the slope associated with a particular independent variable—may themselves change over time. There is no generally satisfactory method for detecting such changes, although it may help to plot residuals against time.
- The performance of a method like GLS estimation may depend crucially on getting the error-generating process right. If the process is not first-order autoregressive, for example, basing estimation on this process may cause more harm than good. Real social processes do not, in any event, unfold according to autoregressive processes which, at best, produce reasonable descriptive summaries. We have to be careful not to construe time series regression models too literally.

Three further cautionary notes are in order:

- Because time series data often manifest strong trends, the independent variables in a time series regression can be strongly collinear.[25] This problem can be compounded when time itself (i.e., the regressor $X_t = t$) is included in the regression to capture a linear trend. The general rationale for employing time as a regressor is to control statistically for omitted factors that change smoothly with time and that are correlated with the independent variables included in the model.
- The models discussed in this section assume contemporaneous effects. That is, all of the variables in the model are measured at time t. It is sometimes reasonable to suppose, however, that the effect of an independent variable (say, X_1) will occur after an interval of time has elapsed, for example, after one time period. The time series regression model would then take the form

$$Y_t = \alpha + \beta_1 X_{t-1,1} + \beta_2 X_{t2} + \cdots + \beta_k X_{tk} + \varepsilon_t$$

  Aside from the loss of the first observation (because $X_{t-1,1}$ is unavailable for $t - 1 = 0$), specifications of this form pose no new problems. If, however, we do not know in advance that the effect of X_1 is lagged some specific number of time periods, and rather want to consider effects at several lags, then autocorrelation of X_1 can induce serious collinearity. Special techniques of estimation (called "distributed lags"[26]) exist to deal with this situation, but these methods require that we know in advance something about the form of the lagged effects of X_1 on Y.
- Finally, the methods of this section are generally inappropriate (in the presence of autocorrelated errors) when the *dependent variable* appears as a lagged effect on the right-hand side of the model, as in

$$Y_t = \alpha + \beta Y_{t-1} + \beta_1 X_{t1} + \cdots + \beta_k X_{tk} + \varepsilon_t$$

---

[23] See Chapter 11.

[24] The bootstrapping methods of Section 16.1 can prove helpful in this context.

[25] See, for example, the regression for Canadian women's labor-force participation described in Exercises 13.5, 13.9, and 14.7.

[26] See, for example, the discussion of distributed lags in Judge et al. (1985, Chapters 9 and 10).

There are several practical and theoretical difficulties that limit the effectiveness and range of application of EGLS estimation.

EXERCISES

14.1 *Generalized least squares: For the linear model $\mathbf{y} = \mathbf{X}\boldsymbol{\beta} + \boldsymbol{\varepsilon}$ with $\boldsymbol{\varepsilon} \sim N_n(\mathbf{0}, \boldsymbol{\Sigma}_{\varepsilon\varepsilon})$, where the error covariance matrix $\boldsymbol{\Sigma}_{\varepsilon\varepsilon}$ is known:

(a) Show that the log likelihood for the model is

$$\log_e L(\boldsymbol{\beta}) = -\frac{n}{2}\log_e 2\pi - \frac{1}{2}\log_e(\det \boldsymbol{\Sigma}_{\varepsilon\varepsilon}) - \frac{1}{2}(\mathbf{y} - \mathbf{X}\boldsymbol{\beta})'\boldsymbol{\Sigma}_{\varepsilon\varepsilon}^{-1}(\mathbf{y} - \mathbf{X}\boldsymbol{\beta})$$

(*Hint*: Use the formula for the multivariate normal distribution.)

(b) Show that the maximum-likelihood estimator of $\boldsymbol{\beta}$ is

$$\mathbf{b}_{\text{GLS}} = (\mathbf{X}'\boldsymbol{\Sigma}_{\varepsilon\varepsilon}^{-1}\mathbf{X})^{-1}\mathbf{X}'\boldsymbol{\Sigma}_{\varepsilon\varepsilon}^{-1}\mathbf{y}$$

and that its sampling variance is

$$V(\mathbf{b}_{\text{GLS}}) = (\mathbf{X}'\boldsymbol{\Sigma}_{\varepsilon\varepsilon}^{-1}\mathbf{X})^{-1}$$

(c) Prove the Gauss-Markov theorem for the GLS estimator. That is, show that under the assumptions $E(\boldsymbol{\varepsilon}) = \mathbf{0}$ and $V(\boldsymbol{\varepsilon}) = \boldsymbol{\Sigma}_{\varepsilon\varepsilon}$, $\mathbf{b}_{\text{GLS}}$ is the minimum-variance linear unbiased estimator of $\boldsymbol{\beta}$. (*Hints*: See Section 9.3.2 and Exercise 12.5.)

14.2 Autocorrelated errors and OLS estimation: Assume that $\mathbf{y} = \mathbf{X}\boldsymbol{\beta} + \boldsymbol{\varepsilon}$ and that the errors follow a first-order autoregressive process, $\varepsilon_t = \rho\varepsilon_{t-1} + \nu_t$, where $\nu_t \sim N(0, \sigma_\nu^2)$.

(a) Show that the OLS estimator $\mathbf{b} = (\mathbf{X}'\mathbf{X})^{-1}\mathbf{X}'\mathbf{y}$ is unbiased despite the autocorrelation of the errors. (*Hint*: Recall the proof that $E(\mathbf{b}) = \boldsymbol{\beta}$ from Section 9.3.)

(b) Show that the variance of the OLS estimator is

$$V(\mathbf{b}) = (\mathbf{X}'\mathbf{X})^{-1}\mathbf{X}'\boldsymbol{\Sigma}_{\varepsilon\varepsilon}\mathbf{X}(\mathbf{X}'\mathbf{X})^{-1}$$

where $\boldsymbol{\Sigma}_{\varepsilon\varepsilon}$ is given by Equation 14.5.

(c) Suppose that we fit the simple-regression model $Y_t = \alpha + \beta x_t + \varepsilon_t$; that the x-values are 1, 2, …, 10; that the errors follow a first-order autoregressive process; and that the variance of the random shocks ν_t is 1. [Recall that the variance of the errors is $\sigma_\varepsilon^2 = \sigma_\nu^2/(1 - \rho^2)$.] Calculate (1) the *true* sampling variance of the OLS estimator; (2) the sampling variance of the OLS estimator according to the usual formula, $\sigma_\varepsilon^2(\mathbf{X}'\mathbf{X})^{-1}$, appropriate when the errors are independent; and (3) the sampling variance of the GLS estimator, using the formula

$$V(\mathbf{b}_{\mathrm{GLS}}) = (\mathbf{X}'\mathbf{\Sigma}_{\varepsilon\varepsilon}^{-1}\mathbf{X})^{-1}$$

Pay particular attention to the sampling variance of the estimators of the slope β. What do you conclude from these calculations? Perform the calculations under the following three circumstances:

(i) $\rho = 0$
(ii) $\rho = .5$
(iii) $\rho = .9$

(d) *Now suppose that the first observation is dropped, and that we perform the regression in (c) using $t = 2, \ldots, 10$. Working with the transformed scores $x_t^* = x_t - \rho x_{t-1}$, and again employing the three different values of ρ, find the sampling variance of the resulting estimator of $\boldsymbol{\beta}$. How does the efficiency of this estimator compare with that of the true GLS estimator? With the OLS estimator? What would happen if there were many more observations?

(e) Repeat (c) [and (d)], but with $x_t = (t-5)^2$ for $t = 1, 2, \ldots, 9$.

14.3 *Show that the appropriate GLS transformation matrix for first-order autoregressive errors is

$$\mathbf{\Gamma} = \begin{bmatrix} \sqrt{1-\rho^2} & 0 & 0 & \cdots & 0 & 0 \\ -\rho & 1 & 0 & \cdots & 0 & 0 \\ 0 & -\rho & 1 & \cdots & 0 & 0 \\ \vdots & \vdots & \vdots & \ddots & \vdots & \vdots \\ 0 & 0 & 0 & \cdots & -\rho & 1 \end{bmatrix}$$

[*Hints*: First show that

$$\frac{1}{1-\rho^2}\begin{bmatrix} 1 & -\rho & 0 & \cdots & 0 & 0 \\ -\rho & 1+\rho^2 & -\rho & \cdots & 0 & 0 \\ 0 & -\rho & 1+\rho^2 & \cdots & 0 & 0 \\ \vdots & \vdots & \vdots & \ddots & \vdots & \vdots \\ 0 & 0 & 0 & \cdots & 1+\rho^2 & -\rho \\ 0 & 0 & 0 & \cdots & -\rho & 1 \end{bmatrix}$$

is the inverse of

$$\mathbf{P} = \begin{bmatrix} 1 & \rho & \rho^2 & \cdots & \rho^{n-1} \\ \rho & 1 & \rho & \cdots & \rho^{n-2} \\ \rho^2 & \rho & 1 & \cdots & \rho^{n-3} \\ \vdots & \vdots & \vdots & \ddots & \vdots \\ \rho^{n-1} & \rho^{n-2} & \rho^{n-3} & \cdots & 1 \end{bmatrix}$$

Then show that $\mathbf{P}^{-1} = [1/(1-\rho^2)]\mathbf{\Gamma}'\mathbf{\Gamma}$. The constant $1/(1-\rho^2)$ can be ignored in forming the square-root matrix $\mathbf{\Gamma}$. Why?]

14.4 *Maximum-likelihood estimation with first-order autoregressive errors: Assume that $\mathbf{y} = \mathbf{X}\boldsymbol{\beta} + \boldsymbol{\varepsilon}$ and that the errors follow a first-order autoregressive process, but that the autoregressive parameter ρ and the variance of the random shocks σ_v^2 are unknown.

(a) Show that the log likelihood under this model can be written in the following form:

$$\log_e L(\boldsymbol{\beta}, \rho, \sigma_v^2) = -\frac{n}{2}\log_e 2\pi - \frac{n}{2}\log_e \sigma_v^2 + \frac{1}{2}\log_e(1-\rho^2) - \frac{1}{2\sigma_v^2}(\mathbf{y}^* - \mathbf{X}^*\boldsymbol{\beta})'(\mathbf{y}^* - \mathbf{X}^*\boldsymbol{\beta})$$

where $\mathbf{y}^* = \mathbf{\Gamma}\mathbf{y}$ and $\mathbf{X}^* = \mathbf{\Gamma}\mathbf{X}$. [*Hints*: Start with Equation 14.1 for the log likelihood of the general model with error covariance matrix $\mathbf{\Sigma}_{\varepsilon\varepsilon}$. Then use $\mathbf{\Sigma}_{\varepsilon\varepsilon} = (1/\sigma_v^2)\mathbf{\Gamma}'\mathbf{\Gamma}$, noting that $\det\mathbf{\Sigma}_{\varepsilon\varepsilon} = (1/\sigma_v^2)^n(\det\mathbf{\Gamma})^2$ and that $\det\mathbf{\Gamma} = \sqrt{1-\rho^2}$.]

(b) Although there are more sophisticated approaches, the maximum-likelihood estimate of $\boldsymbol{\beta}$ can be found by searching over candidate values of ρ (i.e., $-1 < \rho < 1$), because—conditional on the value of ρ—the MLE of $\boldsymbol{\beta}$ can be calculated by generalized least squares. We can simply pick the estimate $\hat{\rho}$ and the associated $\hat{\boldsymbol{\beta}}$ that yields the biggest log likelihood. Apply this approach to Fox and Hartnagel's regression for female theft-conviction rates. (The data are given in `hartnagl.dat`.) How do the maximum-likelihood estimates compare with the OLS and GLS estimates shown in Table 14.1?

(c) Estimated asymptotic standard errors for $\hat{\boldsymbol{\beta}}$, $\hat{\rho}$, and $\hat{\sigma}_v^2$ can be obtained in the usual manner from the inverse of the information matrix. Find $\mathscr{V}(\hat{\boldsymbol{\beta}}, \hat{\rho}, \hat{\sigma}_v^2)$ and calculate $\hat{\mathscr{V}}(\hat{\boldsymbol{\beta}}, \hat{\rho}, \hat{\sigma}_v^2)$ for Fox and Hartnagel's regression. Is $\hat{\rho}$ statistically significant?

14.5 Show that when n is large, the Durbin-Watson statistic D is approximately equal to $2(1 - r_1)$, where r_1 is the lag-one autocorrelation of the OLS residuals. (*Hint*: When n is large, $\sum_{t=1}^{n} E_t^2 \simeq \sum_{t=2}^{n} E_t^2 \simeq \sum_{t=2}^{n} E_{t-1}^2$.)

14.6 Fox and Hartnagel also regressed women's conviction rates for *all* indictable (i.e., serious) offenses on the total fertility rate, women's labor-force-participation rate, postsecondary-degree rate, and men's indictable-offense conviction rate. Apply the methods of this section to this regression. The data are in `hartnagl.dat`.

14.7 Recall from Exercises 13.5 and 13.9 B. Fox's time series regression of Canadian women's labor-force-participation rate on several factors thought to influence this variable. Employing the subset of independent variables selected in Exercise 13.9, apply the methods of this section to diagnose, and—if it appears desirable—correct, serially correlated errors in B. Fox's regression. The data are in Table 13.3 and `bfox.dat`.

14.2 Nonlinear Regression

As I have explained, there is a distinction between independent variables and regressors.[27] The general linear model is linear in the regressors but not necessarily in the independent variables that generate these regressors.

In analysis of variance, for example, the independent variables are qualitative and do not appear directly in the model; a single polytomous independent variable gives rise to several dummy regressors. Likewise, in polynomial regression, a single quantitative independent variable generates several regressors (e.g., linear, quadratic, and cubic terms). Interaction regressors are functions of two or more independent variables. We can also transform the *dependent* variable prior to formulating the linear model.

In its least restrictive form, then, we can write the general linear model as

$$f(Y_i) = \beta_0 f_0(\mathbf{x}_i') + \beta_1 f_1(\mathbf{x}_i') + \cdots + \beta_p f_p(\mathbf{x}_i') + \varepsilon_i$$
$$Y_i' = \beta_0 X_{i0}' + \beta_1 X_{i1}' + \cdots + \beta_p X_{ip}' + \varepsilon_i$$

where

- Y_i is the dependent variable for the ith observation;
- $\mathbf{x}_i'$ is a $1 \times k$ vector of (not necessarily quantitative) independent variables;
- $\beta_0, \beta_1, \ldots, \beta_p$ are parameters to estimate;
- the ε_i are independent and normally distributed errors, with zero expectations and constant variance; and
- the functions $f(\cdot), f_0(\cdot), \ldots, f_p(\cdot)$ do not involve unknown parameters.

For the least-squares estimates of the β's to be unique, the regressors $X_0', \ldots, X_p'$ must be linearly independent. If, as is usually the case, the model includes the constant regressor, then $X_{i0}' \equiv f_0(\mathbf{x}_i') = 1$. An X_j' can be a function of more than one independent variable, encompassing models such as

$$Y = \beta_0 + \beta_1 X_1 + \beta_2 X_2 + \beta_3 X_1^2 + \beta_4 X_2^2 + \beta_5 X_1 X_2 + \varepsilon \qquad [14.12]$$

In an application, the functions $f(\cdot), f_0(\cdot), \ldots, f_p(\cdot)$ may be suggested by prior theoretical considerations, or by examination of the data, as when we transform an independent variable to linearize its relationship to Y.

> Nonlinear regression models that are linear in the parameters, for example the quadratic regression model $Y = \beta_0 + \beta_1 X_1 + \beta_2 X_2 + \beta_3 X_1^2 + \beta_4 X_2^2 + \beta_5 X_1 X_2 + \varepsilon$, can be fit by linear least squares.

---

[27] See Chapters 7 and 8.

14.2.1 Polynomial Regression

Polynomial regression is an important form of nonlinear regression that is accommodated by the general linear model. We have already seen how a quadratic function can be employed to model a U-shaped (or inverted U-shaped) relationship, and how a cubic function can be used to model a relationship in which the direction of curvature changes.[28] More generally, a polynomial of order p can have $p-1$ "bends." Polynomials of degree greater than 3 are rarely employed in data analysis.

Polynomial regressors are particularly useful when a quantitative independent variable is discrete. As we know,[29] we can capture any—potentially nonlinear—partial relationship between Y and X_j by constructing $m-1$ dummy regressors to represent the m distinct values of X_j. The powers $X_j, X_j^2, \ldots, X_j^{m-1}$ provide an alternative—and hence equivalent—basis for the subspace spanned by the dummy regressors.

We can then "step down" through the powers $X_j^{m-1}, X_j^{m-2}, \ldots, X_j^2, X_j^1$, testing the contribution of each term to the model, omitting the term if it proves unnecessary, and refitting the model. We stop dropping terms when one proves to be important. Thus, if the model includes a cubic term, it will also generally include the lower-order quadratic and linear terms. Even if the relationship between Y and X_j is nonlinear, it is usually possible to represent this relationship with a polynomial[30] of degree less than $m-1$.

Polynomials in two or more independent variables can be used to model interactions between quantitative independent variables. Consider, for example, the full quadratic model in two independent variables given in Equation 14.12 above. As illustrated in Figure 14.6(a), this model represents a curved surface relating $E(Y)$ to X_1 and X_2. Of course, certain specific characteristics of the regression surface—such as direction of curvature and monotonicity—depend on the parameters of the model and on the range of values[31] for X_1 and X_2.

The essential structure of nonlinear models is often clarified by differentiating the model with respect to each independent variable.[32] Differentiating Equation 14.12,

$$\frac{\partial E(Y)}{\partial X_1} = \beta_1 + 2\beta_3 X_1 + \beta_5 X_2$$

$$\frac{\partial E(Y)}{\partial X_2} = \beta_2 + 2\beta_4 X_2 + \beta_5 X_1$$

The slope of the partial relationship between Y and X_1, therefore, depends not only on the level of X_1, but also on the specific value at which X_2 is held constant—indicating that X_1 and X_2 interact in affecting Y. Moreover, the shape of the partial relationship between Y and X_1 is quadratic, fixing the value of X_2. Because of the symmetry of the model, similar statements apply to the partial relationship between Y and X_2, holding X_1 constant.[33]

---

[28] See, for example, the discussion of the Canadian occupational prestige data in Section 12.3.

[29] See Section 12.4.1.

[30] Also see Exercise 14.9 for a discussion of orthogonal polynomial contrasts.

[31] See Exercise 14.10.

[32] See Exercise 14.8.

[33] An illustrative application of Equation 14.12 appears in Exercise 14.11.

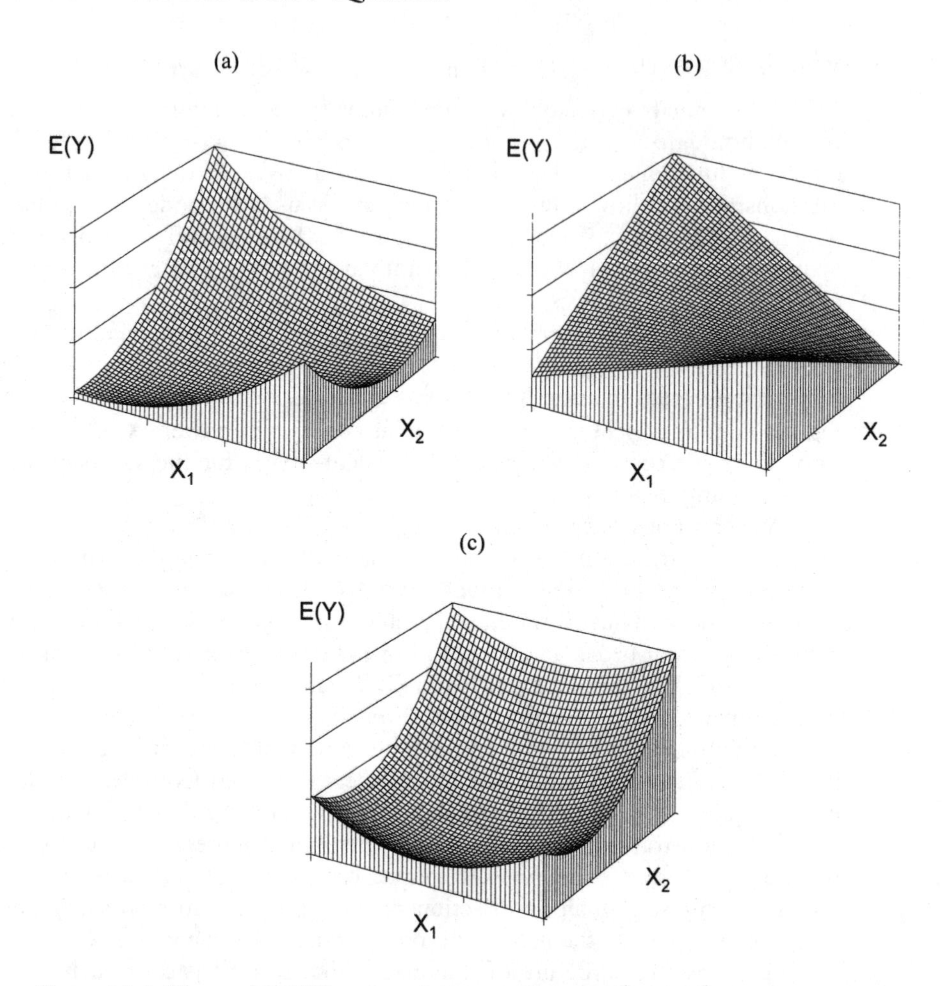

Figure 14.6. The model $E(Y) = \beta_0 + \beta_1 X_1 + \beta_2 X_2 + \beta_3 X_1^2 + \beta_4 X_2^2 + \beta_5 X_1 X_2$, shown in (*a*), represents a curved surface in which the quadratic partial relationship of Y to X_1 changes with the value of X_2 (and the quadratic partial relationship of Y to X_2 changes with the value of X_1). The model $E(Y) = \beta_0 + \beta_1 X_1 + \beta_2 X_2 + \beta_3 X_1 X_2$ in (*b*) represents a curved surface in which the slope of the linear partial relationship of Y to X_1 changes with the value of X_2. The model $E(Y) = \beta_0 + \beta_1 X_1 + \beta_2 X_2 + \beta_3 X_1^2 + \beta_4 X_2^2$ in (*c*) represents a curved surface in which the quadratic partial relationship of Y to X_1 is the same at different levels of X_2.

In contrast, although the model

$$Y = \beta_0 + \beta_1 X_1 + \beta_2 X_2 + \beta_3 X_1 X_2 + \varepsilon$$

also represents a curved surface [an illustration appears in Figure 14.6(*b*)], the slices of this surface in the direction of each independent variable, holding the other constant, are linear:

$$\frac{\partial E(Y)}{\partial X_1} = \beta_1 + \beta_3 X_2$$

$$\frac{\partial E(Y)}{\partial X_2} = \beta_2 + \beta_3 X_1$$

Thus, for example, the slope of the relationship between Y and X_1 is different at different levels of X_2, but at each fixed level of X_2, the relationship between Y and X_1 is linear.

Finally, the model

$$Y = \beta_0 + \beta_1 X_1 + \beta_2 X_2 + \beta_3 X_1^2 + \beta_4 X_2^2 + \varepsilon$$

[illustrated in Figure 14.6(*c*)] represents a curved surface in which the quadratic partial relationship between Y and each of the independent variables is invariant across the levels of the other independent variable:

$$\frac{\partial E(Y)}{\partial X_1} = \beta_1 + \beta_3 X_1$$

$$\frac{\partial E(Y)}{\partial X_2} = \beta_2 + \beta_4 X_2$$

14.2.2 Transformable Nonlinearity

As explained in the previous section, linear statistical models effectively encompass models that are linear in the parameters, even if they are nonlinear in the variables. The forms of nonlinear relationships that can be expressed as linear models are, therefore, very diverse. In certain circumstances, however, theory dictates that we fit models that are nonlinear in their parameters. This is a relatively rare necessity in the social sciences, primarily because our theories are seldom mathematically concrete, although nonlinear models arise in some areas of demography, economics, and psychology, and occasionally in sociology, political science, and other social sciences.

Some models that are nonlinear in the parameters can be transformed into linear models, and, consequently, can be fit to data by linear least squares. A model of this type is the so-called "gravity model" of migration, employed in human geography (see Abler, et al., 1971, pp. 221–233). Let Y_{ij} represent the number of migrants moving from city i to city j; let D_{ij} represent the geographical distance between these cities; and let P_i and P_j represent their respective populations.

The gravity model of migration is built in rough analogy to the Newtonian formula for gravitational attraction between two objects, where population plays the role of mass and migration the role of gravity. The analogy is far from perfect, in part because gravitational attraction is symmetric, but there are two migration streams of generally different sizes between a pair of cities: one from city i to city j, and the other from j to i.

The gravity model is given by the equation

$$\begin{aligned} Y_{ij} &= \alpha \frac{P_i^{\beta} P_j^{\gamma}}{D_{ij}^{\delta}} \varepsilon_{ij} \\ &= \tilde{Y}_{ij} \varepsilon_{ij} \end{aligned} \qquad [14.13]$$

where α, β, γ, and δ are unknown parameters to be estimated from the data; and ε_{ij} is a necessarily positive multiplicative error term that reflects the imperfect determination of migration by distance and population size. When ε_{ij} is

1, Y_{ij} is equal to its "predicted" value $\tilde{Y}_{ij}$, given by the systematic part of the model; when ε_{ij} is less than 1, Y_{ij} is smaller than $\tilde{Y}_{ij}$; and when ε_{ij} is greater than 1, Y_{ij} exceeds $\tilde{Y}_{ij}$. I shall say more about the error presently. Because of the multiplicative form of the gravity model, $\tilde{Y}_{ij}$ is not $E(Y_{ij})$.

Although the gravity model (Equation 14.13) is nonlinear in its parameters, it can be transformed into a linear equation by taking logs:[34]

$$\begin{aligned} \log Y_{ij} &= \log \alpha + \beta \log P_i + \gamma \log P_j - \delta \log D_{ij} + \log \varepsilon_{ij} \\ Y'_{ij} &= \alpha' + \beta P'_i + \gamma P'_j + \delta D'_{ij} + \varepsilon'_{ij} \end{aligned} \qquad [14.14]$$

where

$$\begin{aligned} \alpha' &\equiv \log \alpha \\ P'_i &\equiv \log P_i \\ P'_j &\equiv \log P_j \\ D'_{ij} &\equiv -\log D_{ij} \\ \varepsilon'_{ij} &\equiv \log \varepsilon_{ij} \end{aligned}$$

If we can make the usual linear-model assumptions about the transformed errors ε'_{ij}, then we are justified in fitting the transformed model (Equation 14.14) by linear least squares. In the gravity model, it is probably unrealistic to assume that the transformed errors are independent, because individual cities are involved in many different migration streams. A particularly attractive city, for example, might have positive errors for each of its in-migration streams and negative errors for each of its out-migration streams. We could, however, adapt the model to account for these effects.[35]

Our ability to linearize the model given in Equation 14.13 by a log transformation depends on the multiplicative errors in this model. The multiplicative error specifies that the general magnitude of the difference between Y_{ij} and $\tilde{Y}_{ij}$ is proportional to the size of the latter: The model tends to make larger absolute errors in predicting large migration streams than in predicting small ones. This assumption appears reasonable here. In most cases, we should prefer to specify a form of error—additive or multiplicative—that leads to a simple statistical analysis—supposing, of course, that the specification is reasonable. A subsequent analysis of residuals permits us to subject these assumptions to scrutiny.

Another form of multiplicative model is

$$\begin{aligned} Y_i &= \alpha \exp(\beta_1 X_{i1}) \exp(\beta_2 X_{i2}) \cdots \exp(\beta_k X_{ik}) \varepsilon_i \\ &= \alpha \exp(\beta_1 X_{i1} + \beta_2 X_{i2} + \cdots + \beta_k X_{ik}) \varepsilon_i \end{aligned} \qquad [14.15]$$

---

[34] The log transformation requires that Y_{ij}, α, P_i, P_j, D_{ij}, and ε_{ij} are all positive, as is the case for the gravity model of migration.

[35] See Exercise 14.12 for an illustrative application of the gravity model.

where $e \simeq 2.718$ is the base of the natural logarithms. Taking logs produces the linear equation

$$Y_i' = \alpha' + \beta_1 X_{i1} + \beta_2 X_{i2} + \cdots + \beta_k X_{ik} + \varepsilon_i'$$

with

$$Y_i' \equiv \log_e Y_i$$

$$\alpha' \equiv \log_e \alpha$$

$$\varepsilon_i' \equiv \log_e \varepsilon_i$$

In Equation 14.15, the impact on Y of increasing X_j by one unit is proportional to the level of Y. The effect of rescaling Y by taking logs is to eliminate interaction among the X's. A similar result is, at times, achievable empirically through other power transformations of Y.

> Some nonlinear models—for example, the multiplicative gravity model of migration—can be rendered linear by a transformation.

Multiplicative models provide the most common instance of transformable nonlinearity, but there are also other models to which this approach is applicable. Consider, for example, the model

$$Y_i = \frac{1}{\alpha + \beta X_i + \varepsilon_i} \qquad [14.16]$$

where ε_i is a random error satisfying the standard assumptions. Then, if we take $Y_i' \equiv 1/Y_i$, we can rewrite the model as the linear equation $Y_i' = \alpha + \beta X_i + \varepsilon_i$. This model is illustrated in Figure 14.7 (for a positive β and positive values of X).

14.2.3 Nonlinear Least Squares

Models that are nonlinear in the parameters and that cannot be rendered linear by a transformation are called *essentially nonlinear*. The *general nonlinear model* is given by the equation

$$Y_i = f\left(\underset{(p\times 1)}{\boldsymbol{\beta}},\ \underset{(1\times k)}{\mathbf{x}_i'}\right) + \varepsilon_i \qquad [14.17]$$

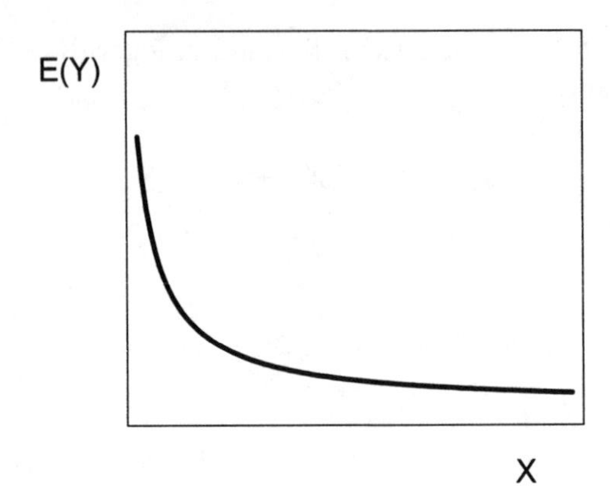

Figure 14.7. The model $E(Y) = 1/(\alpha + \beta X_i + \varepsilon_i)$, for $X > 0$ and $\beta > 0$.

in which

- Y_i is the dependent-variable value for the ith of n observations;
- $\boldsymbol{\beta}$ is a vector of p parameters to be estimated from the data;
- $\mathbf{x}_i'$ is a row vector of scores for observation i on the k independent variables (some of which may be qualitative); and
- ε_i is the error for the ith observation.

It is convenient to write the model in matrix form for the full sample of n observations as

$$\underset{(n\times 1)}{\mathbf{y}} = \mathbf{f}(\underset{(p\times 1)}{\boldsymbol{\beta}}, \underset{(n\times k)}{\mathbf{X}}) + \underset{(n\times 1)}{\boldsymbol{\varepsilon}}$$

I shall assume, as in the general linear model,[36] that $\boldsymbol{\varepsilon} \sim N_n(\mathbf{0}, \sigma_\varepsilon^2 \mathbf{I}_n)$.

An illustrative essentially nonlinear model is the logistic population-growth model (Shryock and Siegel, 1973, pp. 382–385):

$$Y_i = \frac{\beta_1}{1 + \exp(\beta_2 + \beta_3 X_i)} + \varepsilon_i \qquad [14.18]$$

where Y_i is population size, and X_i is time; for equally spaced observations, it is usual to take $X_i = i - 1$. Because the logistic growth model is fit to time series data, the assumption of independent errors is problematic: It may well be the case that the errors are *autocorrelated*—that is, that errors close in time tend to be similar. The additive form of the error is also questionable here, for errors may well grow larger in magnitude as population size increases. Despite these potential difficulties, the logistic population-growth model can provide a useful, if gross and preliminary, representation of the data.[37]

---

[36] For multiplicative errors, we can put the model in the form of Equation 14.17 by taking logs.

[37] See Exercise 14.14.

Under the assumption of independent and normally distributed errors, with zero expectations and common variance, the general nonlinear model (Equation 14.17) has likelihood

$$L(\boldsymbol{\beta}, \sigma_\varepsilon^2) = \frac{1}{(2\pi\sigma_\varepsilon^2)^{n/2}} \exp\left\{-\frac{\sum_{i=1}^n [Y_i - f(\boldsymbol{\beta}, \mathbf{x}_i')]^2}{2\sigma_\varepsilon^2}\right\}$$
$$= \frac{1}{(2\pi\sigma_\varepsilon^2)^{n/2}} \exp\left[-\frac{1}{2\sigma_\varepsilon^2} S(\boldsymbol{\beta})\right]$$

where $S(\boldsymbol{\beta})$ is the sum-of-squares function

$$S(\boldsymbol{\beta}) \equiv \sum_{i=1}^{n} [Y_i - f(\boldsymbol{\beta}, \mathbf{x}_i')]^2$$

As for the general linear model, we therefore maximize the likelihood by minimizing the sum of squared errors $S(\boldsymbol{\beta})$.

To derive estimating equations for the nonlinear model, we need to differentiate $S(\boldsymbol{\beta})$, obtaining

$$\frac{\partial S(\boldsymbol{\beta})}{\partial \boldsymbol{\beta}} = -2 \sum [Y_i - f(\boldsymbol{\beta}, \mathbf{x}_i')] \frac{\partial f(\boldsymbol{\beta}, \mathbf{x}_i')}{\partial \boldsymbol{\beta}}$$

Setting these partial derivatives to $\mathbf{0}$, and replacing the unknown parameters $\boldsymbol{\beta}$ with the estimator $\mathbf{b}$, produces the nonlinear least-squares estimating equations. It is convenient to write the estimating equations in matrix form as

$$[\mathbf{F}(\mathbf{b}, \mathbf{X})]' [\mathbf{y} - \mathbf{f}(\mathbf{b}, \mathbf{x})] = \mathbf{0} \qquad [14.19]$$

where $\underset{(n \times p)}{\mathbf{F}} (\mathbf{b}, \mathbf{X})$ is the matrix of derivatives, with i, jth entry

$$F_{ij} \equiv \frac{\partial f(\mathbf{b}, \mathbf{x}_i')}{\partial B_j}$$

The solution $\mathbf{b}$ of Equation 14.19 is the maximum-likelihood estimate of $\boldsymbol{\beta}$. If there is more than one root to the estimating equations, then we choose the solution that minimizes the sum of squares $S(\mathbf{b})$.

Nonlinear models of the form $Y_i = f(\boldsymbol{\beta}, \mathbf{x}_i') + \varepsilon_i$ can be estimated by nonlinear least squares, finding the value of $\mathbf{b}$ that minimizes $S(\mathbf{b}) = \sum_{i=1}^n [Y_i - f(\mathbf{b}, \mathbf{x}_i')]^2$.

Because the estimating equations (Equation 14.19) arising from a nonlinear model are, in general, themselves nonlinear, their solution is often difficult. It is, for this reason, unusual to obtain nonlinear least-squares estimates by explicitly solving the estimating equations. Instead, it is more common to work directly with the sum-of-squares function.

There are several practical methods for obtaining nonlinear least-squares estimates. I shall pursue in some detail a technique called *steepest descent.* Although the method of steepest descent usually performs poorly relative to alternative procedures, the rationale of the method is simple. Furthermore, many general aspects of nonlinear least-squares calculations can be explained clearly for the steepest-descent procedure. Because of the practical limitations of steepest descent, however, I shall also briefly describe two superior procedures—the *Gauss-Newton method* and the *Marquardt method*—without developing their rationales. Further discussion of nonlinear least squares can be found in the sources listed at the end of the chapter.

The method of steepest descent, like other methods for calculating nonlinear least-squares estimates, begins with a vector $\mathbf{b}^{(0)}$ of initial estimates. These initial estimates can be obtained in a variety of ways. We can, for example, choose p "typical" observations, substitute their values into the model given in Equation 14.17, and solve the resulting system of p nonlinear equations for the p parameters.

Alternatively, we can select a set of reasonable trial values for each parameter, find the residual sum of squares for every combination of trial values, and pick as initial estimates the combination associated with the smallest residual sum of squares. It is sometimes possible to choose initial estimates on the basis of prior research, hypothesis, or substantive knowledge of the process being modeled.

It is unfortunate that the choice of starting values for the parameter estimates may prove consequential: Iterative methods such as steepest descent generally converge to a solution more quickly for initial values that are close to the final values; and, even more importantly, the sum-of-squares function $S(\mathbf{b})$ may have local minima different from the global minimum (as illustrated in Figure 14.8).

Let us denote the gradient (i.e., derivative) vector for the sum-of-squares function as

$$\mathbf{d}(\mathbf{b}) = \frac{\partial S(\mathbf{b})}{\partial \mathbf{b}}$$

The vector $\mathbf{d}(\mathbf{b}^{(0)})$ gives the direction of maximum increase of the sum-of-squares function from the initial point $\{\mathbf{b}^{(0)}, S(\mathbf{b}^{(0)})\}$; the negative of this vector, $-\mathbf{d}(\mathbf{b}^{(0)})$, therefore, gives the direction of steepest descent. Figure 14.8 illustrates these relations for the particularly simple case of one parameter, where we can move either left or right from the initial estimate $B^{(0)}$.

If we move in the direction of steepest descent, then we can find a new estimated parameter vector

$$\mathbf{b}^{(1)} = \mathbf{b}^{(0)} - M_0\mathbf{d}(\mathbf{b}^{(0)})$$

for which $S(\mathbf{b}^{(1)}) < S(\mathbf{b}^{(0)})$: Because $S(\mathbf{b})$ is, by definition, decreasing in the direction of steepest descent, we can always choose a scalar constant M_0 small

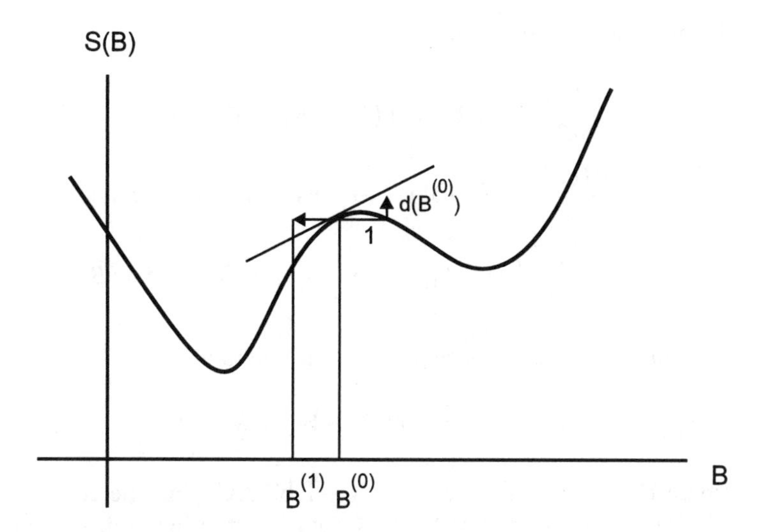

Figure 14.8. The method of steepest descent for one parameter, B. Because the slope of the sum-of-squares function $S(B)$ is positive above the initial estimate $B^{(0)}$, the first step is to the left, to $B^{(1)}$.

enough to improve the residual sum of squares. We can, for instance, first try $M_0 = 1$; if this choice does not lead to a decrease in $S(\mathbf{b})$, then we can take $M_0 = \frac{1}{2}$; and so on.

Our new estimate $\mathbf{b}^{(1)}$ can then be improved in the same manner, by finding

$$\mathbf{b}^{(2)} = \mathbf{b}^{(1)} - M_1\mathbf{d}(\mathbf{b}^{(1)})$$

so that $S(\mathbf{b}^{(2)}) < S(\mathbf{b}^{(1)})$. This procedure continues iteratively until it converges on a solution $\mathbf{b}$—that is, until the changes in $S(\mathbf{b}^{(l)})$ and $\mathbf{b}^{(l)}$ from one iteration to the next are negligible. In practice, however, the method of steepest descent often converges painfully slowly, and at times falls prey to other computational difficulties.

At each iteration, we need to compute the gradient vector $\mathbf{d}(\mathbf{b})$ for the current value of $\mathbf{b} = \mathbf{b}^{(l)}$. From our previous work in this section, we have

$$-\mathbf{d}(\mathbf{b}) = 2[\mathbf{F}(\mathbf{b}, \mathbf{X})]'[\mathbf{y} - \mathbf{f}(\mathbf{b}, \mathbf{x})] \qquad [14.20]$$

$$= 2\sum_{i=1}^{n}\left[\frac{\partial f(\mathbf{b}, \mathbf{x}_i')}{\partial \mathbf{b}}\right][Y_i - f(\mathbf{b}, \mathbf{x}_i')]$$

The partial derivatives $\partial f(\mathbf{b}, \mathbf{x}_i')/\partial B_j$ either can be supplied analytically (which is preferable) or can be evaluated numerically [i.e., approximated by finding the slope of $f(\mathbf{b}, \mathbf{x}_i')$ in a small interval around the current value of B_j]. For example, for the logistic growth model (Equation 14.18) discussed earlier in this section, the analytic derivatives are

$$\frac{\partial f(\mathbf{b}, X_i)}{\partial B_1} = [1 + \exp(B_2 + B_3 X_i)]^{-1}$$
$$\frac{\partial f(\mathbf{b}, X_i)}{\partial B_2} = -B_1[1 + \exp(B_2 + B_3 X_i)]^{-2} \exp(B_2 + B_3 X_i)$$
$$\frac{\partial f(\mathbf{b}, X_i)}{\partial B_3} = -B_1[1 + \exp(B_2 + B_3 X_i)]^{-2} \exp(B_2 + B_3 X_i) X_i$$

In the method of steepest descent, we take

$$\mathbf{b}^{(l+1)} = \mathbf{b}^{(l)} + M_l \mathbf{F}_l' \mathbf{e}^{(l)}$$

where $\mathbf{F}_l \equiv \mathbf{F}(\mathbf{b}^{(l)}, \mathbf{X})$ and $\mathbf{e}^{(l)} = \mathbf{y} - \mathbf{f}(\mathbf{b}^{(l)}, \mathbf{X})$ (and the constant 2 in Equation 14.20 is absorbed into M_l). The Gauss-Newton method, in contrast, calculates

$$\mathbf{b}^{(l+1)} = \mathbf{b}^{(l)} + M_l(\mathbf{F}_l'\mathbf{F}_l)^{-1}\mathbf{F}_l'\mathbf{e}^{(l)}$$

As for steepest descent, the step size M_l is selected so that $S(\mathbf{b}^{(l+1)}) < S(\mathbf{b}^{(l)})$; we first try $M_l = 1$, then $M_l = \frac{1}{2}$, and so on. The direction chosen in the Gauss-Newton procedure is based on a first-order Taylor series expansion of $S(\mathbf{b})$ around $S(\mathbf{b}^{(l)})$.

In the Marquardt procedure,

$$\mathbf{b}^{(l+1)} = \mathbf{b}^{(l)} + (\mathbf{F}_l'\mathbf{F}_l + M_l\mathbf{I}_p)^{-1}\mathbf{F}_l'\mathbf{e}^{(l)}$$

Initially, M_0 is set to some small number, such as 10^{-8}. If $S(\mathbf{b}^{(l+1)}) < S(\mathbf{b}^{(l)})$, then we accept the new value of $\mathbf{b}^{(l+1)}$ and proceed to the next iteration, with $M_{l+1} = M_l/10$; if, however, $S(\mathbf{b}^{(l+1)}) > S(\mathbf{b}^{(l)})$, then we increase M_l by a factor of 10 and try again. When M is small, the Marquardt procedure is similar to Gauss-Newton; as M grows larger, Marquardt approaches steepest descent. Marquardt's method is thus an adaptive compromise between the other two approaches.

Estimated asymptotic sampling covariances for the parameter estimates can be obtained by the maximum-likelihood approach and are given by[38]

$$\widehat{\mathscr{V}(\mathbf{b})} = S_E^2 \left\{[\mathbf{F}(\mathbf{b}, \mathbf{X})]' \, \mathbf{F}(\mathbf{b}, \mathbf{X})\right\}^{-1} \qquad [14.21]$$

We can estimate the error variance from the residuals $\mathbf{e} = \mathbf{y} - \mathbf{f}(\mathbf{b}, \mathbf{X})$, according to the formula[39]

$$S_E^2 = \frac{\mathbf{e}'\mathbf{e}}{n - p}$$

[38] See Bard (1974, pp. 176–179).

[39] Alternatively, we can use the maximum-likelihood estimator of the error variance, without the correction for "degrees of freedom," $\hat{\sigma}_\varepsilon^2 = \mathbf{e}'\mathbf{e}/n$.

Note the similarity of Equation 14.21 to the familiar linear least-squares result, $\widehat{V(\mathbf{b})} = S_E^2(\mathbf{X}'\mathbf{X})^{-1}$. Indeed, $\mathbf{F}(\mathbf{b}, \mathbf{X}) = \mathbf{X}$ for the linear model $\mathbf{y} = \mathbf{X}\boldsymbol{\beta} + \boldsymbol{\varepsilon}$.

> Iterative methods for finding the nonlinear least-squares estimates include the method of steepest descent and, more practically, the Gauss-Newton and Marquardt methods.

Decennial population data for the United States appear in Table 14.2 for the period from 1790 to 1990; the data are plotted in Figure 14.9(*a*). Let us fit the logistic growth model (Equation 14.18) to these data using nonlinear least squares.

The parameter β_1 of the logistic growth model gives the asymptote that expected population approaches as time increases. In 1990, when $Y = 248.710$ (million), population did not appear to be near an asymptote; so as not to extrapolate too far beyond the data, I shall arbitrarily set $B_1^{(0)} = 350$. At time $X_1 = 0$, we have

$$Y_1 = \frac{\beta_1}{1 + e^{\beta_2 + \beta_3 0}} + \varepsilon_1 \qquad [14.22]$$

Ignoring the error, using $B_1^{(0)} = 350$, and substituting the observed value of $Y_1 = 3.929$ into Equation 14.22, we get $\exp(B_2^{(0)}) = (350/3.929) - 1$, or $B_2^{(0)} = \log_e 88.081 = 4.478 \simeq 4.5$. At time $X_2 = 1$,

$$Y_2 = \frac{\beta_1}{1 + \exp(\beta_2 + \beta_3 1)} + \varepsilon_2$$

TABLE 14.2 Population of the United States, in Millions, 1790–1990

Year	*Population*	*Year*	*Population*
1790	3.929	1900	75.995
1800	5.308	1910	91.972
1810	7.240	1920	105.711
1820	9.638	1930	122.775
1830	12.866	1940	131.669
1840	17.069	1950	150.697
1850	23.192	1960	179.323
1860	31.443	1970	203.302
1870	39.818	1980	226.542
1880	50.156	1990	248.710
1890	62.948		

Source of Data: United States (1994).

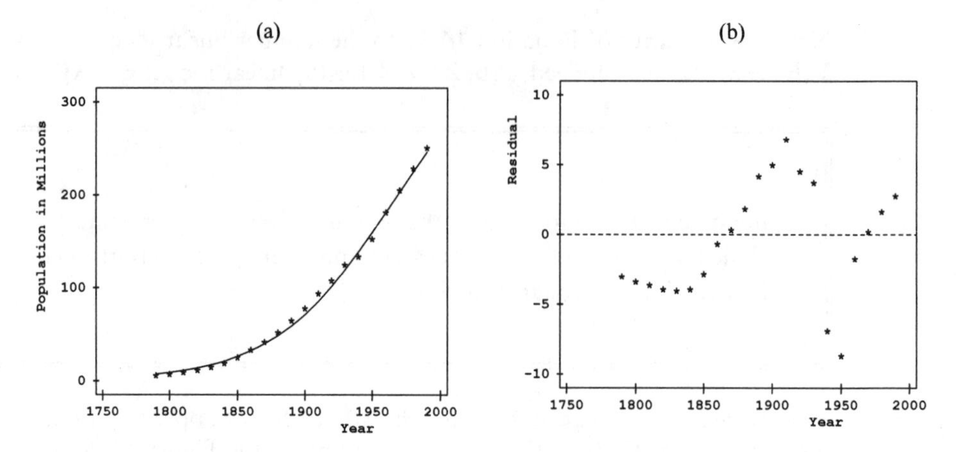

Figure 14.9. Panel (*a*) shows the population of the United States from 1790 through 1990; the line represents the fitted logistic growth model. Residuals from the logistic growth model are plotted against time in panel (*b*).

Again ignoring the error, and making appropriate substitutions, $\exp(4.5+B_3^{(0)}) = (350/5.308) - 1$, or $B_3^{(0)} = \log_e 64.938 - 4.5 = -0.327 \simeq -0.3$.

The Gauss-Newton iterations based on these start values are shown in Table 14.3. Estimated asymptotic standard errors for the coefficients also appear in this table. Although the logistic model captures the major trend in U.S. population growth, the residuals from the least-squares fit [plotted against time in Figure 14.9(*b*)] suggest that the error variance is not constant and that the residuals are autocorrelated.[40] Notice as well the large drop in the residual for 1940, and the large increase for 1960 (when Alaska and Hawaii were first included in the population count).

EXERCISES

14.8 Interpreting effects in nonlinear models (based on Stolzenberg, 1979): For simplicity, disregard the error and let Y represent the systematic part of the dependent variable. Suppose that Y is a function of two independent variables, $Y = f(X_1, X_2)$.

- The *metric effect* of X_1 on Y is defined as the partial derivative $\partial Y/\partial X_1$.
- The *effect of proportional change* in X_1 on Y is defined as $X_1(\partial Y/\partial X_1)$.
- The *instantaneous rate of return* of Y with respect to X_1 is $(\partial Y/\partial X_1)/Y$.
- The *point elasticity* of Y with respect to X_1 is $(\partial Y/\partial X_1)(X_1/Y)$.

Find each of these four measures of the effect of X_1 in each of the following models. Which measure yields the simplest result in each case?

[40] See Exercise 14.13.

TABLE 14.3 Gauss-Newton Iterations for the Logistic Growth Model Fit to the U.S. Population Data. Estimated asymptotic standard errors are given below the final coefficient estimates.

Iteration	*Residual Sum of Squares*	*Coefficients* B_1	B_2	B_3
0	13,007.48	350.0	4.5	−0.3
1	609.57	351.81	3.8405	−0.22706
⋮				
6	356.40	389.17	3.9903	−0.22662
Final	356.40	389.17	3.9903	−0.22662
Standard error		30.81	0.0703	0.01086

How can the several measures be interpreted? How would you fit each model to data, assuming convenient forms for the errors (e.g., additive errors for models a, b, and c)?

(a) $Y = \alpha + \beta_1 X_1 + \beta_2 X_2$.
(b) $Y = \alpha + \beta_1 X_1 + \beta_2 X_1^2 + \beta_3 X_2$.
(c) $Y = \alpha + \beta_1 X_1 + \beta_2 X_2 + \beta_3 X_1 X_2$.
(d) $Y = \exp(\alpha + \beta_1 X_1 + \beta_2 X_2)$.
(e) $Y = \alpha X_1^{\beta_1} X_2^{\beta_2}$.

14.9 *Orthogonal polynomial contrasts: The polynomial regressors X, $X^2, \ldots, X^{m-1}$ generated to represent a quantitative, discrete X with values $1, 2, \ldots, m$ are substantially correlated. It is convenient (but by no means essential) to remove these correlations. Suppose that there are an equal number of observations in the different levels of X, so that it suffices to make the columns of the row basis of the model matrix for X orthogonal. Working with the row basis, begin by subtracting the mean from X, calling the result X^*. *Centering* X in this manner makes X^* orthogonal to the constant regressor **1**. (Why?) X^2 can be made orthogonal to the constant and X^* by projecting the X^2 vector onto the subspace generated by **1** and X^*; call the residual from this projection X^{*2}. The remaining columns $X^{*3}, \ldots, X^{*m-1}$ of the new row basis are formed in a similar manner, each orthogonal to the preceding ones.

(a) Show that the orthogonal polynomial contrasts $\mathbf{1}, X^*, \ldots, X^{*m-1}$ span the same subspace as the original polynomial regressors $\mathbf{1}, X, \ldots, X^{m-1}$.
(b) Show that the incremental sum of squares for each orthogonal contrast X^*, $X^{*2}, \ldots, X^{*m-1}$ is the same as the step-down sum of squares for the corresponding regressor among the original (correlated) polynomial terms, $X^{m-1}, \ldots, X^2, X$. (*Hint*: Remember that X^*, $X^{*2}, \ldots, X^{*m-1}$ are uncorrelated.)
(c) What then is the advantage of orthogonal polynomial contrasts?

14.10 Working with the full quadratic regression model

$$E(Y) = \beta_0 + \beta_1 X_1 + \beta_2 X_2 + \beta_3 X_1^2 + \beta_4 X_2^2 + \beta_5 X_1 X_2$$

and a three-dimensional graphics program, draw pictures of the regression surface (similar to Figure 14.6) for various values of the parameters $\beta_1, \ldots, \beta_5$ and various values of the independent variables X_1 and X_2. You will derive a better impression of the flexibility of this model if you experiment freely, but the following suggestions may prove useful:

- Try both positive and negative values of the parameters.
- Try cases in which there are both positive and negative X's, as well as cases in which the X's are all positive.
- Try setting some of the parameters to 0.

14.11 The data in `ginzberg.dat` (collected by Ginzberg) are analyzed by Monette (1990). The data are for a group of 82 psychiatric patients hospitalized for depression. The dependent variable in the dataset is the patient's score on the Beck scale, a widely used measure of depression. The independent variables are "simplicity" (measuring the degree to which the patient "sees the world in black and white") and "fatalism." (These three variables have been adjusted for other independent variables that can influence depression.) Using the full quadratic regression model

$$Y = \beta_0 + \beta_1 X_1 + \beta_2 X_2 + \beta_3 X_1^2 + \beta_4 X_2^2 + \beta_5 X_1 X_2 + \varepsilon$$

regress the Beck-scale scores on simplicity and fatalism. Are the quadratic and product terms needed here? If possible, graph the data and the fitted regression surface in three dimensions. Do you see any problems with the data? In particular, are there any apparently influential observations? What do standard regression diagnostics for influential observations show?

14.12 Table 14.4 reports interprovincial migration in Canada for the period 1966 to 1971. Also shown in this table are the 1966 and 1971 provincial populations. Table 14.5 gives road distances among the major cities in the 10 provinces. Averaging the 1966 and 1971 population figures, fit the gravity model of migration (Equation 14.13) to the interprovincial migration data. (These tables are given in raveled form in `migratn.dat`.) Display the residuals from the fitted model in a 10×10 table. Can you account for the pattern of residuals? How might the model be modified to provide a more satisfactory fit to the data? (*Hint*: Use dummy regressors to incorporate province effects.)

14.13 Calculate the autocorrelation for the residuals from the logistic growth model fit to the U.S. population data. Recalling the discussion of autocorrelated errors in *linear* regression (in the previous section), does autocorrelation appear to be a serious problem here?

14.14 Using nonlinear least squares, refit the logistic growth model to the U.S. population data (given in Table 14.2) assuming multiplicative rather than

TABLE 14.4 Canadian Interprovincial Migration and Population for the Period 1966–1971

1971 Residence	*1966 Residence*				
	NFLD	PEI	NS	NB	QUE
NFLD		255	2,380	1,140	2,145
PEI	340		1,975	1,310	755
NS	3,340	2,185		8,310	6,090
NB	1,740	1,335	7,635		9,315
QUE	2,235	635	4,350	7,905	
ONT	17,860	3,570	25,730	18,550	99,430
MAN	680	265	1,655	1,355	4,330
SASK	280	125	620	495	1,570
ALTA	805	505	3,300	2,150	7,750
BC	1,455	600	6,075	3,115	16,740
1966 Population	493,396	108,535	756,039	616,788	5,780,845
1971 Population	522,104	111,641	788,960	534,557	6,027,764

1971 Residence	*1966 Residence*				
	ONT	MAN	SASK	ALTA	BC
NFLD	6,295	215	185	425	425
PEI	3,060	400	95	185	330
NS	18,805	1,825	840	2,000	2,490
NB	12,455	1,405	480	1,130	1,195
QUE	48,370	4,630	1,515	3,305	4,740
ONT		23,785	11,805	17,655	21,205
MAN	18,245		16,365	7,190	6,310
SASK	6,845	9,425		10,580	6,090
ALTA	23,550	17,410	41,910		27,765
BC	47,395	26,910	29,920	58,915	
1966 Population	6,960,870	963,066	955,344	1,463,203	1,873,674
1971 Population	7,703,106	988,247	926,242	1,627,874	2,184,621

Sources of Data: Canada (1971, Vol. 1, Part 2, Table 32; 1972, p. 1369).

additive errors:

$$Y_i = \frac{\beta_1}{1 + \exp(\beta_2 + \beta_3 X_i)} \varepsilon_i$$

Which form of the model appears more adequate for these data?

14.15 Table 14.6 gives the population of Canada for decennial censuses between 1851 and 1991. Fit the logistic growth model to these data, assuming (a) additive errors (Equation 14.18), and (b) multiplicative errors (as in the previous exercise). Compare the results of the two fits. Be sure to graph the data and the fit of the models; also graph the residuals from the models.

TABLE 14.5 Road Distances in Miles Among Major Canadian Cities

City	NFLD	PEI	NS	NB	QUE
St. John, NFLD	0	924	952	1119	1641
Charlottetown, PEI	924	0	164	252	774
Halifax, NS	952	164	0	310	832
Fredericton, NB	1119	252	310	0	522
Montreal, QUE	1641	774	832	522	0
Toronto, ONT	1996	1129	1187	877	355
Winnipeg, MAN	3159	2293	2351	2041	1519
Regina, SASK	3542	2675	2733	2423	1901
Edmonton, ALTA	4059	3192	3250	2940	2418
Vancouver, BC	4838	3972	4029	3719	3197

City	ONT	MAN	SASK	ALTA	BC
St. John, NFLD	1996	3159	3542	4059	4838
Charlottetown, PEI	1129	2293	2675	3192	3972
Halifax, NS	1187	2351	2733	3250	4029
Fredericton, NB	877	2041	2423	2940	3719
Montreal, QUE	355	1519	1901	2418	3197
Toronto, ONT	0	1380	1763	2281	3059
Winnipeg, MAN	1380	0	382	899	1679
Regina, SASK	1763	382	0	517	1297
Edmonton, ALTA	2281	899	517	0	987
Vancouver, BC	3059	1679	1297	987	0

Source of Data: Canada (1962).

14.16 Recall the Box-Tidwell regression model

$$Y_i = \alpha + \beta_1 X_{i1}^{\gamma_1} + \cdots + \beta_k X_{ik}^{\gamma_k} + \varepsilon_i$$

In Section 12.5.2, I described a procedure for fitting this model that relies on constructed variables. I applied this procedure to the Canadian occupational prestige data (in `prestige.dat`) to fit the model

$$P = \alpha + \beta_S S_i^{\gamma_S} + \beta_I I_i^{\gamma_I} + \beta_{W1} W + \beta_{W2} W^2 + \varepsilon$$

TABLE 14.6 Population of Canada, in Millions, 1851–1991

Year	*Population*	*Year*	*Population*
1851	2.436	1931	10.377
1861	3.230	1941	11.507
1871	3.689	1951	13.648
1881	4.325	1961	17.780
1891	4.833	1971	21.046
1901	5.371	1981	23.774
1911	7.207	1991	26.429
1921	8.788		

Sources of Data: Urquhart and Buckley (1965, p. 1369) and Canada (1993, Table 3.2).

where P is prestige, S is education, I is income, and W is the percentage of women in the occupation. Fit this model to the data by general nonlinear least squares. Are there any advantages to using general nonlinear least squares in place of Box and Tidwell's procedure? Any disadvantages? Can the two approaches be combined?

14.3 Robust Regression

As I have explained, the efficiency of least-squares regression is seriously impaired by heavy-tailed error distributions; in particular, least squares is vulnerable to outlying observations at high-leverage points.[41] One response to this problem is to employ diagnostics for high-leverage, influential, and outlying data; if unusual data are discovered, then these can be corrected, removed, or otherwise accommodated.

Robust estimation is an alternative approach to outliers and the heavy-tailed error distributions that tend to generate them. Properly formulated, robust estimators are almost as efficient as least squares when the error distribution is normal, and much more efficient when the errors are heavy tailed. Robust estimators hold their efficiency well because they are resistant to outliers. Rather than simply discarding discrepant data, however, robust estimation (as we shall see) downweights them.

Most of this section is devoted to a particular strategy of robust estimation, termed *M-estimation*, due originally to Huber (1964). I shall also briefly explain another approach to robust estimation called *bounded-influence regression.*

14.3.1 *M*-Estimation

Estimating Location

Although our proper interest is in robust regression of linear models, it is helpful initially to narrow our focus to a simpler setting: robust estimation of *location*—that is, estimation of the center of a distribution. Let us, then, begin our exploration of robust estimation with the minimal linear model

$$Y_i = \mu + \varepsilon_i$$

where the observations Y_i (and hence the errors ε_i) are independently sampled from some symmetric distribution with center μ.

If the distribution from which the observations are drawn is normal, then the sample mean $\hat{\mu} = \overline{Y}$ is the maximally efficient estimator of μ, producing the fitted model

$$Y_i = \overline{Y} + E_i$$

---

[41] See Chapter 11.

The mean minimizes the least-squares *objective function*:

$$\sum_{i=1}^{n} \rho_{\text{LS}}(E_i) = \sum_{i=1}^{n} \rho_{\text{LS}}(Y_i - \hat{\mu}) \equiv \sum_{i=1}^{n} (Y_i - \hat{\mu})^2$$

The mean, however, is very sensitive to outliers, as is simply demonstrated: I drew a sample of six observations from the unit-normal distribution, obtaining

$$\begin{array}{lll} Y_1 = -0.068 & Y_2 = -1.282 & Y_3 = 0.013 \\ Y_4 = 0.141 & Y_5 = -0.980 & Y_6 = 1.263 \end{array}$$

The mean of these six values is $\overline{Y} = -0.152$. Now imagine adding a seventh observation, Y_7, allowing it to take on all possible values from -10 to $+10$ (or, with greater imagination, from $-\infty$ to $+\infty$). The result, called the *influence function* of the mean, is graphed in Figure 14.10(*a*). It is apparent from this figure that as the discrepant seventh observation grows more extreme, the sample mean chases it.

The shape of the influence function for the mean follows from the derivative of the least-squares objective function with respect to E:

$$\psi_{\text{LS}}(E) \equiv \rho'_{\text{LS}}(E) = 2E$$

Influence, therefore, is proportional to the residual E. It is convenient to redefine the least-squares objective function as $\rho_{\text{LS}}(E) \equiv \frac{1}{2}E^2$, so that $\psi_{\text{LS}}(E) = E$.

Now consider the sample median as an estimator of μ. The median minimizes the *least-absolute-values* (LAV) objective function:[42]

$$\sum_{i=1}^{n} \rho_{\text{LAV}}(E_i) = \sum_{i=1}^{n} \rho_{\text{LAV}}(Y_i - \hat{\mu}) \equiv \sum_{i=1}^{n} |Y_i - \hat{\mu}|$$

As a result, the median is much more resistant than the mean to outliers. The influence function of the median for the illustrative sample is shown in Figure 14.10(*b*). In contrast to the mean, the influence of a discrepant observation on the median is *bounded*. Once again, the derivative of the objective function gives the shape of the influence function:[43]

$$\psi_{\text{LAV}}(E) \equiv \rho'_{\text{LAV}}(E) = \begin{cases} 1 & \text{for } E > 0 \\ 0 & \text{for } E = 0 \\ -1 & \text{for } E < 0 \end{cases}$$

Although the median is more resistant than the mean to outliers, it is less efficient than the mean if the distribution of Y is normal. When $Y \sim N(\mu, \sigma^2)$, the sampling variance of the mean is σ^2/n, while the variance of the median is

[42] See Exercise 14.17.

[43] Strictly speaking, the derivative of ρ_{LAV} is undefined at $E = 0$, but setting $\psi_{\text{LAV}}(0) \equiv 0$ is convenient.

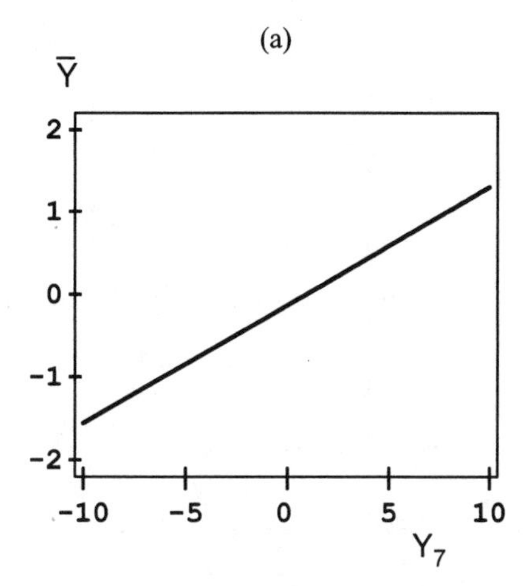

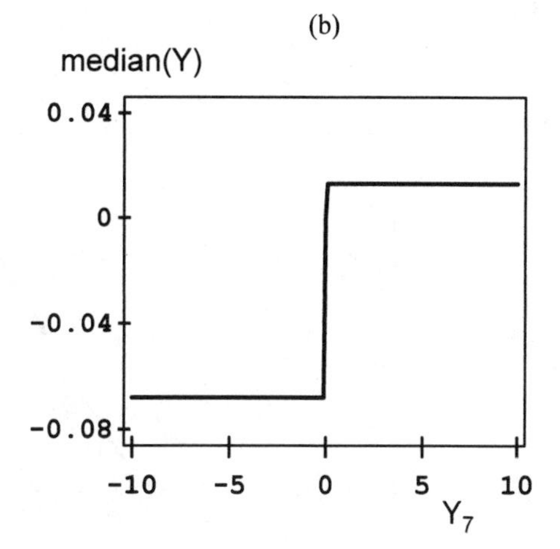

Figure 14.10. The influence functions for the mean (a) and median (b) for the sample $Y_1 = -0.068$, $Y_2 = -1.282$, $Y_3 = 0.013$, $Y_4 = 0.141$, $Y_5 = -0.980$, $Y_6 = 1.263$. The influence function for the median is bounded, while that for the mean is not.

$\pi\sigma^2/2n$—that is, $\pi/2 \simeq 1.57$ times as large as for the mean. Other objective functions combine resistance to outliers with greater robustness of efficiency. Estimators that can be expressed as minimizing an objective function $\sum_{i=1}^{n} \rho(E)$ are called M-estimators.[44]

Two common choices of objective functions are the *Huber* and *biweight* (or *bisquare*) functions:

- The Huber objective function is a compromise between least squares and least absolute values, behaving like least squares in the center and like least absolute values in the tails:

$$\rho_H(E) \equiv \begin{cases} \frac{1}{2}E^2 & \text{for } |E| \leq k \\ k|E| - \frac{1}{2}k^2 & \text{for } |E| > k \end{cases}$$

The Huber objective function ρ_H and its derivative, the influence function ψ_H, are graphed in Figure 14.11:[45]

$$\psi_H(E) = \begin{cases} k & \text{for } E > k \\ E & \text{for } |E| \leq k \\ -k & \text{for } E < -k \end{cases}$$

The value k, which defines the center and tails, is called a *tuning constant.*

It is most natural to express the tuning constant as a multiple of the *scale* (i.e., the spread) of the variable Y, that is, to take $k = cS$, where S is a measure of scale. The sample standard deviation is a poor measure of scale in this context,

[44] Estimators that can be written in this form can be thought of as generalizations of maximum-likelihood estimators, hence the term "M"-estimator. The maximum-likelihood estimator is produced by taking $\rho_{\text{ML}}(y-\mu) \equiv -\log_e p(y-\mu)$ for an appropriate probability or probability density function $p(\cdot)$.

[45] My terminology here is loose, but convenient: Strictly speaking, the ψ-function is not the influence function, but it has the same shape as the influence function.

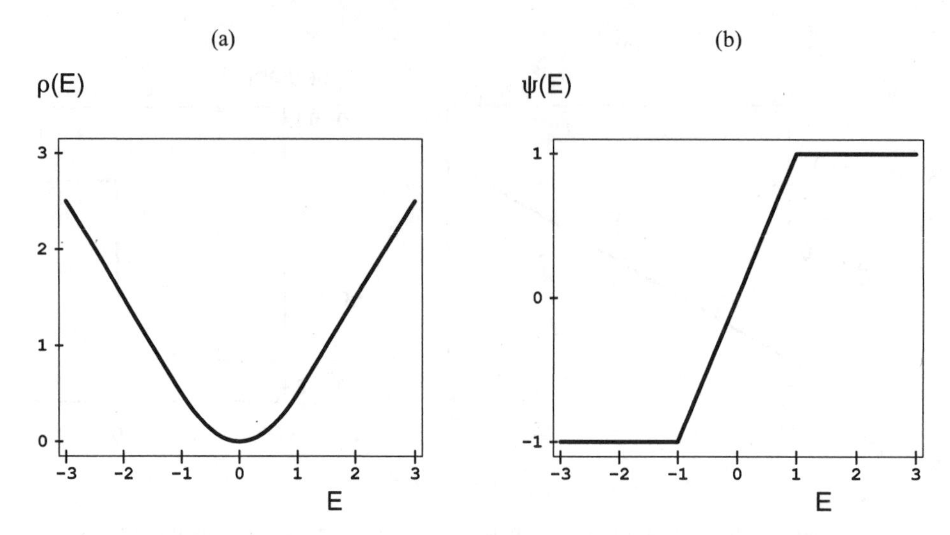

Figure 14.11. Huber objective function ρ_H (a) and "influence function" ψ_H (b). To calibrate these graphs, the tuning constant is set to $k = 1$.

because it is even more affected than the mean by outliers. A common robust measure of scale is the *median absolute deviation*:

$$\text{MAD} \equiv \text{median}|Y_i - \hat{\mu}|$$

The estimate $\hat{\mu}$ can be taken, at least initially, as the median value of Y. We can then define $S \equiv \text{MAD}/0.6745$, which ensures that S estimates the standard deviation σ when the population is normal. Using $k = 1.345S$ (i.e., $1.345/0.6745 \simeq 2$ MADs) produces 95% efficiency relative to the sample mean when the population is normal, along with substantial resistance to outliers when it is not. A smaller tuning constant can be employed for more resistance.

- The biweight objective function levels off at very large residuals:[46]

$$\rho_{\text{BW}}(E) \equiv \begin{cases} \dfrac{k^2}{6}\left\{1 - \left[1 - \left(\dfrac{E}{k}\right)^2\right]^3\right\} & \text{for } |E| \le k \\[2ex] \dfrac{k^2}{6} & \text{for } |E| > k \end{cases}$$

The influence function for the biweight estimator, therefore, "redescends" to 0, completely discounting observations that are sufficiently discrepant:

$$\psi_{\text{BW}}(E) = \begin{cases} E\left[1 - \left(\dfrac{E}{k}\right)^2\right]^2 & \text{for } |E| \le k \\[2ex] 0 & \text{for } |E| > k \end{cases}$$

The functions ρ_{BW} and ψ_{BW} are graphed in Figure 14.12. Using $k = 4.685S$ (i.e., $4.685/0.6745 \simeq 7$ MADs) produces 95% efficiency when sampling from a normal population.

---

[46] The term "bisquare" applies to the ψ-function and to the weight function (hence "biweight"), to be introduced presently—not to the objective function.

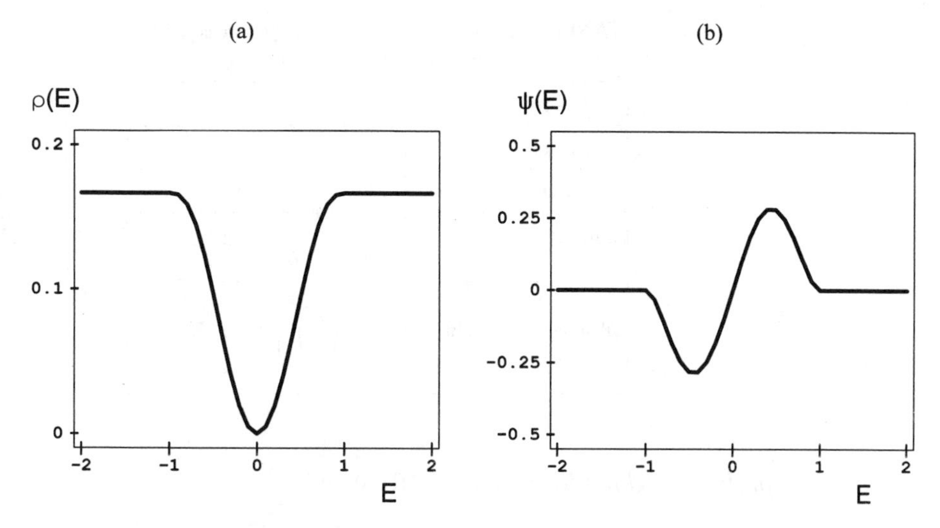

Figure 14.12. Biweight objective function ρ_{BW} (*a*) and "influence function" ψ_{BW} (*b*). To calibrate these graphs, the tuning constant is set to $k = 1$. The influence function returns to 0 when $|E|$ is large.

> Robust *M*-estimators of location, for the parameter μ in the simple model $Y_i = \mu + \varepsilon_i$, minimize the objective function $\sum_{i=1}^{n} \rho(E_i) = \sum_{i=1}^{n} \rho(Y_i - \hat{\mu})$, selecting $\rho(\cdot)$ so that the estimator is relatively unaffected by outlying values. Two common choices of objective function are the Huber and the bisquare. The sensitivity of an *M*-estimator to individual observations is expressed by the influence function of the estimator, which has the same shape as the derivative of the objective function, $\psi(E) \equiv \rho'(E)$.

Calculation of *M*-estimators usually requires an iterative procedure (although iteration is not necessary for the mean and median which, as we have seen, fit into the *M*-estimation framework). An estimating equation for $\hat{\mu}$ is obtained by setting the derivative of the objective function (with respect to $\hat{\mu}$) to 0, obtaining

$$\sum_{i=1}^{n} \psi(Y_i - \hat{\mu}) = 0 \qquad [14.23]$$

There are several general approaches to solving the estimating equation (Equation 14.23); probably the most straightforward, and the simplest to implement computationally, is iteratively to reweight the mean—a special case of

TABLE 14.7 Weight Functions $w(E) \equiv \psi(E)/E$ for Several M-Estimators

Estimator	*Weight Function* $w(E)$
Least squares	1
Least absolute values	$1/\lvert E\rvert$ (for $E \neq 0$)
Huber	1 for $\lvert E\rvert \le k$ $k/\lvert E\rvert$ for $\lvert E\rvert > k$
Bisquare (biweight)	$\left[1-\left(\frac{E}{k}\right)^2\right]^2$ for $\lvert E\rvert \le k$ 0 for $\lvert E\rvert > k$

iteratively reweighted least squares (IRLS):

1. Define the *weight function* $w(E) \equiv \psi(E)/E$. Then the estimating equation becomes

$$\sum_{i=1}^{n}(Y_i - \hat{\mu})\, w_i = 0 \qquad [14.24]$$

where

$$w_i \equiv w\,(Y_i - \hat{\mu})$$

The solution of Equation 14.24 is the weighted mean

$$\hat{\mu} = \frac{\sum w_i Y_i}{\sum w_i}$$

The weight functions corresponding to the least-squares, LAV, Huber, and bisquare objective functions are shown in Table 14.7 and graphed in Figure 14.13. The least-squares weight function accords equal weight to each observation, while the bisquare gives zero weight to observations that are sufficiently outlying; the LAV and Huber weight functions descend toward 0 but never quite reach it.

2. Select an initial estimate $\hat{\mu}^{(0)}$, such as the median of the Y-values.[47] Using $\hat{\mu}^{(0)}$, calculate an initial estimate of scale $S^{(0)}$ and initial weights $w_i^{(0)} = w(Y_i - \hat{\mu}^{(0)})$. Set the iteration counter $l = 0$. The scale is required to calculate the tuning constant $k = cS$ (for prespecified c).
3. At each iteration l, calculate $\hat{\mu}^{(l)} = \sum w_i^{(l-1)} Y_i / \sum w_i^{(l-1)}$. Stop when the change in $\hat{\mu}^{(l)}$ is negligible from one iteration to the next.

[47] Because the estimating equation for redescending M-estimators, such as the bisquare, can have more than one root, the selection of an initial estimate might be consequential.

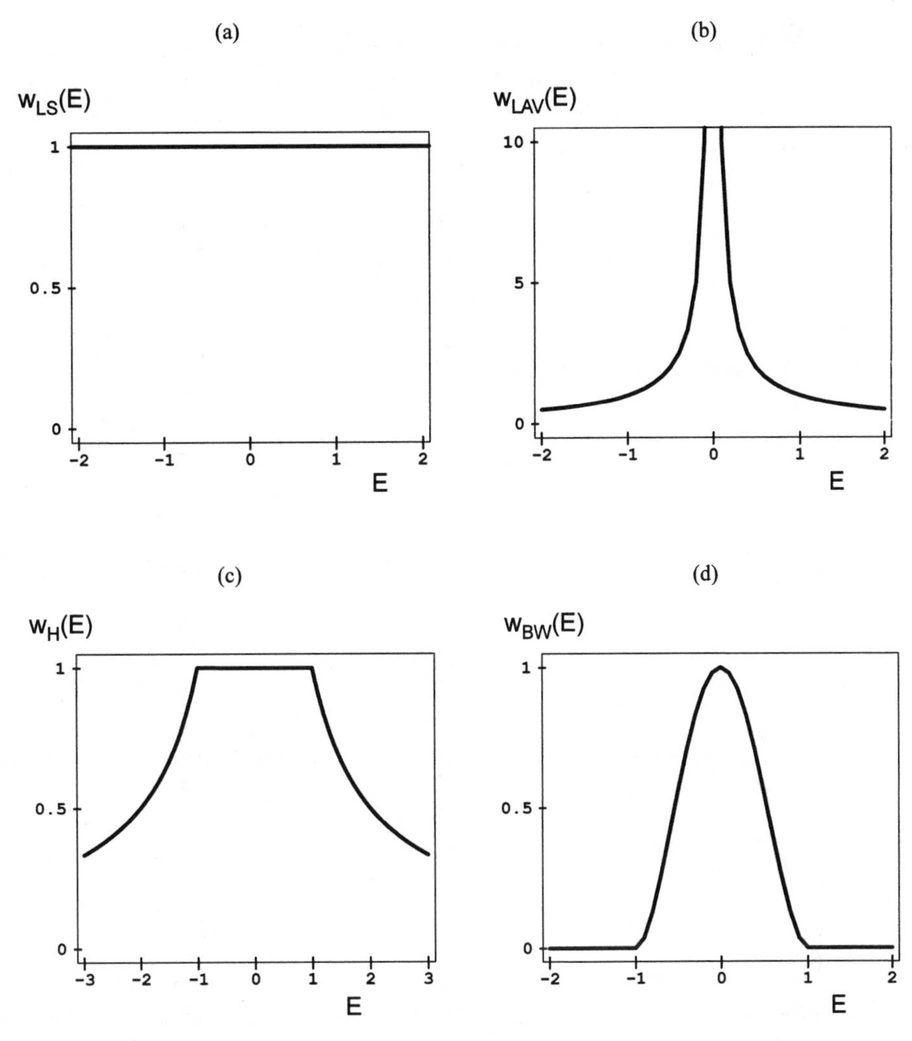

Figure 14.13. Weight functions $w(E)$ for the least-squares (a), least-absolute-values (b), Huber (c), and biweight (d) estimators. The tuning constants for the Huber and biweight estimators are taken as $k = 1$.

An estimating equation for $\hat{\mu}$ is obtained by setting the derivative of the objective function (with respect to $\hat{\mu}$) to 0, obtaining $\sum_{i=1}^{n} \psi(Y_i - \hat{\mu}) = 0$. The simplest procedure for solving this estimating equation is by iteratively reweighted means. Defining the weight function as $w(E) \equiv \psi(E)/E$, the estimating equation becomes $\sum_{i=1}^{n} (Y_i - \hat{\mu})\, w_i = 0$, from which $\hat{\mu} = \sum w_i Y_i / \sum w_i$. Starting with an initial estimate $\hat{\mu}^{(0)}$, initial weights are calculated, and the value of $\hat{\mu}$ is updated. This procedure continues iteratively until the value of $\hat{\mu}$ converges.

M-Estimation in Regression

With the exception of one significant caveat, to be addressed in the next section, the generalization of M-estimators to regression is immediate. We now wish to estimate the linear model

$$\begin{aligned} Y_i &= \alpha + \beta_1 X_{i1} + \cdots + \beta_k X_{ik} + \varepsilon_i \\ &= \underset{(1\times k+1)}{\mathbf{x}_i'} \underset{(k+1\times 1)}{\boldsymbol{\beta}} + \varepsilon_i \end{aligned}$$

The estimated model is

$$\begin{aligned} Y_i &= A + B_1 X_{i1} + \cdots + B_k X_{ik} + E_i \\ &= \mathbf{x}_i'\mathbf{b} + \varepsilon_i \end{aligned}$$

The general M-estimator minimizes the objective function

$$\sum_{i=1}^{n} \rho(E_i) = \sum_{i=1}^{n} \rho(Y_i - \mathbf{x}_i'\mathbf{b})$$

Differentiating the objective function, and setting the derivative to 0, produces

$$\sum_{i=1}^{n} \psi(Y_i - \mathbf{x}_i'\mathbf{b})\mathbf{x}_i = \mathbf{0}$$

which is a system of $k+1$ estimating equations in the $k+1$ elements of $\mathbf{b}$. Using the weight function $w(E) \equiv \psi(E)/E$, and letting $w_i \equiv w(E_i)$, the estimating equations become

$$\sum_{i=1}^{n} w_i(Y_i - \mathbf{x}_i'\mathbf{b})\mathbf{x}_i = \mathbf{0}$$

The solution to these estimating equations minimizes the weighted sum of squares[48] $\sum w_i^2 E_i^2$.

Because the weights depend on the residuals, the estimated coefficients depend on the weights, and the residuals depend on the estimated coefficients, an iterative solution is required. The IRLS algorithm for regression is as follows:

1. Select initial estimates $\mathbf{b}^{(0)}$ and set the iteration counter $l = 0$. Using the initial estimates, find residuals $E_i^{(0)} = Y_i - \mathbf{x}_i'\mathbf{b}^{(0)}$, and from these, calculate the estimated scale of the residuals $S^{(0)}$ and the weights $w_i^{(0)} = w(E_i^{(0)})$.

[48] Cf. the discussion of weighted-least-squares regression in Section 12.2.2.

2. At each iteration l, solve the estimating equations using the current weights, minimizing $\sum w_i^{(l-1)} E_i^2$ to obtain $\mathbf{b}^{(l)}$. The solution is conveniently expressed as

$$\mathbf{b}^{(l)} = (\mathbf{X}'\mathbf{W}\mathbf{X})^{-1}\mathbf{X}'\mathbf{W}\mathbf{y}$$

where the model matrix $\underset{(n \times k+1)}{\mathbf{X}}$ has $\mathbf{x}_i'$ as its ith row, and $\underset{(n \times n)}{\mathbf{W}} \equiv \text{diag}\{w_i^{(l-1)}\}$. Continue until[49] $\mathbf{b}^{(l)} - \mathbf{b}^{(l-1)} \simeq \mathbf{0}$.

The asymptotic covariance matrix of the M-estimator is given by

$$\mathcal{V}(\mathbf{b}) = \frac{E(\psi^2)}{[E(\psi')]^2}(\mathbf{X}'\mathbf{X})^{-1}$$

Using $\sum [\psi(E_i)]^2 / n$ to estimate $E(\psi^2)$ and $[\sum \psi'(E_i)/n]^2$ to estimate $[E(\psi')]^2$ produces the estimated asymptotic covariance matrix $\widehat{\mathcal{V}(\mathbf{b})}$. Research suggests, however, that these sampling variances are not to be trusted unless the sample size is large.[50]

> M-estimation for the regression model $Y_i = \mathbf{x}_i'\boldsymbol{\beta} + \varepsilon_i$ is a direct extension of M-estimation of location: We seek to minimize an objective function of the regression residuals, $\sum_{i=1}^{n} \rho(E_i) = \sum_{i=1}^{n} \rho(Y_i - \mathbf{x}_i'\mathbf{b})$. Differentiating the objective function, and setting the derivatives to 0, produces the estimating equations $\sum_{i=1}^{n} \psi(Y_i - \mathbf{x}_i'\mathbf{b})\mathbf{x}_i = \mathbf{0}$, which generally require iterative solution.

To illustrate M-estimation, recall Duncan's regression of occupational prestige on income and education. In our previous analysis of these data,[51] we discovered two influential observations: *ministers* and *railroad conductors*. Another observation, *reporters*, has a relatively large residual but is not influential; still another observation, *railroad engineers*, is at a high-leverage point, but is not discrepant. Table 14.8 summarizes the results of estimating Duncan's regression using four M-estimators, including ordinary least squares. (The least-squares estimates obtained after deleting *ministers* and *railroad conductors* are also shown for comparison.) The three robust estimators produce quite similar results, with a larger income coefficient and smaller education coefficient than least squares. The redescending bisquare estimator is most different from least squares.

---

[49] As in the location problem, it is possible that the estimating equations for a redescending estimator have more than one root. If you use the bisquare estimator, for example, it is prudent to pick a good start value, such as provided by the Huber estimator.

[50] See Li (1985, pp. 300–301). For an alternative approach that may have better small-sample properties, see Street et al. (1988).

[51] See Section 11.6.

TABLE 14.8 M-Estimates for Duncan's Regression of Occupational Prestige on Income and Education, for 45 U.S. Occupations. The estimator marked "Least squares*" omits ministers and railroad conductors

	Coefficient		
Estimator	*Constant*	*Income*	*Education*
Least squares	−6.065	0.5987	0.5458
Least squares*	−6.409	0.8674	0.3322
Least absolute values	−6.408	0.7477	0.4587
Huber	−7.289	0.7104	0.4819
Bisquare (biweight)	−7.480	0.8184	0.4040

Figure 14.14 shows the final weights for the bisquare estimator applied to Duncan's data. *Ministers*, *railroad conductors*, and *reporters* have very small weights. Rather than simply regarding robust regression as a procedure for automatically downweighting outliers, the method can often be used effectively, as here, to identify outlying observations.

14.3.2 Bounded-Influence Regression

The flies in the ointment of M-estimation in regression are high-leverage outliers. In the location problem, M-estimators such as the Huber and the bisquare bound the influence of individual discrepant observations, but this is not the case in regression—if we admit the possibility of X-values with high leverage. High-leverage observations can force small residuals even when these observations depart from the pattern of the rest of the data.[52]

A key concept in assessing influence is the *breakdown point* of an estimator: The breakdown point is the fraction of arbitrarily "bad" data that the estimator can tolerate without itself being affected to an arbitrarily large extent. In the location problem, for example, the mean has a breakdown point of 0, because a single bad observation can change the mean by an arbitrary amount. The median, in contrast, has a breakdown point of 50%, because fully half the data can be bad without causing the median to become completely unstuck.[53] It is disquieting that in regression analysis all M-estimators have breakdown points of 0.

There are regression estimators, however, that have breakdown points of nearly 50%. One such *bounded-influence* estimator is *least-trimmed-squares* (LTS) regression.

Return to the fitted regression model $Y_i = \mathbf{x}_i'\mathbf{b} + E_i$, ordering the squared residuals from smallest to largest:[54] $(E^2)_{(1)}, (E^2)_{(2)}, \ldots, (E^2)_{(n)}$. Then select $\mathbf{b}$ to minimize the sum of the smaller half of the squared residuals—that is,

$$\sum_{i=1}^{m} (E^2)_{(i)}$$

[52] For an illustration of this phenomenon, see Exercise 14.19.

[53] See Exercise 14.18.

[54] Lest the notation appear confusing, note that it is the *squared* residuals E_i^2 that are ordered from smallest to largest, not the residuals E_i.

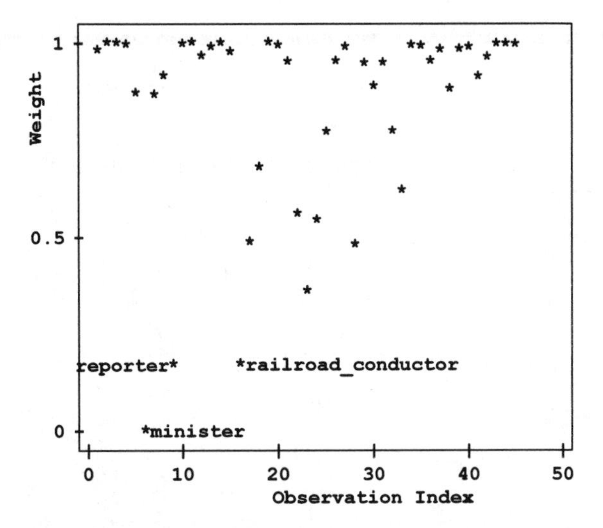

Figure 14.14. Final weights for the bisquare estimator applied to Duncan's regression of occupational prestige on income and education.

for $m = [n/2] + 1$ (where the square brackets indicate rounding down to the next smallest integer).

> Unlike the M-estimator of location, the M-estimator in regression is vulnerable to high-leverage observations. Bounded-influence estimators limit the effect of high-leverage observations. One such bounded-influence estimator is least trimmed squares, which selects the regression coefficients to minimize the smaller half of the squared residuals, $\sum_{i=1}^{m}(E^2)_{(i)}$ (where $m = [n/2] + 1$).

The LTS criterion is easily stated, but the LTS estimate is not so easily computed.[55] Moreover, LTS and other bounded-influence estimators are not a panacea for linear-model estimation, because they can give unreasonable results for some data configurations.[56] In the case of Duncan's occupational prestige regression, the LTS estimates are similar to the bisquare estimates:

$$\widehat{\text{Prestige}} = -5.097 + 0.8003 \times \text{Income} + 0.4027 \times \text{Education}$$

A final caution concerning robust regression: Robust estimation is not a substitute for close examination of the data. Although robust estimators can cope with heavy-tailed error distributions and outliers, they cannot correct nonlinearity, for example.

[55] See, for example, Rousseeuw and Leroy (1987).

[56] See Stefanski (1991).

EXERCISES

14.17 *Prove that the median minimizes the least-absolute-values objective function:

$$\sum_{i=1}^{n} \rho_{\text{LAV}}(E_i) = \sum_{i=1}^{n} |Y_i - \hat{\mu}|$$

14.18 Consider the contrived dataset

$$Y_1 = -0.068 \qquad Y_2 = -1.282 \qquad Y_3 = 0.013 \qquad Y_4 = 0.141$$
$$Y_5 = -0.980$$

(an adaptation of the data used to construct Figure 14.10). Show that more than two values must be changed in order to influence the median of the five values to an arbitrary degree. (Try, for example, to make the first two values progressively and simultaneously larger, graphing the median of the altered dataset against the common value of Y_1 and Y_2; then do the same for the first three observations.)

14.19 The following contrived dataset (discussed in Chapter 3) is from Anscombe (1973):

X	Y
10	7.46
8	6.77
13	12.74
9	7.11
11	7.81
14	8.84
6	6.08
4	5.39
12	8.15
7	6.42
5	5.73

(a) Graph the data and confirm that the third observation is an outlier. Find the least-squares regression of Y on X, and plot the least-squares line on the graph.

(b) Fit a robust regression to the data, using the bisquare or Huber M-estimator. Plot the fitted regression line on the graph. Is the robust regression affected by the outlier?

(c) Omitting the third observation {13, 12.74}, the line through the rest of the data has the equation $Y = 4 + 0.345X$, and the residual of the third observation from this line is 4.24. (Verify these facts.) Generate equally discrepant observations at X-values of 23 and 33, by substituting these values successively into the equation $Y = 4 + 0.345X + 4.24$. Call the resulting Y-values

Y_3' and Y_3''. Redo parts (a) and (b) replacing the third observation with the point $\{23, Y_3'\}$. Then replace the third observation with the point $\{33, Y_3''\}$. What happens?

(d) Repeat part (c) using a bounded-influence estimator.

14.20 Apply one or more robust estimators to each of the following models, comparing the results with least squares. Use the robust-regression weights to identify outlying observations.

(a) Davis's regression of reported on measured weight (Section 11.1 and `davis.dat`).

(b) Davis's regression of measured on reported weight (Section 11.1 and `davis.dat`).

(c) Chirot and Ragin's regression for data on the 1907 Romanian peasant rebellion (Exercise 11.2, Table 11.1, and `chirot.dat`).

(d) Sahlins's regression of acres per gardener on consumers per gardener for the households in Mazulu village (Exercise 5.6, Table 2.1, and `sahlins.dat`).

(e) Moore and Krupat's two-way analysis of variance of conformity by authoritarianism and partner's status (Section 8.2, Table 8.3, and `moore.dat`).

(f) Anscombe's regression of state education expenditures on income, proportion under 18, and proportion urban (Exercise 5.14, Table 5.1, and `anscombe.dat`).

14.4 Nonparametric Regression

This section continues the treatment of nonparametric regression initiated in Chapter 2, employing techniques—such as weighted-least-squares regression—that we have developed in the interim. Although there are many sophisticated methods of nonparametric regression, I shall limit the presentation here to two: *lowess* smoothers for scatterplots, and *additive regression* models.

14.4.1 Smoothing Scatterplots by Lowess

Lowess—an acronym for locally weighted scatterplot smoother—is a direct extension of the locally weighted averages described in Section 2.3.[57] The method produces a smoothed fitted value $\hat{Y}$ corresponding to any X-value in the range of the data (where Y and X are used generically to denote the vertical and horizontal variables in a scatterplot).[58] Often the X-values for which the smooth is computed are simply the data values x_i, and the description of lowess in this section will suppose that this is the case.

To find smoothed values, the lowess procedure fits n polynomial regressions to the data, one for each observation i, emphasizing the points with X-values that are near x_i. It usually suffices to fit a local regression *line*—that is, a local polynomial of degree 1. The lowess procedure is illustrated in Figure 14.15, which examines the relationship between prestige and income in the Canadian occupational prestige data.[59] Lowess is computationally intensive and requires

---

[57] Lowess (also called *loess*) is due to Cleveland (1979).

[58] We already used lowess in Section 4.3 to smooth plots of a dependent variable against a single independent variable, and in Section 12.3.1 to smooth partial-residual plots.

[59] Cf. Figure 2.7, which illustrates the procedure of locally weighted averaging.

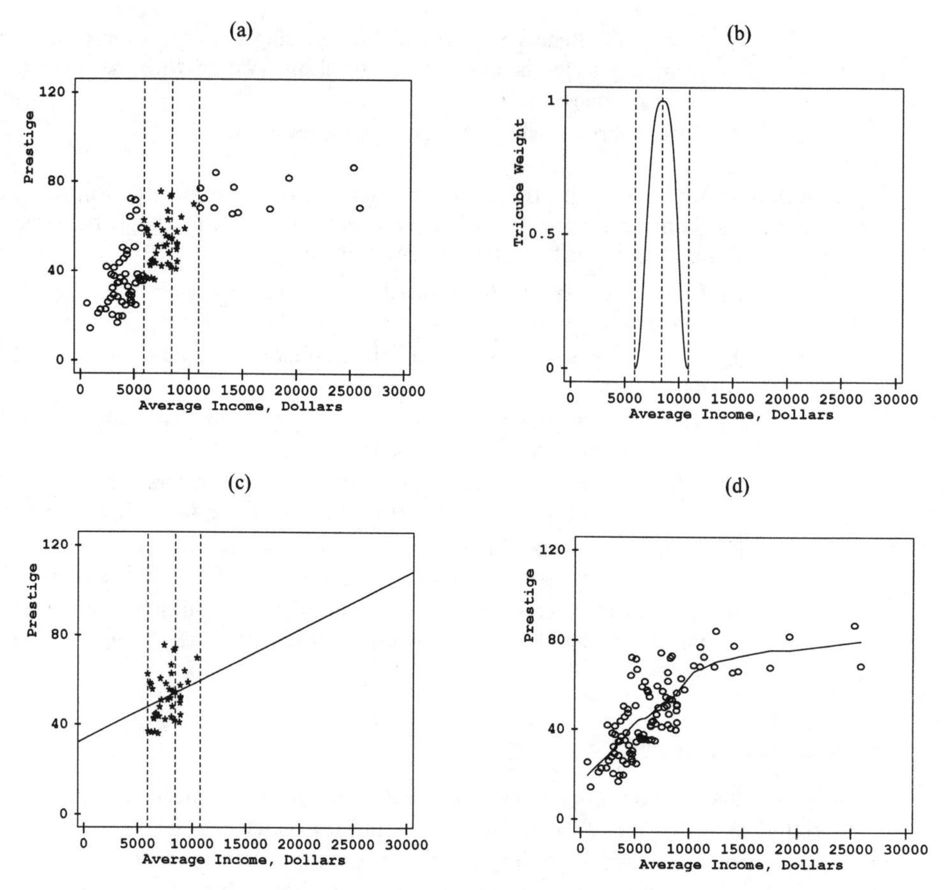

Figure 14.15. Calculating the lowess regression for the data on prestige and income of Canadian occupations. Panels (a), (b), and (c) illustrate the calculation of the locally weighted regression at $x_{(80)} = 8403$: In panel (a), a window is defined enclosing the 40 nearest neighbors of $x_{(80)}$ (corresponding to a span of $s \simeq .4$); panel (b) shows the tricube weight function centered at $x_{(80)}$; panel (c) shows the locally weighted regression line; the fitted value $\hat{Y}_{80}$ is given by the intersection of the local regression line with the vertical line through $x_{(80)}$. Panel (d) shows all 102 fitted values (one from each locally weighted regression) connected by a line. The lowess smooth looks a little rough, suggesting that we might try a larger span. Compare this figure with Figure 2.8, which shows locally weighted averages.

Source: Adapted with permission from Chambers et al. (1983, Figure 4.14).

special computer software for its implementation, but such programs are easy to write and are increasingly common.

The lowess algorithm proceeds as follows:

1. *Choose the span.* Select a fraction of the data $0 < s \leq 1$ (called the *span* of the smoother) to include in each fit, corresponding to $m \equiv [s \times n]$ data values, where the square brackets denote rounding to the nearest integer. Often $s = \frac{1}{2}$ or $s = \frac{2}{3}$ works well. Larger values of s produce smoother results. Some approaches to selecting the span are described below.

2. *Locally weighted regressions.* For each $i = 1, 2, \ldots, n$, select the m values of X closest to x_i, denoted $x_{i1}, x_{i2}, \ldots, x_{im}$ [see Figure 14.15(a)]. The window half-width for observation i is then the distance to the farthest x_{ij}; that is, $h_i \equiv \max_{j=1}^{m} |x_{ij} - x_i|$.

   a. *Calculate weights.* For each of the m observations in the window, compute the weight

   $$w_{ij} \equiv w_T\left(\frac{x_{ij} - x_i}{h_i}\right)$$

   where $w_T(\cdot)$ is the *tricube* weight function:

   $$w_T(z) = \begin{cases} (1 - |z|^3)^3 & \text{for } |z| < 1 \\ 0 & \text{for } |z| \geq 1 \end{cases}$$

   [Here, z simply stands in for the argument to the tricube function— that is, $(x_{ij} - x_i)/h_i$.] Thus, the weights gradually descend to 0 as X_{ij} approaches the boundaries of the window, and attain their maximum value of 1 at the center of the window, x_i [see Figure 14.15(b)].

   b. *Local WLS fit.* Having computed the weights, fit the local regression equation

   $$Y_{ij} = A_i + B_{i1}x_{ij} + B_{i2}x_{ij}^2 + \cdots + B_{ik}x_{ij}^k + E_{ij}$$

   to minimize $\sum_{j=1}^{m} w_{ij}^2 E_{ij}^2$ (i.e., by weighted least squares). Setting the degree of the local polynomial to $k = 1$ fits a local regression line [see Figure 14.15 (c)]. If the relationship between Y and X changes direction quickly, then it may be preferable to use $k = 2$ (a local quadratic).

   c. *Fitted value.* Compute the fitted value

   $$\hat{Y}_i = A_i + B_{i1}x_i + B_{i2}x_i^2 + \cdots + B_{ik}x_i^k$$

   One regression equation is fit, and one fitted value is calculated, for each $i = 1, \ldots, n$. Connecting these fitted values produces the nonparametric regression smooth [see Figure 14.15(d)].

3. *Robustness weights.* The lowess smooth can be made resistant to outliers by adapting the weighting approach of M-estimation. Having calculated the lowess fitted values $\hat{Y}_i$, find the residuals $E_i = Y_i - \hat{Y}_i$. Then, using a weight function $w(\cdot)$ such as the bisquare,[60] calculate "robustness" weights for each observation, $v_i \equiv w(E_i)$. Repeat step 2, using compound weights $v_j w_{ij}$ in the local weighted-least-squares regressions. This procedure can be iterated until the $\hat{Y}_i$ converge, but usually one or two iterations will suffice.

---

[60] See the previous section.

Lowess—nonparametric locally weighted simple regression—is an effective scatterplot smoother. The lowess procedure fits n polynomial regressions, one for each observation i, emphasizing points with X-values in the neighborhood of x_i: Using span s (the fraction of the data defining each neighborhood), weights are calculated for each observation j as $w_{ij} = w_T[(x_{ij} - x_i)/h_i]$, where $w_T(\cdot)$ is the tricube weight function, and h_i is the half-width of the neighborhood surrounding x_i. Observations beyond the span of the smoother, therefore, get weights of 0. The weights are then used to fit the local regression equation by weighted least squares, minimizing $\sum w_{ij}^2 E_{ij}^2$. The smoothed value $\hat{Y}_i$ is calculated by substituting x_i into the fitted local regression equation. The degree of the local polynomial is usually taken as 1 or 2. The lowess smooth can be made resistant to outliers by adapting the weighting approach of M-estimation, associating a robustness weight $v_i = w(Y_i - \hat{Y}_i)$ with each observation [where $w(\cdot)$ is, for example, the biweight function]. The local regressions are then refit, using the compound weights $v_j w_{ij}$.

Selecting the Span

The lowess smoother estimates the nonparametric regression model

$$Y_i = f(x_i) + \varepsilon_i$$

Let us assume that the errors ε_i are identically and independently distributed with zero expectations and common variance σ_ε^2. As in least-squares regression, if X is random, then we shall further assume that X and ε are independent.

Selecting the span in lowess smoothing involves a trade-off of bias against variance: The larger the value of s, the greater the bias and the smaller the variance of $\hat{Y}$ as an estimate of the true regression function $f(x)$. One general approach is to pick s to minimize an estimate of the mean-squared error of $\hat{Y}$—that is, the sum of squared bias and variance.

We could, for example, implement this strategy by using as a criterion the *cross-validation mean-squared prediction error*:[61]

$$\text{CV}(s) = \frac{1}{n}\sum_{i=1}^{n}\left(Y_i - \hat{Y}_i^{(-i)}\right)^2$$

where $\hat{Y}_i^{(-i)}$ is the fitted value evaluated at $X = x_i$, for a locally weighted regression that *omits* the ith observation. The notation CV(s) stresses the dependence

[61] The mean-squared prediction error is, more precisely, MSE + σ_ε^2, but because the term σ_ε^2 is constant, it does not affect the comparison of CV(s) for different spans s.

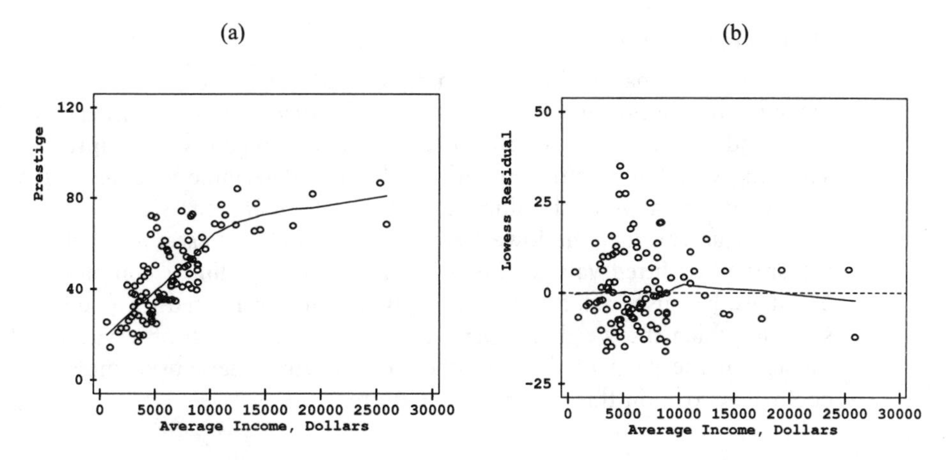

Figure 14.16. Panel (*a*) shows a lowess regression of prestige on income for the Canadian occupational prestige data. The residuals from the lowess fit are smoothed in panel (*b*). In both cases, the span of the smoother is $s = \frac{1}{2}$.

of the prediction error on the choice of s. We select s to make $\mathrm{CV}(s)$ as small as possible.

It is usually sufficient, however, to select the span visually: Begin with a generally reasonable value of s, such as $\frac{1}{2}$ or $\frac{2}{3}$. If the resulting lowess curve looks too rough, try increasing the value of s. This procedure is most effective in a statistical computing environment that updates a plot of the lowess smooth as the value of s is changed. If the lowess curve looks smooth, then try decreasing s until the curve begins to appear rough.

It sometimes helps to smooth the residuals $E_i = Y_i - \hat{Y}_i$ from a tentative lowess fit. If the systematic relationship of Y to X has been adequately captured by $\hat{Y}$, then the smoothed residuals should be unrelated to X (i.e., the smooth should be an approximately horizontal line at $E = 0$). If, alternatively, there is a systematic relationship between the residuals and X, then the regression of Y on X has been over-smoothed. An example, for the regression of prestige on income, appears in Figure 14.16. The smooth of the residuals [in Figure 14.16(*b*)] is close to 0, except for large values of income, where data are sparse because of the skew of the income distribution.

The framework of estimation also explains why local regression is generally superior to local averaging. Suppose that we fit a local regression line (i.e., a local polynomial of degree 1) to the data in the neighborhood of x_i. If the underlying regression function is nearly linear in this neighborhood, then both the local average and the local regression will produce nearly unbiased estimates of $f(x_i)$; the variance of the local regression estimate will be smaller, however.[62] When $f(X)$ is substantially nonlinear in the neighborhood of x_i, fitting a local polynomial regression will produce less bias than using a local average or local regression line. Finally, lowess does not suffer from the flattening that afflicts local averaging near the smallest and largest values[63] of X.

---

[62] See Exercise 14.21.

[63] See Section 2.3.

Statistical Inference[64]

Our principal goal in smoothing a scatterplot is to produce a visual summary of the relationship between Y and X. Formal statistical inference often is of secondary interest. It is nevertheless useful (1) to gain some impression of the sampling variability of the smooth; and (2) to determine whether anything has been gained relative to the simpler linear least-squares fit.

Let us consider the lowess smoother *without* robustness weights. Each of the smoothed fitted values $\hat{Y}_i$ can be expressed as a linear combination of the Y-values, because each fitted value derives from a weighted-least-squares regression, in which the weights depend only on the X-values; observations that fall outside of the span of the smoother for the ith observation implicitly receive coefficients of 0 in this linear combination:

$$\hat{Y}_i = \sum_{j=1}^{n} s_{ij} Y_j$$

Collecting the coefficients $[s_{i1}, s_{i2}, \ldots, s_{in}]$ as the ith row of the matrix $\mathbf{S}$,

$$\underset{(n\times 1)}{\hat{\mathbf{y}}} = \underset{(n\times n)}{\mathbf{S}}\underset{(n\times 1)}{\mathbf{y}}$$

The variance of the fitted values follows directly:

$$V(\hat{\mathbf{y}}) = \mathbf{S}V(\mathbf{y})\mathbf{S}' = \sigma_\varepsilon^2 \mathbf{S}\mathbf{S}' \qquad [14.25]$$

To apply Equation 14.25, we require an estimator of the error variance σ_ε^2, and for that we need to find the degrees of freedom for the residual sum of squares, $\sum E_i^2 = \sum (Y_i - \hat{Y}_i)^2$. There are several ways to proceed, each exploiting an analogy to linear least-squares regression.

The simplest approach is as follows: In linear least squares, the hat matrix $\mathbf{H} \equiv \mathbf{X}(\mathbf{X}'\mathbf{X})^{-1}\mathbf{X}'$ transforms the observed Y's to the fitted values, $\hat{\mathbf{y}} = \mathbf{H}\mathbf{y}$, and the residuals are $\mathbf{e} = (\mathbf{I}_n - \mathbf{H})\mathbf{y}$. The number of parameters in the model is $\text{trace}(\mathbf{H}) = k + 1$, and the residual degrees of freedom are $\text{trace}(\mathbf{I}_n - \mathbf{H}) = n - (k+1)$.

By analogy, the "equivalent" number of parameters for the lowess smooth is $\text{trace}(\mathbf{S})$, and the residual degrees of freedom are

$$df_{\text{res}} = \text{trace}(\mathbf{I}_n - \mathbf{S}) = n - \text{trace}(\mathbf{S})$$

The estimated error variance is, therefore,

$$S_E^2 \equiv \frac{\sum E_i^2}{df_{\text{res}}}$$

[64] The results in this section are developed in greater detail in Hastie and Tibshirani (1990, Chapter 3). Also see Cleveland et al. (1992).

Using this result, $\hat{Y}_i \pm 2S_E\sqrt{v_{ii}}$ (where v_{ii} is the ith diagonal entry of $\mathbf{SS}'$) gives a pointwise approximate 95% confidence envelope around the lowess smooth.[65]

> The lowess fitted values can be expressed as a linear transformation of the observed values, $\hat{\mathbf{y}} = \mathbf{Sy}$. Then the covariance matrix of the fitted values is estimated as $\widehat{V(\hat{\mathbf{y}})} = S_E^2\mathbf{SS}'$, where $S_E^2 = \mathbf{e}'\mathbf{e}/[n-\text{trace}(\mathbf{S})]$, and $\mathbf{e} = \mathbf{y} - \hat{\mathbf{y}}$. This result can be used to put a pointwise 95% confidence envelope around the lowess smooth; trace(**S**) is interpreted as the equivalent number of parameters employed by the lowess smoother.

An example from the Canadian occupational prestige data appears in Figure 14.17, selecting a span of $\frac{2}{3}$ for the lowess regression of prestige on income. The equivalent number of parameters for this fit is trace(**S**) $= 4.668 \simeq 5$ (i.e., about the same as a fourth-degree polynomial regression). Note that at the right, where (due to the skew of income) data are sparse, the confidence envelope is wide.

We can, finally, derive a rough F-test for nonlinearity in analogy to the incremental F-test for linear models. In the present context, the natural null model is the linear regression $Y = \alpha + \beta x + \varepsilon$, with residual sum of squares RSS_0 and degrees of freedom $n - 2$. Let RSS_1 denote the residual sum of squares for the lowess fit. Then, to test the null hypothesis of linearity, we have

$$F_0 = \frac{\dfrac{\text{RSS}_0 - \text{RSS}_1}{\text{trace}(\mathbf{S}) - 2}}{\dfrac{\text{RSS}_1}{n - \text{trace}(\mathbf{S})}}$$

with trace(**S**) $- 2$ and $n -$ trace(**S**) degrees of freedom. A similar test can be computed to compare a lowess smooth with a more complex parametric model (such as a polynomial regression), or to compare two smooths, one with more equivalent parameters than the other.

For the linear regression of prestige on income, $\text{RSS}_0 = 14{,}616.18$; and, for the lowess smoother with span $s = 2/3$, $\text{RSS}_1 = 11{,}962.81$. Recall that $n = 102$. Comparing the two fits,

$$F_0 = \frac{(14{,}616.18 - 11{,}962.81)/(4.668 - 2)}{11{,}962.81/(102 - 4.668)} = 8.09$$

with 2.668 and 97.332 *df*, for which $p = .0001$.

[65] Alternatives to trace(**S**) are trace(**SS**′) and trace(2**S** − **SS**′). These alternatives coincide for linear least-squares regression, substituting the hat matrix **H** for the smoother matrix **S** (see Exercise 14.22), but are generally more difficult to compute, because calculating trace(**S**) requires finding only the diagonal entries of **S**.

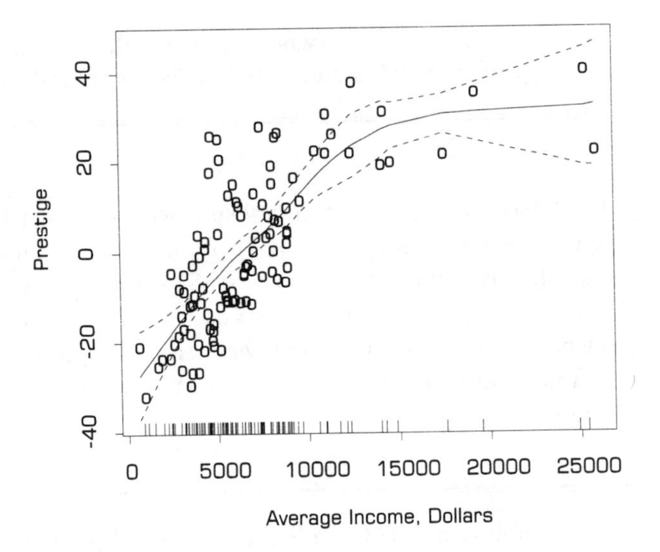

Figure 14.17. Lowess regression of prestige on income, for the Canadian occupational prestige data. The broken lines give an approximate 95% confidence envelope for the lowess smooth. The span of the smoother is $s = \frac{2}{3}$. The lines at the bottom of the graph are a one-dimensional scatterplot for income. Note that the software used to produce this plot rescaled the vertical axis to a mean of 0.

In contrast, when prestige is regressed on the cube root of income,[66] the residual sum of squares is $\text{RSS}_0 = 12{,}725.15$. Comparing this model with the lowess fit, we get $F_0 = 2.32$, with 2.668 and 97.332 *df*, for which $p = .087$: Transforming income substantially improves the fit of the linear model.

14.4.2 Additive Regression Models

The nonparametric additive regression model is

$$Y_i = \alpha + f_1(x_{i1}) + f_2(x_{i2}) + \cdots + f_k(x_{ik}) + \varepsilon_i \tag{14.26}$$

where the f_j are functions that we wish to estimate from the data, and the errors are identically distributed with constant variance and zero expectations.[67] The additive model is more restrictive than the *general* nonparametric multiple-regression model

$$Y_i = f(x_{i1}, x_{i2}, \ldots, x_{ik}) + \varepsilon_i \tag{14.27}$$

and therefore requires some motivation.

It is, as a formal matter, relatively straightforward to extend nonparametric-regression estimators such as lowess to the model given in Equation 14.27. The

[66] See Section 4.3 for the determination of this transformation of income.

[67] I shall at the end of this section consider *semiparametric* regression models, in which the functional form of some of the terms is specified in advance, and models in which some of the terms are functions of more than one independent variable.

principal theoretical problem is to define a multivariate neighborhood around the focal point $\mathbf{x}_i' = [x_{i1}, x_{i2}, \ldots, x_{ik}]$ in the predictor space, to be used in calculating the ith fitted value: If, as is typical, the X_j's are measured on different scales, then the simple Euclidean distance $||\mathbf{x}_i' - \mathbf{x}_j'||$ between $\mathbf{x}_i'$ and the predictors for another observation $\mathbf{x}_j'$ is not meaningful.

We can, however, standardize the X's in some manner prior to calculating distances.[68] Then, having found the m observations with X-values closest to $\mathbf{x}_i'$, we calculate a locally weighted linear or quadratic multiple regression of Y on the X's in the neighborhood of $\mathbf{x}_i'$, to produce the lowess fitted value $\hat{Y}_i$.

Although lowess multiple regression is easy to describe (and to compute), there are two serious practical difficulties:

1. *The "curse of dimensionality"*. As explained in Section 2.2, as the number of independent variables k increases, the number of points in the neighborhood of each $\mathbf{x}_i$ rapidly declines. Suppose, for example, that there are two X's and that the data values are uniformly distributed in the $\{X_1, X_2\}$ plane—the most favorable case. If we define a span of $\frac{1}{2}$ along each dimension, then about $\frac{1}{4}$ of the data values will be included in each neighborhood; more generally, for k uniformly distributed predictors, the proportion of data values in each neighborhood will be about $1/2^k$. The variance of the fitted values, therefore, increases rapidly with the number of predictors. We could, of course, increase the marginal span of the smooth along each dimension, but then the regressions would not really be local, and the bias of the fitted values as estimates of $f(x_{i1}, x_{i2}, \ldots, x_{ik})$ would increase.
2. *Interpretation*. When k exceeds 2, it is difficult to visualize the nonparametric fit to the data. Even for $k = 2$, a three-dimensional perspective plot, or two-dimensional slices of the fitted three-dimensional surface, may not convey a satisfactory impression of the fit.[69]

Fitting the Additive Regression Model

If the X's were independent, then we could fit the additive nonparametric regression (Equation 14.26) by smoothing Y separately against each X_j—just as in linear regression with uncorrelated X's, where the simple regressions of Y on each X_j sum to the multiple regression. Suppose, however, that the X's are related, but that we know all of the *partial-regression functions* f_j except for f_1. We could form the *partial residual*[70]

$$\begin{aligned} Y_{(1)i} &\equiv Y_i - [f_2(x_{i2}) + \cdots + f_k(x_{ik})] \\ &= \alpha + f_1(x_{i1}) + \varepsilon_i \end{aligned}$$

Leaving aside, for the moment, the determination of the regression constant α, smoothing $Y_{(1)}$ against X_1 produces an estimate of f_1.

Of course, in real applications, *all* of the partial-regression functions are unknown, but the notion of smoothing the partial residuals suggests the follow-

[68] See, for example, Cleveland (1993) for this and other details of lowess estimation in multiple regression.

[69] See Cleveland et al. (1992) and Cleveland (1993, Chapter 4 and 5), however.

[70] Cf. the treatment of partial residuals in linear regression in Section 11.6.

ing iterative procedure, called *backfitting*:

1. Find initial estimates $\widehat{f}_j^{(0)}$ of the partial-regression functions, for example, by least-squares linear regression of Y on the X's. Typically, the partial-regression functions are evaluated at the observed X-values, so
$$\widehat{f}_{ij}^{(0)} \equiv \widehat{f}_j^{(0)}(x_{ij}) = B_j x_{ij}$$
I shall refer to the values $\widehat{f}_{ij}$ as the "estimated partial-regression functions."[71] Set the iteration counter $l = 0$. To account for the constant, set $A \equiv \widehat{\alpha} = \overline{Y}$, and center the initial estimates $\widehat{f}_{ij}^{(0)}$ by subtracting the mean $\overline{\widehat{f}}_j^{(0)}$ from each value. This ensures that the residuals $E_i^{(0)}$ have a mean of 0. Repeat this centering step at each iteration.[72]
2. At iteration l, cycle through the predictors $j = 1, \ldots, k$, calculating partial residuals using the most recent values of the other partial-regression functions, and smoothing the partial residuals to update the current regression function. That is,
$$\begin{aligned}
\widehat{f}_{i1}^{(l)} &= \text{lowess}[Y_{(1)i}^{(l)} \text{ on } x_{i1}] \\
&= \mathbf{S}_1\{Y_i - [f_2^{(l-1)}(x_{i2}) + \cdots + f_k^{(l-1)}(x_{ik})]\} \\
\widehat{f}_{i2}^{(l)} &= \text{lowess}[Y_{(2)i}^{(l)} \text{ on } x_{i2}] \\
&= \mathbf{S}_2\{Y_i - [f_1^{(l)}(x_{i1}) + f_3^{(l-1)}(x_{i3}) + \cdots + f_k^{(l-1)}(x_{ik})]\} \\
&\ \ \vdots \\
\widehat{f}_{ik}^{(l)} &= \text{lowess}[Y_{(k)i}^{(l)} \text{ on } x_{ik}] \\
&= \mathbf{S}_k\{Y_i - [f_1^{(l)}(x_{i1}) + \cdots + f_{k-1}^{(l)}(x_{i,k-1})]\}
\end{aligned}$$
where (as in Section 14.4.1), the lowess transformation matrix $\underset{(n\times n)}{\mathbf{S}_j}$ depends only on the configuration of values x_{ij} for the jth predictor.
3. Repeat step 2 until the partial regression functions $\widehat{f}_{ij}^{(l)}$ converge. Applied to a linear model, the backfitting algorithm—performing least-squares simple regression at each step—produces the usual least-squares multiple-regression estimates.[73]

The backfitting algorithm implicitly solves the following set of estimating equations:

$$\begin{bmatrix} 1 & \mathbf{0}' & \mathbf{0}' & \cdots & \mathbf{0}' \\ \mathbf{0} & \mathbf{I}_n & \mathbf{S}_1 & \cdots & \mathbf{S}_1 \\ \mathbf{0} & \mathbf{S}_2 & \mathbf{I}_n & \cdots & \mathbf{S}_2 \\ \vdots & \vdots & \vdots & \ddots & \vdots \\ \mathbf{0} & \mathbf{S}_k & \mathbf{S}_k & \cdots & \mathbf{I}_n \end{bmatrix} \begin{bmatrix} A \\ \widehat{\mathbf{f}}_1 \\ \widehat{\mathbf{f}}_2 \\ \vdots \\ \widehat{\mathbf{f}}_k \end{bmatrix} = \begin{bmatrix} (1/n)\mathbf{1}_n'\mathbf{y} \\ \mathbf{S}_1\mathbf{y} \\ \mathbf{S}_2\mathbf{y} \\ \vdots \\ \mathbf{S}_k\mathbf{y} \end{bmatrix} \qquad [14.28]$$
$$\underset{[(kn+1)\times(kn+1)]}{\mathbf{S}}\ \underset{[(kn+1)\times 1]}{\widehat{\mathbf{f}}} = \underset{[(kn+1)\times n]}{\mathbf{Q}}\ \underset{(n\times 1)}{\mathbf{y}}$$

---

[71] More precisely, $\widehat{f}_{ij}$ is the estimated partial-regression function $\widehat{f}_j(X_j)$ evaluated at $X_j = x_{ij}$.

[72] As Hastie and Tibshirani (1990, p. 115) point out, we can automatically center the partial-regression functions by replacing the smoothing matrix $\mathbf{S}_j$ with the smoothing-and-centering matrix $(\mathbf{I}_n - (1/n)\mathbf{1}_n\mathbf{1}_n')\mathbf{S}_j$ (see Exercise 14.23). The discussion of the estimating equations that follows supposes that $\mathbf{S}_j$ already incorporates this adjustment.

[73] See Exercise 14.24.

where $\hat{\mathbf{f}}_j \equiv [\hat{f}_{1j}, \hat{f}_{2j}, \ldots, \hat{f}_{nj}]'$ are the estimates of the jth partial-regression function, and $\mathbf{y}$ is the dependent-variable vector. The first equation simply sets $A = \overline{Y}$. The remaining matrix equations each have the form

$$\hat{\mathbf{f}}_j + \mathbf{S}_j \sum_{k \neq j} \hat{\mathbf{f}}_k = \mathbf{S}_j \mathbf{y}$$

Rearranging terms, the fitted partial-regression function is the smoothed partial residual, as required:

$$\hat{\mathbf{f}}_j = \mathbf{S}_j \left(\mathbf{y} - \sum_{k \neq j} \hat{\mathbf{f}}_k \right)$$

The estimating equations (Equation 14.28) are a system of $kn + 1$ linear equations in an equal number of unknowns. In principle—barring the analog of perfect collinearity—these equations can be solved directly for the estimates of the partial-regression functions (along with A), but the large number of equations makes direct solution impractical.

> Because of the intrinsic sparseness of high-dimensional data, and because of the difficulty of visualizing general high-dimensional surfaces, it is usually impractical to extend scatterplot smoothers such as lowess to multiple-regression analysis. The nonparametric additive regression model $Y_i = \alpha + f_1(x_{i1}) + f_2(x_{i2}) + \cdots + f_k(x_{ik}) + \varepsilon_i$ is more practical. This model can be fit to data by iteratively smoothing the partial-residual plot for each independent variable in turn, a procedure called backfitting. Iteration proceeds until the estimated partial-regression functions $\widehat{f_j}(x_{ij})$ converge.

Statistical Inference

It is apparent from the solution of the estimating equations,

$$\hat{\mathbf{f}} = \mathbf{S}^{-1}\mathbf{Q}\mathbf{y} = \mathbf{R}\mathbf{y}$$

(where $\mathbf{R} \equiv \mathbf{S}^{-1}\mathbf{Q}$), that each partial-regression function is a linear transformation of the Y-values:[74] $\hat{\mathbf{f}}_j = \mathbf{R}_j \mathbf{y}$. Thus, the covariance matrix for the partial fit is

$$V(\hat{\mathbf{f}}_j) = \mathbf{R}_j V(\mathbf{y}) \mathbf{R}_j' = \sigma_\varepsilon^2 \mathbf{R}_j \mathbf{R}_j'$$

[74] Note that, in general, $\mathbf{R}_j$ (which transforms $\mathbf{y}$ to $\hat{\mathbf{f}}_j$) is different from $\mathbf{S}_j$ (which transforms the partial residual $\mathbf{y} - \sum_{k \neq j} \hat{\mathbf{f}}_k$ to $\hat{\mathbf{f}}_j$).

There are, however, two obstacles to applying this result:

1. The transformation matrix $\mathbf{R}_j$ is not easy to calculate. A rough solution[75] is to substitute $\mathbf{S}_j$. Moreover, if we just want the variances of the elements of $\hat{\mathbf{f}}_j$ (to calculate a confidence envelope for the partial fit), then we need only compute the diagonal elements of $\mathbf{S}_j\mathbf{S}_j'$.
2. We require an estimate of σ_ε^2. A rough solution is to take trace$(\mathbf{S}_j) - 1$ as the degrees of freedom (i.e., the equivalent number of parameters) for the jth fitted partial regression, and to calculate the residual degrees of freedom as[76]

$$df_{\text{res}} = n - \sum_{j=1}^{k}[\text{trace}(\mathbf{S}_j) - 1] - 1$$

Then it is natural to compute the error variance estimate $S_E^2 \equiv \hat{\sigma}_\varepsilon^2 = \sum E_i^2 / df_{\text{res}}$.

At convergence, each estimated partial-regression function can be written as a linear transformation of the Y-values, $\hat{\mathbf{f}}_j = \mathbf{R}_j\mathbf{y}$, where $\hat{\mathbf{f}}_j = [\widehat{f_j(x_{1j})}, \ldots, \widehat{f_j(x_{nj})}]'$. Then $V(\hat{\mathbf{f}}_j) = \sigma_\varepsilon^2\mathbf{R}_j\mathbf{R}_j'$. Because $\mathbf{R}_j$ is difficult to calculate, we use the smoother matrix $\mathbf{S}_j$ in its place. A rough estimate of the error variance σ_ε^2 is $S_E^2 = \mathbf{e}'\mathbf{e}/df_{\text{res}}$, using $df_{\text{res}} = n - \sum_{j=1}^{k}[\text{trace}(\mathbf{S}_j) - 1] - 1$. A rough pointwise confidence envelope for the fitted partial-regression function follows from these results.

Approximate incremental F-tests follow from these results. To test the contribution of the jth predictor, for example, we can refit the model with this predictor omitted. Let RSS_1 denote the residual sum of squares for the full model, and RSS_0 the residual sum of squares for the null model. The incremental F-statistic is then

$$F_0 = \frac{\dfrac{\text{RSS}_0 - \text{RSS}_1}{\text{trace}(\mathbf{S}_j) - 1}}{\dfrac{\text{RSS}_1}{df_{\text{res}}}}$$

with trace$(\mathbf{S}_j) - 1$ and df_{res} degrees of freedom.

---

[75] More sophisticated approaches are described in Hastie and Tibshirani (1990: Ch. 5).

[76] Subtracting 1 from the degrees of freedom for each term reflects the constraint that the partial regression function sums to 0; subtracting 1 in calculating the residual degrees of freedom reflects fitting the regression constant.

> F-tests for lowess and for additive regression models can be formulated in analogy to incremental F-tests in linear models.

Semiparametric Models

It is simple to modify the additive regression model (Equation 14.26) so that some of the independent variables (say the last $k - q$) are fit parametrically:

$$Y_i = \alpha + f_1(x_{i1}) + f_2(x_{i2}) + \cdots + f_q(x_{iq}) \\ + \beta_{q+1}x_{i,\,q+1} + \cdots + \beta_k x_{ik} + \varepsilon_i$$

This approach is natural, for example, when some of the X's are dummy regressors, representing categorical independent variables, but other sorts of prespecified functions—such as polynomial regressors—could be accommodated as well. Moreover, some of $x_{q+1}, \ldots, x_k$ (or $x_1, \ldots, x_q$) could be interaction regressors, representing products of other regressors. Finally, some of the nonparametric partial-regression functions could be functions of more than one independent variable—for example, $f_{12}(x_{i1}, x_{i2})$—representing nonparametric-interaction regression surfaces. This last possibility is, however, limited by the practical problems attending general nonparametric regression.

Semiparametric models can be used to test for nonlinearity in an independent variable, say X_1. Fit two models: (1) the full additive regression model (Equation 14.26); and (2) a semiparametric model that has a linear term in X_1—that is,

$$Y_i = \alpha + \beta_1 x_{i1} + f_2(x_{i2}) + \cdots + f_k(x_{ik}) + \varepsilon_i$$

An incremental F-test comparing the two models, with trace$(\mathbf{S}_j) - 2$ and df_{res} degrees of freedom, tests for nonlinearity in the partial relationship between Y and X_1.

Table 14.9 shows the results of fitting a number of different models to the Canadian occupational prestige data, with income, education, and percentage of women as the independent variables. Several types of models appear in the table: the general additive regression model with all three independent variables; additive models in which one of the independent variables is omitted; semiparametric additive models in which one independent variable is entered linearly; and a linear model (from Section 12.3) in which income is log-transformed, education enters as a cubic polynomial, and percentage of women enters as a quadratic. The partial fits for the general additive model are graphed with approximate 95% confidence envelopes in Figure 14.18.

Using the building blocks in Table 14.9, a variety of incremental F-tests are reported in Table 14.10. In each case, the estimated error variance in the denominator is from model 1: $S_E^2 = 4077.281/92.043 = 44.298$. The results here are consistent with our analysis of the Canadian occupational prestige data

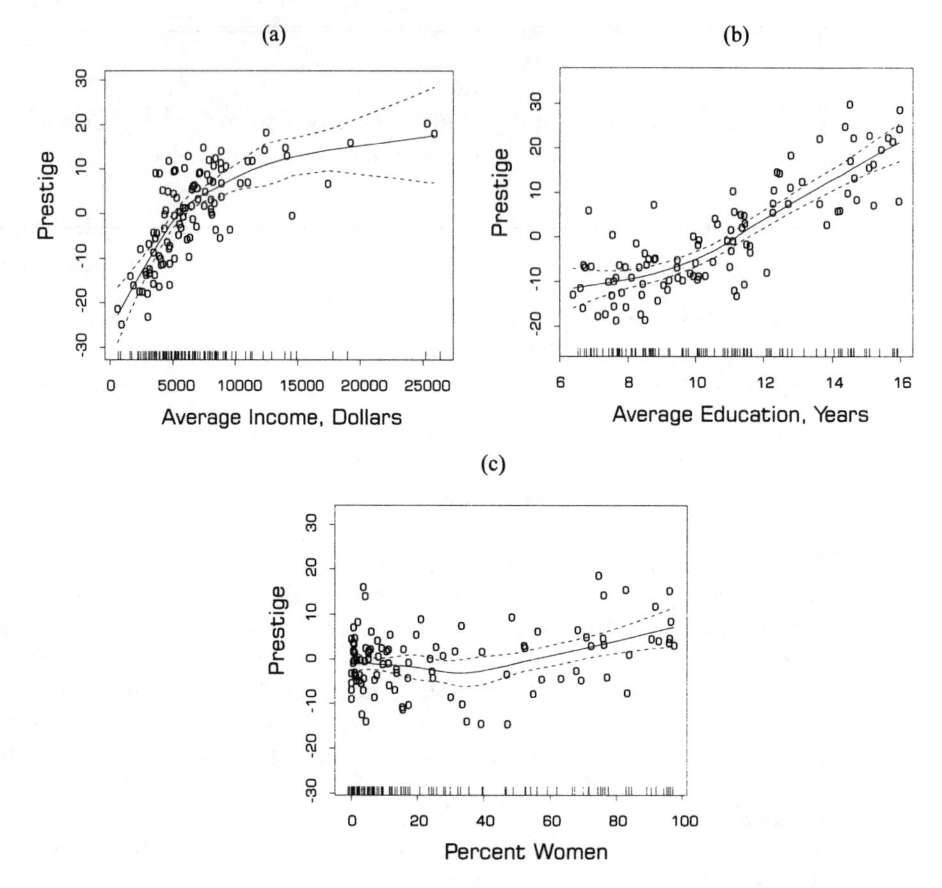

Figure 14.18. Partial-regression fits for the additive nonparametric regression of prestige on (*a*) income, (*b*) education, and (*c*) percentage of women. Each panel shows the partial lowess smooth for the corresponding independent variable; the broken lines give an approximate 95% confidence envelope for the fit; and a one-dimensional scatterplot of the independent variable appears at the bottom of the graph. The span for each lowess smooth is $\frac{2}{3}$. The units of the prestige scale are the same in the three plots to facilitate comparisons among them.

in Chapter 12:

- There is very strong evidence that income and education affect occupational prestige.
- There is somewhat weaker evidence that the percentage of women in the occupations affects prestige.
- There is strong evidence that the effect of income on prestige is nonlinear.
- There is weaker evidence that the effects of education and percentage of women are nonlinear.
- The linear model that we used in Section 12.3, log-transforming income and fitting polynomials in education and percentage of women, fits the data about as well as the additive nonparametric regression model. Indeed, the fits in Figure 14.18 are similar to the partial-residual plots that we inspected in Section 12.3.1.

TABLE 14.9 Residual Sums of Squares and Degrees of Freedom for Models Fit to the Canadian Occupational Prestige Data

Model	df_{res}	*RSS*
1. General additive model	92.043	4077.281
2. Income omitted	95.711	6372.336
3. Income linear	94.711	5376.048
4. Education omitted	94.603	8047.085
5. Education linear	93.603	4509.291
6. Percentage of women omitted	94.772	4632.606
7. Percentage of women linear	93.772	4426.762
8. Linear model from Section 12.3	95	4291.778

TABLE 14.10 Incremental *F*-Tests for Hypotheses Concerning the Canadian Occupational Prestige Data

Source	*Models Contrasted*	*df*	*SS*	*F*	*p*
Income	$2-1$	3.668	2295.055	14.125	$\ll .00001$
Income nonlinear	$3-1$	2.668	1298.767	10.989	$\ll .00001$
Education	$4-1$	2.560	3969.804	35.006	$\ll .00001$
Education nonlinear	$5-1$	1.560	432.010	6.251	.0057
Percentage of women	$6-1$	2.729	555.325	4.594	.0062
Percentage of women nonlinear	$7-1$	1.729	349.481	4.563	.017
Linear model from Section 12.3	$8-1$	2.957	214.497	1.637	.19

EXERCISES

14.21 Suppose that the regression of Y on X is linear in an interval centered at x_0, and that the X-values of the observations in the interval are evenly spaced. Show that both the local linear regression in the interval *and* the local average of Y in the interval are unbiased estimators of $E(Y_0)$ at the center of the interval. Explain why the fitted value $\hat{Y}_0$ from the local regression has smaller variance than the local average. What happens to the unbias of the local regression and local average estimators when the observations are *not* evenly spaced in the interval? (Cf. Exercise 2.1.)

14.22 *As I have pointed out, the smoother matrix $\mathbf{S}$ is analogous to the hat matrix $\mathbf{H}$ in linear least-squares regression. Show that the three definitions of degrees of freedom coincide for linear least squares: That is,

$$\text{trace}(\mathbf{H}) = \text{trace}(\mathbf{HH}') = \text{trace}(2\mathbf{H} - \mathbf{HH}') = n - k - 1$$

14.23 Show that premultiplying the $n \times n$ smoother matrix $\mathbf{S}_j$ by $\mathbf{I}_n - (1/n)\mathbf{1}_n'\mathbf{1}_n$ automatically centers the partial-regression function $\hat{f}_{ij}$.

14.24 Demonstrate by using the Canadian occupational prestige data (in `prestige.dat`) that applying the backfitting algorithm to linear regression produces the usual linear least-squares estimates.

14.25 In each of the following cases, use lowess to smooth the scatterplot relating the dependent variable to each independent variable. Then fit the full additive regression model to the data. Compare the fit of the additive regression model with the linear least-squares fit. Test the linearity of each term in the regression.

(a) Duncan's regression of occupational prestige on income and education (Table 3.2 and `duncan.dat`).

(b) Angell's regression of the moral integration of cities on ethnic heterogeneity and geographic mobility (Table 2.3 and `angell.dat`).

(c) Anscombe's regression of state education expenditures on income, proportion under 18, and proportion urban (Table 5.1 and `anscombe.dat`).

14.26 Why isn't it sensible to fit a nonparametric regression smoother to a component of an additive regression model represented by a set of dummy regressors? Why can we use a semiparametric model in this context?

14.5 Summary

Time Series Regression and Generalized Least Squares

- In time series data, a single individual is tracked over many time periods or points of time. It is not generally reasonable to suppose that the errors in a time series regression are independent.
- To capture serial dependence among the errors in the regression model $\mathbf{y} = \mathbf{X}\boldsymbol{\beta} + \boldsymbol{\varepsilon}$, we drop the assumption that the errors are independent of one another; instead, we assume $\boldsymbol{\varepsilon} \sim N_n(\mathbf{0}, \boldsymbol{\Sigma}_{\varepsilon\varepsilon})$, where nonzero off-diagonal entries in the error covariance matrix $\boldsymbol{\Sigma}_{\varepsilon\varepsilon}$ correspond to correlated errors.
- If the error covariance matrix $\boldsymbol{\Sigma}_{\varepsilon\varepsilon}$ is known, then the maximum-likelihood estimator of $\boldsymbol{\beta}$ is the generalized least-squares estimator

$$\mathbf{b}_{\text{GLS}} = (\mathbf{X}'\boldsymbol{\Sigma}_{\varepsilon\varepsilon}^{-1}\mathbf{X})^{-1}\mathbf{X}'\boldsymbol{\Sigma}_{\varepsilon\varepsilon}^{-1}\mathbf{y}$$

  The sampling variance-covariance matrix of $\mathbf{b}_{\text{GLS}}$ is

$$V(\mathbf{b}_{\text{GLS}}) = (\mathbf{X}'\boldsymbol{\Sigma}_{\varepsilon\varepsilon}^{-1}\mathbf{X})^{-1}$$

  The generalized least-squares estimator can also be expressed as the OLS estimator $(\mathbf{X}^{*\prime}\mathbf{X}^*)^{-1}\mathbf{X}^{*\prime}\mathbf{y}^*$ for the transformed variables $\mathbf{X}^* \equiv \boldsymbol{\Gamma}\mathbf{X}$ and $\mathbf{y}^* \equiv \boldsymbol{\Gamma}\mathbf{y}$, where the transformation matrix $\boldsymbol{\Gamma}$ is a square root of $\boldsymbol{\Sigma}_{\varepsilon\varepsilon}^{-1}$.
- When, more realistically, the error covariance matrix $\boldsymbol{\Sigma}_{\varepsilon\varepsilon}$ is unknown, we need to estimate its contents along with the regression coefficients $\boldsymbol{\beta}$. Without restricting its form, however, $\boldsymbol{\Sigma}_{\varepsilon\varepsilon}$ contains too many unique elements to estimate directly. Supposing that the errors are generated by a stationary time series process reduces the number of independent parameters in $\boldsymbol{\Sigma}_{\varepsilon\varepsilon}$ to n, including the error variance σ_ε^2 and the autocorrelations at various lags, $\rho_1, \ldots, \rho_{n-1}$.

- To reduce the number of parameters in $\boldsymbol{\Sigma}_{\varepsilon\varepsilon}$ further, we can adopt a particular time-series model for the errors. The most commonly employed such model is the first-order autoregressive process $\varepsilon_t = \rho\varepsilon_{t-1} + \nu_t$, where $|\rho| < 1$ and the random shocks ν_t are independently distributed as $N(0, \sigma_\nu^2)$. Under this specification, two errors, ε_t and ε_{t+s}, separated by s time periods have covariance $\rho^s\sigma_\varepsilon^2$. The variance of the regression errors is $\sigma_\varepsilon^2 = \sigma_\nu^2/(1-\rho^2)$.
- To apply GLS estimation for first-order autoregressive errors, we can first estimate the autocorrelation of the errors from the sample autocorrelation of the OLS residuals:

$$\hat{\rho} = r_1 = \frac{\sum_{t=2}^{n} E_t E_{t-1}}{\sum_{t=1}^{n} E_t^2}$$

  We can then use $\hat{\rho}$ to form an estimate of the correlation matrix of the errors or to transform y and **X**. Except for the first observation, these transformations take a particularly simple form: $Y_t^* = Y_t - \hat{\rho}Y_{t-1}$ and $X_{tj}^* = X_{tj} - \hat{\rho}X_{t-1,j}$. This procedure is called empirical GLS estimation.
- To diagnose serial correlation in the errors, we can examine the correlogram of the OLS residuals, calculating the serial correlations

$$r_s = \frac{\sum_{t=s+1}^{n} E_t E_{t-s}}{\sum_{t=1}^{n} E_t^2}$$

  for a number of lags $s = 1, 2, \ldots, m$. If the serial dependence of the residuals is well described by a first-order autoregressive process, then these serial correlations should decay exponentially toward 0. The Durbin-Watson statistic

$$D \equiv \frac{\sum_{t=2}^{n}(E_t - E_{t-1})^2}{\sum_{t=1}^{n} E_t^2} \simeq 2(1 - r_1)$$

  can be used to test for serial correlation of the errors.
- There are several practical and theoretical difficulties that limit the effectiveness and range of application of EGLS estimation.

Nonlinear Regression

- Nonlinear regression models that are linear in the parameters, for example, the quadratic regression model

$$Y = \beta_0 + \beta_1 X_1 + \beta_2 X_2 + \beta_3 X_1^2 + \beta_4 X_2^2 + \beta_5 X_1 X_2 + \varepsilon$$

  can be fit by linear least squares.
- Some nonlinear models can be rendered linear by a transformation. For example, the multiplicative gravity model of migration

$$Y_{ij} = \alpha \frac{P_i^\beta P_j^\gamma}{D_{ij}^\delta} \varepsilon_{ij}$$

  (where Y_{ij} is the number of migrants moving from location i to location j; P_i is the population at location i; D_{ij} is the distance separating the two locations; and ε_{ij} is a multiplicative error term), can be linearized by taking logs.

- More generally, nonlinear models of the form $Y_i = f(\boldsymbol{\beta}, \mathbf{x}_i') + \varepsilon_i$ (in which $\boldsymbol{\beta}$ is a vector of p parameters to be estimated, and $\mathbf{x}_i'$ is a vector of independent-variable values) can be estimated by nonlinear least squares, finding the value of $\mathbf{b}$ that minimizes

$$S(\mathbf{b}) = \sum_{i=1}^{n} E_i^2 = \sum_{i=1}^{n} [Y_i - f(\mathbf{b}, \mathbf{x}_i')]^2$$

Iterative methods for finding the nonlinear least-squares estimates include the method of steepest descent and, more practically, the Gauss-Newton and Marquardt methods. Estimated asymptotic covariances for the coefficients are given by

$$\widehat{\mathcal{V}(\mathbf{b})} = S_E^2 \left\{[\mathbf{F}(\mathbf{b}, \mathbf{X})]' \mathbf{F}(\mathbf{b}, \mathbf{X})\right\}^{-1}$$

where $\mathbf{F}(\mathbf{b}, \mathbf{X})$ is the matrix of derivatives, with i, jth entry $\partial f(\mathbf{b}, \mathbf{x}_i')/\partial B_j$, and $S_E^2 = \sum E_i^2/(n-p)$ is the estimated error variance. An example of an essentially nonlinear model, which requires nonlinear least squares, is the logistic population-growth model

$$Y_i = \frac{\beta_1}{1 + \exp(\beta_2 + \beta_3 X_i)} + \varepsilon_i$$

where Y_i is population size, and X_i is time.

Robust Regression

- Robust M-estimators of location, for the parameter μ in the simple model $Y_i = \mu + \varepsilon_i$, minimize the objective function

$$\sum_{i=1}^{n} \rho(E_i) = \sum_{i=1}^{n} \rho(Y_i - \hat{\mu})$$

selecting $\rho(\cdot)$ so that the estimator is relatively unaffected by outlying values. Two common choices of objective function are the Huber and the bisquare.

- The sensitivity of an M-estimator to individual observations is expressed by the influence function of the estimator, which has the same shape as the derivative of the objective function, $\psi(E) \equiv \rho'(E)$.
- An estimating equation for $\hat{\mu}$ is obtained by setting the derivative of the objective function (with respect to $\hat{\mu}$) to 0, obtaining $\sum_{i=1}^{n} \psi(Y_i - \hat{\mu}) = 0$. The simplest procedure for solving this estimating equation is by iteratively reweighted means. Defining the weight function as $w(E) \equiv \psi(E)/E$, the estimating equation becomes $\sum_{i=1}^{n} (Y_i - \hat{\mu}) w_i = 0$, from which $\hat{\mu} = \sum w_i Y_i / \sum w_i$. Starting with an initial estimate $\hat{\mu}^{(0)}$, initial weights are calculated, and the value of $\hat{\mu}$ is updated. This procedure continues iteratively until the value of $\hat{\mu}$ converges.
- M-estimation for the regression model $Y_i = \mathbf{x}_i'\boldsymbol{\beta} + \varepsilon_i$ is a direct extension of M-estimation of location: We seek to minimize an objective function of the regression residuals,

$$\sum_{i=1}^{n} \rho(E_i) = \sum_{i=1}^{n} \rho(Y_i - \mathbf{x}_i'\mathbf{b})$$

Differentiating the objective function, and setting the derivatives to 0, produces the estimating equations

$$\sum_{i=1}^{n} \psi(Y_i - \mathbf{x}_i'\mathbf{b})\mathbf{x}_i' = 0$$

which generally require iterative solution.

- Using the weight function, the estimating equations can be written as

$$\sum_{i=1}^{n} w_i(Y_i - \mathbf{x}_i'\mathbf{b})\mathbf{x}_i' = 0$$

The solution of the estimating equations then follows by weighted least squares:

$$\mathbf{b} = (\mathbf{X}'\mathbf{W}\mathbf{X})^{-1}\mathbf{X}'\mathbf{W}\mathbf{y}$$

where $\mathbf{W}$ is the diagonal matrix of weights. The method of iteratively reweighted least squares starts with initial estimates $\mathbf{b}^{(0)}$; calculates initial residuals from these estimates; and calculates initial weights from the residuals. The weights are used to update the parameter estimates, and the procedure is iterated until it converges.

- Unlike the M-estimator of location, the M-estimator in regression is vulnerable to high-leverage observations. Bounded-influence estimators limit the effect of high-leverage observations. One such bounded-influence estimator is least trimmed squares, which selects the regression coefficients to minimize the smaller half of the squared residuals, $\sum_{i=1}^{m}(E^2)_{(i)}$ (where $m = [n/2] + 1$).

Nonparametric Regression

- Lowess—nonparametric locally weighted simple regression—is an effective scatterplot smoother. The lowess procedure fits n polynomial regressions, one for each observation i, emphasizing points with X-values in the neighborhood of x_i: Using span s (the fraction of the data defining each neighborhood), weights are calculated for each observation j as $w_{ij} = w_T[(x_{ij} - x_i)/h_i]$, where $w_T(\cdot)$ is the tricube weight function, and h_i is the half-width of the neighborhood surrounding x_i. Observations beyond the span of the smoother, therefore, get weights of 0. The weights are then used to fit the local regression equation

$$Y_{ij} = A_i + B_{i1}x_{ij} + B_{i2}x_{ij}^2 + \cdots + B_{ik}x_{ij}^k + E_{ij}$$

by weighted least squares, minimizing $\sum w_{ij}^2 E_{ij}^2$. The smoothed value $\hat{Y}_i$ is calculated by substituting x_i into the fitted local regression equation. The degree of the local polynomial is usually taken as 1 or 2.

- The lowess smooth can be made resistant to outliers by adapting the weighting approach of M-estimation, associating a robustness weight $v_i = w(Y_i - \hat{Y}_i)$ with each observation [where $w(\cdot)$ is, for example, the biweight function]. The local regressions are then refit, using the compound weights $v_j w_{ij}$.
- The lowess fitted values can be expressed as a linear transformation of the observed values, $\hat{\mathbf{y}} = \mathbf{S}\mathbf{y}$. Then the covariance matrix of the fitted values is estimated as $\widehat{V(\hat{\mathbf{y}})} = S_E^2\mathbf{S}\mathbf{S}'$, where $S_E^2 = \mathbf{e}'\mathbf{e}/[n - \text{trace}(\mathbf{S})]$, and $\mathbf{e} = \mathbf{y} - \hat{\mathbf{y}}$. This result can be used to put a pointwise 95% confidence envelope around the lowess smooth; trace(S) is interpreted as the equivalent number of parameters employed by the lowess smoother.

- Because of the intrinsic sparseness of high-dimensional data, and because of the difficulty of visualizing general high-dimensional surfaces, it is usually impractical to extend scatterplot smoothers such as lowess to multiple-regression analysis. The nonparametric additive regression model

$$Y_i = \alpha + f_1(x_{i1}) + f_2(x_{i2}) + \cdots + f_k(x_{ik}) + \varepsilon_i$$

  is more practical. This model can be fit to data by iteratively smoothing the partial-residual plot for each independent variable in turn, a procedure called backfitting. Iteration proceeds until the estimated partial-regression functions $\widehat{f}_j(x_{ij})$ converge.
- At convergence, each estimated partial-regression function can be written as a linear transformation of the Y-values, $\hat{\mathbf{f}}_j = \mathbf{R}_j\mathbf{y}$, where $\hat{\mathbf{f}}_j = [\widehat{f}_j(x_{1j}), \ldots, \widehat{f}_j(x_{nj})]'$. Then $V(\hat{\mathbf{f}}_j) = \sigma_\varepsilon^2 \mathbf{R}_j\mathbf{R}_j'$. Because $\mathbf{R}_j$ is difficult to calculate, we use the smoother matrix $\mathbf{S}_j$ in its place. A rough estimate of the error variance σ_ε^2 is $S_E^2 = \mathbf{e}'\mathbf{e}/df_{\text{res}}$, using

$$df_{\text{res}} = n - \sum_{j=1}^{k}[\text{trace}(\mathbf{S}_j) - 1] - 1$$

  A rough pointwise confidence envelope for the fitted partial-regression function follows from these results.
- F-tests for lowess and for additive regression models can be formulated in analogy to incremental F-tests in linear models.

14.6 Recommended Reading

Although I have endeavored to present more than a *superficial* introduction to time series regression, nonlinear models, robust regression, and nonparametric regression, the treatment here remains an introduction to these topics.

- Time series analysis—including, but not restricted to, time series regression—is a deep and rich topic, well beyond the scope of the discussion in this chapter. A good, relatively brief, general introduction to the subject may be found in Chatfield (1989). Most econometric texts include some treatment of time series regression. The emphasis is typically on a formal understanding of statistical models and methods of estimation rather than on the use of these techniques in data analysis. Wonnacott and Wonnacott (1979), for example, present an insightful, relatively elementary treatment of time series regression and generalized least squares. Judge et al.'s (1985) presentation of the subject is more encyclopedic, with many references to the literature. An extensive treatment also appears in Harvey (1990).
- Further discussion of nonlinear least squares can be found in many sources, including Gallant (1975), Draper and Smith (1981, Chapter 10), and Bard (1974), in order of increasing detail and difficulty.
- In a volume on robust and exploratory methods, edited by Hoaglin et al. (1983), Goodall presents a high-quality, readable treatment of M-estimators of location. A fine paper by Li on M-estimators for regression appears in a companion volume (Hoaglin et al., 1985). Another good source on M-estimators is Wu (1985). Rousseeuw and Leroy's (1987) book on robust regression and outlier detection emphasizes bounded-influence, high-breakdown estimators.

- Hastie and Tibshirani's (1990) text on generalized additive models includes a wealth of valuable material. Most of the book is leisurely paced and broadly accessible, with many effective examples. Generalized additive models include the additive regression models discussed in this chapter, but also extend the methodology to models such as logistic regression.[77] As a preliminary to generalized additive models, Hastie and Tibshirani include a fine treatment of scatterplot smoothing. A briefer presentation by Hastie of generalized additive models appears in an edited book (Chambers and Hastie, 1992) on statistical modeling in the S computing environment. This book also includes a paper by Cleveland et al. on local regression models.[78] Cleveland's (1993) text on data visualization presents information on lowess smoothing in two and more dimensions, while Härdle (1991) gives an overview of nonparametric regression, with a stress on kernel smoothers for bivariate scatterplots. Additional details may be found in Fan and Gijbels (1996) and in Simonoff (1996).

---

[77] See Chapter 15 for a treatment of logistic regression and generalized *linear* models.

[78] Figures 14.17 and 14.18 were produced using S.

15

Logit and Probit Models

All of the statistical models described in previous chapters are for quantitative dependent variables. It is unnecessary to document the prevalence of qualitative, or nominal, data in the social sciences. In developing the general linear model, I introduced qualitative *independent* variables through the device of coding dummy-variable regressors.[1] There is no reason that qualitative variables should not also appear as dependent variables, affected by other variables, both qualitative and quantitative.

This chapter deals primarily with logit models for qualitative dependent variables, although some related models are also briefly described. The first section of the chapter describes logit and probit models for dichotomous dependent variables. The second section develops similar statistical models for polytomous dependent variables, including ordered categories. The third section discusses the application of logit models to contingency tables, where the independent variables, as well as the dependent variable, are discrete. The fourth and final section briefly describes generalized linear models, a broad synthesis that includes logit, probit, and linear models, among others.

15.1 Models for Dichotomous Data

Logit and probit models express a qualitative dependent variable as a function of several independent variables, much in the manner of the general linear

[1] See Chapters 7 and 8.

model. To understand why these models are required, let us begin by examining a representative problem, attempting to apply linear regression to it. The difficulties that are encountered point the way to more satisfactory statistical models for qualitative data.

In September of 1988, 15 years after the coup of 1973, the people of Chile voted in a plebiscite to decide the future of the military government headed by General Augusto Pinochet. A "yes" vote would represent eight more years of military rule; a "no" vote would set in motion a process to return the country to civilian government. Of course, the no side won the plebiscite, by a clear if not overwhelming margin.

Six months before the plebiscite, the independent research center FLACSO/Chile conducted a national survey of 2700 randomly selected Chilean voters.[2] Of these individuals, 868 said that they were planning to vote yes, and 889 said that they were planning to vote no. Of the remainder, 558 said that they were undecided, 187 said that they planned to abstain, and 168 did not answer the question. I shall look initially only at those who expressed a preference.[3]

Figure 15.1 plots voting intention against a measure of support for the status quo. As seems natural, voting intention appears as a dummy variable, coded 1 for yes, 0 for no. As we shall see presently, this coding makes sense in the context of a dichotomous dependent variable. Because many points would otherwise be overplotted, voting intention is jittered in the figure (although not in the calculations that follow). Support for the status quo is a scale formed from a number of questions about political, social, and economic policies: High scores represent general support for the policies of the miliary regime. (For the moment, disregard the lines plotted in this figure.)

We are used to thinking of a regression as a conditional average. Does this interpretation make sense when the dependent variable is dichotomous? After all, an average between 0 and 1 represents a "score" for the dummy dependent variable that cannot be realized by any individual. In the population, the conditional average $E(Y|x_i)$ is simply the proportion of 1's among those individuals who share the value x_i for the independent variable—the conditional probability π_i of sampling a "yes" in this group; that is,

$$\pi_i \equiv \Pr(Y_i) \equiv \Pr(Y = 1|X = x_i)$$

and, thus,

$$E(Y|x_i) = \pi_i(1) + (1 - \pi_i)(0) = \pi_i \qquad [15.1]$$

---

[2] FLACSO is an acronym for La Facultad Latino-americano des Ciensias Sociales, a respected institution that conducts social research and trains graduate students in several Latin-American countries. During the Chilean military dictatorship, FLACSO/Chile was associated with the opposition to the military government. I worked on the analysis of the survey described here as part of a joint project between FLACSO in Santiago, Chile, and the Centre for Research on Latin America and the Caribbean at York University, Toronto.

[3] It is, of course, difficult to know how to interpret ambiguous responses such as "undecided." It is tempting to infer that respondents were afraid to state their opinions, but there is other evidence from the survey that this is not the case. Few respondents, for example, uniformly refused to answer sensitive political questions, and the survey interviewers reported little resistance to the survey.

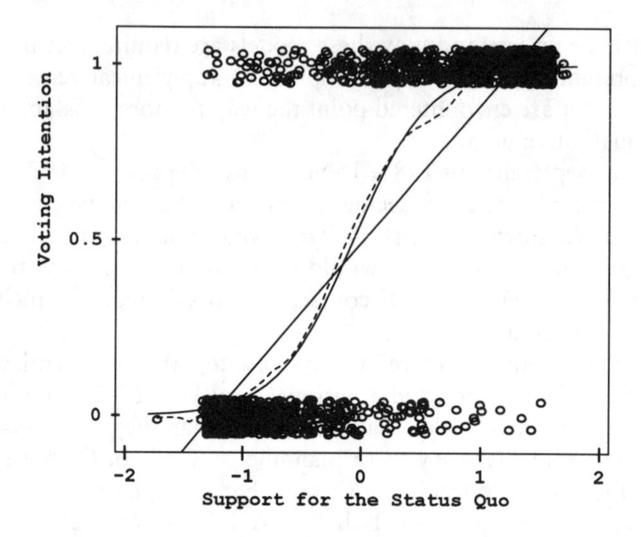

Figure 15.1. Scatterplot of voting intention (1 represents "yes," 0 represents "no") by a scale of support for the status quo, for a sample of Chilean voters surveyed prior to the 1988 plebiscite. The points are jittered vertically to minimize overplotting. The solid straight line shows the linear least-squares fit; the solid curved line shows the fit of the logistic-regression model (described in the next section); the broken line represents a lowess nonparametric regression.

If X is discrete, then in a sample we can calculate the conditional proportion for Y at each value of X. The collection of these conditional proportions represents the sample nonparametric regression of the dichotomous Y on X. In the present example, X is continuous, but we can nevertheless resort to strategies such as local averaging or local regression, as illustrated in Figure 15.1. At low levels of support for the status quo, the conditional proportion of yes responses is close to 0; at high levels, it is close to 1; and in between, the nonparametric regression curve smoothly approaches 0 and 1 in a gentle S-shaped pattern.

15.1.1 The Linear-Probability Model

Although nonparametric regression works here, it would be useful to capture the dependency of Y on X as a simple function. To do so will be particularly helpful when we introduce additional independent variables. As a first effort, let us try linear regression with the usual assumptions:

$$Y_i = \alpha + \beta X_i + \varepsilon_i \qquad [15.2]$$

where $\varepsilon_i \sim N(0, \sigma_\varepsilon^2)$, and ε_i and ε_j are independent for $i \neq j$. If X is random, then we assume that it is independent of ε.

Under Equation 15.2, $E(Y_i) = \alpha + \beta X_i$, and so, using Equation 15.1,

$$\pi_i = \alpha + \beta X_i$$

For this reason, the linear-regression model applied to a dummy dependent variable is called the *linear-probability model*. This model is untenable, but its failure will point the way toward more adequate specifications:

- Because Y_i can take on only the values 0 and 1, the error ε_i is dichotomous as well—and, hence, is not normally distributed, as Equation 15.2 assumes: If $Y_i = 1$, which occurs with probability π_i, then

$$\varepsilon_i = 1 - E(Y_i) = 1 - (\alpha + \beta X_i) = 1 - \pi_i$$

  Alternatively, if $Y_i = 0$, which occurs with probability $1 - \pi_i$, then

$$\varepsilon_i = 0 - E(Y_i) = 0 - (\alpha + \beta X_i) = 0 - \pi_i = -\pi_i$$

  Because of the central-limit theorem, however, the assumption of normality is not critical to least-squares estimation of the normal-probability model, as long as the sample size is sufficiently large.
- The variance of ε cannot be constant, as we can readily demonstrate: If the assumption of linearity holds over the range of the data, then $E(\varepsilon_i) = 0$. Using the relations just noted,

$$V(\varepsilon_i) = \pi_i(1 - \pi_i)^2 + (1 - \pi_i)(-\pi_i)^2 = \pi_i(1 - \pi_i)$$

  The heteroscedasticity of the errors bodes ill for ordinary-least-squares estimation of the linear probability model, but only if the probabilities π_i get close to 0 or 1.[4] Goldberger (1964, pp. 248–250) has proposed a correction for heteroscedasticity employing weighted least squares.[5] Because the variances $V(\varepsilon_i)$ depend on the π_i, however, which, in turn, are functions of the unknown parameters α and β , we require preliminary estimates for the model in order to define weights. Goldberger obtains ad hoc estimates from a preliminary OLS regression; that is, he takes $\widehat{V(\varepsilon_i)} = \widehat{Y}_i(1 - \widehat{Y}_i)$. The fitted values from an OLS regression are not constrained to the interval $[0, 1]$, and so some of these "variances" may be negative.
- This last remark suggests the most serious problem with the linear-probability model: The assumption that $E(\varepsilon_i) = 0$—that is, the assumption of linearity—is only tenable over a limited range of X-values. If the range of the X's is sufficiently broad, then the linear specification cannot confine π to the unit interval $[0, 1]$. It makes no sense, of course, to interpret a number outside of the unit interval as a probability. This difficulty is illustrated in Figure 15.1, in which the least-squares line fit to the Chilean plebiscite data produces fitted probabilities below 0 at low levels and above 1 at high levels of support for the status quo.

Dummy *regressor* variables do not cause comparable difficulties because the general linear model makes no distributional assumptions about the regressors (other than independence from the errors). Nevertheless, for values of π not too close to 0 or 1, the linear-probability model estimated by least squares frequently provides results similar to those produced by the more generally adequate methods described in the remainder of this chapter.

---

[4] See Exercise 15.1. Remember, however, that it is the *conditional* probability, not the marginal probability, of Y that is at issue: The overall proportion of 1's can be near .5 (as in the Chilean plebiscite data), and yet the conditional proportion can still get very close to 0 or 1.

[5] See Section 12.2.2 for a discussion of weighted-least-squares estimation.

> It is problematic to apply least-squares linear regression to a dichotomous dependent variable: The errors cannot be normally distributed and cannot have constant variance. Even more fundamentally, the linear specification does not confine the probability for the response to the unit interval.

One solution to the problems of the linear-probability model—though, as we shall see, not a good solution—is simply to constrain π to the unit interval while retaining the linear relationship between π and X within this interval:

$$\pi = \begin{cases} 0 & \text{for } 0 > \alpha + \beta X \\ \alpha + \beta X & \text{for } 0 \leq \alpha + \beta X \leq 1 \\ 1 & \text{for } \alpha + \beta X > 1 \end{cases} \qquad [15.3]$$

The *constrained linear-probability* model is shown in Figure 15.2. Although it cannot be dismissed on logical grounds, this model has certain unattractive features:

- The critical issue in estimating the linear-probability model is identifying the X-values at which π reaches 0 and 1, because the line $\pi = \alpha + \beta X$ is determined by these two points. As a consequence, estimation of the model is inherently unstable. Assume, for example, that the relationship between π and X is positive. Then, as illustrated in Figure 15.2, our estimate of the point at which $\pi = 0$ is determined by the leftmost X for which $Y = 1$; likewise, our estimate of the point at which $\pi = 1$ is determined by the rightmost X for which $Y = 0$. Both of these points are extreme values, which will vary substantially from sample to sample, and will tend to be more extreme in large samples than in small ones.

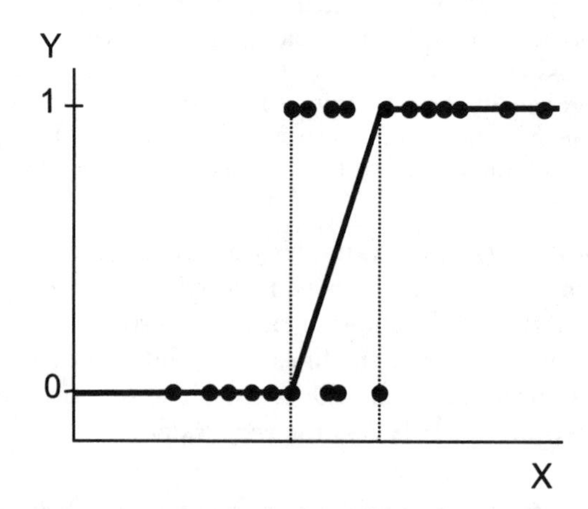

Figure 15.2. The constrained linear probability model. The estimate of the line $\pi = \alpha + \beta X$ is determined by the leftmost 1 and the rightmost 0, as shown by the vertical lines.

- It is much more difficult to estimate the constrained linear-probability model when there are several X's.
- Most fundamentally, the abrupt changes in slope at $\pi = 0$ and $\pi = 1$ are unreasonable. A smoother relationship between π and X (as characterizes the nonparametric regression in Figure 15.1) is more generally sensible.

15.1.2 Transformations of π: Logit and Probit Models

A central difficulty of the unconstrained linear-probability model is its inability to ensure that π stays between 0 and 1. What we require to correct this problem is a positive monotone (i.e., nondecreasing) function that maps the "linear predictor" $\eta = \alpha + \beta X$ into the unit interval. A transformation of this type will allow us to retain the fundamentally linear structure of the model while avoiding the contradiction of probabilities below 0 or above 1. Any cumulative probability distribution function (CDF) meets this requirement. That is, we can respecify the model as

$$\pi_i = P(\eta_i) = P(\alpha + \beta X_i) \qquad [15.4]$$

where the CDF $P(\cdot)$ is selected in advance, and α and β are then parameters to be estimated.

If we choose $P(\cdot)$ as the cumulative rectangular distribution, for example, then we obtain the constrained linear-probability model (Equation 15.3). An a priori reasonable $P(\cdot)$ should be both smooth and symmetric, and should approach $\pi = 0$ and $\pi = 1$ as asymptotes.[6] Moreover, it is advantageous if $P(\cdot)$ is strictly increasing, for then the transformation (Equation 15.4) is one to one, permitting us to rewrite the model as

$$P^{-1}(\pi_i) = \eta_i = \alpha + \beta X_i \qquad [15.5]$$

where $P^{-1}(\cdot)$ is the inverse of the CDF $P(\cdot)$. Thus, we have a linear (Equation 15.5) for a transformation of π, or—equivalently—a nonlinear model (Equation 15.4) for π itself.

The transformation $P(\cdot)$ is often chosen as the CDF of the unit-normal distribution

$$\Phi(z) = \frac{1}{\sqrt{2\pi}} \int_{-\infty}^{z} \exp\left(-\frac{1}{2}Z^2\right) dZ \qquad [15.6]$$

or, even more commonly, of the *logistic distribution*

$$\Lambda(z) = \frac{1}{1 + e^{-z}} \qquad [15.7]$$

In these equations, $\pi \simeq 3.141$ and $e \simeq 2.718$ are the familiar mathematical constants.

[6] This is not to say, however, that $P(\cdot)$ needs to be symmetric in every case, just that symmetric $P(\cdot)$'s are more appropriate *in general.* For an example of an asymmetric choice of $P(\cdot)$, see the discussion of the complementary log-log transformation in Section 15.4.

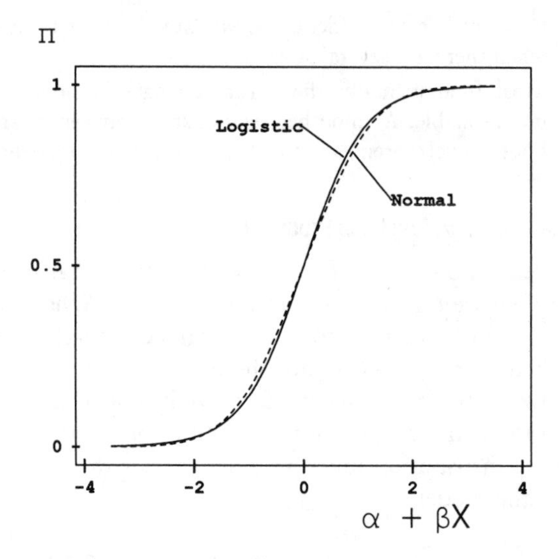

Figure 15.3. Once their variances are equated, the cumulative logistic and cumulative normal distributions—used here to transform $\alpha + \beta X$ to the unit interval—are virtually indistinguishable.

- Using the normal distribution $\Phi(\cdot)$ yields the *linear probit model*:

$$\begin{aligned}\pi_i &= \Phi(\alpha + \beta X_i)\\ &= \frac{1}{\sqrt{2\pi}} \int_{-\infty}^{\alpha+\beta X_i} \exp\left(-\frac{1}{2}Z^2\right) dZ\end{aligned}$$

- Using the logistic distribution $\Lambda(\cdot)$ produces the *linear logistic-regression* or *linear logit model*:

$$\begin{aligned}\pi_i &= \Lambda(\alpha + \beta X_i)\\ &= \frac{1}{1 + \exp[-(\alpha + \beta X_i)]}\end{aligned} \qquad [15.8]$$

Once their variances are equated—the logistic distribution has variance $\pi^2/3$—the logit and probit transformations are so similar that it is not possible, in practice, to distinguish between them, as is apparent in Figure 15.3. It is also clear from this figure that either function is nearly linear over much of its range, say between about $\pi = .2$ and $\pi = .8$. This is why the linear-probability model produces results similar to the logit and probit models,[7] except for extreme values of π_i.

Despite their essential similarity, there are two practical advantages of the logit model compared with the probit model:

1. The equation of the logistic CDF (Equation 15.7) is very simple, while the normal CDF (Equation 15.6) involves an unevaluated integral. This difference is trivial for dichotomous data, because very good closed-form approximations to

[7] See, for example Figure 15.4.

the normal CDF are available, but for polytomous data, where we shall require the *multivariate* logistic or normal distribution, the disadvantage of the probit model is much more acute.[8]

2. The inverse linearizing transformation for the logit model, $\Lambda^{-1}(\pi)$, is directly interpretable as a *log odds*, while the inverse transformation $\Phi^{-1}(\pi)$ does not have a direct interpretation. Rearranging Equation 15.8, we get

$$\frac{\pi_i}{1-\pi_i} = \exp(\alpha + \beta X_i) \qquad [15.9]$$

The ratio $\pi_i/(1-\pi_i)$ is the *odds* that $Y_i = 1$ (e.g., the odds of voting "yes"), an expression of relative chances familiar in gambling. Unlike the probability scale, odds are unbounded above (though bounded below by 0). Taking the log of both sides of Equation 15.9 produces

$$\log_e \frac{\pi_i}{1-\pi_i} = \alpha + \beta X_i$$

The inverse transformation $\Lambda^{-1}(\pi) = \log_e[\pi/(1-\pi)]$, called the *logit* of π, is therefore the log of the odds that Y is 1 rather than 0. As the following table shows, if the odds are "even"—that is, equal to 1, corresponding to $\pi = .5$—then the logit is 0. The logit is symmetric around 0, and unbounded both above and below, making the logit a good candidate for the dependent-variable side of a linear model.

Probability π	*Odds* $\frac{\pi}{1-\pi}$	*Logit* $\log_e \frac{\pi}{1-\pi}$
.01	$1/99 = 0.0101$	-4.60
.05	$5/95 = 0.0526$	-2.94
.10	$1/9 = 0.1111$	-2.20
.30	$3/7 = 0.4286$	-0.85
.50	$5/5 = 1$	0.00
.70	$7/3 = 2.333$	0.85
.90	$9/1 = 9$	2.20
.95	$95/5 = 19$	2.94
.99	$99/1 = 99$	4.60

The logit model is a linear, additive model for the log odds, but (from Equation 15.9) it is also a multiplicative model for the odds:

$$\begin{aligned}\frac{\pi_i}{1-\pi_i} &= \exp(\alpha + \beta X_i) = \exp(\alpha)\exp(\beta X_i)\\ &= \exp(\alpha)[\exp(\beta)]^{X_i}\end{aligned}$$

So, increasing X by 1 changes the logit by β and multiplies the odds by e^{β}. For example, if $\beta = 2$, then increasing X by 1 increases the odds by a factor of $e^2 \simeq 2.718^2 = 7.389$.

Still another way of understanding the parameter β in the logit model is to consider the slope of the relationship between π and X, given by Equa-

[8] See Section 15.2.1.

tion 15.8. Because this relationship is nonlinear, the slope is not constant; the slope is $\beta\pi(1-\pi)$, and hence is at a maximum when $\pi = \frac{1}{2}$, where the slope is $\beta\frac{1}{2}(1-\frac{1}{2}) = \beta/4$, as illustrated in the following table:[9]

π	$\beta\pi(1-\pi)$
.01	$\beta \times .0099$
.05	$\beta \times .0475$
.10	$\beta \times .09$
.20	$\beta \times .16$
.50	$\beta \times .25$
.80	$\beta \times .16$
.90	$\beta \times .09$
.95	$\beta \times .0475$
.99	$\beta \times .0099$

Notice that the slope of the relationship between π and X does not change very much between $\pi = .2$ and $\pi = .8$, reflecting the near linearity of the logistic curve in this range.

The least-squares line fit to the Chilean plebescite data in Figure 15.1, for example, has the equation

$$\widehat{\pi}_{\text{yes}} = 0.492 + 0.394 \times \text{Status quo} \qquad [15.10]$$

As I have pointed out, this line is a poor summary of the data. The logistic-regression model, fit by the method of maximum likelihood (to be developed presently), has the equation

$$\log_e \frac{\widehat{\pi}_{\text{yes}}}{\widehat{\pi}_{\text{no}}} = 0.215 + 3.21 \times \text{Status quo}$$

As is apparent from Figure 15.1, the logit model produces a much more adequate summary of the data, one that is very close to the nonparametric regression. Increasing support for the status quo by one unit multiplies the odds of voting yes by $e^{3.21} = 24.8$. Put alternatively, the slope of the relationship between the fitted probability of voting yes and support for the status quo at $\widehat{\pi}_{\text{yes}} = .5$ is $3.21/4 = 0.80$. Compare this value with the slope ($B = 0.39$) from the linear least-squares regression in Equation 15.10.[10]

15.1.3 An Unobserved-Variable Formulation

An alternative derivation of the logit or probit model posits an underlying regression for a continuous but unobservable dependent variable ξ (representing, e.g., the "propensity" to vote yes), scaled so that

$$Y_i = \begin{cases} 0 & \text{when } \xi_i \le 0 \\ 1 & \text{when } \xi_i > 0 \end{cases} \qquad [15.11]$$

---

[9] See Exercise 15.2.

[10] As I have explained, the slope for the logit model is not constant: It is steepest at $\pi = .5$ and flattens out as π approaches 0 and 1. The linear probability model, therefore, will agree more closely with the logit model when the response probabilities do not (as here) attain extreme values.

That is, when ξ crosses 0, the observed discrete response Y changes from "no" to "yes." The latent variable ξ is assumed to be a linear function of the independent variable X and the (usual) unobservable error variable ε:

$$\xi_i = \alpha + \beta X_i - \varepsilon_i \qquad [15.12]$$

(It is notationally convenient here to *subtract* ε.) We want to estimate the parameters α and β, but cannot proceed by least-squares regression of ξ on X because the latent dependent variable (unlike Y) is not observed.

Using Equations 15.11 and 15.12,

$$\begin{aligned}\pi_i \equiv \Pr(Y_i = 1) &= \Pr(\xi_i > 0) = \Pr(\alpha + \beta X_i - \varepsilon_i > 0)\\ &= \Pr(\varepsilon_i < \alpha + \beta X_i)\end{aligned}$$

If the errors are independently distributed according to the unit-normal distribution, $\varepsilon_i \sim N(0, 1)$, then

$$\pi_i = \Pr(\varepsilon_i < \alpha + \beta X_i) = \Phi(\alpha + \beta X_i)$$

which is the probit model.[11] Alternatively, if the ε_i follow the similar logistic distribution, then we get the logit model

$$\pi_i = \Pr(\varepsilon_i < \alpha + \beta X_i) = \Lambda(\alpha + \beta X_i)$$

We shall have occasion to return to the unobserved-variable formulation of logit and probit models when we consider models for ordinal categorical data.[12]

15.1.4 Logit and Probit Models for Multiple Regression

Generalizing the logit and probit models to several independent variables is straightforward. All we require is a linear predictor that is a function of several regressors. For the logit model,

$$\begin{aligned}\pi_i = \Lambda(\eta_i) &= \Lambda(\alpha + \beta_1 X_{i1} + \beta_2 X_{i2} + \cdots + \beta_k X_{ik}) \qquad [15.13]\\ &= \frac{1}{1 + \exp[-(\alpha + \beta_1 X_{i1} + \beta_2 X_{i2} + \cdots + \beta_k X_{ik})]}\end{aligned}$$

---

[11] The variance of the errors is set conveniently to 1. This choice is legitimate since we have not yet fixed the unit of measurement of the latent variable ξ. The location of the ξ scale was implicitly fixed by setting 0 as the point at which the observable response changes from no to yes. You may be uncomfortable assuming that the errors for an unobservable dependent variable are normally distributed, because we cannot check the assumption by examining residuals, for example. In fact, in most instances we can ensure that the error distribution has any form we please by transforming ξ to make the assumption true. We cannot, however, simultaneously ensure that the true regression is linear. If the latent-variable regression is not linear, then the probit model will not adequately capture the relationship between the dichotomous Y and X.

[12] See Section 15.2.3.

or, equivalently,

$$\log_e \frac{\pi_i}{1-\pi_i} = \alpha + \beta_1 X_{i1} + \beta_2 X_{i2} + \cdots + \beta_k X_{ik}$$

For the probit model,

$$\pi_i = \Phi(\eta_i) = \Phi(\alpha + \beta_1 X_{i1} + \beta_2 X_{i2} + \cdots + \beta_k X_{ik})$$

Moreover, the X's can be as general as in the general linear model, including, for example:

- quantitative independent variables;
- transformations of quantitative independent variables;
- polynomial regressors formed from quantitative independent variables;
- dummy regressors representing qualitative independent variables; and
- interaction regressors.

Interpretation of the partial-regression coefficients in the general logit model (Equation 15.13) is similar to the interpretation of the slope in the logit simple-regression model, with the additional provision of holding other independent variables in the model constant. For example, expressing the model in terms of odds,

$$\frac{\pi_i}{1-\pi_i} = \exp(\alpha + \beta_1 X_{i1} + \cdots + \beta_k X_{ik})$$
$$= \exp(\alpha)[\exp(\beta_1)]^{X_{i1}} \cdots [\exp(\beta_k)]^{X_{ik}}$$

Thus, $\exp(\beta_j)$ is the multiplicative effect on the odds of increasing X_j by 1, holding the other X's constant. Similarly, $\beta_j/4$ is the slope of the logistic regression surface in the direction of X_j at $\pi = .5$.

More adequate specifications transform the linear predictor $\eta_i = \alpha + \beta_1 X_{i1} + \cdots + \beta_k X_{ik}$ smoothly to the unit interval, using a cumulative probability distribution function $P(\cdot)$. Two such specifications are the probit and the logit models, which use the normal and logistic CDFs, respectively. Although these models are very similar, the logit model is simpler to interpret, because it can be written as a linear model for the log odds: $\log_e[\pi_i/(1-\pi_i)] = \alpha + \beta_1 X_{i1} + \cdots + \beta_k X_{ik}$.

The general linear logit and probit models can be fit to data by the method of maximum likelihood. I shall concentrate here on outlining maximum-likelihood estimation for the logit model. Details are given in the next section.

Recall that the dependent variable Y_i takes on the two values 1 and 0 with probabilities π_i and $1 - \pi_i$, respectively. Using a mathematical "trick," the probability distribution for Y_i can be compactly represented as a single equation:[13]

$$p(y_i) \equiv \Pr(Y_i = y_i) = \pi_i^{y_i}(1 - \pi_i)^{1-y_i}$$

where y_i can be 0 or 1. Now consider a particular sample of n independent observations, $y_1, y_2, \ldots, y_n$ (comprising a specific sequence of 0's and 1's). Because the observations are independent, the joint probability for the data is the product of the marginal probabilities:

$$\begin{aligned} p(y_1, y_2, \ldots, y_n) &= p(y_1)p(y_2)\cdots p(y_n) \\ &= \prod_{i=1}^{n} p(y_i) \\ &= \prod_{i=1}^{n} \pi_i^{y_i}(1 - \pi_i)^{1-y_i} \\ &= \prod_{i=1}^{n} \left(\frac{\pi_i}{1 - \pi_i}\right)^{y_i}(1 - \pi_i) \end{aligned} \qquad [15.14]$$

From the general logit model (Equation 15.13),

$$\frac{\pi_i}{1 - \pi_i} = \exp(\alpha + \beta_1 X_{i1} + \cdots + \beta_k X_{ik})$$

and (after some manipulation)[14]

$$1 - \pi_i = \frac{1}{1 + \exp(\alpha + \beta_1 X_{i1} + \cdots + \beta_k X_{ik})}$$

Substituting these results into Equation 15.14 expresses the probability of the data in terms of the parameters of the logit model:

$$\begin{aligned} &p(y_1, y_2, \ldots, y_n) \\ &\quad = \prod_{i=1}^{n} [\exp(\alpha + \beta_1 X_{i1} + \cdots + \beta_k X_{ik})]^{y_i} \left(\frac{1}{1 + \exp(\alpha + \beta_1 X_{i1} + \cdots + \beta_k X_{ik})}\right) \end{aligned}$$

Thinking of this equation as a function of the parameters, and treating the data $(y_1, y_2, \ldots, y_n)$ as fixed, produces the likelihood function, $L(\alpha, \beta_1, \ldots, \beta_k)$. The values of $\alpha, \beta_1, \ldots, \beta_k$ that maximize $L(\alpha, \beta_1, \ldots, \beta_k)$ are the maximum-likelihood estimates $A, B_1, \ldots, B_k$.

Hypothesis tests and confidence intervals follow from general procedures for statistical inference in maximum-likelihood estimation. For an individual coefficient, it is most convenient to test the hypothesis H_0: $\beta_j = \beta_j^{(0)}$ by calculating

[13] See Exercise 15.3.

[14] See Exercise 15.4.

the Wald statistic

$$Z_0 = \frac{B_j - \beta_j^{(0)}}{\widehat{\mathrm{ASE}}(B_j)}$$

where $\widehat{\mathrm{ASE}}(B_j)$ is the estimated asymptotic standard error of B_j. To test the most common hypothesis, H_0: $\beta_j = 0$, we simply divide the estimated coefficient by its standard error, to compute $Z_0 = B_j/\widehat{\mathrm{ASE}}(B_j)$; these tests are analogous to t-tests for individual coefficients in the general linear model. The test statistic Z_0 follows an asymptotic unit-normal distribution under the null hypothesis, an approximation that should be reasonably accurate unless the sample size is small. Similarly, an asymptotic $100(1-a)\%$ confidence interval for β_j is given by

$$\beta_j = B_j \pm z_{a/2}\widehat{\mathrm{ASE}}(B_j)$$

where $z_{a/2}$ is the value from $Z \sim N(0, 1)$ with probability $a/2$ to the right. Wald tests and joint confidence regions for several coefficients can be formulated from the estimated asymptotic variances and covariances of the coefficients.[15]

It is also possible to formulate a likelihood-ratio test for the hypothesis that several coefficients are simultaneously 0, H_0: $\beta_1 = \cdots = \beta_q = 0$. We proceed, as in least-squares regression, by fitting two models to the data: The full model (model 1)

$$\mathrm{logit}(\pi) = \alpha + \beta_1 X_1 + \cdots + \beta_q X_q + \beta_{q+1} X_{q+1} + \cdots + \beta_k X_k$$

and the null model (model 0)

$$\begin{aligned}\mathrm{logit}(\pi) &= \alpha + 0X_1 + \cdots + 0X_q + \beta_{q+1} X_{q+1} + \cdots + \beta_k X_k \\ &= \alpha + \beta_{q+1} X_{q+1} + \cdots + \beta_k X_k\end{aligned}$$

Each model produces a maximized likelihood: L_1 for the full model, L_0 for the null model. Because the null model is a specialization of the full model, $L_1 \geq L_0$. The generalized likelihood-ratio test statistic for the null hypothesis is

$$G_0^2 = 2(\log_e L_1 - \log_e L_0)$$

Under the null hypothesis, this test statistic has an asymptotic chi-square distribution with q degrees of freedom. By extension, a test of the omnibus null hypothesis H_0: $\beta_1 = \cdots = \beta_k = 0$ is obtained by specifying a null model that includes only the constant, $\mathrm{logit}(\pi) = \alpha$.

An analog to the multiple-correlation coefficient can also be obtained from the log likelihood. The maximized log likelihood for the fitted model can be written as[16]

$$\log_e L = \sum_{i=1}^{n} [y_i \log_e P_i + (1 - y_i)\log_e(1 - P_i)]$$

---

[15] See Section 15.1.5.

[16] See Exercise 15.5.

where P_i is the fitted probability[17] that $Y_i = 1$, that is,

$$P_i = \frac{1}{1 + \exp[-(A + B_1 X_{i1} + \cdots + B_k X_{ik})]}$$

Thus, if the fitted model can perfectly predict the Y-values ($P_i = 1$ whenever $y_i = 1$, and $P_i = 0$ whenever $y_i = 0$), then $\log_e L = 0$ (i.e., the maximized likelihood is $L = 1$).[18] To the extent that predictions are less than perfect, $\log_e L < 0$ (and $0 < L < 1$).

By comparing $\log_e L_0$ for the model containing only the constant with $\log_e L_1$ for the full model, we can measure the degree to which using the independent variables improves the predictability of Y. The quantity $G^2 \equiv -2 \log_e L$, called the *deviance* under the model, is a generalization of the residual sum of squares for a linear model.[19] Thus,

$$\begin{aligned} R^2 &\equiv 1 - \frac{G_1^2}{G_0^2} \\ &= 1 - \frac{\log_e L_1}{\log_e L_0} \end{aligned}$$

is analogous to R^2 for a linear model.

> The dichotomous logit model can be fit to data by the method of maximum likelihood. Wald tests and likelihood-ratio tests for the coefficients of the model parallel t-tests and F-tests for the general linear model. The deviance for the model, defined as $G^2 = -2 \times$ the maximized log likelihood, is analogous to the residual sum of squares for a linear model.

To illustrate logistic regression, I shall use data drawn from a sample survey of the Canadian population, conducted in 1977. We shall examine how the labor-force participation of young married women (21 to 30 years of age) is related to the presence of children in their households, to their husbands' income, and to the region of the country in which they reside. Labor-force participation is treated dichotomously here: working versus not working outside of the home.[20]

---

[17] Of course, in a particular sample, Y_i is either 0 or 1, so we can interpret this probability as the population proportion of individuals sharing the ith person's characteristics for whom Y is 1. Other interpretations are also possible, but this is the most straightforward.

[18] Because, for the logit model, π never quite reaches 0 or 1, the predictions cannot be perfect, although they can approach perfection in the limit.

[19] See Exercise 15.6.

[20] See Section 15.2 and Exercise 15.10 for a continuation of this example, distinguishing part-time from full-time work.

TABLE 15.1 Deviances ($-2 \times$ Log Likelihood) for Several Models Fit to the Women's Labor-Force-Participation Data. The following code is used for terms in the models: C constant; I husband's income; K presence of children; R region. The column labeled $k+1$ gives the number of regressors in the model, including the constant

Model	*Terms*	$k+1$	*Deviance*
0	C	1	356.16
1	$C, I, K, R, I \times K$	8	316.54
2	C, I, K, R	7	317.30
3	$C, I, K, I \times K$	4	319.12
4	C, I, R	6	347.86
5	C, K, R	6	322.44

Presence of children and husband's income are expected to be negatively related to working outside of the home.

Husband's annual income (I), measured in thousands of dollars, was determined by subtracting each woman's income from her reported family income. Presence of children is represented by a dummy variable (K) coded 1 if minor children are present in the household and 0 otherwise. Because husband's income might well have a greater effect among women without children, an interaction regressor ($I \times K$) is included in the model. Finally, four dummy variables ($R_1 - R_4$) are employed to represent the five regions of Canada (the Atlantic provinces, Quebec, Ontario, the prairie provinces, and British Columbia, with British Columbia as the baseline).

Several models were fit to the women's labor-force data to provide likelihood-ratio tests for the terms in the logit model; the deviance ($-2 \times$ log likelihood) and the degrees of freedom for each of these models are shown in Table 15.1. An "analysis-of-deviance" table showing the several likelihood-ratio tests appears in Table 15.2. The first row of this table reports an omnibus test for all of the terms in the model. The interaction of husband's income and presence of children, and the region effects, are not statistically significant. There are statistically significant main effects, however, for both husband's income and presence of children.

Deleting the nonsignificant interaction and region effects, the fitted logit model is as follows, with estimated asymptotic standard errors shown beneath

TABLE 15.2 Analysis-of-Deviance Table for Terms in the Logit Model Fit to the Women's Labor-Force-Participation Data.

Source	*Models Contrasted*	*df*	G_0^2	*p*
$I, K, R, I \times K$	$0-1$	7	39.62	$< .0001$
$I \times K$	$2-1$	1	0.76	.38
I (husband's income)	$5-2$	1	5.14	.023
K (presence of children)	$4-2$	1	30.56	$\ll .0001$
R (region)	$3-1$	4	2.58	.63

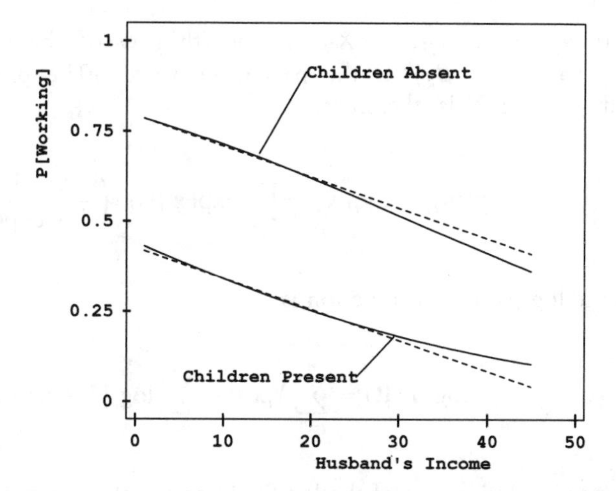

Figure 15.4. Fitted probability of young married women working outside the home, as a function of husband's income and presence of children. The solid line shows the logit model fit by maximum likelihood; the broken line shows the linear least-squares fit.

the coefficients:

$$\begin{aligned} \text{logit}\,(P_W) = {} & 1.336 \;-\; 0.04231I \;-\; 1.576K \\ & (0.384) \;\; (0.01978) \;\;\; (0.292) \end{aligned} \qquad [15.15]$$

$$R^2 = 1 - \frac{319.74}{356.16} = .102$$

where P_W is the fitted probability of working outside of the home, and the R^2 is calculated from the deviance for the model, 319.74. The fit is graphed in Figure 15.4, along with the fit of the linear-probability model estimated by least squares:

$$P_W = 0.7939 - 0.008538I - 0.3674K$$

In this case, the two models produce reasonably similar descriptions of the data. Note that the income slope for the linear-probability model is close to the income slope of the fitted logistic-regression surface near $P_W = .5$: $-0.04231/4 = -0.0106$. The coefficients for presence of children in the two models are also similar: $-1.576/4 = -0.394$.

15.1.5 Estimating the Linear Logit Model*

In this section, I shall develop the details of maximum-likelihood estimation for the general linear logit model (Equation 15.13). It is convenient to rewrite the model in vector form as

$$\pi_i = \frac{1}{1 + \exp(-\mathbf{x}_i'\boldsymbol{\beta})}$$

where $\mathbf{x}_i' = (1, X_{i1}, \ldots, X_{ik})$ is the ith row of the model matrix $\mathbf{X}$, and $\boldsymbol{\beta} = (\alpha, \beta_1, \ldots, \beta_k)'$ is the parameter vector. The probability of the data conditional on $\mathbf{X}$ is, therefore,

$$p(y_1, \ldots, y_n|\mathbf{X}) = \prod_{i=1}^{n} [\exp(\mathbf{x}_i'\boldsymbol{\beta})]^{y_i} \left(\frac{1}{1 + \exp(\mathbf{x}_i'\boldsymbol{\beta})} \right)$$

and the log-likelihood function is

$$\log_e L(\boldsymbol{\beta}) = \sum_{i=1}^{n} Y_i \mathbf{x}_i'\boldsymbol{\beta} - \sum_{i=1}^{n} \log_e[1 + \exp(\mathbf{x}_i'\boldsymbol{\beta})]$$

The partial derivatives of the log likelihood with respect to $\boldsymbol{\beta}$ are

$$\begin{aligned} \frac{\partial \log_e L(\boldsymbol{\beta})}{\partial \boldsymbol{\beta}} &= \sum_{i=1}^{n} Y_i \mathbf{x}_i - \sum_{i=1}^{n} \left(\frac{\exp(\mathbf{x}_i'\boldsymbol{\beta})}{1 + \exp(\mathbf{x}_i'\boldsymbol{\beta})} \right) \mathbf{x}_i \\ &= \sum_{i=1}^{n} Y_i \mathbf{x}_i - \sum_{i=1}^{n} \left(\frac{1}{1 + \exp(-\mathbf{x}_i'\boldsymbol{\beta})} \right) \mathbf{x}_i \end{aligned} \qquad [15.16]$$

Setting the vector of partial derivatives to $\mathbf{0}$ to maximize the likelihood yields estimating equations

$$\sum_{i=1}^{n} \left(\frac{1}{1 + \exp(-\mathbf{x}_i'\mathbf{b})} \right) \mathbf{x}_i = \sum_{i=1}^{n} Y_i \mathbf{x}_i \qquad [15.17]$$

where $\mathbf{b} = (A, B_1, \ldots, B_k)'$ is the vector of maximum-likelihood estimates.

The estimating equations (Equation 15.17) have the following intuitive justification:

$$P_i \equiv \frac{1}{1 + \exp(-\mathbf{x}_i'\mathbf{b})}$$

is the fitted probability for observation i (i.e., the estimated value of π_i). The estimating equations, therefore, set the "fitted sum" $\sum P_i \mathbf{x}_i$ equal to the corresponding observed sum $\sum Y_i \mathbf{x}_i$. In matrix form, we can write the estimating equations as $\mathbf{X}'\mathbf{p} = \mathbf{X}'\mathbf{y}$, where $\mathbf{p} = (P_1, \ldots, P_n)'$ is the vector of fitted values. Note the essential similarity to the least-squares estimating equations $\mathbf{X}'\mathbf{X}\mathbf{b} = \mathbf{X}'\mathbf{y}$, which can be written $\mathbf{X}'\widehat{\mathbf{y}} = \mathbf{X}'\mathbf{y}$.

Because $\mathbf{b}$ is a maximum-likelihood estimator, its estimated asymptotic covariance matrix can be obtained from the inverse of the information matrix

$$\mathscr{I}(\boldsymbol{\beta}) = -E\left[\frac{\partial^2 \log_e L(\boldsymbol{\beta})}{\partial \boldsymbol{\beta}\, \partial \boldsymbol{\beta}'} \right]$$

evaluated at $\boldsymbol{\beta} = \mathbf{b}$. Differentiating Equation 15.16 and making the appropriate substitutions,[21]

$$\begin{aligned}\widehat{\mathscr{V}(\mathbf{b})} &= \left[\sum_{i=1}^{n} \frac{\exp(-\mathbf{x}_i'\mathbf{b})}{[1+\exp(-\mathbf{x}_i'\mathbf{b})]^2}\mathbf{x}_i\mathbf{x}_i'\right]^{-1} \\ &= \left[\sum_{i=1}^{n} P_i(1-P_i)\mathbf{x}_i\mathbf{x}_i'\right]^{-1} \\ &= (\mathbf{X}'\mathbf{V}\mathbf{X})^{-1}\end{aligned}$$

where $\mathbf{V} = \text{diag}\{P_i(1 - P_i)\}$ contains the estimated variances of the Y_i's. The square roots of the diagonal entries of $\widehat{\mathscr{V}(\mathbf{b})}$ are the estimated asymptotic standard errors, which can be used, as described in the previous section, for inferences about individual parameters of the logit model.

As for the linear model estimated by least squares,[22] general linear hypotheses about the parameters of the logit model can be formulated as $H_0\colon \mathbf{L}\boldsymbol{\beta} = \mathbf{c}$, where $\mathbf{L}$ is a $(q \times k + 1)$ hypothesis matrix of rank $q \leq k + 1$, and $\mathbf{c}$ is a $q \times 1$ vector of fixed elements, typically 0.[23] Then the Wald statistic

$$Z_0^2 = (\mathbf{Lb} - \mathbf{c})'\left[\mathbf{L}\widehat{\mathscr{V}(\mathbf{b})}\mathbf{L}'\right]^{-1}(\mathbf{Lb} - \mathbf{c})$$

follows an asymptotic chi-square distribution with q degrees of freedom under the hypothesis H_0. For example, to test the omnibus hypothesis $H_0\colon \beta_1 = \cdots = \beta_k = 0$, we take

$$\underset{(k\times k+1)}{\mathbf{L}} = \begin{bmatrix} 0 & 1 & 0 & \cdots & 0 \\ 0 & 0 & 1 & \cdots & 0 \\ \vdots & \vdots & \vdots & \ddots & \vdots \\ 0 & 0 & 0 & \cdots & 1 \end{bmatrix}$$

and $\mathbf{c} = \underset{(k\times 1)}{\mathbf{0}}$.

Likewise, the asymptotic $100(1 - a)$% joint confidence region for a subset of q parameters $\boldsymbol{\beta}_1$ takes the form

$$(\mathbf{b}_1 - \boldsymbol{\beta}_1)'\mathbf{V}_{11}^{-1}(\mathbf{b}_1 - \boldsymbol{\beta}_1) \leq \chi^2_{q,\,a}$$

Here, $\mathbf{V}_{11}$ is the $(q \times q)$ submatrix of $\widehat{\mathscr{V}(\mathbf{b})}$ that pertains to the estimates $\mathbf{b}_1$, and $\chi^2_{q,\,a}$ is the critical value of the chi-square distribution for q degrees of freedom with probability a to the right.

Unlike the normal equations for a linear model, the logit-model estimating equations (Equation 15.17) are nonlinear functions of $\mathbf{b}$ and, therefore, require iterative solution. One common approach to solving the estimating equations is

---

[21] See Exercise 15.7.

[22] See Section 9.4.3.

[23] See Chapter 9.

the *Newton-Raphson method*, which can be described as follows:

1. Select initial estimates $\mathbf{b}_0$; a particularly simple choice is $\mathbf{b}_0 = 0$.
2. At each iteration $l + 1$, compute new estimates

$$\mathbf{b}_{l+1} = \mathbf{b}_l + (\mathbf{X}'\mathbf{V}_l\mathbf{X})^{-1}\mathbf{X}'(\mathbf{y} - \mathbf{p}_l) \quad [15.18]$$

where $\mathbf{p}_l \equiv \{1/[1+\exp(-\mathbf{x}_i'\mathbf{b}_l)]\}$ is the vector of fitted values from the previous iteration and $\mathbf{V}_l \equiv \text{diag}\{P_{li}(1 - P_{li})\}$.
3. Iterations continue until $\mathbf{b}_{l+1} \simeq \mathbf{b}_l$ to the desired degree of accuracy.

When convergence takes place,

$$(\mathbf{X}'\mathbf{V}_l\mathbf{X})^{-1}\mathbf{X}'(\mathbf{y} - \mathbf{p}_l) \simeq 0$$

and thus the estimating equations $\mathbf{X}'\mathbf{p} = \mathbf{X}'\mathbf{y}$ are approximately satisfied. Conversely, if the fitted sums $\mathbf{X}'\mathbf{p}_l$ are very different from the observed sums $\mathbf{X}'\mathbf{y}$, then there will be a large adjustment in $\mathbf{b}$ at the next iteration. The Newton-Raphson procedure produces the estimated asymptotic covariance matrix of the coefficients $(\mathbf{X}'\mathbf{V}\mathbf{X})^{-1}$ as a by-product.

Now imagine complete convergence of the Newton-Raphson procedure to the maximum-likelihood estimator $\mathbf{b}$. From Equation 15.18, we have

$$\mathbf{b} = \mathbf{b} + (\mathbf{X}'\mathbf{V}\mathbf{X})^{-1}\mathbf{X}'(\mathbf{y} - \mathbf{p})$$

which we can rewrite as

$$\mathbf{b} = (\mathbf{X}'\mathbf{V}\mathbf{X})^{-1}\mathbf{X}'\mathbf{V}\mathbf{y}^*$$

where[24]

$$\mathbf{y}^* \equiv \mathbf{X}\mathbf{b} + \mathbf{V}^{-1}(\mathbf{y} - \mathbf{p})$$

These formulas suggest an analogy between maximum-likelihood estimation of the linear logit model and weighted-least-squares regression. The analogy is the basis of an alternative method for calculating the maximum-likelihood estimates called *iterative weighted least squares* (IWLS):[25]

1. Again, select arbitrary initial values $\mathbf{b}_0$.
2. At each iteration l, calculate fitted values $\mathbf{p}_l \equiv \{1/[1 + \exp(-\mathbf{x}_i'\mathbf{b}_l)]\}$, the variance matrix $\mathbf{V}_l \equiv \text{diag}\{P_{li}(1 - P_{li})\}$, and "pseudo-dependent-variable values" $\mathbf{y}_l^* \equiv \mathbf{X}\mathbf{b}_l + \mathbf{V}_l^{-1}(\mathbf{y} - \mathbf{p}_l)$.
3. Calculate updated estimates by weighted-least-squares regression of the pseudo-values on the X's, using the current variance matrix for weights:

$$\mathbf{b}_{l+1} = (\mathbf{X}'\mathbf{V}_l\mathbf{X})^{-1}\mathbf{X}'\mathbf{V}_l\mathbf{y}_l^*$$

4. Repeat steps 2 and 3 until the coefficients converge.

---

[24] See Exercise 15.8.

[25] This method is also called *iteratively reweighted least squares* (IRLS). See Section 12.2.2 for an explanation of weighted-least-squares estimation.

15.1.6 Diagnostics for Logit Models*

As for linear models fit by least squares, it is desirable to assess the adequacy of a fitted linear logit model. Many of the diagnostic methods described in Chapters 11 and 12 can be extended to logit models by exploiting the analogy between logit models and weighted-least-squares regression developed in the preceding section. The pivotal contribution in this area is Pregibon (1981), and many of the results presented here derive from this source.

Fitting a logit model to dichotomous data by maximum likelihood is intrinsically more complex than fitting a linear model by least squares. One consequence of this relative complexity is that it becomes more difficult to assess how the fit is affected—both potentially and actually—by variations in the data. The discrete, binary character of the dependent variable renders many direct analogs of graphical diagnostics for linear least squares difficult to interpret. The reader should be cautioned that the parallels between the methods described in this section and those of Chapters 11 and 12 are less than perfect. Nevertheless, diagnostic methods for logit models are valuable data-analytic tools, and are likely to be improved and extended in the future.

> Diagnostics for linear models—including studentized residuals, hat values, influence diagnostics, partial-residual plots, partial-regression plots, and constructed-variable plots—can be extended to the logit model.

Residuals in Logit Model

There are several different ways of defining residuals for logit models. Most straightforwardly, we can take $E_i \equiv Y_i - P_i$, where, we recall, $Y_i = 0$ or 1 is the dichotomous dependent variable, and $P_i = 1/[1 + \exp(-\mathbf{x}_i'\mathbf{b})]$ is the fitted probability for the ith observation.

It is also possible to define residuals by developing parallels to the residual sum of squares in linear models. I shall pursue two such possibilities, both based on chi-square-like statistics. The *goodness-of-fit* or *Pearson statistic*

$$Z^2 \equiv \sum_{i=1}^{n} \frac{E_i^2}{P_i(1 - P_i)}$$

is a measure of the overall lack of fit of the model, large values indicating a poor fit—that is, an inability to match Y-values of 0 and 1 with low and high fitted probabilities, respectively. Although it might appear that Z^2 should follow an asymptotic chi-square distribution with $n - k$ degrees of freedom, this is not the case, because the number of components of the statistic grows with the sample size n. The ith component of Z^2,

$$Z_i \equiv \frac{E_i}{\sqrt{P_i(1 - P_i)}}$$

is an indicator of lack of fit for observation i.

A statistic similar to Z^2, but based on the likelihood of the fitted model, is the *deviance*,[26]

$$G^2 \equiv -2\log_e L(\mathbf{b}) = -2\sum [Y_i \log_e P_i + (1 - Y_i)\log_e(1 - P_i)]$$

which implicitly contrasts the fitted model with a model that predicts each observation perfectly (and hence for which $L = 1$ and $\log L = 0$). Like the goodness-of-fit statistic Z^2, the deviance G^2 is not appropriate for a formal likelihood-ratio test of the model, because it does not follow an asymptotic chi-square distribution. The ith component of the deviance is

$$G_i \equiv \pm\sqrt{-2\,[Y_i \log_e P_i + (1 - Y_i)\log_e(1 - P_i)]}$$

where the sign is chosen to agree with that of E_i. The discreteness of the dependent variable Y, and hence of the residuals E for specific combinations of X-values, makes the Z_i and G_i difficult to examine; in particular, the Z_i and G_i are not normally distributed. They are, nevertheless, useful building blocks for other diagnostics.

In Section 15.1.5, I demonstrated an analogy between logistic regression and weighted-least-squares regression, according to which $\mathbf{y}^* \equiv \mathbf{Xb} + \mathbf{V}^{-1}\mathbf{e}$ plays the role of the "dependent variable," and where $\mathbf{V} = \text{diag}\{P_i(1-P_i)\}$. Landwehr, et al. (1980) suggest that the Y_i^* may be thought of as *pseudo-observations* on the logit scale. Thus,

$$\widehat{Y}_i^* \equiv A + B_1 X_{i1} + \cdots + B_k X_{ik}$$

is the logit fitted value, and

$$E_i^* \equiv \frac{E_i}{P_i(1 - P_i)}$$

is the logit residual.

Residual and Partial-Residual Plots

Nonlinearity on the logit scale can sometimes be detected by plotting the logit residuals E_i^* against individual X's. Following Landwehr, et al. (1980), *logit partial residuals* for the jth regressor are defined by

$$\mathbf{e}_j^* \equiv \mathbf{e}^* + B_j \mathbf{x}_j$$

where $\mathbf{e}^*$ is the vector of logit residuals, $\mathbf{x}_j$ is the jth *column* of the model matrix $\mathbf{X}$, and B_j is the estimated coefficient for the jth regressor. Because of discreteness, plots of logit residuals and partial residuals are typically difficult to examine. Landwehr, et al. (1980) suggest enhancing these plots with a nonresistant smoother (such as lowess without robustness weighting).

[26] See Section 15.1.4.

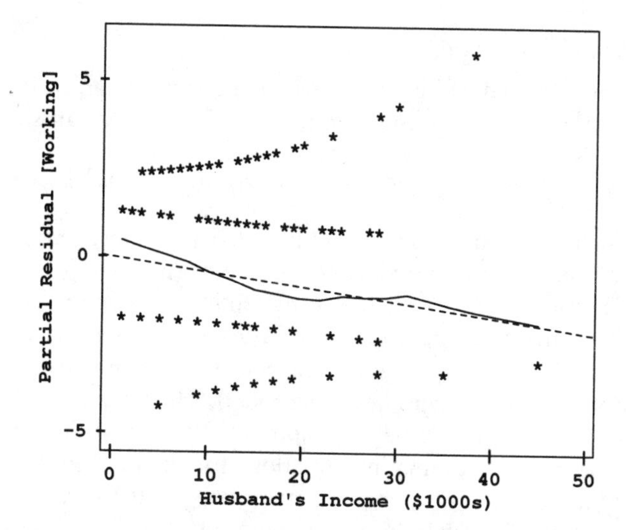

Figure 15.5. Partial-residual plot for husband's income, for the logistic-regression model fit to the women's labor-force-participation data. The broken line gives the logit fit; the solid line shows a lowess smooth of the plot. Note the four bands due to the four combinations of values of the dichotomous dependent variable and the dichotomous independent variable presence of children. Because husband's income is also discrete, many points are overplotted.

An illustrative partial-residual plot for husband's income in the women's labor-force-participation model is shown in Figure 15.5. The four bands that appear in the plot are due to the discreteness of the response and of the other independent variable in the model, presence of children: At each level of husband's income only four partial residuals can occur, corresponding to the four combinations of values of presence of children and labor-force participation.[27] Although there is a suggestion of nonlinearity in the plot, I was unable to improve the fit of the model.

Hat Values and the Hat Matrix

Pregibon (1981) shows that

$$\mathbf{H} = \mathbf{V}^{1/2}\mathbf{X}(\mathbf{X}'\mathbf{V}\mathbf{X})^{-1}\mathbf{X}'\mathbf{V}^{1/2}$$

is, in many respects, analogous to the hat matrix in least-squares regression.[28] For instance, $\mathbf{z} = (\mathbf{I}_n - \mathbf{H})\mathbf{z}$ [while in least-squares regression $\mathbf{e} = (\mathbf{I}_n - \mathbf{H})\mathbf{e}$]. Furthermore, $\mathbf{H}$ is symmetric and idempotent; the diagonal values $h_i \equiv h_{ii}$ are bounded by 0 and 1; and the average value of h_i is $\bar{h} = (k+1)/n$. The h_i are often useful for detecting high-leverage observations. Note that unlike in linear models, however, the hat values in logistic regression depend on the dependent-variable scores (through the fitted probabilities $\mathbf{p}$, which determine $\mathbf{V}$) as well as on the model matrix $\mathbf{X}$.

---

[27] Because husband's income is also discrete—measured to the nearest $1000—many of the points are overplotted. Separating the points by jittering is not really helpful here: The bands in the plot make interpretation difficult in any event.

[28] Also see Exercise 15.9.

Studentized Residuals

Studentized residuals for logistic regression can be defined by dedicating a parameter to the ith observation, producing the "mean-shift" outlier model

$$\text{logit}(\pi_j) = \alpha + \beta_1 X_{j1} + \cdots + \beta_k X_{jk} + \gamma D_j \qquad [15.19]$$

in which the dummy variable D is coded 1 for observation i and 0 for all other observations. This procedure is equivalent to deleting the ith observation from the sample. If L represents the maximized likelihood for the original model, and $L_{(-i)}$ the maximized likelihood for the model given in Equation 15.19, then $G^2_{(-i)} \equiv 2\log_e(L_{(-i)}/L)$ is asymptotically distributed as χ^2 with 1 degree of freedom under the hypothesis H_0: $\gamma = 0$. Thus, $G_{(-i)}$ (with sign chosen to agree with that of E_i) follows an asymptotic unit-normal distribution.

Because the estimating equations for the logit model are nonlinear, to calculate the $G^2_{(-i)}$ exactly requires iteratively refitting the model for each observation, a substantial burden of computation. Pregibon (1981) shows that a good approximation is available that does not require refitting the model:

$$G^2_{(-i)} \simeq G_i^2 + \frac{Z_i^2 h_i}{1 - h_i} \qquad [15.20]$$

Using the deviance residual G_i in place of the Pearson residual Z_i in Equation 15.20, and taking the square root, produces the approximate studentized residual

$$G_i^* \equiv \frac{G_i}{\sqrt{1 - h_i}}$$

Normal-probability plots of G_i^* are useful for revealing outliers, although interpretative problems can be caused by the discreteness of the dependent variable.[29] A Bonferroni outlier test can be based on the largest absolute G_i^*.

Influence Diagnostics

In Chapter 11, I developed the topic of influence in linear models by examining how the deletion of an observation affects the least-squares regression coefficients. Analogously, for the linear logit model, we can define $\mathbf{d}_i \equiv \mathbf{b} - \mathbf{b}_{(-i)}$, where $\mathbf{b}_{(-i)}$ represents the maximum-likelihood estimator of $\boldsymbol{\beta}$ omitting observation i. To find $\mathbf{b}_{(-i)}$ requires refitting the model, but an approximation, $\mathbf{b}^*_{(-i)}$, can be obtained by performing a single Newton-Raphson iteration starting at $\mathbf{b}_0 = \mathbf{b}$. Pregibon (1981) proves that

$$\mathbf{d}_i^* \equiv \mathbf{b} - \mathbf{b}^*_{(-i)} = (\mathbf{X}'\mathbf{V}'\mathbf{X})^{-1}\mathbf{x}_i \frac{E_i}{1 - h_i} \qquad [15.21]$$

where $\mathbf{x}_i$ is the ith row of the model matrix $\mathbf{X}$, written as a column vector. (A similar one-step approximation provides the basis for Equation 15.20.)

---

[29] Landwehr, et al. (1980) suggest a bootstrapping procedure to construct an empirical confidence envelope for the normal-probability plot, similar to Atkinson's procedure, described in Section 12.1, for studentized residuals from linear models.

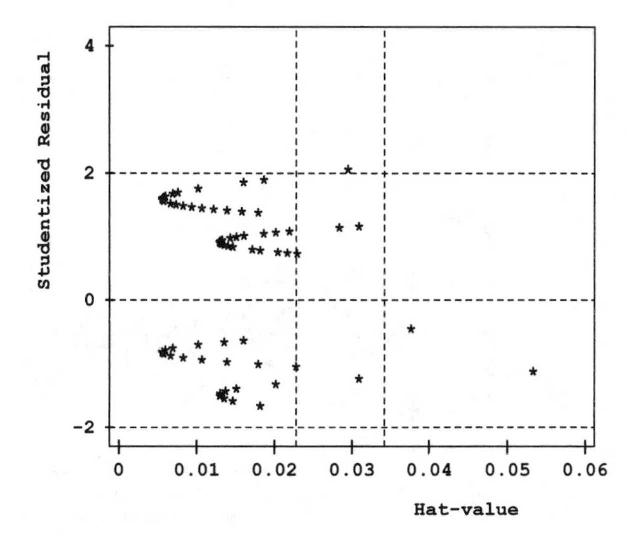

Figure 15.6. Plot of studentized residuals versus hat values for the logit model fit to the women's labor-force-participation data. Vertical lines are drawn at twice and three times the average hat value. Many points are overplotted.

An analog of Cook's influence statistic follows from the Wald test statistic for the hypothesis H_0: $\boldsymbol{\beta} = \boldsymbol{\beta}_0$:

$$\begin{aligned} Z_0^2 &= (\mathbf{b} - \boldsymbol{\beta}_0)' \widehat{\mathscr{V}(\mathbf{b})}^{-1} (\mathbf{b} - \boldsymbol{\beta}_0) \\ &= (\mathbf{b} - \boldsymbol{\beta}_0)' \mathbf{X}'\mathbf{V}\mathbf{X} (\mathbf{b} - \boldsymbol{\beta}_0) \end{aligned}$$

We can use Z_0^2 for the "hypothesis" H_0: $\boldsymbol{\beta} = \mathbf{b}^*_{(-i)}$ as a scale-invariant scalar index of the distance between $\mathbf{b}^*_{(-i)}$ and $\mathbf{b}$. Substituting from Equation 15.21, we obtain

$$\begin{aligned} D_i^* &\equiv (\mathbf{b} - \mathbf{b}^*_{(-i)})' \mathbf{X}'\mathbf{V}\mathbf{X} (\mathbf{b} - \mathbf{b}^*_{(-i)}) \\ &= \frac{E_i^2}{(1 - h_i)^2} \mathbf{x}_i' (\mathbf{X}'\mathbf{V}'\mathbf{X})^{-1} \mathbf{x}_i \\ &= \frac{Z_i^2 h_i}{(1 - h_i)^2} \end{aligned}$$

As in linear models, therefore, the influence of an observation is an increasing function of its residual and its leverage.

Figure 15.6 plots studentized residuals versus hat values for the logistic regression model fit to the women's labor-force-participation data. None of the studentized residuals is particularly large, but there are some hat values that exceed the rule-of-thumb cutoffs $2\bar{h}$ and $3\bar{h}$. The strange pattern in the plot is due to the dichotomous dependent variable and the dichotomous regressor for presence of children. Because the other independent variable—husband's income—is also discrete, many points are overplotted.

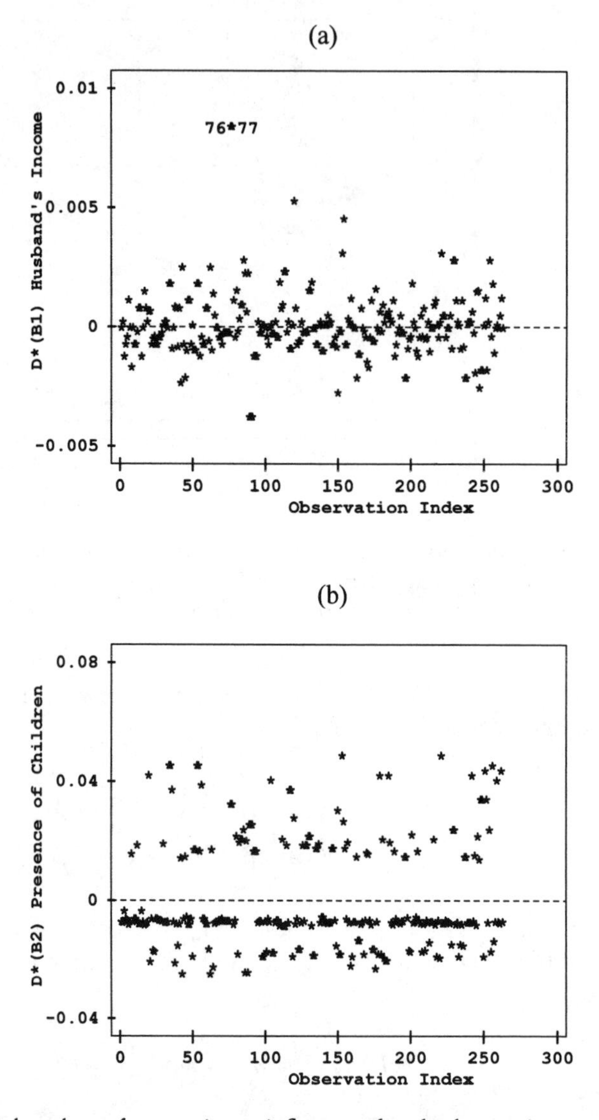

Figure 15.7. Index plots of approximate influence of each observation on the coefficients of husband's income and presence of children.

The largest Cook's distances by far ($D = 0.199$) belong to observations 76 and 77: two women whose husbands have substantial incomes ($38,000), and who have children at home, but who nevertheless work outside of the home. The index plots in Figure 15.7(a) and (b) show approximate influence on the coefficients for husband's income and presence of children, respectively: Observations 76 and 77 appear to be raising the husband's income coefficient, pushing it toward 0; none of the observations appears to have much impact on the presence-of-children coefficient. Omitting observations 76 and 77 changes the husband's income coefficient by about 50%, producing the modified fit

$$\begin{aligned} \text{logit}(P_W) = \; &1.608 - 0.06035I - 1.647K \\ &(0.405) \quad (0.02119) \quad (0.298) \end{aligned}$$

Partial-Regression Plot

Wang (1985) has suggested a logit-model analog to the partial-regression plot.[30] Let $Z_i^{(1)}$ represent the Pearson residuals for the logistic regression of Y_i on $X_{i2}, \ldots, X_{ik}$ (i.e., excluding X_{i1}). Similarly, let $X_i^{(1)}$ represent the residuals from the weighted-least-squares regression of X_{i1} on $X_{i2}, \ldots, X_{ik}$, using weights $V_i^{(1)} = P_i^{(1)}(1 - P_i^{(1)})$, where the $P_i^{(1)}$ are the fitted probabilities obtained from the logistic regression model for Y deleting X_1. Wang (1985) shows that

$$\mathbf{x}^{(1)} = (\mathbf{I}_n - \mathbf{H}_{(1)})\mathbf{V}_{(1)}^{1/2}\mathbf{x}_1$$

where $\mathbf{x}^{(1)} \equiv \{X_i^{(1)}\}$; $\mathbf{H}_{(1)}$ is the hat matrix for the logistic regression deleting X_1; and

$$\mathbf{V}_{(1)}^{1/2} \equiv \text{diag}\left\{\sqrt{P_i^{(1)}\left(1 - P_i^{(1)}\right)}\right\}$$

The residuals $Z_i^{(1)}$ are plotted against $X_i^{(1)}$. A disadvantage of this procedure is that, unlike the partial-regression plot in least-squares regression, it requires literally recomputing the maximum-likelihood fit for the model deleting each regressor in turn.

Constructed-Variable Plot for Transforming an X

In a later paper, Wang (1987) directly generalizes Atkinson's constructed-variable plot for the power transformation of an independent variable in least-squares regression. We are contemplating replacing X_1 with $X_1^{(\lambda)}$, where

$$X_1^{(\lambda)} = \begin{cases} \dfrac{X^\lambda - 1}{\lambda} & \text{for } \lambda \neq 0 \\ \log_e X & \text{for } \lambda = 0 \end{cases}$$

As in least-squares regression, we form the constructed variable $X_1 \log_e X_1$ and add it to the logistic regression. A test for the coefficient of the constructed variable is a test of the need to transform X_1. The suggested transformation, again as in least-squares regression, is

$$\widetilde{\lambda}_1 = 1 + \frac{D_1}{B_1}$$

where D_1 is the coefficient of the constructed variable in the *auxiliary* regression, and B_1 is the coefficient of X_1 in the *original* regression. A partial-regression plot for the constructed variable $X_1 \log_e X_1$ shows leverage and influence on the decision to transform X_1.

---

[30] Wang's work applies to the wider class of generalized linear models, not just to logit models. See Section 15.4 for an introduction to generalized linear models.

EXERCISES

15.1 Nonconstant error variance in the linear-probability model: Make a table showing the variance of the error $V(\varepsilon) = \pi(1-\pi)$ for the following values of π:

$$.001, .01, .05, .1, .3, .5., .7, .9, .95, .99, .999$$

When is the heteroscedasticity problem serious?

15.2 *Show that the slope of the logistic-regression curve, $\pi = 1/\left[1 + e^{-(\alpha+\beta X)}\right]$, can be written $\beta\pi(1-\pi)$. (*Hint*: Differentiate π with respect to X, and then substitute for expressions that equal π and $1-\pi$.)

15.3 Substitute first $y_i = 0$ and then $y_i = 1$ into the expression

$$p(y_i) \equiv \Pr(Y_i = y_i) = \pi_i^{y_i}(1-\pi_i)^{1-y_i}$$

to show that this equation captures $p(0) = 1 - \pi_i$ and $p(1) = \pi_i$.

15.4 *Show that, for the logit multiple-regression model,

$$\pi_i = \frac{1}{1 + \exp[-(\alpha + \beta_1 X_{i1} + \beta_2 X_{i2} + \cdots + \beta_k X_{ik})]}$$

the probability that $Y_i = 0$ can be written

$$1 - \pi_i = \frac{1}{1 + \exp(\alpha + \beta_1 X_{i1} + \cdots + \beta_k X_{ik})}$$

15.5 *Show that the maximized likelihood for the fitted logit model can be written as

$$\log_e L = \sum_{i=1}^{n} [y_i \log_e P_i + (1 - y_i)\log_e(1 - P_i)]$$

where

$$P_i = \frac{1}{1 + \exp[-(A + B_1 X_{i1} + \cdots + B_k X_{ik})]}$$

is the fitted probability that $Y_i = 1$. [*Hint*: Use $p(y_i) = \pi_i^{y_i}(1-\pi_i)^{1-y_i}$.]

15.6 *Deviance in least-squares regression: The log likelihood for the linear regression model with normal errors can be written as

$$\log_e L(\alpha, \beta_1, \ldots, \beta_k, \sigma_\varepsilon^2) = -\frac{n}{2}\log_e\left(2\pi\sigma_\varepsilon^2\right) - \frac{\sum_{i=1}^{n} \varepsilon_i^2}{2\sigma_\varepsilon^2}$$

where $\varepsilon_i = Y_i - (\alpha + \beta_1 X_{i1} + \cdots + \beta_k X_{ik})$ (see Section 9.3.3). Let l represent the maximized log likelihood, treated as a function of the regression coefficients $\alpha, \beta_1, \ldots, \beta_k$ but not of the error variance σ_ε^2, which is regarded as a "nuisance parameter." Let $l' = -(n/2)\log_e(2\pi\sigma_\varepsilon^2)$ represent the log likelihood for a model that fits the data perfectly (i.e., for which all $\varepsilon_i = 0$). Then the deviance is defined as $-2\sigma_\varepsilon^2(l - l')$. Show that, by this definition, the deviance for the normal linear model is just the residual sum of squares. [For the logit model, there is no nuisance parameter, and $l' = 0$; the deviance for this model is, therefore, $-2 \log_e L$, as stated in the text. See McCullagh and Nelder (1989, Chapter 2) for further discussion of the deviance.]

15.7 *Evaluate the information matrix for the logit model:

$$\mathscr{I}(\boldsymbol{\beta}) = -E\left[\frac{\partial^2 \log_e L(\boldsymbol{\beta})}{\partial \boldsymbol{\beta}\, \partial \boldsymbol{\beta}'}\right]$$

and show that the estimated asymptotic covariance matrix of the coefficients is

$$\widehat{\mathscr{V}(\mathbf{b})} = \left[\sum_{i=1}^{n} \frac{\exp(-\mathbf{x}_i'\mathbf{b})}{[1 + \exp(-\mathbf{x}_i'\mathbf{b})]^2} \mathbf{x}_i\mathbf{x}_i'\right]^{-1}$$

15.8 *Show that the maximum-likelihood estimator for the logit model can be written as

$$\mathbf{b} = (\mathbf{X}'\mathbf{V}\mathbf{X})^{-1}\mathbf{X}'\mathbf{V}\mathbf{y}^*$$

where

$$\mathbf{y}^* \equiv \mathbf{X}\mathbf{b} + \mathbf{V}^{-1}(\mathbf{y} - \mathbf{p})$$

(*Hint*: Simply multiply out the equation.)

15.9 *Demonstrate the following properties of the hat matrix

$$\mathbf{H} = \mathbf{V}^{1/2}\mathbf{X}(\mathbf{X}'\mathbf{V}\mathbf{X})^{-1}\mathbf{X}'\mathbf{V}^{1/2}$$

in logistic regression: (a) $\mathbf{H}$ is symmetric; (b) $\mathbf{H}$ is idempotent; (c) $\mathbf{z} = (\mathbf{I}_n - \mathbf{H})\mathbf{z}$; (d) $0 \leq h_i \leq 1$; (e) $\sum h_i = k + 1$.

15.10 Restricting attention to women working outside of the home, fit a dichotomous logit model to the Canadian women's labor-force data (in `womenlf.dat`) in which working full time versus part time is the dependent variable. Include terms for husband's income, presence of children, the interaction between husband's income and presence of children, and region. Construct an analysis-of-deviance table for this model similar to Table 15.2, and a graph of a final model fit to the data similar to Figure 15.4. Then fit the linear-probability model to the data by least squares. Compare the fit of the linear-probability to the logit model.

15.11 Greene and Shaffer (1992) analyzed decisions by the Canadian Federal Court of Appeal on cases filed by refugee applicants who had been turned down by the Immigration and Refugee Board. (Greene and Shaffer's study is described briefly in Section 1.2.) Restricting our attention to the 10 (of 12) judges who were present on the Court during the entire period of the study, and to countries of origin that produced at least 20 appeals during this period, we shall elaborate Greene and Shaffer's analysis using a logistic regression. The dependent variable is whether or not leave was granted to appeal the decision of the Refugee Board. We shall examine a random subsample of cases for which an independent expert rated the merit of the case. (The judge does not decide whether the applicant is granted refugee status; if the case has any merit, an appeal should be granted.) Aside from the expert's rating of merit (yes or no), and the identity of the judge, the logistic regression will include the following independent variables: the language in which the appeal was filed (English or French); the location in which the original application for refugee status was filed (Toronto, Montreal, other); and the applicant's nation of origin. Nation of origin will be handled in two ways: (1) as a factor with 17 categories; and (2) using the logit of the success rates for all cases from the applicant's nation decided during the period of the study [i.e., $\log_e$(number of leaves granted/number of leaves denied)]. The principal object of the analysis is to determine whether the substantial differences among the judges in their rates of granting leave to appeal can be explained by differences in characteristics of the cases. Recall that the cases were assigned to judges not at random, but on a rotating basis. The data for this exercise are in `greene.dat`.

(a) Fit a logit model to the data, regressing the judge's decision on judge, the expert's rating, language, location, and the logit of the success rate for cases originating in the applicant's nation. Be sure to test for differences among the judges, using either a Wald test or a likelihood-ratio test (or both). Also perform tests for the other independent variables.

(b) Fit an alternative model, employing dummy variables for nation of origin. Compare the two models to test the linearity (on the logit scale) of the partial effect of national success rate. (Also see Exercise 15.17.)

(c) *Adjusted proportions: Fit a final model to the data, including the independent variables found to be important in part (a). Calculate the adjusted logit (i.e., the fitted logit) for each judge, setting other independent variables to their sample means. Then translate these logits to the probability scale to obtain adjusted proportions. The adjusted proportions are analogous to adjusted means in linear models (see Exercises 7.5, 7.10, 8.8, and 8.12). Construct a contingency table relating judge to decision, and calculate the sample proportions of leave granted by each judge. Finally, compare these raw sample proportions, showing the marginal relationship between judge and decision, with the adjusted proportions, showing the partial relationship controlling for other factors. What do you conclude?

15.2 Models for Polytomous Data

A substantial limitation of the logit and probit models of the previous section is that they apply only to dichotomous dependent variables. In the Chilean plebiscite data, for example, many of the voters surveyed indicated that they were undecided, and some said that they planned to abstain or refused to reveal their voting intentions.[31] Polytomous data of this sort are, of course, common,

[31] See Exercise 15.14.

and it is desirable to model them in a natural manner—not simply to ignore some of the categories (e.g., restricting attention to those who responded "yes" or "no") or to combine categories arbitrarily to produce a dichotomy.

In this section, I shall describe three general approaches to modeling polytomous data:[32]

1. modeling the polytomy directly as a set of unordered categories, using a generalization of the dichotomous logit model;
2. constructing a set of nested dichotomies from the polytomy, fitting an independent logit or probit model to each dichotomy; and
3. extending the unobserved-variable interpretation of the dichotomous logit and probit models to ordered polytomies.

15.2.1 The Polytomous Logit Model

It is possible to generalize the dichotomous logit model to a polytomy by employing the multivariate logistic distribution. This approach has the advantage of treating the categories of the polytomy in a nonarbitrary, symmetric manner (but the disadvantage that the analysis is relatively complex).[33]

Suppose that the dependent variable Y can take on any of m qualitative values, which, for convenience, we number $1, 2, \ldots, m$. To anticipate the example employed in this section, a married woman can (1) work full time, (2) work part time, or (3) not work outside of the home. Although the categories of Y are numbered, we do not, in general, attribute ordinal properties to these numbers: They are simply category labels. Let π_{ij} denote the probability that the ith observation falls in the jth category of the dependent variable; that is, $\pi_{ij} \equiv \Pr(Y_i = j)$, for $j = 1, \ldots, m$.

We have available k regressors, $X_1, \ldots, X_k$, on which the π_{ij} depend. More specifically, suppose that this dependence can be modeled using the *multivariate logistic distribution*:

$$\pi_{ij} = \frac{\exp(\gamma_{0j} + \gamma_{1j}X_{i1} + \cdots + \gamma_{kj}X_{ik})}{1 + \sum_{l=1}^{m-1} \exp(\gamma_{0l} + \gamma_{1l}X_{i1} + \cdots + \gamma_{kl}X_{ik})} \quad \text{for } j = 1, \ldots, m-1 \qquad [15.22]$$

$$\pi_{im} = 1 - \sum_{l=1}^{m-1} \pi_{ij} \quad \text{(for category } m\text{)}$$

There is, then, one set of parameters, $\gamma_{0j}, \gamma_{1j}, \ldots, \gamma_{kj}$, for each dependent-variable category but the last. The last category (i.e., category m) functions as a type of baseline. The use of a baseline category is one way of avoiding redundant parameters because of the restriction, reflected in the second part of

---

[32] Some of the references at the end of the chapter describe additional statistical models for polytomous data.

[33] A similar probit model based on the multivariate-normal distribution is also possible in principle, but it is numerically less tractable because of the necessity of evaluating a multivariate integral.

Equation 15.22, that the response category probabilities for each observation must sum to 1:[34] $\sum_{j=1}^{m} \pi_{ij} = 1$. The denominator of π_{ij} in the first line of the equation imposes this restriction.

Some algebraic manipulation of Equation 15.22 produces[35]

$$\log_e \frac{\pi_{ij}}{\pi_{im}} = \gamma_{0j} + \gamma_{1j}X_{i1} + \cdots + \gamma_{kj}X_{ik} \quad \text{for } j = 1, \ldots, m-1$$

The regression coefficients, therefore, affect the log odds of membership in category j versus the baseline category. It is also possible to form the log odds of membership in *any* pair of categories j and j':

$$\begin{aligned} \log_e \frac{\pi_{ij}}{\pi_{ij'}} &= \log_e \left(\frac{\pi_{ij}/\pi_{im}}{\pi_{ij'}/\pi_{im}} \right) \\ &= \log_e \frac{\pi_{ij}}{\pi_{im}} - \log_e \frac{\pi_{ij'}}{\pi_{im}} \\ &= (\gamma_{0j} - \gamma_{0j'}) + (\gamma_{1j} - \gamma_{1j'})X_{i1} + \cdots + (\gamma_{kj} - \gamma_{kj'})X_{ik} \end{aligned}$$

Thus, the regression coefficients for the logit between any pair of categories are the differences between corresponding coefficients for the logit between each of the categories and the baseline category.

To gain further insight into the polytomous logit model, suppose that the model is specialized to a dichotomous dependent variable. Then, $m = 2$, and

$$\log_e \frac{\pi_{i1}}{\pi_{i2}} = \log_e \frac{\pi_{i1}}{1 - \pi_{i1}} = \gamma_{01} + \gamma_{11}X_{i1} + \cdots + \gamma_{k1}X_{ik}$$

When it is applied to a dichotomy, the polytomous logit model is, therefore, identical to the dichotomous logit model of the previous section.

The maximum-likelihood fit of the polytomous logit model to the women's labor-force-participation data is as follows, treating not working outside of the home as the baseline category:

$$\log_e \frac{P_{\text{FT}}}{P_{\text{NW}}} = \underset{(0.484)}{1.983} - \underset{(0.02809)}{0.09723} \times \text{Income} - \underset{(0.362)}{2.559} \times \text{Children}$$

$$\log_e \frac{P_{\text{PT}}}{P_{\text{NW}}} = \underset{(0.592)}{-1.432} + \underset{(0.02345)}{0.00689} \times \text{Income} + \underset{(0.4690)}{0.0215} \times \text{Children}$$

---

[34] An alternative is to treat the categories symmetrically:

$$\pi_{ij} = \frac{\exp(\gamma_{0j} + \gamma_{1j}X_{i1} + \cdots + \gamma_{kj}X_{ik})}{\sum_{l=1}^{m} \exp(\gamma_{0l} + \gamma_{1l}X_{i1} + \cdots + \gamma_{kl}X_{ik})}$$

but to impose a linear restriction, analogous to a sigma constraint, on the parameters. This approach produces somewhat more difficult computations, however, and has no real advantages.

[35] See Exercise 15.12.

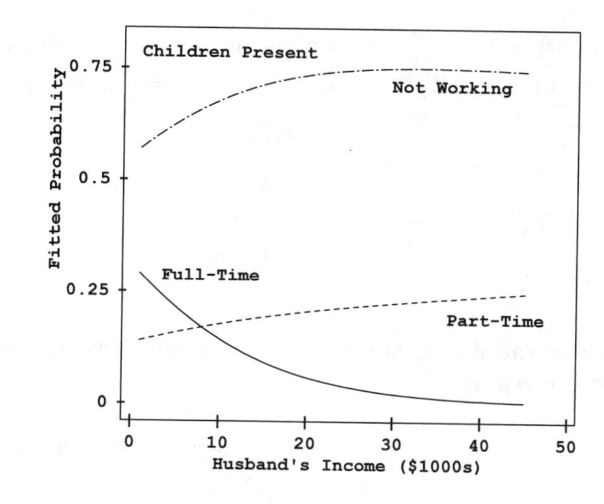

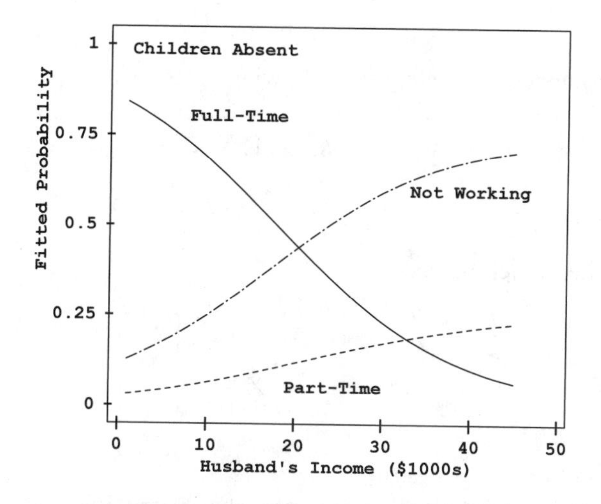

Figure 15.8. Fitted probabilities for the polytomous logit model, showing women's labor-force participation as a function of husband's income and presence of children. The upper panel is for children present, the lower panel for children absent.

Here, P_{FT} is the fitted probability of working full time, P_{PT} the fitted probability of working part time, and P_{NW} the fitted probability of not working outside of the home. High husband's income and the presence of children, therefore, decrease the odds of working full time relative to not working, but make virtually no difference to the odds of working part time as opposed to not working. The fitted probabilities are graphed in Figure 15.8.

*Details of Estimation**

To fit the model given in Equation 15.22 to data, we may again invoke the maximum-likelihood method. First, note that each Y_i takes on its possible values $1, 2, \ldots, m$ with probabilities $\pi_{i1}, \pi_{i2}, \ldots, \pi_{im}$; as mentioned above,

$\Pr(Y_i = m) = 1 - \sum_{j=1}^{m-1} \pi_{ij}$. Following Nerlove and Press (1973), let us define indicator variables $W_{i1}, \ldots, W_{im}$ so that $W_{ij} = 1$ if $Y_i = j$, and $W_{ij} = 0$ if $Y_i \neq j$; thus,

$$
\begin{aligned}
p(y_i) &= \pi_{i1}^{w_{i1}} \pi_{i2}^{w_{i2}} \cdots \pi_{im}^{w_{im}} \\
&= \prod_{j=1}^{m} \pi_{ij}^{w_{ij}}
\end{aligned}
$$

If the observations are sampled independently, then their joint probability distribution is given by

$$
\begin{aligned}
p(y_1, \ldots, y_n) &= p(y_1) \times \cdots \times p(y_n) \\
&= \prod_{i=1}^{n} \prod_{j=1}^{m} \pi_{ij}^{w_{ij}}
\end{aligned}
$$

For compactness, define the following vectors:

$$
\begin{aligned}
\mathbf{x}_i' &\equiv (1, X_{i1}, \ldots, X_{ik}) \\
\boldsymbol{\gamma}_j &\equiv (\gamma_{0j}, \gamma_{1j}, \ldots, \gamma_{kj})'
\end{aligned}
$$

and the model matrix

$$
\underset{(n \times k+1)}{\mathbf{X}} \equiv \begin{bmatrix} \mathbf{x}_1' \\ \mathbf{x}_2' \\ \vdots \\ \mathbf{x}_n' \end{bmatrix}
$$

It is convenient to impose the restriction $\sum_{j=1}^{m} \pi_{ij} = 1$ by setting $\boldsymbol{\gamma}_m = 0$ (making category m the baseline, as explained previously). Then, employing Equation 15.22,

$$
p(y_1, \ldots, y_n | \mathbf{X}) = \prod_{i=1}^{n} \prod_{j=1}^{m} \left(\frac{\exp(\mathbf{x}_i' \boldsymbol{\gamma}_j)}{1 + \sum_{l=1}^{m-1} \exp(\mathbf{x}_i' \boldsymbol{\gamma}_l)} \right)^{w_{ij}} \quad [15.23]
$$

and the log likelihood is

$$
\begin{aligned}
\log_e L(\boldsymbol{\gamma}_1, \ldots, \boldsymbol{\gamma}_{m-1}) &= \sum_{i=1}^{n} \sum_{j=1}^{m} W_{ij} \left[\mathbf{x}_i' \boldsymbol{\gamma}_j - \log_e \left(1 + \sum_{l=1}^{m-1} \exp(\mathbf{x}_i' \boldsymbol{\gamma}_l) \right) \right] \\
&= \sum_{i=1}^{n} \sum_{j=1}^{m-1} W_{ij} \mathbf{x}_i' \boldsymbol{\gamma}_j - \sum_{i=1}^{n} \log_e \left(1 + \sum_{l=1}^{m-1} \exp(\mathbf{x}_i' \boldsymbol{\gamma}_l) \right)
\end{aligned}
$$

because $\sum_{j=1}^{m} W_{ij} = 1$ and $\boldsymbol{\gamma}_m = 0$; setting $\boldsymbol{\gamma}_m = 0$ accounts for the 1 in the denominator of Equation 15.23, because $\exp(\mathbf{x}_i' 0) = 1$.

Differentiating the log likelihood with respect to the parameters, and setting the partial derivatives to 0, produces the nonlinear estimating equations:[36]

$$\sum_{i=1}^{n} W_{ij}\mathbf{x}_i = \sum_{i=1}^{n} \mathbf{x}_i \frac{\exp(\mathbf{x}_i'\mathbf{c}_j)}{1+\sum_{l=1}^{m-1}\exp(\mathbf{x}_i'\mathbf{c}_l)} \quad \text{for } j=1,\ldots,m-1 \qquad [15.24]$$
$$= \sum_{i=1}^{n} P_{ij}\mathbf{x}_i$$

where $\mathbf{c}_j \equiv \widehat{\boldsymbol{\gamma}}_j$ are the maximum-likelihood estimators, and

$$P_{ij} \equiv \frac{\exp(\mathbf{x}_i'\mathbf{c}_j)}{1+\sum_{l=1}^{m-1}\exp(\mathbf{x}_i'\mathbf{c}_l)}$$

are the fitted probabilities. As in the dichotomous logit model, the maximum-likelihood estimator sets observed sums equal to fitted sums. The estimating equations (Equation 15.24) are nonlinear and, therefore, require iterative solution, for example, by the Gauss-Newton method.

Let us stack up all of the parameters in a large vector:

$$\underset{[(m-1)(k+1)\times 1]}{\boldsymbol{\gamma}} \equiv (\boldsymbol{\gamma}_1', \ldots, \boldsymbol{\gamma}_{m-1}')'$$

The information matrix is[37]

$$\underset{[(m-1)(k+1)\times(m-1)(k+1)]}{\mathcal{I}(\boldsymbol{\gamma})} = \begin{bmatrix} \mathcal{I}_{11} & \mathcal{I}_{12} & \cdots & \mathcal{I}_{1,m-1} \\ \mathcal{I}_{21} & \mathcal{I}_{22} & \cdots & \mathcal{I}_{2,m-1} \\ \vdots & \vdots & \ddots & \vdots \\ \mathcal{I}_{m-1,1} & \mathcal{I}_{m-1,2} & \cdots & \mathcal{I}_{m-1,m-1} \end{bmatrix}$$

where

$$\underset{[(k+1)\times(k+1)]}{\mathcal{I}_{jj}} = -E\left[\frac{\partial^2 \log_e L(\boldsymbol{\gamma})}{\partial\boldsymbol{\gamma}_j \partial\boldsymbol{\gamma}_j'}\right] \qquad [15.25]$$
$$= \sum_{i=1}^{n} \frac{\mathbf{x}_i\mathbf{x}_i' \exp(\mathbf{x}_i'\boldsymbol{\gamma}_j)[1+\sum_{l=1}^{m-1}\exp(\mathbf{x}_i'\boldsymbol{\gamma}_l) - \exp(\mathbf{x}_i'\boldsymbol{\gamma}_j)]}{[1+\sum_{l=1}^{m-1}\exp(\mathbf{x}_i'\boldsymbol{\gamma}_l)]^2}$$

and

$$\underset{[(k+1)\times(k+1)]}{\mathcal{I}_{jj'}} = -E\left[\frac{\partial^2 \log_e L(\boldsymbol{\gamma})}{\partial\boldsymbol{\gamma}_j \, \partial\boldsymbol{\gamma}_{j'}'}\right] \qquad [15.26]$$
$$= -\sum_{i=1}^{n} \frac{\mathbf{x}_i\mathbf{x}_i' \exp[\mathbf{x}_i'(\boldsymbol{\gamma}_{j'}+\boldsymbol{\gamma}_j)]}{[1+\sum_{l=1}^{m-1}\exp(\mathbf{x}_i'\boldsymbol{\gamma}_l)]^2}$$

The estimated asymptotic covariance matrix of $\mathbf{c} \equiv (\mathbf{c}_1', \ldots, \mathbf{c}_{m-1}')'$ is obtained from the inverse of the information matrix, replacing $\boldsymbol{\gamma}$ with $\mathbf{c}$.

[36] See Exercise 15.13.

[37] See Exercise 15.13.

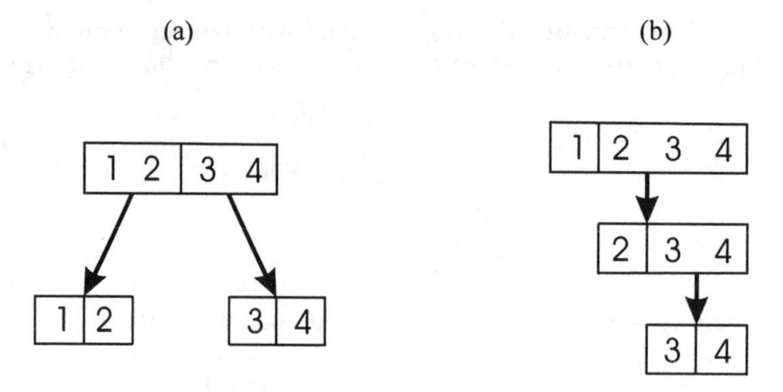

Figure 15.9. Alternative sets of nested dichotomies [(a) and (b)] for a four-category polytomous dependent variable.

15.2.2 Nested Dichotomies

Perhaps the simplest approach to polytomous data—because it employs the already-familiar dichotomous logit or probit model—is to fit separate models to each of a set of dichotomies derived from the polytomy. These dichotomies are constructed so that the likelihood for the polytomous dependent variable is the product of the likelihoods for the dichotomies—that is, the models are statistically independent even though they are fitted to data from the same sample. The likelihood is separable in this manner if the set of dichotomies is *nested*.[38] Although the system of nested dichotomies comprises a model for the polytomy, and although this model often produces fitted probabilities that are very similar to those produced by the polytomous logit model of the previous section, the two models are not equivalent.

A nested set of $m - 1$ dichotomies is produced from an m-category polytomy by successive binary partitions of the categories of the polytomy. Two examples for a four-category variable are shown in Figure 15.9. In part (a) of this figure, the dichotomies are $\{12, 34\}$, $\{1, 2\}$, and $\{3, 4\}$; in part (b), the nested dichotomies are $\{1, 234\}$, $\{2, 34\}$, and $\{3, 4\}$. This simple—and abstract—example illustrates a key property of nested dichotomies: The set of nested dichotomies selected to represent a polytomy is not unique. Because the results of the analysis and their interpretation depend on the set of nested dichotomies that is selected, this approach to polytomous data is reasonable only when a particular choice of dichotomies is substantively compelling. If the dichotomies are purely arbitrary, or if alternative sets of dichotomies are equally reasonable and interesting, then nested dichotomies should probably not be used to analyze the data.

Nested dichotomies are an especially attractive approach when the categories of the polytomy represent ordered progress through the stages of a process. Imagine, for example, that the categories in Figure 15.9(b) represent adults' attained level of education: (1) less than high school; (2) high-school graduate; (3) some postsecondary; (4) postsecondary degree. Because individuals normally progress through these categories in sequence, the dichotomy $\{1, 234)$ represents

[38] A proof of this property of nested dichotomies will be given presently.

the completion of high school; {2, 34} the continuation to postsecondary education, conditional on high-school graduation; and {3, 4} the completion of a degree conditional on undertaking a postsecondary education.[39]

In the case of the women's labor-force data, the following system of nested dichotomies appears reasonable, if not completely compelling:

1. Not working outside the home versus working: {Not Working, (Full Time, Part Time)}.
2. Among those working outside the home, working part time versus full time: {Part Time, Full Time}.

A dichotomous logit model was fit to the first of these dichotomies in Section 15.1.4, where we discovered that the effects of region and the interaction between husband's income and presence of children are both small. A final model,[40] including only main effects for husband's income and presence of children, is

$$\log_e \frac{P_{\text{PT}} + P_{\text{FT}}}{P_{\text{NW}}} = \underset{(0.384)}{1.336} - \underset{(0.01978)}{0.04231I} - \underset{(0.292)}{1.576K} \qquad [15.27]$$

$$R^2 = .102$$

where P_{PT} is the fitted probability of working part time, P_{FT} is the fitted probability of working full time, and P_{NW} is the fitted probability of not working outside of the home.

Similar results are obtained for the full-time versus part-time dichotomy, where region effects and the income-by-children interaction are also negligible.[41] The fit for the second dichotomy is

$$\log_e \frac{P_{\text{FT}}}{P_{\text{PT}}} = \underset{(0.767)}{3.478} - \underset{(0.0391)}{0.1073I} - \underset{(0.541)}{2.651K}$$

$$R^2 = .276$$

and thus the impact of both independent variables is greater on the odds of working full time versus part time than on the odds of working versus not working. In this instance, the general polytomous logit model (fit in Section 15.2.1) is more revealing of the structure of the data: The general polytomous logit model shows that the independent variables have little effect on working part time versus not working outside of the home; this result is not apparent in the models fit to the nested dichotomies.

Because the nested dichotomies are independent, we can pool Wald or likelihood-ratio test statistics across the two models: For example, the likelihood-ratio test statistic for the effect of income on the working versus not-working dichotomy is $G_0^2 = 5.14$, on 1 degree of freedom.[42] The likelihood-ratio test

---

[39] Fienberg (1980, pp. 110–116) terms ratios of odds formed from nested dichotomies "continuation ratios."

[40] This model is reproduced from Equation 15.15.

[41] See Exercise 15.10.

statistic for the effect of income on the full-time versus part-time dichotomy is $G_0^2 = 12.98$, also on 1 degree of freedom. Consequently, $G_0^2 = 5.14 + 12.98 = 18.12$ on 2 degrees of freedom (for which $p = .0001$) tests the effect of income on the three-category dependent variable.

*Why Nested Dichotomies Are Independent**

For simplicity, I shall demonstrate the independence of the nested dichotomies $\{12, 3\}$ and $\{1, 2\}$. By repeated application, this result applies generally to any system of nested dichotomies. Let W_{i1}, W_{i2}, and W_{i3} be dummy variables indicating whether the polytomous dependent variable Y_i is 1, 2, or 3. For example, $W_{i1} = 1$ if $Y_i = 1$, and 0 otherwise. Let Y_i' be a dummy variable representing the first dichotomy, $\{12, 3\}$: That is, $Y_i' = 1$ when $Y_i = 1$ or 2, and $Y_i' = 0$ when $Y_i = 3$. Likewise, let Y_i'' be a dummy variable representing the second dichotomy, $\{1, 2\}$: $Y_i'' = 1$ when $Y_i = 1$, and $Y_i'' = 0$ when $Y_i = 2$; Y_i'' is undefined when $Y_i = 3$. We need to show that $p(y_i) = p(y_i')p(y_i'')$. [To form this product, we adopt the convention that $p(y_i'') \equiv 1$ when $Y_i = 3$.]

The probability distribution of Y_i' is given by

$$\begin{aligned} p(y_i') &= (\pi_{i1} + \pi_{i2})^{y_i'} \pi_{i3}^{1-y_i'} \\ &= (\pi_{i1} + \pi_{i2})^{w_{i1}+w_{i2}} \pi_{i3}^{w_{i3}} \end{aligned} \qquad [15.28]$$

where $\pi_{ij} \equiv \Pr(Y_i = j)$ for $j = 1, 2, 3$. To derive the probability distribution of Y_i'', note that

$$\begin{aligned} \Pr(Y_i'' = 1) &= \Pr(Y_i = 1 | Y_i \neq 3) = \frac{\pi_{i1}}{\pi_{i1} + \pi_{i2}} \\ \Pr(Y_i'' = 0) &= \Pr(Y_i = 2 | Y_i \neq 3) = \frac{\pi_{i2}}{\pi_{i1} + \pi_{i2}} \end{aligned}$$

and, thus,

$$\begin{aligned} p(y_i'') &= \left(\frac{\pi_{i1}}{\pi_{i1} + \pi_{i2}}\right)^{y_i''} \left(\frac{\pi_{i2}}{\pi_{i1} + \pi_{i2}}\right)^{1-y_i''} \\ &= \left(\frac{\pi_{i1}}{\pi_{i1} + \pi_{i2}}\right)^{w_{i1}} \left(\frac{\pi_{i2}}{\pi_{i1} + \pi_{i2}}\right)^{w_{i2}} \end{aligned} \qquad [15.29]$$

Multiplying Equation 15.28 by Euation 15.29 produces

$$p(y_i')p(y_i'') = \pi_{i1}^{w_{i1}} \pi_{i2}^{w_{i2}} \pi_{i3}^{w_{i3}} = p(y_i)$$

which is the required result.

Because the dichotomies Y' and Y'' are independent, it is legitimate to combine models for these dichotomies to form a model for the polytomy Y. Likewise, we can sum likelihood-ratio or Wald test statistics for the two dichotomies.

---

[42] See Section 15.1.4.

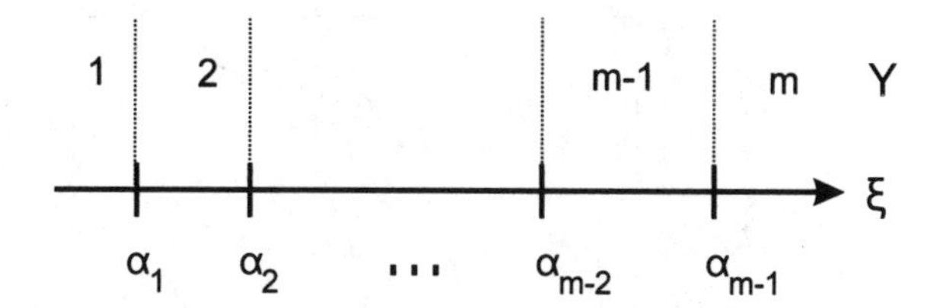

Figure 15.10. The boundaries $\alpha_1 < \alpha_2 < \cdots < \alpha_{m-1}$ divide the latent continuum ξ into m regions, corresponding to the values of the observable variable Y.

15.2.3 Ordered Logit and Probit Models

Imagine (as in Section 15.1.3) that there is a latent (i.e., unobservable) variable ξ that is a linear function of the X's plus a random error:

$$\xi_i = \alpha + \beta_1 X_{i1} + \cdots + \beta_k X_{ik} + \varepsilon_i$$

Now, however, suppose that instead of dividing the range of ξ into two regions to produce a dichotomous response, the range of ξ is dissected by $m-1$ boundaries into m regions. Denoting the boundaries by $\alpha_1 < \alpha_2 < \cdots < \alpha_{m-1}$, and the resulting response by Y, we observe

$$Y_i = \begin{cases} 1 & \text{if } \xi_i \leq \alpha_1 \\ 2 & \text{if } \alpha_1 < \xi_i \leq \alpha_2 \\ \vdots & \\ m-1 & \text{if } \alpha_{m-2} < \xi_i \leq \alpha_{m-1} \\ m & \text{if } \alpha_{m-1} < \xi_i \end{cases} \qquad [15.30]$$

The boundaries, regions, and corresponding values of ξ and Y are represented graphically in Figure 15.10.

Using Equation 15.30, we can determine the cumulative probability distribution of Y:

$$\begin{aligned} \Pr(Y_i \leq j) &= \Pr(\xi_i \leq \alpha_j) \\ &= \Pr(\alpha + \beta_1 X_{i1} + \cdots + \beta_k X_{ik} + \varepsilon_i \leq \alpha_j) \\ &= \Pr(\varepsilon_i \leq \alpha_j - \alpha - \beta_1 X_{i1} - \cdots - \beta_k X_{ik}) \end{aligned}$$

If the errors ε_i are independently distributed according to the standard normal distribution, then we obtain the ordered probit model.[43] If the errors follow the similar logistic distribution, then we get the ordered logit model. In the latter event,

$$\begin{aligned} \text{logit}\,[\Pr(Y_i \leq j)] &= \log_e \frac{\Pr(Y_i \leq j)}{\Pr(Y_i > j)} \\ &= \alpha_j - \alpha - \beta_1 X_{i1} - \cdots - \beta_k X_{ik} \end{aligned}$$

---

[43] As in the dichotomous case, we conveniently fix the error variance to 1 to set the scale of the latent variable ξ.

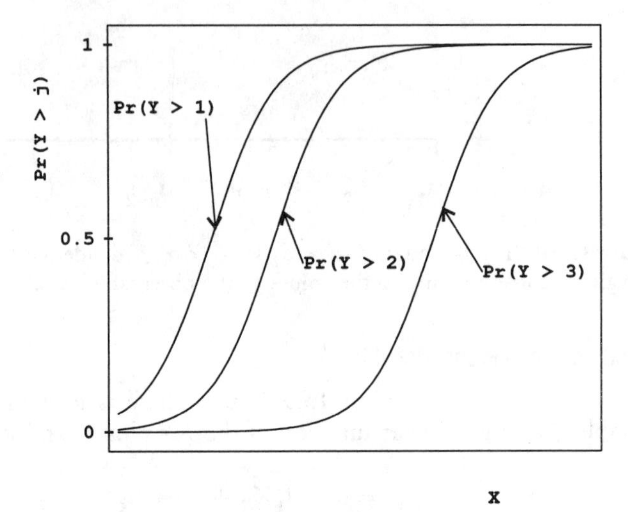

Figure 15.11. The proportional-odds model for four response categories and a single independent variable X. The logistic regression curves are parallel.

Source: Adapted from Agresti (1990, Figure 9.1), *Categorical Data Analysis*. Copyright ©1990 John Wiley & Sons, Inc. Reprinted by permission of John Wiley & Sons, Inc.

Equivalently,

$$\begin{aligned} \text{logit}\,[\Pr(Y_i > j)] &= \log_e \frac{\Pr(Y_i > j)}{\Pr(Y_i \leq j)} \\ &= (\alpha - \alpha_j) + \beta_1 X_{i1} + \cdots + \beta_k X_{ik} \end{aligned} \quad [15.31]$$

for $j = 1, 2, \ldots, m - 1$.

The logits in Equation 15.31 are for cumulative categories—at each point contrasting categories above category j with category j and below. The slopes for each of these regression equations are identical; the equations differ only in their intercepts. The logistic regression surfaces are, therefore, parallel to each other, as illustrated in Figure 15.11 for $m = 4$ response categories and a single X. Put another way, for a fixed set of X's, any two different cumulative log odds (i.e., logits)—say, at categories j and j'—differ only by the constant $(\alpha_j - \alpha_{j'})$. The odds, therefore, are proportional to one another, and for this reason, Equation 15.31 is called the *proportional-odds model.*

There are $(k + 1) + (m - 1) = k + m$ parameters to estimate in the proportional-odds model, including the regression coefficients $\alpha, \beta_1, \ldots, \beta_k$ and the category boundaries $\alpha_1, \ldots, \alpha_{m-1}$. Note, however, that there is an extra parameter in the regression equations (Equation 15.31), because each equation has its own constant, $-\alpha_j$, along with the common constant α. A simple solution is to set $\alpha = 0$ (and to absorb the negative sign in α_j), producing[44]

$$\text{logit}\,[\Pr(Y_i > j)] = \alpha_j + \beta_1 X_{i1} + \cdots + \beta_k X_{ik}$$

[44] Setting $\alpha = 0$ implicitly establishes the origin of the latent variable ξ (just as fixing the error variance establishes its unit of measurement). An alternative would be to fix one of the boundaries to 0. These choices are arbitrary and inconsequential.

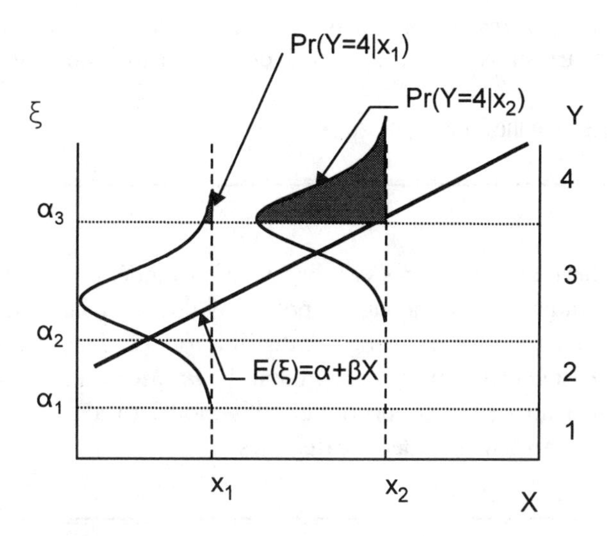

Figure 15.12. The proportional-odds model for four response categories and a single independent variable X. The latent dependent variable ξ has a linear regression on X. The latent continuum ξ appears at the left of the graph, the observable response Y at the right. The conditional logistic distribution of the latent variable is shown for two values of the independent variable, x_1 and x_2. The shaded area in each distribution gives the conditional probability that $Y = 4$.

Source: Adapted from Agresti (1990, Figure 9.2), *Categorical Data Analysis*. Copyright ©1990 John Wiley & Sons, Inc. Reprinted by permission of John Wiley & Sons, Inc.

Figure 15.12 illustrates the proportional-odds model for $m = 4$ response categories and a single X. The conditional distribution of the latent variable ξ is shown for two representative values of the independent variable, x_1 [where $\Pr(Y > 3) = \Pr(Y = 4)$ is small] and x_2 [where $\Pr(Y = 4)$ is about .5]. McCullagh (1980) explains how Equation 15.31 can be fit by the method of maximum likelihood (and discusses alternatives to the proportional-odds model).

Applying the proportional-odds model to the women's labor-force data produces the following results:

$$\log_e \frac{P_{\text{PT}} + P_{\text{FT}}}{P_{\text{NW}}} = \underset{(0.377)}{1.852} - \underset{(0.0194)}{0.0539} \times \text{Income} - \underset{(0.280)}{1.972} \times \text{Children}$$

$$\log_e \frac{P_{\text{FT}}}{P_{\text{NW}} + P_{\text{PT}}} = \underset{(0.3619)}{0.9409} - \underset{(0.0194)}{0.0539} \times \text{Income} - \underset{(0.280)}{1.972} \times \text{Children}$$

Thus, the "propensity" to work outside of the home declines with husband's income and presence of children. The coefficients for income and children are identical in the equations for the logit of working versus not working and the logit of working full time versus working part time, but the intercepts differ, reflecting the assumption of proportional odds. As it turns out, however, proportional odds are not reasonable for these data:[45] A score test of the assumption yields a chi-square statistic of 18.6 with 2 degrees of freedom, for which $p = .0001$.

15.2.4 Comparison of the Three Approaches

Several approaches can be taken to modeling polytomous data, including: (1) modeling the polytomy directly using a logit model based on the multivariate logistic distribution; (2) constructing a set of $m-1$ nested dichotomies to represent the m categories of the polytomy; and (3) fitting the proportional-odds model to a polytomous dependent variable with ordered categories.

The three approaches to modeling polytomous data—the polytomous logit model, logit models for nested dichotomies, and the proportional-odds model—address different sets of log odds, corresponding to different dichotomies constructed from the polytomy. Consider, for example, the ordered polytomy $\{1, 2, 3, 4\}$—representing, say, four ordered educational categories:

- Treating category 1 as the baseline, the coefficients of the polytomous logit model apply directly to the dichotomies $\{1, 2\}$, $\{1, 3\}$, and $\{1, 4\}$, and indirectly to any pair of categories.
- Forming continuation dichotomies (one of several possibilities), the nested-dichotomies approach models $\{1, 234\}$, $\{2, 34\}$, and $\{3, 4\}$.
- The proportional-odds model applies to the dichotomies $\{1, 234\}$, $\{12, 34\}$, and $\{123, 4\}$, imposing the restriction that only the intercepts of the three regression equations differ.

Which of these models is most appropriate depends partly on the structure of the data and partly on our interest in them. For the present example, I would prefer the continuation dichotomies (because of their straightforward interpretation), unless the more parsimonious proportional-odds model fits the data well.

EXERCISES

15.12 *Show that the polytomous logit model of Equation 15.22 can be written in the form

$$\log_e \frac{\pi_{ij}}{\pi_{im}} = \gamma_{0j} + \gamma_{1j}X_{i1} + \cdots + \gamma_{kj}X_{ik} \quad \text{for } j = 1, \ldots, m-1$$

[45] Compare the equation for $\log_e[(P_{\mathrm{PT}} + P_{\mathrm{FT}})/P_{\mathrm{NW}}]$ in the proportional-odds model with the unconstrained fit for the same logit in Equation 15.27.

15.13 *Derive the estimating equations (Equation 15.24) and the information matrix (Equations 15.25 and 15.26) for the polytomous logit model.

15.14 Analyze the Chilean plebiscite data (in `chile.dat`) by:

(a) fitting a polytomous logit model to the data, distinguishing the categories (i) yes, (ii) no, (iii) abstain, (iv) undecided, and (v) refused to respond;

(b) fitting a sequence of logit models to the following nested dichotomies: {(i, ii, iii, iv), v}; {(i, ii, iii), iv}; {(i, ii), iii}; and {i, ii};

(c) fitting the proportional-odds model, using the categories (ii), [(iii) or (iv)], (i)—that is omitting those who refused to respond, and combining the categories abstain and undecided into a middle category.

Which of these approaches do you prefer? Why?

15.3 Discrete Independent Variables and Contingency Tables

When the independent variables—as well as the dependent variable—are discrete, the joint sample distribution of the variables defines a contingency table of counts: Each cell of the table records the number of observations possessing a particular combination of characteristics. An example, drawn from *The American Voter* (Campbell et al., 1960), a classical study of electoral behavior, appears in Table 15.3. This table, based on data from a sample survey conducted after the 1956 U. S. presidential election, relates voting turnout in the election to strength of partisan preference (classified as weak, medium, or strong) and perceived closeness of the election (one-sided or close).

The last column of Table 15.3 gives the *empirical logit* for the dependent variable,

$$\log_e \frac{\text{Proportion voting}}{\text{Proportion not voting}}$$

TABLE 15.3 Voter Turnout by Perceived Closeness of the Election and Intensity of Partisan Preference, for the 1956 U.S. Presidential Election. Frequency counts are shown in the body of the table.

		Turnout		*Logit*
Perceived Closeness	*Intensity of Preference*	*Voted*	*Did Not Vote*	$\log_e \frac{\textit{Voted}}{\textit{Did Not Vote}}$
One-sided	Weak	91	39	0.847
	Medium	121	49	0.904
	Strong	64	24	0.981
Close	Weak	214	87	0.900
	Medium	284	76	1.318
	Strong	201	25	2.084

Source of Data: Campbell et al. (1960, Table 5–3).

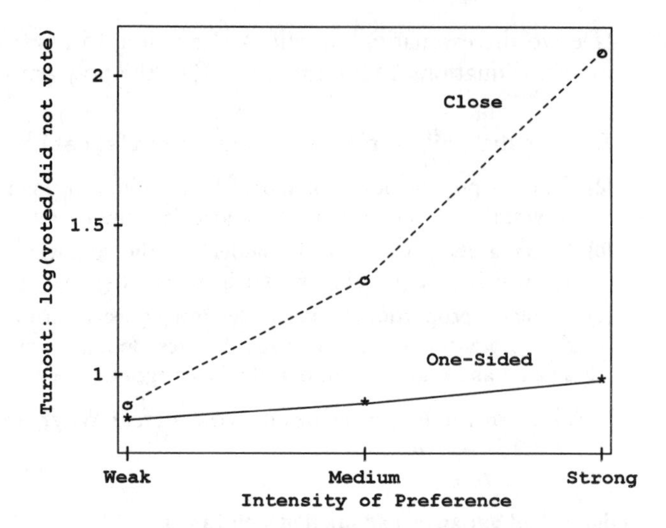

Figure 15.13. Empirical logits for voter turnout by intensity of partisan preference and perceived closeness of the election, for the 1956 U.S. presidential election.

for each of the six combinations of categories of the independent variables.[46] For example,

$$\text{logit}\,(\text{voted} \mid \text{one-sided, weak preference}) = \log_e \frac{91/130}{39/130} = \log_e \frac{91}{39} = 0.847$$

Because the conditional proportions voting and not voting share the same denominator, the empirical logit can also be written as

$$\log_e \frac{\text{Number voting}}{\text{Number not voting}}$$

The empirical logits from Table 15.3 are graphed in Figure 15.13, much in the manner of profiles of cell means for a two-way analysis of variance.[47] Perceived closeness of the election and intensity of preference appear to interact in affecting turnout: Turnout increases with increasing intensity of preference, but only if the election is perceived to be close. Those with medium or strong preference who perceive the election to be close are more likely to vote than those who perceive the election to be one-sided; this difference is greater among those with strong than with medium partisan preference.

The methods of this chapter are fully appropriate for tabular data. When, as in Table 15.3, the independent variables are qualitative or ordinal, it is natural

[46] This calculation will fail if there is a zero frequency in the table because, in this event, the proportion voting or not voting for some combination of independent-variable values will be 0. A simple remedy is to add 0.5 to each of the cell frequencies. Adding 0.5 to each count also serves to reduce the bias of the sample logit as an estimator of the corresponding population logit. See Cox and Snell (1989, pp. 31–32).

[47] See Section 8.2.1.

to use logit or probit models that are analogous to analysis-of-variance models. Treating perceived closeness of the election as the "row" factor and intensity of partisan preference as the "column" factor, for example, yields the model

$$\text{logit}\, \pi_{jk} = \mu + \alpha_j + \beta_k + \gamma_{jk} \tag{15.32}$$

where

- π_{jk} is the conditional probability of voting in combination of categories j of perceived closeness and k of preference (i.e., in cell jk of the independent-variable table);
- μ is the general level of turnout in the population;
- α_j is the main effect on turnout of membership in the jth category of perceived closeness;
- β_k is the main effect on turnout of membership in the kth category of preference; and
- γ_{jk} is the interaction effect on turnout of simultaneous membership in categories j of perceived closeness and k of preference.

When all of the variables—independent as well as dependent—are discrete, their joint distribution defines a contingency table of frequency counts. It is natural to employ logit models that are analogous to analysis-of-variance models to analyze contingency tables.

Under the usual sigma constraints, Equation 15.32 leads to deviation-coded regressors, as in the analysis of variance. Adapting the SS(·) notation of Chapter 8,[48] likelihood-ratio tests for main effects and interactions can then be constructed in close analogy to the incremental F-tests for the two-way ANOVA model. Deviances under several models for the *American Voter* data are shown in Table 15.4, and the "analysis-of-deviance" table for these data is given in Table 15.5. The log-likelihood-ratio statistic for testing H_0: all $\gamma_{jk} = 0$, for example, is

$$\begin{aligned} G_0^2(\gamma|\alpha, \beta) &= G^2(\alpha, \beta) - G^2(\alpha, \beta, \gamma) \\ &= 1363.552 - 1356.434 \\ &= 7.118 \end{aligned}$$

with $6-4 = 2$ degrees of freedom, for which $p = .028$. The interaction discerned in Figure 15.13 is, therefore, statistically significant, but not overwhelmingly so.

[48] In Chapter 8, we used SS(·) to denote the *regression* sum of squares for a model including certain terms. Because the deviance is analogous to the *residual* sum of squares, we need to take differences of deviances in the opposite order.

TABLE 15.4 Deviances for Models Fit to the *American Voter* Data. Terms: α Perceived Closeness, β Intensity of Preference, γ Closeness $\times$ Preference Interaction. The column labeled $k+1$ gives the number of parameters in the model, including the constant μ.

Model	*Terms*	$k+1$	*Deviance*: G^2
1	α, β, γ	6	1356.434
2	α, β	4	1363.552
3	α, γ	4	1368.042
4	β, γ	5	1368.554
5	α	2	1382.658
6	β	3	1371.838

15.3.1 The Binomial Logit Model*

Although the models for dichotomous and polytomous dependent variables described in this chapter can be directly applied to tabular data, there is some advantage in reformulating these models to take direct account of the replication of combinations of independent-variable values. In analyzing dichotomous data, for example, we previously treated each observation individually, so the dummy dependent variable Y_i takes on either the value 0 or the value 1.

Suppose, instead, that we group all of the n_i observations that share the specific combination of independent-variable values $\mathbf{x}_i' = (x_{i1}, x_{i2}, \ldots, x_{ik})$. Let Y_i count the number of these observations that fall in the first of the two categories of the dependent variable; we arbitrarily term these observations "successes." The count Y_i can take on any integer value between 0 and n_i. Let m denote the number of *distinct combinations* of the independent variables (e.g., $m = 6$ in Table 15.3).

As in our previous development of the dichotomous logit model, let π_i represent $\Pr(\text{success}|\mathbf{x}_i)$. Then the success count Y_i follows the binomial

TABLE 15.5 Analysis-of-Deviance Table for the *American Voter* Data, Showing Alternative Likelihood-Ratio Tests for the Main Effects of Perceived Closeness of the Election and Intensity of Partisan Preference.

Source	*Models Contrasted*	*df*	G_0^2	*p*
Perceived closeness		1		
$\alpha\|\beta$	6 − 2		8.286	.0040
$\alpha\|\beta, \gamma$	4 − 1		12.120	.0005
Intensity of preference		2		
$\beta\|\alpha$	5 − 2		19.106	<.0001
$\beta\|\alpha, \gamma$	3 − 1		11.608	.0030
Closeness × Preference		2		
$\gamma\|\alpha, \beta$	2 − 1		7.118	.028

distribution:

$$p(y_i) = \binom{n_i}{y_i} \pi_i^{y_i}(1 - \pi_i)^{n_i - y_i} \qquad [15.33]$$
$$= \binom{n_i}{y_i} \left(\frac{\pi_i}{1 - \pi_i}\right)^{y_i} (1 - \pi_i)^{n_i}$$

To distinguish grouped dichotomous data from ungrouped data, I shall refer to the former as *binomial data* and the latter as *binary data*.[49] Polytomous data can be handled in a similar manner, employing the multinomial distribution.[50]

Suppose, next, that the dependence of the response probabilities π_i on the independent variables is well described by the logit model

$$\log_e \frac{\pi_i}{1 - \pi_i} = \mathbf{x}_i'\boldsymbol{\beta}$$

Substituting this model into Equation 15.33, the likelihood for the parameters is

$$L(\boldsymbol{\beta}) = \prod_{i=1}^{m} \binom{n_i}{y_i} [\exp(\mathbf{x}_i'\boldsymbol{\beta})]^{y_i} \left(\frac{1}{1 + \exp(\mathbf{x}_i'\boldsymbol{\beta})}\right)^{n_i}$$

Maximizing the likelihood leads to precisely the same maximum-likelihood estimators, coefficient standard errors, and statistical tests as the binary logit model of Section 15.1.5.[51] The binomial logit model nevertheless has the following advantages:

- Because we deal with m binomial observations rather than the larger $n = \sum_{i=1}^{m} n_i$ binary observations, computations for the binomial logit model are more efficient, particularly when the n_i are large.
- The overall deviance for the binomial logit model, $-2 \log_e L(\mathbf{b})$, implicitly contrasts the model with a *saturated* model that has one parameter for each of the m combinations of independent-variable values (e.g., the full two-way "ANOVA" model with main effects and interactions fit in the previous section to the *American Voter* data). The saturated model necessarily recovers the m empirical logits perfectly, and, consequently, has a likelihood of 1 and a log likelihood of 0. The deviance for a less-than-saturated model, therefore, provides a likelihood-ratio test, on $m - k - 1$ degrees of freedom, of the hypothesis that the functional form of the model is correct.[52] In contrast, the deviance for the binary logit model cannot be used for a statistical test, because the degrees of freedom $n - k - 1$ (unlike $m - k - 1$) grow as the sample size n grows.
- As long as the frequencies n_i are not very small, many diagnostics are much better behaved for the cells of the binomial logit model than for individual binary observations. For example, the individual components of the deviance for the

---

[49] Binary data can be thought of as a limiting case of binomial data, for which all $n_i = 1$.

[50] See Exercise 15.16.

[51] See Exercise 15.15.

[52] See, for example, Exercise 15.17. This test is analogous to the test for "lack of fit" in a linear model with a discrete independent variable described in Section 12.4.1.

binomial logit model,

$$G_i \equiv \pm\sqrt{-2\left[Y_i \log_e \frac{n_i P_i}{Y_i} + (n_i - Y_i)\log_e \frac{n_i(1-P_i)}{n_i - Y_i}\right]}$$

can be compared with the unit-normal distribution to locate outlying cells. Here $P_i = 1/[1 + \exp(-\mathbf{x}_i'\mathbf{b})]$ is the fitted probability of "success" for cell i, and, therefore, $\widehat{Y}_i = n_i P_i$ is the expected number of "successes" in this cell. The sign of G_i is selected to agree with that of the simple cell residual, $E_i = Y_i - \widehat{Y}_i$. The other logistic regression diagnostics described in Section 15.1.6 can also be adapted to binomial data.[53]

> Although the binary logit model can be applied to tables in which the dependent variable is dichotomous, it is also possible to use the equivalent binomial logit model; the binomial logit model is based on the frequency counts of "successes" and "failures" for each combination of independent-variable values. When it is applicable, the binomial logit model offers several advantages, including efficient computation, a test of the fit of the model based on its deviance, and better-behaved diagnostics.

EXERCISES

15.15 *Derive the maximum-likelihood estimating equations for the binomial logit model. Show that this model produces the same estimated coefficients as the dichotomous (binary) logit model of Section 15.1. (*Hint*: Compare the log likelihood for the binomial model with the log likelihood for the binary model; by separating individual observations sharing a common set of X-values, show that the former log likelihood is equal to the latter, except for a constant factor. This constant is irrelevant because it does not influence the maximum-likelihood estimator; moreover, the constant disappears in likelihood-ratio tests.)

15.16 *Use the multinomial distribution (see Appendix D, Section D.2.2) to specify a polytomous logit model for discrete independent variables (analogous to the binomial logit model), where combinations of independent-variable values are replicated. Derive the likelihood under the model, and the maximum-likelihood estimating equations.

[53] See, for example, Collett (1991, Chapter 5).

TABLE 15.6 Perceived Need by Income.

	Perceived Need	
Family Income	*No*	*Yes*
\$0	8	5
\$1–\$1,999	17	16
\$2,000–\$3,999	88	76
\$4,000–\$5,999	125	108
\$6,000–\$7,999	134	75
\$8,000–\$9,999	130	94
\$10,000–\$11,999	168	79
\$12,000–\$13,999	178	65
\$14,000–\$15,999	240	60
\$16,000–\$17,999	141	45
\$18,000–\$19,999	160	45
\$20,000–\$24,999	299	71
\$25,000–\$29,999	199	29
\$30,000–\$39,999	162	23
\$40,000–\$49,999	61	7
\$50,000–\$74,999	36	3
\$75,000–\$99,999	4	0
\$100,000 or more	7	1

Source of Data: These data were collected as part of the Social Change in Canada Project, directed by T. Atkinson, M. Ornstein, and H. Stevenson of York University. The research was supported by SSHRCC Grant S75-0332. The data were made available by the Institute for Social Research of York University. Neither the principal investigators nor the disseminating archive are responsible for the interpretations presented here.

15.17 The data shown in Table 15.6 are drawn from the same 1977 survey that produced the Canadian women's labor-force data employed in this chapter. Respondents to the survey are classified by family income (represented by 18 categories) and by their responses to the question, "During the past year, have there been any major things you or your family *really* needed to buy but have not been able to afford?" I shall refer to this second variable as "perceived need."

(a) Calculate the empirical logits, $\log_e(\text{Yes/No})$, for perceived need within income categories, and plot these logits against family income. Because there is a zero frequency, you should add 0.5 to each frequency prior to calculating the logits. Does the relationship between perceived need and income appear to be roughly linear on the logit scale? How would you characterize the relationship between perceived need and income? In drawing the graph, use the category midpoints, expressed in thousands of dollars, as income scores, employing an arbitrary figure (say, 125) for the final, open, category.

(b) Using the binomial logit model, fit two models to the data: (i) assuming a linear relationship between the logit of perceived need and income; and (ii) coding dummy regressors for the income categories. Show the fitted line from (i) on the graph drawn in part (a). Then perform a likelihood-ratio test for nonlinearity on the logit scale.

15.18 Harris, et al. (1978) conducted a social-psychological experiment that examined reactions to invasions of personal space. The research took place

TABLE 15.7 Reactions to Invasions of Personal Space.

(1) Density	*(2) Sex of Subject*	*(3) Sex of Intruder*	*(4) Response*	
			Yes	*No*
Low	Male	Male	18	1
		Female	15	8
	Female	Male	17	5
		Female	12	7
High	Male	Male	13	6
		Female	16	4
	Female	Male	10	9
		Female	14	6

Source of Data: Harris, et al. (1978, Table 1).

in a field setting provided by a public escalator. The primary results of the study were presented in the form of a contingency table (Table 15.7). Three of the variables in the table were design or independent variables: (1) density of people on the escalator, rated as either high or low; (2) the sex of the subject; and (3) the sex of the intruder. The fourth variable was a dichotomous response or dependent variable: whether or not the subject reacted in some manner to the intrusion.

The authors analyzed the data by separately examining the two-way (partial) tables relating density to response within combinations of categories of the other two variables. Because there is a statistically significant relationship between density and response in only one of the four partial tables, the authors concluded that " 'males in the present study were more likely to react to a personal space invasion under low-density conditions than high-density conditions, but only when the intruder was another male. Density had no effect on responses by female subjects' " (Harris, et al., 1978, pp. 352–353). The implication is that there is a three-way interaction among the independent variables in determining response.

(a) Calculate the dependent-variable odds within combinations of independent-variable categories. Compute and graph the log odds (as in Figure 15.13), commenting on the results.

(b) Construct an analysis-of-deviance table, testing the various interactions and main effects of the independent variables on response. Do these tests square with the descriptive findings in part (a)?

(c) On the basis of the tests in part (b), fit a final logit model that incorporates only those effects shown to be important (and, of course, effects marginal to them). Using the parameter estimates for the model, calculate and graph the fitted logits.

(d) Test for independence between density and response separately in each of the four partial tables. (You may either fit a logit model to each table or perform a traditional Pearson chi-square test of independence.) Do you obtain the results reported by Harris, et al.?

(e) Do the results of your logit analysis support the authors' conclusions [replicated in part (d)]? Which analysis to you prefer? Why? (Cf. Fox, 1979.)

TABLE 15.8 Electoral Participation in the American South.

(1) Race	*(2) County Percent Nonwhite*	*(3) State Suffrage Law*	*(4) Frequency of Voting* *All or Most Elections*	*Some Elections*	*Never*
White	< 30	Restrictive	108	58	64
		Moderate	190	74	108
	> 30	Restrictive	46	18	21
		Moderate	41	10	16
Black	< 30	Restrictive	3	5	5
		Moderate	15	8	26
	> 30	Restrictive	0	6	76
		Moderate	3	10	27

Source of Data: Campbell et al., (1960, Table 11-5).

15.19 The data in Table 15.8, drawn from *The American Voter*, are for a sample of respondents residing in the southern United States. All of the southern states had more or less restrictive election laws, aimed primarily at limiting the political participation of African Americans. The table relates respondents' frequency of voting to race, the racial composition of the respondent's county of residence, and the restrictiveness of the state suffrage law.

Analyze the data in this table using:

(a) the polytomous (multinomial) logit model;

(b) dichotomous (binomial) logit models for the nested dichotomies {Never, (Some, All or Most)} and {Some, All or Most};

(c) the proportional-odds model.

15.4 Generalized Linear Models*

Generalized linear models, due to Nelder and Wedderburn (1972), represent a grand synthesis that includes, among others, most of the statistical models discussed in this book. In addition to substantially expanding the range of application of traditional linear models, generalized linear models provide a unified computational basis for fitting a very broad class of statistical models to data. An extensive discussion of generalized linear models is beyond the scope of this book, but I would like, nevertheless, to outline the topic.

The generalized linear model consists of three components:

1. A *random component*, in the form of a dependent variable Y_i, which, conditional on the independent variables, follows one of the distributions in the exponential family: the normal, Poisson, binomial, gamma, or inverse-Gaussian distributions.
2. A *linear predictor*

$$\eta_i = \alpha + \beta_1 X_{i1} + \cdots + \beta_k X_{ik}$$

on which the dependent variable depends.

3. A *link function* $L(\cdot)$ that transforms the expectation of the dependent variable $\mu_i \equiv E(Y_i)$ to the linear predictor η_i. Standard link functions include:
   - the identity link: $L(\mu_i) = \mu_i$;
   - the log link: $L(\mu_i) = \log_e \mu_i$;
   - the inverse link: $L(\mu_i) = 1/\mu_i$;
   - the square-root link: $L(\mu_i) = \sqrt{\mu_i}$;
   - the logit link: $L(\pi_i) = \log_e [\pi_i/(1 - \pi_i)]$;
   - the probit link: $L(\pi_i) = \Phi(\pi_i)$, where $\Phi(\cdot)$ is the CDF of the unit-normal distribution; and
   - the complementary log-log link: $L(\pi_i) = \log_e [-\log_e(1 - \pi_i)]$.

Generalized linear models represent a grand synthesis that includes, but is not restricted to, linear models with normally distributed errors, and logit and probit models for dichotomous data. Generalized linear models consist of three components: (1) a random component, representing the conditional distribution of the dependent variable, selected from the family of exponential distributions; (2) a linear predictor—a linear function of regressors on which the dependent variable depends; and (3) a linearizing link function that transforms the expectation of the dependent variable to the linear predictor.

The last three link functions—the logit, probit, and complementary log-log—are intended for binomial data, where μ_i represents the expected number of "successes" in n_i binomial trials, and where, therefore, $\pi_i = \mu_i/n_i$. The logit and probit links are familiar from the earlier sections of this chapter. The complementary log-log link, graphed in Figure 15.14, approaches $\pi = 0$ more gradually than $\pi = 1$, and is more appropriate than the logit or probit link when the data behave asymmetrically in this manner.[54] Combining the identity link with the normal distribution produces the general linear model.

A particularly important generalized linear model that we have not previously encountered is the *Poisson regression model*, which pairs the Poisson distribution with the log link. This model is often appropriate when the dependent variable is a count.[55]

The Poisson regression model can also be applied to a contingency table of counts, treating each of the classifications generating the table as a "factor," and the cell counts as the "dependent variable." For example, for two classifications, the model takes the ANOVA-like form:

$$\log_e \mu_{jk} = \mu + \alpha_j + \beta_k + \gamma_{jk} \qquad [15.34]$$

[54] If, alternatively, the data approach $\pi = 1$ more gradually than $\pi = 0$, then the definitions of "success" and "failure" can simply be interchanged.

[55] Exercise 15.20, for example, applies the Poisson regression model to Ornstein's interlocking-directorate data (discussed in Section 12.2).

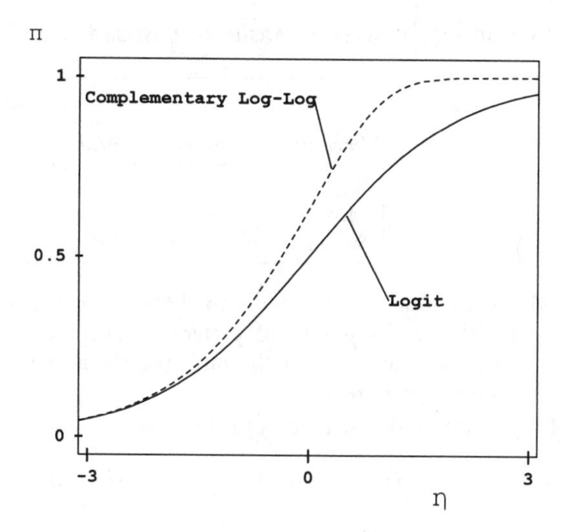

Figure 15.14. The asymmetric complementary log-log link function compared with the symmetric logit link. The complementary log-log link approaches 0 more gradually than 1.

where the parameters α_j, β_k, and γ_{jk} follow the usual sigma constraints, and where the $\log_e \mu_{jk}$ represent log expected cell counts. Equation 15.34 is called the *log-linear model* for the two-way contingency table.[56]

Maximum-likelihood estimates for generalized linear models can be obtained by iterative weighted least squares, in much the same manner as for logit models.[57] Indeed, most of the results of this chapter, including the material presented here on diagnostics, can be extended directly to generalized linear models.

EXERCISES

15.20 Apply the Poisson regression model to Ornstein's interlocking-directorate data (discussed in Section 12.2 and located in `ornstein.dat`), regressing number of interlocks on the square root of assets, and dummy variables for sector and nation of control.

15.21 The following table collapses the *American Voter* data given in Table 15.3 over perceived closeness of the election, showing the marginal

[56] See Exercise 15.21. Sociologists often employ log-linear models to analyze tabular data. When there is a dependent variable in the table, however, these models are equivalent to the logit model for tabular data discussed in the previous section—and the logit model, because it explicitly takes account of the dependent variable, can be interpreted more simply. In tables with no dependent variable, and in tables that have special structure (such as mobility tables), log-linear models are of more direct interest. See, for example, Fienberg (1980).

[57] See, for example, McCullagh and Nelder (1989, Chapter 2).

relationship between intensity of partisan preference and turnout:

	Voter Turnout	
Intensity of Preference	*Voted*	*Did Not Vote*
Weak	305	126
Medium	405	125
Strong	265	49

(a) Fit a binomial logit model to the table, treating turnout as the dependent variable and intensity of preference as a three-category factor. Perform a likelihood-ratio test of the null hypothesis that turnout is unrelated to intensity of preference.

(b) Fit two log-linear models to the table:

$$\log_e \mu_{jk} = \mu + \alpha_j + \beta_k + \gamma_{jk} \qquad [15.35]$$

and

$$\log_e \mu_{jk} = \mu + \alpha_j + \beta_k \qquad [15.36]$$

Because the γ_{jk} represent associations between the two variables in the contingency table, the model given in Equation 15.36 is a null model specifying that the two variables are independent. Show that the estimated expected frequencies under this model (i.e., the fitted values) imply the independence of the two variables. (*Hint*: Show that the expected cell frequencies $\widehat{\mu}_{jk}$ are each the product of the corresponding row and column sums, divided by the total frequency.) Equation 15.35, in contrast, is a "saturated" model, capable of capturing any pattern of cell frequencies—there are as many independent parameters in the model as cells in the table. Show that the estimated expected frequencies under Equation 15.35 are identical to the observed frequencies in the table. Perform a likelihood-ratio test for the hypothesis of independence by comparing Equation 15.35 with Equation 15.36. Does this test produce the same result as that in part (a)? (Note: The likelihood-ratio chi-square test for independence here is similar to, but not identical to, the traditional Pearson chi-square test for independence in a two-way table.)

15.5 Summary

- It is problematic to apply least-squares linear regression to a dichotomous dependent variable: The errors cannot be normally distributed and cannot have constant variance. Even more fundamentally, the linear specification does not confine the probability for the response to the unit interval.
- More adequate specifications transform the linear predictor $\eta_i = \alpha + \beta_1 X_{i1} + \cdots + \beta_k X_{ik}$ smoothly to the unit interval, using a cumulative probability distribution function $P(\cdot)$. Two such specifications are the probit and the logit models, which use the normal and logistic CDFs, respectively. Although these models are very similar, the logit model is simpler to interpret, because it can be written as a linear model for the log odds:

$$\log_e \frac{\pi_i}{1 - \pi_i} = \alpha + \beta_1 X_{i1} + \cdots + \beta_k X_{ik}$$

- The dichotomous logit model can be fit to data by the method of maximum likelihood. Wald tests and likelihood-ratio tests for the coefficients of the model parallel t-tests and F-tests for the general linear model. The deviance for the model, defined as $G^2 = -2 \times$ the maximized log likelihood, is analogous to the residual sum of squares for a linear model.
- Diagnostics for linear models—including studentized residuals, hat values, influence diagnostics, partial-residual plots, partial-regression plots, and constructed-variable plots—can be extended to the logit model.
- Several approaches can be taken to modeling polytomous data, including:
  1. modeling the polytomy directly using a logit model based on the multivariate logistic distribution;
  2. constructing a set of $m-1$ nested dichotomies to represent the m categories of the polytomy; and
  3. fitting the proportional-odds model to a polytomous dependent variable with ordered categories.
- When all of the variables—independent as well as dependent— are discrete, their joint distribution defines a contingency table of frequency counts. It is natural to employ logit models that are analogous to analysis-of-variance models to analyze contingency tables. Although the binary logit model can be applied to tables in which the dependent variable is dichotomous, it is also possible to use the equivalent binomial logit model; the binomial logit model is based on the frequency counts of "successes" and "failures" for each combination of independent-variable values. When it is applicable, the binomial logit model offers several advantages, including efficient computation, a test of the fit of the model based on its deviance, and better-behaved diagnostics.
- Generalized linear models represent a grand synthesis that includes, but is not restricted to, linear models with normally distributed errors, and logit and probit models for dichotomous data. Generalized linear models consist of three components:

  1. a random component, representing the conditional distribution of the dependent variable, selected from the family of exponential distributions;
  2. a linear predictor—a linear function of regressors on which the dependent variable depends; and
  3. a linearizing link function that transforms the expectation of the dependent variable to the linear predictor.

  The Poisson regression model, which is often applicable when the dependent variable represents a count, combines the Poisson distribution with the log link.

15.6 Recommended Reading

The topics introduced in this chapter could easily be expanded to fill several books, and there is a large literature—both in journals and texts—dealing with logit and related models for qualitative data, with generalized linear models, and with the analysis of contingency tables.

- Agresti (1990) presents an excellent and comprehensive overview of statistical methods for qualitative data. The emphasis is on logit and log-linear models for contingency tables, but there is also some treatment of logistic regression models and other topics.
- Fienberg's (1980) widely read text on the analysis of contingency tables provides an accessible and lucid introduction to log-linear models and related subjects, such as logit models and models for ordered categories.
- The second edition of Cox and Snell's (1989) classic text concentrates on logit models for dichotomous data but also includes some discussion of polytomous nominal and ordinal data.
- Collett (1991) also focuses on the binary and binomial logit models. The book is noteworthy for its extensive review of diagnostic methods for logit models. In texts that deal primarily with diagnostics for linear models fit by least squares, Cook and Weisberg (1982) and Atkinson (1985) also present some material that extends these methods to logistic regression and other generalized linear models.
- McCullagh and Nelder (1989), the "bible" of generalized linear models, is a rich and interesting—if generally difficult—text. Dobson (1983) presents a much briefer overview of generalized linear models at a more moderate level of statistical sophistication. Aitkin et al.'s (1989) text, geared to the statistical computer package GLIM for fitting generalized linear models, is still more accessible.

16

Assessing Sampling Variation: Bootstrapping and Cross-Validation

This chapter introduces two modern techniques for assessing sampling variation: *Bootstrapping*, discussed in the first section of the chapter, allows us to compute estimated standard errors, confidence intervals, and hypothesis tests without making strong assumptions about the distribution of the data, and without having to derive asymptotic results. The method has a simple rationale, is easy to implement, and is broadly applicable, but it requires very intensive computation.

Cross-validation, which is described in the second section of the chapter, provides a simple basis for honest statistical inference when—as is typically the case in a careful investigation—we need to examine the data in order to formulate a descriptively adequate statistical model. In cross-validation, the data are divided at random into two parts: One part is used for data exploration and model formulation; the second part is used to validate the model, thus preserving the integrity of statistical inferences.

16.1 Bootstrapping

Bootstrapping is a nonparametric approach to statistical inference that substitutes computation for more traditional distributional assumptions and asymptotic results.[1] Bootstrapping offers a number of advantages:

- The bootstrap is quite general, although there are some cases in which it fails.

[1] The term "bootstrapping," coined by Efron, refers to using the sample to learn about the sampling distribution of a statistic without reference to external assumptions—as in "pulling oneself up by one's bootstraps."

- Because it does not require distributional assumptions (such as normally distributed errors), the bootstrap can provide more accurate inferences when the data are not well behaved or when the sample size is small.
- It is possible to apply the bootstrap to statistics with sampling distributions that are difficult to derive, even asymptotically.
- It is relatively simple to apply the bootstrap to complex data-collection plans (such as stratified and clustered samples).

> Bootstrapping is a broadly applicable, nonparametric approach to statistical inference that substitutes intensive computation for more traditional distributional assumptions and asymptotic results. The bootstrap can be used to derive accurate standard errors, confidence intervals, and hypothesis tests for most statistics.

16.1.1 Bootstrapping Basics

My principal aim is to explain how to bootstrap regression analysis, linear models, and so on, but the topic is best introduced in a simpler context: Suppose that we draw an independent random sample from a large population.[2] For concreteness and simplicity, imagine that we sample four working married couples, determining in each case the husband's and wife's income, as recorded in Table 16.1. I shall focus on the difference in incomes between husbands and wives, denoted Y_i for the ith couple.

We want to estimate the mean difference in income between husbands and wives in the population. Please bear with me as I review some basic statistical theory: A point estimate of this population mean difference μ is the sample mean, $\overline{Y} = \sum Y_i/n = (6-3+5+3)/4 = 2.75$. Elementary statistical theory tells us that the standard deviation of the sampling distribution of sample means (i.e., the standard error of the mean) is $\text{SE}(\overline{Y}) = \sigma/\sqrt{n}$, where σ is the population standard deviation of Y.

If we knew σ, and if Y were normally distributed, then a 95% confidence interval for μ would be

$$\mu = \overline{Y} \pm 1.96 \frac{\sigma}{\sqrt{n}}$$

[2] In an *independent random sample*, each element of the population can be selected more than once. In a *simple random sample*, in contrast, once an element is selected into the sample, it is removed from the population, so that sampling is done "without replacement." When the population is very large in comparison with the sample, the distinction between independent and simple random sampling becomes inconsequential.

TABLE 16.1 Contrived "Sample" of Four Married Couples, Showing Husbands' and Wives' Incomes in Thousands of Dollars.

Observation	*Husband's Income*	*Wife's Income*	*Difference* Y_i
1	24	18	6
2	14	17	−3
3	40	35	5
4	44	41	3

where $z_{.025} = 1.96$ is the standard normal value with a probability of .025 to the right. If Y is not normally distributed in the population, then this result applies asymptotically. Of course, the asymptotics are cold comfort when $n = 4$.

In a real application, we do not know σ. The standard unbiased estimator of σ is $S = \sqrt{\sum(Y_i - \overline{Y})^2/(n-1)}$, from which an estimate of the standard error of the mean is $\widehat{\mathrm{SE}}(\overline{Y}) = S/\sqrt{n}$. If the population is normally distributed, then we can take account of the added uncertainty associated with estimating the standard error by substituting the heavier-tailed t-distribution for the normal distribution, producing the 95% confidence interval

$$\mu = \overline{Y} \pm t_{n-1,.025} \frac{S}{\sqrt{n}}$$

Here, $t_{n-1,.025}$ is the critical value of t with $n - 1$ degrees of freedom and a right-tail probability of .025.

In the present case, $S = 4.031$, $\widehat{\mathrm{SE}}(\overline{Y}) = 4.031/\sqrt{4} = 2.015$, and $t_{3,.025} = 4.30$. The 95% confidence interval for the population mean is thus

$$\mu = 2.75 \pm 4.30 \times 2.015 = 2.75 \pm 8.66$$

or, equivalently,

$$-5.91 < \mu < 11.41$$

As one would expect, this confidence interval—which is based on only four observations—is very wide and includes 0. It is, unfortunately, hard to be sure that the population is reasonably close to normally distributed when we have such a small sample, and so the t-interval may not be accurate.[3]

Bootstrapping begins by using the distribution of data values in the sample (here, $Y_1 = 6$, $Y_2 = -3$, $Y_3 = 5$, $Y_4 = 3$) to estimate the distribution of Y in the

[3] To say that a confidence interval is "accurate" means that it has the stated coverage. That is, a 95% confidence interval is accurate if it is constructed according to a procedure that encloses the population mean in 95% of samples.

population.[4] That is, we define the random variable Y^* with distribution[5]

y^*	$p^*(y^*)$
6	.25
−3	.25
5	.25
3	.25

Notice that

$$E^*(Y^*) = \sum_{\text{all } y^*} y^* p(y^*) = 2.75 = \overline{Y}$$

and

$$V^*(Y^*) = \sum [y^* - E^*(Y^*)]^2 p(y^*)$$
$$= 12.187 = \frac{3}{4}S^2 = \frac{n-1}{n}S^2$$

Thus, the expectation of Y^* is just the sample mean of Y, and the variance of Y^* is [except for the factor $(n - 1)/n$, which is trivial in larger samples] the sample variance of Y.

We next mimic sampling from the original population by treating the sample as if it were the population, enumerating all possible samples of size $n = 4$ from the probability distribution of Y^*. In the present case, each bootstrap sample selects four values *with replacement* from among the four values of the original sample. There are, therefore, $4^4 = 256$ different bootstrap samples,[6] each selected with probability 1/256. A few of the 256 samples are shown in Table 16.2. Because the four observations in each bootstrap sample are chosen with replacement, particular bootstrap samples usually have repeated observations from the original sample: Indeed, of the illustrative bootstrap samples shown in Table 16.2, only sample 100 does *not* have repeated observations.

Let us denote the bth bootstrap sample[7] as $\mathbf{y}_b^* = [Y_{b1}^*, Y_{b2}^*, Y_{b3}^*, Y_{b4}^*]'$, or, more generally, $\mathbf{y}_b^* = [Y_{b1}^*, Y_{b2}^*, \ldots, Y_{bn}^*]'$, where $b = 1, 2, \ldots, n^n$. For each

[4] An alternative would be to resample from a distribution given by a nonparametric density estimate (see, e.g., Silverman and Young, 1987). Typically, however, little, if anything, is gained by using a more complex estimate of the population distribution. Moreover, the simpler method explained here generalizes more readily to more complex situations in which the population is multivariate or not simply characterized by a distribution.

[5] The asterisks on $p^*(\cdot)$, E^*, and V^* remind us that this probability distribution, expectation, and variance are conditional upon the specific sample in hand. If we were to select another sample, the values of Y_1, Y_2, Y_3, and Y_4, would change, and—along with them—the probability distribution of Y^*, its expectation, and variance.

[6] Many of the 256 samples have the same elements, but in different order—for example, [6, 3, 5, 3] and [3, 5, 6, 3]. We could enumerate the unique samples without respect to order and find the probability of each, but it is simpler to work with the 256 orderings, because each ordering has equal probability.

[7] If vector notation is unfamiliar, then think of $\mathbf{y}_b^*$ simply as a list of the bootstrap observations Y_{bi}^* for sample b.

TABLE 16.2 A Few of the 256 Bootstrap Samples for the Dataset [6, −3, 5, 3], and the Corresponding Bootstrap Means, $\overline{Y}_b^*$

Bootstrap Sample b	Y_{b1}^*	Y_{b2}^*	Y_{b3}^*	Y_{b4}^*	$\overline{Y}_b^*$
1	6	6	6	6	6.00
2	6	6	6	−3	3.75
3	6	6	6	5	5.75
⋮	⋮				⋮
100	−3	5	6	3	2.75
101	−3	5	−3	6	1.25
⋮	⋮				⋮
255	−3	3	3	5	3.50
256	3	3	3	3	3.00

such bootstrap sample, we calculate the mean, $\overline{Y}_b^* = (\sum_{i=1}^n Y_{bi}^*)/n$. The sampling distribution of the 256 bootstrap means is shown in Figure 16.1.

The mean of the 256 bootstrap sample means is just the original sample mean, $\overline{Y} = 2.75$. The standard error (standard deviation) of the bootstrap means is

$$\begin{aligned} \text{SE}^*(\overline{Y}^*) &= \sqrt{\frac{\sum_{b=1}^{n^n}(\overline{Y}_b^* - \overline{Y})^2}{n^n}} \\ &= 1.745 \end{aligned}$$

We divide here by n^n rather than by $n^n - 1$ because the distribution of the 256 bootstrap sample means (Figure 16.1) is known, not estimated. The standard error of the bootstrap means is nearly equal to the usual estimate of the standard error of the sample mean; the slight slippage is due to the factor $\sqrt{n/(n-1)}$, which is usually negligible (though not when $n = 4$):[8]

$$\begin{aligned} \widehat{\text{SE}}(\overline{Y}) &= \sqrt{\frac{n}{n-1}}\text{SE}^*(\overline{Y}^*) \\ 2.015 &= \sqrt{\frac{4}{3}} \times 1.745 \end{aligned}$$

This precise relationship between the usual formula for standard error and the bootstrap standard error is peculiar to linear statistics like the mean. For the mean, then, the bootstrap standard error is just a more complicated way to calculate what we already know, but

- bootstrapping might still provide more accurate confidence intervals, as I shall explain presently; and

[8] See Exercise 16.1.

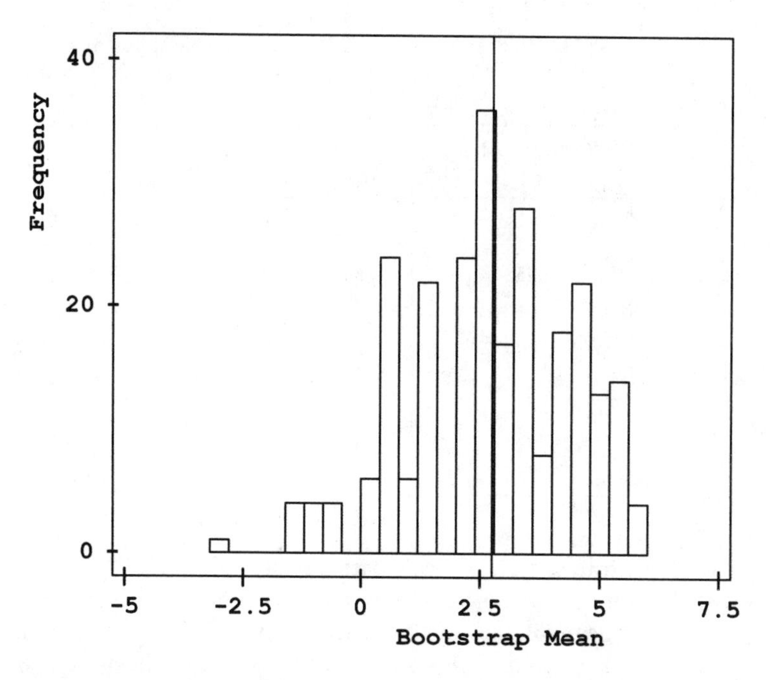

Figure 16.1. Histogram of the 256 bootstrap means from the sample [6, −3, 5, 3]. The vertical line gives the mean of the original sample, $\overline{Y} = 2.75$, which is also the mean of the 256 bootstrap means.

- bootstrapping can be applied to nonlinear statistics for which we do not have standard-error formulas, or for which only asymptotic standard errors are available.

Bootstrapping exploits the following central analogy:

The population is to the sample as the sample is to the bootstrap samples.

Consequently,

- the bootstrap observations Y_{bi}^* are analogous to the original observations Y_i;
- the bootstrap mean $\overline{Y}_b^*$ is analogous to the mean of the original sample $\overline{Y}$;
- the mean of the original sample $\overline{Y}$ is analogous to the (unknown) population mean μ; and
- the distribution of the bootstrap sample means is analogous to the (unknown) sampling distribution of means for samples of size n drawn from the original population.

Bootstrapping uses the sample data to estimate relevant characteristics of the population. The sampling distribution of a statistic is then constructed empirically by resampling from the sample. The resampling procedure is designed to parallel the process by which sample observations were drawn from the population. For example, if the data represent an independent random sample of size n, then each bootstrap sample selects n observations with replacement from the original sample. The key bootstrap analogy is: *The population is to the sample as the sample is to the bootstrap samples.*

The bootstrapping calculations that we have undertaken thus far depend on very small sample size, because the number of bootstrap samples (n^n) quickly becomes unmanageable: Even for samples as small as $n = 10$ it is impractical to enumerate all of the bootstrap samples. Consider the "data" shown in Table 16.3, an extension of the previous example. The mean and standard deviation of the differences in income Y are $\overline{Y} = 4.6$ and $S = 5.948$. Thus, the estimated standard error of the sample mean is $\widehat{\text{SE}}(\overline{Y}) = 5.948/\sqrt{10} = 1.881$.

Although we cannot (as a practical matter) enumerate all of the 10^{10} bootstrap samples, it is easy to draw at random a large number of bootstrap samples. To estimate the standard error of a statistic (here, the mean), 100 or 200 bootstrap samples should be more than sufficient. To find a confidence interval, we shall need a larger number of bootstrap samples, say 1000 or 2000.

A practical bootstrapping procedure, therefore, is as follows:

1. Let r denote the number of *bootstrap replications*—that is, the number of bootstrap samples to be selected.

TABLE 16.3 Contrived "Sample" of 10 Married Couples, Showing Husbands' and Wives' Incomes in Thousands of Dollars

Observation	*Husband's Income*	*Wife's Income*	*Difference* Y_i
1	24	18	6
2	14	17	−3
3	40	35	5
4	44	41	3
5	24	18	6
6	19	9	10
7	21	10	11
8	22	30	−8
9	30	23	7
10	24	15	9

TABLE 16.4 A Few of the $r = 2000$ Bootstrap Samples Drawn From the Dataset [6, −3, 5, 3, 6, 10, 11, −8, 7, 9], and the Corresponding Bootstrap Means, $\overline{Y}_b^*$

b	Y_{b1}^*	Y_{b2}^*	Y_{b3}^*	Y_{b4}^*	Y_{b5}^*	Y_{b6}^*	Y_{b7}^*	Y_{b8}^*	Y_{b9}^*	$Y_{b,10}^*$	$\overline{Y}_b^*$
1	6	10	6	5	−8	9	9	6	11	3	5.7
2	9	9	7	7	3	3	−3	−3	−8	6	3.0
3	9	−3	6	5	10	6	10	10	10	6	6.9
⋮	⋮										⋮
1999	6	9	6	3	11	6	6	7	3	9	6.6
2000	7	6	7	3	10	6	9	3	10	6	6.7

2. For each bootstrap sample $b = 1, \ldots, r$, randomly draw n observations Y_{b1}^*, $Y_{b2}^*, \ldots, Y_{bn}^*$ with replacement from among the n sample values, and calculate the bootstrap sample mean, $\overline{Y}_b^* = \sum Y_{bi}^*/n$.
3. From the r bootstrap samples, *estimate* the standard error of the bootstrap means[9]

$$\widehat{\mathrm{SE}}^*(\overline{Y}^*) = \sqrt{\frac{\sum_{b=1}^{r}(\overline{Y}_b^* - \overline{\overline{Y}}^*)^2}{r-1}}$$

where $\overline{\overline{Y}}^* = \sum_{b=1}^{r} \overline{Y}_b^*/r$ is the mean of the bootstrap means. We can, if we wish, "correct" $\widehat{\mathrm{SE}}^*(\overline{Y}^*)$ for degrees of freedom, multiplying by $\sqrt{n/(n-1)}$.

To illustrate this procedure, I drew $r = 2000$ bootstrap samples, each of size $n = 10$, from the "data" given in Table 16.3, calculating the mean $\overline{Y}_b^*$ for each sample. A few of the 2000 bootstrap replications are shown in Table 16.4, and the distribution of bootstrap means is graphed in Figure 16.2.

We know from statistical theory that were we to enumerate all of the 10^{10} bootstrap samples (or, alternatively, to sample infinitely from the population of bootstrap samples), the average bootstrap mean would be $E^*(\overline{Y}^*) = \overline{Y} = 4.6$, and the standard deviation of the bootstrap means would be $\mathrm{SE}^*(\overline{Y}^*) = \widehat{\mathrm{SE}}(\overline{Y})\sqrt{(n-1)/n} = 1.881\sqrt{9/10} = 1.784$. For the 2000 bootstrap samples that I selected, $\overline{\overline{Y}}^* = 4.693$ and $\widehat{\mathrm{SE}}(\overline{Y}^*) = 1.750$—both quite close to the theoretical values.

[9] It is important to distinguish between the "ideal" bootstrap estimate of standard error, $\mathrm{SE}^*(\overline{Y}^*)$, which is based on *all* n^n bootstrap samples, and the estimate of this quantity, $\widehat{\mathrm{SE}}^*(\overline{Y}^*)$, which is based on r randomly selected bootstrap samples. By making r large enough, we seek to ensure that $\widehat{\mathrm{SE}}^*(\overline{Y}^*)$ is close to $\mathrm{SE}^*(\overline{Y}^*)$. Even $\mathrm{SE}^*(\overline{Y}^*) = \widehat{\mathrm{SE}}(\overline{Y})$ is an imperfect estimate [of the true standard error of the sample mean, $\mathrm{SE}(\overline{Y})$], however, because it is based on a particular sample of size n drawn from the original population.

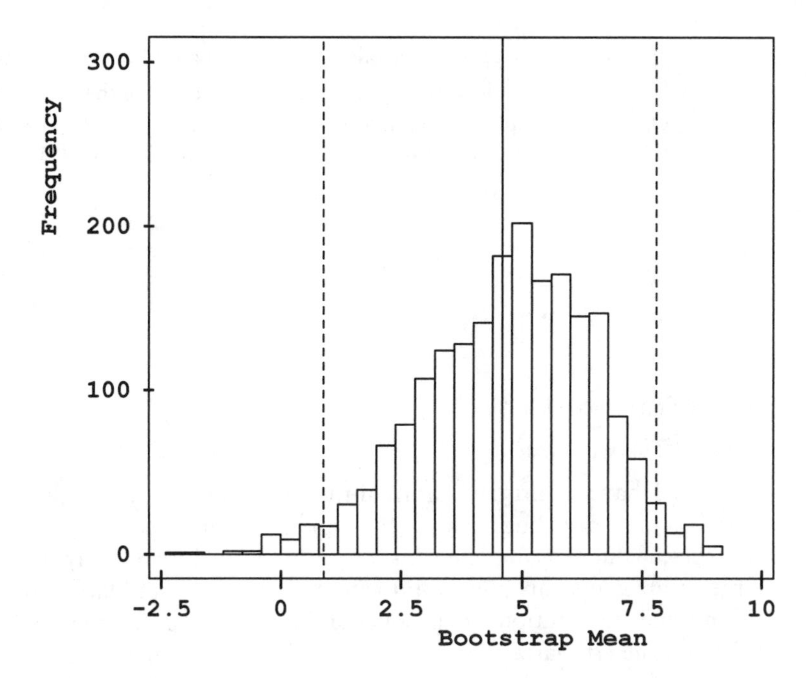

Figure 16.2. Histogram of $r = 2000$ bootstrap means, produced by resampling from the "sample" [6, −3, 5, 3, 6, 10, 11, −8, 7, 9]. The solid vertical line gives the sample mean, $\overline{Y} = 4.6$; the broken vertical lines give the boundaries of the 95% percentile confidence interval for the population mean μ based on the 2000 bootstrap samples. The procedure for constructing this confidence interval is described in the next section.

The bootstrapping procedure described in this section can be generalized to derive the empirical sampling distribution for an estimator $\hat{\theta}$ of the parameter θ:

1. Specify the data-collection scheme $\mathscr{S}$ that gives rise to the observed sample when applied to the population:[10]

$$\mathscr{S}(\text{Population}) \Longrightarrow \text{Sample}$$

   The estimator $\hat{\theta}$ is some function $s(\cdot)$ of the observed sample. In the preceding example, the data-collection procedure is independent random sampling from a large population.
2. Using the observed sample data as a "stand-in" for the population, replicate the data-collection procedure, producing r bootstrap samples:

$$\mathscr{S}(\text{Sample}) \begin{cases} \Longrightarrow & \text{Bootstrap sample}_1 \\ \Longrightarrow & \text{Bootstrap sample}_2 \\ & \vdots \\ \Longrightarrow & \text{Bootstrap sample}_r \end{cases}$$

[10] The "population" can be real—the population of working married couples—or hypothetical—the population of conceivable replications of an experiment. What is important in the present context is that the sampling procedure can be described concretely.

3. For each bootstrap sample, calculate the estimate $\hat{\theta}_b^* = s(\text{Bootstrap sample}_b)$.
4. Use the distribution of the $\hat{\theta}_b^*$'s to estimate properties of the sampling distribution of $\hat{\theta}$. For example, the bootstrap estimate of the standard error of $\hat{\theta}$ is $\widehat{\text{SE}}^*(\hat{\theta}^*)$ (i.e., the standard deviation of the r bootstrap replications $\hat{\theta}_b^*$):[11]

$$\widehat{\text{SE}}^*(\hat{\theta}^*) \equiv \sqrt{\frac{\sum_{b=1}^{r}(\hat{\theta}_b^* - \overline{\theta}^*)^2}{r-1}}$$

where $\overline{\theta}^* \equiv \sum_{b=1}^{r} \hat{\theta}_b^*/r$.

16.1.2 Bootstrap Confidence Intervals

Normal-Theory Intervals

As I have mentioned, normal-theory confidence intervals for means are based on the t-distribution when the population variance of Y is unknown. Most statistics, including sample means, are asymptotically normally distributed, so in large samples we can use the bootstrap estimate of standard error, along with the normal distribution, to produce a $100(1-a)\%$ confidence interval for θ based on the estimator $\hat{\theta}$:

$$\theta = \hat{\theta} \pm z_{a/2}\widehat{\text{SE}}^*(\hat{\theta}^*) \qquad [16.1]$$

In Equation 16.1, $z_{a/2}$ is the standard normal value with probability $a/2$ to the right.

This approach will work well if the bootstrap sampling distribution of the estimator is approximately normal, so it is advisable to examine a normal quantile comparison plot of the bootstrap distribution. There is no advantage to calculating normal-theory bootstrap confidence intervals for linear statistics like the mean, for the ideal bootstrap standard error of the statistic and the estimated standard error based directly on the sample coincide. Using bootstrap resampling in this setting just makes for extra work, and introduces an additional small random component into estimated standard errors.

> Having produced r bootstrap replicates $\hat{\theta}_b^*$ of an estimator $\hat{\theta}$, the bootstrap estimate of standard error is the standard deviation of the bootstrap replicates: $\widehat{\text{SE}}^*(\hat{\theta}^*) = \sqrt{\sum_{b=1}^{r}(\hat{\theta}_b^* - \overline{\theta}^*)^2/(r-1)}$, where $\overline{\theta}^*$ is the mean of the $\hat{\theta}_b^*$. In large samples, where we can rely on the normality of $\hat{\theta}$, a 95% confidence interval for θ is given by $\hat{\theta} \pm 1.96\widehat{\text{SE}}^*(\hat{\theta}^*)$.

[11] We may want to apply the correction factor $\sqrt{n/(n-1)}$.

Percentile Intervals

Another very simple approach is to use the quantiles of the bootstrap sampling distribution of the estimator to establish the end points of a confidence interval nonparametrically. Let $\hat{\theta}^*_{(b)}$ represent the ordered bootstrap estimates, and suppose that we want to construct a $(100 - a)\%$ confidence interval. If the number of bootstrap replications r is large (as it should be to construct a percentile interval), then the $a/2$ and $1 - a/2$ quantiles of $\hat{\theta}^*_b$ are approximately $\hat{\theta}^*_{(\text{lower})}$ and $\hat{\theta}^*_{(\text{upper})}$, where lower $= ra/2$ and upper $= r(1 - a/2)$. If lower and upper are not integers, then we can interpolate between adjacent ordered values $\hat{\theta}^*_{(b)}$, or round off to the nearest integer.

> A nonparametric confidence interval for θ can be constructed from the quantiles of the bootstrap sampling distribution of $\hat{\theta}^*$. The 95% percentile interval is $\hat{\theta}^*_{(\text{lower})} < \theta < \hat{\theta}^*_{(\text{upper})}$, where the $\hat{\theta}^*_{(b)}$ are the ordered bootstrap replicates, lower $= .025 \times r$, and upper $= .975 \times r$.

A 95% confidence interval for the $r = 2000$ resampled means in Figure 16.2, for example, is constructed as follows:

$$\begin{aligned}
\text{lower} &= 2000(.05/2) = 50 \\
\text{upper} &= 2000(1 - .05/2) = 1950 \\
\overline{Y}^*_{(50)} &= 0.9 \\
\overline{Y}^*_{(1950)} &= 7.8 \\
0.9 &< \mu < 7.8
\end{aligned}$$

The end points of this interval are marked in Figure 16.2. Because of the skew of the bootstrap distribution, the percentile interval is not quite symmetric around $\overline{Y} = 4.6$. For comparison, the standard t-interval for the mean of the original sample of 10 observations is

$$\begin{aligned}
\mu &= \overline{Y} \pm t_{9,\,.025}\widehat{\text{SE}}(\overline{Y}) \\
&= 4.6 \pm 2.262 \times 1.881 \\
&= 4.6 \pm 4.255 \\
0.345 &< \mu < 8.855
\end{aligned}$$

In this case, the standard interval is a bit wider than the percentile interval, especially at the top.

*Improved Bootstrap Intervals**

I shall briefly describe an adjustment to percentile intervals that improves their accuracy.[12] As before, we want to produce a $100(1 - a)\%$ confidence interval for θ having computed the sample estimate $\hat{\theta}$, and bootstrap replicates $\hat{\theta}_b^*$, $b = 1, \ldots, r$. We require $z_{a/2}$, the unit-normal value with probability $a/2$ to the right, and two "correction factors," Z and A, defined in the following manner:

- Calculate

$$Z \equiv \Phi^{-1}\left[\frac{\overset{r}{\underset{b=1}{\#}}(\hat{\theta}_b^* < \hat{\theta})}{r}\right]$$

  where $\Phi^{-1}(\cdot)$ is the inverse of the standard normal distribution function, and $\#(\hat{\theta}_b^* < \hat{\theta})/r$ is the proportion of bootstrap replicates below the estimate $\hat{\theta}$. If the bootstrap sampling distribution is symmetric, and if $\hat{\theta}$ is unbiased, then this proportion will be close to .5, and the "correction factor" Z will be close to 0.
- Let $\hat{\theta}_{(-i)}$ represent the value of $\hat{\theta}$ produced when the ith observation is deleted from the sample;[13] there are n of these quantities. Let $\overline{\theta}$ represent the average of the $\hat{\theta}_{(-i)}$; that is, $\overline{\theta} \equiv \sum_{i=1}^{n} \hat{\theta}_{(-i)}/n$. Then calculate

$$A \equiv \frac{\sum_{i=1}^{n}(\hat{\theta}_{(-i)} - \overline{\theta})^3}{6[\sum_{i=1}^{n}(\hat{\theta}_{(-i)} - \overline{\theta})^2]^{3/2}} \qquad [16.2]$$

With the correction factors Z and A in hand, compute

$$A_1 \equiv \Phi\left[Z + \frac{Z - z_{a/2}}{1 - A(Z - z_{a/2})}\right]$$

$$A_1 \equiv \Phi\left[Z + \frac{Z + z_{a/2}}{1 - A(Z + z_{a/2})}\right]$$

where $\Phi(\cdot)$ is the cumulative standard normal distribution function. Notice that when the correction factors Z and A are both 0, $A_1 = \Phi(-z_{a/2}) = a/2$, and $A_2 = \Phi(z_{a/2}) = 1 - a/2$. The values A_1 and A_2 are used to locate the end points of the corrected percentile confidence interval. In particular, the corrected interval is

$$\hat{\theta}^*_{(\text{lower}^*)} < \theta < \hat{\theta}^*_{(\text{upper}^*)}$$

where $\text{lower}^* = rA_1$ and $\text{upper}^* = rA_2$.

---

[12] The interval described here is called a "bias-corrected, accelerated" (or BC_a) percentile interval. Details can be found in Efron and Tibshirani (1993, Chapter 14); also see Stine (1990) for a discussion of different procedures for constructing bootstrap confidence intervals.

[13] The $\hat{\theta}_{(-i)}$ are called the *jackknife values* of the statistic $\hat{\theta}$. The jackknife values can also be used as an alternative to the bootstrap to find a nonparametric confidence interval for θ. See Exercise 16.2.

> The lower and upper bounds of percentile confidence intervals can be corrected to improve the accuracy of these intervals.

Applying this procedure to the "data" in Table 16.3, we have $z_{.05/2} = 1.96$ for a 95% confidence interval. There are 861 bootstrapped means below $\overline{Y} = 4.6$, and so $Z = \Phi^{-1}(861/2000) = -0.1747$. The $\overline{Y}_{(-i)}$ are 4.444, 5.444, ..., 4.111; the mean of these values[14] is $\overline{\overline{Y}} = \overline{Y} = 4.6$, and (from Equation 16.2), $A = -0.05630$. Using the correction factors,

$$A_1 = -0.1747 + \frac{-0.1747 - 1.96}{1 - [-.05630(-0.1747 - 1.96)]} = .0046$$

$$A_2 = -0.1747 + \frac{-0.1747 + 1.96}{1 - [-.05630(-0.1747 + 1.96)]} = .9261$$

Multiplying by r, we have $2000 \times .0046 \simeq 9$ and $2000 \times .9261 \simeq 1852$, from which

$$\overline{Y}^*_{(9)} < \mu < \overline{Y}^*_{(1852)} \qquad [16.3]$$
$$-0.2 < \mu < 7.0$$

Unlike the other confidence intervals that we have calculated for the "sample" of 10 differences in income between husbands and wives, the interval given in Equation 16.3 includes 0.

16.1.3 Bootstrapping Regression Models

The procedures of the previous section can be easily extended to regression, linear models, and so on. The most straightforward approach is to collect the dependent-variable value and regressors for each observation

$$\mathbf{z}'_i \equiv [Y_i, X_{i1}, \ldots, X_{ik}]$$

Then the observations $\mathbf{z}'_1, \mathbf{z}'_2, \ldots, \mathbf{z}'_n$ can be resampled, and the regression estimator computed for each of the resulting bootstrap samples, $\mathbf{z}^{*\prime}_{b1}, \mathbf{z}^{*\prime}_{b2}, \ldots, \mathbf{z}^{*\prime}_{bn}$, producing r sets of bootstrap regression coefficients, $\mathbf{b}^*_b = [A^*_b, B^*_{b1}, \ldots, B^*_{bk}]'$. The methods of the previous section can be applied to compute standard errors or confidence intervals for the regression estimates.

Directly resampling the observations $\mathbf{z}'_i$ implicitly treats the regressors X as random, rather than fixed. We may want to treat X as fixed (if, for example,

[14] The average of the jackknifed estimates is not, in general, the same as the estimate calculated for the full sample, but this *is* the case for the jackknifed sample means. See Exercise 16.2.

the data derive from an experimental design):

1. Estimate the regression coefficients $A, B_1, \ldots, B_k$ for the original sample, and calculate the fitted value and residual for each observation:
$$\hat{Y}_i = A + B_1 x_{i1} + \cdots + B_k x_{ik}$$
$$E_i = Y_i - \hat{Y}_i$$
2. Select bootstrap samples of the *residuals*, $\mathbf{e}_b^* = [E_{b1}^*, E_{b2}^*, \ldots, E_{bn}^*]'$ and, from these, calculate bootstrapped Y-values, $\mathbf{y}_b^* = [Y_{b1}^*, Y_{b2}^*, \ldots, Y_{bn}^*]'$, where $Y_{bi}^* = \hat{Y}_i + E_{bi}^*$.
3. Regress the bootstrapped Y-values on the *fixed* X-values to obtain bootstrap regression coefficients.
   *If, for example, estimates are calculated by least-squares regression, then $\mathbf{b}_b^* = (\mathbf{X}'\mathbf{X})^{-1}\mathbf{X}'\mathbf{y}_b^*$ for $b = 1, \ldots, r$.
4. The resampled $\mathbf{b}_b^* = [A_b^*, B_{b1}^*, \ldots, B_{bk}^*]'$ can be used in the usual manner to construct bootstrap standard errors and confidence intervals for the regression coefficients.

Bootstrapping with fixed X draws an analogy between the fitted value $\hat{Y}$ in the sample and the conditional expectation of Y in the population, and between the residual E in the sample and the error ε in the population. Although no assumption is made about the shape of the error distribution, the bootstrapping procedure, by constructing the Y_{bi}^* according to the linear model, implicitly assumes that the functional form of the model is correct.

Furthermore, by resampling residuals and randomly reattaching them to fitted values, the procedure implicitly assumes that the errors are identically distributed. If, for example, the true errors have nonconstant variance, then this property will *not* be reflected in the resampled residuals. Likewise, the unique impact of a high-leverage outlier will be lost to the resampling.[15]

> Regression models and similar statistical models can be bootstrapped by (1) treating the regressors as random, and selecting bootstrap samples directly from the observations $\mathbf{z}_i' = [Y_i, X_{i1}, \ldots, X_{ik}]$; or (2) treating the regressors as fixed, and resampling from the residuals E_i of the fitted regression model. In the latter instance, bootstrap observations are constructed as $Y_{bi}^* = \hat{Y}_i + E_{bi}^*$, where the $\hat{Y}_i$ are the fitted values from the original regression, and the E_{bi}^* are the resampled residuals for the bth bootstrap sample. In each bootstrap sample, the Y_{bi}^* are then regressed on the original X's. A disadvantage of fixed-X resampling is that the procedure implicitly assumes that the regression model fit to the data is correct, and that the errors are identically distributed.

[15] For these reasons, random resampling may be preferable even if the X-values are best conceived as fixed. See Exercise 16.3.

TABLE 16.5 Statistics for $r = 2000$ Bootstrapped Huber Regressions Applied to Duncan's Occupational Prestige Data. The values Z and A are the adjustment factors for the bootstrap percentile confidence intervals. Three bootstrap confidence intervals are shown for each coefficient

	Coefficient		
	Constant	*Income*	*Education*
Average Bootstrap Estimate	−7.033	0.6916	0.4921
Bootstrap standard error	3.069	0.1741	0.1374
Z	−0.08386	−0.02125	−0.00875
A	0.02145	−0.04274	0.03412
Normal-theory interval	(−13.304, −1.274)	(0.5363, 0.8845)	(0.3445, 0.6193)
Percentile interval	(−13.199, −0.845)	(0.3193, 1.0211)	(0.2204, 0.7825)
Adjusted percentile interval	(−13.447, −1.179)	(0.2816, 0.9855)	(0.2551, 0.8112)

To illustrate bootstrapping regression coefficients, I shall use Duncan's regression of occupational prestige on the income and educational levels of 45 U.S. occupations.[16] The Huber M-estimator applied to Duncan's regression produces the following fit, with estimated asymptotic standard errors shown in parentheses beneath each coefficient:[17]

$$\widehat{\text{Prestige}} = \underset{(3.588)}{-7.289} + \underset{(0.1005)}{0.7104}\ \text{Income} + \underset{(0.0825)}{0.4819}\ \text{Education}$$

Using random resampling, I drew $r = 2000$ bootstrap samples, calculating the Huber estimator for each bootstrap sample. The results of this computationally intensive procedure are summarized in Table 16.5. The bootstrapped regression coefficients for income and education are graphed in Figure 16.3(a) and (b), along with the percentile confidence intervals for these coefficients. Figure 16.3(c) shows a scatterplot of the bootstrapped coefficients for income and education, which gives a sense of the covariation of the two estimates; it is clear that the income and education coefficients are substantially negatively correlated.

The bootstrap standard errors of the income and education coefficients are substantially larger than the estimated asymptotic standard errors, underscoring the inadequacy of the latter in small samples. The simple normal-theory confidence intervals based on the bootstrap standard errors (and formed as the estimated coefficients ± 1.96 standard errors) are quite different from the percentile intervals for the income and education coefficients; the percentile intervals are very similar, however, to the adjusted percentile intervals (because the adjustment factors, Z and A, are small). Comparing the average bootstrap coefficients $\overline{A}^*$, $\overline{B}_1^*$, and $\overline{B}_2^*$ with the corresponding estimates A, B_1, and B_2 suggests that there is little, if any, bias in the Huber estimates.[18]

---

[16] These data were discussed in Chapter 5, and at several other points in this text.

[17] M-estimation is a method of robust regression described in Section 14.3.

[18] For the use of the bootstrap to estimate bias, see Exercise 16.4.

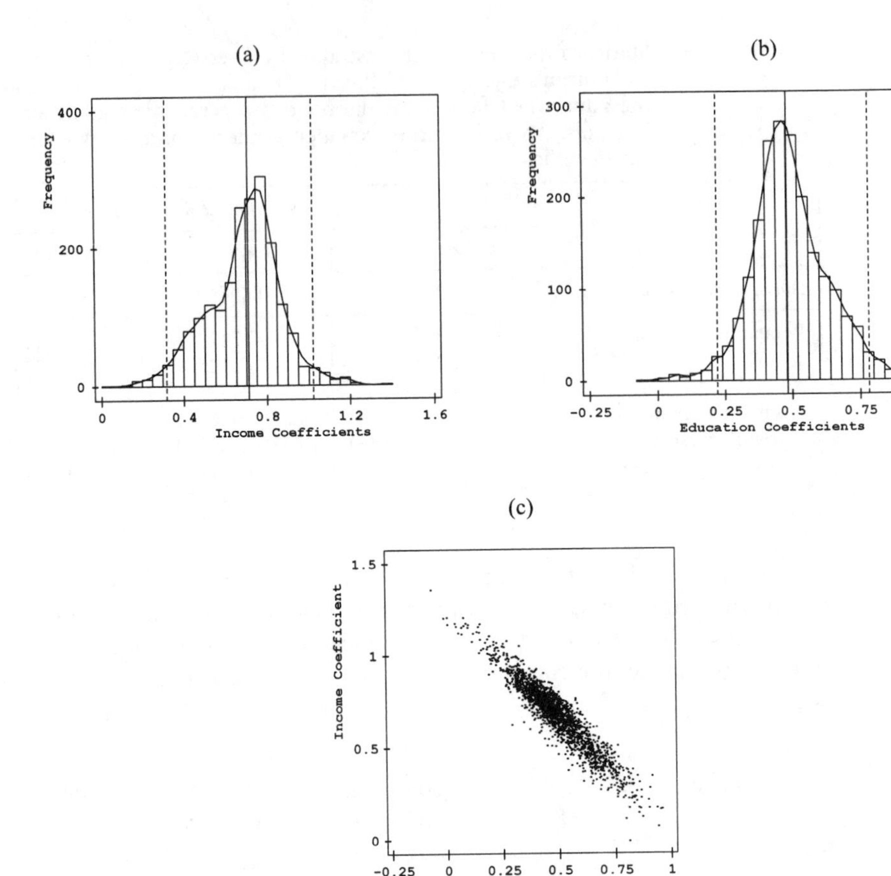

Figure 16.3. Panels (a) and (b) show histograms and kernel density estimates for the $r = 2000$ bootstrap replicates of the income and education coefficients in Duncan's occupational prestige regression. The regression model was fit by M-estimation using the Huber weight function. Panel (c) shows a scatterplot of the income and education coefficients for the 2000 bootstrap samples.

16.1.4 Bootstrap Hypothesis Tests*

In addition to providing standard-error estimates and confidence intervals, the bootstrap can also be used to test statistical hypotheses. The application of the bootstrap to hypothesis testing is more or less obvious for individual coefficients, because a bootstrap confidence interval can be used to test the hypothesis that the corresponding parameter is equal to any specific value (typically 0 for a regression coefficient).

More generally, let $T \equiv t(\mathbf{z})$ represent a test statistic, written as a function of the sample $\mathbf{z}$. The contents of $\mathbf{z}$ vary by context. In regression analysis, for example, $\mathbf{z}$ is the $n \times k+1$ matrix $[\mathbf{y}, \mathbf{X}]$ containing the dependent variable and the regressors.

For concreteness, suppose that T is the Wald-like test statistic for the omnibus null hypothesis H_0: $\beta_1 = \cdots = \beta_k = 0$ in a robust regression, calculated using the estimated asymptotic covariance matrix for the regression coefficients. That is, let $\underset{(k\times k)}{\mathbf{V}_{11}}$ contain the rows and columns of $\widehat{\mathscr{V}(\mathbf{b})}$ that pertain to the k slope coefficients $\mathbf{b}_1 = [B_1, \ldots, B_k]'$. We can write the null hypothesis as H_0: $\boldsymbol{\beta}_1 = \mathbf{0}$. Then the test statistic is,

$$T = \mathbf{b}_1'\mathbf{V}_{11}\mathbf{b}_1$$

We could compare the obtained value of this statistic to the quantiles of χ_k^2, but we are loath to do so because we do not trust the asymptotics. We can, instead, construct the sampling distribution of the test statistic nonparametrically, using the bootstrap.

Let $T_b^* \equiv t(\mathbf{z}_b^*)$ represent the test statistic calculated for the bth bootstrap sample, $\mathbf{z}_b^*$. We have to be careful to draw a proper analogy here: Because the original-sample estimates play the role of the regression parameters in the bootstrap "population' (i.e., the original sample), the bootstrap analog of the null hypothesis—to be used with each bootstrap sample—is H_0: $\beta_1 = B_1, \ldots, \beta_k = B_k$. The bootstrapped test statistic is, therefore,

$$T_b^* = (\mathbf{b}_{b1}^* - \mathbf{b}_1)'\mathbf{V}_{b,11}^*(\mathbf{b}_{b1}^* - \mathbf{b}_1)$$

Having obtained r bootstrap replications of the test statistic, the bootstrap estimate of the p-value for H_0 is simply[19]

$$\hat{p}^* = \frac{\#_{b=1}^r(T_b^* \geq T)}{r}$$

Bootstrap hypothesis tests proceed by constructing an empirical sampling distribution for the test statistic. If T represents the test statistic computed for the original sample, and T_b^* is the test statistic for the bth of r bootstrap samples, then the p-value for the test is $\#(T_b^* \geq T)/r$.

[19] There is a subtle point here: We use the sample estimate $\mathbf{b}_1$ in place of the hypothesized parameter $\boldsymbol{\beta}_1^{(0)}$ to calculate the bootstrapped test statistic T_b^* *regardless* of the hypothesis that we are testing—because in the central bootstrap analogy $\mathbf{b}_1$ stands in for $\boldsymbol{\beta}_1$ (and the bootstrapped sampling distribution of the test statistic is computed under the assumption that the hypothesis is true). See Exercise 16.5 for an application of this test to Duncan's regression.

16.1.5 Bootstrapping Complex Sampling Designs

One of the great virtues of the bootstrap is that it can be applied in a natural manner to more complex sampling designs. If, for example, the population is divided into S strata, with n_s observations drawn from stratum s, then bootstrap samples can be constructed by resampling n_s observations with replacement from the sth stratum. Likewise, if observations are drawn into the sample in clusters rather than individually, then the bootstrap should resample clusters. We can still calculate estimates and test statistics in the usual manner, using the bootstrap to assess sampling variation in place of the standard formulas, which are appropriate for independent random samples.[20]

When different observations are selected for the sample with unequal probabilities, it is common to take account of this fact by differentially weighting the observations in inverse proportion to their probability of selection.[21] Thus, for example, in calculating the (weighted) sample mean of a variable Y, we take

$$\overline{Y}^{(w)} = \frac{\sum_{i=1}^{n} w_i Y_i}{\sum_{i=1}^{n} w_i}$$

and to calculate the (weighted) correlation of X and Y, we take

$$r_{XY}^{(w)} = \frac{\sum w_i(X_i - \overline{X})(Y_i - \overline{Y})}{\sqrt{[\sum w_i(X_i - \overline{X})^2][\sum w_i(Y_i - \overline{Y})^2]}}$$

Other statistical formulas can be adjusted analogously.[22]

The case weights are often scaled so that $\sum w_i = n$, but simply to incorporate the weights in the usual formulas for standard errors does not produce correct results. Once more, the bootstrap provides a straightforward solution: Draw bootstrap samples in which the probability of inclusion is proportional to the probability of inclusion in the original sample, and calculate bootstrap replicates of the statistics of interest using the case weights.

The essential "trick" of using the bootstrap in these (and other) instances is to resample from the data in the same way as the original sample was drawn from the population. Statistics are calculated for each bootstrap replication in the same manner as for the original sample.

> The bootstrap can be applied to complex sampling designs (involving, e.g., stratification, clustering, and case-weighting) by resampling from the sample data in the same manner as the original sample was selected from the population.

---

[20] An application appears in Exercise 16.8.

[21] These "case weights" are to be distinguished from the weights used in weighted-least-squares regression (see Section 12.2.2).

[22] See Exercise 16.6.

Social scientists frequently analyze data from complex sampling designs as if they originate from independent random samples (even though there are often nonnegligible dependencies among the observations), or employ ad hoc adjustments (e.g., by weighting). A tacit defense of common practice is that to take account of the dependencies in complex sampling designs is too difficult. The bootstrap provides a simple solution.[23]

16.1.6 Concluding Remarks

If the bootstrap is so simple, and of such broad application, why isn't it used more in the social sciences? Beyond the problem of lack of familiarity (which surely can be remedied), there are, I believe, three serious obstacles to increased use of the bootstrap:

1. Common practice—such as relying on asymptotic results in small samples or treating dependent data as if they were independent—usually understates sampling variation, and makes results look stronger than they really are. Researchers are understandably reluctant to report honest standard errors when the usual calculations indicate greater precision. It is best, however, not to fool yourself, regardless of what you think about fooling others.
2. Although the conceptual basis of the bootstrap is intuitively simple, and although the calculations are straightforward, to apply the bootstrap, it is necessary to write or find suitable statistical software. There is some bootstrapping software available, but the nature of the bootstrap—which adapts resampling to the data-collection plan and statistics employed in an investigation—apparently precludes full generality, and makes it difficult to use traditional statistical computer packages. After all, researchers are not tediously going to draw 2000 samples from their data unless a computer program can fully automate the process. This impediment is much less acute in programmable statistical computing environments.[24]
3. Even with good software, the bootstrap is computationally intensive. This barrier to bootstrapping is more apparent than real, however. Computational speed is central to the exploratory stages of data analysis: When the outcome of one of many small steps immediately affects the next, rapid results are important. This is why a responsive computing environment is particularly useful for regression diagnostics, for example. It is not nearly as important to calculate standard errors and p-values quickly. With powerful, yet relatively inexpensive, desktop computers, there is nothing to preclude the machine from cranking away unattended for a few hours. The time and effort involved in a bootstrap calculation are typically small compared with the totality of a research investigation—and are a small price to pay for accurate and realistic inference.

---

[23] More traditional approaches to estimating sampling variation in complex survey samples may be found, for example, in Kish (1965).

[24] See, for example, the bootstrapping software for the S statistical-computing environment described by Efron and Tibshirani (1993, Appendix).

EXERCISES

16.1 *Show that the mean of the n^n bootstrap means is the sample mean

$$E^*(\overline{Y}^*) = \frac{\sum_{b=1}^{n^n} \overline{Y}_b^*}{n^n} = \overline{Y}$$

and that the standard deviation (standard error) of the bootstrap means is

$$SE^*(\overline{Y}^*) = \sqrt{\frac{\sum_{b=1}^{n^n}(\overline{Y}_b^* - \overline{Y})^2}{n^n}} = \frac{S}{\sqrt{n-1}}$$

where $S = \sqrt{\sum_{i=1}^{n}(Y_i - \overline{Y})^2/(n-1)}$ is the sample standard deviation. (*Hint*: Exploit the fact that the mean is a linear function of the observations.)

16.2 *The jackknife: The "jackknife" is an alternative to the bootstrap that requires less computation, but that often does not perform as well, and is not quite as general. Efron and Tibshirani (1993, Chapter 11) show that the jackknife is an approximation to the bootstrap. Here is a brief description of the jackknife for the estimator $\hat{\theta}$ of a parameter θ:

1. Divide the sample into m independent groups. In most instances (unless the sample size is very large), we take $m = n$, in which case each observation constitutes a "group." If the data originate from a cluster sample, then the observations in a cluster should be kept together.
2. Recalculate the estimator omitting the jth group, $j = 1, \ldots, m$, denoting the resulting value of the estimator as $\hat{\theta}_{(-j)}$. The *pseudo-value* associated with the jth group is defined as $\hat{\theta}_j^* \equiv \hat{\theta} - \hat{\theta}_{(-j)}$.
3. The average of the pseudo-values, $\hat{\theta}^* \equiv (\sum_{j=1}^{m} \hat{\theta}_j^*)/m$ is the jackknifed estimate of θ. A jackknifed $100(1-a)\%$ confidence interval for θ is given by

$$\theta = \hat{\theta}^* \pm t_{a/2,m-1}\frac{S^*}{\sqrt{n}}$$

where $t_{a/2,m-1}$ is the critical value of t with probability $a/2$ to the right for $m-1$ degrees of freedom; and $S^* \equiv \sqrt{\sum_{j=1}^{m}(\hat{\theta}_j^* - \hat{\theta}^*)^2/(m-1)}$ is the standard deviation of the pseudo-values.

(a) Show that when the jackknife procedure is applied to the mean with $m = n$, the pseudo-values are just the original observations, $\hat{\theta}_i^* = Y_i$; the jackknifed estimate $\hat{\theta}^*$ is, therefore, the sample mean $\overline{Y}$; and the jackknifed confidence interval is the same as the usual confidence interval. Demonstrate these results numerically for the contrived "data" in Table 16.3. (These results are peculiar to linear statistics like the mean.)

(b) Find jackknifed confidence intervals for the Huber M-estimator of Duncan's regression of occupational prestige on income and education. Compare these intervals with the bootstrap and normal-theory intervals given in Table 16.5. (The data are in Table 3.2 and `duncan.dat`.)

16.3 Random versus fixed resampling in regression:

(a) Recall (from Chapter 2), Davis's data (in `davis.dat`) on measured and reported weight for 101 women engaged in regular exercise. Bootstrap the least-squares regression of reported weight on measured weight, drawing $r = 1000$ bootstrap samples using (i) random-X resampling and (ii) fixed-X resampling. In each case, plot a histogram (and, if you wish, a density estimate) of the 1000 bootstrap slopes, and calculate the bootstrap estimate of standard error for the slope. How does the influential outlier in this regression affect random resampling? How does it affect fixed resampling?

(b) Randomly construct a dataset of 100 observations according to the regression model $Y_i = 5 + 2x_i + \varepsilon_i$, where $x_i = 1, 2, \ldots, 100$, and the errors are independent (but seriously heteroscedastic) with $\varepsilon_i \sim N(0, X_i^2)$. As in (a), bootstrap the least-squares regression of Y on x, using (i) random resampling and (ii) fixed resampling. In each case, plot the bootstrap distribution of the slope coefficient, and calculate the bootstrap estimate of standard error for this coefficient. Compare the results for random and fixed resampling. For a few of the bootstrap samples, plot the least-squares residuals against the fitted values. How do these plots differ for fixed versus random resampling?

(c) Why might random resampling be preferred in these contexts, even if (as is *not* the case for Davis's data) the X-values are best conceived as fixed.

16.4 Bootstrap estimates of bias: The bootstrap can be used to estimate the bias of an estimator $\hat{\theta}$ of a parameter θ, simply by comparing the mean of the bootstrap distribution $\overline{\theta}^*$ (which stands in for the expectation of the estimator) with the sample estimate $\hat{\theta}$ (which stands in for the parameter); that is, $\widehat{\text{bias}} = \overline{\theta}^* - \hat{\theta}$. [Further discussion and more sophisticated methods are described in Efron and Tibshirani (1993, Chapter 10).] Employ this approach to estimate the bias of the maximum-likelihood estimator of the variance, $\hat{\sigma}^2 = \sum(Y_i - \overline{Y})^2/n$ for a sample of $n = 10$ observations drawn from the normal distribution $N(0, 100)$. Use $r = 500$ bootstrap replications. How close is the bootstrap bias estimate to the theoretical value $-\sigma^2/n = -100/10 = -10$?

16.5 *Test the omnibus null hypothesis H_0: $\beta_1 = \beta_2 = 0$ for the Huber M-estimator in Duncan's regression of occupational prestige on income and education.

(a) Base the test on the estimated asymptotic covariance matrix of the coefficients.

(b) Use the bootstrap approach described in Section 16.1.4.

16.6 Case weights:

(a) *Show how case weights can be used to "adjust" the usual formulas for the least-squares coefficients and their covariance matrix. How do these case-weighted formulas compare with those for weighted-least-squares regression (discussed in Section 12.2.2)?

(b) Using data from a sample survey that employed disproportional sampling and for which case weights are supplied, estimate a least-squares regression (i) ignoring the case weights; (ii) using the case weights to estimate both the regression coefficients and their standard errors (rescaling the case weights, if necessary, so that they sum to the sample size); and (iii) using the case weights but estimating coefficient standard errors with the bootstrap. Compare the estimates and standard errors obtained in (i), (ii), and (iii).

16.7 *Bootstrapping time-series regression: Bootstrapping can be adapted to time series regression but, as in the case of fixed-X resampling, the procedure makes strong use of the model fit to the data—in particular, the manner in which serial dependency in the data is modeled. Suppose (as in Section 14.1) that the errors in the linear model $\mathbf{y} = \mathbf{X}\boldsymbol{\beta} + \boldsymbol{\varepsilon}$ follow a first-order autoregressive process, $\varepsilon_i = \rho\varepsilon_{i-1} + \upsilon_i$; the υ_i are independently and identically distributed with zero expectations and common variance σ_υ^2. Suppose further that we use an empirical GLS procedure to obtain estimates $\hat{\rho}$ and $\hat{\boldsymbol{\beta}}$. From the residuals $\mathbf{e} = \mathbf{y} - \mathbf{X}\hat{\boldsymbol{\beta}}$, we can estimate υ_i as $V_i = E_i - \hat{\rho}E_{i-1}$ for $i = 2, \ldots, n$; by convention, we take $V_1 = E_1$. Then, for each bootstrap replication, we sample n values with replacement from the V_i—call them $V_{b1}^*, V_{b2}^*, \ldots, V_{bn}^*$. Using these values, we construct residuals $E_{b1}^* = V_{b1}^*$, and $E_{bi}^* = \hat{\rho}E_{b,i-1}^* + V_{bi}^*$ for $i = 2, \ldots, n$; and from these residuals and the original fitted values $\hat{Y}_i = \mathbf{x}_i'\hat{\boldsymbol{\beta}}$, we construct bootstrapped Y-values, $Y_{bi}^* = \hat{Y}_i + E_{bi}^*$. The Y_{bi}^* are used along with the original $\mathbf{x}_i'$ to obtain a bootstrap replication $\hat{\boldsymbol{\beta}}_b^*$ of the pseudo-GLS coefficient estimates. (Why are the $\mathbf{x}_i'$ treated as fixed?) Employ this procedure to estimate the standard errors of the coefficient estimates in the pseudo-GLS regression for the Canadian women's crime-rate data (discussed in Section 14.1 and located in `hartnagl.dat`). Compare the bootstrap estimates of standard errors with the usual estimated asymptotic standard errors. Which standard-error estimates do you prefer? Why?

16.8 In Exercise 15.11, you fit a logistic regression model to Greene's refugee-appeals data (in `greene.dat`). Recall, however, that these cases were assigned to judges not at random, but in "batches" on a rotating basis. Treating the "batch" as a cluster of related observations, bootstrap the final logistic regression model fit in part (c) of Exercise 15.11.

(a) Compare the bootstrap standard errors of the regression coefficients with the usual estimated asymptotic standard errors.

(b) *Perform a bootstrap hypothesis test for differences among judges. Does the result square with the usual Wald or likelihood-ratio test?

(c) Which inferences would a court likely find more convincing: those based on the usual asymptotic theory or those based on the bootstrap?

16.2 Cross-Validation

Cross-validation is a general term describing several related methods that share a common property: Part of the data (called the "training" or "exploratory" subsample) is used to select or estimate a statistical model, which is then evaluated using the other part of the data (the "validatory" or "confirmatory" subsample). In some applications of this very simple—but powerful—idea, the roles of training and validatory subsamples are interchanged or rotated.[25] It makes semantic sense to restrict the term *cross*-validation to these applications, but the term is used more broadly, and, indeed, in this section I shall employ a form of cross-validation in which there is just one exploratory subsample and one validatory subsample.

I have stressed the importance of descriptive adequacy in statistical modeling, and—in support of this goal—I have described a variety of methods for screening data, and evaluating and, if necessary, modifying statistical models. This process of data exploration, model fitting, model criticism, and model respecification is typically iterative, requiring several failed attempts before an adequate description of the data is achieved. In the process, variables may be dropped from the model; terms such as interactions may be incorporated or deleted; variables may be transformed; and unusual data may be corrected, deleted, or otherwise accommodated.

The outcome should be a model that more accurately reflects the principal characteristics of the data at hand, but the risk of iterative modeling is that we shall capitalize on chance—overfitting the data and overstating the strength of our results. It is obviously problematic to employ the same data both to explore and to validate a statistical model, but the apparent alternative of analyzing data blindly, simply to preserve the "purity" of classical statistical inference, is surely worse.

An ideal solution to this dilemma would be to collect new data with which to validate a model, but this solution is often impractical: Lack of funds or other constraints may preclude the collection of new data; and, in certain circumstances—for example, the examination of historical records—it is impossible even in principle to collect new data.

Cross-validation simulates the collection of new data by randomly dividing the data that we have in hand into two, possibly equal, parts—the first part to be used for exploration and model formulation, the second for model validation, formal estimation, and testing. This is such a simple idea that it hardly requires detailed explanation. Perhaps the only subtle point is that the division of the data into exploratory and validatory subsamples can exploit the sampling structure of the data. If, for example, the data are collected in clusters, then the observations from each cluster can be randomly apportioned between the two subsamples.

---

[25] This approach is particularly common when we intend to use a statistical model for prediction. To assess the quality of predictions using the sample data on which the model was calibrated likely overstates its performance. Cross-validation can be employed to assess the predictive performance of the model more realistically. In Section 14.4.1, for example, we used the cross-validation mean-squared prediction error to help select the span s of a nonparametric-regression estimator:

$$\mathrm{CV}(s) = \frac{1}{n}(Y_i - \hat{Y}_i^{(-i)})^2$$

where $\hat{Y}_i^{(-i)}$ is the fitted value of the ith observation computed using a regression that omits this observation. In this case, the training subsample includes the remaining $n - 1$ observations, and the validation subsample consists of observation i alone. The procedure is repeated n times, for $i = 1, \ldots, n$—putting the "cross" in "cross-validation."

> When the same data are employed for selecting a statistical model and for drawing statistical inferences based on the model, the integrity of the inferences is compromised. Cross-validation is a general strategy for protecting the accuracy of statistical inference when—as is typically the case—it is not possible to collect new data with which to validate the model. In cross-validation, the data at hand are divided at random into two subsamples: an exploratory subsample, which is used to select a statistical model for the data; and a validatory subsample, which is used for formal statistical inference.

16.2.1 An Illustration

To illustrate cross-validation, I shall present an abbreviated account of some research that I conducted on the Canadian refugee determination process, employing data that were collected and described by Greene and Shaffer (1992).[26] Greene and Shaffer's data pertain to decisions of the Canadian Federal Court of Appeal on cases filed by claimants who were refused refugee status by the Immigration and Refugee Board. The Court either grants or denies leave to appeal the board's decision, and a single judge (rather than the usual tribunal) hears the request for leave to appeal.

During the period of the study, the 10 judges who adjudicated these cases varied widely in the rates with which they approved requests for leave to appeal negative decisions of the Refugee Board—with approval rates ranging from 13% to 56% of cases. The cases are not assigned randomly to judges, however, but rather are heard on a rotating basis. Although it seems unlikely that this procedure would introduce systematic differences into the leave requests processed by different judges, it is conceivable that this is the case. The Crown, therefore, contended, in defending the fairness of the current procedure, that it was insufficient simply to demonstrate large and statistically significant differences among the rates of approval for the 10 judges.

To determine whether systematic differences among the cases heard by the judges could account for differences in their judgments, I controlled statistically for several factors that—it was suggested—might influence the decisions, including:

1. the rate of success of leave applications from the applicant's country;
2. whether or not this country was identified as a "high refugee producing" country;
3. the region of the world in which the applicant's country is located (Latin America, Europe, Africa, the Middle East, or Asia and the Pacific Islands); and
4. the date of the applicant's case.

[26] Also see Section 1.2 and Exercises 15.11 and 16.8. The analysis of the refugee data reported in this section uses a larger subset of cases.

These independent variables were included in a logistic regression, along with dummy variables identifying the judge who heard the case. The dependent variable was whether or not leave to appeal was granted. Prior to constructing the logistic regression model, the roughly 800 cases meeting the criteria for inclusion in the study were randomly divided into exploratory and validatory subsamples. The data in the exploratory subsample were carefully examined, and several variations of the analysis were undertaken. For example, the date of the case was treated both as a quantitative variable with a linear effect and categorically, divided into five quarters (the period of the study was slightly in excess of one year).

Two of the models fit to the data in the exploratory subsample are shown in Table 16.6. Model 1 contains two independent variables—national rate of success and judge—that are highly statistically significant; high-refugee country has a small and nonsignificant coefficient; and the region and time-period effects are relatively small but approach statistical significance. Examination of the region coefficients suggested that applicants from Latin America might be treated differently; likewise, the resolution of time period into polynomial components suggested a possible linear effect of time. Model 2 is a final model for the exploratory subsample, incorporating a dummy regressor for Latin America and a linear trend over the five time periods, both of which appear to be statistically significant.

The last columns of Table 16.6 (for model 3) show the result of refitting model 2 to the data from the validation subsample. The national rate of success and judge both are still highly statistically significant, but neither the coefficient for Latin America nor the linear time trend proves to be statistically significant. That these latter two coefficients appeared to be statistically significant in the exploratory subsample illustrates the risk of selecting and testing a model on the same data. Most notably, differences among judges were essentially the same before and after controlling for the other independent variables in the analysis.

TABLE 16.6 Wald Tests for Terms in the Linear-Logit Model for the Canadian Refugee Data, Exploratory and Validatory Subsamples; Z^2 Is the Wald Test Statistic

		Subsample					
		Exploratory				*Validatory*	
		Model 1		*Model 2*		*Model 3*	
Term	*df*	Z^2	p	Z^2	p	Z^2	p
Country success	1	14.55	.0001	15.72	.0001	25.64	< .0001
High-refugee country	1	0.09	.77				
Region	4	6.47	.17				
Latin America	1	4.44	.035	4.98	.026	0.58	.45
Time period	4	9.24	.055				
Linear	1	6.46	.011	5.74	.017	0.98	.32
Quadratic	1	1.73	.19				
Cubic	1	0.16	.69				
Quartic	1	0.19	.67				
Judge	9	29.75	.0005	29.67	.0005	37.45	< .0001

16.2.2 Concluding Remarks

Like the bootstrap, cross-validation is a good, simple, broadly applicable procedure that is rarely used in practice. I believe that researchers rebel against the idea of dividing their data in half. In very small samples, division of the data is usually not practical. Even in samples of moderate size, however (such as the refugee-appeal data discussed in the previous section), halving the sample size makes it more difficult to find "statistically significant" results.

Yet, if statistical inference is to be more than an incantation to be spoken over the data, it is necessary to conduct research honestly. This is not to say that procedures of inference cannot be approximate—simplifying abstraction of some sort is unavoidable—but substantial errors of inference are easy to introduce when the same data are used both to formulate and to test a statistical model.

Cross-validation is not a panacea for these problems, but it goes a long way toward solving them. Issues such as variable selection and choice of transformation are neatly handled by cross-validation. Problems such as influential data are less easily dealt with, because these problems are particular to specific observations. That we locate an outlier in the exploratory subsample, for example, does not imply that an outlier is present in the validatory subsample. The reverse could be true as well, of course. We can, however, use the distribution of residuals in the exploratory subsample to help us decide whether to use a method of estimation in the validatory subsample that is resistant to unusual data.

EXERCISE

16.9 Using a dataset and initial statistical model of your choice, randomly divide the data into two subsamples. Use the first (exploratory) subsample to "fine-tune" the statistical model—transforming variables where necessary, eliminating independent variables that appear to be unimportant, and so on. Then fit the final model to the second (validation) subsample. How do the results for the two subsamples compare?

16.3 Summary

- Bootstrapping is a broadly applicable, nonparametric approach to statistical inference that substitutes intensive computation for more traditional distributional assumptions and asymptotic results. The bootstrap can be used to derive accurate standard errors, confidence intervals, and hypothesis tests for most statistics.
- Bootstrapping uses the sample data to estimate relevant characteristics of the population. The sampling distribution of a statistic is then constructed empirically by resampling from the sample. The resampling procedure is designed to parallel the process by which sample observations were drawn from the population. For example, if the data represent an independent random sample of size

n, then each bootstrap sample selects n observations with replacement from the original sample. The key bootstrap analogy is: *The population is to the sample as the sample is to the bootstrap samples.*

- Having produced r bootstrap replicates $\hat{\theta}_b^*$ of an estimator $\hat{\theta}$, the bootstrap estimate of standard error is the standard deviation of the bootstrap replicates:

$$\widehat{\text{SE}}^*(\hat{\theta}^*) = \sqrt{\frac{\sum_{b=1}^{r}(\hat{\theta}_b^* - \overline{\theta}^*)^2}{r-1}}$$

where $\overline{\theta}^*$ is the mean of the $\hat{\theta}_b^*$. In large samples, where we can rely on the normality of $\hat{\theta}$, a 95% confidence interval for θ is given by $\hat{\theta} \pm 1.96\widehat{\text{SE}}^*(\hat{\theta}^*)$.

- A nonparametric confidence interval for θ can be constructed from the quantiles of the bootstrap sampling distribution of $\hat{\theta}^*$. The 95% percentile interval is $\hat{\theta}^*_{(\text{lower})} < \theta < \hat{\theta}^*_{(\text{upper})}$, where the $\hat{\theta}^*_{(b)}$ are the ordered bootstrap replicates, lower $= .025 \times r$, and upper $= .975 \times r$.
- The lower and upper bounds of percentile confidence intervals can be corrected to improve the accuracy of these intervals.
- Regression models and similar statistical models can be bootstrapped by (1) treating the regressors as random, and selecting bootstrap samples directly from the observations $\mathbf{z}_i' = [Y_i, X_{i1}, \ldots, X_{ik}]$; or (2) treating the regressors as fixed, and resampling from the residuals E_i of the fitted regression model. In the latter instance, bootstrap observations are constructed as $Y_{bi}^* = \hat{Y}_i + E_{bi}^*$, where the $\hat{Y}_i$ are the fitted values from the original regression, and the E_{bi}^* are the resampled residuals for the bth bootstrap sample. In each bootstrap sample, the Y_{bi}^* are then regressed on the original X's. A disadvantage of fixed-X resampling is that the procedure implicitly assumes that the regression model fit to the data is correct, and that the errors are identically distributed.
- Bootstrap hypothesis tests proceed by constructing an empirical sampling distribution for the test statistic. If T represents the test statistic computed for the original sample, and T_b^* is the test statistic for the bth of r bootstrap samples, then the p-value for the test is $\#(T_b^* \geq T)/r$.
- The bootstrap can be applied to complex sampling designs (involving, e.g., stratification, clustering, and case weighting) by resampling from the sample data in the same manner as the original sample was selected from the population.
- When the same data are employed for selecting a statistical model and for drawing statistical inferences based on the model, the integrity of the inferences is compromised. Cross-validation is a general strategy for protecting the accuracy of statistical inference when—as is typically the case—it is not possible to collect new data with which to validate the model. In cross-validation, the data at hand are divided at random into two subsamples: an exploratory subsample, which is used to select a statistical model for the data; and a validatory subsample, which is used for formal statistical inference.

16.4 Recommended Reading

- Bootstrapping is a rich topic; the presentation in this chapter has stressed computational procedures at the expense of a detailed account of statistical properties and limitations. Although Efron and Tibshirani's (1993) book on the bootstrap contains some relatively advanced material, most of the exposition requires only modest statistical background and is eminently readable. Another good, briefer source is Stine's (1990) paper, which includes a fine discussion of the rationale of bootstrap confidence intervals. Young's (1994) paper, and the commentary that follows it, focus on practical difficulties in applying the bootstrap.

- Cross-validation is an essentially simple idea. An overview is given by Bailey, et al. (1989). Some more detail can be found in Mosteller and Tukey (1968, 1977).

Appendix A

Notation

Specific notation is introduced at various points in the appendices and chapters. Throughout the text, I adhere to the following general conventions, with few exceptions. [Examples are shown in brackets.]

- Known scalar constants (including subscripts) are represented by lowercase italic letters [a, b, x_i, x_1^*].
- Observable scalar random variables are represented by uppercase italic letters [X, Y_i, B_0'] or, if the names contain more than one character, by roman letters, the first of which is uppercase [RegSS, RSS_0]. Where it is necessary to make the distinction, *specific values* of random variables are represented as constants [x, y_i, b_0'].
- Scalar parameters are represented by lowercase Greek letters [α, β, β_j^*, γ_2]. Their estimators are generally denoted by "corresponding" italic characters [A, B, B_j^*, C_2], or by Greek letters with diacritics [$\hat{\alpha}$, $\tilde{\beta}$]. See the Greek alphabet in Table A.1.
- Unobservable scalar random variables are also represented by lowercase Greek letters [ε_i].
- Vectors and matrices are represented by boldface characters—lowercase for vectors [$\mathbf{x}_1$, $\boldsymbol{\beta}$], uppercase for matrices [$\mathbf{X}$, $\boldsymbol{\Sigma}_{12}$]. Roman letters are used for constants and observable random variables [$\mathbf{y}$, $\mathbf{x}_1$, $\mathbf{X}$]. Greek letters are used for parameters and unobservable random variables [$\boldsymbol{\beta}$, $\boldsymbol{\Sigma}_{12}$, $\boldsymbol{\varepsilon}$]. It is occasionally convenient to show the order of a matrix or vector below the matrix [$\underset{(n\times 1)}{\boldsymbol{\varepsilon}}$, $\underset{(n\times k+1)}{\mathbf{X}}$]. The order of an identity matrix is given by a subscript [$\mathbf{I}_n$]. A zero ma-

TABLE A.1 The Greek Alphabet With Roman "Equivalents"

Greek Letter			*Roman Equivalent*	
Lowercase	*Uppercase*		*Phonetic*	*Other*
α	A	alpha	a	
β	B	beta	b	
γ	Γ	gamma	g, n	c
δ, ∂	Δ	delta	d	
ε	E	epsilon	e	
ζ	Z	zeta	z	
η	H	eta	e	
θ	Θ	theta	th	
ι	I	iota	i	
κ	K	kappa	k	
λ	Λ	lambda	l	
μ	M	mu	m	
ν	N	nu	n	
ξ	Ξ	xi	x	
o	O	omicron	o	
π	Π	pi	p	
ρ	P	rho	r	
σ	Σ	sigma	s	
τ	T	tau	t	
υ	Υ	upsilon	y, u	
ϕ	Φ	phi	ph	
χ	X	chi	ch	x
ψ	Ψ	psi	ps	
ω	Ω	omega	o	w

trix or vector is represented by a boldface 0 [$\mathbf{0}$]; a vector of 1's is represented by a boldface 1, possibly subscripted with its number of elements [$\mathbf{1}_n$]. Vectors are column vectors, unless they are explicitly transposed [column: $\mathbf{x}$; row: $\mathbf{x}'$].

- Diacritics and symbols such as * (asterisk) and ′ (prime) are used freely as modifiers to denote alternative forms [$\mathbf{X}^*$, β', $\tilde{\varepsilon}$].
- The symbol $\equiv$ can be read as "is defined by," or "is equal to by definition" [$\bar{X} \equiv (\sum X_i)/n$].
- The symbol $\sim$ means "is distributed as" [$\varepsilon_i \sim N(0, \sigma_\varepsilon^2)$].
- The operator $E(\ )$ denotes the expectation of a scalar, vector, or matrix random variable [$E(Y_i)$, $E(\boldsymbol{\varepsilon})$, $E(\mathbf{X})$].
- The operator $V(\ )$ denotes the variance of a scalar random variable or the variance-covariance matrix of a vector random variable [$V(\varepsilon_i)$, $V(\mathbf{b})$].
- Estimated variances or variance-covariance matrices are indicated by a circumflex ("hat") placed over these expressions [$\widehat{V(\varepsilon_i)}$, $\widehat{V(\mathbf{b})}$].
- The operator $C(\ )$ gives the covariance of two scalar random variables or the covariance matrix of two vector random variables [$C(X, Y)$, $C(\mathbf{x}_i, \boldsymbol{\varepsilon})$].
- The operators $\mathscr{E}(\ )$ and $\mathscr{V}(\ )$ denote asymptotic expectation and variance, respectively. Their usage is similar to that of $E(\ )$ and $V(\ )$ [$\mathscr{E}(B)$, $\mathscr{V}(\hat{\boldsymbol{\beta}})$, $\widehat{\mathscr{V}(B)}$].
- Probability limits are specified by plim [plim $b = \beta$].
- Standard mathematical functions are shown in lowercase [cos W, trace($\mathbf{A}$)]. The base of the log function is always specified explicitly, unless it is irrelevant [$\log_e L$]. The exponential function $\exp(x)$ represents e^x.

- The summation sign $\sum$ is used to denote continued addition [$\sum_{i=1}^{n} X_i \equiv X_1 + X_2 + \cdots + X_n$]. Often, the range of the index is suppressed if it is clear from the context [$\sum_i X_i$], and the index may be suppressed as well [$\sum X_i$]. The symbol $\prod$ similarly indicates continued multiplication [$\prod_{i=1}^{n} p(Y_i) \equiv p(Y_1) \times p(Y_2) \times \cdots \times p(Y_n)$]. The symbol # indicates a count [$\#_{i=1}^{n}(T_b^* \geq T)$].
- To avoid awkward and repetitive phrasing in the statement of definitions and results, the words "if" and "when" are understood to mean "if and only if," unless explicitly indicated to the contrary. Terms are generally set in *italics* when they are introduced. ["Two vectors are *orthogonal* if their inner product is 0."]

Appendix B

Vector Geometry*

It is, I suppose, natural that linear algebra is the algebra of linear statistical models. By providing a graphical representation of linear algebra, vector geometry enlists our visual intuition in the effort to understand the properties and structure of linear models.[1] This appendix (along with the starred sections of the text) assumes some familiarity with matrices and linear algebra. Some basic references are given at the end of the appendix.

Considered algebraically, vectors are one-column (or one-row) matrices. Vectors also have the following geometric interpretation: The vector $\mathbf{x} = (x_1, x_2, \ldots, x_n)'$ is represented as a directed line segment extending from the origin of an n-dimensional Cartesian coordinate space to the point defined by the entries (called the *coordinates*) of the vector. Some examples of geometric vectors in two- and three-dimensional space are shown in Figure B.1.

B.1 Basic Operations

The basic arithmetic operations defined for vectors have simple geometric interpretations. To add two vectors $\mathbf{x}_1$ and $\mathbf{x}_2$ is, in effect, to place the "tail" of one at the tip of the other. When a vector is shifted from the origin in this manner, it retains its length and orientation (the angles that it makes with respect to the coordinate axes); length and

[1] See Chapter 10.

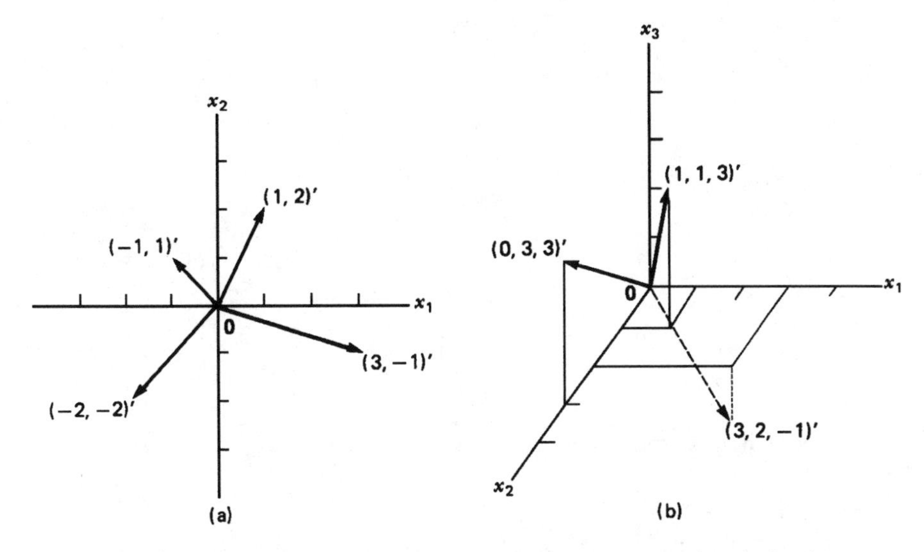

Figure B.1. Examples of geometric vectors in (*a*) two-dimensional and (*b*) three-dimensional space. Each vector is a directed line segment from the origin (0) to the point given by the entries of the vector.

orientation serve to define a vector uniquely. The operation of vector addition, illustrated in two dimensions in Figure B.2, is equivalent to completing a parallelogram in which $\mathbf{x}_1$ and $\mathbf{x}_2$ are two adjacent sides; the vector sum is the diagonal of the parallelogram, starting at the origin.

As shown in Figure B.3, the difference $\mathbf{x}_1 - \mathbf{x}_2$ is a vector whose length and orientation are obtained by proceeding from the tip of $\mathbf{x}_2$ to the tip of $\mathbf{x}_1$. Likewise, $\mathbf{x}_2 - \mathbf{x}_1$ proceeds from $\mathbf{x}_1$ to $\mathbf{x}_2$.

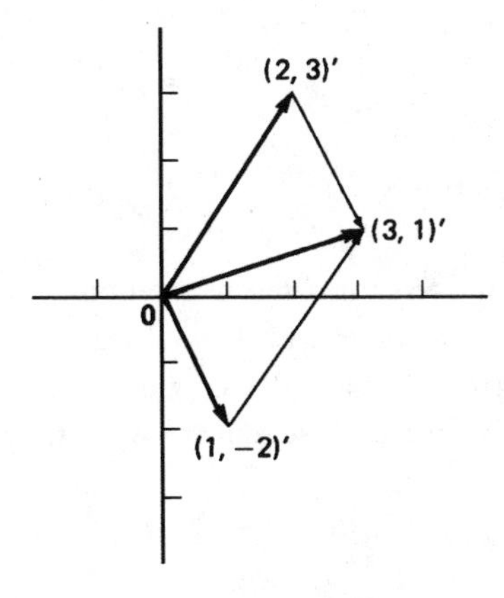

Figure B.2. Vectors are added by placing the "tail" of one on the tip of the other and completing the parallelogram. The sum is the diagonal of the parallelogram starting at the origin.

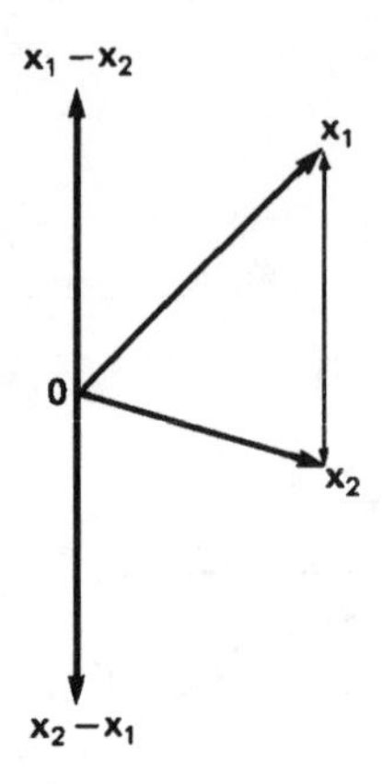

Figure B.3. Vector differences $\mathbf{x}_1 - \mathbf{x}_2$ and $\mathbf{x}_2 - \mathbf{x}_1$.

The length of a vector $\mathbf{x}$, denoted $||\mathbf{x}||$, is the square root of its sum of squared coordinates:

$$||\mathbf{x}|| = \sqrt{\sum_{i=1}^{n} x_i^2}$$

This result follows from the Pythagorean theorem in two dimensions, as shown in Figure B.4(a). The result can be extended one dimension at a time to higher-dimensional coordinate spaces, as shown for a three-dimensional space in Figure B.4(b). The distance between two vectors $\mathbf{x}_1$ and $\mathbf{x}_2$, defined as the distance separating their tips, is given by $||\mathbf{x}_1 - \mathbf{x}_2|| = ||\mathbf{x}_2 - \mathbf{x}_1||$ (see Figure B.3).

The product $a\mathbf{x}$ of a scalar (i.e., individual number) a and a vector $\mathbf{x}$ is a vector of length $|a| \times ||\mathbf{x}||$, as is readily verified:

$$||a\mathbf{x}|| = \sqrt{\sum (ax_i)^2}$$

$$= \sqrt{a^2 \sum x_i^2}$$

$$= |a| \times ||\mathbf{x}||$$

If the scalar a is positive, then the orientation of $a\mathbf{x}$ is the same as that of $\mathbf{x}$; if a is negative, then $a\mathbf{x}$ is *collinear* with (i.e., along the same line as) $\mathbf{x}$ but in the opposite direction. The negative $-\mathbf{x} = (-1)\mathbf{x}$ of $\mathbf{x}$ is, therefore, a vector of the same length as $\mathbf{x}$ but of opposite orientation. These results are illustrated for two dimensions in Figure B.5.

B.2 Vector Spaces and Subspaces

The *vector space of dimension* n is the set of all vectors $\mathbf{x} = (x_1, x_2, \ldots, x_n)'$; the coordinates x_i may be any real numbers. The vector space of dimension 1 is, therefore, the real line; the vector space of dimension 2 is the plane; and so on.

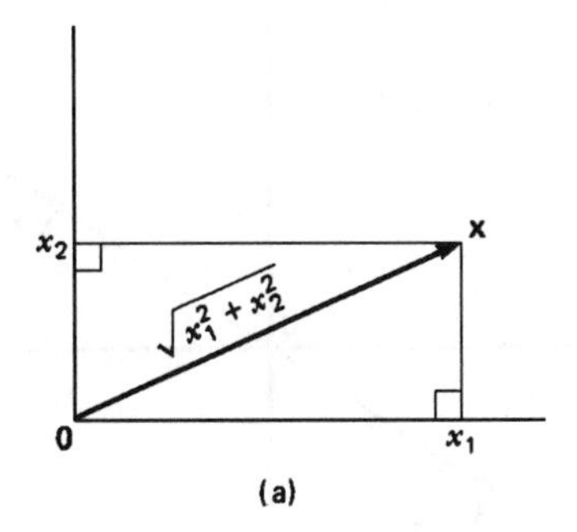

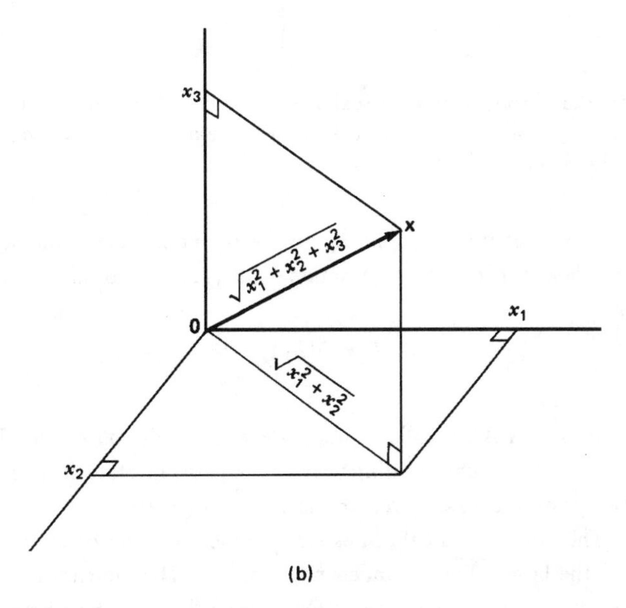

Figure B.4. The length of a vector is the square root of its sum of squared coordinates, $||\mathbf{x}|| = \sqrt{\sum_{i=1}^{n} x_i^2}$. This result is illustrated in (*a*) two and (*b*) three dimensions.

The *subspace* of the *n*-dimensional vector space that is *generated* by a set of *k* vectors $\{\mathbf{x}_1, \mathbf{x}_2, \ldots, \mathbf{x}_k\}$ is the subset of vectors $\mathbf{y}$ in the space that can be expressed as linear combinations of the generating set:[2]

$$\mathbf{y} = a_1\mathbf{x}_1 + a_2\mathbf{x}_2 + \cdots + a_k\mathbf{x}_k$$

The set of vectors $\{\mathbf{x}_1, \mathbf{x}_2, \ldots, \mathbf{x}_k\}$ is said to *span* the subspace that it generates.

A set of vectors $\{\mathbf{x}_1, \mathbf{x}_2, \ldots, \mathbf{x}_k\}$ is *linearly independent* if no vector in the set can be expressed as a linear combination of other vectors:

$$\mathbf{x}_j = a_1\mathbf{x}_1 + \cdots + a_{j-1}\mathbf{x}_{j-1} + a_{j+1}\mathbf{x}_{j+1} + \cdots + a_k\mathbf{x}_k \qquad \text{[B.1]}$$

---

[2] Notice that each of $\mathbf{x}_1, \mathbf{x}_2, \ldots, \mathbf{x}_k$ is a vector, with *n* coordinates; that is, $\{\mathbf{x}_1, \mathbf{x}_2, \ldots, \mathbf{x}_k\}$ is a set of *k* vectors, *not* a vector with *k* coordinates.

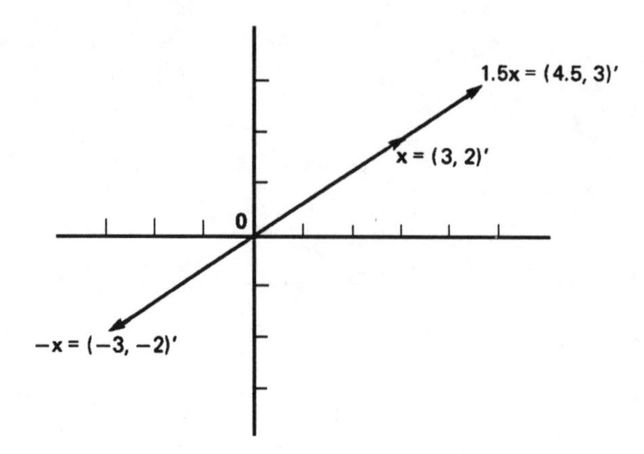

Figure B.5. Product $a\mathbf{x}$ of a scalar and a vector, illustrated in two dimensions. The vector $a\mathbf{x}$ is collinear with $\mathbf{x}$; it is in the same direction as $\mathbf{x}$ if $a > 0$, and in the opposite direction from $\mathbf{x}$ if $a < 0$.

(where some of the constants a_l can be 0). Equivalently, the set of vectors is linearly independent if there are no constants $b_1, b_2, \ldots, b_k$, not all 0, for which

$$b_1\mathbf{x}_1 + b_2\mathbf{x}_2 + \cdots + b_k\mathbf{x}_k = \underset{(n\times 1)}{\mathbf{0}} \qquad \text{[B.2]}$$

Equation B.1 or B.2 is called a linear *dependency* or *collinearity*. If these equations hold, then the vectors comprise a *linearly dependent* set. Note that the zero vector is linearly dependent on every other vector, inasmuch as $\mathbf{0} = 0\mathbf{x}$.

The *dimension* of the subspace spanned by a set of vectors is the number of vectors in the largest linearly independent subset. The dimension of the subspace spanned by $\{\mathbf{x}_1, \mathbf{x}_2, \ldots, \mathbf{x}_k\}$ cannot, therefore, exceed the smaller of k and n. These relations are illustrated for a vector space of dimension $n = 3$ in Figure B.6. Figure B.6(a) shows the one-dimensional subspace (i.e., the line) generated by a single nonzero vector $\mathbf{x}$; Figure B.6(b) shows the one-dimensional subspace generated by two collinear vectors $\mathbf{x}_1$ and $\mathbf{x}_2$; Figure B.6(c) shows the two-dimensional subspace (the plane) generated by two linearly independent vectors $\mathbf{x}_1$ and $\mathbf{x}_2$; and Figure B.6(d) shows the plane generated by three linearly dependent vectors $\mathbf{x}_1$, $\mathbf{x}_2$, and $\mathbf{x}_3$, no two of which are collinear. (In this last case, any one of the vectors lies in the plane generated by the other two.)

A linearly independent set of vectors $\{\mathbf{x}_1, \mathbf{x}_2, \ldots, \mathbf{x}_k\}$—such as $\{\mathbf{x}\}$ in Figure B.6(a) or $\{\mathbf{x}_1, \mathbf{x}_2\}$ in Figure B.6(c)—is said to provide a *basis* for the subspace that it spans. Any vector $\mathbf{y}$ in this subspace can be written *uniquely* as a linear combination of the basis vectors:

$$\mathbf{y} = c_1\mathbf{x}_1 + c_2\mathbf{x}_2 + \cdots + c_k\mathbf{x}_k$$

The constants $c_1, c_2, \ldots, c_k$ are called the *coordinates of* $\mathbf{y}$ *with respect to the basis* $\{\mathbf{x}_1, \mathbf{x}_2, \ldots, \mathbf{x}_k\}$.

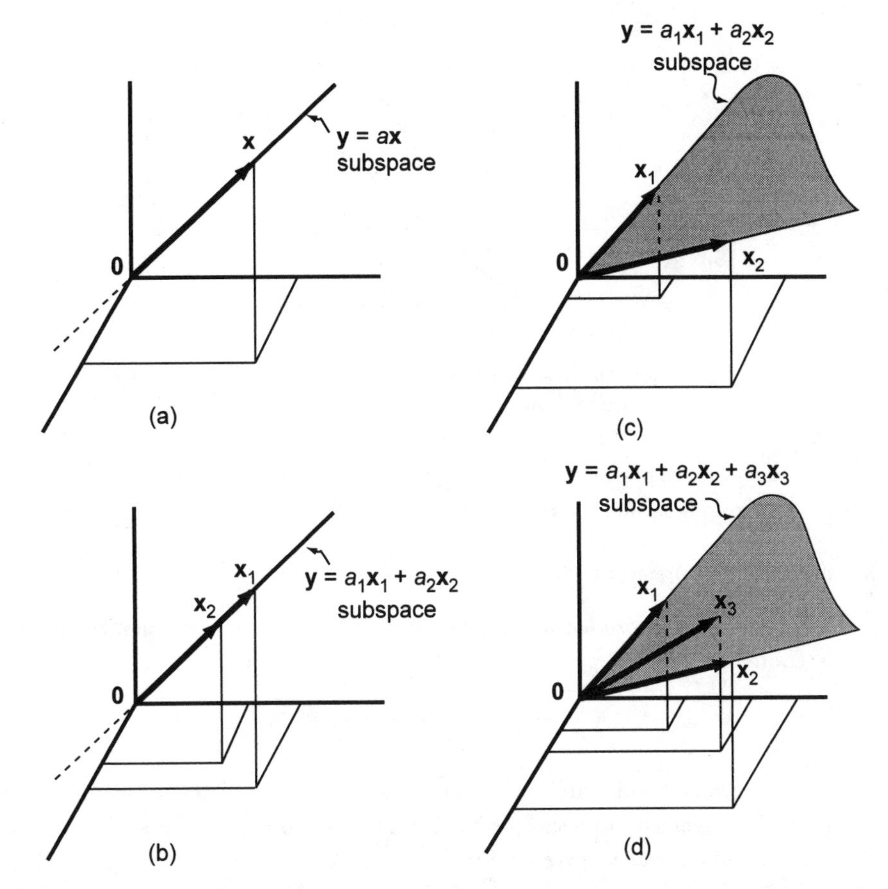

Figure B.6. Subspaces generated by sets of vectors in three-dimensional space. (*a*) One nonzero vector generates a one-dimensional subspace (a line). (*b*) Two collinear vectors also generate a one-dimensional subspace. (*c*) Two linearly independent vectors generate a two-dimensional subspace (a plane). (*d*) Three linearly dependent vectors, two of which are linearly independent, generate a two-dimensional subspace. The planes in (*c*) and (*d*) extend infinitely; they are drawn between x_1 and x_2 only for clarity.

The coordinates of a vector with respect to a basis for a two-dimensional subspace can be found geometrically by the parallelogram rule of vector addition, as illustrated in Figure B.7. Finding coordinates algebraically entails the solution of a system of linear simultaneous equations in which the c_j's are the unknowns:

$$\begin{aligned}\underset{(n\times 1)}{\mathbf{y}} &= c_1\mathbf{x}_1 + c_2\mathbf{x}_2 + \cdots + c_k\mathbf{x}_k \\ &= [\mathbf{x}_1, \mathbf{x}_2, \ldots, \mathbf{x}_k]\begin{bmatrix} c_1 \\ c_2 \\ \vdots \\ c_k \end{bmatrix} \\ &= \underset{(n\times k)}{\mathbf{X}}\underset{(k\times 1)}{\mathbf{c}}\end{aligned}$$

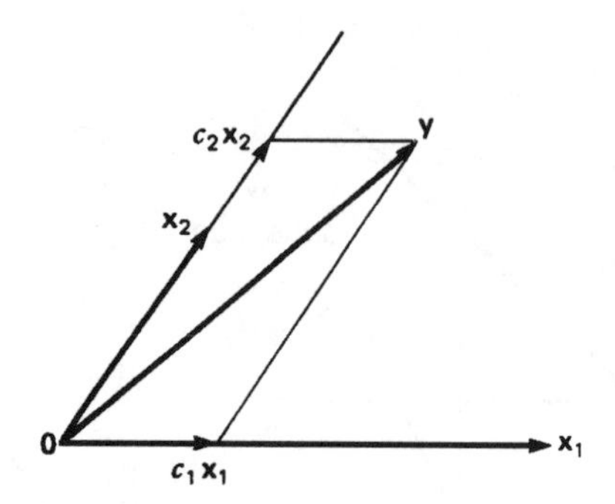

Figure B.7. The coordinates of y with respect to the basis $\{\mathbf{x}_1, \mathbf{x}_2\}$ of a two-dimensional subspace can be found from the parallelogram rule of vector addition.

When the vectors in $\{\mathbf{x}_1, \mathbf{x}_2, \ldots, \mathbf{x}_k\}$ are linearly independent, the matrix $\mathbf{X}$ is of full column rank k, and the equations have a unique solution.[3]

B.3 Orthogonality and Orthogonal Projections

The *inner product* (or *dot product*) of two vectors is the sum of products of their coordinates:

$$\mathbf{x} \cdot \mathbf{y} \equiv \sum_{i=1}^{n} x_i y_i$$

Two vectors $\mathbf{x}$ and $\mathbf{y}$ are *orthogonal* (i.e., perpendicular) if their inner product is 0. The essential geometry of vector orthogonality is shown in Figure B.8. Although $\mathbf{x}$ and $\mathbf{y}$ lie in an n-dimensional space (and therefore cannot, in general, be visualized directly), they span a subspace of dimension 2 which, by convention, we make the plane of the paper.[4] When $\mathbf{x}$ and $\mathbf{y}$ are orthogonal [as in Figure B.8(a)], the two right triangles with vertices $(\mathbf{0}, \mathbf{x}, \mathbf{x}+\mathbf{y})$ and $(\mathbf{0}, \mathbf{x}, \mathbf{x}-\mathbf{y})$ are congruent; consequently, $||\mathbf{x}+\mathbf{y}|| = ||\mathbf{x}-\mathbf{y}||$. Because the squared length of a vector is the inner product of the vector with itself ($\mathbf{x} \cdot \mathbf{x} = \sum x_i^2$), we have

$$(\mathbf{x}+\mathbf{y}) \cdot (\mathbf{x}+\mathbf{y}) = (\mathbf{x}-\mathbf{y}) \cdot (\mathbf{x}-\mathbf{y})$$

$$\mathbf{x} \cdot \mathbf{x} + 2\mathbf{x} \cdot \mathbf{y} + \mathbf{y} \cdot \mathbf{y} = \mathbf{x} \cdot \mathbf{x} - 2\mathbf{x} \cdot \mathbf{y} + \mathbf{y} \cdot \mathbf{y}$$

$$4\mathbf{x} \cdot \mathbf{y} = 0$$

$$\mathbf{x} \cdot \mathbf{y} = 0$$

When, in contrast, $\mathbf{x}$ and $\mathbf{y}$ are not orthogonal [as in Figure B.8(b)], then $||\mathbf{x}+\mathbf{y}|| \neq ||\mathbf{x}-\mathbf{y}||$, and $\mathbf{x} \cdot \mathbf{y} \neq 0$.

---

[3] The *column space* of a matrix is the vector space generated by its columns. The *row space* of a matrix is similarly defined. The *rank* of a matrix is the dimension of its row or column space (which, it turns out, have the same dimension). Matrix rank is intimately related to the solution of systems of linear simultaneous equations. If the topic is unfamiliar, then consult one of the references on linear algebra given at the end of this appendix.

[4] I often use this device in applying vector geometry to statistical problems, where the subspace of interest can often be confined to two or three dimensions, even though the dimension of the full vector space is typically equal to the sample size n.

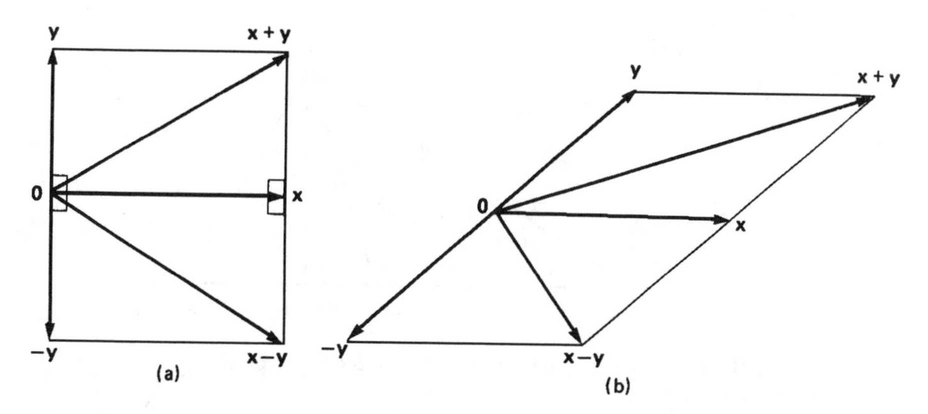

Figure B.8. When two vectors **x** and **y** are orthogonal, as in (*a*), their inner product $\mathbf{x} \cdot \mathbf{y}$ is 0. When the vectors are not orthogonal, as in (*b*), their inner product is nonzero.

The definition of orthogonality can be extended to matrices in the following manner: The matrix $\underset{(n \times k)}{\mathbf{X}}$ is orthogonal if each pair of its columns is orthogonal—that is, if $\mathbf{X}'\mathbf{X}$ is diagonal.[5] The matrix **X** is *orthonormal* if $\mathbf{X}'\mathbf{X} = \mathbf{I}$.

The *orthogonal projection* of one vector **y** onto another vector **x** is a scalar multiple $\hat{\mathbf{y}} = b\mathbf{x}$ of **x** such that $(\mathbf{y} - \hat{\mathbf{y}})$ is orthogonal to **x**. The geometry of orthogonal projection is illustrated in Figure B.9. By the Pythagorean theorem (see Figure B.10), $\hat{\mathbf{y}}$ is the point along the line spanned by **x** that is closest to **y**. To find b, we note that

$$\mathbf{x} \cdot (\mathbf{y} - \hat{\mathbf{y}}) = \mathbf{x} \cdot (\mathbf{y} - b\mathbf{x}) = 0$$

Thus, $\mathbf{x} \cdot \mathbf{y} - b\mathbf{x} \cdot \mathbf{x} = 0$ and $b = (\mathbf{x} \cdot \mathbf{y})/(\mathbf{x} \cdot \mathbf{x})$.

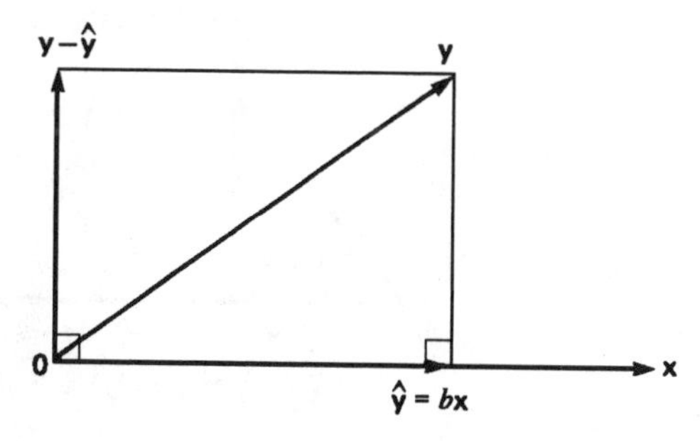

Figure B.9. The orthogonal projection $\hat{\mathbf{y}} = b\mathbf{x}$ of **y** onto **x**.

The orthogonal projection of **y** onto **x** can be used to determine the angle w separating two vectors, by finding its cosine. I shall distinguish between two cases:[6] In

---

[5] The i, jth entry of $\mathbf{X}'\mathbf{X}$ is $\mathbf{x}_i'\mathbf{x}_j = \mathbf{x}_i \cdot \mathbf{x}_j$, where $\mathbf{x}_i$ and $\mathbf{x}_j$ are, respectively, the ith and jth columns of **X**. The ith diagonal entry of $\mathbf{X}'\mathbf{X}$ is likewise $\mathbf{x}_i'\mathbf{x}_i = \mathbf{x}_i \cdot \mathbf{x}_i$.

[6] By convention, we examine the smaller of the two angles separating a pair of vectors, and, therefore, never encounter angles that exceed 180°. Call the smaller angle w; then the larger angle is $360 - w$. Our convention is of no consequence because $\cos(360 - w) = \cos w$.

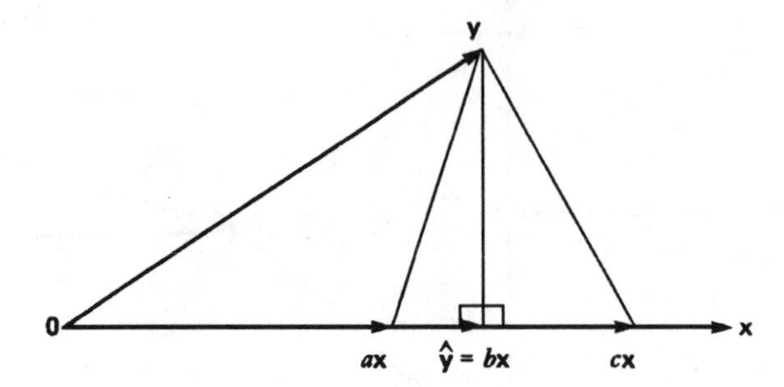

Figure B.10. The orthogonal projection $\hat{\mathbf{y}} = b\mathbf{x}$ is the point along the line spanned by **x** that is closest to **y**.

Figure B.11(a), the angle separating the vectors is between 0° and 90°; in Figure B.11(b), the angle is between 90° and 180°. In the first instance,

$$\cos w = \frac{||\hat{\mathbf{y}}||}{||\mathbf{y}||} = \frac{b||\mathbf{x}||}{||\mathbf{y}||} = \frac{\mathbf{x} \cdot \mathbf{y}}{||\mathbf{x}||^2} \times \frac{||\mathbf{x}||}{||\mathbf{y}||} = \frac{\mathbf{x} \cdot \mathbf{y}}{||\mathbf{x}|| \times ||\mathbf{y}||}$$

and, likewise, in the second instance,

$$\cos w = -\frac{||\hat{\mathbf{y}}||}{||\mathbf{y}||} = \frac{b||\mathbf{x}||}{||\mathbf{y}||} = \frac{\mathbf{x} \cdot \mathbf{y}}{||\mathbf{x}|| \times ||\mathbf{y}||}$$

In both instances, the sign of b for the orthogonal projection of **y** onto **x** correctly reflects the sign of $\cos w$.

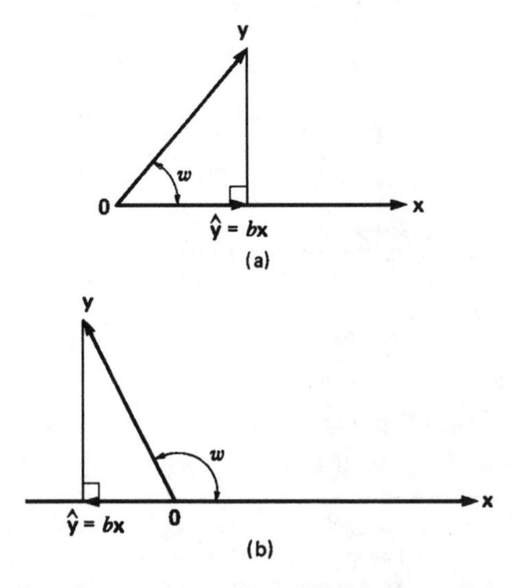

Figure B.11. The angle w separating two vectors, **x** and **y**: (a) $0° < w < 90°$; (b) $90° < w < 180°$.

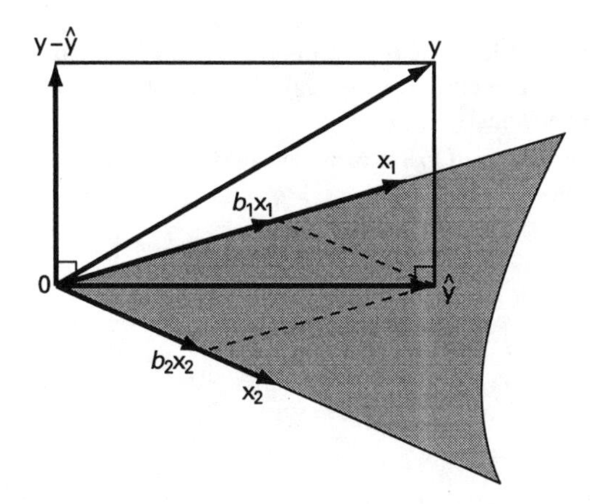

Figure B.12. The orthogonal projection $\hat{\mathbf{y}}$ of $\mathbf{y}$ onto the subspace (plane) spanned by $\mathbf{x}_1$ and $\mathbf{x}_2$.

The orthogonal projection of a vector $\mathbf{y}$ onto the subspace spanned by a set of vectors $\{\mathbf{x}_1, \mathbf{x}_2, \ldots, \mathbf{x}_k\}$ is the vector

$$\hat{\mathbf{y}} = b_1\mathbf{x}_1 + b_2\mathbf{x}_2 + \cdots + b_k\mathbf{x}_k$$

formed as a linear combination of the $\mathbf{x}_j$'s such that $(\mathbf{y} - \hat{\mathbf{y}})$ is orthogonal to each and every vector $\mathbf{x}_j$ in the set. The geometry of orthogonal projection for $k = 2$ is illustrated in Figure B.12. The vector $\hat{\mathbf{y}}$ is the point closest to $\mathbf{y}$ in the subspace spanned by the $\mathbf{x}_j$'s.

Placing the constants b_j into a vector $\mathbf{b}$, and gathering the vectors $\mathbf{x}_j$ into an $(n \times k)$ matrix $\mathbf{X} \equiv [\mathbf{x}_1, \mathbf{x}_2, \ldots, \mathbf{x}_k]$, we have $\hat{\mathbf{y}} = \mathbf{Xb}$. By the definition of an orthogonal projection,

$$\mathbf{x}_j \cdot (\mathbf{y} - \hat{\mathbf{y}}) = \mathbf{x}_j \cdot (\mathbf{y} - \mathbf{Xb}) = 0 \quad \text{for } j = 1, \ldots, k$$

Equivalently, $\mathbf{X}'(\mathbf{y} - \mathbf{Xb}) = 0$, or $\mathbf{X}'\mathbf{y} = \mathbf{X}'\mathbf{Xb}$. We can solve this matrix equation uniquely for $\mathbf{b}$ as long as $\mathbf{X}'\mathbf{X}$ is nonsingular, in which case $\mathbf{b} = (\mathbf{X}'\mathbf{X})^{-1}\mathbf{X}'\mathbf{y}$. The matrix $\mathbf{X}'\mathbf{X}$ is nonsingular if $\{\mathbf{x}_1, \mathbf{x}_2, \ldots, \mathbf{x}_k\}$ is a linearly independent set of vectors, providing a basis for the subspace that it generates; otherwise, $\mathbf{b}$ is not unique.

B.4 Recommended Reading

There is a plethora of books on linear algebra and matrices. Most presentations develop the fundamental properties of vector spaces, but often, unfortunately, without explicit visual representation.

- Several matrix texts, including Healy (1986), Graybill (1983), Searle (1982), and Green and Carroll (1976), focus specifically on statistical applications. The last of these sources has a strongly geometric orientation.

- Davis (1965), who presents a particularly lucid and simple treatment of matrix algebra, includes some material on vector geometry (limited, however, to two dimensions).
- Namboodiri (1984) provides a compact introduction to matrix algebra (but not to vector geometry).

Appendix C

Multivariable and Matrix Differential Calculus*

Multivariable differential calculus—the topic of this brief appendix—finds frequent application in statistics, particularly to optimization problems (e.g., *least* squares, *maximum* likelihood). The essential ideas of multivariable calculus are straightforward extensions of calculus of a single independent variable, but the topic is frequently omitted from introductory treatments of calculus—hence, its inclusion here. The material in this appendix assumes familiarity with the basics of differential calculus.

C.1 Partial Derivatives

Consider a function $y = f(x_1, x_2, \ldots, x_n)$ of several independent variables.[1] The *partial derivative* of y with respect to a particular x_i is the derivative of $f(x_1, x_2, \ldots, x_n)$ treating the other x's constant. To distinguish it from the ordinary derivative dy/dx, the standard notation for the partial derivative uses Greek deltas in place of *d*'s: $\partial y/\partial x_i$.

For example, for the function

$$y = f(x_1, x_2) \equiv x_1^2 + 3x_1x_2^2 + x_2^3 + 6$$

the partial derivatives with respect to x_1 and x_2 are

$$\frac{\partial y}{\partial x_1} = 2x_1 + 3x_2^2 + 0 + 0 = 2x_1 + 3x_2^2$$

$$\frac{\partial y}{\partial x_2} = 0 + 6x_1x_2 + 3x_2^2 + 0 = 6x_1x_2 + 3x_2^2$$

[1] I am using the term "independent variable" here in the mathematical, rather than the statistical, sense.

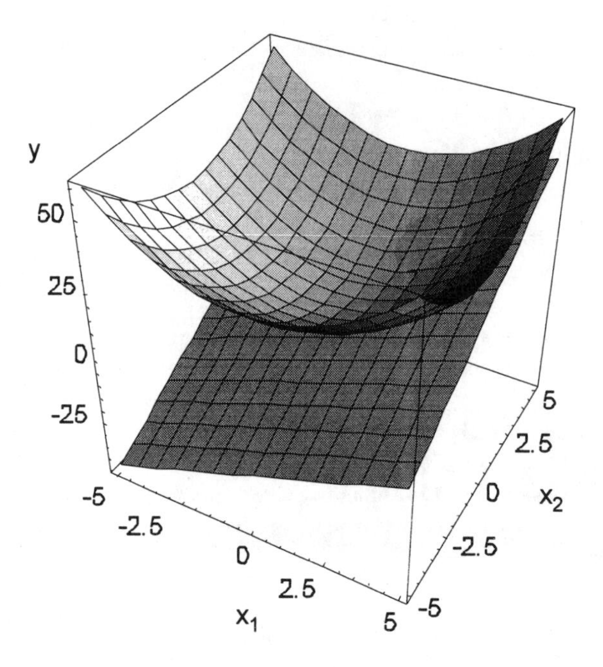

Figure C.1. The function $y = f(x_1, x_2) = x_1^2 + x_1x_2 + x_2^2 + 10$, showing the tangent plane at $x_1 = 1$, $x_2 = 2$.

The "trick" in partial differentiation with respect to x_i is to treat all of the other x's as constants (i.e., literally to hold other x's constant). Thus, when we differentiate with respect to x_1, terms such as x_2^2 and x_2^3 are constants.

The partial derivative $\partial f(x_1, x_2, \ldots, x_n)/\partial x_1$ gives the slope of the tangent hyperplane to the function $f(x_1, x_2, \ldots, x_n)$ in the direction of x_1. For example, the tangent plane to the function $f(x_1, x_2) \equiv x_1^2 + x_1x_2 + x_2^2 + 10$ above the pair of values $x_1 = 1$, $x_2 = 2$ is shown in Figure C.1.

At a local minimum or maximum, the slope of the tangent hyperplane is 0 in all directions. Consequently, to minimize or maximize a function of several variables, we need to differentiate the function with respect to each variable, set the partial derivatives to 0, and solve the resulting set of simultaneous equations.[2]

Let us, for example, find the values of x_1 and x_2 that minimize the function $y = f(x_1, x_2) \equiv x_1^2 + x_1x_2 + x_2^2 + 10$. Differentiating,

$$\frac{\partial y}{\partial x_1} = 2x_1 + x_2$$

$$\frac{\partial y}{\partial x_2} = x_1 + 2x_2$$

Setting these partial derivatives to 0 produces the unique solution $x_1 = 0$, $x_2 = 0$. In this case, the solution is particularly simple because the partial derivatives are linear functions of x_1 and x_2. The value of the function at its minimum is $y = 0^2 + (0 \times 0) + 0^2 + 10 = 10$.

[2] I shall explain in Section C.3 how to distinguish maxima from minima.

The slopes of the tangent plane above the pair of values $x_1 = 1$, $x_2 = 2$, illustrated in Figure C.1, are $\partial y/\partial x_1 = 2(1) + 2 = 4$, and $\partial y/\partial x_2 = 1 + 2(2) = 5$.

C.2 Lagrange Multipliers for Constrained Optimization

The method of Lagrange multipliers[3] permits us to optimize a function of the form $y = f(x_1, x_2, \ldots, x_n)$ subject to a constraint of the form $g(x_1, x_2, \ldots, x_n) = 0$. The method, in effect, incorporates the constraint into the set of partial derivatives.

Here is a simple example: Minimize $y = f(x_1, x_2) \equiv x_1^2 + x_2^2$ subject to the restriction that $x_1 + x_2 = 1$. (In the absence of this restriction, it is obvious that $x_1 = x_2 = 0$ minimizes the function.) To solve this constrained minimization problem:

1. Rewrite the constraint in the required form, $g(x_1, x_2, \ldots, x_n) = 0$. That is, $x_1 + x_2 - 1 = 0$.
2. Construct a new function incorporating the constraint. In the general case, this function takes the form

$$h(x_1, x_2, \ldots, x_n, L) \equiv f(x_1, x_2, \ldots, x_n) - L \times g(x_1, x_2, \ldots, x_n)$$

   The new independent variable L is called a *Lagrange multiplier*. For the example,

$$h(x_1, x_2, L) \equiv x_1^2 + x_2^2 - L(x_1 + x_2 - 1)$$

3. Find the values of $x_1, x_2, \ldots, x_n$ that (along with L) optimize the function $h(x_1, x_2, \ldots, x_n, L)$. That is, differentiate $h(x_1, x_2, \ldots, x_n, L)$ with respect to each of $x_1, x_2, \ldots, x_n$ and L; set the $n+1$ partial derivatives to 0; and solve the resulting system of simultaneous equations for $x_1, x_2, \ldots, x_n$ and L. For the example,

$$\frac{\partial h(x_1, x_2, L)}{\partial x_1} = 2x_1 - L$$
$$\frac{\partial h(x_1, x_2, L)}{\partial x_2} = 2x_2 - L$$
$$\frac{\partial h(x_1, x_2, L)}{\partial L} = -x_1 - x_2 + 1$$

   Notice that the partial derivative with respect to L, when equated to 0, reproduces the constraint $x_1 + x_2 = 1$. Consequently, whatever solutions satisfy the equations produced by setting the partial derivatives to 0, necessarily satisfy the constraint. In this case, there is only one solution: $x_1 = x_2 = 0.5$ (and $L = 1$).

The method of Lagrange multipliers easily extends to handle several restrictions, by introducing a separate Lagrange multiplier for each restriction.

C.3 Differential Calculus in Matrix Form

The function $y = f(x_1, x_2, \ldots, x_n)$ of the independent variables $x_1, x_2, \ldots, x_n$ can be written as the function $y = f(\mathbf{x})$ of the vector $\mathbf{x} = [x_1, x_2, \ldots, x_n]'$. The *vector partial derivative* of y with respect to $\mathbf{x}$ is defined as the column vector of partial derivatives of

[3] The method is named after the 18th-century French mathematician J. L. Lagrange.

y with respect to each of the entries of $\mathbf{x}$:

$$\frac{\partial y}{\partial \mathbf{x}} \equiv \left[\frac{\partial y}{\partial x_1}, \frac{\partial y}{\partial x_2}, \ldots, \frac{\partial y}{\partial x_n}\right]'$$

If, therefore, y is a linear function of $\mathbf{x}$,

$$y = \underset{(1\times n)(n\times 1)}{\mathbf{a}' \quad \mathbf{x}} = a_1x_1 + a_2x_2 + \cdots + a_nx_n$$

then $\partial y/\partial x_i = a_i$, and $\partial y/\partial \mathbf{x} = \mathbf{a}$. Alternatively, suppose that y is a *quadratic form* in $\mathbf{x}$,

$$y = \underset{(1\times n)(n\times n)(n\times 1)}{\mathbf{x}' \quad \mathbf{A} \quad \mathbf{x}}$$

where the matrix $\mathbf{A}$ is symmetric. Expanding the matrix product gives us

$$y = a_{11}x_1^2 + a_{22}x_2^2 + \cdots + a_{nn}x_n^2 + 2a_{12}x_1x_2 + \cdots + 2a_{1n}x_1x_n + \cdots + 2a_{n-1,\,n}x_{n-1}x_n$$

and, thus,

$$\frac{\partial y}{\partial x_i} = 2(a_{i1}x_1 + a_{i2}x_2 + \cdots + a_{in}x_n) = 2\mathbf{a}_i'\mathbf{x}$$

where $\mathbf{a}_i'$ represents the ith row of $\mathbf{A}$. Placing these partial derivatives in a vector produces $\partial y/\partial \mathbf{x} = 2\mathbf{A}\mathbf{x}$. The vector partial derivatives of linear and quadratic functions are strikingly similar to the analogous scalar derivatives of functions of one variable: $d(ax)/dx = a$ and $d(ax^2)/dx = 2ax$.

The so-called Hessian matrix of second-order partial derivatives of the function $y = f(\mathbf{x})$ is defined in the following manner:

$$\frac{\partial^2 y}{\partial \mathbf{x}\,\partial \mathbf{x}'} \equiv \begin{bmatrix} \frac{\partial^2 y}{\partial x_1^2} & \frac{\partial^2 y}{\partial x_1\,\partial x_2} & \cdots & \frac{\partial^2 y}{\partial x_1\,\partial x_n} \\ \frac{\partial^2 y}{\partial x_2\,\partial x_1} & \frac{\partial^2 y}{\partial x_2^2} & \cdots & \frac{\partial^2 y}{\partial x_2\,\partial x_n} \\ \vdots & \vdots & \ddots & \vdots \\ \frac{\partial^2 y}{\partial x_n\,\partial x_1} & \frac{\partial^2 y}{\partial x_n\,\partial x_2} & \cdots & \frac{\partial^2 y}{\partial x_n^2} \end{bmatrix}$$

For instance, $\partial^2(\mathbf{x}'\mathbf{A}\mathbf{x})/\partial \mathbf{x}\,\partial \mathbf{x}' = 2\mathbf{A}$, for a symmetric matrix $\mathbf{A}$.

To minimize a function $y = f(\mathbf{x})$ of several variables, we can set the vector partial derivative to $\mathbf{0}$, $\partial y/\partial \mathbf{x} = \mathbf{0}$, and solve the resulting set of simultaneous equations for $\mathbf{x}$, obtaining a solution $\mathbf{x}^*$. This solution represents a (local) minimum of the function in question if the Hessian matrix evaluated at $\mathbf{x} = \mathbf{x}^*$ is positive definite. The solution represents a maximum if the Hessian is negative definite.[4] Again, there is a strong parallel

[4] The square matrix $\mathbf{H}$ is *positive definite* if $\mathbf{x}'\mathbf{H}\mathbf{x} > 0$ for any nonzero vector $\mathbf{x}$. A positive-definite Hessian is a sufficient but not necessary condition for a minimum. Likewise, the square matrix $\mathbf{H}$ is *negative definite* if $\mathbf{x}'\mathbf{H}\mathbf{x} < 0$ for any nonzero vector $\mathbf{x}$; a negative-definite Hessian is a sufficient but not necessary condition for a maximum.

with the scalar results for a single x: Recall that the second derivative d^2y/dx^2 is positive at a minimum and negative at a maximum.

I showed earlier that the function $y = f(x_1, x_2) = x_1^2 + x_1x_2 + x_2^2 + 10$ has a *stationary point* (i.e., a point at which the partial derivatives are 0) at $x_1 = x_2 = 0.5$. The second-order partial derivatives of this function are

$$\frac{\partial^2 y}{\partial x_1\,\partial x_2} = \frac{\partial^2 y}{\partial x_2\,\partial x_1} = 1$$

$$\frac{\partial^2 y}{\partial x_1^2} = \frac{\partial^2 y}{\partial x_2^2} = 2$$

The Hessian evaluated at $x_1 = x_2 = 0.5$ (or, indeed, at any point), is, therefore,

$$\begin{bmatrix} \dfrac{\partial^2 y}{\partial x_1^2} & \dfrac{\partial^2 y}{\partial x_1\,\partial x_2} \\ \dfrac{\partial^2 y}{\partial x_2\,\partial x_1} & \dfrac{\partial^2 y}{\partial x_2^2} \end{bmatrix} = \begin{bmatrix} 2 & 1 \\ 1 & 2 \end{bmatrix}$$

This matrix is clearly positive definite, verifying that the value $y = 10$ at $x_1 = x_2 = 0.5$ is a minimum of $f(x_1, x_2)$.

Appendix D

Probability and Estimation

The purpose of this appendix is to outline basic results in probability and statistical inference that are employed principally in the starred parts of the text. Material in the unstarred portions of this appendix is, however, used occasionally in the unstarred parts of the text. For good reason, elementary statistics courses—particularly in the social sciences—often provide only the barest introduction to probability and the theory of estimation. After a certain point, however, some background in these topics is necessary.

In Section D.1, I review concepts in elementary probability theory. Sections D.2 and D.3 briefly describe several probability distributions that are of particular importance in the study of linear and related models. Section D.4 outlines asymptotic distribution theory, which we shall occasionally require to determine properties of statistical estimators, a subject that is taken up in Section D.5. The concluding section of the appendix, Section D.6, develops the broadly applicable and centrally important method of maximum-likelihood estimation. Taken together, the sections of this appendix provide a "crash course" in some of the basics of mathematical statistics.

D.1 Elementary Probability Theory

D.1.1 Basic Definitions

In probability theory, an *experiment* is a repeatable procedure for making an observation; an *outcome* is a possible observation resulting from an experiment; and the *sample space* of the experiment is the set of all possible outcomes. Any specific *realization* of the experiment produces a particular outcome in the sample space. Sample spaces may be discrete and finite, discrete and infinite (i.e., countably infinite[1]), or continuous.

If, for example, we flip a coin twice and record on each flip whether the coin shows heads (H) or tails (T), then the sample space of the experiment is discrete and finite, consisting of the outcomes $S = \{HH, HT, TH, TT\}$. If, alternatively, we flip a coin repeatedly until a head appears, and record the number of flips required to obtain this result, then the sample space is discrete and infinite, consisting of the positive integers, $S = \{1, 2, 3, \ldots\}$. If we burn a light bulb until it fails, recording the burning time in hours and fractions of an hour, then the sample space of the experiment is continuous and consists of all positive real numbers (not bothering to specify an upper limit for the life of a bulb): $S = \{x\colon x > 0\}$. In this section, I shall limit consideration to discrete, finite sample spaces.

An *event* is a subset of the sample space of an experiment—that is, a set of outcomes. An event is said to occur in a realization of the experiment if one of its constituent outcomes occurs. For example, for $S = \{HH, HT, TH, TT\}$, the event $E \equiv \{HH, HT\}$, representing a head on the first flip of the coin, occurs if we obtain either the outcome HH or the outcome HT.

Let $S = \{o_1, o_2, \ldots, o_n\}$ be the sample space of an experiment; let $O_1 \equiv \{o_1\}$, $O_2 \equiv \{o_2\}, \ldots, O_n \equiv \{o_n\}$ be the *simple events*, each consisting of one of the outcomes; and let the event $E = \{o_a, o_b, \ldots, o_m\}$ be any subset[2] of S. *Probabilities* are numbers assigned to events in a manner consistent with the following axioms (rules):

P1. $\Pr(E) \geq 0$: The probability of an event is nonnegative.
P2. $\Pr(E) = \Pr(O_a) + \Pr(O_b) + \cdots + \Pr(O_m)$: The probability of an event is the sum of probabilities of its constituent outcomes.
P3. $\Pr(S) = 1; \Pr(\emptyset) = 0$, where $\emptyset$ is the *empty event*, which contains no outcomes: The sample space is exhaustive—some outcome must occur.

Suppose, for example, that all outcomes in the sample space $S = \{HH, HT, TH, TT\}$ are equally likely,[3] so that

$$\Pr(HH) = \Pr(HT) = \Pr(TH) = \Pr(TT) = .25$$

Then, for $E \equiv \{HH, HT\}$, $\Pr(E) = .25 + .25 = .5$.

In classical statistics, the perspective adopted in most applications of statistics (and in this book), probabilities are interpreted as long-run proportions. Thus, if the probability of an event is $\frac{1}{2}$, then the event will occur approximately half the time when the experiment is repeated many times, and the approximation is expected to improve as the number of repetitions increases.

---

[1] To say that a set is countably infinite means that a one-to-one relationship can be established between the elements of the set and the natural numbers $0, 1, 2, \ldots$.

[2] The subscripts $a, b, \ldots, m$ are each (different) numbers between 1 and n.

[3] Equally likely outcomes produce a simple example—and correspond to a "fair" coin "fairly" flipped—but any assignment of probabilities to outcomes that sums to 1 is consistent with the axioms.

A number of important relations can be defined among events. The *intersection* of two events, E_1 and E_2, denoted $E_1 \cap E_2$, contains all outcomes common to the two; $\Pr(E_1 \cap E_2)$ is thus the probability that *both* E_1 *and* E_2 occur simultaneously. If $E_1 \cap E_2 = \emptyset$, then E_1 and E_2 are said to be *disjoint* or *mutually exclusive*. By extension, the intersection of many events $E_1 \cap E_2 \cap \cdots \cap E_k$ contains all outcomes that are members of each and every event. Consider, for example, the events $E_1 \equiv \{HH, HT\}$ (a head on the first trial), $E_2 \equiv \{HH, TH\}$ (a head on the second trial), and $E_3 \equiv \{TH, TT\}$ (a tail on the first trial). Then $E_1 \cap E_2 = \{HH\}$, $E_1 \cap E_3 = \emptyset$, and $E_2 \cap E_3 = \{TH\}$.

The *union* of two events $E_1 \cup E_2$ contains all outcomes that are in either or both events, $\Pr(E_1 \cup E_2)$ is the probability that E_1 occurs *or* that E_2 occurs (or that *both* occur). The union of several events $E_1 \cup E_2 \cup \cdots \cup E_k$ contains all outcomes that are in one or more of the events. If these events are disjoint, then

$$\Pr(E_1 \cup E_2 \cup \cdots \cup E_k) = \sum_{i=1}^{k} \Pr(E_i)$$

otherwise

$$\Pr(E_1 \cup E_2 \cup \cdots \cup E_k) < \sum_{i=1}^{k} \Pr(E_i)$$

(because some outcomes contribute more than once when the probabilities are summed). For two events,

$$\Pr(E_1 \cup E_2) = \Pr(E_1) + \Pr(E_2) - \Pr(E_1 \cap E_2)$$

Subtracting the intersection corrects for double counting. To extend the previous example, assuming equally likely outcomes,

$$\begin{aligned}
\Pr(E_1 \cup E_3) &= \Pr(HH, HT, TH, TT) = 1 \\
&= \Pr(E_1) + \Pr(E_3) \\
&= .5 + .5 \\
\Pr(E_1 \cup E_2) &= \Pr(HH, HT, TH) = .75 \\
&= \Pr(E_1) + \Pr(E_2) - \Pr(E_1 \cap E_2) \\
&= .5 + .5 - .25
\end{aligned}$$

The *conditional probability of* E_2 *given* E_1 is

$$\Pr(E_2|E_1) \equiv \frac{\Pr(E_1 \cap E_2)}{\Pr(E_1)}$$

The conditional probability is interpreted as the probability that E_2 will occur if E_1 is known to have occurred. Two events are *independent*[4] if $\Pr(E_1 \cap E_2) = \Pr(E_1)\Pr(E_2)$. Independence of E_1 and E_2 implies that $\Pr(E_1) = \Pr(E_1|E_2)$ and that $\Pr(E_2) = \Pr(E_2|E_1)$.

[4] Independence is different from disjointness. If two events are disjoint, then they cannot occur together, and they are, therefore, *dependent*.

More generally, a set of events $\{E_1, E_2, \ldots, E_k\}$ is independent if, for every subset $\{E_a, E_b, \ldots, E_m\}$ containing two or more of the events,

$$\Pr(E_a \cap E_b \cap \cdots \cap E_m) = \Pr(E_a)\Pr(E_b)\cdots\Pr(E_m)$$

Appealing once more to our example, the probability of a head on the second trial (E_2) given a head on the first trial (E_1) is

$$\begin{aligned} \Pr(E_2|E_1) &= \frac{\Pr(E_1 \cap E_2)}{\Pr(E_1)} \\ &= \frac{.25}{.5} = .5 \\ &= \Pr(E_2) \end{aligned}$$

Likewise, $\Pr(E_1 \cap E_2) = .25 = \Pr(E_1)\Pr(E_2) = .5 \times .5$. The events E_1 and E_2 are, therefore, independent.

The *difference* between two events $E_1 - E_2$ contains all outcomes in the first event that are not in the second. The difference $\overline{E} \equiv S - E$ is called the *complement* of the event E. Note that $\Pr(\overline{E}) = 1 - \Pr(E)$. From the example, where $E_1 \equiv \{HH, HT\}$ with all outcomes equally likely, $\Pr(\overline{E}_1) = \Pr(TH, TT) = .5 = 1 - .5$.

Let $E \equiv E_1 \cap E_2 \cap \cdots E_k$. Then $\overline{E} = \overline{E}_1 \cup \overline{E}_2 \cup \cdots \cup \overline{E}_k$. Applying previous results,

$$\begin{aligned} \Pr(E_1 \cap E_2 \cap \cdots \cap E_k) &= \Pr(E) = 1 - \Pr(\overline{E}) \\ &\geq 1 - \sum_{i=1}^{k} \Pr(\overline{E}_i) \end{aligned} \qquad \text{[D.1]}$$

Suppose that all of the events $E_1, E_2, \ldots, E_k$ have equal probabilities, say $\Pr(E_i) = 1 - b$ [so that $\Pr(\overline{E}_i) = b$]. Then

$$\begin{aligned} \Pr(E_1 \cap E_2 \cap \cdots \cap E_k) &\equiv 1 - a \\ &\geq 1 - kb \end{aligned} \qquad \text{[D.2]}$$

Equation D.2 and the more general Equation D.1 are called *Bonferroni inequalities*.

Equation D.2 has the following use in simultaneous statistical inference: Suppose that b is the Type I error rate for each of k nonindependent statistical tests. Let a represent the combined Type I error rate for the k tests—that is, the probability of falsely rejecting at least one of k true null hypotheses. Then $a \leq kb$. For instance, if we test 20 true statistical hypotheses, each at a significance level of .01, then the probability of rejecting at least one hypothesis is at most $20 \times .01 = .20$ (i.e., no more than one chance in five)—a sober reminder that "data dredging" can prove seriously misleading.

D.1.2 Random Variables

A *random variable* is a function that assigns a number to each outcome of the sample space of an experiment. For the sample space $S = \{HH, HT, TH, TT\}$, introduced earlier, a random variable X that counts the number of heads in an outcome is defined as follows:

Outcome	Value x of X
HH	2
HT	1
TH	1
TT	0

If, as in this example, X is a discrete random variable, then we write $p(x)$ for $\Pr(X = x)$, where the uppercase letter X represents the random variable, while the lowercase letter x denotes a *particular value* of the variable. The probabilities $p(x)$ for all values of X comprise the *probability distribution* of the random variable. If, for example, each of the four outcomes of the coin-flipping experiment has probability .25, then the probability distribution of the number of heads is

x	$p(x)$
0	.25
1	.50
2	.25
sum	1.00

The *cumulative distribution function* (*CDF*) of a random variable X, written $P(x)$, gives the probability of observing a value of the variable that is less than or equal to a particular value:

$$P(x) \equiv \Pr(X \leq x) = \sum_{x' \leq x} p(x')$$

For the example,

x	$P(x)$
0	.25
1	.75
2	1.00

Random variables defined on continuous sample spaces may themselves be continuous. We still take $P(x)$ as $\Pr(X \leq x)$, but it generally becomes meaningless to refer to the probability of observing individual values of X. The *probability density function* $p(x)$

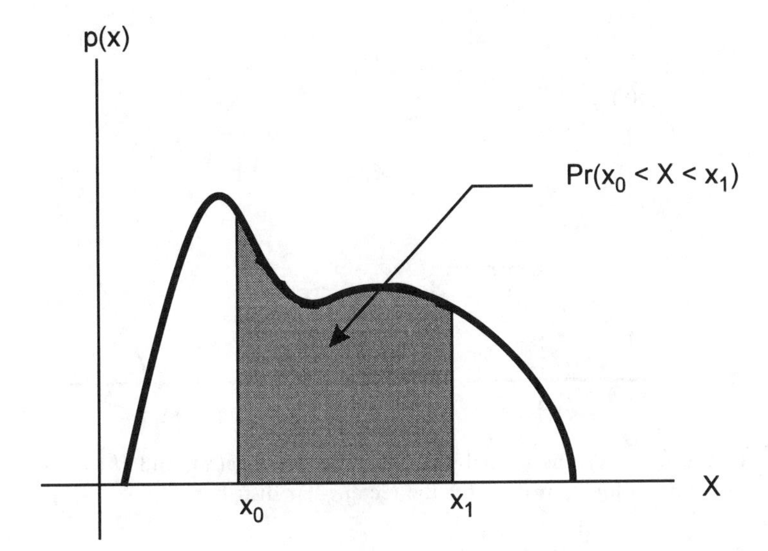

Figure D.1. Areas under the probability density function $p(x)$ are interpreted as probabilities.

is, nevertheless, the continuous analog of the discrete probability distribution, defining $p(x) \equiv dP(x)/dx$. Reversing this relation,[5] $P(x) = \int_{-\infty}^{x} p(x)\,dx$; and

$$\Pr(x_0 \leq x \leq x_1) = P(x_1) - P(x_0) = \int_{x_0}^{x_1} p(x)\,dx$$

Thus, as illustrated in Figure D.1, areas under the density function are interpreted as probabilities.

A particularly simple continuous probability distribution is the *rectangular distribution*:

$$p(x) = \begin{cases} 0 & a > x \\ \dfrac{1}{b-a} & a \leq x \leq b \\ 0 & x > b \end{cases}$$

This density function is pictured in Figure D.2(a), and the corresponding cumulative distribution function is shown in Figure D.2(b). The total area under a density function must be 1; here,

$$\int_{-\infty}^{\infty} p(x)dx = \int_{a}^{b} p(x)dx = \frac{1}{b-a}(b-a) = 1$$

[5] If you are unfamiliar with integral calculus, do not be too concerned: The principal point to understand is that *areas* under the density curve $p(x)$ are interpreted as probabilities, and that the *height* of the CDF $P(x)$ gives the probability of observing values of X less than or equal to the value x. The integral sign $\int$ is the continuous analog of a sum, and represents the area under a curve.

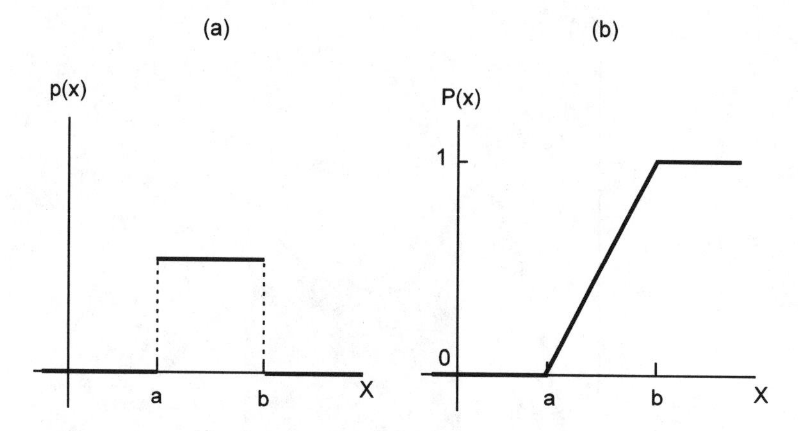

Figure D.2. (*a*) The probability density function $p(x)$, and (*b*) the cumulative distribution function $P(x)$ for the rectangular distribution.

Two fundamental properties of a random variable are its *expected value* (or *mean*) and its *variance*.[6] The expected value specifies the center of the probability distribution of the random variable (in the same sense as the mean of a set of scores specifies the center of their distribution), while the variance indicates how spread out the distribution is around its expectation. The expectation is interpretable as the mean score of the random variable that would be observed over many repetitions of the experiment, while the variance is the mean-squared distance between the scores and their expectation.

In the discrete case, the expectation of a random variable X, symbolized by $E(X)$ or μ_X, is given by

$$E(X) \equiv \sum_{\text{all } x} xp(x)$$

The analogous formula for the continuous case is

$$E(X) \equiv \int_{-\infty}^{\infty} xp(x)\, dx$$

The variance of a random variable X, written $V(X)$ or σ_X^2, is defined as $E[(X - \mu_X)^2]$. Thus, in the discrete case,

$$V(X) \equiv \sum_{\text{all } x} (x - \mu_X)^2 p(x)$$

while, in the continuous case,

$$V(X) \equiv \int_{-\infty}^{\infty} (x - \mu_X)^2 p(x)\, dx$$

[6] The expectation and variance are undefined for some random variables, a possibility that I shall ignore here.

The variance is expressed in the squared units of the random variable (e.g., "squared number of heads"), but the standard deviation $\sigma \equiv +\sqrt{\sigma^2}$ is measured in the same units as the variable.

For our example,

x	$p(x)$	$xp(x)$	$x-\mu$	$(x-\mu)^2 p(x)$
0	.25	0.00	−1	0.25
1	.50	0.50	0	0.00
2	.25	0.50	1	0.25
sum	1.00	$\mu = 1.00$		$\sigma^2 = 0.50$

Thus, $E(X) = 1$, $V(X) = 0.5$, and $\sigma = \sqrt{0.5} \simeq 0.707$.

The *joint probability distribution* of two discrete random variables X_1 and X_2 gives the probability of simultaneously observing any pair of values for the two variables. We write $p_{12}(x_1, x_2)$ for $\Pr(X_1 = x_1$ and $X_2 = x_2)$; it is usually unambiguous to drop the subscript on p, simply writing $p(x_1, x_2)$. The *joint probability density* $p(x_1, x_2)$ of two continuous random variables is defined analogously. Extension to the joint probability or joint probability density $p(x_1, x_2, \ldots, x_n)$ of several random variables is straightforward.

To distinguish it from the joint probability distribution, we call $p_1(x_1)$ the marginal probability distribution or marginal probability density for X_1. Note that $p_1(x_1) = \sum_{x_2} p(x_1, x_2)$ or $p_1(x_1) = \int_{-\infty}^{\infty} p(x_1, x_2)\, dx_2$. We usually drop the subscript, to write $p(x_1)$.

In the fair coin-flipping experiment, for example, let X_1 count the number of heads, and let $X_2 = 1$ if both coins are the same and 0 if they are different:

Outcome	Pr	x_1	x_2
HH	.25	2	1
HT	.25	1	0
TH	.25	1	0
TT	.25	0	1

The joint and marginal distributions for X_1 and X_2 are as follows:

$p(x_1, x_2)$			
	x_2		
x_1	0	1	$p(x_1)$
0	0	.25	.25
1	.50	0	.50
2	0	.25	.25
$p(x_2)$	.50	.50	1.00

The *conditional* probability or probability density of X_1 given X_2 is

$$p_{1|2}(x_1|x_2) = \frac{p_{12}(x_1, x_2)}{p_1(x_2)}$$

As usual, it is generally convenient to drop the subscript, writing $p(x_1|x_2)$. For our example, $p(x_1|x_2)$ is

$p(x_1\|x_2)$		
	x_2	
x_1	0	1
0	0	.5
1	1.0	0
2	0	.5
sum	1.0	1.0

The conditional expectation of X_1 given $X_2 = x_2$—written $E_{1|2}(X_1|x_2)$ or, more compactly, $E(X_1|x_2)$—is found from the conditional distribution $p_{1|2}(x_1|x_2)$, as is the conditional variance of X_1 given $X_2 = x_2$, written $V_{1|2}(X_1|x_2)$ or $V(X_1|x_2)$. Using the illustrative conditional distributions $p(x_1|x_2)$,

$$E(X_1|0) = 0(0) + 1(1) + 0(2) = 1$$
$$V(X_1|0) = 0(0-1)^2 + 1(1-1)^2 + 0(2-1)^2 = 0$$
$$E(X_1|1) = .5(0) + 0(1) + .5(2) = 1$$
$$V(X_1|1) = .5(0-1)^2 + 0(1-1)^2 + .5(2-1)^2 = 1$$

The random variables X_1 and X_2 are said to be *independent* if $p(x_1) = p(x_1|x_2)$ for all values of X_1 and X_2; that is, when X_1 and X_2 are independent, the conditional and marginal distributions of X_1 are identical. Equivalent conditions for independence are $p(x_2) = p(x_2|x_1)$ and $p(x_1, x_2) = p(x_1)p(x_2)$: When X_1 and X_2 are independent, their joint probability or probability density is the product of their marginal probabilities or densities. In our example, it is clear that X_1 and X_2 are *not* independent. More generally, the set of n random variables $\{X_1, X_2, \ldots, X_n\}$ is independent if for every subset $\{X_a, X_b, \ldots, X_m\}$ of size $m = 2$ or larger,

$$p(x_a, x_b, \ldots, x_m) = p(x_a)p(x_b)\cdots p(x_m)$$

The *covariance* of two random variables is a measure of their *linear* dependence:

$$C(X_1, X_2) = \sigma_{12} \equiv E[(X_1 - \mu_1)(X_2 - \mu_2)]$$

When large values of X_1 are associated with large values of X_2 (and, conversely, small values with small values), the covariance is positive; when large values of X_1 are associated with small values of X_2 (and vice versa), the covariance is negative. The covariance is

0 otherwise, for instance—but not exclusively—when the random variables are independent. In our previous example, X_1 and X_2 are not independent, but σ_{12} is nevertheless 0 (as the reader can verify). The covariance of a variable with itself is its variance: $C(X, X) = V(X)$.

The *correlation* $\rho_{12} \equiv \sigma_{12}/\sigma_1\sigma_2$ between two random variables X_1 and X_2 is a normalized version of the covariance. The smallest possible value of the correlation, $\rho = -1$, is indicative of a perfect inverse linear relationship between the random variables, while the largest value, $\rho = 1$, is indicative of a perfect direct linear relationship; $\rho = 0$ corresponds to a covariance of 0 and indicates the absence of a linear relationship.

*Vector Random Variables**

It is often convenient to write a collection of random variables as a *vector random variable*: for example, $\underset{(n\times1)}{\mathbf{x}} = [X_1, X_2, \ldots, X_n]'$. The expectation of a vector random variable is simply the vector of expectations of its elements:

$$E(\mathbf{x}) = \boldsymbol{\mu}_x \equiv [E(X_1), E(X_2), \ldots, E(X_n)]'$$

The *variance-covariance matrix* of a vector random variable $\mathbf{x}$ is defined by analogy with the scalar variance as

$$V(\mathbf{x}) = \underset{(n\times n)}{\boldsymbol{\Sigma}_{xx}} \equiv E[(\mathbf{x} - \boldsymbol{\mu}_x)(\mathbf{x} - \boldsymbol{\mu}_x)'] = \begin{bmatrix} \sigma_1^2 & \sigma_{12} & \cdots & \sigma_{1n} \\ \sigma_{21} & \sigma_2^2 & \cdots & \sigma_{2n} \\ \vdots & \vdots & \ddots & \vdots \\ \sigma_{n1} & \sigma_{n2} & \cdots & \sigma_n^2 \end{bmatrix}$$

The diagonal entries of $V(\mathbf{x})$ are the variances of the X's, and the off-diagonal entries are their covariances. The variance-covariance matrix $V(\mathbf{x})$ is symmetric and positive semi-definite.[7] The *covariance matrix* of two vector random variables $\underset{(n\times1)}{\mathbf{x}}$ and $\underset{(m\times1)}{\mathbf{y}}$ is

$$C(\mathbf{x}, \mathbf{y}) = \underset{(n\times m)}{\boldsymbol{\Sigma}_{xy}} \equiv E[(\mathbf{x} - \boldsymbol{\mu}_x)(\mathbf{y} - \boldsymbol{\mu}_y)']$$

and consists of the covariances between pairs of X's and Y's.

---

[7] An order-n square matrix $\mathbf{A}$ is *positive semidefinite* if $\mathbf{x}'\mathbf{A}\mathbf{x} \geq 0$ for every vector $\mathbf{x} \neq \mathbf{0}$. A square matrix $\mathbf{A}$ is *positive definite* if $\mathbf{x}'\mathbf{A}\mathbf{x} > 0$ for all nonzero $\mathbf{x}$. All of the eigenvalues of a positive-semidefinite matrix are nonnegative; all of the eigenvalues of a positive-definite matrix are positive.

D.1.3 Transformations of Random Variables

Suppose that the random variable Y is a linear function $a+bX$ of a discrete random variable X, which has expectation μ_X and variance σ_X^2. Then

$$\begin{aligned}E(Y) = \mu_Y &= \sum_x (a+bx)p(x) \\ &= a\sum p(x) + b\sum xp(x) \\ &= a + b\mu_X\end{aligned}$$

and (employing this property of the expectation operator)

$$\begin{aligned}V(Y) = E[(Y-\mu_Y)^2] &= E\{[(a+bX)-(a+b\mu_X)]^2\} \\ &= b^2E[(X-\mu_X)^2] = b^2\sigma_X^2\end{aligned}$$

Now, let Y be a linear function $a_1X_1 + a_2X_2$ of two discrete random variables X_1 and X_2, with expectations μ_1 and μ_2, variances σ_1^2 and σ_2^2, and covariance σ_{12}. Then

$$\begin{aligned}E(Y) = \mu_Y &= \sum_{x_1}\sum_{x_2}(a_1x_1 + a_2x_2)p(x_1, x_2) \\ &= \sum_{x_1}\sum_{x_2} a_1x_1p(x_1,x_2) + \sum_{x_1}\sum_{x_2} a_2x_2p(x_1,x_2) \\ &= a_1\sum_{x_1} x_1p(x_1) + a_2\sum_{x_2} x_2p(x_2) \\ &= a_1\mu_1 + a_2\mu_2\end{aligned}$$

and

$$\begin{aligned}V(Y) &= E[(Y-\mu_Y)^2] \\ &= E\{[(a_1X_1 + a_2X_2) - (a_1\mu_1 + a_2\mu_2)]^2\} \\ &= a_1^2E[(X_1-\mu_1)^2] + a_2^2E[(X_2-\mu_2)^2] \\ &\quad +2a_1a_2E[(X_1-\mu_1)(X_2-\mu_2)] \\ &= a_1^2\sigma_1^2 + a_2^2\sigma_2^2 + 2a_1a_2\sigma_{12}\end{aligned}$$

When X_1 and X_2 are independent and, consequently, $\sigma_{12} = 0$, this expression simplifies to $V(Y) = a_1^2\sigma_1^2 + a_2^2\sigma_2^2$.

Although I have developed these rules for discrete random variables, they apply equally to the continuous case. For instance, if $Y = a + bX$ is a linear function of the continuous random variable X, then[8]

$$\begin{aligned}E(Y) &= \int_{-\infty}^{\infty}(a+bx)p(x)\,dx \\ &= a\int_{-\infty}^{\infty}p(x)\,dx + b\int_{-\infty}^{\infty}xp(x)\,dx \\ &= a + bE(X)\end{aligned}$$

---

[8] If you are unfamiliar with calculus, then simply think of the integral $\int$ as the continuous analog of the sum $\sum$.

*Transformations of Vector Random Variables**

These results generalize to vector random variables in the following manner: Let $\underset{(m\times 1)}{\mathbf{y}}$ be a linear transformation $\underset{(m\times n)}{\mathbf{A}}\underset{(n\times 1)}{\mathbf{x}}$ of the vector random variable $\mathbf{x}$, which has expectation $E(\mathbf{x}) = \boldsymbol{\mu}_x$ and variance-covariance matrix $V(\mathbf{x}) = \boldsymbol{\Sigma}_{xx}$. Then it can be shown (in a manner analogous to the scalar proofs given previously) that

$$E(\mathbf{y}) = \boldsymbol{\mu}_y = \mathbf{A}\boldsymbol{\mu}_x$$

$$V(\mathbf{y}) = \boldsymbol{\Sigma}_{yy} = \mathbf{A}\boldsymbol{\Sigma}_{xx}\mathbf{A}'$$

If the entries of $\mathbf{x}$ are pairwise independent, then all of the off-diagonal entries of $\boldsymbol{\Sigma}_{xx}$ are 0, and the variance of each entry of $\mathbf{y}$ takes a particularly simple form:

$$\sigma^2_{Y_i} = \sum_{j=1}^{n} a^2_{ij}\sigma^2_{X_j}$$

At times, when $\mathbf{y} = f(\mathbf{x})$, we need to know not only $E(\mathbf{y})$ and $V(\mathbf{y})$, but also the probability distribution of $\mathbf{y}$. Indeed, the transformation $f(\cdot)$ may be nonlinear. Suppose that there is the same number of elements n in $\mathbf{y}$ and $\mathbf{x}$; that the function f is differentiable; and that f is one to one over the domain of $\mathbf{x}$-values under consideration (i.e., there is a unique pairing of $\mathbf{x}$-values and $\mathbf{y}$-values). This last property implies that we can write the reverse transformation $\mathbf{x} = f^{-1}(\mathbf{y})$. The probability density for $\mathbf{y}$ is given by

$$p(\mathbf{y}) = p(\mathbf{x})\left|\det\left(\frac{\partial \mathbf{x}}{\partial \mathbf{y}}\right)\right| = p(\mathbf{x})\left|\det\left(\frac{\partial \mathbf{y}}{\partial \mathbf{x}}\right)\right|^{-1}$$

where $|\det(\partial\mathbf{x}/\partial\mathbf{y})|$, called the *Jacobian* of the transformation, is the absolute value of the $(n \times n)$ determinant

$$\begin{vmatrix} \dfrac{\partial X_1}{\partial Y_1} & \cdots & \dfrac{\partial X_n}{\partial Y_1} \\ \vdots & \ddots & \vdots \\ \dfrac{\partial X_1}{\partial Y_n} & \cdots & \dfrac{\partial X_n}{\partial Y_n} \end{vmatrix}$$

and $|\det(\partial\mathbf{y}/\partial\mathbf{x})|$ is similarly defined.

D.2 Discrete Distributions: The Binomial Distribution and Its Relatives*

In this section, I define three important discrete distributions: the binomial distribution; its generalization, the multinomial distribution; and the Poisson distribution, which can be construed as an approximation to the binomial.

D.2.1 The Binomial Distribution

The coin-flipping experiment described at the beginning of Section D.1.2 gives rise to a binomial random variable that counts the number of heads in two flips of a fair coin. To extend this example, let the random variable X count the number of heads in n independent flips of a coin. Let π denote the probability (not necessarily .5) of obtaining a head on any given flip; then $1 - \pi$ is the probability of obtaining a tail. The probability of observing exactly x heads and $n - x$ tails [i.e., $\Pr(X = x)$] is given by the *binomial distribution*:

$$p(x) = \binom{n}{x} \pi^x (1 - \pi)^{n-x} \qquad \text{[D.3]}$$

where x is any integer between 0 and n, inclusive; the factor $\pi^x(1 - \pi)^{n-x}$ is the probability of observing x heads and $n - x$ tails in a *particular* arrangement; and $\binom{n}{x} \equiv n!/[x!(n - x)!]$, called the *binomial coefficient*, is the number of *different* arrangements of x heads and $n - x$ tails.

The expectation of the binomial random variable X is $E(X) = n\pi$, and its variance is $V(X) = n\pi(1 - \pi)$. Figure D.3 shows the binomial distribution for $n = 10$ and $\pi = .7$.

D.2.2 The Multinomial Distribution

Imagine n repeated, independent trials of a process that on each trial can give rise to one of k different categories of outcomes. Let the random variable X_i count the number of outcomes in category i. Let π_i denote the probability of obtaining an outcome in category i on any given trial. Then $\sum_{i=1}^{k} \pi_i = 1$ and $\sum_{i=1}^{k} X_i = n$.

Suppose, for instance, that we toss a die n times, letting X_1 count the number of 1's, X_2 the number of 2's, ..., X_6 the number of 6's. Then $k = 6$, and π_1 is the probability

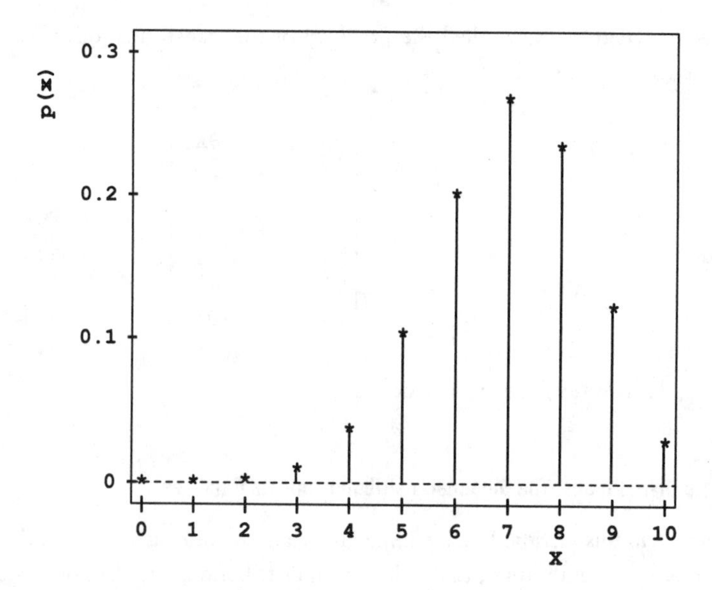

Figure D.3. The binomial distribution for $n = 10$ and $\pi = .7$.

of obtaining a 1 on any toss, π_2 is the probability of obtaining a 2, and so on. If the die is "fair," then $\pi_1 = \pi_2 = \cdots = \pi_6 = 1/6$.

Returning to the general case, the vector random variable $\mathbf{x} \equiv [X_1, X_2, \ldots, X_k]'$ follows the *multinomial distribution*

$$p(\mathbf{x}) = p(x_1, x_2, \ldots, x_k) = \frac{n!}{x_1!x_2!\cdots x_k!}\pi_1^{x_1}\pi_2^{x_2}\cdots\pi_k^{x_k}$$

The rationale for this formula is similar to that of the binomial: $\pi_1^{x_1}\pi_2^{x_2}\cdots\pi_k^{x_k}$ gives the probability of obtaining x_1 outcomes in category 1, x_2 in category 2, and so on, in a particular arrangement; and $n!/(x_1!x_2!\cdots x_k!)$ counts the number of different arrangements. Finally, if $k = 2$, then $x_2 = n - x_1$, and the multinomial distribution reduces to the binomial distribution of Equation D.3.

D.2.3 The Poisson Distribution

The 19th century French mathematician S. Poisson introduced the distribution that bears his name as an approximation to the binomial. The approximation is accurate when n is large and π is small, and when the product of the two, $\lambda \equiv n\pi$, is neither large nor small. The Poisson distribution is

$$p(x) = \frac{\lambda^x e^{-\lambda}}{x!} \quad \text{for } x = 1, 2, 3, \ldots$$

Although the domain of X is all positive integers, the approximation works because $p(x) \simeq 0$ when x is sufficiently large. (Here, $e \simeq 2.718$ is the mathematical constant.)

The Poisson distribution arises naturally in several other contexts. Suppose, for example, that we observe a process that randomly produces events of a particular kind (such as births or auto accidents), counting the number of events X that occur in a fixed time interval. This count follows a Poisson distribution if the following conditions hold:

- Although the particular time points at which the events occur are random, the *rate* of occurrence is fixed during the interval of observation
- If we focus attention on a sufficiently small subinterval of length s, then the probability of observing one event in that subinterval is proportional to its length, λs, and the probability of observing more than one event is negligible. In this context, it is natural to think of the parameter λ of the Poisson distribution as the *rate of occurrence* of the event.
- The occurrence of events in nonoverlapping subintervals is independent.

The expectation of a Poisson random variable is $E(X) = \lambda$, and its variance is also $V(X) = \lambda$. Figure D.4 illustrates the Poisson distribution with rate parameter $\lambda = 5$ (implying that, on average, five events occur during the fixed period of observation).

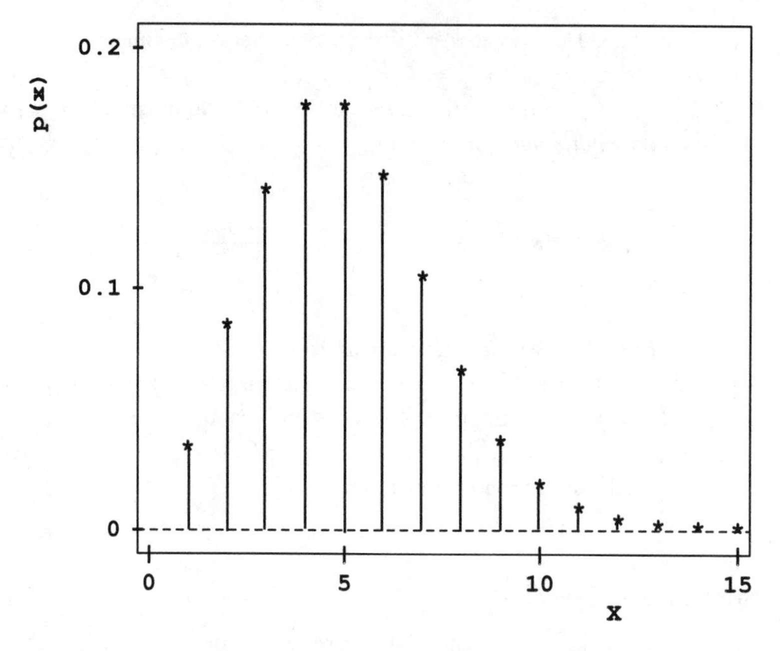

Figure D.4. The Poisson distribution with rate parameter $\lambda = 5$.

D.3 Continuous Distributions: The Normal Distribution and Its Relatives

In this section, I describe five families of continuous random variables that play central roles in the development of linear statistical models: the univariate normal, chi-square, t-, and F-distributions, and the multivariate-normal distribution. Despite the relatively complex formulas defining these distributions, I have left most of the section unstarred, because some familiarity with the normal, chi-square, t-, and F-distributions is important to understanding statistical inference in linear models.[9]

D.3.1 The Normal Distribution

A *normally distributed* (or *Gaussian*) random variable X has probability density function

$$p(x) = \frac{1}{\sigma\sqrt{2\pi}} \exp\left[-\frac{(x-\mu)^2}{2\sigma^2}\right]$$

where the parameters of the distribution μ and σ^2 are, respectively, the mean and variance of X. There is, therefore, a different normal distribution for each choice of μ and σ^2; several examples are shown in Figure D.5. I frequently use the abbreviated notation $X \sim N(\mu, \sigma^2)$, meaning that X is normally distributed with expectation μ and variance σ^2.

[9] You may, if you wish, skip the formulas in favor of the graphs and verbal descriptions of the several distributions.

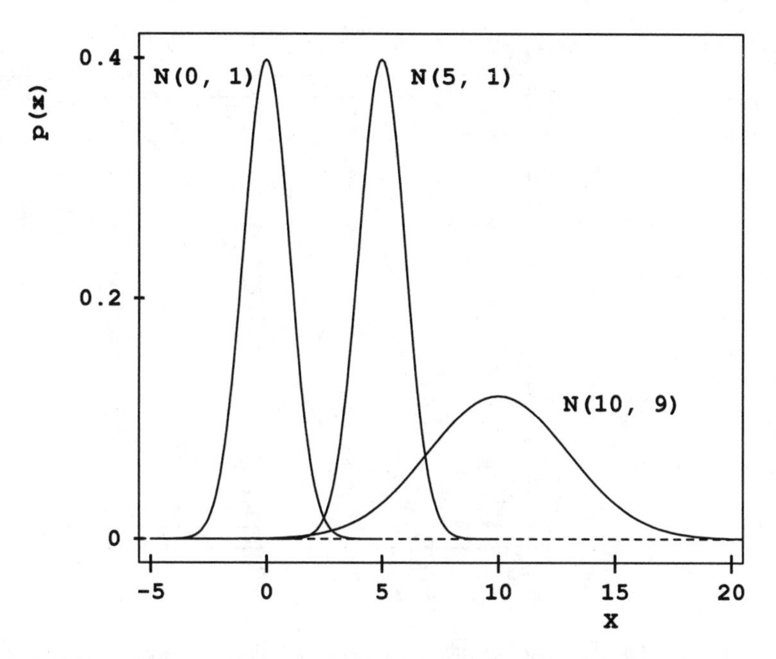

Figure D.5. Normal density functions: $N(0, 1)$, $N(5, 1)$, and $N(10, 9)$.

Of particular importance is the *unit-normal* random variable $Z \sim N(0, 1)$, with density function

$$\phi(z) = \frac{1}{\sqrt{2\pi}} \exp(-z^2/2)$$

The CDF of the unit-normal distribution, $\Phi(z)$, is shown in Figure D.6.

D.3.2 The Chi-Square (χ^2) Distribution

If $Z_1, Z_2, \ldots, Z_n$ are independently distributed unit-normal random variables, then

$$X^2 \equiv Z_1^2 + Z_2^2 + \cdots + Z_n^2$$

follows a *chi-square distribution* with n degrees of freedom, abbreviated χ_n^2. The probability density function of the chi-square variable is

$$P(x^2) = \frac{1}{2^{n/2}\Gamma(\frac{n}{2})} (x^2)^{(n-2)/2} \exp(-x^2/2)$$

where $\Gamma(\cdot)$ is the *gamma function*

$$\Gamma\left(\frac{n}{2}\right) \equiv \begin{cases} \left(\frac{n}{2} - 1\right)! & \text{for } n \text{ even} \\ \left(\frac{n}{2} - 1\right)\left(\frac{n}{2} - 2\right)\cdots\left(\frac{3}{2}\right)\left(\frac{1}{2}\right)\sqrt{\pi} & \text{for } n \text{ odd} \end{cases}$$

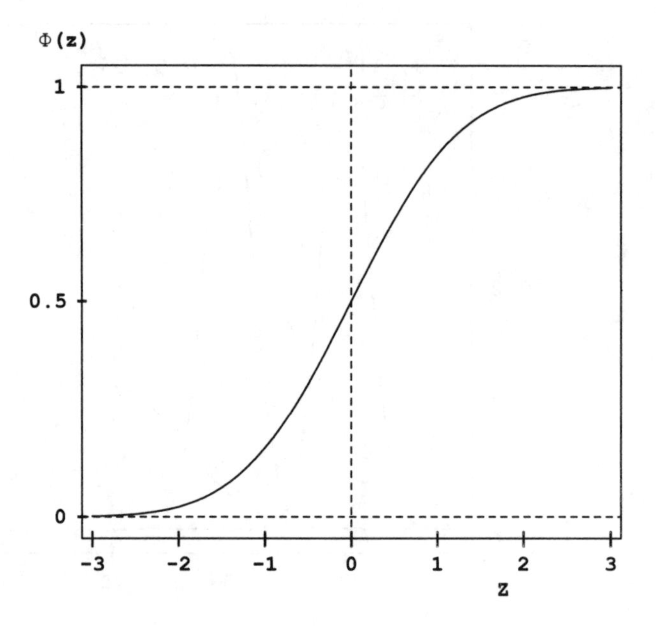

Figure D.6. The CDF of the unit-normal distribution, $\Phi(z)$.

The expectation and variance of a chi-square random variable are $E(X^2) = n$, and $V(X^2) = 2n$. Several chi-square distributions are graphed in Figure D.7.

D.3.3 The t-Distribution

If Z follows a unit-normal distribution, and X^2 independently follows a chi-square distribution with n degrees of freedom, then

$$t \equiv \frac{Z}{\sqrt{\dfrac{X^2}{n}}}$$

is a t random variable with n degrees of freedom,[10] abbreviated t_n. The probability density function of t is

$$p(t) = \frac{\Gamma\left(\dfrac{n+1}{2}\right)}{\sqrt{\pi n}\,\Gamma\left(\dfrac{n}{2}\right)} \times \frac{1}{\left(1+\dfrac{t^2}{n}\right)^{(n+1)/2}} \qquad \text{[D.4]}$$

From the symmetry of this formula around $t = 0$, it is clear that[11] $E(t) = 0$. It can be shown that $V(t) = n/(n-2)$, for $n > 2$; thus, the variance of t is large for small degrees of freedom, and approaches 1 as n increases.

---

[10] I write a lowercase t for the random variable in deference to nearly universal usage.

[11] When $n = 1$, the expectation $E(t)$ does not exist, but the median and mode of t are still 0; t_1 is called the *Cauchy distribution*.

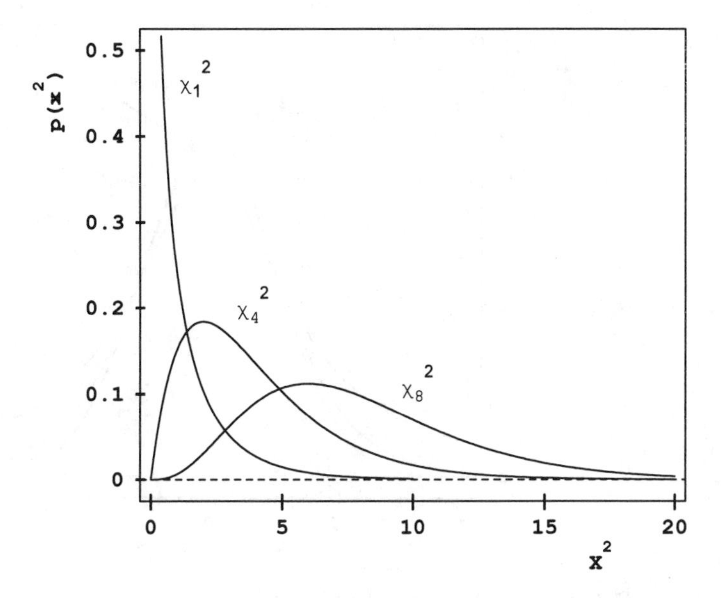

Figure D.7. Chi-square density functions: χ_1^2, χ_4^2, and χ_8^2.

Several t-distributions are shown in Figure D.8. As degrees of freedom grow, the t-distribution more and more closely approximates the unit-normal distribution, and in the limit, $t_\infty = N(0, 1)$. The normal approximation to the t-distribution is quite close for n as small as 30.

D.3.4 The *F*-Distribution

Let X_1^2 and X_2^2 be independently distributed chi-square variables with n_1 and n_2 degrees of freedom, respectively. Then

$$F \equiv \frac{X_1^2/n_1}{X_2^2/n_2}$$

follows an *F-distribution* with n_1 numerator degrees of freedom and n_2 denominator degrees of freedom, abbreviated F_{n_1, n_2}. The probability density for F is

$$p(f) = \frac{\Gamma\left(\frac{n_1+n_2}{2}\right)}{\Gamma\left(\frac{n_1}{2}\right)\Gamma\left(\frac{n_2}{2}\right)} \left(\frac{n_1}{n_2}\right)^{n_1/2} f^{(n_1-2)/2} \left(1+\frac{n_1}{n_2}f\right)^{-(n_1+n_2)/2} \qquad \text{[D.5]}$$

Comparing Equations D.4 and D.5, it can be shown that $t_n^2 = F_{1,n}$. As n_2 grows larger, F_{n_1,n_2} approaches $\chi_{n_1}^2/n_1$ and, in the limit, $F_{n,\infty} = \chi_n^2/n$.

For $n_2 > 2$, the expectation of F is $E(F) = n_2/(n_2 - 2)$, which is approximately 1 for large values of n_2. For $n_2 > 4$,

$$V(F) = \frac{2n_2^2(n_1 + n_2 - 2)}{n_1(n_2 - 2)^2(n_2 - 4)}$$

Figure D.9 shows several F probability density functions.

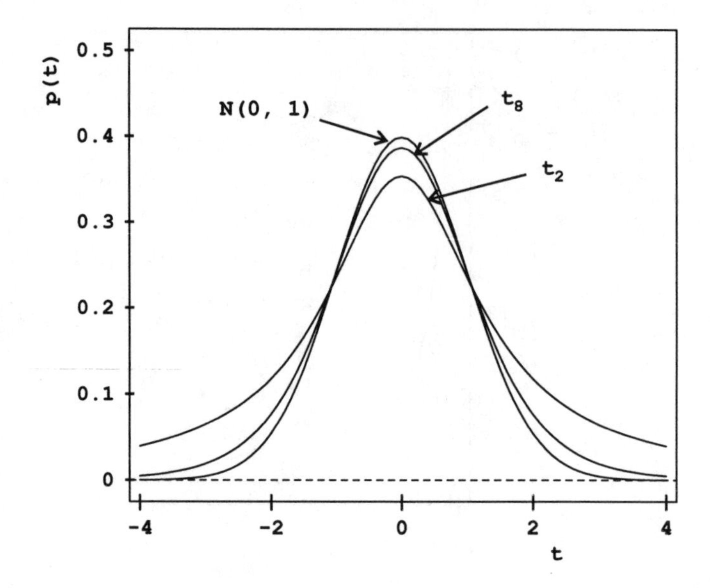

Figure D.8. t density functions: t_2, t_8, and $N(0,\,1) = t_\infty$.

D.3.5 The Multivariate-Normal Distribution*

The joint probability density for a *multivariate-normal* vector random variable $\mathbf{x} = [X_1, X_2, \ldots, X_n]'$ with mean vector $\boldsymbol{\mu}$ and positive-definite variance-covariance matrix $\boldsymbol{\Sigma}$ is given by

$$p(\mathbf{x}) = \frac{1}{(2\pi)^{n/2}\sqrt{\det \boldsymbol{\Sigma}}} \exp\left[-\frac{1}{2}(\mathbf{x}-\boldsymbol{\mu})'\boldsymbol{\Sigma}^{-1}(\mathbf{x}-\boldsymbol{\mu})\right]$$

which we abbreviate as $\mathbf{x} \sim N_n(\boldsymbol{\mu}, \boldsymbol{\Sigma})$.

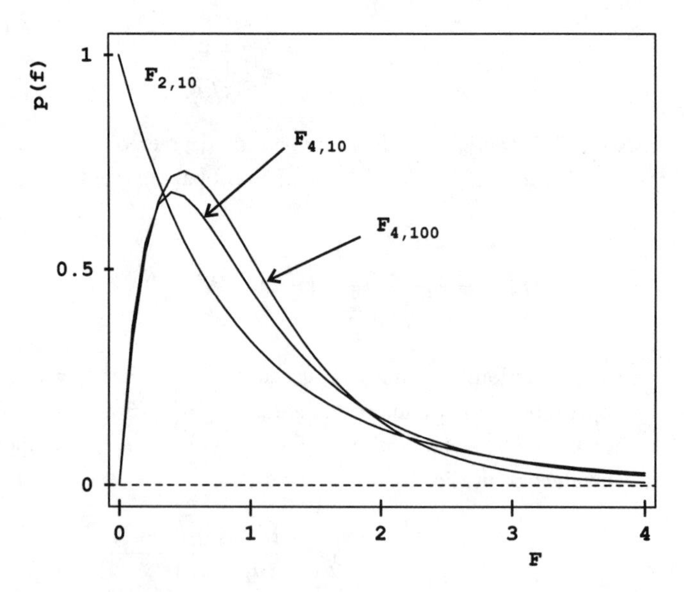

Figure D.9. F density functions: $F_{2,\,10}$, $F_{4,\,10}$, and $F_{4,\,100}$.

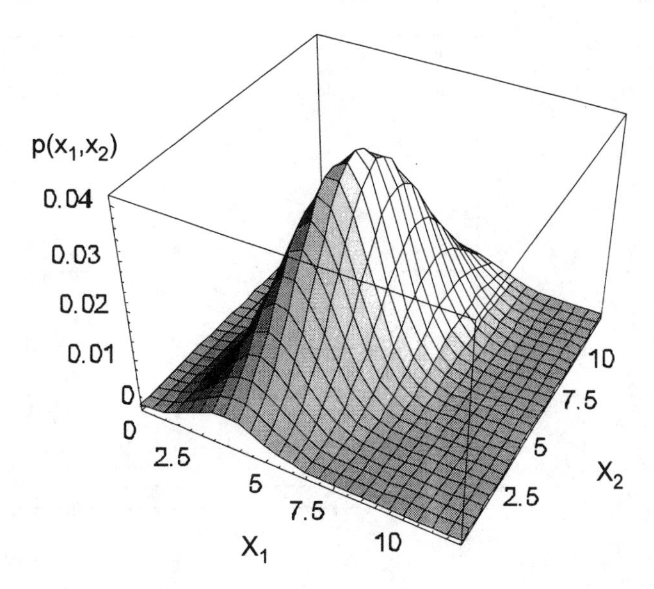

Figure D.10. The bivariate-normal density function for $\mu_1 = 5$, $\mu_2 = 6$, $\sigma_1 = 1.5$, $\sigma_2 = 3$, and $\sigma_{12} = 2.25$. The slices of the density surface are normal both in the direction of X_1 and in the direction of X_2.

If $\mathbf{x}$ is multivariately normally distributed, then the marginal distribution of each of its components is univariate normal,[12] $X_i \sim N(\mu_i, \sigma_i^2)$. Furthermore, if $\mathbf{x} \sim N_n(\boldsymbol{\mu}, \boldsymbol{\Sigma})$ and

$$\underset{(m\times 1)}{\mathbf{y}} = \underset{(m\times n)(n\times 1)}{\mathbf{A}\ \mathbf{x}}$$

is a linear transformation of $\mathbf{x}$ with $\text{rank}(\mathbf{A}) = m \leq n$, then $\mathbf{y} \sim N_m(\mathbf{A}\boldsymbol{\mu}, \mathbf{A}\boldsymbol{\Sigma}A')$. We say that a vector random variable $\mathbf{x}$ follows a *singular normal distribution* if the covariance matrix $\boldsymbol{\Sigma}$ of $\mathbf{x}$ is singular, but if a maximal linearly independent subset of $\mathbf{x}$ is multivariately normally distributed.

A bivariate-normal density function for $\mu_1 = 5$, $\mu_2 = 6$, $\sigma_1 = 1.5$, $\sigma_2 = 3$, and $\rho_{12} = .5$ [i.e., $\sigma_{12} = (.5)(1.5)(3) = 2.25$] is depicted in Figure D.10.

D.4 Asymptotic Distribution Theory: An Introduction*

Partly because it is at times difficult to determine the small-sample properties of statistical estimators, it is of interest to investigate how an estimator behaves as the sample size grows. *Asymptotic distribution theory* provides tools for this investigation. I shall merely outline the theory here: More complete accounts are available in many sources, including some of the references at the end of this appendix.

---

[12] The converse is not true: Each X_i can be univariately normally distributed without $\mathbf{x}$ being multivariate normal.

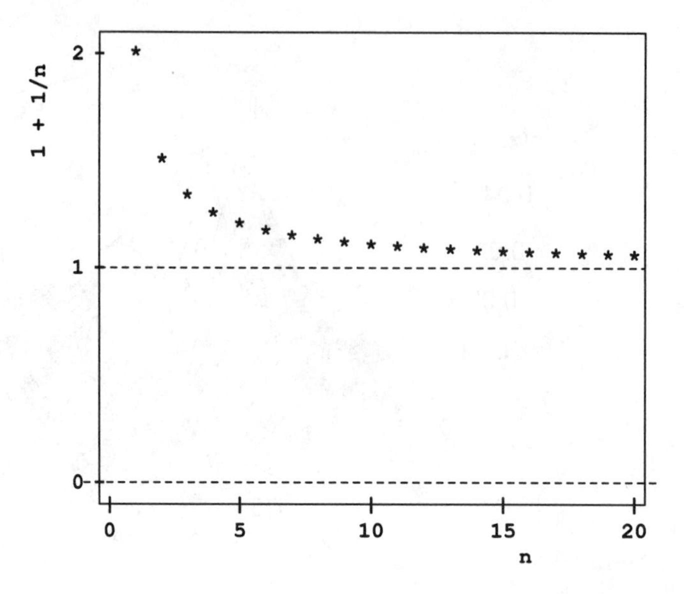

Figure D.11. The first 20 values of the sequence $a_n = 1 + 1/n$, which has the limit $a = 1$.

D.4.1 Probability Limits

Although asymptotic distribution theory applies to sequences of random variables, it is necessary first to consider the *nonstochastic*[13] *infinite sequence* $\{a_1, a_2, \ldots, a_n, \ldots\}$. As the reader is likely aware, this sequence has a *limit* a when, given any positive number ε, no matter how small, there is a positive integer $n(\varepsilon)$ such that $|a_n - a| < \varepsilon$ for all $n > n(\varepsilon)$. In words: a_n can be made arbitrarily close to a by picking n sufficiently large.[14] To describe this state of affairs compactly, we write $\lim_{n\to\infty} a_n = a$. If, for example, $a_n = 1 + 1/n$, then $\lim_{n\to\infty} a_n = 1$; this sequence and its limit are graphed in Figure D.11.

Consider now a sequence of random variables $\{X_1, X_2, \ldots, X_n, \ldots\}$. In a typical statistical application, X is some estimator and n is the size of the sample from which the estimator is calculated. Let $p_n \equiv \Pr(|X_n - a| < \delta)$, where a is a constant and δ is a small positive number. Think of p_n as the probability that X_n is close to a. Suppose that the *nonstochastic* sequence of probabilities $\{p_1, p_2, \ldots, p_n, \ldots\}$ approaches a limit of 1;[15] that is, $\lim_{n\to\infty} \Pr(|X_n - a| < \delta) = 1$. Then, as n grows, the random variable X_n concentrates more and more of its probability in a small region around a, a situation that is illustrated in Figure D.12. If this result holds regardless of how small δ is, then

---

[13] By "nonstochastic" I mean that each a_n is a fixed number rather than a random variable.

[14] The notation $n(\varepsilon)$ stresses that the required value of n depends on the selected criterion ε.

[15] To say that $\{p_1, p_2, \ldots, p_n, \ldots\}$ is a nonstochastic sequence is only apparently contradictory: Although these probabilities are based on random variables, the probabilities themselves are each specific numbers—such as, .6, .9, and so forth.

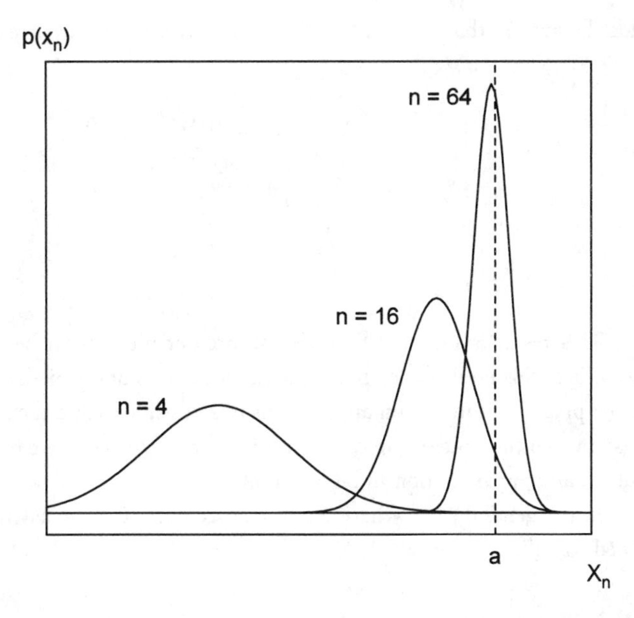

Figure D.12. plim $X_n = a$: As n grows, the distribution of X_n concentrates more and more of its probability in a small region around a.

we say that a is the *probability limit* of X_n, denoted plim $X_n = a$. We generally drop the subscript n to write the even more compact expression, plim $X = a$.

Probability limits have the following very useful property: If plim $X = a$, and if $Y = f(X)$ is some continuous function of X, then plim $Y = f(a)$. Likewise, if plim $X = a$, plim $Y = b$, and $Z = f(X, Y)$ is a continuous function of X and Y, then plim $Z = f(a, b)$.

D.4.2 Asymptotic Expectation and Variance

We return to the sequence of random variables $\{X_1, X_2, \ldots, X_n, \ldots\}$. Let μ_n denote the expectation of X_n. Then $\{\mu_1, \mu_2, \ldots, \mu_n, \ldots\}$ is a nonstochastic sequence. If this sequence approaches a limit μ, then we call μ the *asymptotic expectation* of X, also written $\mathcal{E}(X)$.

Although it seems natural to define an asymptotic variance analogously as the limit of the sequence of variances, this definition is not satisfactory because (as the following example illustrates) $\lim_{n\to\infty} V(X_n)$ is 0 in most interesting cases. Suppose that we calculate the mean $\overline{X}_n$ for a sample of size n drawn from a population with mean μ and variance σ^2. We know, from elementary statistics, that $E(\overline{X}_n) = \mu$ and that

$$V(\overline{X}_n) \equiv E[(\overline{X}_n - \mu)^2] = \frac{\sigma^2}{n}$$

Consequently, $\lim_{n\to\infty} V(\overline{X}_n) = 0$. Inserting the factor $\sqrt{n}$ within the square, however, produces the expectation $E\{[\sqrt{n}(\overline{X}_n - \mu)]^2\} = \sigma^2$. Dividing by n and taking the limit yields the answer that we want, defining the *asymptotic variance* of the sample mean:

$$\mathscr{V}(\overline{X}) \equiv \lim_{n\to\infty} \frac{1}{n} E\{[\sqrt{n}(\overline{X}_n - \mu)]^2\}$$
$$= \frac{1}{n}\mathscr{E}\{[\sqrt{n}(\overline{X}_n - \mu)]^2\}$$
$$= \frac{\sigma^2}{n}$$

This result is uninteresting for the present illustration because $\mathscr{V}(\overline{X}) = V(\overline{X})$—indeed, it is this equivalence that motivated the definition of the asymptotic variance in the first place—but in certain applications it is possible to find the asymptotic variance of a statistic when the finite-sample variance is intractable. Then we can apply the asymptotic result as an approximation in large samples.

In the general case, where X_n has expectation μ_n, the asymptotic variance of X is defined to be[16]

$$\mathscr{V}(X) \equiv \frac{1}{n}\mathscr{E}\{[\sqrt{n}(X_n - \mu_n)]^2\} \quad \text{[D.6]}$$

D.4.3 Asymptotic Distribution

Let $\{P_1, P_2, \ldots, P_n, \ldots\}$ represent the CDFs of a sequence of random variables $\{X_1, X_2, \ldots, X_n, \ldots\}$. The CDF of X converges to the *asymptotic distribution* P if, given any positive number ε, however small, we can find a sufficiently large $n(\varepsilon)$ such that $|P_n(x) - P(x)| < \varepsilon$ for all $n > n(\varepsilon)$ and for all values x of the random variable. A familiar illustration is provided by the *central-limit theorem*, which (in one of its versions) states that the mean of a set of independent and identically distributed random variables with finite expectations and variances follows an approximate normal distribution, the approximation improving as the number of random variables increases.

The results of this section extend straightforwardly to vectors and matrices: We say the plim $\underset{(m\times 1)}{\mathbf{x}} = \underset{(m\times 1)}{\mathbf{a}}$ when plim $X_i = a_i$ for $i = 1, 2, \ldots, m$. Likewise, plim $\underset{(m\times p)}{\mathbf{X}} = \underset{(m\times p)}{\mathbf{A}}$ means that plim $X_{ij} = a_{ij}$ for all i and j. The asymptotic expectation of the vector random variable $\underset{(m\times 1)}{\mathbf{x}}$ is defined as the vector of asymptotic expectations of its elements, $\boldsymbol{\mu} = \mathscr{E}(\mathbf{x}) \equiv [\mathscr{E}(X_1), \mathscr{E}(X_2), \ldots, \mathscr{E}(X_m)]'$. The asymptotic variance-covariance matrix of $\mathbf{x}$ is given by

$$\mathscr{V}(\mathbf{x}) \equiv \frac{1}{n}\mathscr{E}\{[\sqrt{n}(\mathbf{x}_n - \boldsymbol{\mu}_n)][\sqrt{n}(\mathbf{x}_n - \boldsymbol{\mu}_n)]'\}$$

---

[16] It is generally preferable to define asymptotic expectation and variance in terms of the asymptotic distribution (see the next section), because the sequences used for this purpose here do not exist in all cases (see Theil, 1971, pp. 375–376; also see McCallum, 1973). My use of the symbols $\mathscr{E}(\cdot)$ and $\mathscr{V}(\cdot)$ for asymptotic expectation and variance is not standard: The reader should be aware that these symbols are sometimes used in place of $E(\cdot)$ and $V(\cdot)$ to denote *ordinary* expectation and variance.

D.5 Properties of Estimators[17]

An *estimator* is a sample statistic (i.e., a function of the observations of a sample) used to estimate an unknown population parameter. Because its value varies from one sample to the next, an estimator is a random variable. An *estimate* is the value of an estimator for a particular sample. The probability distribution of an estimator is called its *sampling distribution*; the variance of this distribution is called the *sampling variance* of the estimator; and the standard deviation of the sampling distribution is often called the *standard error* of the estimator.

D.5.1 Bias

An estimator A of the parameter α is *unbiased* if $E(A) = \alpha$. The difference $E(A) - \alpha$ (which, of course, is 0 for an unbiased estimator) is the *bias* of A.

Suppose, for example, that we draw n independent observations X_i from a population with mean μ and variance σ^2. Then the sample mean $\overline{X} \equiv \sum X_i/n$ is an unbiased estimator of μ, while

$$S_*^2 \equiv \frac{\sum(X_i - \overline{X})^2}{n} \qquad \text{[D.7]}$$

is a biased estimator of σ^2, because $E(S_*^2) = [(n-1)/n]\sigma^2$; the bias of S_*^2 is, therefore, $-\sigma^2/n$. Sampling distributions of unbiased and biased estimators are illustrated in Figure D.13.

*Asymptotic Bias**

The *asymptotic bias* of an estimator A of α is $\mathscr{E}(A) - \alpha$, and the estimator is *asymptotically unbiased* if $\mathscr{E}(A) = \alpha$. Thus, S_*^2 is asymptotically unbiased, because its bias $-\sigma^2/n \to 0$ as $n \to \infty$.

D.5.2 Mean-Squared Error and Efficiency

To say that an estimator is unbiased means that its average value over repeated samples is equal to the parameter being estimated. This is clearly a desirable property for an estimator to possess, but it is cold comfort if the estimator does not provide estimates that are close to the parameter: In forming the expectation, large negative estimation errors for some samples could offset large positive errors for others.

The *mean-squared error* (MSE) of an estimator A of the parameter α is literally the average squared difference between the estimator and the parameter: $\text{MSE}(A) \equiv E[(A - \alpha)^2]$. The *efficiency* of an estimator is inversely proportional to its mean-squared error. We generally prefer a more efficient estimator to a less efficient one.

---

[17] Most of the material in this and the following section can be traced to a remarkable, seminal paper on estimation by Fisher (1922).

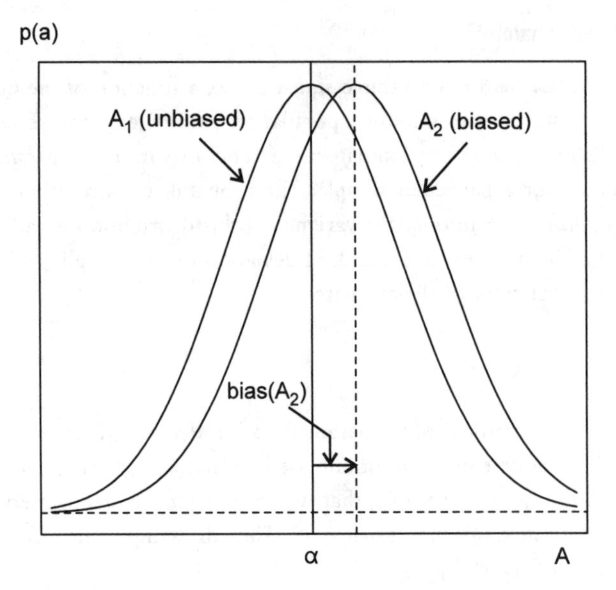

Figure D.13. The estimator A_1 is an unbiased estimator of α because $E(A_1) = \alpha$; the estimator A_2 has a positive bias, because $E(A_2) > \alpha$.

The mean-squared error of an unbiased estimator is simply its sampling variance, because $E(A) = \alpha$. For a biased estimator,

$$\begin{aligned}\text{MSE}(A) &= E[(A-\alpha)^2] \\ &= E\{[A - E(A) + E(A) - \alpha]^2\} \\ &= E\{[A - E(A)]^2\} + [E(A) - \alpha]^2 \\ &\qquad + 2[E(A) - E(A)][E(A) - \alpha] \\ &= V(A) + [\text{bias}(A)]^2 + 0\end{aligned}$$

The efficiency of an estimator increases, therefore, as its sampling variance and bias decline. In comparing two estimators, an advantage in sampling variance can more than offset a disadvantage due to bias, as illustrated in Figure D.14.

*Asymptotic Efficiency**

Asymptotic efficiency is inversely proportional to *asymptotic mean-squared error (AMSE)* which, in turn, is the sum of asymptotic variance and squared asymptotic bias.

D.5.3 Consistency*

An estimator A of the parameter α is *consistent* if $\text{plim}\, A = \alpha$. A sufficient (but not necessary[18]) condition for consistency is that an estimator be asymptotically unbiased

[18] There are cases in which $\text{plim}\, A = \alpha$, but the variance and asymptotic expectation of A do not exist. See Johnston (1972, p. 272) for an example.

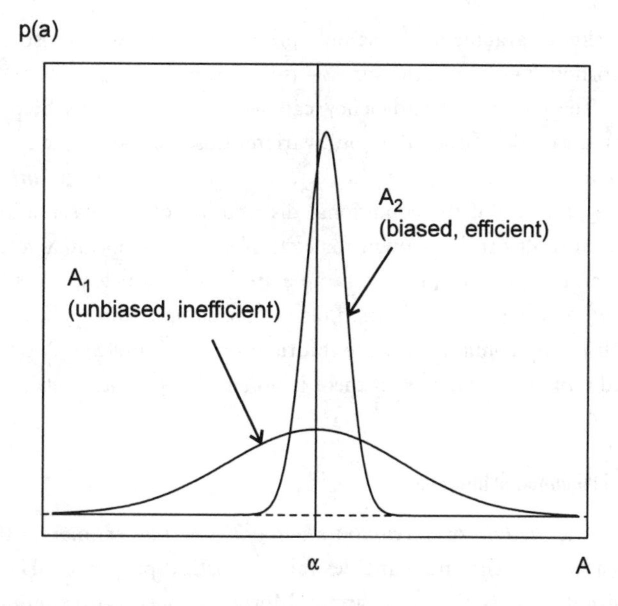

Figure D.14. Relative efficiency of estimators: Even though it is biased, A_2 is a more efficient estimator of α than the unbiased estimator A_1, because the smaller variance of A_2 more than compensates for its small bias.

and that the sampling variance of the estimator approach 0 as n increases; this condition implies that the mean-squared error of the estimator approaches a limit of 0. Figure D.12 illustrates consistency, if we construe X as an estimator of a. The estimator S_*^2 given in Equation (D.7) is a consistent estimator of the population variance σ^2 even though it is biased in finite samples.

D.5.4 Sufficiency*

Sufficiency is a more difficult property to grasp than unbias, efficiency, or consistency. A statistic S based on a sample of observations is *sufficient* for the parameter α if the statistic exhausts all of the information about α that is present in the sample. More formally, suppose that the observations $X_1, X_2, \ldots, X_n$ are drawn from a probability distribution with parameter α, and let the statistic $S \equiv f(X_1, X_2, \ldots, X_n)$. Then S is a sufficient statistic for α if the probability distribution of the observations *conditional* on S, that is, $p(x_1, x_2, \ldots, x_n|S)$, does not depend on α. The sufficient statistic S need not be an estimator of α.

To illustrate the idea of sufficiency, suppose that n observations are independently sampled, and that each observation X_i takes on the value 1 with probability[19] π and the value 0 with probability $1 - \pi$. I shall demonstrate that the sample sum $S \equiv \sum_{i=1}^{n} X_i$ is a sufficient statistic for π. If we know the value s of S, then there are $\binom{n}{s}$ different possible arrangements of the s 1's and $n - s$ 0's, each with probability $1/\binom{n}{s}$. Because this probability does not depend on the parameter π, the statistic S is sufficient for π.

[19] The Greek letter π is used because the probability cannot be directly observed. Since π is a probability, it is a number between 0 and 1—*not* the mathematical constant $\simeq 3.1416$.

By a similar argument, the sample proportion $P \equiv S/n$ is also a sufficient statistic. The proportion P—but not the sum S—is an estimator of π.

The concept of sufficiency can be extended to sets of parameters and statistics: Given a sample of (possibly multivariate) observations $\mathbf{x}_1, \mathbf{x}_2, \ldots, \mathbf{x}_n$, a vector of statistics $\mathbf{s} = [S_1, S_2, \ldots, S_p]' \equiv f(\mathbf{x}_1, \mathbf{x}_2, \ldots, \mathbf{x}_n)$ is *jointly sufficient* for the parameters $\boldsymbol{\alpha} = [\alpha_1, \alpha_2, \ldots, \alpha_k]'$ if the conditional distribution of the observations given $\mathbf{s}$ does not depend on $\boldsymbol{\alpha}$. It can be shown, for example, that the mean $\overline{X}$ and variance S^2 calculated from an independent random sample are jointly sufficient statistics for the parameters μ and σ^2 of a normal distribution (as are the sample sum $\sum X_i$ and sum of squares $\sum X_i^2$, which jointly contain the same information as $\overline{X}$ and S^2). A set of sufficient statistics is called *minimally sufficient* if there is no smaller sufficient set.

D.6 Maximum-Likelihood Estimation

The *method of maximum likelihood* provides estimators that have both a reasonable intuitive basis and many desirable statistical properties. The method is very broadly applicable and is simple to apply. Moreover, once a maximum-likelihood estimator is derived, the general theory of maximum-likelihood estimation provides standard errors, statistical tests, and other results useful for statistical inference. A disadvantage of the method, however, is that it frequently requires strong assumptions about the structure of the data.

Let us first consider a simple example: Suppose that we want to estimate the probability π of getting a head on flipping a particular coin. We flip the coin "independently" 10 times (i.e., we sample $n = 10$ flips), obtaining the following result: $HHTHHHTTHH$. The probability of obtaining this sequence—in advance of collecting the data—is a function of the unknown parameter π:

$$\begin{aligned}
\Pr(\text{data}|\text{parameter}) &= \Pr(HHTHHHTTHH|\pi) \\
&= \pi\pi(1-\pi)\pi\pi\pi(1-\pi)(1-\pi)\pi\pi \\
&= \pi^7(1-\pi)^3
\end{aligned}$$

The data for our particular sample are *fixed*, however: We have already collected them. The parameter π also has a fixed value, but this value is unknown, and so we can let it vary in our imagination between 0 and 1, treating the probability of the observed data as a function of π. This function is called the likelihood function:

$$\begin{aligned}
L(\text{parameter}|\text{data}) &= L(\pi|HHTHHHTTHH) \\
&= \pi^7(1-\pi)^3
\end{aligned}$$

The probability function and the likelihood function are the same equation, but the probability function is a function of the data with the value of the parameter fixed, while the likelihood function is a function of the parameter with the data fixed.

Here are some representative values[20] of the likelihood for different values of π:

π	$L(\pi\|\text{data}) = \pi^7(1-\pi)^3$
0.0	0.0
.1	.0000000729
.2	.00000655
.3	.0000750
.4	.000354
.5	.000977
.6	.00179
.7	.00222
.8	.00168
.9	.000478
1.0	0.0

The full likelihood function is graphed in Figure D.15. Although each value of $L(\pi|\text{data})$ is a notional probability, the function $L(\pi|\text{data})$ is not a probability distribution or a density function: It does not integrate to 1, for example. In the present instance, the probability of obtaining the sample of data that we have in hand, *HHTHHHTTHH*, is small regardless of the true value of π. This is usually the case: Unless the sample is very small, *any specific* sample result—including the one that is realized—will have low probability.

Nevertheless, the likelihood contains useful information about the unknown parameter π. For example, π *cannot* be 0 or 1, because if it were either of these values, then the observed data could not have been obtained. Reversing this reasoning, the value of π that is most supported by the data is the one for which the likelihood is largest. This value is the *maximum-likelihood estimate* (MLE), denoted $\hat{\pi}$. Here, $\hat{\pi} = .7$, which is just the sample proportion of heads, 7/10.

*Generalization of the Example**

More generally, for n independent flips of the coin, producing a particular sequence that includes x heads and $n - x$ tails,

$$L(\pi|\text{data}) = \Pr(\text{data}|\pi) = \pi^x(1-\pi)^{n-x}$$

We want the value of π that maximizes $L(\pi|\text{data})$, which we often abbreviate $L(\pi)$. As is typically the case, it is simpler—and equivalent—to find the value of π that maximizes the log of the likelihood, here

$$\log_e L(\pi) = x \log_e \pi + (n-x)\log_e(1-\pi) \tag{D.8}$$

[20] The likelihood is a continuous function of π for values of π between 0 and 1. This contrasts, in the present case, with the probability function, because there is a finite number of possible samples, $2^{10} = 1024$.

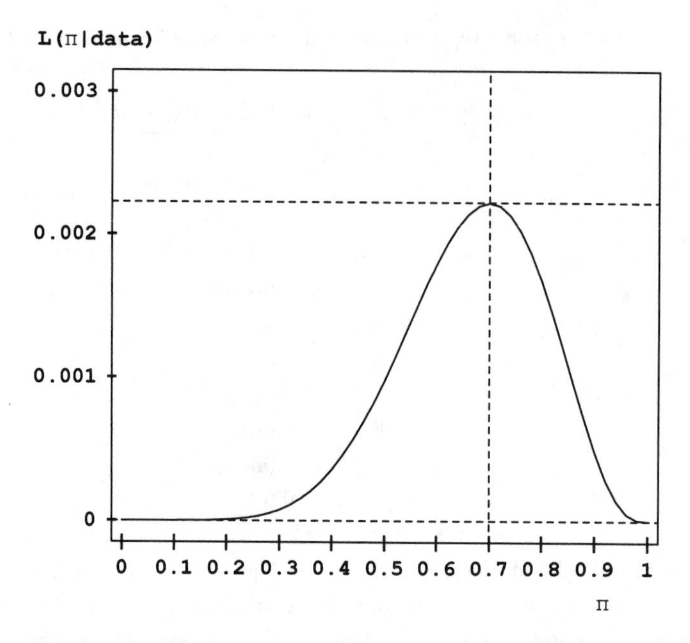

Figure D.15. The likelihood function $L(\pi|HHTHHHTTHH) = \pi^7(1-\pi)^3$.

Differentiating $\log_e L(\pi)$ with respect to π produces

$$\begin{aligned}\frac{d\log_e L(\pi)}{d\pi} &= \frac{x}{\pi} + (n-x)\frac{1}{1-\pi}(-1) \\ &= \frac{x}{\pi} - \frac{n-x}{1-\pi}\end{aligned}$$

Setting the derivative to 0 and solving produces the MLE which, as before, is the sample proportion x/n. The maximum-likelihood *estimator* is $\hat{\pi} = X/n$. To avoid this slightly awkward substitution of estimator for estimate in the last step, we usually replace x by X in the log likelihood function (Equation D.8).

D.6.1 Properties of Maximum-Likelihood Estimators*

Under very broad conditions, maximum-likelihood estimators have the following general properties (see the references at the end of this appendix):

- Maximum-likelihood estimators are consistent.
- They are asymptotically unbiased, although they may be biased in finite samples.
- They are asymptotically efficient—no asymptotically unbiased estimator has a smaller asymptotic variance.
- They are asymptotically normally distributed.
- If there is a sufficient statistic for a parameter, then the maximum-likelihood estimator of the parameter is a function of a sufficient statistic.
- The asymptotic sampling variance of the MLE $\hat{\alpha}$ of a parameter α can be obtained from the second derivative of the log likelihood:

$$\mathscr{V}(\hat{\alpha}) = \frac{1}{-E\left[\dfrac{d^2 \log_e L(\alpha)}{d\alpha^2}\right]} \qquad \text{[D.9]}$$

The denominator of $\mathcal{V}(\hat{\alpha})$ is called the *Fisher information*

$$\mathcal{I}(\alpha) \equiv -E\left[\frac{d^2 \log_e L(\alpha)}{d\alpha^2}\right]$$

In practice, we substitute the MLE $\hat{\alpha}$ into Equation D.9 to obtain an *estimate* of the asymptotic sampling variance, $\widehat{\mathcal{V}(\hat{\alpha})}$.

- $L(\hat{\alpha})$ is the value of the likelihood function at the MLE $\hat{\alpha}$, while $L(\alpha)$ is the likelihood for the true (but generally unknown) parameter α. The *log-likelihood-ratio statistic*

$$G^2 \equiv -2 \log_e \frac{L(\alpha)}{L(\hat{\alpha})} = 2[\log_e L(\hat{\alpha}) - \log_e L(\alpha)]$$

follows an asymptotic chi-square distribution with 1 degree of freedom. Because, by definition, the MLE maximizes the likelihood for our particular sample, the value of the likelihood at the true parameter value α is generally smaller than at the MLE $\hat{\alpha}$ (unless, by good fortune, $\hat{\alpha}$ and α happen to coincide).

Establishing these results is well beyond the scope of this appendix, but the results do make some intuitive sense. For example, if the log likelihood has a sharp peak, then the MLE is clearly differentiated from nearby values. Under these circumstances, the second derivative of the log likelihood is a large negative number; there is a lot of "information" in the data concerning the value of the parameter; and the sampling variance of the MLE is small. If, in contrast, the log likelihood is relatively flat at its maximum, then alternative estimates quite different from the MLE are nearly as good as the MLE; there is little information in the data concerning the value of the parameter; and the sampling variance of the MLE is large.

D.6.2 Statistical Inference: Wald, Likelihood-Ratio, and Score Tests

The properties of maximum-likelihood estimators described in the previous section lead directly to three common and general procedures—called the *Wald test*, the *likelihood-ratio test*, and the *score test*[21]—for testing the statistical hypothesis H_0: $\alpha = \alpha_0$. The Wald and likelihood-ratio tests can be "turned around" to produce confidence intervals for α.

Wald test. Relying on the asymptotic[22] normality of the MLE $\hat{\alpha}$, we calculate the test statistic

$$Z_0 \equiv \frac{\hat{\alpha} - \alpha_0}{\sqrt{\widehat{\mathcal{V}(\hat{\alpha})}}}$$

which is asymptotically distributed as $N(0, 1)$ under H_0.

---

[21] The score test is sometimes called the *Lagrange-multiplier test*.

[22] Asymptotic results apply approximately, with the approximation growing more accurate as the sample size n gets larger.

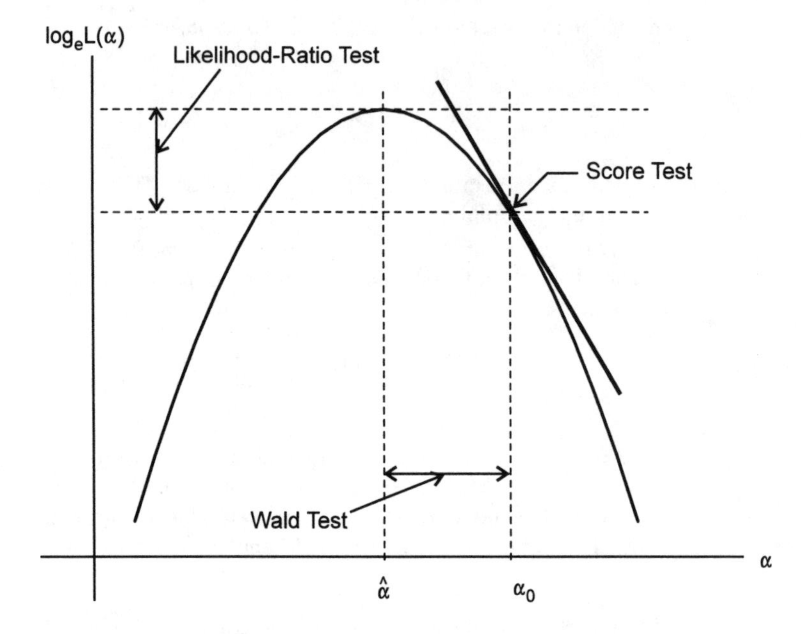

Figure D.16. Tests of the hypothesis H_0: $\alpha = \alpha_0$: The likelihood-ratio test compares $\log_e L(\hat{\alpha})$ with $\log_e L(\alpha_0)$; the Wald test compares $\hat{\alpha}$ with α_0; and the score test examines the slope of $\log_e L(\alpha)$ at $\alpha = \alpha_0$.

Likelihood-ratio test. Employing the log-likelihood ratio, the test statistic

$$G_0^2 \equiv -2 \log_e \frac{L(\alpha_0)}{L(\hat{\alpha})} = 2[\log_e L(\hat{\alpha}) - \log_e L(\alpha_0)]$$

is asymptotically distributed as χ_1^2 under H_0.

Score test. The "score" $S(\alpha)$ is the slope of the log likelihood at a particular value[23] of α. At the MLE, the score is 0: $S(\hat{\alpha}) = 0$. It can be shown that the *score statistic*

$$S_0 \equiv \frac{S(\alpha_0)}{\sqrt{\mathcal{I}(\alpha_0)}}$$

is asymptotically distributed as $N(0, 1)$ under H_0.

Unless the log likelihood is quadratic, the three test statistics can produce somewhat different results in specific samples, although the three tests are asymptotically equivalent. In certain contexts, the score test has the practical advantage of not requiring the computation of the MLE $\hat{\alpha}$ (because S_0 depends only on the null value α_0, which is specified in H_0). Figure D.16 shows the relationship among the three test statistics, and clarifies the intuitive rationale of each: The Wald test measures the distance between $\hat{\alpha}$ and α_0, using the standard error to calibrate this distance. If $\hat{\alpha}$ is far from α_0, for example, then doubt is cast on H_0. The likelihood-ratio test measures the distance between $\log_e L(\hat{\alpha})$

[23] *That is, $S(\alpha) \equiv d \log_e L(\alpha)/d\alpha$.

and $\log_e L(\alpha_0)$; if $\log_e L(\hat{\alpha})$ is much larger than $\log_e L(\alpha_0)$, then H_0 is probably wrong. Finally, the score test statistic measures the slope of log likelihood at α_0; if this slope is very steep, then we are probably far from the peak of the likelihood function, casting doubt on H_0.

*An Illustration**

It is instructive to apply these results to our previous example, in which we sought to estimate the probability π of obtaining a head from a coin based on a sample of n flips. Recall that the MLE of π is the sample proportion $\hat{\pi} = X/n$, where X counts the number of heads in the sample. The second derivative of the log likelihood (Equation D.8) is

$$\frac{d^2 \log_e L(\pi)}{d\pi^2} = -\frac{X}{\pi^2} - \left[-\frac{n-X}{(1-\pi)^2}(-1)\right]$$
$$= \frac{-X + 2\pi X - n\pi^2}{\pi^2(1-\pi)^2}$$

Noting that $E(X) = n\pi$, the information is

$$\mathcal{I}(\pi) = \frac{-n\pi + 2n\pi^2 - n\pi^2}{-\pi^2(1-\pi^2)} = \frac{n}{\pi(1-\pi)}$$

and the asymptotic variance of $\hat{\pi}$ is $\mathcal{V}(\hat{\pi}) = [\mathcal{I}(\pi)]^{-1} = \pi(1-\pi)/n$, a familiar result.[24] The *estimated* asymptotic sampling variance is $\widehat{\mathcal{V}(\hat{\pi})} = \hat{\pi}(1-\hat{\pi})/n$.

For our sample of $n = 10$ flips with $X = 7$ heads, $\widehat{\mathcal{V}(\hat{\pi})} = (.7 \times .3)/10 = 0.0210$, and a 95% asymptotic confidence interval for π based on the Wald statistic is

$$\pi = .7 \pm 1.96 \times \sqrt{0.0210} = .7 \pm .290$$

Alternatively, to test the hypothesis that the coin is fair, H_0: $\pi = .5$, we can calculate the Wald test statistic

$$Z_0 = \frac{.7 - .5}{\sqrt{0.0210}} = 1.38$$

which corresponds to a two-tail p-value [from $N(0, 1)$] of .168.

The log likelihood, recall, is

$$\log_e L(\pi) = X \log_e \pi + (n - X)\log_e(1-\pi)$$
$$= 7\log_e \pi + 3\log_e(1-\pi)$$

Using this equation,

$$\log_e L(\hat{\pi}) = \log_e L(.7) = 7\log_e .7 + 3\log_e .3 = -6.1086$$
$$\log_e L(\pi_0) = \log_e L(.5) = 7\log_e .5 + 3\log_e .5 = -6.9315$$

[24] In this case, the asymptotic variance coincides with the exact, finite-sample variance of $\hat{\pi}$.

The likelihood-ratio test statistic for H_0 is, therefore,

$$G_0^2 = 2[-6.1086 - (-6.9315)] = 1.646$$

which corresponds to a p-value (from χ_1^2) of .199.

Finally, for the score test,

$$S(\pi) = \frac{d \log_e L(\pi)}{d\pi} = \frac{X}{\pi} - \frac{n - X}{1 - \pi}$$

from which $S(\pi_0) = 7/.5 - 3/.5 = 8$. Evaluating the information at π_0 produces $\mathcal{I}(\pi_0) = \mathcal{I}(.5) = 10/(.5 \times .5) = 40$. The score statistic is, therefore,

$$S_0 = \frac{S(\pi_0)}{\sqrt{\mathcal{I}(\pi_0)}} = \frac{8}{\sqrt{40}} = 1.265$$

for which the two-tail p-value [from $N(0, 1)$] is .206.

The three tests are in reasonable agreement, but all are quite inaccurate! An exact test, using the null binomial distribution of X (the number of heads),

$$p(x) = \binom{10}{x} .5^x .5^{10-x} = \binom{10}{x} .5^{10}$$

yields a two-tail p-value of .3438 [corresponding to $\Pr(X \leq 3 \text{ or } X \geq 7)$]. We must be careful in applying asymptotic results to small samples.

D.6.3 Several Parameters*

The maximum-likelihood method can be generalized to simultaneous estimation of several parameters. Let $p(\underset{(n \times m)}{\mathbf{X}} \mid \underset{(k \times 1)}{\boldsymbol{\alpha}})$ represent the probability or probability density for n possibly multivariate observations $\mathbf{X}$ ($m \geq 1$) which depend on k independent parameters[25] $\boldsymbol{\alpha}$. The likelihood $L(\boldsymbol{\alpha}) \equiv L(\boldsymbol{\alpha}|\mathbf{X})$ is a function of the parameters $\boldsymbol{\alpha}$, and we seek the values $\hat{\boldsymbol{\alpha}}$ that maximize this function. As before, it is generally more convenient to work with $\log_e L(\boldsymbol{\alpha})$ in place of $L(\boldsymbol{\alpha})$. To maximize the likelihood, we find the vector partial derivative $\partial \log_e L(\boldsymbol{\alpha})/\partial \boldsymbol{\alpha}$, set this derivative to $\mathbf{0}$, and solve the resulting matrix equation for $\hat{\boldsymbol{\alpha}}$. If there is more than one root, then we choose the solution that produces the largest likelihood.

As in the case of a single parameter, the maximum-likelihood estimator is consistent, asymptotically unbiased, asymptotically efficient, asymptotically normal (but now *multivariate* normal), and based on sufficient statistics. The asymptotic variance-covariance matrix of the MLE is

$$\underset{(k \times k)}{\mathcal{V}(\hat{\boldsymbol{\alpha}})} = \left\{ -E\left[\frac{\partial^2 \log_e L(\boldsymbol{\alpha})}{\partial \boldsymbol{\alpha}\, \partial \boldsymbol{\alpha}'} \right] \right\}^{-1} \qquad \text{[D.10]}$$

---

[25] To say that the parameters are independent means that the value of none can be obtained from the values of the others. If there is a dependency among the parameters, then the redundant parameter can simply be replaced by a function of other parameters.

The matrix in braces in Equation D.10 is called the *Fisher information matrix*, $\mathcal{I}(\boldsymbol{\alpha})$ (not to be confused with the identity matrix **I**). Notice how the formulas for several parameters closely parallel those for one parameter.

Generalizations of the score and Wald tests follow directly. The Wald statistic for H_0: $\boldsymbol{\alpha} = \boldsymbol{\alpha}_0$ is

$$Z_0^2 \equiv (\hat{\boldsymbol{\alpha}} - \boldsymbol{\alpha}_0)' \widehat{\mathcal{V}(\hat{\boldsymbol{\alpha}})}^{-1} (\hat{\boldsymbol{\alpha}} - \boldsymbol{\alpha}_0)$$

The score vector is $S(\boldsymbol{\alpha}) \equiv \partial \log_e L(\boldsymbol{\alpha})/\partial\boldsymbol{\alpha}$; and the score statistic is

$$S_0^2 \equiv S(\boldsymbol{\alpha}_0)' \mathcal{I}(\boldsymbol{\alpha}_0)^{-1} S(\boldsymbol{\alpha}_0)$$

The likelihood-ratio test also generalizes straightforwardly:

$$G_0^2 \equiv -2 \log_e \left[\frac{L(\boldsymbol{\alpha}_0)}{L(\hat{\boldsymbol{\alpha}})} \right]$$

All three test statistics are asymptotically distributed as χ_k^2 under H_0.

Each of these tests can be adapted to more complex hypotheses. Suppose, for example, that we wish to test the hypothesis H_0 that p of the k elements of $\boldsymbol{\alpha}$ are equal to particular values. Let $L(\hat{\boldsymbol{\alpha}}_0)$ represent the maximized likelihood under the constraint represented by the hypothesis (i.e., setting the p parameters equal to their hypothesized values, but leaving the other parameters free to be estimated); $L(\hat{\boldsymbol{\alpha}})$ represents the globally maximized likelihood when the constraint is relaxed. Then, under the hypothesis H_0,

$$G_0^2 \equiv -2 \log_e \left[\frac{L(\hat{\boldsymbol{\alpha}}_0)}{L(\hat{\boldsymbol{\alpha}})} \right]$$

has an asymptotic chi-square distribution with p degrees of freedom.

The following example (adapted from Theil, 1971, pp. 389–390) illustrates these results: A sample of n independent observations X_i is drawn from a normally distributed population with unknown mean μ and variance σ^2. We want to estimate μ and σ^2. The likelihood function is

$$\begin{aligned} L(\mu, \sigma^2) &= \prod_{i=1}^{n} \frac{1}{\sigma\sqrt{2\pi}} \exp\left[-\frac{(X_i - \mu)^2}{2\sigma^2} \right] \\ &= (2\pi\sigma^2)^{-n/2} \exp\left[-\frac{1}{2\sigma^2} \sum_{i=1}^{n} (X_i - \mu)^2 \right] \end{aligned}$$

and the log likelihood is

$$\log_e L(\mu, \sigma^2) = -\frac{n}{2} \log_e 2\pi - \frac{n}{2} \log \sigma^2 - \frac{1}{2\sigma^2} \sum (X_i - \mu)^2$$

with partial derivatives

$$\begin{aligned} \frac{\partial \log_e L(\mu, \sigma^2)}{\partial \mu} &= \frac{1}{\sigma^2} \sum (X_i - \mu) \\ \frac{\log_e L(\mu, \sigma^2)}{\partial \sigma^2} &= -\frac{n}{2\sigma^2} + \frac{1}{2\sigma^4} \sum (X_i - \mu)^2 \end{aligned}$$

Setting the partial derivatives to 0 and solving simultaneously for the maximum-likelihood estimators of μ and σ^2 produces

$$\hat{\mu} = \frac{\sum X_i}{n} = \overline{X}$$

$$\hat{\sigma}^2 = \frac{\sum(X_i - \overline{X})^2}{n}$$

The matrix of second derivatives of the log likelihood is

$$\begin{bmatrix} \dfrac{\partial^2 \log_e L}{\partial \mu^2} & \dfrac{\partial^2 \log_e L}{\partial \mu\, \partial \sigma^2} \\ \dfrac{\partial^2 \log_e L}{\partial \sigma^2\, \partial \mu} & \dfrac{\partial^2 \log_e L}{\partial (\sigma^2)^2} \end{bmatrix} = \begin{bmatrix} -\dfrac{n}{\sigma^2} & -\dfrac{1}{\sigma^4}\sum(X_i - \mu) \\ -\dfrac{1}{\sigma^4}\sum(X_i - \mu) & \dfrac{n}{2\sigma^4} - \dfrac{n}{4\sigma^6}\sum(X_i - \mu)^2 \end{bmatrix}$$

Taking expectations, noting that $E(X_i - \mu) = 0$ and that $E[(X_i - \mu)^2] = \sigma^2$, produces the negative of the information matrix:

$$-\mathscr{I}(\mu, \sigma^2) = \begin{bmatrix} -\dfrac{n}{\sigma^2} & 0 \\ 0 & -\dfrac{n}{2\sigma^4} \end{bmatrix}$$

The asymptotic variance-covariance matrix of the maximum-likelihood estimators is, as usual, the inverse of the information matrix:

$$\mathscr{V}(\hat{\mu}, \hat{\sigma}^2) = [\mathscr{I}(\mu, \sigma^2)]^{-1} = \begin{bmatrix} \dfrac{\sigma^2}{n} & 0 \\ 0 & \dfrac{2\sigma^4}{n} \end{bmatrix}$$

The result for the sampling variance of $\hat{\mu} = \overline{X}$ is the usual one (σ^2/n). The MLE of σ^2 is biased but consistent (and, indeed, is the estimator S_*^2 given previously in Equation D.7).

D.7 Recommended Reading

Almost any introductory text in mathematical statistics, and many econometric texts, cover the subject matter of this appendix more formally and in greater detail. A relatively compact summary appears in Zellner (1983). Wonnacott and Wonnacott (1990) present insightful treatments of many of these topics at a much lower level of mathematical sophistication; I particularly recommend this source if you found the unstarred parts of this appendix too terse. A good, relatively accessible discussion of asymptotic distribution theory appears in Theil (1971, Chapter 8). A general treatment of Wald, likelihood-ratio, and score tests can be found in Engle (1984).

References

Abler, R., Adams, J. S., and Gould, P. (1971). *Spatial Organization: The Geographer's View of the World*. Prentice-Hall, Englewood Cliffs, NJ.

Achen, C. H. (1982). *Interpreting and Using Regression*. Sage, Beverly Hills, CA.

Agresti, A. (1990). *Categorical Data Analysis*. Wiley, New York.

Aitkin, M., Anderson, D., Francis, B., and Hinde, J. (1989). *Statistical Modeling in GLIM*. Clarendon Press, Oxford.

Andrews, D. F. (1979). The robustness of residual displays. In Launer, R. L. and Wilkenson, G. N., editors, *Robustness in Statistics*, pages 19–32. Academic Press, New York.

Angell, R. C. (1951). The moral integration of American cities. *American Journal of Sociology*, 57 (part 2):1–140.

Anscombe, F. J. (1960). Rejection of outliers [with commentary]. *Technometrics*, 2:123–166.

Anscombe, F. J. (1961). Examination of residuals. *Proceedings of the Fourth Berkeley Symposium on Mathematical Statistics and Probability*, 1:1–36.

Anscombe, F. J. (1973). Graphs in statistical analysis. *American Statistician*, 27:17–22.

Anscombe, F. J. (1981). *Computing in Statistical Science Through APL*. Springer-Verlag, New York.

Anscombe, F. J. and Tukey, J. W. (1963). The examination and analysis of residuals. *Technometrics*, 5:141–160.

Atkinson, A. C. (1985). *Plots, Transformations, and Regression: An Introduction to Graphical Methods of Diagnostic Regression Analysis*. Clarendon Press, Oxford.

Bailey, R. A., Harding, S. A., and Smith, G. L. (1989). Cross-validation. In Kotz, S. and Johnson, N. L., editors, *Encyclopedia of Statistical Sciences, Supplement Volume*, pages 39–44. Wiley, New York.

Bard, Y. (1974). *Nonlinear Parameter Estimation*. Academic Press, New York.

Barnett, V. and Lewis, T. (1994). *Outliers in Statistical Data*. Wiley, New York, third edition.

Bartlett, M. S. (1937). Properties of sufficiency and statistical tests. *Proceedings of the Royal Society, A*, 160:268–282.

Becker, R. A., Chambers, J. M., and Wilks, A. R. (1988). *The New S Language: A Programming Environment for Data Analysis and Graphics*. Wadsworth, Pacific Grove, CA.

Beckman, R. J. and Cook, R. D. (1983). Outliers. *Technometrics*, 25:119–163.

Beckman, R. J. and Trussell, H. J. (1974). The distribution of an arbitrary studentized residual and the effects of updating in multiple regression. *Journal of the American Statistical Association*, 69:199–201.

Belsley, D. A. (1984). Demeaning condition diagnostics through centering [with commentary]. *Americn Statistician*, 38:73–93.

Belsley, D. A., Kuh, E., and Welsch, R. E. (1980). *Regression Diagnostics: Identifying Influential Data and Sources of Collinearity*. Wiley, New York.

Blau, P. M. and Duncan, O. D. (1967). *The American Occupational Structure*. Wiley, New York.

Bollen, K. (1989). *Structural Equations With Latent Variables*. Wiley, New York.

Box, G. E. P. and Cox, D. R. (1964). An analysis of transformations. *Journal of the Royal Statistical Society, B*, 26:211–252.

Box, G. E. P. and Tidwell, P. W. (1962). Transformation of the independent variables. *Technometrics*, 4:531–550.

Breusch, T. S. and Pagan, A. R. (1979). A simple test for heteroscedasticity and random coefficient variation. *Econometrica*, 47:1287–1294.

Campbell, A., Converse, P. E., Miller, W. E., and Stokes, D. E. (1960). *The American Voter*. Wiley, New York.

Campbell, D. T. and Stanley, J. C. (1963). *Experimental and Quasi-Experimental Designs for Research*. Rand McNally, Chicago.

Canada (1962). *Major Roads (Map)*. Department of Mines and Technical Surveys, Ottawa.

Canada (1971). *Census of Canada*. Statistics Canada, Ottawa.

Canada (1972). *Canada Year Book 1972*. Statistics Canada, Ottawa.

Canada (1993). *Canada Year Book 1994*. Statistics Canada, Ottawa.

Chambers, J. M., Cleveland, W. S., Kleiner, B., and Tukey, P. A. (1983). *Graphical Methods for Data Analysis*. Wadsworth, Belmont, CA.

Chambers, J. M. and Hastie, T. J., editors (1992). *Statistical Models in S*. Wadsworth and Brooks/Cole, Pacific Grove, CA.

Chatfield, C. (1989). *Analysis of Time Series: An Introduction*. Chapman and Hall, London, fourth edition.

Chatterjee, S. and Hadi, A. S. (1988). *Sensitivity Analysis in Linear Regression*. Wiley, New York.

Chatterjee, S. and Price, B. (1991). *Regression Analysis by Example*. Wiley, New York, second edition.

Chirot, D. and Ragin, C. (1975). The market, tradition and peasant rebellion: The case of Romania in 1907. *American Sociological Review*, 40:428–444.

Cleveland, W. S. (1979). Robust locally weighted regression and smoothing scatterplots. *Journal of the American Statistical Association*, 74:829–836.

Cleveland, W. S. (1993). *Visualizing Data*. Hobart Press, Summit, NJ.

Cleveland, W. S. (1994). *The Elements of Graphing Data*. Hobart Press, Summit, NJ, revised edition.

Cleveland, W. S., Grosse, E., and Shyu, W. M. (1992). Local regression models. In Chambers, J. M. and Hastie, T. J., editors, *Statistical Models in S*, pages 309–376. Wadsworth and Brooks/Cole, Pacific Grove, CA.

Collett, D. (1991). *Modelling Binary Data*. Chapman and Hall, London.

Conover, W. J., Johnson, M. E., and Johnson, M. M. (1981). A comparative study of tests for homogeneity of variances, with applications to the outer continental shelf bidding data. *Technometrics*, 23:351–361.

Cook, R. D. (1977). Detection of influential observations in linear regression. *Technometrics*, 19:15–18.

Cook, R. D. (1993). Exploring partial residual plots. *Technometrics*, 35:351–362.

Cook, R. D. (1994). On the interpretation of regression plots. *Journal of the American Statistical Association*, 89:177–189.

Cook, R. D. (1996). Added-variable plots and curvature in linear regression. *Technometrics*, 38:275–278.

Cook, R. D. and Weisberg, S. (1980). Characterizations of an empirical influence function for detecting influential cases in regression. *Technometrics*, 22:495–508.

Cook, R. D. and Weisberg, S. (1982). *Residuals and Influence in Regression*. Chapman and Hall, New York.

Cook, R. D. and Weisberg, S. (1983). Diagnostics for heteroscedasticity in regression. *Biometrika*, 70:1–10.

Cook, R. D. and Weisberg, S. (1989). Regression diagnostics with dynamic graphics [with commentary]. *Technometrics*, 31:277–311.

Cook, R. D. and Weisberg, S. (1994). *An Introduction to Regression Graphics*. Wiley, New York.

Coombs, C. H., Dawes, R. M., and Tversky, A. (1970). *Mathematical Psychology: An Elementary Introduction*. Prentice-Hall, Englewood Cliffs, NJ.

Cox, D. R. and Snell, E. J. (1989). *Analysis of Binary Data*. Chapman and Hall, London, second edition.

Davis, C. (1990). Body image and weight preoccupation: A comparison between exercising and non-exercising women. *Appetite*, 15:13–21.

Davis, P. J. (1965). *Mathematics of Matrices: A First Book of Matrix Theory and Linear Algebra*. Blaisdell, New York.

Dempster, A. P. (1969). *Elements of Continuous Multivariate Analysis*. Addison-Wesley, Reading MA.

Dobson, A. J. (1983). *An Introduction to Statistical Modelling*. Chapman and Hall, London.

Draper, N. R. and Smith, H. (1981). *Applied Regression Analysis*. Wiley, New York, second edition.

Draper, N. R. and Van Nostrand, R. C. (1979). Ridge regression and James-Stein estimators: Review and comments. *Technometrics*, 21:451–466.

Duan, N. and Li, K. C. (1991). Slicing regression: A link-free regression method. *Annals of Statistics*, 19:505–530.

Duncan, O. D. (1961). A socioeconomic index for all occupations. In Reiss, Jr., A. J., editor, *Occupations and Social Status*, pages 109–138. Free Press, New York.

Duncan, O. D. (1975). *Introduction to Structural Equation Models*. Academic Press, New York.

Duncan, O. D. (1984). *Notes on Social Measurement: Historical and Critical*. Russell Sage Foundation, New York.

Durbin, J. and Watson, G. S. (1950). Testing for serial correlation in least squares regression I. *Biometrika*, 37:409–428.

Durbin, J. and Watson, G. S. (1951). Testing for serial correlation in least squares regression II. *Biometrika*, 38:159–178.

Efron, B. and Tibshirani, R. J. (1993). *An Introduction to the Bootstrap*. Chapman and Hall, New York.

Engle, R. F. (1984). Wald, likelihood ratio, and Lagrange multiplier tests in econometrics. In Griliches, Z. and Intriligator, M. D., editors, *Handbook of Econometrics*, volume II, pages 775–879. North-Holland, Amsterdam.

Ericksen, E. P., Kadane, J. B., and Tukey, J. W. (1989). Adjusting the 1980 census of population and housing. *Journal of the American Statistical Association*, 84:927–944.

Fan, J. and Gijbels, I. (1996). *Local Polynomial Modelling and Its Applications*. Chapman and Hall, London.

Fienberg, S. E. (1980). *The Analysis of Cross-Classified Categorical Data*. MIT Press, Cambridge, MA, second edition.

Fisher, R. A. (1922). On the mathematical foundations of theoretical statistics. *Philosophical Transactions of the Royal Society of London, A*, 222:309–368.

Fisher, R. A. (1925). *Statistical Methods for Research Workers*. Oliver and Boyd, Edinburgh.

Fox, B. (1980). *Women's Domestic Labour and Their Involvement in Wage Work: Twentieth-Century Changes in the Reproduction of Daily Life*. Ph.D. thesis, University of Alberta.

Fox, J. (1979). Comment on Harris, Luginbuhl, and Fishbein: Loglinear and logit models for tabular experimental data. *Social Psychology Quarterly*, 42:431–433.

Fox, J. (1984). *Linear Statistical Models and Related Methods: With Applications to Social Research*. Wiley, New York.

Fox, J. (1991). *Regression Diagnostics: An Introduction*. Sage, Newbury Park, CA.

Fox, J. (1992). Statistical graphics. In Borgatta, E. F. and Borgatta, M. L., editors, *Encyclopedia of Sociology*, volume 4, pages 2054–2073. Macmillan, New York.

Fox, J. and Guyer, M. (1978). "Public" choice and cooperation in *n*-person prisoner's dilemma. *Journal of Conflict Resolution*, 22:469–481.

Fox, J. and Hartnagel, T. F. (1979). Changing social roles and female crime in Canada: A time series analysis. *Canadian Review of Sociology and Anthropology*, 16:96–104.

Fox, J. and Monette, G. (1992). Generalized collinearity diagnostics. *Journal of the American Statistical Association*, 87:178–183.

Fox, J. and Suschnigg, C. (1989). A note on gender and the prestige of occupations. *Canadian Journal of Sociology*, 14:353–360.

Francis, I. (1973). Comparison of several analysis of variance programs. *Journal of the American Statistical Association*, 68:860–865.

Freedman, D., Pisani, R., Purves, R., and Adhikari, A. (1991). *Statistics*. Norton, New York, second edition.

Freedman, D. A. (1987). As others see us: A case study in path analysis [with commentary]. *Journal of Educational Statistics*, 12:101–223.

Freedman, J. L. (1975). *Crowding and Behavior*. Viking Press, New York.

Friendly, M. (1991). *SAS System for Statistical Graphics*. SAS Institute, Cary, NC.

Friendly, M. and Franklin, P. (1980). Interactive presentation in multitrial free recall. *Memory and Cognition*, 8:265–270.

Furnival, G. M. and Wilson, R. W. (1974). Regression by leaps and bounds. *Technometrics*, 16:499–511.

Gallant, A. R. (1975). Nonlinear regression. *American Statistican*, 29:73–81.

Goldberger, A. S. (1964). *Econometric Theory*. Wiley, New York.

Goldberger, A. S. (1973). Structural equation models: An overview. In Goldberger, A. S. and Duncan, O. D., editors, *Structural Equation Models in the Social Sciences*, pages 1–18. Seminar Press, New York.

Goodall, C. (1983). *M*-estimators of location: An outline of the theory. In Hoaglin, D. C., Mosteller, F., and Tukey, J. W., editors, *Understanding Robust and Exploratory Data Analysis*, pages 339–403. Wiley, New York.

Gould, S. J. (1989). *Wonderful Life: The Burgess Shale and the Nature of History*. Norton, New York.

Graybill, F. A. (1983). *Introduction to Matrices With Applications in Statistics*. Wadsworth, Belmont, CA, second edition.

Green, P. E. and Carroll, J. D. (1978). *Mathematical Tools for Applied Multivariate Analysis*. Academic Press, New York.

Greene, I. and Shaffer, P. (1992). Leave to appeal and leave to commence judicial review in Canada's refugee-determination system: Is the process fair? *International Journal of Refugee Law*, 4:71–83.

Härdle, W. (1991). *Smoothing Techniques: With Implementation in S*. Springer-Verlag, New York.

Harris, B., Luginbuhl, J. E. R., and Fishbein, J. E. (1978). Density and personal space in a field setting. *Social Psychology*, 41:350–353.

Harvey, A. (1990). *The Econometric Analysis of Time Series*. MIT Press, Cambridge, MA, second edition.

Hastie, T. J. (1992). Generalized additive models. In Chambers, J. M. and Hastie, T. J., editors, *Statistical Models in S*, pages 249–307. Wadsworth and Brooks/Cole, Pacific Grove, CA.

Hastie, T. J. and Tibshirani, R. J. (1990). *Generalized Additive Models*. Chapman and Hall, London.

Healy, M. J. R. (1986). *Matrices for Statistics*. Clarendon Press, Oxford.

Hernandez, F. and Johnson, R. A. (1980). The large sample behavior of transformations to normality. *Journal of the American Statistical Association*, 75:855–861.

Hoaglin, D. C., Mosteller, F., and Tukey, J. W., editors (1983). *Understanding Robust and Exploratory Data Analysis*. Wiley, New York.

Hoaglin, D. C., Mosteller, F., and Tukey, J. W., editors (1985). *Exploring Data Tables, Trends, and Shapes*. Wiley, New York.

Hoaglin, D. C. and Welsch, R. E. (1978). The hat matrix in regression and ANOVA. *American Statistician*, 32:17–22.

Hocking, R. R. (1976). The analysis and selection of variables in linear regression. *Biometrics*, 32:1–49.

Hocking, R. R. (1985). *The Analysis of Linear Models*. Brooks/Cole, Monterey, CA.

Hocking, R. R. and Speed, F. M. (1975). The full rank analysis of some linear model problems. *Journal of the American Statistical Association*, 70:706–712.

Hoerl, A. E. and Kennard, R. W. (1970a). Ridge regression: Biased estimation for nonorthogonal problems. *Technometrics*, 12:55–67.

Hoerl, A. E. and Kennard, R. W. (1970b). Ridge regression: Applications to nonorthogonal problems. *Technometrics*, 12:69–82.

Huber, P. J. (1964). Robust estimation of a location parameter. *Annals of Mathematical Statistics*, 35:73–101.

Hurvich, C. M. and Tsai, C.-L. (1990). The impact of model selection on inference in linear regression. *American Statistician*, 44:214–217.

Johnston, J. (1972). *Econometric Methods*. McGraw-Hill, New York, second edition.

Judge, G. G., Griffiths, W. E., Hill, E. C., Lütkepohl, H., and Lee, T.-C. (1985). *The Theory and Practice of Econometrics*. Wiley, New York, second edition.

Kish, L. (1965). *Survey Sampling*. Wiley, New York.

Kish, L. (1987). *Statistical Design for Research*. Wiley, New York.

Kmenta, J. (1986). *Elements of Econometrics*. Macmillan, New York, second edition.

Koch, G. G. and Gillings, D. B. (1983). Inference, design based vs. model based. In Kotz, S. and Johnson, N. L., editors, *Encyclopedia of Statistical Sciences, Volume 4*, pages 84–88. Wiley, New York.

Landwehr, J. M., Pregibon, D., and Shoemaker, A. C. (1980). Some graphical procedures for studying a logistic regression fit. In *Proceedings of the Business and Economic Statistics Section, American Statistical Association*, pages 15–20.

Leinhardt, S. and Wasserman, S. S. (1978). Exploratory data analysis: An introduction to selected methods. In Schuessler, K. F., editor, *Sociological Methodology 1979*, pages 311–365. Jossey-Bass, San Francisco.

Li, G. (1985). Robust regression. In Hoaglin, D. C., Mosteller, F., and Tukey, J. W., editors, *Exploring Data Tables, Trends, and Shapes*, pages 281–343. Wiley, New York.

Mallows, C. L. (1973). Some comments on C_p. *Technometrics*, 15:661–676.

Mallows, C. L. (1986). Augmented partial residuals. *Technometrics*, 28:313–319.

Mandel, J. (1982). Use of the singular value decomposition in regression analysis. *American Statistician*, 36:15–24.

Manski, C. (1991). Regression. *Journal of Economic Literature*, 29:34–50.

McCallum, B. T. (1973). A note concerning asymptotic covariance expressions. *Econometrica*, 41:581–583.

McCullagh, P. (1980). Regression models for ordinal data [with commentary]. *Journal of the Royal Statistical Society, B*, 42:109–142.

McCullagh, P. and Nelder, J. A. (1989). *Generalized Linear Models*. Chapman and Hall, London, second edition.

Monette, G. (1990). Geometry of multiple regression and 3-D graphics. In Fox, J. and Long, J. S., editors, *Modern Methods of Data Analysis*, pages 209–256. Sage, Newbury Park, CA.

Moore, D. S. (1995). *The Basic Practice of Statistics*. Freeman, New York.

Moore, D. S. and McCabe, G. P. (1993). *Introduction to the Practice of Statistics*. Freeman, New York, second edition.

Moore, Jr., J. C. and Krupat, E. (1971). Relationship between source status, authoritarianism, and conformity in a social setting. *Sociometry*, 34:122–134.

Morrison, D. F. (1976). *Multivariate Statistical Methods*. McGraw-Hill, New York, second edition.

Mosteller, F. and Tukey, J. W. (1968). Data analysis, including statistics. In Lindzey, G. and Aronson, E., editors, *The Handbook of Social Psychology, Volume Two: Research Methods*, pages 80–203. Addison-Wesley, Reading, MA, second edition.

Mosteller, F. and Tukey, J. W. (1977). *Data Analysis and Regression*. Addison-Wesley, Reading, MA.

Namboodiri, K. (1984). *Matrix Algebra: An Introduction*. Sage, Beverly Hills, CA.

Nelder, J. A. (1976). Letter to the editor. *American Statistician*, 30:103.

Nelder, J. A. (1977). A reformulation of linear models [with commentary]. *Journal of the Royal Statistical Society, A*, 140:48–76.

Nelder, J. A. and Wedderburn, R. W. M. (1972). Generalized linear models. *Journal of the Royal Statistical Society, A*, 135:370–384.

Nerlove, M. and Press, S. J. (1973). *Univariate and Multivariate Log-Linear and Logistic Models*. RAND Corporation, Santa Monica, CA.

Obenchain, R. L. (1977). Classical *F*-tests and confidence intervals for ridge regression. *Technometrics*, 19:429–439.

Ornstein, M. D. (1976). The boards and executives of the largest Canadian corporations: Size, composition, and interlocks. *Canadian Journal of Sociology*, 1:411–437.

Ornstein, M. D. (1983). *Accounting for Gender Differences in Job Income in Canada: Results From a 1981 Survey*. Labour Canada, Ottawa.

Pecknold, J. C., McClure, D. J., Appeltauer, L., Wrzesinski, L., and Allan, T. (1982). Treatment of anxiety using fenobam (a nonbenzodiazepine) in a double-blind standard (diazepam) placebo-controlled study. *Journal of Clinical Psychopharmacology*, 2:129–133.

Pregibon, D. (1981). Logistic regression diagnostics. *Annals of Statistics*, 9:705–724.

Putter, J. (1967). Orthonormal bases of error spaces and their use for investigating the normality and variance of residuals. *Journal of the American Statistical Association*, 62:1022–1036.

Rao, C. R. (1973). *Linear Statistical Inference and Its Applications*. Wiley, New York, second edition.

Riordan, C. A., Quigley-Fernandez, B., and Tedeschi, J. T. (1982). Some variables affecting changes in interpersonal attraction. *Journal of Experimental Social Psychology*, 18:358–374.

Robey, B., Shea, M. A., Rutstein, O., and Morris, L. (1992). The reproductive revolution: New survey findings. Technical Report M-11, Population Reports.

Rousseeuw, P. J. and Leroy, A. M. (1987). *Robust Regression and Outlier Detection*. Wiley, New York.

Sahlins, M. (1972). *Stone Age Economics*. Aldine, New York.

Sall, J. (1990). Leverage plots for general linear hypotheses. *American Statistician*, 44:308–315.

Scheffé, H. (1959). *The Analysis of Variance*. Wiley, New York.

Scudder, T. (1962). *The Ecology of the Gwembe Tonga*. Manchester University Press, Manchester.

Searle, S. R. (1971). *Linear Models*. Wiley, New York.

Searle, S. R. (1982). *Matrix Algebra Useful for Statistics*. Wiley, New York.

Searle, S. R. (1987). *Linear Models for Unbalanced Data*. Wiley, New York.

Searle, S. R., Speed, F. M., and Henderson, H. V. (1981). Some computational and model equivalences in analysis of variance of unequal-subclass-numbers data. *American Statistician*, 35:16–33.

Searle, S. R., Speed, F. M., and Milliken, G. A. (1980). Population marginal means in the linear model: An alternative to least squares means. *American Statistician*, 34:216–221.

Seber, G. A. F. (1977). *Linear Regression Analysis*. Wiley, New York.

Shryock, H. S., Siegel, J. S., and Associates (1973). *The Methods and Materials of Demography (Two Volumes)*. U.S. Bureau of the Census, Washington, DC.

Silverman, B. W. (1986). *Density Estimation for Statistics and Data Analysis*. Chapman and Hall, London.

Silverman, B. W. and Young, G. A. (1987). The bootstrap: To smooth or not to smooth? *Biometrika*, 74:469–479.

Simonoff, J. S. (1996). *Smoothing Methods in Statistics*. Springer-Verlag, New York.

Speed, F. M. and Hocking, R. R. (1976). The use of the $R(\ )$-notation with unbalanced data. *American Statistician*, 30:30–33.

Speed, F. M., Hocking, R. R., and Hackney, O. P. (1978). Methods of analysis of linear models with unbalanced data. *Journal of the American Statistical Association*, 73:105–112.

Speed, F. M. and Monlezun, C. J. (1979). Exact F tests for the method of unweighted means in a 2^k experiment. *American Statistician*, 33:15–18.

Spence, I. and Lewandowsky, S. (1990). Graphical perception. In Fox, J. and Long, J. S., editors, *Modern Methods of Data Analysis*, pages 13–57. Sage, Newbury Park, CA.

Stefanski, L. A. (1991). A note on high-breakdown estimators. *Statistics and Probability Letters*, 11:353–358.

Steinhorst, R. K. (1982). Resolving current controversies in analysis of variance. *American Statistician*, 36:138–139.

Stine, R. (1990). An introduction to bootstrap methods: Examples and ideas. In Fox, J. and Long, J. S., editors, *Modern Methods of Data Analysis*, pages 325–373. Sage, Newbury Park, CA.

Stine, R. A. (1995). Graphical interpretation of variance inflation factors. *American Statistician*, 49:53–56.

Stine, R. A. and Fox, J., editors (1996). *Statistical Computing Environments for Social Research*. Sage, Thousand Oaks, CA.

Stolzenberg, R. M. (1979). The measurement and decomposition of causal effects in nonlinear and nonadditive models. In Schuessler, K. F., editor, *Sociological Methodology 1980*, pages 459–488. Jossey-Bass, San Francisco.

Stone, M. (1987). *Coordinate-Free Multivariable Statistics: An Illustrated Geometric Progression from Halmos to Gauss and Bayes*. Clarendon Press, Oxford.

Street, J. O., Carroll, R. J., and Ruppert, D. (1988). A note on computing robust regression estimates via iteratively reweighted least squares. *American Statistician*, 42:152–154.

Theil, H. (1971). *Principles of Econometrics*. Wiley, New York.

Thompson, M. E. (1988). Superpopulation models. In Kotz, S. and Johnson, N. L., editors, *Encyclopedia of Statistical Sciences, Volume 9*, pages 93–99. Wiley, New York.

Tierney, L. (1990). *Lisp-Stat: An Object-Oriented Environment for Statistical Computing and Dynamic Graphics*. Wiley, New York.

Tierney, L. (1995). Data analysis using Lisp-Stat. *Sociological Methods and Research*, 23:329–351.

Torgerson, W. S. (1958). *Theory and Methods of Scaling*. Wiley, New York.

Tufte, E. R. (1983). *The Visual Display of Quantitative Information*. Graphics Press, Cheshire, CT.

Tukey, J. W. (1972). Some graphic and semigraphic displays. In Bancroft, T. A., editor, *Statistical Papers in Honor of George W. Snedecor*, pages 293–316. Iowa State University Press, Ames, IA.

Tukey, J. W. (1977). *Exploratory Data Analysis*. Addison-Wesley, Reading, MA.

United States (1994). *Statistical Abstract of the United States: 1994*. U.S. Bureau of the Census, Washington, DC.

Urquhart, M. C. and Buckley, K. A. H., editors (1965). *Historical Statistics of Canada*. Macmillan, Toronto.

Velleman, P. F. and Hoaglin, D. C. (1981). *Applications, Basics, and Computing of Exploratory Data Analysis*. Duxbury, Boston.

Velleman, P. F. and Welsch, R. E. (1981). Efficient computing of regression diagnostics. *American Statistician*, 35:234–241.

Vinod, H. D. (1978). A survey of ridge regression and related techniques for improvements over ordinary least squares. *Review of Economics and Statistics*, 60:121–131.

Vinod, H. D. and Ullah, A. (1981). *Recent Advances in Regression Methods*. Dekker, New York.

Wang, P. C. (1985). Adding a variable in generalized linear models. *Technometrics*, 27:273–276.

Wang, P. C. (1987). Residual plots for detecting nonlinearity in generalized linear models. *Technometrics*, 29:435–438.

Weisberg, S. (1985). *Applied Linear Regression*. Wiley, New York, second edition.

White, H. (1980). A heteroscedasticity-consistent covariance matrix estimator and a direct test for heteroscedasticity. *Econometrica*, 38:817–838.

Wonnacott, R. J. and Wonnacott, T. H. (1979). *Econometrics*. Wiley, New York, second edition.

Wonnacott, T. H. and Wonnacott, R. J. (1990). *Introductory Statistics*. Wiley, New York, fifth edition.

Wu, L. L. (1985). Robust *M*-estimation of location and regression. In Tuma, N. B., editor, *Sociological Methodology 1985*, pages 316–388. Jossey-Bass, San Francisco.

Yates, F. (1934). The analysis of multiple classifications with unequal numbers in the different classes. *Journal of the American Statistical Association*, 29:51–66.

Young, G. A. (1994). Bootstrap: More than a stab in the dark? [with commentary]. *Statistical Science*, 9:382–415.

Zellner, A. (1983). Statistical theory and econometrics. In Griliches, Z. and Intriligator, M. D., editors, *Handbook of Econometrics, Volume 1*, pages 67–178. North-Holland, Amsterdam.

Author Index

Subject Index

Dataset Index

Names of files containing datasets are given in brackets.

About the Author

John Fox is Professor of Sociology at McMaster University in Hamilton, Canada. He was previously Professor of Sociology and of Mathematics and Statistics at York University in Toronto, where he also directed the Statistical Consulting Service at the Institute for Social Research. Professor Fox earned a Ph.D. in Sociology from the University of Michigan in 1972. He has taught many workshops on statistical topics, at such places as the summer program of the Inter-University Consortium for Political and Social Research and the annual meetings of the American Sociological Association. His recent and current work includes research on statistical methods and on Canadian society. He is author of *Regression Diagnostics* (Sage, 1991) and co-editor (with Robert Stine) of *Statistical Computing Environments for Social Research* (Sage, 1996).